Table II (*continued*)

z	0	1	2	3	4	5	6	7	8	9
.0	.5000	.5040	.5080	.5120	.5160	.5199	.5239	.5279	.5319	.5359
.1	.5398	.5438	.5478	.5517	.5557	.5596	.5636	.5675	.5714	.5753
.2	.5793	.5832	.5871	.5910	.5948	.5987	.6026	.6064	.6103	.6141
.3	.6179	.6217	.6255	.6293	.6331	.6368	.6406	.6443	.6480	.6517
.4	.6554	.6591	.6628	.6664	.6700	.6736	.6772	.6808	.6844	.6879
.5	.6915	.6950	.6985	.7019	.7054	.7088	.7123	.7157	.7190	.7224
.6	.7257	.7291	.7324	.7357	.7389	.7422	.7454	.7486	.7517	.7549
.7	.7580	.7611	.7642	.7673	.7703	.7734	.7764	.7794	.7823	.7852
.8	.7881	.7910	.7939	.7967	.7995	.8023	.8051	.8078	.8106	.8133
.9	.8159	.8186	.8212	.8238	.8264	.8289	.8315	.8340	.8365	.8389
1.0	.8413	.8438	.8461	.8485	.8508	.8531	.8554	.8577	.8599	.8621
1.1	.8643	.8665	.8686	.8708	.8729	.8749	.8770	.8790	.8810	.8830
1.2	.8849	.8869	.8888	.8907	.8925	.8944	.8962	.8980	.8997	.9015
1.3	.9032	.9049	.9066	.9082	.9099	.9115	.9131	.9147	.9162	.9177
1.4	.9192	.9207	.9222	.9236	.9251	.9265	.9278	.9292	.9306	.9319
1.5	.9332	.9345	.9357	.9370	.9382	.9384	.9406	.9418	.9430	.9441
1.6	.9452	.9463	.9474	.9484	.9495	.9505	.9515	.9525	.9535	.9545
1.7	.9554	.9564	.9573	.9582	.9591	.9599	.9608	.9616	.9625	.9633
1.8	.9641	.9648	.9656	.9664	.9671	.9678	.9686	.9693	.9700	.9706
1.9	.9713	.9719	.9726	.9732	.9738	.9744	.9750	.9756	.9762	.9767
2.0	.9772	.9778	.9783	.9788	.9793	.9798	.9803	.9808	.9812	.9817
2.1	.9821	.9826	.9830	.9834	.9838	.9842	.9846	.9850	.9854	.9857
2.2	.9861	.9864	.9868	.9871	.9874	.9878	.9881	.9884	.9887	.9890
2.3	.9893	.9896	.9898	.9901	.9904	.9906	.9909	.9911	.9913	.9916
2.4	.9918	.9920	.9922	.9925	.9927	.9929	.9931	.9932	.9934	.9936
2.5	.9938	.9940	.9941	.9943	.9945	.9946	.9948	.9949	.9951	.9952
2.6	.9953	.9955	.9956	.9957	.9959	.9960	.9961	.9962	.9963	.9964
2.7	.9965	.9966	.9967	.9968	.9969	.9970	.9971	.9972	.9973	.9974
2.8	.9974	.9975	.9976	.9977	.9977	.9978	.9979	.9979	.9980	.9981
2.9	.9981	.9982	.9982	.9983	.9984	.9984	.9985	.9985	.9986	.9986
3.	.9987	.9990	.9993	.9995	.9997	.9998	.9998	.9999	.9999	1.0000

Notes: 1. Enter table at Z, read out $P(Z \leq z)$, the shaded area.

2. For a general normal X, enter table at $z = (x - \mu)/\sigma$ to read $P(X \leq x)$.

3. Entries opposite 3 are for 3.0, 3.1, 3.2 . . . , 3.9.

4. For $z \geq 4$, $P(Z > z) = P(Z < -z) \doteq \dfrac{1}{\sigma\sqrt{2\pi}} e^{-z^2/2}$.

Statistics:
Theory and Methods

Statistics:
Theory and Methods

Donald A. Berry

Bernard W. Lindgren

UNIVERSITY OF MINNESOTA

BROOKS/COLE PUBLISHING COMPANY
PACIFIC GROVE, CALIFORNIA

Brooks/Cole Publishing Company

A Division of Wadsworth, Inc.

© 1990 by Wadsworth, Inc., Belmont, California 94002. All rights reserved. No part of this book may be reproduced, stored in a retrieval system, or transcribed, in any form or by any means—electronic, mechanical, photocopying, recording, or otherwise—without the prior written permission of the publisher, Brooks/Cole Publishing Company, Pacific Grove, California 93950, a division of Wadsworth, Inc.

Printed in the United States of America

10   9   8   7   6   5   4   3   2   1

Library of Congress Cataloging-in-Publication Data

Berry, Donald A.

    Statistics : theory and methods / Donald A. Berry, Bernard W. Lindgren.

       p.   cm.

      ISBN 0-534-09942-4

      1. Statistics.   I. Lindgren, B. W. (Bernard W.)   II. Title.

QA276.12.B48   1990

519.5—dc20   89-48611

                                                CIP

Sponsoring Editor: *John Kimmel*

Marketing Representative: *Cathy Twiss*

Editorial Assistant: *Mary Ann Zuzow*

Production Editor: *Penelope Sky*

Production Assistant: *Barbara Kimmel*

Manuscript Editor: *David Hoyt*

Interior and Cover Design: *Lisa Thompson*

Art Coordinators: *Lisa Torri and Cloyce Wall*

Interior Illustration: *John Foster*

Typesetting: *Polyglot Compositors*

Printing and Binding: *Arcata Graphics/Fairfield*

Preface

This text is designed for a two-quarter or two-semester course in statistical theory and methodology. The effective application of statistical methods requires an understanding of the theory behind them. So a first course in statistics for mathematically prepared students should have a solid component of theory. On the other hand, statistical theory has no reason for existence without applications. Indeed, most students find the theory dull without the associated methods and applications to real-life problems.

The student need not have had a previous course in statistics but should have a background in differential and integral calculus. However, because these mathematical tools may be rusty or untried, we proceed rather carefully, especially in the early chapters. Occasionally we may exceed the students' knowledge of calculus, but we do so as gently as possible, motivating the mathematics by an immediate application to statistics.

In the course of the exposition we give as "examples" specific instances of the general discussion to illustrate an area of application. To provide examples of solved problems we include problems with solutions before the regular problem set at the end of each chapter. In both cases, we've grouped problems according to chapter sections, each group corresponding roughly to the material of a class lecture. An asterisk beside a problem number indicates that the answer is in the back of the book.

Many of our examples and problems use real data from real situations, drawn from a number of different fields of application. Many are in the area of the health sciences, because health issues are of broad general interest, and because the experimental designs involved are often of the simpler types that we deal with in the text.

Although statistical theory involves mathematics, this is a statistics text, not a mathematics text. We do not take the "definition–theorem–proof" approach. Indeed, calculus alone is not a sufficient prerequisite for a rigorous development of probability or for proofs of theorems on sufficiency and large-sample properties. We do use mathematical arguments, but in some places we say, "it can be shown . . . ," and provide a reference.

The coverage is fairly standard: distribution theory, principles of inference, and the basic methodology of one- and two-sample inference, with brief introductions to the areas of linear models, categorical data analysis, and Bayesian inference. We emphasize the importance of nonparametric methods and have integrated them into the sections on inference rather than set them apart as optional.

Our approach to inference is for the most part traditional, but we take pains to point out the questions that these methods do and do not address. We emphasize the distinction between gathering statistical evidence and making decisions. In testing hypotheses we emphasize P-values, because researchers generally use P-values rather than fixed type I error rates.

Our tables are more extensive than is customary. The traditional .05 and .01 critical values were originally used simply because of limited computing power. Now any microcomputer can calculate P-values, and we have used this capability to construct tables of tail-areas for the distributions we use most often (t, chi-square, F). We also provide the usual tables of percentiles.

In our development of probability, we take independence of *variables* (rather than events) as basic, because statistical applications of probability focus on the independence of variables. Independence for several events is then easy to define in terms of their indicator variables. We introduce joint distributions early, because even conditioning one event on another is really a bivariate matter. Also, regarding binomial and hypergeometric variables as sums of several Bernoulli variables greatly simplifies their treatment.

No contemporary discussion of statistical inference is complete without some mention of the Bayesian approach. Every method of inference involves a subjective element, but Bayesian inference takes explicit account of subjectivity, in a framework in which hypotheses can be assigned probabilities. We introduce this approach in our last chapter along with related ideas of statistical decision theory.

We gratefully acknowledge the help of many students in spotting errors and checking answers; special thanks to Ruth Meyers, Lukas Makris, David Bjostad, Lee Li-shya, and Bret Larget for help in researching material for examples and in preparing problem solutions. We also thank the following reviewers: David Birkes, Oregon State University; James Hutton, Bucknell University; Mark Pinsky, Northwestern University; Françoise Seillier-Moiseiwitsch, University of North Carolina, Chapel Hill; Gerald L. Sievers, Western Michigan University; Furman Smith, Auburn University at Montgomery; and Mohamed Tahir, Temple University.

Donald A. Berry
Bernard W. Lindgren

Contents

C H A P T E R 1   Probability   5

1.1　Sample Spaces and Events　5
1.2　Event Algebra　7
1.3　Probability for Experiments with Symmetries　12
1.4　Composition of Experiments　15
1.5　Sampling at Random　22
1.6　Binomial and Multinomial Coefficients　25
1.7　Discrete Probability Distributions　26
　　　Chapter Perspective　30
　　　Solved Problems　30
　　　Problems　39

C H A P T E R 2   Discrete Random Variables   46

2.1　Probability Functions for Discrete Random Variables　47
2.2　Joint Distributions　51
2.3　Conditional Probability　55
2.4　Bayes' Theorem　60
2.5　Independent Random Variables　65
2.6　Exchangeable Random Variables　70
　　　Chapter Perspective　72
　　　Solved Problems　73
　　　Problems　85

C H A P T E R 3   Averages   93

3.1　The Mean　94
3.2　Expected Value of a Function of Random Variables　99

3.3   Variability   104
3.4   Covariance and Correlation   108
3.5   Sums of Random Variables   113
3.6   Probability Generating Functions   116
      Chapter Perspective   121
      Solved Problems   122
      Problems   130

C H A P T E R 4   **Bernoulli and Related Variables**   **135**

4.1   Sampling Bernoulli Populations   135
4.2   The Binomial Distribution   139
4.3   Hypergeometric Distributions   143
4.4   Inverse Sampling   147
4.5   Approximating Binomial Probabilities   151
4.6   Poisson Distributions   155
4.7   Sample Proportions and the Law of Large Numbers   160
4.8   Multinomial Distributions   162
4.9   Applying the p.g.f.   164
      Chapter Perspective   166
      Solved Problems   167
      Problems   179

C H A P T E R 5   **Continuous Random Variables**   **186**

5.1   The Distribution Function   187
5.2   Density and the Probability Element   195
5.3   The Median and Other Percentiles   204
5.4   Expected Value   208
5.5   Average Deviations   213
5.6   Several Variables   217
5.7   Covariance and Correlation   226
5.8   Independence   229
5.9   Conditional Distributions   233
5.10 Moment Generating Functions   236
      Chapter Perspective   242
      Solved Problems   243
      Problems   260

C H A P T E R 6   **Families of Continuous Distributions**   **269**

6.1   Normal Distributions   269
6.2   Exponential Distributions   275

6.3   Gamma Distributions   278
6.4   Beta Distributions   282
6.5   Chi-Square Distributions   284
6.6   *F*- and *t*-Distributions   286
6.7   Distributions for Reliability   290
6.8   Bivariate Normal Distributions   294
      Chapter Perspective   299
      Solved Problems   299
      Problems   311

C H A P T E R 7   Organizing and Describing Data   317

7.1   Frequency Distributions   317
7.2   Data on Continuous Variables   322
7.3   Order Statistics   329
7.4   The Sample Mean   333
7.5   Some Measures of Dispersion   337
7.6   Correlation   343
      Chapter Perspective   347
      Solved Problems   349
      Problems   357

C H A P T E R 8   Samples, Statistics, and Sampling Distributions   364

8.1   Random Sampling   365
8.2   Likelihood   368
8.3   Sufficient Statistics   372
8.4   Sampling Distributions   375
8.5   Simulating Sample Distributions   378
8.6   Order Statistics   381
8.7   Moments of Sample Means and Proportions   386
8.8   Sampling Distributions for Large Samples   390
8.9   The m.g.f. of the Sample Mean   395
8.10 Independence of Mean and Variance   397
      Chapter Perspective   399
      Solved Problems   399
      Problems   407

C H A P T E R 9   Estimation   412

9.1   Errors in Estimation   413
9.2   Determining Sample Size   417

9.3   Large-Sample Confidence Intervals   420
9.4   When σ Is Unknown—Small n 423
9.5   Pivotal Quantities   426
9.6   Estimating a Mean Difference   427
9.7   Estimating Variability   430
9.8   The Method of Moments   432
9.9   Maximum Likelihood Estimation   433
9.10   Consistency   436
9.11   Efficiency   439
       Chapter Perspective   444
       Solved Problems   445
       Problems   452

C H A P T E R 10   Significance Testing   459

10.1   Hypotheses   461
10.2   Assessing the Evidence   462
10.3   One-Sample Z-Tests   468
10.4 t-Tests   472
10.5   Some Nonparametric Tests   474
10.6   Testing a Population Variance   480
       Chapter Perspective   481
       Solved Problems   482
       Problems   486

C H A P T E R 11   Tests as Decision Rules   491

11.1   Rejection Regions and Errors   492
11.2   The Power Function   496
11.3   Choosing a Sample Size   498
11.4   Quality Control   500
11.5   Most Powerful Tests   503
11.6   Likelihood Ratio Tests   509
       Chapter Perspective   512
       Solved Problems   512
       Problems   518

C H A P T E R 12   Comparing Two Populations   523

12.1   Treatment Effects   523
12.2   Large-Sample Comparison of Means   527
12.3   Comparing Proportions   529
12.4   Two-Sample t-Tests   533

12.5   Two-Sample Nonparametric Tests   536
12.6   Paired Comparisons   541
12.7   Comparing Variances   543
        Chapter Perspective   545
        Solved Problems   545
        Problems   551

C H A P T E R **13**   **Goodness of Fit   557**

13.1   Fitting a Distribution with Two Categories   557
13.2   The Chi-Square Test   559
13.3   Tests Based on c.d.f.s   564
13.4   Testing Normality   570
13.5   Testing Independence   574
13.6   A Likelihood Ratio Test for Goodness of Fit   579
13.7   Testing Homogeneity Using Paired Data   582
        Chapter Perspective   585
        Solved Problems   586
        Problems   592

C H A P T E R **14**   **Analysis of Variance   600**

14.1   One-Way ANOVA: The Method   600
14.2   One-Way ANOVA: The Theory   605
14.3   Multiple Comparisons   610
14.4   Two-Way ANOVA   613
        Chapter Perspective   618
        Solved Problems   618
        Problems   623

C H A P T E R **15**   **Regression   629**

15.1   Models with a Single Explanatory Variable   630
15.2   Least Squares   633
15.3   Distribution of the Least-Squares Estimators   638
15.4   Inference for Regression Parameters   642
15.5   Predicting Y from X 645
15.6   The Regression Effect   650
15.7   Multiple Regression   654
        Chapter Perspective   660
        Solved Problems   660
        Problems   664

C H A P T E R **16**   **Bayesian Methods and Making Decisions**   **671**

16.1 Assessing Prior Probabilities   672
16.2 Using Bayes' Theorem to Update Probabilities   675
16.3 Loss Functions   681
16.4 Bayesian Point Estimates   684
16.5 Probability of an Interval   686
16.6 Probability of the Null Hypothesis   688
16.7 Hypothesis Testing as Decision Making   691
16.8 Bayesian Predictive Distributions   693
Chapter Perspective   696
Solved Problems   696
Problems   701

APPENDIX 1   Tables   707
APPENDIX 2   Answers to Selected Problems   752
Index   761

Statistics:
Theory and Methods

Prologue

An article in the *British Medical Journal*[1] reported on a survey of 13 consecutive issues of that esteemed periodical. The survey found that of the 77 papers in those issues, 62 used statistical analyses. Of these, 32 had statistical errors of one kind or another, and 18 had fairly serious faults. The authors of the article venture the opinion that less respected journals probably show an even greater frequency of errors.

We want to make two points from this survey. First, *statistics* is more than a name for facts and figures about sports or the economy. It is a discipline that is an integral part of scientific research. Second, its methods are much misunderstood and much misused. Indeed, some of the criticisms leveled by Gore et al. can be challenged—not all statisticians agree on what is correct. One purpose of this text is to give you a critical attitude toward applications of statistical techniques, along with an appreciation of their utility.

To set the stage for the step-by-step development, we offer here some accounts of situations and problems whose treatments involve probability and/or statistical ideas. These are given to whet your appetite. We describe some of the challenges but postpone any solutions until we've developed appropriate tools.

You will see from these examples that statistical ideas and methods are not restricted to medical or even to scientific applications. Business decisions, industrial production plans, governmental policies, and research in education and agriculture all rely heavily on statistical analyses.

Example **a**   **A Smoking Survey**
"Cigarette smoking is at its lowest point in 20 years among high school students in Santa Maria," according to a poll taken in driver's education classes at Santa Maria

---

[1] S. M. Gore, I. G. Jones, and E. C. Rytter, "Misuse of statistical methods: Critical assessment of articles in BMJ from January to March 1976," *Br. Med. J. 1* (1977), 85–87.

High.[2] The teacher has been conducting polls on drug and alcohol use for the past 20 years, and his surveys are viewed as a barometer of trends among students in Santa Maria.

Of 173 students polled in 1986, only 16% responded that they smoked cigarettes. In a similar survey taken in 1985, 28% smoked. Other results showed that 56% drank alcohol, compared to 57% in 1985; 10% used cocaine when available, compared to 9% the previous year; 4% used LSD; and 6% abused pills.

These matters are of general concern. Are these sample percentages typical of all high school students? Or of high school students in California, or even in Santa Maria? The teacher thought the poll was "very, very accurate," because the voting was done behind closed doors. Practically all students take driver's education, so he believed that the poll included a good cross-section of teenagers.   ■

Example b   **Atomic Tests and Leukemia**

In the summer of 1957, in a desert area of Nevada, the United States carried out an above-ground atomic explosion code-named "Smoky." About 3,000 military men and civilians were "invited" to watch. Some 20 years later, a compensation claim by one of the participants, a leukemia victim, prompted the U.S. Centers for Disease Control (CDC) to conduct a search for Smoky participants.

The overall rate of leukemia for men of ages similar to those of the Smoky participants is 1 in 1500. The CDC tracked down 450 individuals and found that there were 8 cases of leukemia among them. Is this convincing evidence that participation in Smoky was a cause of the disease? If so, the claim should be allowed. On the other hand, it is *possible* that 450 people selected "at random" would include 8 who had leukemia. We need to know the extent of this possibility before we can decide whether Smoky was the culprit.   ■

Example c   **Corneal Thickness and Glaucoma**

A study[3] was conducted to determine whether glaucoma can be linked with other eye defects—in particular, with increased corneal thickness. Clearly, one would need to know about the corneal thickness of normal eyes as well as that of glaucomatous eyes. One way to get such information is to measure the corneal thickness of some glaucomatous eyes and some normal eyes. Another possibility, the one used in this study, is to use subjects with glaucoma in one eye but not in the other. The study used eight subjects, and the corneal thickness measurements are shown in Table a.

---

[2] Reported in the Santa Barbara (Calif.) *News-Press*, Jan. 13, 1987.
[3] Reported by N. Ehlers, "On Corneal Thickness and Intraocular Pressure, II", *Acta Ophthalmologica 48* (1970), 1107–1112.

Table a

Patient	Glaucomatous eye	Contralateral eye
1	488	484
2	478	478
3	480	492
4	426	444
5	440	436
6	410	398
7	458	464
8	460	476

An inspection of these numbers does not indicate a clear relationship. In the text of this book, we will develop some precise tools for analyzing such data. ■

Example d   The data in Table b are from the report of an inertia welding experiment.[4] To make a weld, the operator stops a rotating part by forcing it into contact with a stationary part. The resulting friction generates heat that produces a hot-pressure weld. Seven welds were made using four different velocities of rotation (the controlled variable x). Each weld was then tested for breaking strength (the response Y), with results as shown in Table b.

Table b

Velocity (x) $(10^2$ $ft/min)$	Breaking strength (Y) (ksi)
2.00	89
2.5	97, 91
2.75	98
3.00	100, 104, 97

The data seem to indicate a relationship between velocity and breaking strength of the resulting weld, but since different breaking strengths can occur for the same velocity of rotation, the relationship is not perfect. In particular, there may be random errors that influence the results. How can we describe or "model" the relationship in the presence of such errors? ■

These examples involve some aspects of statistical problems generally. One is that responses and measurements are **variable**. One source of variability is the existence of individual differences, reflecting "randomness" in a population of

---

[4] G. E. P. Box, W. G. Hunter, and J. S. Hunter, *Statistics for Experimenters* (New York: Wiley, 1978), p. 473.

subjects or experimental units. Measurements involve variability when the measurement process itself cannot be performed in exactly the same way one time as another, so that possible measured values constitute a kind of **population** from which one draws with each measurement.

The target population in an investigation embodies some underlying truth or status or experimental mechanism that results in variability in the data. The reason for collecting data is to learn something about this target population. In order to draw conclusions about the population, we set up mathematical models that explicitly allow for randomness. The fundamental problem of statistics is to decide which of the various possible models for a population best represent it, in light of the data in a sample. Probability plays an essential role in representing randomness, so we begin our study of statistics with probability, the subject of Chapter 1.

Probability

Example **1a**

A Case of Discrimination?

Dr. Benjamin Spock, a well-known pediatrician and author, was tried for conspiracy in 1968 for his encouragement and aid to draft resisters during the Vietnam conflict. He was convicted by an all-male jury. His lawyers felt that since he was revered by millions of mothers, Spock might have had a better chance for acquittal had the jury included women.

One argument presented in the appeal of the verdict was that women were grossly underrepresented in the list of potential jurors from which the jury was drawn. In one 30-month period, the lists of the judge in the Spock trial included just 14.4% women—86 out of 597. In that same period, the jury lists for the other six judges in the same court included 29% women. Neither percentage is close to the proportion of women in the area (over 50%), but at the very least we could expect the methods of selection of Spock's judge to be no less fair than those of the other judges.

Suppose the names on Spock's judge's list were taken at random from the same population as those on the other judges' lists (with 29% women). What are the chances that a list of 597, randomly selected from that population, would include no more than 14.9% women? ■

To address the question posed in this example, we'd need to calculate "chances" or **probabilities**. Typically, the phenomena we'll be describing involve uncertainties; the factors that influence the results cannot be controlled—or even identified in some cases. We can discuss chances meaningfully if we make assumptions about the experiment. These define a probability **model**, which includes these three main ingredients: a sample space, events, and probabilities of events.

1.1 Sample Spaces and Events

A phenomenon whose outcome is uncertain is an **experiment of chance**. A model for it should take into account all possible outcomes.

Definition

> The **sample space** Ω in a model for an experiment of chance is the set of all outcomes considered to be possible.

Example **1.1a**   **Sample Space for Selecting Jurors**

To obtain potential jurors, names are often selected "at random" from lists of registered voters (as in Example 1a). In any one selection, any name on the list might be drawn. The sample space Ω is just the set of all voters.   ∎

In this chapter, we consider only **discrete** sample spaces. A discrete sample space is one whose outcomes are finite in number or, if they are infinite in number, are at most countable—can be listed in sequence (as can the integers, for example). When defining a set by listing its outcomes, we use braces around the list. For example, $\{a, b\}$ is the set whose only two elements are the outcomes a and b.

Example **1.1b**   **Sample Spaces for Coin-Tossing**

Consider the toss of two coins. Sometimes two people each toss a coin to see who pays for coffee. Either they match or they don't match. The sample space consisting of these two outcomes may be adequate for modeling the coffee decision:

$$\Omega = \{\text{same, different}\}.$$

In other situations, one might be interested in how many heads show. If so, $\Omega = \{0, 1, 2\}$ is an appropriate sample space. It is also adequate if one is only interested in whether the coins match. However, if one is interested in which coin falls which way, neither of these sample spaces suffices; we need an even more detailed sample space:

$$\Omega = \{HH, HT, TH, TT\}. \tag{1}$$

("HT" means that the first coin falls heads and the second falls tails.)

Different purposes may suggest different sample spaces, even when the physical act is the same. The sample space shown in (1) for the coin toss is the most detailed, and it serves the purpose in each of the situations described.   ∎

The sample spaces in the preceding examples are finite, but countably infinite sample spaces are useful in some cases, as in the next example.

Example **1.1c**   **An Infinite Sample Space**

Consider a stream of potential customers entering a gift shop. Starting at an arbitrary point in time, we count how many people come in before one makes a purchase. This number is a positive integer. However, we can't be certain that a purchase will be made within any specified finite number of potential customers, so we take Ω to be infinite: $\Omega = \{1, 2, 3, \ldots\}$.   ∎

Events

Usually, there are subsets of the sample space that are of special interest. Any subset of the sample space is called an **event**. An event can be defined by giving either

(1) a list of its outcomes or (2) a descriptive phrase or condition that characterizes its outcomes. The symbol for an event will usually be a capital letter, possibly with a subscript.

An individual outcome ω, considered as a set $\{\omega\}$ whose only member is ω, is an event. Another special kind of event is the whole sample space Ω.

Definition

> An **event** is a set of outcomes—a **subset** of the sample space Ω. An event E is said to occur if any one of its outcomes occurs.

Example 1.1d

Events Defined by Bets in Roulette

When a ball is rolled on a spinning roulette wheel, it eventually comes to rest in one of 38 compartments numbered 0, 00, 1, 2, . . . , 36. The 0 and 00 are colored green, and half of the numbers from 1 to 36 are red and the rest black. The set of 38 compartments in which the ball may land is an obvious choice for the sample space:

$$\Omega = \{0, 00, 1, 2, \ldots, 36\}.$$

One can bet on any individual number or on various sets of more than one number. These sets are events. For example, "odd" is an event; a bet on "odd" wins if any odd number turns up:

$$\text{odd} = \{1, 3, \ldots, 35\}.$$

Other bets are defined by the casino: "black" (which wins if any black number turns up, "first dozen" (which wins if the outcome is any of the integers in the event $\{1, 2, \ldots, 12\}$), and so on. ∎

1.2 Event Algebra

The algebra of sets or events deals with relationships among events and with combinations of events.

Example 1.2a

Dealing a Card

The natural sample space for the selection of a card from a standard deck of playing cards[1] is the set of the 52 cards in the deck. Some events in this sample space that might be of interest are:

R: The card is red.

H: The card is a heart.

F: The card is a face card.

3: The card is a three.

---

[1] A standard deck consists of 52 cards, 13 in each of four suits: spades, hearts, diamonds, and clubs. The 26 hearts and diamonds are red; the rest, black. The 13 denominations in each suit are 2, 3, . . . , 9, 10, J (jack), Q (queen), K (king), and A (Ace). The jack, queen, and king are "face cards."

There are some obvious relationships among these events. For instance, every outcome of H is also in R, but no outcome is in both F and 3.

Combinations of conditions R, H, F, and 3 might be of interest. For instance, consider these:

The card is a red face card (that is, both red and a face card).

The card is not a 3.

The card is a heart or a face card.

Each statement is a condition that defines a set of outcomes—an event in the sample space. They can be thought of as combinations or operations involving R, H, F, and 3.   ■

Inclusion, **equality**, and **complementation** are basic notions in the algebra of events:

(i) E is **included** in F: $E \subset F$, if and only if every outcome in E is also in F.
(ii) E and F are **equal**: $E = F$, if and only if $E \subset F$ *and* $F \subset E$.
(iii) The **complement** of E is the set E^c of all outcomes not in E. (Alternative notations for E^c are E' and $\bar{E}$.)

Venn
Diagrams

As in the algebra of sets, representations by means of **Venn diagrams** can be helpful in understanding the relationships among events and combinations of events. A Venn diagram represents the sample space as a set of points in the plane, usually bounded and connected, such as a large rectangle. An event is then a subset of this rectangle. Figure 1.1 shows a sample space and these events:

E: points within the circle,

F: points within the triangle, and

G: points within the smaller rectangle.

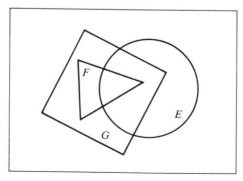

Figure 1.1

In Figure 1.1, F is a subset of G: $F \subset G$. However, E and F are not ordered by inclusion: F is not a subset of E, and E is not a subset of F. The complement of E is the big rectangle with the circle representing E deleted.

In terms of conditions that define events, the relation $E \subset F$ means that condition E *implies* condition F. The complement of E is defined by the *denial* of the condition that defines E.

The sample space Ω is itself an event. Its "complement"—the denial of Ω—is a condition satisfied by no outcomes. Any condition satisfied by *no* outcomes in the sample space will be called an event, denoted by $\varnothing$ and referred to as the **empty set**. Clearly,

The Empty Set

$$\Omega^c = \varnothing, \qquad \text{and} \qquad \varnothing^c = \Omega.$$

Events that have outcomes in common **intersect**, and the outcomes they have in common make up their **intersection** (as in the intersection of two streets). When an outcome is described as lying in *either* of two events (or in both) it is said to be in their **union**.

The **intersection** of events E and F, denoted by EF, is the set of all points in *both* E and F. The **union** of events E and F, denoted by $E \cup F$, is the set of outcomes contained either in E or in F (or in both).

Like the operations of multiplication and addition of numbers, the operations of set intersection and union are **commutative**:

$$EF = FE, \qquad \text{and} \qquad E \cup F = F \cup E.$$

Properties of Intersections and Unions

They are also **associative**:

$$(EF)G = E(FG) = EFG,$$

and

$$(E \cup F) \cup G = E \cup (F \cup G) = E \cup F \cup G.$$

Thus, we have unambiguous definitions for intersections and unions of three events. These are special cases of general definitions: The union of any collection of events, countable or not, is the set of all outcomes that are in at least one of the events, and the intersection of any collection of events is the set of all outcomes that are in every event.

Two events that do not intersect are said to be **disjoint** or **mutually exclusive**. In the Venn diagram of Figure 1.2, events E (circle) and G (rectangle) are disjoint: $EG = \varnothing$. Events E and F (triangle) intersect, and their intersection EF is the shaded region.

Figure 1.3 gives a Venn diagram illustrating set union. The shaded region defines the union of the events E (circle), F (triangle), and G (rectangle).

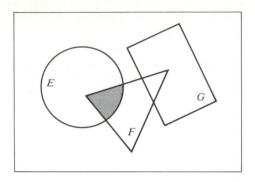

Figure 1.2

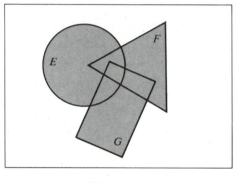

Figure 1.3

Example 1.2b   Consider the sample space of Example 1.2a—the deck of cards. With obvious notation for king, queen, jack, the four suits (spades, hearts, diamonds, clubs), and the two colors (black and red), we have these relations:

$$H \subset R \qquad \text{(all hearts are red)}$$
$$F3 = \varnothing \qquad \text{(no card is both 3 and a face card)}$$
$$B = R^c \qquad \text{(cards that are not red are black)}$$
$$SB = S \qquad \text{(cards that are black and spades are spades)}$$
$$B = S \cup C \qquad \text{(black cards are spades or clubs)}$$
$$R = D \cup H \qquad \text{(red cards are diamonds or hearts)}$$
$$F = K \cup Q \cup J \qquad \text{(face cards are kings, queens, or jacks)}$$
$$A = BA \cup RA \qquad \text{(aces are red aces or black aces)}$$
$$\Omega = B \cup R = S \cup H \cup D \cup C = F \cup F^c.$$

∎

In translating verbal statements into mathematical ones, watch for the key words *and*, *or*, *implies*, and *not*:[2]

Word/Symbol Equivalents:

Condition *E and* condition *F*: *EF*.
Condition *E or* condition *F*: $E \cup F$.
Condition *E implies* condition *F*: $E \subset F$.
Not condition *E*: E^c.

When both intersections and unions are used in a single expression, we need a convention for the order in which operations are to be carried out. The convention we use is the same as for ordinary arithmetic of numbers, with union corresponding to addition and intersection to multiplication. Parentheses play the usual role. Thus, for example, $(A \cup B)(C \cup D)$ means the intersection of the events $A \cup B$ and $C \cup D$. However, $A \cup BC \cup D$ means the union of the three events A, BC, and D, which is ordinarily not the same as the event $(A \cup B)(C \cup D)$.

Distributive Laws

Ordinary numbers obey the **distributive law**: $a(b + c) = ab + ac$. The operations of event union and event intersection obey a similar law. Set intersection is distributive over set union: For any events A, B, C,

$$A(B \cup C) = AB \cup AC. \tag{1}$$

This is evident in a Venn diagram, provided the diagram is sufficiently general, as in Figure 1.4. The event represented by both sides of the equation is shown as a shaded region.

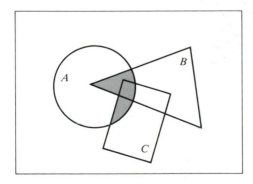

Figure 1.4

---

[2] English usage can be misleading in this regard. For example, "the set of hearts and diamonds" means the set of red cards—cards that are either hearts *or* diamonds. The word "and" is used in the sense that if we put together the hearts and the diamonds, we get all the red cards—the set of cards that are hearts or diamonds.

A "dual" distributive law also holds: Interchanging operations of union and intersection in Equation (1), we have

$$A \cup BC = (A \cup B)(A \cup C). \tag{2}$$

(See Problem 6(f) at the end of this chapter.) This says that the operation of union is distributive over intersection.

"Proofs by picture" can be convincing in simple cases such as Equations (1) and (2). However, in more complicated situations, the picture one draws may have some special characteristic that makes a generally false statement true for that picture. The **Proving Set Equality**   sure way to establish a set equality is to show that each side is a subset of the other. Problem 6(f) at the end of the chapter asks you to use this method of proof to show (2). We'll illustrate the method to establish (1) in Solved Problem D at the end of the chapter.

1.3   Probability for Experiments with Symmetries

We now complete the specification of a probability model by assigning probabilities to events. Probabilities give numerical expression to the intuitive notion of the relative likelihoods of events. Thus, if A is more likely to occur than B, the probability of A should be greater than the probability of B.

Assigning probabilities can be difficult. Indeed, any particular assignment is subject to revision in the light of experience—a statistical problem.

There is an important class of experiments for which a particular way of assigning probabilities is quite compelling and may be generally agreed on as suitable. These are experiments with finite sample spaces, in which there is no reason to think that any one outcome is more likely than another. We think of the outcomes as **equally likely** and therefore assign the same probability to each one. This probability could be any number, but common practice takes it to be $1/N$, where N is the number of outcomes in the sample space.

Suppose $\Omega = \{\omega_1, \omega_2, \ldots, \omega_N\}$. If the outcomes ω_i are "equally likely," we say that each has probability $1/N$, and write

$$P(\omega_i) = \frac{1}{N}, \qquad i = 1, 2, \ldots, N. \tag{1}$$

Experiments with Equally Likely Outcomes   There are a number of familiar experiments in which the possible outcomes are commonly assumed to be equally likely: the roll of a die, the toss of a coin, the spin of a wheel of fortune, the selection of a ticket in a lottery, etc. Because of the symmetry, we usually see no reason to favor one outcome as more likely than another.

The tickets in a lottery are sold to finance the prize and are commonly called *chances.* A single ticket has "one chance in *N*" of winning; one who holds *k* of the *N* tickets has "*k* chances in *N*" of winning. Two people who hold equal numbers of tickets have the same chance of winning.

Lottery
Models

We'll frequently use the random selection of a lottery ticket as a model for an experiment with equally likely outcomes. The sample space is the set of lottery tickets, and each ticket has probability $1/N$ of being drawn. A selection in which outcomes are equally likely is termed **random**. Thus, the phrases *random selection* and *select at random* mean that we assume the ticket has probability $1/N$— "one chance in *N*"—of being drawn.

Definition

> An individual is drawn **at random** from a population if all individuals in the population are equally likely.

Probability
of An Event

Someone with *k* tickets is said to have "*k* chances in *N*" of having a ticket drawn—of holding a winning ticket. Those *k* tickets constitute an event, and the probability of the event is defined to be k/N, the sum of the *k* probabilities $1/N$ of the individual tickets.

Definition

> Suppose an event *E* consists of *k* outcomes in a sample space with *N* equally likely outcomes. The **probability** of *E* is
>
> $$P(E) = \frac{k}{N}. \tag{2}$$
>
> The **odds on** *E* are *k* to $N - k$; the **odds against** *E* are $N - k$ to *k*.

Example **1.3a**

In Example 1.2b, we defined events for the selection of a card from a deck of playing cards. When the selection is random, each card has probability 1/52. From Equation (2), the probability of the event *S* (the card is a spade) is the number of spades divided by 52: $P(S) = 13/52$. The probability that the card is a face card is $P(F) = 12/52$; and so on.   ∎

Condition (2) is appropriate only when the outcomes are equally likely, but we may adopt this model for making calculations even when we think it is only approximately correct. For example, after several shuffles of a deck of cards, the assumption of equally likely outcomes in the next deal may not be quite correct. However, the easy calculations based on the assumption of equal likelihood may be adequate, and these are all we can do, not knowing the correct model.

As defined by Equation (2), the probability of an event has some important properties. First, $0 \le k \le N$, so $0 \le k/n \le 1$. Second, the sample space Ω has probability $N/N = 1$; its complement is the empty set $\varnothing$, with probability $0/N = 0$.

Third, if E contains k and F contains m outcomes *and* $EF = \emptyset$, then $E \cup F$ contains $k + m$ outcomes, so

$$P(E \cup F) = \frac{k + m}{N} = \frac{k}{N} + \frac{m}{N} = P(E) + P(F).$$

These properties will be adopted as axioms in the more general setting to be described in Section 1.7.

Properties of $P(E)$:

   (i) $0 \le P(E) \le 1$, for any event E.
  (ii) $P(\Omega) = 1$.
 (iii) $P(E \cup F) = P(E) + P(F)$ whenever $EF = \emptyset$.

Properties (i)–(iii) imply many other useful properties. For instance, additivity for any finite collection of pairwise disjoint events follows by induction from (iii)— see Equation (4) below. Also, applying (iii) with $F = E^c$, we obtain

$$1 = P(\Omega) = P(E \cup E^c) = P(E) + P(E^c),$$

or

$$P(E^c) = 1 - P(E). \tag{3}$$

This is useful for finding the probability of E when it is easier to calculate the probability of E^c. More generally, for disjoint events $E_1, \ldots, E_k$, (iii) implies

$$P(E_1 \cup \cdots \cup E_k) = P(E_1) + \cdots P(E_k). \tag{4}$$

Any event E is disjoint from its complement, but together they fill out the sample space: $\Omega = E \cup E^c$. "Multiplying" through by an event F (see Equation (1) in Section 1.2) yields

$$F = FE \cup FE^c.$$

But FE and FE^c are also disjoint, so we may apply (iii) to obtain

$$P(F) = P(FE) + P(FE^c). \tag{5}$$

Partition of Ω

We extend Equation (5) as follows. A collection of events $E_1, \ldots, E_k$ that are mutually disjoint and fill out Ω: $E_1 \cup \cdots \cup E_k = \Omega$ is a **partition** of Ω. A partition of Ω automatically partitions any event in the sample space: The events FE_i are disjoint because the E_i are disjoint; and by the distributive law,

$$F = FE_1 \cup FE_2 \cup \cdots \cup FE_k.$$

(See Figure 1.5.)

From Equation (4), it then follows that for any partition $\{E_i\}$,

$$P(F) = P(FE_1) + \cdots + P(FE_k). \tag{6}$$

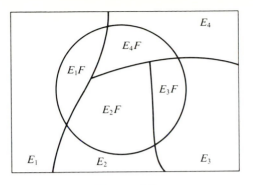

Figure 1.5

Law of Total
Probability

This is called the **law of total probability**. The following example illustrates Equations (6) for $k = 2, 3$.

Example **1.3b**

Survey Sampling

In a political poll, an individual is selected at random from a population of voters. Some events of interest include these:

M: Male, R: Republican,

F: Female, D: Democrat,

I: Independent.

Since $M \cup F = \Omega$, it follows from Equation (5) that $P(R) = P(RM) + P(RF)$. Similarly, the events R, D, and I are mutually disjoint and together account for the whole sample space:

$R \cup D \cup I = \Omega$.

"Multiplying" through by M, we obtain

$M = RM \cup DM \cup IM$.

This says that a male is Republican, Democrat, or Independent. Since RM, DM, and IM are mutually disjoint, we may apply Equation (3):

$P(M) = P(RM) + P(DM) + P(IM)$. ∎

Using Equation (2) to find the probability of an event E requires counting the outcomes in Ω and the outcomes in E. The counting process can often be simplified by breaking down the experiment into a sequence of two or more simpler experiments. This brings us to the topic of the next section.

1.4

Composition of Experiments

Calculating probabilities for experiments with equally likely outcomes involves **counting**. Enumerating all possibilities can be quite tedious, and it is easy to miss some of them. In many important cases, we can break the experiment down into subexperiments that have relatively few outcomes. The following example illustrates how decomposition can be exploited for counting outcomes.

Example 1.4a   **Throwing Two Dice**

Suppose we throw two dice, one green and one red. In many games ("craps" and Monopoly, for example), the payoff or the next move is determined by the total number of points showing on the two dice. The sample space

$$\Omega = \{2, 3, 4, \ldots, 12\}$$

is adequate for this purpose. However, anyone experienced in throwing dice will tell you that these outcomes are not equally likely; a 7, for example, is much easier to get than a 12. There is a more detailed sample space, in which intuition suggests that outcomes are equally likely.

The experiment of throwing the two dice consists of two simpler experiments— tossing the green die and tossing the red die. For this composite experiment, we can take the outcomes to be the ordered pairs (i, j), where i is the number of points on the green die, and j the number on the red one. For each of the six possibilities for i, there are six possibilities for j, making a total of 36 for the pair. Using the simplified notation "$i\ j$" for the pair (i, j), we can list the outcomes as follows:

$$
\begin{array}{cccccc}
1\ 1 & 1\ 2 & 1\ 3 & 1\ 4 & 1\ 5 & 1\ 6 \\
2\ 1 & 2\ 2 & 2\ 3 & 2\ 4 & 2\ 5 & 2\ 6 \\
3\ 1 & 3\ 2 & 3\ 3 & 3\ 4 & 3\ 5 & 3\ 6 \\
4\ 1 & 4\ 2 & 4\ 3 & 4\ 4 & 4\ 5 & 4\ 6 \\
5\ 1 & 5\ 2 & 5\ 3 & 5\ 4 & 5\ 5 & 5\ 6 \\
6\ 1 & 6\ 2 & 6\ 3 & 6\ 4 & 6\ 5 & 6\ 6 \\
\end{array}
$$

With pairs listed in this rectangular array, it is clear that there are 6×6 or 36 possible pairs.

Suppose the 36 pairs are equally likely. Then, since there are six pairs with sum 7, $P(7) = \frac{6}{36}$. There is only one pair with sum 12: $P(12) = \frac{1}{36}$.   ∎

The multiplication idea illustrated in this example clearly extends, by induction, to any finite number of subexperiments:

Multiplication Principle:

If a composite experiment $\mathscr{E}$ consists of experiments $\mathscr{E}_1, \mathscr{E}_2, \ldots, \mathscr{E}_k$, where the sample space of $\mathscr{E}_i$ contains n_i outcomes, the number of outcome sequences in $\mathscr{E}$ is

$$n_1 n_2 \cdots n_k.$$

Example **1.4b**   **Three Coin Tosses**
In a sequence of three tosses of a coin, there are two outcomes at each toss, heads (H) and tails (T). The number of outcomes sequences is $2 \cdot 2 \cdot 2 = 8$:

$$HHH, HHT, HTH, THH, TTH, THT, HTT, TTT.$$

The number 8 is evident in a **tree diagram**, in which the choices at each stage are represented as branches. Figure 1.6 shows the tree diagram for the toss of three coins.

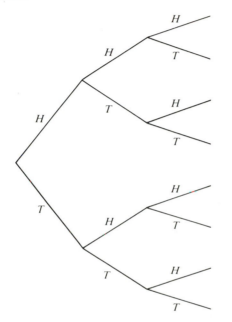

Figure 1.6 Tree diagram for three coins

The same count (8) is correct when three coins are tossed simultaneously, provided the coins are distinguished (for example, a nickel, a penny, and a dime) so that a "sequence" of outcomes is defined.

In Examples 1.4a and 1.4b, the sample space for $\mathscr{E}_i$ does not depend on the outcomes of the other subexperiments. However, the multiplication principle applies provided the *number* of outcomes in $\mathscr{E}_i$ (not necessarily the outcomes themselves) is the same for every outcome of $\mathscr{E}_{i-1}$, for $i = 2, \ldots, k$. In the next example, the sample space for $\mathscr{E}_2$ depends on the outcome of $\mathscr{E}_1$, but the *number* of outcomes in $\mathscr{E}_2$ does not.

Example **1.4c**   **Counting Sequences of Selections**
Consider a bowl with four chips, numbered 1, 2, 3, 4. Suppose we select three chips from a bowl in sequence. The first chip is any of the four chips; no matter which one

was drawn, there are three left for the second selection. After the first two selections, there are two left. The number of possible sequences is therefore $4 \cdot 3 \cdot 2 = 24$. Figure 1.7 shows these 24 sequences in a tree diagram.

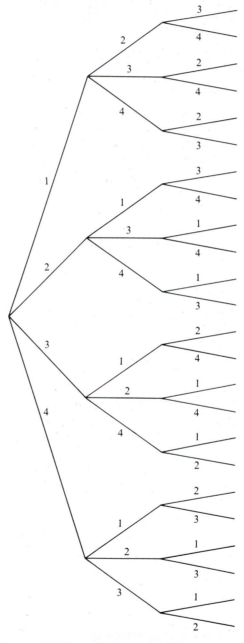

Figure 1.7 Tree diagram for Example 1.4c   ■

> In selecting n objects one at a time from a group of N objects, the number of possible sequences is
>
> $$N(N - 1) \cdots (N - n + 1) = (N)_n$$
>
> This is called the number of **permutations** of N things taken n at a time.

[A notation for $(N)_n$ used on calculators is $_N P_n$.]

In the special case $n = N$, *all* members of the population are included in the sample. Complete sampling is seldom done, but applying Equation (1) in this case

Permutations

gives a count of the number of possible arrangements of a set of distinct objects:

$$(N)_N = N(N - 1)(N - 2) \cdots 3 \cdot 2 \cdot 1 = N!. \tag{2}$$

The $N!$ arrangements are distinguishable provided that the objects being arranged are themselves distinguishable. When some of the objects are identical, the number of arrangements that can be distinguished is less than $N!$, as in the following example.

Example 1.4d

Counting Patterns

A child has six blocks—three red (R), two blue (B), and one white (W). In how many distinct ways can these be arranged in a row?

The total number of permutations of six blocks is 6! or 720. But consider this arrangement: $R\ W\ R\ B\ B\ R$. The two B's and three R's could be rearranged while your back is turned in any of $2! \times 3! = 12$ ways, and you would not know the difference—the pattern would be unchanged. Similarly, there are 12 such rearrangements possible for each pattern. So the number of distinct *patterns* is $\frac{720}{12}$ or 60. ∎

In the relatively simple setting of this last example, the 60 patterns could be listed (and counted) with a little patience. However, in making such a list, it's hard to be sure of including every possibility unless you know the total count. The method used in Example 1.4a illustrates the general scheme:

> The number of distinct sequences of N objects, m_1 alike of type 1, m_2 alike of type 2, . . . , m_k alike of type k (where $\Sigma m_i = N$) is
>
> $$\binom{N}{m_1, m_2, \ldots, m_k} = \frac{N!}{m_1! m_2! \cdots m_k!}. \tag{3}$$

When there are just two types of objects or individuals in the population, we have the important special case of Equation (3), in which $k = 2$. The number of distinct sequences of m objects of one type and $N - m$ of the other is

$$\binom{N}{m, N - m} = \frac{N!}{m!(N - m)!} = \frac{N(N - 1) \cdots (N - m + 1)}{m(m - 1) \cdots 3 \cdot 2 \cdot 1}.$$

Since $m + n = N$, fixing m determines n. So a more convenient (and more common) notation drops one or the other:

$$\binom{N}{m, N - m} = \binom{N}{m} = \binom{N}{N - m} = \frac{N!}{m!(N - m)!}. \tag{4}$$

Example 1.4e   **Arranging Two Types of Objects**

Three blocks have the letter A and two have the letter B. In how many ways can these five blocks be arranged? The numbers are small enough that we can actually make a list of all possible patterns:

  AAABB, AABAB, AABBA, ABAAB, ABABA, ABBAA, BAAAB, BAABA, BABAA, BBAAA.

Using the scheme we just developed, we can obtain the count 10 without having to make a list:

$$\binom{5}{3} = \binom{5}{2} = \frac{5 \cdot 4}{2 \cdot 1} = 10.$$

Viewing this problem in another way is instructive. An arrangement of three A's and two B's can be accomplished by choosing two of the five sequence positions for the B's and depositing the A's in the other three—or choosing three positions for the A's and putting the B's in the other two. So the desired arrangement can be accomplished by selecting positions—a selection that is made without regard to order. There are ten possible selections of three sequence positions for the A's from the five possible positions. Written to correspond to the above list of letter sequences, they are as follows:

  123, 124, 125, 134, 135, 145, 234, 235, 245, 345.

(For example, 134 means that we are to put A's in positions 1, 3, and 4.)   ■

Combinations   A selection *without* regard to order is a **combination**. Following the reasoning of the preceding example, we have a simple way of counting combinations. The number of combinations of n things chosen from N is the same as the number of permutations of N objects, n of which are of one type and the rest of another: $\binom{N}{n}$.

Another way of arriving at this number of combinations is to consider forming a sequence in the following two steps. First, select the objects to be used in the sequence; this we can do in $\binom{N}{n}$ ways. Second, arrange the objects that have been selected; this we can do in $n!$ ways. So

$$\binom{N}{n} \cdot n! = N(N - 1) \cdots (N - n + 1) = (N)_n.$$

Dividing by $n!$, we obtain Equation (4) as the number of combinations.

The number of distinct **combinations** of n objects selected from N is

$$\binom{N}{n} = \frac{N!}{n!(N-n)!} = \frac{(N)_n}{(n)_n} = \frac{N(N-1)\cdots(N-n+1)}{n(n-1)\cdots 3\cdot 2\cdot 1}. \tag{5}$$

[An alternative notation for $\binom{N}{n}$ is $_NC_n$.]

Example 1.4f

Counting Poker Hands

Dealing five cards from a well-shuffled deck of 52 playing cards can be thought of as selecting five cards at random. There are $\binom{52}{5}$ or 2,598,960 distinct hands of five cards. How many of these hands contain two aces, two 10's, and a jack?

We find the answer in a sequence of steps: pick two aces from the four aces in the deck, then pick two 10's from the four 10's, and finally pick one of the four jacks. (We could have picked the 10's first, then the jack, and then the aces with the same result.) The number of ways of constructing this hand is then the product of the numbers of ways of carrying out each step:

$$\binom{4}{2}\binom{4}{2}\binom{4}{1} = 6\cdot 6\cdot 4 = 144.$$

A poker hand like the above is said to have "two pair"—two cards in each of two denominations plus one card in a third denomination. To count *all* the two-pair hands, we see how many ways the three denominations can be picked and then multiply by 144 (which applies to any particular choice of three denominations). The two denominations from which the pairs are selected can be picked from the 13 denominations in $\binom{13}{2} = 78$ ways and the denomination for the fifth card in 11 ways, so there are $78\cdot 11 = 858$ choices for the denominations. Then

$$P(\text{two pair}) = \frac{78\cdot 6\cdot 6\cdot 11\cdot 4}{2{,}598{,}960} = \frac{858\cdot 144}{2{,}598{,}960} = .0475. \qquad \blacksquare$$

Partitioning a Set

The quantity $\binom{N}{n}$, the number of combinations of n things, has another interpretation. We've seen that it is exactly the same as $\binom{N}{N-n}$, the number of ways of selecting n or the number of ways of leaving $N-n$ behind (unselected). That is, it counts the number of ways of dividing the N objects into two distinct piles, with n in one pile and $N-n$ in the other. Similarly, the quantity

$$\binom{N}{n_1, n_2, \ldots, n_k} = \frac{N!}{n_1! n_2! \cdots n_k!} \tag{6}$$

is the number of ways of partitioning N objects into k distinct piles, with n_1 in one, n_2 in another, and so on.

Example 1.4g

Partitioning into Three Groups

We want to divide a group of twelve bridge players into three tables, with four at each table. The number of ways of dividing the twelve into three *distinguishable* groups is

$$\binom{12}{4,\,4,\,4} = \frac{12!}{4!4!4!} = 34,650.$$

The same result can be obtained in another way. Identify the tables as Table 1, Table 2, and Table 3. First choose four players from the twelve for Table 1 (495 ways), then four from the remaining eight for Table 2 (70 ways), and then four from the remaining four for Table 3 (1 way). The number of ways of partitioning in this manner is: $495 \cdot 70 \cdot 1 = 34,650$. (See Problem 32 at the end of this chapter.)

Since the groups counted in this way are identified by table number, the count is appropriate only when the bridge tables are distinguishable. If we are concerned only with which players are together at the same table, the count we've made is too high. We'd need to divide it by the number of possible arrangements of the three tables, obtaining $\frac{34,650}{3!} = 5775$ as the number of distinct groupings without regard to ordering of the tables.   ■

1.5   **Sampling at Random**

An important type of composite experiment is sampling one at a time from a population of N individuals. This can be done *with* or *without* replacement. In sampling with replacement, we replace each individual drawn and thoroughly mix the population before the next selection. In sampling without replacement, we set aside the one drawn at each stage and sample at random from those that remain.

Suppose we sample at random, without replacement, until we obtain a sample sequence of length n. The multiplication principle of Section 1.4 says there are $(N)_n$ possible sample sequences. Intuition suggests that if sampling is at random at each step, these sample sequences are equally likely. In the next example, we make this assumption and examine some of its consequences.

Example 1.5a   **Random Selection**
Consider again the bowl of six chips numbered from 1 to 6, as in Example 1.4c, and a sequence of three selections at random, without replacement. There are $6 \cdot 5 \cdot 4 = 120$ possible sample sequences. Assuming these to be equally likely, we can find the probability of any event by counting. For instance, the number of sequences with 2 in the first position is the number of ways of filling the second spot (5) times the number of ways of then filling the third spot (4), or 20. Thus,

$$P(\text{1st chip is 2}) = \frac{20}{120} = \frac{1}{6}.$$

Similar calculations for the other chips show that each possible outcome of the first selection has probability $\frac{1}{6}$—they are equally likely. This agrees with the assumption that the first selection is at random.

Another consequence of the assumption of equally likely sequences is not so obvious: The *second* chip drawn is also equally likely to be any of the six! This is because the number of sequences with 4 in the second spot is the number of ways of filling the remaining two positions, or $5 \cdot 4$, so the probability is again $\frac{20}{120} = \frac{1}{6}$.

This may seem wrong at first glance. To see intuitively why it is correct, realize that in finding the probability that the second chip is 4, you don't know which chip is removed at the first selection. You have to regard all six chips as possible outcomes of the second selection, and there is no reason to prefer one over the other. If you are still not convinced, imagine that you select the first chip with your left hand but don't look at it and then select a second chip with your right hand from those that remain. Is one hand more likely to contain chip 2 than the other? Look at it this way: If you open your right hand first, the chip in the right hand becomes the first chip. (See "exchangeability" in Section 2.6.) ∎

If experiment $\mathscr{E}$ consists of $\mathscr{E}_1$ followed by $\mathscr{E}_2, \ldots,$ followed by $\mathscr{E}_k$, where in each $\mathscr{E}_i$ the n_i possible outcomes are equally likely, then the $n_1 n_2 \cdots n_k$ possible sample sequences of $\mathscr{E}$ are equally likely.

In terms of tree diagrams, with successive nodes (branch points) corresponding to successive experiments, there are the same number of branchings at each of the nodes for a given experiment and equal probabilities for each branch at a node, and the $n_1 \cdots n_k$ paths through the tree are equally likely.

When permutations of N things n at a time are equally likely, it follows that the possible combinations (in which we ignore order) are equally likely.

Example 1.5b   **Sampling Without Replacement**

Again consider six chips in a bowl, and a selection of three, one at a time at random, without replacement. There are 120 equally likely possible sequences, as in the preceding example. Suppose we ignore order and count combinations of three chips; there are 20 of these ("6 choose 3"). Each combination is made up of six sequences. These sequences are equally likely, so the 20 combinations are equally likely.

If the chips are four red and two white, we can find the probability that the selection of three chips includes two red ones and a white. This event does not involve ordering, so we can use the sample space of 20 equally likely combinations. Of these, the number of combinations with two reds and a white is 12:

$$P(2 \text{ red}, 1 \text{ white}) = \frac{\binom{4}{2}\binom{3}{1}}{\binom{6}{3}} = \frac{12}{20} = \frac{3}{5}.$$ ∎

In selecting n objects at random from N, one at a time without replacement, the $\binom{N}{n}$ possible combinations are equally likely.

Next consider sampling *n* objects *with* replacement (and mixing) from a population of *N* objects. The number of possible sample sequences is N^n, and these are equally likely.

Example 1.5c   **Random Integers**

Political pollsters and others survey samples taken from populations of people. One way of obtaining such a sample would be to write each individual's name on a lottery ticket and select tickets at random. Such a scheme is ordinarily infeasible. A more workable method of sampling at random uses *random digits* and assumes the availability of a numbered list of population members. Members are selected from a population of *N* individuals by selecting random integers from 1 to *N*. Random integers are formed as sequences of random digits.

A **random digit** is a digit selected at random from the ten digits: 0, 1, 2, . . . , 9. A **random-digit sequence** is a sequence of digits selected at random without replacement from the ten digits. Such sequences can be generated by sampling at random with replacement from a lottery with tickets numbered from 0 to 9, but there are computer programs that produce sequences of digits that behave, for most practical purposes, like random-digit sequences. (Some programs do a better job of this than others.) Tables of random digits produced by computer programs have been published. Table XIV in Appendix 1 of this book is taken from such a publication. A sequence of random digits is obtained by reading down in a column, or across in a row, or in any other systematic fashion (taking every third digit, or reading diagonally, or whatever).

Random-digit sequences can be used to select at random from a set of integers. For example, to select at random from the integers from 0 to 99, first select a digit at random for the tens place and then a second digit at random for the units place. In this way, we select one of the integers 00, 01, . . . , 99. These 100 possible ordered pairs are equally likely. To obtain a sequence of two-digit integers, we simply group successive random digits in pairs. For instance, the digits in the first row of Table XIV are

  4 2 9 1 6   5 0 1 9 9   2 6 4 3 5   9 7 1 1 7.

The successive pairs of digits in this sequence define a sequence of ten random integers on the range from 00 to 99:

  42, 91, 65, 01, 99, 26, 43, 59, 71, 17.

Similarly, a random integer on the range 000 to 999 is obtained by successive digits in Table XIV in groups of three.

Sequences of random integers represent sampling *with* replacement. However, in taking a sample survey, there is not much point in interviewing the same person twice. This can happen in sampling with replacement, so survey sampling is typically done without replacement. Sampling without replacement can be achieved by simply ignoring an integer that has already been drawn. (This can be very inefficient if the sample size is a substantial fraction of the population size. There are more efficient ways of using random digits than ignoring repetitions, but we won't pursue this issue here.) ■

1.6   Binomial and Multinomial Coefficients

The numbers $\binom{n}{m}$ are called **binomial coefficients**, because they are coefficients in binomial expansions:

Binomial Theorem:

$$(x + y)^n = \sum_{m=0}^{n} \binom{n}{m} x^m y^{n-m}. \tag{1}$$

This theorem from algebra can be proved by mathematical induction on n. That the coefficients are numbers of combinations is not just a coincidence. The following reasoning shows why and provides another way to prove the theorem.

Consider $n = 5$. Every term in the expansion of $(x + y)^5$ product is a product of five factors, each being either the x or the y from the corresponding $(x + y)$. There are 32 such products in the expansion, because there are two choices (use either the x or the y) for each of the five $(x + y)$'s. For example, one product is

$$x \cdot x \cdot y \cdot x \cdot y = x^3 y^2.$$

However, there are several other products equal to $x^3 y^2$, and these "like terms" are combined in Equation (1). There are as many of them as there are ways of choosing three of the five $(x + y)$'s to supply the x, the rest of them supplying the y. So there are $\binom{5}{3} = 10$ terms of the form $x^3 y^2$. This same reasoning applies in general to yield Equation (1).

We can also obtain expansions of powers of a *multinomial*, in terms of "multinomial coefficients." (See Equation A(6) of Section 1.4.) Reasoning as in the binomial case, we obtain the multinomial theorem:

$$(x_1 + x_2 + \cdots + x_k)^n = \sum \binom{n}{m_1, \ldots, m_k} x_1^{m_1} x_2^{m_2} \cdots x_k^{m_k}, \tag{2}$$

where the sum is taken over all possible combinations of the nonnegative integers $m_1, m_2, \ldots, m_k$ that add up to n.

Substituting the value 1 for each x_j in Equation (2) gives

$$\sum \binom{n}{m_1, \ldots, m_k} = (1 + 1 + \cdots + 1)^n = k^n. \tag{3}$$

This says that the total number of partitions of n objects into k distinct groups (of any sizes) is k^n. Alternatively, this result can be reasoned along the lines of the next example, which illustrates Equation (3) when $k = 2$.

Example 1.6a   **Choosing Committees**

How many ways are there of forming a committee using individuals from a group of five when the size of the committee is not specified? Each committee represents a partition of the group into those on the committee and those not on it. The total number possible is the number with no members (the best kind!) plus the number with one member plus the number with two members, and so on:

$$\binom{5}{0} + \binom{5}{1} + \binom{5}{2} + \binom{5}{3} + \binom{5}{4} + \binom{5}{5} = 32.$$

We can also count the number of possible committees by deciding in turn whether or not to assign each individual member to the committee: $2^5 = 32$. (The tree diagram has five branching stages, with two choices at each branch point.) ∎

1.7   Discrete Probability Distributions

We have defined probability for events in a sample space with equally likely outcomes. In many situations, the outcomes in the most natural or convenient sample space are not equally likely.

Example 1.7a   Outcomes with Unequal Probabilities

Suppose a thumbtack is dropped onto a flat surface. It lands either with point up (U) or with point resting on the surface (D). Not having the symmetry of a coin, there is nothing to suggest that U and D are equally likely—or what the odds might be if they are not equal.

In the lottery model for a coin toss, we use equal numbers of tickets marked H and marked T. In the lottery model for the toss of a thumbtack, the appropriate proportions of U-tickets and D-tickets in the lottery are unknown. There is no reason to expect that they are equal. Using experience as the basis for estimating the appropriate proportions is a typical problem in statistical inference, one to be addressed in Chapter 9.

The *probability* of the outcome U is

$$P(U) = \frac{\text{Number of } U\text{-tickets}}{\text{Total number of tickets}}.$$

Also,

$$P(D) = \frac{\text{Number of } D\text{-tickets}}{\text{Total number of tickets}} = 1 - P(U). \qquad ∎$$

Example 1.7b   Unequal Odds in Horse Races

A horse race is an experiment of chance—the outcome is uncertain. Someone who knows nothing about horse racing may regard the horses as being equally likely to win, but a well-educated fan knows that some horses are faster than others.

Consider a race involving just five horses: A, B, C, D, and E. You feel that A is twice as likely to win as B, five times as likely as C and as D, and ten times as likely as E. For you, an appropriate lottery model contains tickets for horses A through E in numbers proportional to $10:5:2:2:1$, respectively. The corresponding probabilities are then these numbers divided by their total, 20. They are given in Table 1.1, listed together with your odds against a horse's winning. The table shows "odds against"

Table 1.1

Horse	Probability	Odds against
A	.50	1–1
B	.25	3–1
C	.10	9–1
D	.10	9–1
E	.05	19–1
	1.00	

Table 1.2

Horse	Track odds against	Payoff (including $2 bet)
A	33–50	$3.32
B	58–25	6.64
C	73–10	16.60
D	73–10	16.60
E	78–5	33.20

because the odds column at a racetrack is given in these terms. However, the "odds" posted at a racetrack are based on track payoff rather than on anyone's probabilities. For example, at a track you might see the "to win" table shown in Table 1.2 (which ignores "place" and "show"). You'd see this table if half of the money bet was placed on A, one-quarter on B, 10% on each of C and D, and 5% on E.

You can calculate the track's take from the "probabilities" corresponding to the "odds against" table. In Table 1.2, these probabilities total

$$\frac{50}{83} + \frac{25}{83} + \frac{10}{83} + \frac{10}{83} + \frac{5}{83} = \frac{100}{83}.$$

The track's take is 1 minus the inverse of this sum:

$$1 - \frac{83}{100} = .17.$$

The track keeps exactly 17% of everything bet—a typical "track take," no matter which horse wins. So winning "in the long run" is possible only for (at most) the very few bettors who are racing experts.   ■

Models with equally likely outcomes require finite sample spaces. More generally, discrete sample spaces can be countably infinite, as in the next example. If an event contains infinitely many outcomes, the "sum" of their probabilities is an infinite sum; such sums (as we learn in calculus) are defined as the limits of finite sums as more and more terms are added. The sum of the probabilities of all the outcomes must, of course, converge to 1.

Example 1.7c   **Waiting for "Heads"**

A fair coin is tossed until heads turns up. This can happen on the first toss (in which case the experiment stops there), but it may not happen until later. The sample space is

$$\Omega = \{H, TH, TTH, TTTH, TTTTH, TTTTTH, \ldots\}.$$

What probabilities should be assigned to these outcomes?

For an ideal toss, the probability of $\{H\}$ is $\frac{1}{2}$. In the model for tossing three coins (or one coin three times) given in Example 1.4d, we found that $P(\{TH\}) = \frac{1}{4}$ and $P(\{TTH\}) = \frac{1}{8}$. Continuing this pattern, we take the probabilities in the present sample space to be

$$\frac{1}{2}, \frac{1}{4}, \frac{1}{8}, \frac{1}{16}, \frac{1}{32}, \frac{1}{64}, \ldots,$$

respectively. These are terms in a geometric sequence with common ratio $\frac{1}{2}$; they sum to 1.

With probabilities of outcomes thus defined, we can find the probability of any event as the sum of the probabilities of its outcomes. For instance,

$$P(\text{at least 3 tosses}) = P(TTH, TTTH, TTTTH, \ldots)$$
$$= \frac{1}{8} + \frac{1}{16} + \frac{1}{32} + \cdots = \frac{1}{4}.$$

We can also calculate this by observing that requiring at least three tosses to get heads is the same as having the first two tosses result in tails:

$$P(\text{at least 3 tosses}) = P(\text{sequence starts } TT) = \frac{1}{4}. \qquad \blacksquare$$

A discrete probability model consists of

(a)   A sample space Ω with at most a countable number of outcomes ω, and

(b)   A nonnegative number $P(\omega)$ assigned to each ω as its probability so that $\Sigma P(\omega) = 1$.

The probability of an event E is defined as the *sum* of the probabilities of the outcomes in E:

$$P(E) = \sum_{\omega \text{ in } E} P(\omega). \qquad (1)$$

In Section 1.3, we derived a number of properties of probability for experiments that have equally likely outcomes [Properties (i)–(iii) and Equations (4)–(6) and (8)]. These properties continue to hold in the general discrete model as a consequence of Equation (1). For example, the additivity property [(iii) of Section 1.3] still holds: When $EF = \emptyset$, the event $E \cup F$ consists precisely of those outcomes that are in E and those that are in F; its probability is then the sum of the probabilities of the outcomes in E and the probabilities of the outcomes in F.

Fundamental Properties of Probability:

For any events E, F in Ω,

   (i) $P(E) \geq 0$.
 (ii) $P(\Omega) = 1$.
(iii) $P(E \cup F) = P(E) + P(F)$,   if $EF = \emptyset$.

Various other useful relations flow logically from Properties (i)–(iii):

Some further properties of probability:

  (iv) $P(E_1 \cup \cdots \cup E_k) = \sum_1^k P(E_i)$
        whenever $E_i E_j = \emptyset$ for $i \neq j$.
   (v) $P(E^c) = 1 - P(E)$.
  (vi) $P(\emptyset) = 0$.
 (vii)   If $E \subset F$, then $P(E) \leq P(F)$.
(viii) $P(E) \leq 1$.
  (ix) $P(E \cup F) = P(E) + P(F) - P(EF)$.

Property (iii) implies (iv) by induction [and is a special case of (iv)]. Property (vi) follows from (v) and (ii), with $E = \Omega$. Property (vii) follows from applying additivity (iii) to the relation $F = E \cup E^c F$, which holds when $E \subset F$, since $P(FE^c)$ is nonnegative. Property (viii) is a special case of (vii) with $F = \Omega$. Proof of Property (ix) is left as an exercise (Solved Problem at the end of the chapter).

    A property that deserves special mention is the law of total probability, Equation (6) of Section 1.3. The proof given in Section 1.3 (page 14) is general, being based on Properties (i)–(iii), which hold in general.

Law of Total Probability:

If $E_1, E_2, \ldots, E_k$ constitute a partition of the sample space, then for any event F,

$$P(F) = P(FE_1) + \cdots + P(FE_k). \tag{2}$$

Chapter Perspective

Random sampling is basic to statistical methodology. This chapter introduced the probability theory we require for developing and analyzing that methodology. In sampling *at random*, we choose each sample member in such a way that all available individuals are equally likely. The notion of *equal likelihood* of outcomes is fundamental in sampling.

When sample space outcomes are equally likely, we calculate probabilities of events by *counting*. We found the multiplication principle useful in counting. You may be frustrated when you find combinatorial arguments difficult to follow or to imitate. Don't give up. In what follows, you will need some experience in simple counting problems to deal effectively with some of the more interesting and important methods, but the important ideas of statistical theory and practice are not combinatorial.

The basic properties of probability given in Section 1.7 are quite general; they will also apply to the continuous case (Chapters 5 and 6). Before taking up the continuous case, we develop the discrete case further in Chapters 2–4.

Solved Problems

Sections 1.1–1.2

A. Give an appropriate sample space for these experiments:
  (i) Tossing a penny and a die together.
  (ii) Selecting a family at random and asking whether they are watching TV and, if so, which channel.
  (iii) Spinning a pointer that has a continuous scale showing the angle (measured from a reference position) at which it stops.

  Solution:
  (i) The most detailed sample space consists of all pairs of an outcome for the penny and an outcome for the die:

  $$\Omega = \{H1, H2, H3, H4, H5, H6, T1, T2, T3, T4, T5, T6\}.$$

  (ii) This depends on the city, and whether the viewer has cable TV. In Minneapolis, ignoring cable:

  $$\Omega = \{\text{None}, 2, 4, 5, 9, 11, 17, 23, 29, 41\}.$$

  (iii) If we measure the angle in degrees, $\Omega = [0, 360)$; if in radians, $\Omega = [0, 2\pi)$. (These assume infinite accuracy. If we round to the nearest integer (for degrees), then $\Omega = \{0, 1, 2, \ldots, 359\}$.)

B. In each of the following cases, reproduce the Venn diagram of Figure 1.8 (where $E = $ set of points in the circle, $F = $ set of points in the triangle) and shade the event:
  (i) $E^c F$ **(ii)** $E \cup F^c$ **(iii)** $E \cup E^c F$ **(iv)** $EF \cup E^c F^c$.

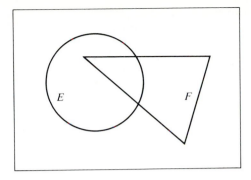

Figure 1.8

Solution: (see Figure 1.9.)

(i) E^cF consists of the points that are in F but not in E.

(ii) $E \cup F^c$ consists of all points except those not in E and in F—that is, those in E^cF. [See (i).]

(iii) $E \cup E^cF$ consists of points that are either in E or in its complement and in F; this describes $E \cup F$.

(iv)   The points in E^cF^c are not in E and not in F; they are outside both the circle and triangle. EF is the intersection of circle and triangle, so the union we want consists of both shaded regions.

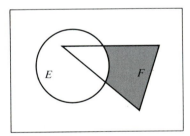

(i)

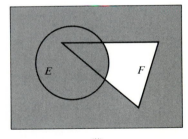

(ii)

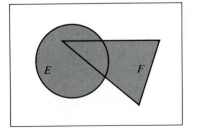

(iii)

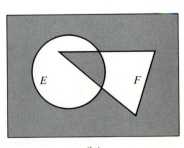

(iv)

Figure 1.9

C. List all possible events in the sample space $\Omega = \{a, b, c\}$, and then, with $E = \{a, b\}$, $F = \{a, c\}$, and $G = \{b, c\}$, list the outcomes in each of the following:

(i) EFG^c (ii) EFG (iii) $E \cup FG$ (vi) $EF^c \cup EF$

Solution:

For each of a, b, c, we decide whether to include that point or not, so there are $8 = 2^3$ ways to form a subset. The subsets are $\emptyset, \{a\}, \{b\}, \{c\}, \{a, b\}, \{b, c\}, \{a, c\}, \{a, b, c\} = \Omega$.

(i) $EF = \{a\}$ and $G^c = \{a\}$, so $EFG^c = \{a\}$.
(ii) No point is in all three, so $EFG = \emptyset$.
(iii) $FG = \{c\}$, so $E \cup FG = \{a, b, c\} = \Omega$.
(iv) $EF^c \cup EF = E(F^c \cup F) = E\Omega = E = \{a, b\}$.
      [Or, $EF^c = \{b\}$, $EF = \{a\}$, and their union is $\{a, b\}$.]

D. Show Equation (1) of Section 1.2: $A(B \cup C) = AB \cup AC$. Show also that set intersection is distributive over finite unions generally.

Solution:

First suppose an outcome ω is in $A(B \cup C)$; then it is both in A and in the union of B and C. Since it is in A and in either B or C, it is in either AB or AC. Hence, it is in the right-hand side (r.h.s) of Equation (1). Conversely, suppose ω is in the r.h.s. of Equation (1); then it is in either AB or AC (or in both). In either case, it must be in A. But if it is in AB, it must be in B, and if it is in AC, it is in C. So ω is in $B \cup C$ and in A. Therefore, it is in the l.h.s. of Equation (1). Since each side is a subset of the other, the equality is established.

To show that $A(B_1 \cup \cdots \cup B_n) = AB_1 \cup \cdots \cup AB_n$, we would use induction. The crucial step is to write

$$A(B_1 \cup \cdots \cup B_n) = A[(B_1 \cup \cdots \cup B_{n-1}) \cup B_n]$$

and apply the distributive law for $n - 1$ and for two terms.

Section 1.3

E. In a draft lottery, each of the 366 dates in a year is marked on a capsule. When a capsule is selected at random from the 366, what is the probability that

(a) it is a December date?
(b) it is not a date in April, May, or June?

Solution:

(a) The 366 dates are equally likely; since there are 31 days in the month of December, the probability is $\frac{31}{366}$.
(b) There are $30 + 31 + 30$ days in April, May, and June, so $366 - 91$ or 275 days are not in these months: $\frac{275}{366}$.

F. When two dice are tossed, we can describe the outcomes as ij, to indicate that the first die shows i and the second, j. The complete list of outcomes was shown in Example 1.4a. Assuming that the 36 outcome pairs are equally likely, find the probability that

(i)   The sum $(i + j)$ is 6.
(ii) The second die shows 5.
(iii) They differ by at least 3.
(iv) The larger of the two numbers is 4.

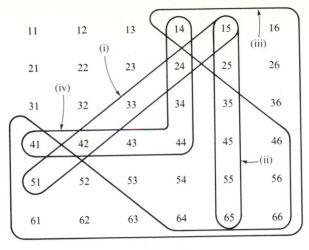

Figure 1.10

Solution: (see Figure 1.10)

(i)   There are five pairs with sum 6 (51, 42, 33, 24, 15), so $\frac{5}{36}$.

(ii) There are six pairs with the second die showing 5, so $\frac{6}{36}$. [Although this is what you would have thought the answer should be, the stated problem defines the model in the sample space of pairs, and P (second die shows 5) $= \frac{1}{6}$ is a consequence.]

(iii) There are 12 pairs with this property, so the probability is $\frac{12}{36}$ or $\frac{1}{3}$.

(iv) There are seven pairs like this, so the probability is $\frac{7}{36}$.

Section 1.4

G. In how many ways can ten distinguishable coins fall?

Solution:

Each coin can fall in two ways. The number of outcomes for ten coins is then $2 \times 2 \times \cdots \times 2 = 2^{10} = 1024$. (This uses the multiplication principle of Section 1.4.)

H. Given a menu with three appetizers, two soups, four salads, six entrees, and five desserts, how many different meals (consisting of appetizer, soup, salad, entree, and dessert) are possible?

Solution:

The multiplication principle gives the number of possibilities as the product of the numbers of choices at the various stages: $3 \cdot 2 \cdot 4 \cdot 6 \cdot 5 = 720$.

I. Count the possible permutations of the letters in the word

(i)   *banana.*   (ii)   *statistics.*

Solution:

(i)   There are 3 *a*'s, 2 *n*'s, and 1 *b*—six letters in all. The number of distinct patterns is

$$\frac{6!}{3!2!1!} = 60.$$

(ii) (similarly) $\dfrac{10!}{3!3!2!1!1!} = 50,400.$

J. How many connecting cables are needed to link each pair of nine offices directly?

Solution:

One cable is needed for each pair of offices, and there are as many pairs as the number of ways of choosing two from nine things: $\binom{9}{2} = 36.$

K. In the Illinois lottery, players must try to match six numbers drawn from the numbers 1 to 44 without replacement. At one point, the anticipated payoff was $32 million. News reports asserted that there are about 7 million possible combinations. Verify this.

Solution:

The six numbers selected from 44 form a *combination*;

$$\binom{44}{6} = \frac{44 \cdot 43 \cdot 42 \cdot 41 \cdot 40 \cdot 39}{6 \cdot 5 \cdot 4 \cdot 3 \cdot 2 \cdot 1} = 7,059,052.$$

L. A baby sitter has three raisin, one chocolate chip, and two peanut butter cookies to give out (one each) to six children. In how many distinct ways can he give them out?

Solution:

Imagine the children coming to him in any particular sequence. Whatever it is, he can hand out the cookies in order, after arranging them in a row. There are 60 ways to arrange them:

$$\frac{6!}{3!2!1!} = 60.$$

Section 1.5

M. In the draft lottery of Problem E, with selections made one at a time and at random, find the probability that the first three selections include exactly two December dates.

Solution:

The number of possible sequences is $366 \cdot 365 \cdot 364$. The two dates in December may be selected first $(31 \cdot 30 \cdot 335)$, or last (same product) or first and last (same product). The probability is then $3(31 \cdot 30 \cdot 335)$ divided by $366 \cdot 365 \cdot 364$, or about .019.

N. Four men hang up their coats at a party and later pick them up randomly. What is the probability that no one gets the right coat?

Solution:

There are four ways the first man can select a coat; after this, three ways the second can select; and so on. A total of $4 \cdot 3 \cdot 2 \cdot 1 = 24$ ways—equally likely, since the selections are random. We can make a list of these and observe, for each, whether or not anyone got the right coat. Suppose the men select in the order A, B, C, D. The coat orders are as follows:

ABCD	ADCB	BACD	CABD	*CDAB	*DCAB
ABDC	ADBC	*BADC	*CADB	*CDBA	*DCBA
ACDB	BCAD	*BDAC	CBAD	*DABC	DBAC
ACBD	*BCDA	BDCA	CBDA	DACB	DBCA

Those marked * have no match of man to coat; there are nine *'s, so the desired probability is $\frac{9}{24} = \frac{3}{8}$. (Note that the systematic listing makes clear why the multiplication principle works.) [Is there a "formula" that would work with n men and coats? The answer is yes—it's provided by an extension of Property (ix) in Section 1.7 to the union of n events (see Problem 54). For,

$$P(\text{no match}) = 1 - P(\text{at least one match})$$

$$= 1 - P(A \text{ gets right coat or } B \text{ does or } C \text{ does or } \dots).$$

The events in the union on the right are not disjoint, so the more complicated formula for the probability of a union is required. To apply it, you need such things as

$$P(A \text{ does and } C \text{ does}) = (n - 2)!/n!.$$

The answer for $n \geq 2$ is

$$\frac{1}{2} - \frac{1}{6} + \frac{1}{24} - \cdots + (-1)^n/n!$$

which converges to $1/e \doteq .368$. Even for as few as four men, the answer $(.375)$ is close to this asymptotic result.]

O. In a sequence of four random digits, what is the probability that the digits are all distinct?

Solution:

To construct a sequence with distinct digits, we choose the first in one of 10, the second in one of 9, the third in one of 8, and the fourth in one of 7 ways: $10 \cdot 9 \cdot 8 \cdot 7$ ways in all. Without the restriction "distinct," there are $10 \cdot 10 \cdot 10 \cdot 10$ ways. Hence,

$$P(\text{all distinct}) = \frac{\#(\text{sequence of distinct digits})}{\#(\text{sequences})} = \frac{9 \cdot 8 \cdot 7}{10 \cdot 10 \cdot 10}$$

or about .504.

P. A committee of four is to be selected at random from a group consisting of ten labor and five management representatives. What is the probability that the committee selected includes

   (i)   Two representatives from labor and two from management?

   (ii)   At least one representative from each group?

   (iii)   The chair of the labor delegation and the chair of the management delegation?

Solution:

There are $\binom{15}{4}$ or 1365 committees possible in all; because they're selected "at random," these are equally likely. So for any given type of committee, count the number of that type and divide by 1365:

   (i)   To form such a committee, we choose two from the ten labor *and* two from the five management people. The number of such selections is $\binom{10}{2} \cdot \binom{5}{2} = 45 \cdot 10 = 450$.

(Note that "and" translates into "times"—think of tree diagrams.) So

$$P(2 \text{ labor, 2 manag.}) = \frac{450}{1365} \doteq .330.$$

(ii)   The complement of the given condition is that they are either all from labor: $\binom{10}{4} = 210$ such committees, *or* all from management: $\binom{5}{4} = 5$ such committees— 215 in all. (The "or" translates as "plus".) So

$$P(\text{at least one of each}) = 1 - P(\text{all labor or all manag.})$$

$$= 1 - \frac{215}{1365} = \frac{1150}{1365} \doteq .842.$$

(iii)   If the two chairs are on the committee, we have only two more to select (from the remaining 13 delegates) to fill out the committee:

$$P(\text{chairs included}) = \binom{13}{2} \bigg/ \binom{15}{4} \doteq .057.$$

Q.   Find the probability that a five-card poker hand has exactly one pair (and three other cards, so that there are four distinct denominations in all).

Solution:

As in Example 1.4f, there are 2,598,960 distinct five-card hands. Assuming a random deal these are equally likely. To form a hand containing exactly one pair we first pick a denomination for the pair: 13 ways. Next we pick two of the four cards from that denomination: 6 ways. We need three more cards, and these must be of three distinct denominations, different from that of the pair:

$$\binom{12}{3} = 220 \text{ ways.}$$

Then we pick a card from each of the three selected denominations: $4 \cdot 4 \cdot 4 = 64$ ways. The number of complete hands with just one pair is then $13 \cdot 6 \cdot 220 \cdot 64 = 1,098,240$, and the probability of such a hand is this number divided by 2,598,960, or about .423. Although there are many ways of reasoning to get a wrong answer, students often come up with this count of the three denominations:

$$\binom{12}{1}\binom{11}{1}\binom{10}{1} = 1320 = 6 \cdot 220.$$

This gives six times the correct count, because it orders the three cards of the nonpaired denominations. Thus, for example, the selections (2, 5, 7) and (5, 2, 7) as the denominations of the odd cards should not be regarded as different. Order in a poker hand doesn't matter, and the denominator 2,598,960 is the number selections of five *without* regard to order.

R.   A group of nine individuals is to be divided at random into three committees of three each.

  (i)   Find the number of distinct ways of doing this, assuming that the three committees are indistinguishable. (All that matters is who is with whom, not the label for the group.)

(ii) Two of the nine are best friends. Find the probability that they end up on the same committee.

Solution:

(i) The number of divisions into distinct groups:

$$\frac{9!}{3!3!3!} = 1680.$$

However, if the three committees are indistinguishable, this is too high by a factor of 3!, the number of permutations of the committees. The desired number is $\frac{1680}{6} = 280$.

(ii) Using the sample space of distinct divisions, 1680 in number, we count divisions with the two friends in committee A. We first fill the one open position in A (7 ways) and then choose three from the remaining six people to be in B (20 ways). The same count would apply if we put the friends in B, or in C. So the number of committees with the two friends on the same committee is $3 \cdot 7 \cdot 20 = 480$. Answer: $\frac{420}{1680} = \frac{1}{4}$.

Alternatively, if we take the denominator to be the number of unordered divisions (280), the numerator is 7 (the number of ways of completing whatever group the two friends are in) times the number of unordered divisions of the remaining six people:

$$\frac{7 \cdot \binom{6}{3}}{2} = 70.$$

Again, the answer is $\frac{1}{4}$. We just have to be careful to use the same kind of outcomes in both numerator and denominator.

There is a much simpler way: Wherever the first friend is assigned, two of the remaining eight slots are in the same group, so the friend has two chances in eight of being assigned to one of them.

Sections 1.6–1.7

S. Write out the expansion of $(x + y + z)^4$.

Solution:

There are 81 terms before collecting like terms. Collecting:

Type of term	Number of that type	Total
x^4, y^4, z^4	1 each	3
$xz^3, x^3z, yz^3, y^3z, xy^3, x^3y$	$\frac{4!}{3!} = 4$ each	24
x^2yz, xy^2z, x^2yz	$\frac{4!}{2!} = 12$ each	36
x^2y^2, x^2z^2, y^2z^2	$\frac{4!}{(2!2!)} = 6$ each	18
	Total:	81

The expansion is then

$$(x + y + z)^4 = x^4 + y^4 + z^4 + 4(xz^3 + x^3z + yz^3 + y^3z + xy^3 + x^3y)$$
$$+ 12(x^2yz + xy^2z + x^2yz) + 6(x^2y^2 + x^2z^2 + y^2z^2).$$

T.　Determine the coefficient of x^8y^6 in the expansion of $(x^2 + y)^{10}$.

Solution:

$x^8y^6 = (x^2)^4y^6$, so the coefficient is $\binom{10}{4} = 210$.

U.　Show: $\begin{pmatrix} n+1 \\ a, b, c \end{pmatrix} = \begin{pmatrix} n \\ a-1, b, c \end{pmatrix} + \begin{pmatrix} n \\ a, b-1, c \end{pmatrix} + \begin{pmatrix} n \\ a, b, c-1 \end{pmatrix}$, where $a + b + c = n + 1$.

Solution:

Write each term on the right in terms of factorials, and take out common factors:

$$\frac{n!}{(r-1)!s!t!} + \frac{n!}{r!(s-1)!t!} + \frac{n!}{r!s!(t-1)!} = \frac{n!}{(r-1)!(s-1)!(t-1)!} \left\{ \frac{1}{st} + \frac{1}{rt} + \frac{1}{rs} \right\}$$
$$= \frac{n!(r+s+t)}{r!s!t!}.$$

Note: This has the following interpretation. Suppose we want to arrange 4 A's, 3 B's, and 2 C's. Take any one of these and start the arrangement with it. If it's an A, there are 8 left, and $\binom{8}{3,3,2}$ ways to finish the arrangement. If a B is chosen as the start, there are $\binom{8}{4,2,2}$ ways, and starting with a C, $\binom{8}{4,3,1}$ ways to finish. Hence

$$\begin{pmatrix} 9 \\ 4, 3, 2 \end{pmatrix} = \begin{pmatrix} 8 \\ 3, 3, 2 \end{pmatrix} + \begin{pmatrix} 8 \\ 4, 2, 2 \end{pmatrix} + \begin{pmatrix} 8 \\ 4, 3, 1 \end{pmatrix}.$$

V.　Suppose the probability of having a tossed thumbtack fall with point up (U) is $P(U) = p$. Consider the following model for the result of tossing four tacks:

Outcome	Probability
DDDD	$(1-p)^4$
UDDD, DUDD, DDUD, or DDDU	$4p(1-p)^3$
UUDD, UDUD, UDDU, DUUD, DUDU, or DDUU	$6p^2(1-p)^2$
DUUU, UDUU, UUDU, or UUUD	$4p^3(1-p)$
UUUU	p^4

(i)　Show that the assigned probabilities sum to 1.

(ii)　Find the probability of at least 2 U's.

Solution:

(i)　We apply the binomial theorem [Equation (1) of Section 1.6] with $x = p$, $y = 1 - p$, and $n = 4$:

$$1 = [p + (1 - p)]^4 = p^4 + 4p^3(1 - p) + 6p^2(1 - p)^2 + 4p(1 - p)^3 + (1 - p)^4.$$

These five terms account for all the given probabilities.

(ii) The sequences involving two or more U's are the last three entries in the probability table. The probability of this event is therefore the sum of the last three terms of the expansion in (i) (or 1 minus the first two terms):

$$p^2(6 - 8p + 3p^2).$$

W. Show Property (ix) of Section 1.7: $P(E \cup F) = P(E) + P(F) - P(EF)$.

Solution:

Since $E \cup F = E \cup E^c F$ and E is disjoint from $E^c F$, additivity of probability over disjoint events implies

$$P(E \cup F) = P(E) + P(E^c F).$$

According to the law of total probability [Equation (2) of Section 1.7],

$$P(F) = P(EF) + P(E^c F).$$

Subtracting the second equation from the first yields the desired result.

Problems

For problems marked with an asterisk, answers are provided in Appendix 2.

Sections 1.1–1.2

*1. Give an appropriate sample space for each of these experiments:
 (a) Tossing a die repeatedly until a six appears.
 (b) Counting the smokers in a class of 30 students.
 (c) Asking the political party preference of a voter selected at random from a population of voters.
 (d) Weighing a student on a scale in the health service.
 (e) Recording the time to failure of a light bulb.

*2. Suppose we toss a nickel and a penny together. Take the sample space to be $\{Hh, Ht, Th, Tt\}$, where H and T refer to the nickel and h and t to the penny.
 (a) List the outcomes in each of these events:

 E: Not both coins show heads.

 F: At least one coin shows heads.

 G: The penny shows heads.

 (b) List the outcomes in each of the following:

 EFG, $\quad E^c \cup F^c$, $\quad (EF)^c$, $\quad E \cup FG$, $\quad EG^c$.

 (c) How many distinct events can be defined on this sample space? (Include $\varnothing$ and Ω in your count.)

3. In the usual sample space for the toss of a six-sided die, consider these events:

 E: The outcome is not 6.

 F: The outcome is an even number.

 G: The outcome is less than 3.

List the outcomes in each of the following:

(a) EG (b) $F \cup G$ (c) EG^c (d) $E^c \cup FG$.

4. In the Venn diagram of Figure 1.11, let E denote the set of points within the circle, F the set within the triangle, and G the set within the rectangle. For each of the following, reproduce the diagram and shade the set of points representing the given event:

(a) EFG (b) FG^c (c) $E \cup FG$ (d) $E^c \cup F^c$ (e) $(E \cup F)(E \cup G)$.

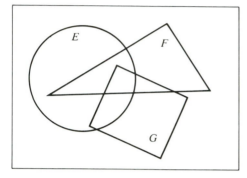

Figure 1.11 Venn diagram for Problem 4

*5. Give each of the following as one of the events E, Ω, or $\varnothing$:

(a) $E \cup \varnothing$ (c) $E \cup \Omega$ (e) EE (g) $E^c \cup E$

(b) $E\varnothing$ (d) $E\Omega$ (f) $E \cup E$ (h) $E^c E$.

6. Show the following:

(a) $EF \subset E \cup F$.

(b) If $E \subset F$, then $EF = E$.

(c) If $E \subset F$ and $F \subset G$, then $E \subset G$.

(d) $(EF)^c = E^c \cup F^c$.

(e) $(E_1 \cup E_2)^c = E_1^c E_2^c$.

(f) $E \cup FG = (E \cup F)(E \cup G)$.

(g) $EF \cup EF^c = E$

(h) $E \cup F = EF \cup EF^c \cup E^c F$

7. In Problem 6, (d) and (e) are special cases of DeMorgan's laws. The intersection of the events in a family $\{E_\alpha\}$ is defined as the set of all outcomes that belong to E_α for every α. The union of the events in a family $\{E_\alpha\}$ is defined as the set of all outcomes each of which is in some one or more of the E_α. Obtain DeMorgan's laws, extending (d) and (e) of Problem 6 to the case of arbitrary collections of events $\{E_\alpha\}$.

Section 1.3

*8. One card is drawn at random from a deck of playing cards. (See Example 1.2a.) Find the probability that the card selected is

(a) not a diamond.

(b) an "honor card" (one of ace, king, queen, jack, 10).

(c) a red honor card.

    (d) red.

    (e) a spade but not an honor card.

    (f) not a black honor card.

9. The 38 equally likely positions on a roulette wheel are numbered 0, 00, 1, 2, . . . , 36. (See Example 1.1d.) The 0 and 00 are green; of the rest, half are black and half red. Determine the probability that the outcome of a spin is

    (a) red.   **(c)** an odd number.

    (b) not black.   **(d)** in the "third dozen" (25, . . . , 36).

∗10. Consider selecting two letters from $\{A, B, C, D, E\}$. The possible pairs are easily listed:

$$AB, \; AC, \; AD, \; AE, \; BC, \; BD, \; BE, \; CD, \; CE, \; DE.$$

Suppose these are equally likely. Find the probability that any particular letter (say, A) is one of those selected.

11. In a population of 100 undergrads, 20 are seniors, 25 are juniors, and 30 are freshmen. Given that 6 of the seniors, 5 of the sophomores, and 3 of the freshmen smoke, and that $\frac{7}{9}$ of the upperclassmen (juniors and seniors) don't smoke, find the probability that a student picked at random is

    (a) a smoker.   **(d)** not a senior and doesn't smoke.

    (b) a sophomore.   **(e)** a freshman or a nonsmoker.

    (c) a junior nonsmoker.   **(f)** a senior or a smoking freshman.

∗12. You are one of ten finalists currently tied for first place in a contest. Three are to be selected at random from the ten to receive the top prizes. What are your chances of receiving a top prize?

13. In an episode of the TV show "All in the Family," Mike claimed that he could identify different brands of cola by taste alone. He was challenged and presented with three glasses, one filled with Coke, one with Pepsi, and one with RC Cola.

    (a) Suppose Mike really was *not* able to discriminate among the three brands of cola so that his matching brands to glasses was completely random. Find the probability that none of his identifications would be correct.

    (b) Find the probability of no matches had there been four brands of cola instead of three.

Section 1.4

∗14. Three roads lead from town A to town B, and four lead from B to C. How many distinct routes can be followed from A to C via B?

15. In how many ways can eight six-sided dice fall? (Imagine them to be of different colors.)

∗16. A sample of 4 individuals is to be selected one at a time from a population of 15 individuals. The names are recorded in the order in which they are drawn. How many distinct sample sequences are possible?

17. In how many ways can five math books and four statistics books be placed on a shelf, with the math books together and the statistics books together?

∗18. A student takes a multiple-choice exam consisting of 20 questions, with four choices per question.

    (a)   How many answer sheets are possible?

    (b)   How many are possible if no two successive responses are the same?

19.   A club with ten members will choose a president, secretary, and treasurer.

    (a)   In how many ways can this be done?

    (b)   One of the ten club members wonders what his chances are of being selected for an office. What are they, if the ways of making the selection are equally likely?

∗20.   In a medical trial, each of ten subjects is assigned to one of two treatments, *A* and *B*, by the toss of a coin.

    (a)   In how many ways can the assignments be made?

    (b)   In how many of these assignments will exactly half of the subjects be given treatment *A* and the other half, *B*?

    (c)   Assuming that the assignments in (a) are equally likely, find the probability that all ten subjects get the same treatment.

21.   Four indistinguishable balls are to be put into three containers placed in a row. How many distinct configurations are possible? (One configuration, for instance, would be to have three balls in container #1, none in #2, and one in #3.)

∗22.   Suppose we have twelve books and a shelf that holds only eight. In how many ways can we select eight of the twelve books *and* arrange them on the shelf?

23.   How many distinct permutations are there of the letters in

    ∗(a)   The word *minimum*?

    (b)   The word *lollipop*?

24.   Find the number of ways three men and four women can be arranged in a row of seven chairs

    ∗(a)   without restriction.

    ∗(b)   if the men sit together.

    ∗(c)   if the men sit on one end and the women on the other.

    (d)   if men and women alternate.

    (e)   if two particular people cannot abide each other and must be separated.

∗25.   If *n* countries in a bloc exchange ambassadors, how many ambassadors are involved?

∗26.   In two-person cribbage, each player is dealt a hand of six cards.

    (a)   How many distinct hands are possible for one player?

    (b)   How many distinct deals—six cards to each player—are possible? (It makes a difference who gets which hand; if the players were to exchange hands in a given deal, their new hands would constitute a different "deal.")

27.   How many committees consisting of two men and two women can be chosen from a group consisting of six men and five women?

∗28.   A delegation of three from the ten members of a city council is to be chosen to attend a convention.

    (a)   In how many ways can a delegation be chosen?

    (b)   In how many ways, if two particular members will not attend together?

    (c)   In how many ways, if two particular members will either both go or neither?

∗29.   In how many ways can one choose a committee from a group of ten people if the size of the committee is not specified?

30. Suppose you are to answer nine out of twelve questions on an exam.
  (a) In how many ways can you choose nine questions to answer?
  (b) In how many ways can you choose if you are required to answer Nos. 1 and 2?
  (c) In how many ways, if you must answer No. 1 or No. 2 but not both?
  (d) In how many ways, if you must answer at least three of the first four questions?

*31. In how many ways can one divide a group of six tennis players into three pairs to enter a doubles tournament?

32. In how many ways can a group of eight individuals be divided into
  (a) two groups, one of three and one of five?
  (b) four teams of two each, each committee with a different task? (Order is important.)
  (c) four teams of two each, all with the same task? (Order is not important.)

33. Show the following:

$$\binom{n}{a,\,b,\,c} = \binom{n}{a}\binom{n-a}{b}, \qquad \text{where } a+b+c = n,$$

and write out a general formula of this type (for dividing n things into groups of m_1, $m_2, \ldots, m_k$).

Section 1.5

*34. A chain of pizza parlors ("Godfather's") once offered chance cards to its customers. Each card had eight numbers arranged in a circle, hidden by large dots that could be scratched away. Exactly three of the numbers were 7's, and a customer who uncovered the 7's by scratching away the dots from three numbers would "win." (The card would be void if more than three numbers were revealed.) Find the probability that a person will uncover the three 7's in scratching out just three of the eight dots.

*35. License plates are issued at random, each having three letters followed by three digits. The initial distribution of plates includes all those on which the first letter is either A or B.
  (a) What is the probability that you get AAA-111? BBQ-279?
  (b) What is the probability that you get a plate on which the three letters are the same and the three digits are the same?

*36. I have ten socks in a drawer; two pair are green, two pair blue, and one pair red. One blue sock has a hole in it.
  (a) Dressing in the dark, I pick two socks at random. What is the probability that they are the same color?
  (b) If I pick three at random, what is the probability that at least two of the three are the same color?
  (c) If I pick four at random, what is the probability that at least two of the four are the same color?
  (d) What is the probability that if I pick four at random, I get the sock with the hole?

37. Four cards are dealt from a standard deck of 52 playing cards, at random (with no replacement). Find the probability that
  (a) they are all from the same suit.

   (b)   they are half red and half black.

   (c)   there is one of each suit.

*38.   A committee of three is picked at random from a group of eight students—two freshmen, two sophomores, two juniors, and two seniors. Find the probability that
   (a)   no freshmen are selected.
   (b)   the committee includes one senior, one junior, and one other.

39.   Four of eight candidates for a council position are members of minority groups. If names are placed on the ballot in a random order, what is the probability that the names of the minority candidates head the list?

*40.   In Toronto, Canada, a family named Kelly bought nine tickets for a dollar each in the Interprovincial Lottery drawing. They won $13,890,588.80. In this lottery, six numbers were drawn at random from numbers 1 to 49 (without replacement). To win, one must pick all six numbers exactly, but order is not important. Calculate the probability that a person with nine different entries would win this lottery.

41.   Explain how you could use Table XIV of the Appendix to
   (a)   call a Bingo game. (Numbers are selected at random and without replacement, from the numbers 1 to 75.)
   (b)   obtain a sequence of random selections from a list with 3,750 names.
   (c)   substitute for the pair of dice in playing any game in which the players throw dice (e.g., Monopoly).

*42.   Find the probability that if we put four distinct balls randomly into ten distinct containers (one at a time), no container gets more than one ball. (Note the connection between this and Solved Problem O.)

43.   Find the probability that no two among ten persons at a party have the same birthday, assuming that all 365^{10} possible sequences of birthdays are equally likely. (*Hint*: Find the count for the numerator as you did in Problem 42, but with ten balls and 365 containers.)

*44.   In comparing a sample of three numbers called X's with a sample of five numbers called Y's, a method we describe in Section 10.5 involves putting all eight numbers in numerical order and looking at the pattern of X's and Y's. Suppose all possible patterns are equally likely.
   (a)   How many patterns of three X's and five Y's are possible?
   (b)   Find the probability that the X's will be together on one end or the other.
   (c)   Find the probability that there will be just one Y to the left of the middle.
   (d)   Find the probability that there will be exactly two Y's to the left of the middle.
   (e)   Find the probability that there will be exactly three Y's to the left of the middle.

*45.   Find the probability that in a five-card poker hand (random selection of five from a standard deck) that
   (a)   three of the cards are of one denomination and the others are of two different denominations ("three of a kind").
   (b)   three are of one denomination; the other two are from a second denomination ("full house").

46.   Eight glasses are filled, four with New Coke and four with Classic Coke. A taster is to select four, the ones he thinks are the Classic Cokes. Suppose the taster really cannot tell

them apart and is only guessing. Find the probability that among the four selected there are at least three Classic Cokes.

Sections 1.6–1.7

*47. Write out the expansion of each of the following:

(a) $(x + y)^5$.

(b) $(x + y + z)^3$.

48. Show that

$$\binom{n}{k} + \binom{n}{k-1} = \binom{n+1}{k}.$$

*49. Use the identity of the preceding problem to obtain the values of $\binom{6}{k}$ from the values of $\binom{5}{k}$ for $k = 0, 1, \ldots, 5$ given in Example 1.6a.

50. Find the coefficient of

*(a) x^3y^5 in the expansion of $(x + y)^8$.

*(b) x^2y^4z in the expansion of $(x + y + z)^7$.

(c) $x^2y^2z^2w^2$ in the expansion of $(x + y + z + w)^8$.

51. In the game of Monopoly, throwing a double (the two dice showing the same number) has special significance. (For instance, a double gets you out of Jail free.)

*(a) Find the probability of a double.

(b) Following the reasoning of Example 1.7c, give a model for the experiment in which the pair of dice is cast repeatedly until a double occurs.

*(c) In (b), find the probability that it takes at least three tosses to get a double.

*52. Given $P(A) = .59$, $P(B) = .30$, $P(AB) = .21$, find

(a) $P(AB^c)$.

(b) $P(A \cup B^c)$.

53. Show that $P(EF) \leq P(E \cup F) \leq P(E) + P(F)$ using the axioms of probability, properties (i)–(iii) of Section 1.7.

54. Apply Property (vii) of Section 1.7 twice to show

$$P(E \cup F \cup G) = P(E) + P(F) + P(G) - P(EF) - P(FG) - P(EG) + P(EFG).$$

[This is an extension of Equation (5) in Section 1.7 to the case of three events. Can you conjecture the extension to n events?]

55. (a) Use induction to show $P(E_1 \cup \cdots \cup E_k) \leq \Sigma P(E_i)$.

(b) Use (a) and DeMorgan's laws (Problem 7) to show

$$P(E_1 E_2 \cdots E_k) \geq 1 - \Sigma P(E_i^c),$$

[This inequality will be useful in connection with multiple comparisons in Section 14.3.]

CHAPTER 2

Discrete Random Variables

Example **2a**

Pets and Recovery

A research study reported at a 1978 meeting of the American Heart Association focused on the possibility that having a pet might contribute to a patient's recovery. Patients were classified according to (1) whether or not they live at least one year after surgery, and (2) whether or not they have a pet.

With regard to whether a randomly selected patient does or does not have a pet, the observation varies from patient to patient. We say it is a *random variable*. Similarly, the observation of whether the patient survives or not is a random variable. The question is whether or not these variables are related. In particular, do patients who have pets live longer than those who don't? ∎

We are often interested in characteristics that vary from one individual to another—sometimes focusing on one characteristic, but also on relationships among several characteristics (as in the example).

Random Variables

A characteristic of the outcome of an experiment of chance is called a **random variable**. Random variables are **numerical** or **categorical**. The possible values of a numerical random variable are numbers; the possible values of a categorical random variable are categories in some scheme of classification.[1] Values of a numerical variable are always ordered; values of a categorical random variable may or may not be ordered.

Numerical characteristics such as distance, time, and mass are usually thought of as numbers on a continuous scale—the possible values cannot be listed. We say that these random variables are **continuous**. When the possible values of a random variable are countable, we say that the variable is **discrete**.

---

[1] A numerical value is a special kind of category, but we reserve the term *categorical* for nonnumerical categories.

Example 2b   A student chosen at random from the undergraduates at a university can be classified in various ways:

Random variable	Values
Sex	Male, Female
Class	Fresh., Soph., Jr., Sr., other
Height (inches)	Numbers on the interval from 30 to 90
College	Arts, Education, Technology, other
Shoe size	3, 3.5, 4, 4.5, ... , 16
No. of siblings	0, 1, 2, 3, ...

Some of these variables are numerical (shoe size, number of siblings); some are categorical with unordered categories (sex, college); and class is categorical but with ordered categories. If measured with infinite precision, a person's height is a number on a continuous scale, but heights measured and recorded to the nearest inch are discrete. Similarly, foot length and width might be considered to be measured on a continuous scale, but shoe sizes are discrete. ■

   In this chapter, we study methods for describing discrete random variables and their interrelationships. Continuous variables will be treated in Chapter 5.

2.1   Probability Functions for Discrete Random Variables

The value of a random variable depends on the outcome of the experiment being observed; in Example 2b, it depends on which student is selected. Thus, a random variable is a *function* of the outcome. We sometimes use function notation to denote a random variable, as in $X(\omega)$, but will often suppress the dependence on the outcome ω and call it X. Ordinarily, we represent random variables using uppercase letters appearing late in the alphabet $(X, Y, U, \ldots)$, perhaps with subscripts; their possible values are then referred to by corresponding lowercase letters $(x, y, u, \ldots)$.

   Each possible value x of a random variable X defines an *event*: the set of sample space outcomes ω assigned the value x by the function $X(\omega)$. The probability of that event in the sample space is simply the sum of the probabilities of its outcomes:[2]

$$P[X(\omega) = x] = P[\text{set of } \omega\text{'s with } X(\omega) = x] = \sum_{X(\omega)=x} P(\omega).$$

Example 2.1a   Selecting a viewer at random from a population of TV viewers is an experiment of chance. Advertisers like to know which of the major commercial TV networks the viewer is watching at a given moment. This is a random variable. Its values have

---

[2] $P[E]$ is the same as $P(E)$. We use square brackets only when there are parentheses in the specification of the event E.

probabilities equal to the corresponding population proportions, since with random sampling all viewers are equally likely to be selected. These probabilities are generally not known, so we give them literal names:

Network	Probability
NBC	p
CBS	q
ABC	r
Other	s

All viewers are accounted for in the listed values, so the probabilities (population proportions) must sum to 1: $p + q + r + s = 1$. ∎

The probability of observing the value x clearly depends on x. It is a *function* of x, called the **probability function (p.f.)** of the random variable X and denoted by $f(x)$. When necessary for clarity, we use a subscript to distinguish the p.f. of X: $f_X(x)$.

Probability function (p.f.) of the discrete random variable $X(\omega)$:

$$f(x) = P[X(\omega) = x] = \sum_{X(\omega) = x} P(\omega).$$

Example 2.1b   **Checking Lot Quality**
It is common practice in industry to decide whether a shipment of articles is acceptable on the basis of a sample taken at random from the shipment. Suppose that a shipment of 20 computer circuit boards includes two that are defective. (In practice, we would not know this.) When three circuit boards are chosen at random (without replacement), the $\binom{20}{3} = 1140$ possible samples of three boards are equally likely. Each of these possible samples is an ω, and the number of defectives in ω is a random variable, $X(\omega)$, with possible values 0, 1, and 2. Since we assumed that the possible samples ω are equally likely, we can find the value of the p.f. $f(x)$ by counting. For example, the probability of 0 defectives ($X = 0$) is

$$f(0) = \frac{\text{Number of samples with 0 defectives}}{\text{Number of possible samples}}$$

$$= \frac{\binom{18}{3}}{\binom{20}{3}} = \frac{136}{190}.$$

In general,

$$f(x) = \frac{\binom{18}{x}\binom{2}{2-x}}{\binom{20}{3}} \qquad \text{for } x = 0, 1, 2.$$

When the list of possible values is short, as in this case, it is easy to give the probability function in table form:

x	0	1	2
$f(x)$	$\dfrac{136}{190}$	$\dfrac{51}{190}$	$\dfrac{3}{190}$

Figure 2.1 gives a graphical representation of this p.f.

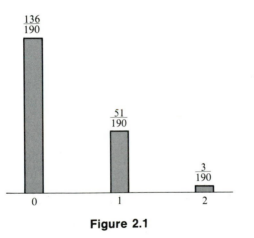

Figure 2.1

Example 2.1c

Waiting for a Success

Example 1.7c considered an experiment in which we toss a coin until we obtain heads. The outcomes ω are the sequences T, TH, TTH, $TTTH$, ... Define $X(\omega)$ to be the number of tails before the first heads. The possible values of X are 0, 1, 2, 3, ... With probabilities assigned to the sequences ω as in Example 1.7c, the p.f. for X is

$$f(x) = \frac{1}{2^{x+1}} \qquad \text{for } x = 0, 1, 2, \ldots.$$

This function is partially shown in Figure 2.2.

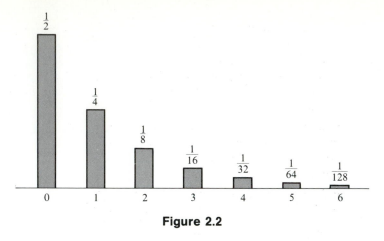

Figure 2.2

Sometimes the outcome ω itself is a random variable of interest: $X(\omega) = \omega$. The probability table for this random variable is simply the probability table for the sample space Ω. The next example is a case in point.

Example 2.1d   The usual sample space for the toss of a die is the set of six faces of the die, identified by number:

$$\Omega = \{1, 2, 3, 4, 5, 6\}.$$

The probability function for $X(\omega) = \omega$ is

$$f_X(x) = \frac{1}{6}, \qquad x = 1, 2, 3, 4, 5, 6.$$

There are other random variables on this sample space. Consider

$$Y(\omega) = \begin{cases} 1 & \text{if } \omega \text{ is odd,} \\ 0 & \text{if } \omega \text{ is even.} \end{cases}$$

The p.f. for Y is $f_Y(y) = \frac{1}{2}$, $\qquad y = 0$ or 1.   ■

In the above examples, we first defined probabilities in a sample space Ω. From these we calculated the probability function of the random variable of interest. The p.f. must be nonnegative, since its values are probabilities. Its values sum to 1, since the events $\{X = x\}$ partition Ω:

Properties of a probability function $f(x)$:

(i) $f(x) \geq 0$ for all x,

(ii) $\sum f(x) = 1.$

In constructing a model for a random variable X, it is often convenient and natural to specify its distribution by assigning probabilities directly in its value-space. For this we use an $f(x)$ that satisfies (i) and (ii) in the box. This amounts to treating the x-axis as Ω, with $f(x)$ as $P(\omega)$. For example, if X = the number of points showing in the throw of a die, we can define the distribution by $f(x) = \frac{1}{6}, x = 1, \ldots, 6$.

A probability function—whether given by table, graph, or formula—shows how probability is distributed among the possible values of a random variable. Thus, we speak of the **probability distribution** or, simply, the **distribution** of the random variable.

2.2   **Joint Distributions**

Discovering and understanding relationships among variables is useful for applications (see Example 2a). Changing one of two related variables may *cause* a change in the other. In such a case, we can control an effect to some extent by controlling the cause. Moreover, as we'll see in Chapter 15, a relationship between variables can be useful in *predicting* one variable from a knowledge of others, even when the relationship is not causal.

To study a relationship between random variables $X(\omega)$ and $Y(\omega)$, we need to consider them together, as a **random vector** (X, Y). The "values" of this random vector are pairs (x, y). The probability distribution of (X, Y) is called **bivariate** and defined by a **joint probability function**:

$$f(x, y) = P[X(\omega) = x, Y(\omega) = y].$$

This is found as a sum of probabilities $P(\omega)$ for all ω such that $X(\omega) = x$ and $Y(\omega) = y$.

Joint probability function for (X, Y):

$$f(x, y) = P(X = x, Y = y), \tag{1}$$

where

(i) $f(x, y) \geq 0$ for all x, y,

(ii) $\displaystyle\sum_{(x, y)} f(x, y) = 1.$

Properties (i) and (ii) parallel corresponding properties of a univariate p.f. $f(x)$. As in the univariate case, they follow from similar properties of $P(\omega)$. Also, as in the univariate case, we may define a bivariate distribution by giving a function $f(x, y)$ satisfying properties (i) and (ii), without reference to an underlying Ω.

Example 2.2a

Class Versus Sex

We cross-classified the 95 students in one of our classes according to sex and class. The entry in each cell is the number of students in that category of the cross-classification:

		$Y = class$			
		Fr.	So.	Jr.	Sr.
$X = sex$	Male	4	26	15	5
	Female	5	20	15	5

Suppose we consider this class of 95 students to be a population and select a student at random. The individual students (ω's) are then equally likely. The probability that a student falls in a particular cell of the two-way table is the *proportion* of students with the pair of characteristics that define that cell. The proportion for a given cell is just the ratio of the cell-frequency to the population size. For instance,

$$f(M, So.) = P(\text{male sophomore}) = \frac{\#(\text{male sophomores})}{\#(\text{students})}.$$

The complete table of joint probabilities is as follows:

		$Y = class$				
		Fr.	So.	Jr.	Sr.	f_Y
$X = sex$	M	$\frac{4}{95}$	$\frac{26}{95}$	$\frac{15}{95}$	$\frac{5}{95}$	$\frac{50}{95}$
	F	$\frac{5}{95}$	$\frac{20}{95}$	$\frac{15}{95}$	$\frac{5}{95}$	$\frac{45}{95}$
	f_X	$\frac{9}{95}$	$\frac{46}{95}$	$\frac{30}{95}$	$\frac{10}{95}$	1

The subtotals of the probabilities in rows and in columns are shown in the right and lower margins. These are probabilities for the individual variables Sex (X) and Class (Y) and define the (univariate) distributions of these variables. For instance, from the law of total probability [Equation (2) in Section 1.7], we have

$$f_X(M) = f(M, Fr.) + f(M, So.) + f(M, Jr.) + f(M, Sr.) = \frac{50}{95},$$

and

$$f_Y(So.) = f(M, So.) + f(F, So.) = \frac{46}{95}.$$

The univariate distributions of X and Y, which appear in the margins of the table of joint probabilities, are called **marginal**. ∎

Marginal p.f.'s:

$$f_X(x) = P(X = x) = \sum_y f(x, y), \tag{2}$$

$$f_Y(y) = P(Y = y) = \sum_x f(x, y). \tag{3}$$

Example 2.2b   **Selecting Chips at Random**

A bowl contains four chips marked 1, 2, 3, and 4. One is selected at random and then another at random from those that remain. Let

$X =$ number on the first chip,

$Y =$ number on the second chip.

The outcomes of the composite experiment are the possible pairs (x, y), where x and y are each one of the numbers 1, 2, 3, 4. Suppose we assume that the 12 pairs (x, y) with $x \neq y$ are equally likely, so that each pair has probability $\frac{1}{12}$:

$$f(x, y) = P(X = x, Y = y) = \frac{1}{12} \quad \text{for } x \neq y.$$

The table of joint probabilities is as follows:

x y	1	2	3	4	$f_Y(y)$
1	0	1/12	1/12	1/12	1/4
2	1/12	0	1/12	1/12	1/4
3	1/12	1/12	0	1/12	1/4
4	1/12	1/12	1/12	0	1/4
$f_X(x)$	1/4	1/4	1/4	1/4	1

The marginal entries give the p.f.'s for X and Y considered individually. The two marginal distributions are the same. ∎

We define joint probability functions for more than two random variables in the same way as for two. In the case of three variables (X, Y, Z),

$$f(x, y, z) = P(X = x, Y = y, Z = z). \tag{4}$$

A three-way table giving these joint probabilities requires three dimensions, but we can show them on a two-dimensional sheet of paper by "layering," a technique that will be illustrated in the next example. Summing entries in a three-way table along one direction, say, X, produces the marginal joint distribution of Y and Z, and we obtain the univariate distribution for any one variable by summing over the other two variables.

Example 2.2c   **Taste Testing**

Problem 13 of Chapter 1 dealt with Mike's attempt to identify three brands of cola in a taste test. We assumed there that Mike really was *not* able to discriminate among the three brands of cola and assigned probability $\frac{1}{6}$ to each possible arrangement. (Think of the glasses of cola lined up in a sequence and his assignment of brand to glass as a permutation of the correct brand sequence.)

The following variable is the **indicator function** for the event C—that Mike correctly identifies the Coke:

$$X_C = \begin{cases} 1 & \text{if he correctly identifies Coke,} \\ 0 & \text{otherwise.} \end{cases}$$

Define X_P and X_R similarly for Pepsi and Royal Crown, respectively. The six possible assignments (the equally likely ω's) are shown in the following table. It also shows values of the indicator variables as well as the values of $Y = X_C + X_P + X_R$, the number of correct identifications.

Actual brand:	C	P	R	X_C	X_P	X_R	Y
Mike's	C	P	R	1	1	1	3
identification:	C	R	P	1	0	0	1
	P	R	C	0	0	0	0
	P	C	R	0	0	1	1
	R	P	C	0	1	0	1
	R	C	P	0	0	0	0

The joint probability function of the X's is actually given in this table, but for variables with more values we'd use a three-way table given in layers. For example, choosing X_R as the layering variable, we have these two-way tables for X_C and X_P:

For $X_R = 1$:   For $X_R = 0$:

		X_P	
		0	1
X_C	0	1/6	0
	1	0	1/6

		X_P	
		0	1
	0	1/3	1/6
	1	1/6	0

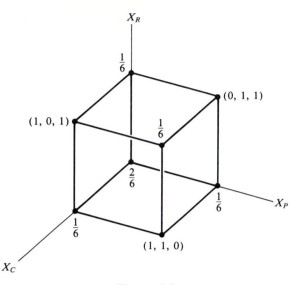

Figure 2.3

Figure 2.3 shows the joint distribution of the three X's, in which we have indicated the joint distribution of (X_C, X_P, X_R) by marking each of the possible triples (x, y, z) with the corresponding probability. The layers referred to above are easy to see in this figure. (We could have used either of the other variables as the layering variable; because of symmetry, the two-way layers in each case look the same as those given above.)

The distribution of Y is apparent in the original table: $f_Y(3) = \frac{1}{6}$, $f_Y(1) = \frac{1}{2}$, and $f_Y(0) = \frac{1}{3}$. ∎

| 2.3 | **Conditional Probability** |

Suppose we have settled on a particular probability model for some experiment of chance. If we then learn something more about the experiment, we may need to revise the model accordingly.

| Example **2.3a** | **Partial Information About a Card** |

When a card is selected at random from a standard deck of playing cards, the probability of a spade is $\frac{1}{4}$. Suppose, however, you learn that the card selected is red. Are the odds that the card is a spade still 1:3? Obviously not. The model for the experiment needs to be updated. All black cards are ruled out, and the appropriate sample space is now just the set of red cards—the subset of the original sample space defined by the event "red." Events in this new sample space must be given new probabilities. ∎

When one learns that the event F has occurred, how does this information change the probability of another event E? Consider the lottery model for the experiment—that is, a random selection of a ticket from a lottery. To say that F has occurred is to say that only the tickets with the property F are now possible. For calculating our new probabilities, what has changed is the available pool of tickets. So if the tickets in F were equally likely *before* learning that the outcome is in F, it is reasonable to view them as still equally likely *after* learning this. However, the denominator of the fractions defining probabilities changes.

When some tickets are removed or otherwise disallowed, the assumption that the remaining tickets are equally likely is our rationale for defining conditional probability.

Conditional probability of an outcome in Ω, given that F has occurred:

$$P(\omega \mid F) = \begin{cases} 0 & \text{if } \omega \text{ is not in } F, \\ \dfrac{P(\omega)}{P(F)} & \text{if } \omega \text{ is in } F, \end{cases} \tag{1}$$

provided $P(F) \neq 0$.

In Section 1.7, we defined probability for any event E as the sum of the probabilities of its outcomes. We now do this with the conditional probabilities:

$$P(E \mid F) = \sum_{\omega \text{ in } E} P(\omega \mid F). \tag{2}$$

Combining Equations (1) and (2), we have:

Conditional probability of E given F:

$$P(E \mid F) = \frac{P(EF)}{P(F)}, \tag{3}$$

provided $P(F) \neq 0$.

Example 2.3b   **Smoking Versus Sex**
Students in a class of 50 were cross-classified according to sex, and according to whether they now smoke (S), used to smoke but quit (Q), or never smoked (N):

		Smoking			
		S	Q	N	
Sex	M	12	8	10	30
	F	14	4	2	20
	Total	26	12	12	50

So the probability that a randomly selected student is a smoker is $\frac{26}{50}$. Suppose we now observe that the student selected is male; this information changes the odds that he is a smoker. Of the 30 males, 12 are smokers. We take the new odds that the student is a smoker to be 12:18. That is, the 30 males in the reduced sample space are still equally likely, and this defines the **conditional probability of** S **given** M: $P(S\,|\,M) = \frac{12}{30}$.

It is instructive to write this last fraction in terms of unconditional probabilities:

$$P(S\,|\,M) = \frac{\#(\text{male smokers})}{\#(\text{males})} = \frac{12}{30} = \frac{12/50}{30/50} = \frac{P(MS)}{P(M)}.$$

Reasoning similarly, we would define the conditional probability that a student is male given that the student smokes:

$$P(M\,|\,S) = \frac{\#(\text{male smokers})}{\#(\text{smokers})} = \frac{12}{26} = \frac{12/50}{26/50} = \frac{P(MS)}{P(S)}. \qquad \blacksquare$$

Conditional probabilities satisfy properties of probability [(i)–(iii) of Section 1.7]. In particular, they are additive over disjoint events:

$$P(E_1 \cup E_2\,|\,F) = P(E_1\,|\,F) + P(E_2\,|\,F), \qquad \text{when } E_1 E_2 = \varnothing.$$

Multiplying through the defining formula for conditional probability [Equation (3)] by $P(F)$, we obtain a formula for the probability of the intersection of two events. This is useful when the conditional probability $P(E\,|\,F)$ is known:

Multiplication Rule:

$$P(EF) = P(E)P(F\,|\,E). \tag{4}$$

The multiplication rule may seem more intuitive than the definition of conditional probability to which it is equivalent. The following example illustrates its use.

Example 2.3c **Two Cards Selected in Succession**
Suppose we select two cards at random, one at a time and without replacement, from a standard deck of playing cards. What is the probability that the first card is the king of hearts and the second the ten of spades?

The outcomes of this composite experiment are sequences of two cards. Since the first selection is random, $P(\text{heart king}) = \frac{1}{52}$. At the second draw, again random, the 51 cards that remain are equally likely, so

$$P(\text{2nd is spade ten}\,|\,\text{first is heart king}) = \frac{1}{51}.$$

So, by the multiplication rule,

$$P(\text{heart king, then spade ten}) = P(\text{1st is heart king})$$
$$\cdot\, P(\text{2nd is spade ten} \mid \text{1st is heart king})$$
$$= \frac{1}{52} \cdot \frac{1}{51} = \frac{1}{2652}.$$

The probability of these same two cards but in reverse order is clearly the same. The probability of this pair *without* regard to order is then $\frac{2}{2652} = \frac{1}{1326}$. This would be the same for any particular pair.

The number of *combinations* of two cards from 52 is $52 \cdot \frac{51}{2} = 1326$. In Section 1.5, we used intuition to argue that combinations obtained by a sequence of random selections without replacement are equally likely. We've shown this to be the case in this example as a result of our definition of conditional probability. ∎

The multiplication rule [Equation (4)] extends easily to more than two events. Given events E, F, and G, we apply (4) to the two events EF and G:

$$P(EFG) = P[(EF)G] = P(EF)P(G \mid EF).$$

Applying the rule once more to obtain $P(EF)$, we have

$$P(EFG) = P(E)P(F \mid E)P(G \mid EF). \tag{5}$$

Repeating this process yields a multiplication rule for the probability of the intersection of any finite collection of events:

Extended multiplication rule:

$$P(E_1 E_2 \cdots E_k) = P(E_1)P(E_2 \mid E_1)P(E_3 \mid E_1 E_2) \cdots P(E_k \mid E_1 E_2 \cdots E_{k-1}) \tag{6}$$

Example 2.3d   **Star Wars**

In a speech reprinted in the *Congressional Record*,[3] we find this paragraph:

> Most of the discussion of the "Star Wars" defense assumes a many-layered defense with three or four distinct layers. The idea behind having several layers is that the total defense can be made nearly perfect in this way, even if the individual layers are less than perfect. For example, if each layer has, say, an 80-percent effectiveness—which means that one in five missiles or warheads will get through—a combination of three such layers will have an overall effectiveness better than 99 percent, which means that no more than one warhead in 100 will reach its target.

Suppose a missile meets the layers in sequence. The writer assumes that it has probability .20 of penetrating the first layer unscathed; that the conditional probability of its penetrating the second layer (given that it got through the first) is

---

[3] June 17, 1985.

.20; and that the conditional probability of its penetrating the third layer (given that it got through the first two) is .20. With E_i denoting the event that the missile gets through layer i,

$$P(E_1 E_2 E_3) = P(E_3 \mid E_1 E_2) P(E_2 \mid E_1) P(E_1) = .2 \times .2 \times .2 = .008.$$

So the probability that the missile is foiled in its mission is $1 - .008$, or 99.2 percent. ∎

Being probabilities, the values of the p.f. of a random variable must be modified in view of partial information about the random variable. Consider the following example.

Example 2.3e

A Conditional Probability Function
In Example 2.1c, we gave a possible probability model for X, the number of tosses of a coin required to get heads: $f(x) = \frac{1}{2}^x$, $x = 1, 2, \ldots$ Let F denote the event $\{X \leq 3\}$—that is, that at most three tosses are needed to get heads. How does the information that F has occurred change the appropriate distribution for X?
The probability of the condition is

$$P(F) = \frac{1}{2} + \frac{1}{4} + \frac{1}{8} = \frac{7}{8}.$$

So, given F, the new probabilities of the possible values (those in F) are the unconditional probabilities divided by $\frac{7}{8}$. Thus, they are in the same proportion as before, in the ratio $4:2:1$. The *conditional* p.f. of X given F is as follows:

x	1	2	3	4	5	$\cdots$
$f(x \mid F)$	$\dfrac{4}{7}$	$\dfrac{2}{7}$	$\dfrac{1}{7}$	0	0	$\cdots$

∎

Conditional p.f.'s

Conditional p.f.'s commonly arise when one learns the value of one of the two variables in a bivariate distribution. Thus, the **conditional p.f.** of X given $Y = y$ is the conditional probability that $X = x$, given $Y = y$. Using f for p.f., we include a vertical bar followed by the value of the conditioning variable to indicate that it is a conditional probability function:

$$f(x \mid y) = P(X = x \mid Y = y) = \frac{f(x, y)}{f_Y(y)} \tag{7}$$

where $f(x, y)$ is the joint probability function of X and Y—a function that is usually different from $f(x \mid y)$.

Example 2.3f

As in Example 2.2b, let X denote the first and Y the second chip in a random selection of two chips taken without replacement from a bowl containing four chips numbered 1, 2, 3, 4. The joint p.f. is given by the following table (repeated from

Example 2.2b):

x / y	1	2	3	4	$f_Y(y)$
1	0	1/12	1/12	1/12	1/4
2	1/12	0	1/12	1/12	1/4
3	1/12	1/12	0	1/12	1/4
4	1/12	1/12	1/12	0	1/4
$f_X(x)$	1/4	1/4	1/4	1/4	1

Given that $Y = 1$, for example, the only outcomes possible are the pairs corresponding to the first row of probabilities. From Equation (7),

$$f(x\,|\,1) = \frac{f(x,\,1)}{f_Y(1)} = \begin{cases} 0 & \text{for } x = 1, \\ 1/3 & \text{for } x = 2,\,3,\,4. \end{cases}$$

These conditional probabilities are proportional to the probabilities in the first row of joint probabilities; division by the row sum ensures that they add to 1. ∎

2.4   Bayes' Theorem

The multiplication rule, Equation (4) of Section 2.3, expresses $P(EF)$ as a product:

$$P(EF) = P(E)P(F\,|\,E).$$

If we simply interchange E and F, the rule says

$$P(EF) = P(F)P(E\,|\,F).$$

So

$$P(E\,|\,F)P(F) = P(F\,|\,E)P(E). \tag{1}$$

This simple observation is the essence of Bayes' theorem, which follows from Equation (1) upon dividing by $P(F)$:

$$P(E\,|\,F) = \frac{P(F\,|\,E)P(E)}{P(F)}. \tag{2}$$

Bayes' theorem is sometimes referred to as the *law of inverse probability* because it expresses $P(E\,|\,F)$ in terms of $P(F\,|\,E)$, with the roles of E and F inverted. If you want to find a conditional probability $P(E\,|\,F)$, the key to deciding whether or not Bayes' theorem can help you is whether or not you know $P(F\,|\,E)$.

In most cases, we can calculate the denominator $P(F)$ in Equation (2) by means of the law of total probability [Equation (2) of Section 1.7]:

$$P(F) = P(FE) + P(FE^c). \tag{3}$$

Applying the multiplication rule [Equation (4) of Section 2.3] to both terms on the r.h.s. of Equation (3), we obtain

$$P(F) = P(F\mid E)P(E) + P(F\mid E^c)P(E^c). \tag{4}$$

The following examples illustrate the calculations.

Example 2.4a

Smoking and Gender

Consider a population of students, of whom 40% are female (F) and 60% male (M). Suppose 15% of the females and 10% of the males smoke. What fraction of the smokers are female? Equivalently, what is the probability that a student selected at random is female, given that the student is a smoker?

Translating into symbols, we know these probabilities (or population proportions):

$$P(F) = .4, \qquad P(M) = .6, \qquad P(S\mid F) = .15, \qquad P(S\mid M) = .10.$$

We want to find $P(F\mid S)$, the "inverse" of $P(S\mid F)$. Using Equations (3) and (4), we can find the proportion of smokers:

$$P(S) = P(SF) + P(SM) = P(S\mid F)P(F) + P(S\mid M)P(M)$$
$$= .15 \times .40 + .10 \times .60 = .06 + .06 = .12.$$

The ratio of female smokers to male smokers is .06 to .06, or 1 to 1, and

$$P(F\mid S) = \frac{P(S\mid F)P(F)}{P(S)} = \frac{.15 \times .40}{.12} = \frac{.06}{.06 + .06} = \frac{1}{2}.$$

Figure 2.4 shows a tree diagram for this example. The main branching is into female and male, in the ratio 4 to 6. A further branching of each of these shows the proportion of smokers (S) and nonsmokers (S^c), the branches being marked with the appropriate *conditional* probability (that is, given that branch). The products of the probabilities along a path from left to right are probabilities for the four possible pairs of conditions: FS, FS^c, MS, and MS^c. The population is thus divided into these four subsets, in the proportions 6, 34, 6, and 54, respectively. The condition that a person selected at random is a smoker defines the union of FS and MS. In this smaller sample space of smokers, which is 12% of the original sample space, the ratio of females to males is 6 to 6.

Still another way to see how Bayes' theorem works is in terms of the joint distribution of the variables smoking (with values S and S^c) and gender (with values M and F). The probability table is as follows:

	S	S^c	
M	.06	.54	.60
F	.06	.34	.40
	.12	.88	1

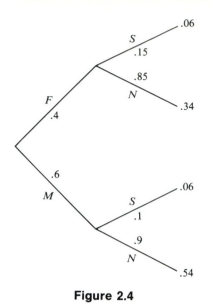

Figure 2.4

The entries in this table are simply the branch labels in the tree diagram of Figure 2.4. Conditional probabilities are proportional to the rows or to the columns (depending on the conditioning variable). ∎

Example **2.4b**   **A Paternity Case**

Mr. G has been accused in a paternity suit of being the father of Ms. H's child. The state introduces expert testimony concerning the genetic makeup of Mr. G, Ms. H, and the child. The evidence is complicated, involving many genetic factors: we limit our discussion to blood group.

Suppose we know Mr. G. is type AB, and Ms. H. is type A. The datum D is that the child is type B. (We now know that the mother is heterozygous AO, since if she were homozygous AA, the child could not be type B.) Let G denote the event that Mr. G is the child's father. A court is interested in determining $P(G|D)$, the probability that Mr. G is the father in view of the data D.

Let $p = P(G)$, the "prior" probability (prior to the data D) that Mr. G is the father. With Equations (2) and (4) combined, Bayes' theorem says:

$$P(G|D) = \frac{P(D|G) \cdot p}{P(D|G) \cdot p + P(D|G^c)(1 - p)}. \tag{5}$$

This is referred to as the *posterior probability* of G, given D.

A reasonable interpretation of p is that it is a juror's personal probability that Mr. G is the father, based on evidence other than the genetic data. Unfortunately, in paternity cases in many states in the U.S. and in other countries, court-appointed pathologists always take p to be $\frac{1}{2}$. They take the fact that a man is accused as reason to say he is as likely to be the father as not! (This assumption has silly implications. For example, if three men are suspected of being the father, their p's would sum to $\frac{3}{2}$.

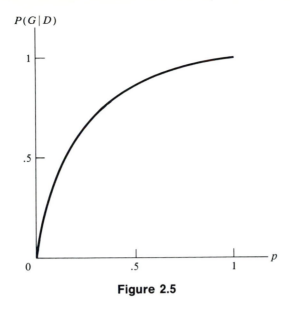

Figure 2.5

Moreover, it can turn out that their *posterior* probabilities (5) add up to more than 1 when there are only two potential fathers.)

The probabilities $P(D|G)$ and $P(D|G^c)$ follow from Mendel's laws of inheritance. (You need not be concerned about the details of the genetics.) The probability that a type AB father and a type AO mother produce the child in question is $\frac{1}{4}$, and this is $P(D|G)$. The probability $P(D|G^c)$ is $\frac{1}{2}$ times the proportion of B alleles in the appropriate gene pool, typically about .08. With these probabilities substituted in Equation (5), the result is

$$P(G|D) = \frac{.25p}{.25p + .04(1 - p)} = \frac{p}{.16 + .84p}.$$

The graph of this function of p is shown in Figure 2.5. With the assumption that $p = .5$, the posterior probability that Mr. G is the father is $P(G|D) \doteq .86$. However, rather than saying, "The probability of paternity is 86%," the expert should show the jury Figure 2.5 and explain how it converts a prior probability p (based on other evidence) to a posterior probability. ∎

General Form of Bayes' Theorem

Suppose we know $P(F|E_i)$ for each of at most countably many events E_i that partition Ω: $\{E_1, E_2, \ldots\}$. (*Partition* was defined in Section 1.3.) We can express the denominator $P(F)$ in Bayes' formula [Equation (2)] using the correspondingly more general form of the law of total probability [from Equation (2) in Section 1.7]:

$$P(F) = P(F|E_1)P(E_1) + P(F|E_2)P(E_2) + \cdots. \tag{6}$$

Then, *given* the occurrence of F, the odds on $E_1, E_2, \ldots$ are proportional to the terms in this decomposition of $P(F)$. This fact is expressed in the following version of

Bayes' theorem:

Bayes' Theorem:

If $\{E_1, E_2, \ldots\}$ is a partition of Ω, then for $i = 1, 2, \ldots,$

$$P(E_i \mid F) = \frac{P(F \mid E_i)P(E_i)}{P(F \mid E_1)P(E_1) + P(F \mid E_2)P(E_2) + \cdots}. \qquad (7)$$

In terms of odds,

$$\frac{P(E_i \mid F)}{P(E_j \mid F)} = \frac{P(F \mid E_i)}{P(F \mid E_j)} \cdot \frac{P(E_i)}{P(E_j)}. \qquad (8)$$

Note that Equation (7) generalizes Equation (5).

Example 2.4c   **Delinquency and Birth Order**

Are delinquent children likely to be youngest in their families? A research study investigated the possible relationship between delinquency and birth order among high school girls. Delinquency (D) was somewhat arbitrarily defined according to answers on a questionnaire. In the group reported in the study, 40% were eldest (E), 30% middle (M), 20% youngest (Y), and 10% only children (O). The proportions classified as delinquent were 5% for an eldest child, 10% for a middle child, 15% for a youngest child, and 20% for an only child. (Actual percentages have been rounded for simplicity.) Thus, we have

$$P(E) = .40, \qquad P(M) = .30, \qquad P(Y) = .20, \qquad P(O) = .10,$$
$$P(D \mid E) = .05, \qquad P(D \mid M) = .10, \qquad P(D \mid Y) = .15, \qquad P(D \mid O) = .20.$$

Given these probabilities, what proportion of the delinquent girls are youngest in their families?

Using the law of total probability [Equation (6)] we find the proportion of delinquents to be

$$P(D) = P(D \mid E)P(E) + P(D \mid M)P(M) + P(D \mid Y)P(Y) + P(D \mid O)P(O)$$
$$= .05 \times .40 + .10 \times .30 + .15 \times .20 + .20 \times .10$$
$$= .02 + .03 + .03 + .02 = .10.$$

So, from Equation (7),

$$P(E \mid D) = \frac{.02}{.02 + .03 + .03 + .02} = .2.$$

Similarly, $P(M \mid D) = .3$, $P(Y \mid D) = .3$, and $P(O \mid D) = .2$. Figure 2.6 shows the computations in terms of a tree diagram. ■

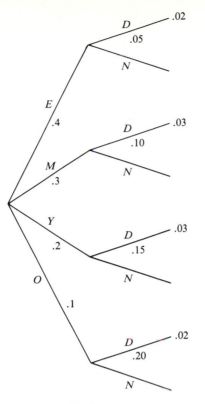

Figure 2.6

2.5   Independent Random Variables

Two random variables are **independent** when the marginal probability function of one variable is the same as the conditional p.f. given any value of the other variable.

Example **2.5a**   **Smoking versus Running**
Each student in one of our statistics classes was classified on the variables $X =$ smoking status and $Y =$ running status. The categories of Y were defined as R, for someone who runs at least five miles a week, and R^c, for all others. The categories of X (as in Example 2.3b) are S (now smokes), Q (used to but quit), and N (never smoked). Here are the results:

| | | \multicolumn{3}{c}{X: smoking} | |
		S	Q	N	
Y: running	R	2	4	16	22
	R^c	6	12	48	66
		8	16	64	88

In a random selection from this class, the joint probabilities for X and Y are as follows (obtained by dividing the above entries by 88):

		S	Q	N	f_Y
	R	$\dfrac{1}{44}$	$\dfrac{2}{44}$	$\dfrac{8}{44}$	$\dfrac{11}{44}$
Y: running	R^c	$\dfrac{3}{44}$	$\dfrac{6}{44}$	$\dfrac{24}{44}$	$\dfrac{33}{44}$
	f_X	$\dfrac{1}{11}$	$\dfrac{2}{11}$	$\dfrac{8}{11}$	1

X: Smoking

Suppose we learn that a student is a runner. Given $Y = R$, the conditional probabilities for X are proportional to the joint probabilities in the first row, in the ratio $1:2:8$. The conditional probabilities for X given $Y = R^c$ are in the ratio $3:6:24$, the same as $1:2:8$. Not surprisingly, the marginal probabilities f_X are in the same ratio:

$$f(S \mid R) = f(S \mid W) = \frac{1}{11} = f_X(S),$$

$$f(Q \mid R) = f(Q \mid W) = \frac{2}{11} = f_X(Q),$$

$$f(N \mid R) = f(N \mid W) = \frac{8}{11} = f_X(N).$$

So the p.f. for the classification according to smoking (X) does not change with the information about running (Y).

Likewise, since the columns in the table are proportional, each one defining odds of $1:3$, we see that

$$f(R \mid S) = f(R \mid Q) = f(R \mid N) = \frac{1}{4} = f_Y(R).$$

The coincidence of conditional with unconditional probabilities (rows or columns) is equivalent to the fact that each joint probability is the product of corresponding marginal probabilities. For example,

$$f(S, R) = \frac{1}{44} = \frac{1}{11} \cdot \frac{11}{44} = f_X(S) \cdot f_Y(R).$$

You should verify that this multiplicative property holds for each entry in the table.

■

When the distribution for X does not change upon learning that $Y = y$:

$$f(x \mid y) = f_X(x) \qquad \text{for all } (x, y), \tag{1}$$

the multiplication rule [Equation (4) of Section 2.3] becomes

$$f(x, y) = f(x \mid y)f_Y(y) = f_X(x)f_Y(y).$$

That is, the joint p.f. is the product of the marginal p.f.'s. We take this as the definition of independence:

Random variables X and Y are **independent** if and only if for every pair (x, y),

$$f(x, y) = f_X(x)f_Y(y). \tag{2}$$

Postulating Independence

Given the joint probabilities in Example 2.5a, verification of Equation (2) shows that X and Y are independent. However, we usually use (2) to *construct* a model— defining joint probabilities from marginal probabilities, assuming independence. Researchers may be willing to assume tentatively that two variables are independent, and this allows them to calculate joint probabilities from marginal probabilities. They seldom know for sure that variables are independent, but assumptions must be made in any scientific endeavor, or science would not advance. Scientists can make calculations under the assumption of independence and, at the same time, worry about the validity of this assumption. They may be able to check their assumptions using experimental results. (We'll show how to check independence in Chapter 13.)

Example 2.5b

ABO Blood Typing

In a certain population, 42% have type O blood, 44% have type A, 10% have type B, and 4% have type AB. Suppose 10% have negative Rh-factor. What proportion of the population are O-negative ?

Conceivably, the proportion of O-negatives is anything from 0 to 10%. However, if blood type and Rh-factor are independent, application of Equation (2) shows that the proportion of O-negatives is $.42 \times .10 = .042$, or just over 4%. Moreover, under this assumption, all joint proportions are products:

		O	*A*	*B*	*AB*	
	+	.378	.396	.090	.036	.900
Rh factor	−	.042	.044	.010	.004	.100
		.420	.440	.100	.040	1

Blood type

The assumption of independence can be checked experimentally by comparing these joint probabilities with corresponding proportions in a sample.   ∎

Two random variables that are not independent are **dependent**. In the dependent case, not all of the joint probabilities factor according to (2). All it takes to show that two variables are dependent is *one* joint probability that is not the product of the corresponding marginal probabilities. In particular, an entry of 0 in a table of joint probabilities is an immediate tip-off: A joint probability $f(x, y)$ cannot be zero *and* equal to the product $f_X(x) \cdot f_Y(y)$ unless at least one factor is zero.

Example 2.5c   In Example 2.2b, we considered a model for two random selections of a chip from a bowl with four chips marked 1, 2, 3, and 4. The joint probabilities $f(x, y)$ are as follows:

y x	1	2	3	4	$f_X(x)$
1	0	$\frac{1}{12}$	$\frac{1}{12}$	$\frac{1}{12}$	$\frac{1}{4}$
2	$\frac{1}{12}$	0	$\frac{1}{12}$	$\frac{1}{12}$	$\frac{1}{4}$
3	$\frac{1}{12}$	$\frac{1}{12}$	0	$\frac{1}{12}$	$\frac{1}{4}$
4	$\frac{1}{12}$	$\frac{1}{12}$	$\frac{1}{12}$	0	$\frac{1}{4}$
$f_Y(y)$	$\frac{1}{4}$	$\frac{1}{4}$	$\frac{1}{4}$	$\frac{1}{4}$	1

The variables X and Y are dependent. To show this, we need find only *one* instance in which factorization fails:

$$f(1, 1) = 0 \neq \left(\frac{1}{4}\right) \cdot \left(\frac{1}{4}\right) = f_X(1)f_Y(1). \qquad \blacksquare$$

Several Variables   We now extend the notion of independence to more than two variables. It is not enough to say that the variables must be independent in pairs. Problem 51 shows that random variables X, Y, and Z can be independent in pairs even though X and $Y + Z$ are dependent. With the following definition of independence, this sort of thing cannot happen. Variables X, Y, and Z are said to be independent provided the joint probability function factors:

$$f(x, y, z) = f_X(x)f_Y(y)f_Z(z) \qquad (3)$$

for all possible values x, y, and z. This implies, in particular, that independence holds for each pair. For example, if X, Y, and Z are independent, then

$$f_{X,Y}(x, y) = \sum_z f(x, y, z) = \sum_z f_X(x)f_Y(y)f_Z(z)$$

$$= f_X(x)f_Y(y) \cdot \sum_z f_Z(z)$$

$$= f_X(x)f_Y(y).$$

The following defines independence of an arbitrary number of random variables:

Random variables $X_1, X_2, \ldots, X_k$ are **independent** when for all $x_1, \ldots, x_k$,

$$f(x_1, x_2, \ldots, x_k) = f_1(x_1) \cdots f_k(x_k), \tag{4}$$

where f_i is the p.f. of X_i.

Just as in the case $k = 3$, this definition implies that any *subset* of a set of k independent variables is a set of independent variables. Moreover, the conditional probabilities for any subset of the k variables, given the values of some of the others, are the same as the unconditional probabilities.

As in the case of two variables, we use the factorization condition (4) in constructing models for several variables that are assumed to be independent.

Example 2.5d   **Rolling Eight Dice**

In the carnival game "Razzle Dazzle," players toss eight dice simultaneously. A player who throws a total of 8 with these eight dice wins immediately. What is the probability of throwing a total of 8?

The outcome of each die is a random variable. If we assume that all six possible values are equally likely, the probability of "1" is $\frac{1}{6}$ for each die. The only way of getting a total of 8 is for every die to show "1." Then, if we assume independence,

$$P(\text{total of } 8) = f(1, 1, \ldots, 1)$$

$$= f_1(1) \cdots f_8(1) = \left(\frac{1}{6}\right)^8$$

$$= \frac{1}{1,679,616}.$$

In this game, the operator claims that since there are 41 possible sums $(8, 9, \ldots, 48)$, the probability of each is $\frac{1}{41}$. (To be continued in Example 3.6c.) ■

Independent Events

In statistical theory, the notion of independent variables is extremely important and fundamental. The notion of independence can be extended to *events*, as follows. For any event E, define the **indicator variable**:

$$X_E = \begin{cases} 1, & \text{if } E \text{ occurs}, \\ 0, & \text{if } E \text{ does not occur}. \end{cases}$$

So $f_{X_E}(1) = P(E)$, and $f_{X_E}(0) = P(E^c)$. Then:

Events $E_1, E_2, \ldots, E_k$ are **independent** if and only if their indicator variables $X_{E_1}, X_{E_2}, \ldots, X_{E_k}$ are independent.

In the case of *two* events E and F, the margins of the joint probability table for their indicator variables are as follows:

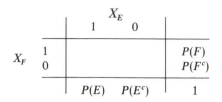

When the indicator variables are independent, the joint probabilities in this table are products. Thus,

$$f(1, 1) = P(EF) = P(E)P(F).$$

Moreover, if this relationship holds, *all* of the entries must factor. For instance,

$$P(EF^c) = P(E) - P(EF) = P(E) - P(E)P(F) = P(E) \cdot P(F^c).$$

So E and F are independent if and only if $P(EF) = P(E)P(F)$. Moreover, E and F are independent if and only if E and F^c are independent, if and only if E^c and F are independent, and if and only if E^c and F^c are independent.

The condition $P(EF) = P(E)P(F)$ is automatically satisfied when $P(F) = 0$, because then $P(EF) = 0$. Thus, any event whose probability is zero is independent of any other event.

The condition $P(EF) = P(E)P(F)$ implies that if $P(F) \neq 0$, $P(E \mid F) = P(E)$; and conversely, $P(E \mid F) = P(E)$ implies that $P(EF) = P(E)P(F)$. Likewise, when $P(E) \neq 0$, $P(F \mid E) = P(F)$ if and only if $P(EF) = P(E)P(F)$. Thus, when events are independent, knowing that one has occurred (or not) does not affect the odds on the other.

In the case of more than two events, Equation (4) implies that the probability of the intersection of independent events factors into the product of their individual probabilities. Thus, when events E, F, and G are independent,

$$P(EFG) = P(E)P(F)P(G).$$

However, the converse is not true. (See Problem 47.) To verify independence of a collection of events in terms of a factorization criterion, we must check to see that the factorization holds for every subcollection of those events.

2.6   ## Exchangeable Random Variables

Consider an experiment with outcome X, a random variable with p.f. $f(x)$. A sequence of n trials of the experiment results in the observations $(X_1, X_2, \ldots, X_n)$, where the p.f. of each X_i is $f(x)$. When $X_1, X_2, \ldots,$ and X_n are independent, we refer to this sequence of observations as a **random sample**. The joint p.f. of the variables in this sample sequence is a product of the marginal p.f.'s:

$$f(x_1, x_2, \ldots, x_n) = f(x_1)f(x_2) \cdots f(x_n) = \prod_{i=1}^{n} f(x_i).$$

The product on the right is a **symmetric** function of the arguments $x_1, \ldots, x_n$: Its value does not change for any rearrangement of the arguments. We conclude that the joint distribution of any subset of the X_i's is the same as that of any other subset of the same number of the X_i's. Thus, not only do the individual X's have the same distribution, but every pair has the same distribution as every other pair, and so on.

In arriving at the last conclusion, the only property of the joint p.f. we used is that it is a symmetric function of the arguments $x_1, \ldots, x_n$. So the conclusion that the marginal p.f.'s of a particular dimension are all the same is valid whenever the joint probability function has this symmetry. Variables whose joint distribution has this symmetry property are said to be **exchangeable**.

Random variables $X_1, X_2, \ldots, X_n$ are **exchangeable** when their joint p.f. is a symmetric function of its arguments. In this case, the k-dimensional marginal distributions are all the same, for $k = 1, \ldots, n - 1$.

Sampling *with* replacement from a finite population yields observations that are exchangeable, but so does sampling *without* replacement. We show this in two simple special cases.

Example 2.6a   **Exchangeability in Sampling Without Replacement**
We return to the joint p.f. of Examples 2.2b and 2.5c for the two chips selected at random without replacement from a bowl with four chips marked 1, 2, 3, and 4:

x \ y	1	2	3	4	$f_X(x)$
1	0	$\frac{1}{12}$	$\frac{1}{12}$	$\frac{1}{12}$	$\frac{1}{4}$
2	$\frac{1}{12}$	0	$\frac{1}{12}$	$\frac{1}{12}$	$\frac{1}{4}$
3	$\frac{1}{12}$	$\frac{1}{12}$	0	$\frac{1}{12}$	$\frac{1}{4}$
4	$\frac{1}{12}$	$\frac{1}{12}$	$\frac{1}{12}$	0	$\frac{1}{4}$
$f_Y(y)$	$\frac{1}{4}$	$\frac{1}{4}$	$\frac{1}{4}$	$\frac{1}{4}$	1

We saw in Example 2.5c that X and Y are not independent. However, the symmetry about the main diagonal of this table means that the joint probability function is symmetric in x and y. Thus, X and Y are exchangeable. (That the marginal distributions are the same is clear from the table.)

Observe that the conditional probabilities for the first chip given the second are the same as the corresponding conditional probabilities for the second given the first. This result is another consequence of the exchangeability of X and Y.   ■

Example 2.6b   **Exchangeability in Poker Hands**

A poker hand is a sample of five cards without replacement. Assuming adequate shuffling (and no cheating), we may consider the successive cards as selected at random from those that remain. Let X_i refer to the denomination $(2, 3, \ldots, Q, K, A)$ of card i, for $i = 1, \ldots, 5$. Using the multiplication rule [Equation (6) of Section 2.3], we find the probability of the sample sequence $(7, 2, J, 4, 7)$ to be

$$f(7, 2, J, 4, 7) = \frac{4}{52} \cdot \frac{4}{51} \cdot \frac{4}{50} \cdot \frac{4}{49} \cdot \frac{3}{48},$$

and for a permutation of this sequence, the same:

$$f(J, 4, 7, 7, 2) = \frac{4}{52} \cdot \frac{4}{51} \cdot \frac{4}{50} \cdot \frac{3}{49} \cdot \frac{4}{48}.$$

Such symmetry occurs for any sequence—the factors in the numerators are simply a rearrangement of the factors for any other permutation.

Symmetry implies that the distribution for the first card drawn is the same as the distribution for each of the other cards. It also implies that the joint distribution of the first two cards drawn is the same as the joint distribution of the third and fifth, or of any other pair.   ■

To see that the sample observations in sampling from a finite population without replacement are exchangeable, it may help to ignore the time or sequencing in the sampling and think of simultaneous selection. Assigning subscripts simultaneously to the observations, it should be clear that the joint distribution of (X_1, X_2) is the same as that of (X_4, X_2), for example.

Chapter Perspective

This chapter has introduced the language of random variables and probability functions, univariate and multivariate. Multivariate distributions are important for two reasons. (1) Researchers are often seeking to understand relationships among variables. (2) Even in the case of a univariate distribution, a sample from it is a set of several variables.

The notion of independence is important in representing the lack of any relationship among variables. Dependence among variables can be exploited for purposes of control or prediction (Chapter 15).

Sampling a population is the basic means of learning about its unknown characteristics. Independence of the observations in a sample makes the sampling process easiest to analyze, as we'll see in Chapter 8. Exchangeability or symmetry of the sample observations has important consequences that apply when sampling is done at random, with or without replacement.

An important new idea of the chapter is that of conditioning—revising a probability model in the light of new information. A probability model for some mechanism defines probabilities of outcomes, but a researcher needs a way of drawing conclusions about the mechanism that produces a set of data, given the data. Bayes's theorem shows us how to reverse the roles of mechanism and data. It is fundamental to the process of learning in that it provides a way of updating what we know in view of new data. (See Chapter 16.)

In Chapter 7, we'll examine ways of organizing and summarizing the data in a sample. We first take up some useful ways of describing unknown aspects of populations (Chapter 3) and then study some important, special classes of discrete random variables (Chapter 4). Continuous variables will be studied in Chapters 5 and 6.

Solved Problems

Sections 2.1–2.2

A. In Example 1.7c, we gave probabilities for the number of tosses of a coin required to obtain heads for the first time. Let X denote this number of tosses. Its p.f. is

$$f(k) = P(X = k) = \left(\frac{1}{2}\right)^k, \qquad k = 1, 2, \ldots$$

(a) Find $P(X > 2)$.
(b) Find $P(X \text{ is odd})$.

Solution:

(a) $P(X > 2) = 1 - f(1) - f(2) = 1 - \dfrac{1}{2} - \dfrac{1}{4} = \dfrac{1}{4}$.

(b) $P(X = 1, 3, \ldots) = \dfrac{1}{2} + \dfrac{1}{8} + \dfrac{1}{32} + \cdots$

$$= \left(\frac{1}{2}\right)\left[1 + \frac{1}{4} + \frac{1}{16} + \cdots\right]$$

The quantity in brackets is a geometric series with sum $1/(1 - \frac{1}{4})$, so

$$P(X = 1, 3, \ldots) = \frac{\frac{1}{2}}{1 - \frac{1}{4}} = \frac{2}{3}.$$

Alternatively, X is odd if *either* we get heads on the first toss *or* we get two tail, and then it takes an odd number of additional tosses to get the first heads:

$$P = P(X = 1, 3, \ldots) = \frac{1}{2} + \frac{1}{4}P.$$

Solving for P, we find $P = \frac{2}{3}$.

B. Let $\omega = (i, j)$ denote an ordered pair of digits ("digit" $= 0, 1, \ldots, 9$), and let $P(\omega) = .01$ for each pair. Find the p.f. of each of the variables $Y = i + j$ and $Z = i - j$.

Solution:
Referring to Table 2.1, which gives $P[(i, j)]$, we see that $P(Y = 0) = P[(0, 0)] = .01$, $P(Y = 1) = P[(0, 1) \text{ or } (1, 0)] = .02$, etc. It is apparent that $P(Y = k) = (k + 1)/100$, up to $k = 9$. After passing the largest diagonal, the number of pairs with a given sum decreases, and $P(Y = k) = (19 - k)/100$ for $k = 10$ through 18.

Table 2.1

j \ i	0	1	2	3	4	5	6	7	8	9
0	.01	.01	.01	.01	.01	.01	.01	.01	.01	.01
1	.01	.01	.01	.01	.01	.01	.01	.01	.01	.01
2	.01	.01	.01	.01	.01	.01	.01	.01	.01	.01
3	.01	.01	.01	.01	.01	.01	.01	.01	.01	.01
4	.01	.01	.01	.01	.01	.01	.01	.01	.01	.01
5	.01	.01	.01	.01	.01	.01	.01	.01	.01	.01
6	.01	.01	.01	.01	.01	.01	.01	.01	.01	.01
7	.01	.01	.01	.01	.01	.01	.01	.01	.01	.01
8	.01	.01	.01	.01	.01	.01	.01	.01	.01	.01
9	.01	.01	.01	.01	.01	.01	.01	.01	.01	.01

The formulas can be combined economically as follows:

$$f_Y(k) = \frac{10 - |k - 9|}{100}, \qquad k = 0, 1, \ldots, 18.$$

We've circled the entries in Table 2.1 for which $Y = 13$. Similarly, differences $i - j$ are constant on diagonals in the other direction, as shown for $k = 7$, where $f_Z(7) = .03$. Thus,

$$f_Z(k) = \frac{10 - |k|}{100}, \qquad k = -9, -8, \ldots, 9.$$

(Observe that Z has the same triangular distribution as Y except that it is shifted 9 units to the left.)

C. Find the probability function of Y, the number of hearts in a random selection of two cards from a standard deck of playing cards.

Solution:
"Random selection" means that the $\binom{52}{2} = 1326$ combinations of two cards are equally likely. To find the probability, we need to count the combinations with 0, 1, and 2 hearts:

None: $\binom{39}{2} = 741$,   One: $\binom{39}{1}\binom{13}{1} = 507$,   Two: $\binom{13}{2} = 78$.

So

$$f(0) = \frac{741}{1326} = \frac{57}{102}, \qquad f(1) = \frac{39}{102}, \qquad f(2) = \frac{6}{102}.$$

D. Suicides in the U.S. in 1975 were classified as shown in Table 2.2, given as percentages for males and percentages for females. Find the following:
(a) The proportion of suicides that were by poison.
(b) The proportion of suicides that were males.
(c) The proportion of suicides that were not males who took poison.
(d) The proportion of suicides that were by males or by poison.

Table 2.2

Category	Male	Female
Firearms, explosives	63.5	35.7
Poison	15.8	43.0
Hanging, strangulation	14.3	11.5
Other	6.3	9.75
Number:	18,595	7,088

Solution:
Frequencies (percentage times column total, rounded) are as follows:

Category	Male	Female	Total
Firearms, explosives	11,810	2,530	14,340
Poison	2,940	3,050	5,990
Hanging, strangulation	2,660	810	3,470
Other	1,170	690	1,860
Total:	18,580	7,080	25,660

There is no point to selecting at random, but it is convenient to use the notation of probability: A population proportion can be interpreted as a probability under the assumption of a random selection from the population.

(a) Proportion who took poison $= P(\text{poison}) = \dfrac{5990}{25660} = .233.$

(b) $P(\text{Male}) = \dfrac{18580}{25660} = .721.$

(c) $1 - P(\text{male \& poison}) = 1 - \dfrac{2940}{25660} = .885.$

(d) $P(\text{poison}) + P(\text{male}) - P(\text{male \& poison}) = .233 + .721 - .115 = .839,$
or: $P(\text{male}) + P(\text{female \& poison}) = .721 + .119 = .839.$

E. In Problem 36 of Chapter 1, we picked three socks at random from a drawer containing ten socks—four blue, four green, two red. Let $G = \#(\text{green})$, $R = \#(\text{red})$ in the selection of three.

- **(a)** Construct a probability table for the joint distribution of G and R.
- **(b)** Give the marginal distributions of G and R.
- **(c)** Find $P(R + G = 2)$.
- **(d)** Find $P(R = G)$.

Solution:

(a) Possible values: 0, 1, 2, 3 for G; 0, 1, 2 for R. However, some pairs are impossible (for example, $G = 2$, $R = 2$). For two of the possible pairs, the probabilities are

$$P(G = 0, R = 1) = \frac{\binom{4}{0}\binom{2}{1}\binom{4}{2}}{\binom{10}{3}} = \frac{1}{10},$$

$$P(G = 1, R = 1) = \frac{\binom{4}{1}\binom{2}{1}\binom{4}{1}}{\binom{10}{3}} = \frac{4}{15}.$$

The complete table of joint probabilities is as follows:

R \\ G	0	1	2	3	f_R
0	1/30	6/30	6/30	1/30	14/30
1	3/30	8/30	3/30	0	14/30
2	1/30	1/30	0	0	2/30
f_G	5/30	15/30	9/30	1/30	1

(b) The marginal distributions are shown in the right and lower margins of the table in (a).

(c) $P(R + G = 2) = \dfrac{1 + 8 + 6}{30} = \dfrac{1}{2}.$

(d) $P(R = G) = \dfrac{1 + 8 + 0}{30} = \dfrac{3}{10}.$ $[R = G = 0, \text{ or } 1, \text{ or } 2.]$

Section 2.3

F. A digit is selected at random from 0, 1, 2, . . . , 9. Consider the following events:

E: The digit is even.

T: The digit is a multiple of 3.

F: The digit is a multiple of 4.

Find   **(a)** $P(E \mid F)$,   **(b)** $P(F \mid E)$,   **(c)** $P(E \mid T)$,   **(d)** $P(T \mid E)$,
   (e) $P(T \mid F)$,   and   **(f)** $P(F \mid T)$.

Solution:

(a) $F \subset E$, so $P(EF) = P(F)$; $P(E \mid F) = \dfrac{P(EF)}{P(F)} = \dfrac{P(F)}{P(F)} = 1$.

(b) $P(F \mid E) = \dfrac{P(EF)}{P(E)} = \dfrac{P(F)}{P(E)} = \dfrac{\frac{2}{10}}{\frac{5}{10}} = \dfrac{2}{5}$.

(c) $P(E \mid T) = \dfrac{P[\{6\}]}{P[\{3, 6, 9\}]} = \dfrac{1}{3}$.

(d) $P(T \mid E) = \dfrac{P[\{6\}]}{P[\{0, 2, 4, 6, 8\}]} = \dfrac{1}{5}$.

(e) $P(T \mid F) = 0$,   since $TF = \varnothing$.

(f) $P(F \mid T) = 0$,   for the same reason as (e).

G. According to life tables used by the National Center for Health Statistics, the proportion of black males born in 1973 who will live to age 20 is .947. The proportion who live to age 65 is .499. On the basis of these proportions, find the probability that a black male who is 20 years old in 1993 will reach age 65.

Solution:

We take the given proportions as probabilities:

$$P(\text{Age} > 20) = .947, \qquad P(\text{Age} > 65) = .499.$$

The desired probability is conditional:

$$P(\text{Age} > 65 \mid \text{Age} > 20) = \frac{P(\text{Age} > 65 \text{ and Age} > 20)}{P(\text{Age} > 20)}.$$

But $\{\text{Age} > 65\} \subset \{\text{Age} > 20\}$, so their intersection is $\{\text{Age} > 65\}$:

$$P(\text{Age} > 65 \mid \text{Age} > 20) = \frac{P(\text{Age} > 65)}{P(\text{Age} > 20)} = \frac{.499}{.947} \doteq .527.$$

H. Referring to the joint distribution in Problem D above, and again interpreting proportions as probabilities, find the following:

(a) $P(\text{poison} \mid \text{male})$ **(d)** $P(\text{female} \mid \text{firearms or explosives})$

(b) $P(\text{male} \mid \text{poison})$ **(e)** $P(\text{firearms or explosives} \mid \text{not poison})$.

(c) $P(\text{female} \mid \text{poison})$

Solution:

(a) This is just the proportion of males who took poison, given as 15.8% or .158. For (b)–(e), refer to the solution to Problem D:

(b) $\dfrac{P(\text{male \& poison})}{P(\text{male})} = \dfrac{.115}{.233} = \dfrac{2940}{5990} = .491$.

(c) Since female is the complement of male, this is 1 minus the answer to (b): .509.

(d) $P(\text{female} \mid \text{firearms or explosives}) = \dfrac{2530/25660}{14340/25660} = .177$.

(e) $P(\text{firearms or explosives} \mid \text{not poison}) = \dfrac{14340/25660}{1 - .233} = .729$.

I. A carton of ten items includes two that are defective and eight that are good. We select two at random, one at a time and without replacement. Find the probability that
   (a) The first is defective and the second is good (D_1 and G_2).
   (b) The first is good and the second is defective (D_2 and G_1).
   (c) One is defective and the other is good.
   (d) Both are defective.
   (e) Neither is defective.

Solution:

(a) $P(D_1G_2) = P(D_1) \cdot P(G_2 \mid D_1) = \dfrac{2}{10} \cdot \dfrac{8}{9} = \dfrac{8}{45}.$

Alternatively, we can count sequences:

$$P(D_1G_2) = \frac{(2)_1(8)_1}{(10)_2} = \frac{2 \cdot 8}{10 \cdot 9}.$$

(b) $P(G_1D_2) = P(G_1) \cdot P(D_2 \mid G_1) = \dfrac{2}{10} \cdot \dfrac{8}{9} = \dfrac{8}{45}.$

(c) $= (a) + (b) = \dfrac{16}{45}.$

(d) $P(D_1D_2) = P(D_1)P(D_2 \mid D_1) = \dfrac{2}{10} \cdot \dfrac{1}{9} = \dfrac{1}{45}.$

Alternatively, we can count combinations:

$$P(D_1D_2) = \binom{2}{2} / \binom{10}{2} = \frac{1}{45}.$$

(e) $P(\text{Both good}) = \dfrac{8}{10} \cdot \dfrac{7}{9}.$

Or, $P(G_1G_2) = 1 - [(c) + (d)]$, since (c), (d), and (e) exhaust the possibilities.

J. Two cards are picked at random from a standard deck. Find the probability that
   (a) Both are hearts, given that both are red.
   (b) Both are face cards, given that they are of the same suit.
   (c) Both are red, given that the second one is red.
   (d) Both are red, given that at least one is red.

Solution:

"At random" tells us that the $\binom{52}{2}$ possible pairs are equally likely. Define these events: $H_i =$ the ith card is a heart; $R_i =$ the ith card is red; and $F_i =$ the ith card is a face card.

(a) $P(\text{both hearts} \mid \text{both red})$

$$= \frac{P(H_1H_2R_1R_2)}{P(R_1R_2)} = \frac{P(H_1H_2)}{P(R_1R_2)} = \frac{\binom{13}{2}/\binom{52}{2}}{\binom{26}{2}/\binom{52}{2}} = \frac{6}{25} = .24.$$

Alternative reasoning: There are $\binom{13}{2} = 78$ pairs of hearts in the reduced sample space of 26 red cards, and $\binom{26}{2} = 325$ pairs altogether (equally likely), so the answer is $\frac{78}{325} = .24$.

(b) $P(\text{both } F \mid \text{same suit})$

$$= \frac{P(F_1 F_2 \text{ and same suit})}{P(\text{same suit})} = \frac{4 \cdot \binom{3}{2}/\binom{52}{2}}{4 \cdot \binom{13}{2}/\binom{52}{2}} = \frac{1}{26} \doteq .04.$$

One could also reason that, whatever the suit, there are 3 pairs of face cards out of $\binom{13}{2} = 78$ equally likely pairs: $\frac{3}{78}$.

(c) $P(R_1 R_2 \mid R_2) = \dfrac{P(R_1 R_2)}{P(R_2)} = \dfrac{(26/52)(25/51)}{1/2} = \dfrac{25}{51} \doteq .49.$

(d) $P(R_1 R_2) = P(R_1^c R_2^c) = \dfrac{26}{52} \cdot \dfrac{25}{51} \doteq \dfrac{25}{102},$

so it follows that

$$P(R_1 R_2 \mid R_1 \cup R_2) = \frac{P(R_1 R_2)}{1 - P(R_1 \cup R_2)} \doteq \frac{25/102}{1 - 25/102} = \frac{25}{77}.$$

[Observe the distinction between the conditioning events in (c) and in (d). In both cases, we know that there is at least one red card, but in (c) we also know that a particular draw produced a red card. In (c), the reduced sample space is $\{R_1 R_2, R_1^c R_2\}$, whereas in (d) it is $\{R_1 R_2, R_1^c R_2, R_1 R_2^c\}$.]

K. In Problem E above, we gave the joint distribution of random variables R (number of red socks) and G (number of green socks) in this table of joint probabilities:

R \ G	0	1	2	3	f_R
0	1/30	6/30	6/30	(1/30)	14/30
1	(3/30	8/30	3/30	0)	(14/30)
2	1/30	1/30	0	0	2/30
f_G	5/30	15/30	9/30	(1/30)	1

Find the conditional p.f. of G given $R = 1$ and the conditional p.f. of R given $G = 3$.

Solution:

The row and column circled in the table correspond to the conditions $R = 1$ and $G = 3$, respectively. In each case, the conditional p.f. has the shape of the cross-section defined by the condition. Thus, for $R = 1$, the joint probabilities are $\frac{3}{30}, \frac{8}{30}, \frac{3}{30}, 0$, and the conditional probabilities are in the same proportion $(3:8:3:0)$. We find them by dividing by their sum so that the conditional probabilities add up to 1:

$$\frac{3}{14}, \frac{8}{14}, \frac{3}{14}, 0.$$

When $G = 3$, the only possible value of R is 0. The values of the (conditional) p.f. are 1, 0, 0 for $R = 0, 1, 2$, respectively.

Section 2.4

L. A laboratory test for detecting a certain disease occasionally gives a false positive or a false negative. Suppose 2% of healthy individuals get a false positive: $P(+ \,|\, H) = .02$, and 2% of diseased individuals get a false negative: $P(- \,|\, D) = .02$. Suppose further that 1 in 1,000 persons has the disease: $P(D) = .001$. Find $P(D \,|\, +)$, the probability that a person who gets a positive reading has the disease.

Solution:

We're after $P(D \,|\, +)$, but we're given $P(+ \,|\, D)$; so Bayes's theorem [Equation (5) or (7) of Section 2.4] should help:

$$P(D \,|\, +) = \frac{P(+ \,|\, D)P(D)}{P(+)} = \frac{P(+ \,|\, D)P(D)}{P(+ \,|\, D)P(D) + P(+ \,|\, H)P(H)}$$

$$= \frac{.98 \times .001}{.98 \times .001 + .02 \times .999} \doteq .047.$$

Why is this probability so small when the test equipment seems to be rather accurate? Essentially because the disease is so rare: A positive reading is about 20 times more likely to occur because of a false reading than because the person has the disease. (The probability that the person has the disease has actually increased 47-fold with the knowledge of the positive reading.) We also see that $P(-) = 1 - P(+) = .97904$, and

$$P(D \,|\, -) = \frac{P(- \,|\, D)P(D)}{P(-)} = \frac{.02 \times .001}{.97904} \doteq .00002.$$

The tree diagram is shown in Figure 2.7.

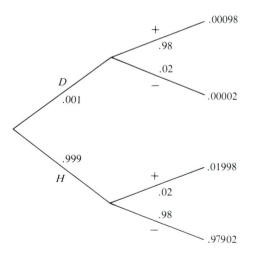

Figure 2.7 Tree diagram for Problem L

M. In a certain high school, the sizes of the sophomore, junior, and senior classes are equal. Of the sophomores, 10% smoke; 15% of the juniors smoke; and 20% of the seniors smoke.

(a) Find the proportion of smokers in the high school.

(b) Find the proportion of sophomores among the smokers in the school.

Solution:

Regarding the given proportions as probabilities, we are given

$$P(\text{soph.}) = P(\text{jun.}) = P(\text{sen.}) = \frac{1}{3},$$

$$P(S \mid \text{soph.}) = .10, \qquad P(S \mid \text{jun.}) = .15, \qquad \text{and} \qquad P(S \mid \text{sen.}) = .20.$$

(a) By the law of total probability [Equation (6) of Section 2.4],

$$P(S) = .10 \times \frac{1}{3} + .15 \times \frac{1}{3} + .20 \times \frac{1}{3} = .15.$$

(b) Bayes' theorem [Equation (7) of Section 2.4] tells us that the proportions of sophomores, juniors, and seniors among the smokers are proportional to the terms of the sum that gives us $P(S)$ in (a). Thus,

$$P(\text{soph.} \mid S) = \frac{P(S \mid \text{soph.})P(\text{soph.})}{P(S)}$$

$$= \frac{.10 \times \frac{1}{3}}{.15}$$

$$= \frac{2}{9}.$$

N. A contestant in the TV show "Let's Make a Deal" was presented with three boxes. One box contained a valuable prize, the key to a Lincoln Continental. She chose box 2. The MC then opened box 1 and showed that it was empty. He claimed that her chances had now gone up from one in three to one in two! Is this correct? (Assume that the MC knew which box contained the key and would not open that box.)

Solution:

We've assumed that the MC will open box 1 if the key is in box 3 and box 3 if the key is in box 1. We need to know what he would do if the key were in the contestant's box; assume that he chooses *randomly* between boxes 1 and 3. Initially,

$$P(\text{key in 1}) = P(\text{key in 2}) = P(\text{key in 3}) = \frac{1}{3}.$$

We are assuming:

$$P(\text{shows 1} \mid \text{key in 1}) = 0, \qquad P(\text{shows 1} \mid \text{key in 3}) = 1,$$

and

$$P(\text{shows 1} \mid \text{key in 2}) = \frac{1}{2}.$$

Then

$$P(\text{key in 2}\,|\,\text{shows 1}) = \frac{P(\text{shows 1}\,|\,\text{key in 2}) \cdot P(\text{key in 2})}{P(\text{shows 1})}.$$

$$= \frac{(1/2)(1/3)}{0 + (1/2)(1/3) + (1)(1/3)} = \frac{1}{3}.$$

This is consistent with the intuitive notion that the MC has given no information in opening a box you know that he knows is empty. The MC (Monte Hall) put it this way in a letter to a statistician, "Oh, and incidentally, after one [box] is seen to be empty, her chances are no longer 50–50 but remain what they were in the first place, one out of three. It just seems to the contestant that one box having been eliminated, she stands a better chance. Not so."

Sections 2.5–2.6

O. You toss a coin and I throw a die. What is the probability that you get heads (H) *and* I throw a six?

Solution:
There is no apparent connection between the toss of the coin and the throw of the die, so we assume that they are independent experiments. Then

$$P(H \text{ and six}) = P(H) \cdot P(\text{six}) = \frac{1}{2} \cdot \frac{1}{6} = \frac{1}{12}.$$

P. Articles are produced by a process whose output is 1% defective. (Take this to mean that as each article is produced, there is one chance in a hundred that it will be defective.) Find the probability that no defectives turn up in a batch of 50 articles, assuming independence of the successive trials.

Solution:
The probability that all articles are good is the *product* of the probabilities for each [by Equation (4) of Section 2.5]:

$$P(\text{all good}) = (.99)(.99) \cdots (.99) = (.99)^{50} \doteq .605.$$

Q. Find the probability that the total number of points thrown with three dice is five.

Solution:
We denote the outcomes of the individual dice by (X, Y, Z) and assume these to be independent. The sequences with a sum of five are (1, 1, 3), (1, 3, 1), (3, 1, 1), (1, 2, 2), (2, 1, 2), and (2, 2, 1). Each of these six sequences has probability $(\frac{1}{6})^3$, so $P(X + Y + Z = 5) = 6/6^3 = \frac{1}{36}$.

R. A random variable X has three possible values: $\{a, b, c\}$. A second variable Y has two possible values: $\{r, s\}$. Their joint distribution is defined by the accompanying table of probabilities.

	a	b	c
r	$\dfrac{1}{4}$	$\dfrac{1}{8}$	$\dfrac{1}{8}$
s	$\dfrac{1}{4}$	0	$\dfrac{1}{4}$

(a) Are the events $X = a$ and $Y = r$ independent?
(b) Are the variables X and Y independent?
(c) Find $P(Y = s \mid X = c)$.
(d) Construct a probability table with the same marginals but in which the variables are independent.

Solution:

(a) Yes. $P(X = a \text{ and } Y = r) = \frac{1}{4} = P(X = a) \cdot P(Y = r)$.
  $[P(X = a)$ is the first column total, $\frac{1}{2}$, and $P(Y = r)$ the first row total, also $\frac{1}{2}$.]

(b) No—the multiplication in (a) does not hold throughout the table. The most obvious failure is the entry 0:

$$P(X = b \text{ and } Y = s) = 0 \neq \frac{1}{8} \cdot \frac{1}{2} = P(X = b) \cdot P(Y = s).$$

(c) $P(Y = s \mid X = c) = \dfrac{P(Y = s \text{ and } X = c)}{P(Y = c)} = \dfrac{1/4}{3/8} = \dfrac{2}{3}.$

(d) With marginal totals filled in, we multiply them to get joint probabilities:

$\dfrac{1}{4}$	$\dfrac{1}{16}$	$\dfrac{3}{16}$	$\dfrac{1}{2}$
$\dfrac{1}{4}$	$\dfrac{1}{16}$	$\dfrac{3}{16}$	$\dfrac{1}{2}$
$\dfrac{1}{2}$	$\dfrac{1}{8}$	$\dfrac{3}{8}$	

S. Suppose $P(E) = .3$ and $P(F) = .5$. Find $P(E \cup F)$ in these two cases: (a) E and F are disjoint. (b) E and F are independent.

Solution:

We apply the addition rule: $P(E \cup F) = P(E) + P(F) - P(EF)$.

(a) If E and F are disjoint, $P(EF) = 0$, and

$$P(E \cup F) = P(E) + P(F) = .3 + .5 = .8.$$

(b) When E and F are independent, the multiplication rule for independent events says that $P(EF) = P(E) \cdot P(F) = .15$. So

$$P(E \cup F) = .3 + .5 - .15 = .65.$$

[It is clear from these two calculations that disjointness and independence are *not* the same! They are actually opposites, in the sense that E and F cannot be both disjoint and independent unless either $P(E)$ or $P(F)$ is 0:

E, F independent implies $P(EF) = P(E) \cdot P(F)$,

E, F disjoint implies $P(EF) = P(\emptyset) = 0.$]

T.   Show that if A, B, and C are independent events, then A is independent of the event $B \cup C$.

Solution:

We need to show that $P[A(B \cup C)] = P(A)P(B \cup C)$. Independence of A, B, and C means that

(i) $P(ABC) = P(A)P(B)P(C)$, and

(ii) $P(AB) = P(A)P(B), \qquad P(BC) = P(B)P(C), \qquad P(AC) = P(A)P(C).$

Then

$$P[A(B \cup C)] = P(AB \cup AC)$$

(intersection is distributive over unions—Equation (1) of Section 1.1)

$$= P(AB) + P(AC) - P[(AB)(AC)]$$

(by the extended addition rule—Property (ix) of Section 1.7)

$$= P(A)P(B) + P(A)P(C) - P(A)P(B)P(C)$$

(from independence, (i)–(ii) above)

$$= P(A)[P(B) + P(C) - P(BC)]$$

(upon factoring out $P(A)$ and using (ii))

$$= P(A)P(B \cup C)$$

(again using the extended addition rule).

U.   Thirteen cards are dealt from a standard deck, one at a time, after thorough shuffling. Find the probability that the eighth card dealt is a heart, given that the fifth card is a heart.

Solution:

In sampling without replacement from a finite population, the outcomes of the various selections are exchangeable, as are all pairs of selections. This means that the desired probability is the same as the probability that the second card is a heart given that the first card is a heart. Since with one heart missing there are 12 hearts among the 51 cards available for the second selection: $P(H_8 \mid H_5) = \frac{12}{51}$.

V.   A bowl contains five white and three black chips. Two chips are selected at random, one at a time. Find the probability that the first is black, given that the second is white,

(a)   If there is replacement and mixing after the first selection.

(b)   If there is no replacement.

Solution:

Let B_1 denote the event that a black is drawn first, W_2 the event that a white is drawn second, and so on.

(a) The selections are independent, so $P(B_1 \mid W_2) = P(B_1) = \frac{3}{8}$.

(b) The selections are exchangeable, so $P(B_1 \mid W_2) = P(B_2 \mid W_1) = \frac{3}{7}$. (With one white gone, there are four white and three black chips left.) Or, we could use Bayes's theorem:

$$P(B_1 \mid W_2) = \frac{P(W_2 \mid B_1)P(B_1)}{P(W_2 \mid B_1)P(B_1) + P(W_2 \mid W_1)P(W_1)} = \frac{(5/7)(3/8)}{(5/7)(3/8) + (4/7)(5/8)} = \frac{3}{7}.$$

Problems

For problems marked with an asterisk, answers are provided in Appendix 2.

Sections 2.1–2.2

*1. Example 1.4a gave the 36 outcomes for the toss of two dice. Let a and b denote the numbers of points on the first die and second die, respectively. Assume the 36 outcomes are equally likely and construct a probability table for the random variable $X(a, b) = a + b$.

*2. Referring to Problem 1, construct the probability table for the random variable $Y(a, b) = a - b$.

3. A carton of 12 eggs includes two that are rotten. Let X denote the number of rotten eggs among three chosen at random from the carton. Give the probability distribution of X. (That is, list the possible values of X and find the probability of each.)

*4. Two chips are drawn without replacement from a bowl containing five chips numbered from 1 to 5. List the ten possible selections of and construct probability tables for these random variables:

(a) $S = $ sum of the numbers on the selected chips.

(b) $D = $ magnitude of their difference (that is, their absolute difference).

(c) $U = $ larger of the two numbers drawn.

*5. A report based on the 1970 census gives the cross-classification of state and local government employees (in thousands) shown in Table 2.3.

Table 2.3

	State	Local	Total
Education	939	3,723	4,662
Highways	295	298	593
Health	443	465	908
Police	47	396	443
Other	604	1,823	2,427
Total	2,328	6,705	9,033

Find the probability that one individual selected at random from these 9,033,000 government employees
  (a)   is in education.
  (b)   is a state employee.
  (c)   is in education or police work.
  (d)   is a state employee not in education.

6.   A bowl contains five chips, numbered 1, 2, 3, 4, and 5. Two are selected, one at a time and at random, *with* replacement. Let X denote the number on the first chip selected and Y the number on the second chip. Construct a table of joint probabilities and use it to find
  (a) $P(Y = 2)$.   **(c)** $P(X = 1, \text{ and } Y = 2 \text{ or } 4)$.
  (b) $P(X + Y = 5)$.   **(d)** $P(|X - Y| = 2)$.

∗7.   We want to know the relationship (if any) between the level of education of a person and his or her opinion about the death penalty proposed for skyjackers. Suppose that the joint probabilities for these two variables are as follows:

		Opinion	
		Favor	*Oppose*
	Grade school	.15	.05
Education:	*High school*	.20	.10
	College	.25	.25

  (a)   Give the probability table for the variable "education."
  (b)   Give the table for the variable "opinion."

8.   Repeat Problem 6 with the change that the second selection is without replacement of the first.

∗9.   Let (X, Y) have the joint distribution defined by the following table of joint probabilities:

		X		
		1	*2*	*3*
	1	.2	0	.4
Y	*2*	0	.3	.1

  (a)   Find the marginal p.f.'s.
  (b)   Find $P(X + Y > 3)$.
  (c)   Find the p.f. of $Z = X + Y$.

10.   Consider again the variables S and D defined in Problem 4.
  (a)   Give the table of joint probabilities.
  (b)   Find the p.f. of the variable $X = S - D$.

Section 2.3

∗11.   A card is selected at random from a standard deck of playing cards. Consider the events

A (ace), *S* (spade), and *B* (black). Find

(a) $P(A|S)$.   (c) $P(S|B)$.   (e) $P(A|B)$.

(b) $P(S|A)$.   (d) $P(B|S)$.   (f) $P(B|A)$.

12. An ordinary die (six faces numbered 1, 2, . . . , 6) is tossed. Find the following, assuming the faces to be equally likely:

   (a) The probability that the toss results in a 4, given that it is even.

   (b) The probability that the outcome is 6, given that it is divisible by 3.

*13. In the game of bridge, each of four players is dealt 13 of the 52 cards.

   (a) A player gets a hand with just one ace. What is the probability that his partner has no aces?

   (b) Another player sees that she and her partner (the player opposite) have the ace and queen of spades but not the king or jack. What is the probability that both the king and jack are held by the player on her left?

14. The probability that a white male born in 1973 will live to age 40 is .97, and the probability that he will live to age 65 is .66. Find the probability that if he reaches age 40, he will live to be 65.

*15. Three cards are selected at random from a standard deck of playing cards. Find the probability

   (a) that all are spades, given that all are black.

   (b) that none is an ace, given that none is a face card. (A face card is a king, queen, or jack.)

   (c) that none is a spade, given that the three cards selected are of different suits.

16. Two chips are selected without replacement from a bowl with five chips numbered 1 to 5, as in Problem 6. Find the probability

   (a) that the first chip is 2, given that the second is 3.

   (b) that the 2 and 3 are chosen (in any order), given that the sum is 5.

   (c) that the first chip is 2, given that the sum is 5.

*17. Suppose the two chips in Problem 16 are selected *with* replacement of the first chip before the second selection. Find the probability that the 4 is drawn both times,

   (a) given that the first one drawn is the 4.

   (b) given that at least one of the two is a 4.

*18. To see a certain executive, I have to get by two secretaries. I figure that my chances are one in five of getting past the outer secretary; and if I do, the chances are one in five that I get past the private secretary. What is the probability (according to my figuring) that I get to see the executive?

19. A card is drawn at random from a deck of playing cards, and then another is drawn at random from those that remain. Find the probability that

   (a) both are hearts.

   (b) the first is not a heart but the second is a heart.

   (c) the second is a heart.

*20. I am to draw either *A* or *B* as my opponent in a certain game. My chances of beating *A* are 4 in 10 and of beating *B*, 8 in 10. If *A* and *B* are equally likely to be my opponent, what is the probability that I win this game?

21. Show that $P(EF|G) = P(E|FG)P(F|G)$ for any events E, F, and G for which the probabilities are defined.

22. Three chips are drawn at random without replacement from a bowl containing four white and six black chips.

(a) Find the probability that the first is black and the second and third are white.

(b) Find the probability that the sample includes exactly one black and two white chips, in two ways: (1) By finding [as in (a)] the probabilities of WBW and WWB, and (2) by assuming that the $\binom{10}{3}$ selections of three chips are equally likely and counting combinations.

*23. Consider this two-stage experiment: First roll a regular four-sided die with outcomes labeled $1, 2, 3$, and 4. Suppose the result of this roll is $X = k$. At the second stage, select k chips at random without replacement from a bowl with two red and two black chips. Let $Y = $ the number of red chips selected.

(a) Construct a tree diagram and label the branches with the appropriate probability or conditional probability.

(b) Construct a table of joint probabilities for (X, Y). (These are the products of the probabilities along the 12 paths.)

(c) Obtain the marginal distribution of Y in the table of (a).

(d) Find the probability that $X = 4$, given that $Y = 2$.

24. Suppose the distribution of rat hairs in a jar of peanut butter is as follows:

Number (ω)	0	1	2	3	4
Probability [$P(\omega)$]	.8	.1	.05	.03	.02

You open a jar and come across a rat hair.

(a) For your jar, find the distribution of the number of rat hairs given this information.

(b) What is the probability that there are more than the one you have found? (That is, what is the probability that there were more than one in the jar, given that there was at least one?)

Section 2.4

*25. Suppliers A and B provide, respectively, 10% and 90% of a certain item that is used in large quantities. Suppose that 2% of those provided by A are defective and 5% of those provided by B are defective. A randomly selected item is found to be defective; find the probability that the item was supplied by A.

26. Of the undergraduates in a certain university, 10% are foreign students, and 20% of the graduate students are foreign students. There are four times as many undergraduates as graduate students. Find the probability that a randomly chosen foreign student is an undergraduate.

27. A letter to the editor of a medical journal includes this argument:

Let us say that the prevalence of stenosis was fivefold that of achalasia ... , i.e., .030 vs. .006. Given the frequency of an absent gastric air bubble found by

Orlando: 50% in achalasia, 17% in stenosis, one could conclude that in a randomly selected group of 1,000 patients, stenosis with an absent air bubble would be more likely than achalasia . . . .[4]

Use Bayes' theorem to reproduce the writer's calculation of odds of about 5 to 3 in favor of stenosis given an absent air bubble.

*28. Evidence in a paternity suit indicates that four particular men are equally likely to have been the father of the child. Blood typing in the ABO system reveals that the phenotypes of mother and child are both type O. The men are found to have blood types as shown in the following table. The table also gives the probability for each man of producing a type O child with the type O mother. (These probabilities are determined from genetic theory together with, in the case of the first two men, observed proportions of blood types in the population.)

Man i	Phenotype	P(type O child | man i & O-mother)
1	A	.431
2	B	.472
3	O	1
4	AB	0

Find the probability that man 1 is the father, given all the evidence.

29. Suppose the numbers of freshmen, sophomores, juniors, and seniors in a certain college are in the proportion 6:5:4:3, respectively. Females make up 15% of the freshmen, 20% of the sophomores, 30% of the juniors, and 35% of the seniors. What fraction of the females are freshmen?

*30. A study on learning disability[5] used a Bayesian analysis to estimate the probability that a child of a parent with a reading disability would also have a reading disability. Notation: FD = father disabled, CD = child disabled, and CN = child normal. Based on sample information, it was estimated that $P(FD|CD) = .29$ and $P(FD|CN) = .04$. The population incidence of child reading disability $P(CD)$ is subject to debate, so the study report gave results for various values of $P(CD)$. Calculate $P(CD|FD)$
   (a) assuming $P(CD) = .05$.
   (b) assuming $P(CD) = .10$.

Sections 2.5–2.6

*31. You are to toss two dice. Assuming independent tosses and "fair" dice (equally likely sides), find the probability that
   (a) both dice show a "six."
   (b) the two dice match.
   (c) the total number of points showing is 4.

---

[4] S. L., Ludwig in *Gastroenterology 76* (1979), 432.
[5] Reported in *Journal of Learning Disabilities 17* (1984), 616–618.

*32.   A baseball player has a batting average of .330. Assume this to be the probability that he gets a hit each time he bats, and that the times at bat are independent. (These assumptions may not be exactly appropriate, but they're not bad.) What is the probability that he gets no hits in four times at bat?

*33.   At one point in the 1978 National League baseball season, Pete Rose had hit in 37 consecutive games. He needed 20 more to break Joe DiMaggio's major-league record of 56 straight games. The Las Vegas odds were 99 to 1 against his doing it. See if you can approximate these odds by calculating along the lines of the preceding problem. Assume independence, an average of .330, and four at-bats per game.

34.   Each card in a standard deck is assigned a number of "points" by the Goren system of bridge bidding: Ace gets 4 points, king 3, queen 2, jack 1, and any other card gets 0. Let X denote the number of points for a card drawn at random from the deck, and let Y be the indicator variable of the event "heart." (That is, $Y = 1$ if the card is a heart, otherwise 0.) Show that X and Y are independent.

*35.   Let Z denote the total number of points (see Problem 34) and W the number of hearts in a bridge hand (a random deal of 13 cards from a standard deck). Are Z and W independent? (*Hint*: Look at the distribution of Z, given $W = 13$.)

36.   A set of Christmas-tree lights has eight bulbs in "series" (all the lights go out if any one bulb fails). If the probability that any given bulb does not burn out during the first 100 hours is .9, what is the probability that the string stays lit for 100 hours? (Assume that the bulbs burn out independently.)

37.   A news item in 1984 told of a female birth, the first daughter born in that family in at least 130 years. A Department of Health statistician said that the probability of such a string of male births is almost zero, and calculated the probability of ten pregnancies resulting in ten boys to be about .00098. How did he obtain this result, and what assumptions are required to get it? (*Note*: Despite the very low probability, the existence of a family with ten male births in a row is not surprising in view of the enormous number of families in the U.S.)

*38.   To play "Russian Roulette," the player places a bullet in one of the six chambers of a revolver, spins the cylinder, points the revolver at his or her head, and pulls the trigger. The probability of surviving one play is $\frac{5}{6}$. (We recommend that any experimentation be done with a simulation: Select one chip at random from a bowl containing six chips, one marked "bullet." Spinning the chamber corresponds to replacing the chip and mixing.) Suppose you plan to play two games of Russian Roulette in succession. Find your chances of survival if

(a)   you spin the cylinder between plays.

(b)   you do not spin the cylinder between plays.

39.   Suppose you toss four ordinary dice independently. Find the probability that there is at least one pair. (For instance, the sequences 2 4 2 1 and 1 3 1 1 each have at least one pair.)

*40.   Find the probability that in a sequence of independent tosses of a die, a "six" turns up before the fourth toss.

41. Random variables X and Y have the joint probabilities given in the following table:

X \ Y	1	2	3
a	$\frac{1}{12}$	$\frac{1}{6}$	$\frac{1}{12}$
b	$\frac{1}{6}$	0	$\frac{1}{6}$
c	0	$\frac{1}{3}$	0

    (a) Are X and Y independent? (Explain.)
    (b) Find $P(Y = 2 \mid X = a)$.
    (c) Find $P(X = a \mid Y = 2)$.
    (d) Find $P(X = a \mid Y \geq 2)$.
    (e) Suppose that U and V are independent and have the same marginal distributions as X and Y. Construct the table of joint probabilities for U and V.

*42. Determine whether any two of the events A, S, and B in Problem 11 are independent.

43. Consider again Problem 11.
    (a) The events B^c and S are disjoint. Are they independent?
    (b) Are A, S, and B independent?

*44. Given: A, B, and C are pairwise independent, and their union is the whole sample space. If $P(A) = .6$, $P(B) = .4$, and $P(C) = .3$, find $P(ABC)$.

45. Show the following:
    (a) If X and Y are independent as defined by Equation (2) of Section 2.5, then
$$f(x \mid y) = f_X(x).$$
    (b) If X, Y, and Z are independent, then $f(x \mid y, z) = f_X(x)$.

46. Suppose events B and C are disjoint. Show that if A is independent of B and independent of C, then A is independent of $B \cup C$.

47. Consider a sample space Ω of five outcomes: $\{\omega_1, \ldots, \omega_5\}$, with probabilities
$$P(\omega_1) = \frac{1}{8}, \qquad P(\omega_2) = P(\omega_3) = P(\omega_4) = \frac{3}{16}.$$

Let $A = \{\omega_1, \omega_2, \omega_3\}$, $B = \{\omega_1, \omega_2, \omega_4\}$, and $C = \{\omega_1, \omega_3, \omega_4\}$. Show that $P(ABC) = P(A)P(B)P(C)$, but that A, B, and C are not pairwise independent.

48. Show that if A, B, C, and D are independent, then AB and $C \cup D$ are independent.

*49. Consider a sequence of random selections, one at a time without replacement, from an urn with eight black and five red balls. Find the probability that
    (a) the fifth one selected is black.
    (b) the fifth one selected is black, given that the seventh one is red.
    (c) The fifth one is black and the seventh one red.

50. Let M denote the number of ω's with $X(\omega) = x$ in a population of N individuals, and consider a sequence of n individuals selected at random without replacement. Let X_i denote the outcome of the ith selection. Show that $P(X_i = x) = M/N$ for all i by finding the number of possible sequences and the number of sequences in which $X_i = x$.

51. Define the joint p.f. of X, Y, Z as follows:

$$f(x, y, z) = \begin{cases} 1/4 & \text{for } (1, 0, 0), (0, 1, 0), (0, 0, 1), \text{ or } (1, 1, 1), \\ 0 & \text{otherwise.} \end{cases}$$

(a) Show that the variables X, Y, and Z are pairwise independent but not independent.

(b) Show that X and $Y + Z$ are not independent.

Averages

Example **3a**

"Let's Make a Deal"

A contestant on the TV game show "Let's Make a Deal" had won $9,000 in prizes and was then offered the option of trading these for whatever lay behind one of three doors. He was to choose one of the doors and would receive the prize behind that door. Behind one door was a $20,000 prize; behind another, a $5,000 prize; and behind the third, a $2,000 prize. Should the contestant take the option or stick with what he had already won?

As far as the contestant was concerned, the arrangement of prizes behind the three doors was random. Thus, he was equally likely to choose the $2,000 door, the $5,000 door, or the $20,000 door. The value of the door he would choose is a random variable. He must compare this random quantity with a nonrandom quantity ($9,000). Specifying a single number to take the place of a random quantity in such a situation lies at the heart of averaging. ■

Numerical variables—temperature, earnings, blood pressure, bowling scores, and the like—are often described using a single number called an *average*. In common usage, *average* suggests little more than *ordinary* or *typical*, so in this sense an *average* value would be one that is somewhere in the middle of the values that can occur—neither very high nor very low relative to the other values. Among various ways of making this notion of average more precise, the most usual is the *arithmetic mean*, which we take up first.

Averaging is also useful in measuring variability and other aspects of a distribution, in measuring the strength of relationships, and in defining *generating functions*—powerful tools for dealing with the sums of random variables that arise in calculating means.

3.1   The Mean

Example 3.1a   **Value of the Random Door**

We return to the game show situation that opened this chapter. The amount X the contestant would receive (in dollars) has this probability function:

x	$f(x)$
2,000	1/3
5,000	1/3
20,000	1/3

To assign a numerical value to this random option, imagine that the distribution of X describes a "game" to be played repeatedly, say, 3,000 times. Using the long-run interpretation of the probabilities, we'd expect that about "one-third of the time" (or about 1,000 times), the contestant would get $2,000, one-third of the time $5,000, and one-third $20,000. In 3,000 trials, he would expect to get about

$$\$2000 \cdot 1000 + \$5000 \cdot 1000 + \$20000 \cdot 1000 = \$27,000,000.$$

Distributing this equally over the 3,000 trials gives the contestant about $9,000 per trial. This amount per trial has been called, traditionally, the "mathematical expectation" for a single play of the game:

$$\$2,000 \times \frac{1}{3} + \$5,000 \times \frac{1}{3} + \$20,000 \times \frac{1}{3} = \$9,000.$$

The contestant's mathematical expectation in playing the game happens to be precisely equal to the value of his nonrandom alternative—what he has already won and can choose to keep. The "expected" reward in choosing a door ($9,000) is not one of the amounts he could win, but it is between the extremes of $2,000 and $20,000.

Despite the apparent equivalence of the two options, some very rational people would distinctly prefer the sure $9,000 and others, the random option. We'll return to this point in Example 3.2a.   ■

The older term *mathematical expectation* has evolved into *expected value*, and a common notation for the expected value of X is $E(X)$. (When there is no possibility of confusion, we may omit the parentheses and write EX.) The terms *mean value*, *mean*, and *average value* are also commonly used to denote this characteristic of a distribution. The symbol μ is commonly used as a generic abbreviation for *mean*. When there is more than one random variable in a given context, we either use a subscript for clarity (μ_X, for example) or use $E(X)$.

> **Expected value** of a numerical random variable X with probability function $f(x)$:
>
> $$E(X) = \sum xf(x), \tag{1}$$
>
> provided the sum is absolutely convergent. This is also called the *mean value* or *average value* of X and is often denoted by μ or μ_X.

Example 3.1b

Expected Number of Correct Guesses

Let Y denote the number of soft-drink brands that Michael gets correct in his test described in Problem 13 of Chapter 1. In Example 2.2c we found the p.f. for Y shown in the accompanying table, which also gives a column of products $yf(y)$ whose sum is the mean value, $E(Y)$:

y	$f(y)$	$yf(y)$
0	2/6	0
1	3/6	3/6
3	1/6	3/6
Sums:	1	$1 = \mu$

So Michael would get one correct "on the average." ∎

Example 3.1c

Blood Types

Consider the distribution of blood types, as given in Example 2.5b:

x	$f(x)$
O	.42
A	.44
B	.10
AB	.04

Blood type is not numerical, so "expected blood type" would not make sense. Only numerical random variables have mean values. ∎

The only ingredients of the formula for expected value are the various possible values x and their probabilities $f(x)$; it is not necessary to refer to the underlying sample space Ω. Nevertheless, the expected value *can* be found in terms of probabilities in Ω:

Expected value calculated from Ω-probabilities:

$$E[X(\omega)] = \sum_{\omega} X(\omega)P(\omega) \qquad (2)$$

This is equivalent to Equation (1).

Example 3.1d **Average Range**

Suppose two chips are selected at random from four chips numbered from 1 to 4. (This is the prototype for sampling at random from finite populations.) The range R is the magnitude of the *difference* between the numbers on the chips that are drawn:

$R(\omega) = $ larger number in $\omega - $ smaller number in ω.

This difference does not depend on the order in which the chips were drawn, so we take the set of $\binom{4}{2} = 6$ possible combinations as the sample space. These six combinations (ω's), their probabilities, and the corresponding R-values are as shown in the following table, which includes a column of products:

ω	$P(\omega)$	$R(\omega)$	$R(\omega)P(\omega)$
1, 2	1/6	1	1/6
1, 3	1/6	2	2/6
1, 4	1/6	3	3/6
2, 3	1/6	1	1/6
2, 4	1/6	2	2/6
3, 4	1/6	1	1/6

From this tabulation, we see that $f_R(1) = \frac{1}{2}$, $f_R(2) = \frac{1}{3}$, and $f_R(3) = \frac{1}{6}$. Using Equation (1), we find

$$E(R) = \sum r f_R(r) = 1 \cdot f_R(1) + 2 \cdot f_R(2) + 3 \cdot f_R(3)$$
$$= 1 \cdot \frac{1}{2} + 2 \cdot \frac{1}{3} + 3 \cdot \frac{1}{6} = \frac{5}{3}.$$

We can also calculate this average directly from the column of products $R(\omega)P(\omega)$, using (2):

$$E(R) = R(1, 2)P(1, 2) + R(1, 3)P(1, 3) + \cdots$$
$$= 1 \cdot \frac{1}{6} + 2 \cdot \frac{1}{6} + 3 \cdot \frac{1}{6} + 1 \cdot \frac{1}{6} + 2 \cdot \frac{1}{6} + 1 \cdot \frac{1}{6} = \frac{5}{3}. \qquad \blacksquare$$

Using (2), we can easily establish some important properties of the mean:

Properties of the expected value:

(i) $E(a) = a$, for any constant a.
(ii) $E(aX) = aE(X)$, for any constant a.
(iii) $E(X + Y) = E(X) + E(Y)$.

These are special cases of

(iv) $E(aX + bY + c) = aE(X) + bE(Y) + c$,
      for any constants a, b, c.

Formula (iv) follows easily from (2):

$$E(aX + bY + c) = \sum [aX(\omega) + bY(\omega) + c]P(\omega)$$
$$= a \sum X(\omega)P(\omega) + b \sum Y(\omega)P(\omega) + c \sum P(\omega)$$
$$= aE(X) + bE(Y) + c.$$

Example 3.1e   **Sum of Two Dice**
In many parlor and gambling games, players toss two dice at each turn. Let X denote the number of points showing on one die and Y the number on the other. The probability functions of X and Y are the same:

$$f_Y(x) = f_X(x) = \frac{1}{6}, \qquad \text{for } x = 1, 2, 3, 4, 5, 6.$$

Applying (1), we see that their mean values are

$$E(X) = E(Y) = (1 + 2 + 3 + 4 + 5 + 6) \cdot \frac{1}{6} = 3.5.$$

The total number of points on the two dice is the sum $X + Y$. According to (iii), the expected total is the sum of the expected values:

$$E(X + Y) = E(X) + E(Y) = 3.5 + 3.5 = 7.$$

Alternatively, we could have first derived the p.f. for $Z = X + Y$ and used (1) to calculate its mean. However, using (iii) is easier.   ∎

Probability as Relative Mass   Calculating a mean value is the same as calculating a center of gravity (c.g.) in physics. If we place masses $m_1, \ldots, m_k$ at $x_1, \ldots, x_k$, respectively, they will balance at

$$\text{c.g.} = \frac{x_1 m_1 + \cdots + x_k m_k}{m_1 + \cdots + m_k} = \sum_{i=1}^{k} x_i \left(\frac{m_i}{\sum m_j} \right).$$

The *relative* masses $m_i/\Sigma\, m_j$ play the same role in this calculation that probabilities play in (1). Like probabilities, they are nonnegative and add up to 1.

The parallel between relative masses and probabilities is important in giving an intuitive feel for an expected value as a balance point. It also provides a way of roughly checking the calculation of a mean value by a visual examination of the probability function. The balance property of the mean $E(X) = \mu$ can be expressed in this way:

The Mean as a Balance Point

$$E(X - \mu) = \sum(x - \mu)f(x) = E(X) - E(\mu) = \mu - \mu = 0. \tag{3}$$

The positive products $(x - \mu)f(x)$ exactly cancel the negative ones.

The balance property shows, incidentally, that if the distribution of X is symmetric about the value $X = a$, then that value is the mean.

If $f(x)$ is **symmetric** about $x = a$—that is, if for all x,

$f(a - x) = f(a + x),$

then (provided EX exists)

$E(X) = a.$

(4)

Example 3.1f

Average of Two Dice, Using Symmetry

The probability function for the total number of points in a toss of two dice is symmetric about 7 (see Problem 1 in Chapter 2):

$$f(2) = f(12) = \frac{1}{36}, \qquad f(3) = f(11) = \frac{2}{36}, \qquad f(4) = f(10) = \frac{3}{36},$$

$$f(5) = f(9) = \frac{4}{36}, \qquad f(6) = f(8) = \frac{5}{36}, \qquad \text{and} \qquad f(7) = \frac{6}{36}.$$

So (4) holds with $a = 7$, which is then the mean (see Example 3.1e). That the distribution balances at 7 is evident from the graph of $f(x)$ in Figure 3.1.   ∎

Conditional distributions, being distributions, may have mean values. Thus, for a conditional distribution with p.f.

$$f(y\,|\,x) = P(Y = y\,|\,X = x),$$

Conditional Mean

the conditional mean of Y given $X = x$ is

$$E(Y\,|\,x) = \sum y \cdot f(y\,|\,x). \tag{5}$$

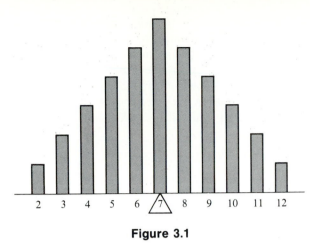

Figure 3.1

Example **3.1g** Let (X, Y) have the distribution defined by the joint probabilities in the following table:

x \ y	0	1	2
1	.3	.2	.1
2	.1	.1	.2

The conditional probabilities for Y given $X = 1$ are in the proportion $3:2:1$, or $\frac{3}{6}, \frac{2}{6}, \frac{1}{6}$. The conditional mean is then

$$E(Y \mid X = 1) = 0 \cdot \frac{3}{6} + 1 \cdot \frac{2}{6} + 2 \cdot \frac{1}{6} = \frac{2}{3}$$

$$= \frac{0 \times .3 + 1 \times .2 + 2 \times .1}{.6}.$$

Similarly,

$$E(Y \mid X = 2) = 0 \cdot \frac{1}{4} + 1 \cdot \frac{1}{4} + 2 \cdot \frac{2}{4} = \frac{5}{4}. \blacksquare$$

3.2 Expected Value of a Function of Random Variables

One is often interested in a *function* of a random variable: $g(X)$. Such transformations are important in a number of settings. The simplest nontrivial transformations are *linear*—the type of transformation we usually encounter in changing from one scale of measurement to another (for example, inches to centimeters or Fahrenheit to Celsius). Nonlinear transformations, such as the square root and logarithm, are useful in data analysis. In particular, they can reduce any adverse effect that large observations have on an analysis. More immediately, we'll encounter the nonlinear functions $g(x) = x^2$ (in defining a variance in Section 3.3) and $h(x) = t^x$ (in the generating functions of Section 3.6).

Any function of a random variable is also a random variable. Thus, if $X(\omega)$ is a random variable and g a function, then $Y(\omega) = g[X(\omega)]$ is also a random variable. In more abbreviated notation, we write $Y = g(X)$.

Finding the expected value of $Y = g(X)$ is a straightforward application of (2) of Section 3.1:

$$E[Y(\omega)] = \sum Y(\omega)P(\omega) = \sum g[X(\omega)] \cdot P(\omega).$$

Alternatively, we may think of the set of possible X-values itself as a sample space, with probabilities given by $f(x)$. Then $Y = g(X)$ can be treated as a random variable defined on this sample space. With this interpretation the mean $E(Y)$ is a weighted average of the values of $g(x)$ at the various possible x-values; the weights are $f(x)$:

Mean value of a function g of a random variable X:

$$E[g(X)] = \sum g(x)f(x). \qquad (1)$$

When $g(x) = x$, this reduces to (1) of Section 3.1.

Finding the mean of $g(X)$ by means of (1) usually involves less work than using the defining formula, $E(Y) = \sum y f_Y(y)$, because (1) avoids the need for finding the p.f. f_Y.

Example **3.2a**   **Utility of the Random Door**

The game show contestant of Example 3a had the option of choosing between keeping \$9,000 already won and a random prospect X with p.f.

$$f(x) = \frac{1}{3}, \qquad \text{for} \qquad x = 2{,}000,\ 5{,}000,\ \text{or}\ 20{,}000.$$

His expected reward from this random prospect is \$9,000, calculated in Example 3.1a. This average equals the \$9,000 he had already won (and could choose to keep). Is he indifferent between the two options? It is quite unlikely that he would be willing to let a coin toss make the decision for him. A rational choice would depend on the contestant's financial circumstances.

Expected
Utility

The usefulness or value of money is a function of amount, but not necessarily proportional to amount. The **utility function** is a subjective, numerical assessment of the usefulness of money for the purpose of comparing alternatives. In utility theory, choices among random prospects are made by maximizing *expected* utility.

In the present instance, some would tend to choose the certain \$9,000, not wanting to risk going down to \$2,000 or \$5,000 even though there is a chance of ending up with \$20,000. Suppose, however, the contestant must pay off a \$20,000 debt the next day and has no other source of funds. The \$9,000 he has with certainty would not be much more useful to him than $2,000. The contestant's utility function might be approximated by the following:

$$u(M) = \begin{cases} 0, & M < 20{,}000, \\ 1, & M \geq 20{,}000. \end{cases}$$

This treats all dollar amounts of $20,000 or more as equally valuable or "useful," and all dollar amounts of less than $20,000 as valueless. The expected utility, from (1), is

$$E[u(X)] = \sum u(x)f(x)$$

$$= [u(2{,}000) + u(5{,}000) + u(20{,}000)] \cdot \frac{1}{3}$$

$$= \frac{0 + 0 + 1}{3} = \frac{1}{3}.$$

The utility of $9,000 is 0, less than $\frac{1}{3}$, so the contestant should take the option of choosing a door. In this simple situation, the conclusion is obvious, because the random prospect offers some chance of reaching the important plateau of $20,000. Problem 10 considers other utility functions for which the optimal choice may be different. ∎

For a given y, the value of the conditional p.f. $f(y|x)$ is a function $g(x)$, to which we can apply (1) to find $E[g(X)]$:

$$E[f(y|X)] = \sum_x f(y|x)f_X(x)$$

$$= \sum_x \frac{f(x, y)}{f_X(x)} \cdot f_X(x)$$

$$= \sum_x f(x, y) = f_Y(y).$$

The conditional mean $E(Y|x)$ is another function $g(x)$, to which we can apply (1) to obtain an **iterated expectation**:

Iterated Expectation

$$E[E(Y|X)] = \sum_x \sum_y y \cdot \frac{f(x, y)}{f_X(x)} \cdot f_X(x) = \sum_y y \left\{ \sum_x f(x, y) \right\} = E(Y).$$

Given random variables (X, Y),

$$E[f(y|X)] = f_Y(y) \qquad \text{for all } y, \tag{2}$$

and

$$E[E(Y|X)] = E(Y). \tag{3}$$

Example 3.2b   Consider again the distribution of Example 3.1g. The conditional probabilities for $Y \mid X$ are shown in the following table:

y	0	1	2	$f_X(x)$
$f(y \mid X = 1)$	1/2	1/3	1/6	.6
$f(y \mid X = 2)$	1/4	1/4	1/2	.4
$f_Y(y)$	.4	.3	.3	1

Weighting the entries $f(y \mid x)$ with $f_X(x)$ and summing, according to (2), yields the unconditional probabilities in the bottom margin. For example,

$$f_Y(0) = \frac{1}{2} \times .6 + \frac{1}{4} \times .4 = .4.$$

To illustrate (3): In Example 3.1g, we found

$$E(Y \mid X = 1) = \frac{2}{3},$$

$$E(Y \mid X = 2) = \frac{5}{4}.$$

Then, using (3), we calculate EY as an iterated expectation:

$$E[E(Y \mid X)] = \frac{2}{3}f_X(1) + \frac{5}{4}f_X(2) = \frac{2}{3}(.6) + \frac{5}{4}(.4) = .9.$$

This value of EY is of course the same as that calculated using the marginal distribution of Y in the usual formula: $\Sigma \, yf(y)$.   ■

Functions of Several Variables

Some important relationships among variables involve functions of more than one random variable. Consider first the case of two random variables X and Y, with probability function $f(x, y)$. A function $g(x, y)$ defines a new random variable $Z = g(X, Y)$. From (2) of Section 3.1, with $\omega = (x, y)$ and $P(\omega) = g(x, y)$, we have

$$E[g(X, Y)] = \sum_{(x, y)} g(x, y)f(x, y). \tag{4}$$

As in the case of a single variable, (4) is usually simpler than first finding $f_Z(z) = P(Z = z)$ and then using the formula $E(Z) = \Sigma \, zf_Z(z)$.

Consider the particular function $g(x, y) = x$. Applying (4), we find

$$E(X) = \sum_x \sum_y xf(x, y) = \sum_x x \sum_y f(x, y) = \sum_x xf_X(x).$$

Of course, this agrees with the defining formula for $E(X)$ given by (1) of Section 3.1.

Example 3.2c   As in Example 3.1e, let X and Y denote the numbers showing when two dice are rolled. This time, as in one setting of the game "Risk," we are interested in the larger of the two numbers: $V = \max\{X, Y\}$. All of the possible pairs are equally likely:

$$f(x, y) = \frac{1}{36}.$$

The values of V for the various possible pairs are as shown in the following array:

y x	1	2	3	4	5	6
1	1	2	3	4	5	6
2	2	2	3	4	5	6
3	3	3	3	4	5	6
4	4	4	4	4	5	6
5	5	5	5	5	5	6
6	6	6	6	6	6	6

Using (4), we would calculate $E(V)$ by multiplying each table entry by $\frac{1}{36}$ and adding. Accumulating terms in the sum that involve like values of V, we obtain

$$E(V) = 1 \cdot \frac{1}{36} + 2 \cdot \frac{3}{36} + 3 \cdot \frac{5}{36} + 4 \cdot \frac{7}{36} + 5 \cdot \frac{9}{36} + 6 \cdot \frac{11}{36} = \frac{161}{36} \doteq 4.47.$$

Alternatively, we can find $E(V)$ by finding the p.f. f_V and applying (1) of Section 3.1. The process of finding f_V is identical to that of accumulating probabilities of "like values" of V in the above calculation.

The graph of the p.f. f_V is given in Figure 3.2. It shows 4.47 to be reasonable as the balance point of the distribution.

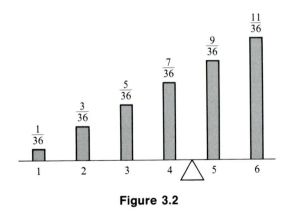

Figure 3.2   ■

Applying (4) to the case of independent random variables yields an important result. Consider the product of functions $g(X)$ and $h(Y)$, where X and Y are independent. The joint probability function of independent variables factors [(2) of

Section 2.5], so

$$E[g(X)h(Y)] = \sum_{(x,y)} g(x)h(y)f(x, y) = \sum_x \sum_y g(x)f_X(x)h(y)f_Y(y)$$
$$= \sum g(x)f_X(x) \sum h(y)f_Y(y) = E[g(X)] \cdot E(h(Y)].$$

When X and Y are independent,

$$E[g(X)h(Y)] = E[g(X)]E[h(Y)]. \tag{5}$$

In view of the factorization of the probability function in the case of several independent variables [(4) in Section 2.5], formula (5) extends to that case in the obvious way.

3.3   Variability

The average value is only one characteristic of a random variable. The "average" temperature in Minneapolis, for example, is far from the whole story. There can be large deviations from the average—it has been as hot as 106°F and as cold as $-34°$F. The amount of variability or dispersion of values about the middle value is important. We describe such variability in terms of deviations from the "middle."

The Mean Deviation

The distance of X from its mean is the absolute deviation $|X - \mu|$, a random variable. We take the *average* of this distance, the **mean absolute deviation**, as a measure of dispersion in the distribution of the random variable X:

$$\text{m.a.d.} = E|X - \mu| = \sum |x - \mu| \cdot f(x). \tag{1}$$

This is also commonly called the *mean deviation*, for simplicity. When the distribution is dispersed over a wide range, there will be large deviations from the mean, and the m.a.d. will tend to be large. When the distribution is concentrated in a narrow range, it will tend to be small.

Example 3.3a

Let X be the number of points that show at a toss of a fair die. The absolute deviations of the possible values of X from its mean, $\mu = 3.5$, are

2.5, 1.5, .5, .5, 1.5, 2.5.

Since the values of X are equally likely, so are these absolute deviations, and the mean deviation is their ordinary average. So we simply add them and divide by 6:

$$\text{m.a.d.} = \frac{9}{6} = 1.5. \quad \blacksquare$$

Variance

A more traditional way to measure variability is to average the *squared* deviations. The average squared deviation is called the **variance** of the distribution.

It is denoted by var X or by σ^2 (or σ_X^2):

$$\text{var } X = E[(X - \mu)^2].\tag{2}$$

An alternative formula for variance is often easier to work with:

$$\text{var } X = E(X^2) - \mu^2.\tag{3}$$

In words, this says that the variance of X is its average square minus the square of its average. To show that formulas (2) and (3) for var X are equivalent we simply expand the squared deviation in (2):

$$(X - \mu)^2 = X^2 - 2\mu X + \mu^2$$

and average term by term:

$$E[(X - \mu)^2] = E(X^2) - 2\mu E(X) + \mu^2.$$

Since $E(X) = \mu$, the last two terms combine to give $-\mu^2$. This establishes (3).

Example 3.3b   Consider again the number of points that show at a toss of a fair die. The variance of this variable X is the weighted sum of the squares of the deviations of its values from the mean, $\mu = \frac{7}{2}$, the weight being the corresponding probability—$\frac{1}{6}$, for each possible value:

$$\sigma^2 = E\left[\left(X - \frac{7}{2}\right)^2\right] = \sum_{i=1}^{6}\left(i - \frac{7}{2}\right)^2 \cdot \frac{1}{6} = \frac{35}{12}.$$

Alternatively, we can calculate this using (3); we first find

$$E(X^2) = \sum_{i=1}^{6} i^2 \cdot \frac{1}{6} = \frac{91}{6}.$$

Then

$$\text{var } X = \frac{91}{6} - \left(\frac{7}{2}\right)^2 = \frac{182}{12} - \frac{147}{12} = \frac{35}{12}. \qquad \blacksquare$$

Variance of a random variable:

$$\text{var } X = E[(X - \mu)^2] = \sigma^2.\tag{2}$$

Alternative formula for calculating σ^2:

$$\text{var } X = E(X^2) - \mu^2.\tag{3}$$

It is clear from (2) that a variance is never negative. If you come up with a *negative* variance when using (3), you have made a mistake in your calculations.

Moments   The average deviation and average squared deviation are examples of **moments**. The rth moment of X (or of probability distribution) about c is defined by analogy with the corresponding moment of a mass distribution:

$$E[(X - c)^r] = \sum (x - c)^r f(x).$$ (4)

The first moment about 0 is the mean or expected value:

$$E(X - 0) = E(X) = \mu.$$

The first moment of X about its mean is 0:

$$E(X - \mu) = 0,$$ (5)

by (3) of Section 3.1.

Moments about 0 [$c = 0$ in (4)] are called, simply, *moments*. Moments about the mean [$c = \mu$ in (4)] are **central moments**.

Parallel Axis Theorem

The alternative formula (3) is a special case of the **parallel axis theorem.**[1] To develop this result, we start with a deviation about the value $x = c$ and add and subtract μ:

$$X - c = (X - \mu) + (\mu - c).$$

We then square the right-hand side and average term by term [using Properties (i)–(iv) of Section 3.1]:

$$E[(X - c)^2] = E[(X - \mu)^2 + (\mu - c)^2 + 2(X - \mu)(\mu - c)]$$
$$= \sigma^2 + (\mu - c)^2 + 2(\mu - c)E(X - \mu).$$

The last term is 0, since $E(X - \mu) = 0$. So the second moment about $x = c$ can be obtained from the second moment about $x = \mu$ by adding $(\mu - c)^2$:

Parallel axis theorem:

$$E[(X - c)^2] = \sigma^2 + (\mu - c)^2.$$ (6)

Clearly, setting $c = \mu$ *minimizes* (6) and yields the variance. The variance is therefore the *smallest* second moment.

Although a variance measures variability, it is awkward to interpret. In particular, the unit of measurement of var X is the square of the unit of X. So a variance is measured in square pounds, square seconds, and the like. A more easily interpreted measure of dispersion is the (positive) square root of the variance. This is a kind of average deviation, called the **standard deviation** (s.d.). Its units are the same

Standard Deviation

---

[1] The terminology is borrowed from physics.

as those of X:

> Standard deviation of a random variable:
> $$\sigma_X = \sqrt{\operatorname{var} X} = \sqrt{E(X^2) - \mu^2}. \tag{7}$$

You may find it helpful to think of the standard deviation as a typical or ordinary deviation from the mean.

The standard deviation (7) and the mean deviation (1) are "typical" deviations, and both have the dimensions of X, but they are usually different. In the early development of statistical methodology, there was considerable controversy as to whether the standard deviation or the mean deviation should be used to measure dispersion. The standard deviation prevailed, and in current statistical practice it still dominates. However, the mean deviation has the important advantage that it is less influenced by the large deviations that often occur in actual populations.

Example 3.3c   Continuing Example 3.3b, we find the standard deviation of the number of points thrown with the toss of a die to be

$$\sigma = \sqrt{\operatorname{var} X} = \sqrt{\frac{35}{12}} \doteq 1.71.$$

We see that σ is not as small as the smallest of the deviations (.5) nor as large as the largest (2.5). This is true generally: σ is a typical deviation. In Example 3.3a, we found the mean deviation to be 1.5, not far from but somewhat smaller than σ. (See Figure 3.3.) ∎

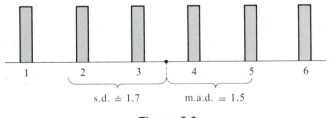

Figure 3.3

The idea of mean (absolute) deviation can help in understanding the standard deviation. As in the preceding example, the mean deviation is usually somewhat smaller than the standard deviation. It is never larger (see Problem 23). It is sometimes easy to guess the value of the mean deviation, and sometimes even easy to calculate it.

It is often necessary or convenient to rescale a measurement by means of a linear transformation: $Y = a + bX$. The moments of this new variable Y are related to those of X. According to Property (iv) of Section 3.1, the mean of Y is obtained by applying the same linear transformation to the mean of X:

$$\mu_Y = E(Y) = a + bE(X).$$

The deviation of Y from its mean involves only the scale factor

$$Y - \mu_Y = a + bX - [a + b\mu_X] = b[X - \mu_X].$$

Hence,

$$\text{var } Y = E[b^2(X - \mu_X)^2] = b^2 \text{ var } X.$$

To find the standard deviation, we take the square root:

$$\sigma_Y = |b|\sigma_X.$$

Moments of a linear function of $Y = a + bX$:

$$\mu_Y = a + b\mu_X,$$
$$\sigma_Y = |b|\sigma_X. \tag{8}$$

3.4   ## Covariance and Correlation

In studying relationships between two variables, a *product moment* is useful—the average product of the deviations of the two variables from their respective means. This is called their **covariance**, calculated using (4) of Section 3.2:

$$\text{cov}(X, Y) = E[(X - \mu_X)(Y - \mu_Y)].$$
$$= \sum (x - \mu_X)(y - \mu_Y)f(x, y). \tag{1}$$

The covariance is an extension of the notion of variance in the sense that it reduces to the variance when $Y = X$:

$$\text{cov}(X, X) = E[(X - \mu_X)(X - \mu_X)] = \text{var } X.$$

We'll sometimes use this notation for covariance:

$$\sigma_{X,Y} = \text{cov}(X, Y).$$

An alternative formula, analogous to the alternative formula for variance [(3) in Section 3.3], can be more convenient for calculations:

$$\text{cov}(X, Y) = E(XY) - \mu_X\mu_Y. \tag{2}$$

This follows upon expanding the product in (1) and collecting terms, just as we derived (3) in Section 3.3. Indeed, with $Y = X$, (2) reduces to (3) of Section 3.3.

Covariance of X and Y:

$$\sigma_{X,Y} = \text{cov}(X, Y) = E[(X - \mu_X)(Y - \mu_Y)]. \tag{1}$$

Alternatively,

$$\text{cov}(X, Y) = E(XY) - \mu_X\mu_Y. \tag{2}$$

Example 3.4a   In Examples 2.2b and 2.5c, we considered selecting two chips at random without replacement from four chips numbered 1, 2, 3, and 4. The joint distribution of the first chip (X) and the second chip (Y) is given by the following table:

y \ x	1	2	3	4	$f_Y(y)$
1	0	1/12	1/12	1/12	1/4
2	1/12	0	1/12	1/12	1/4
3	1/12	1/12	0	1/12	1/4
4	1/12	1/12	1/12	0	1/4
$f_X(x)$	1/4	1/4	1/4	1/4	1

We'll use (2) to find $\text{cov}(X, Y)$. To find the expected product $E(XY)$, we multiply x and y in each of the twelve possible pairs and then multiply these products by their corresponding probabilities (each $\frac{1}{12}$) and add:

$$E(XY) = (2 + 3 + 4 + 2 + 6 + 8 + 3 + 6 + 12 + 4 + 8 + 12)\frac{1}{12} = \frac{70}{12}.$$

From the symmetry of the marginals, it follows that $\mu_X = \mu_Y = \frac{5}{2}$. Using Equation (2), we find

$$\sigma_{X,Y} = \frac{70}{12} - \frac{5}{2} \cdot \frac{5}{2} = -\frac{5}{12}.$$

Problem 27 asks you to verify this using Equation (1).]   ∎

A covariance is a measure of concordance or association of two variables. If X tends to be large when Y is large, and small when Y is small, then the deviations in (1) will tend to be of the same sign and their products positive. The covariance, the sum of these products, will therefore be positive. Similarly, if X tends to be small when Y

is large and large when Y is small, the product of the deviations will tend to be negative, and the covariance will be negative. (This is what happened in the above example.)

Still another possibility is that relatively large X's occur with both small and large Y's. Then positive and negative products will tend to cancel, and the covariance will be close to zero. Two variables X and Y with zero covariance are said to be *uncorrelated*. In particular, if X and Y are *independent*, then they are uncorrelated. To see this, apply (4) of Section 3.2 with $g(x) = x$ and $h(y) = y$:

$$E(XY) = E(X)E(Y). \tag{3}$$

Clearly, (2) shows that in this case the covariance is 0.

Example 3.4b   Consider the distribution for (X, Y) defined by the following table of joint probabilities:

x	2	4	6	$f_Y(y)$
y				
1	1/5	0	1/5	2/5
2	0	1/5	0	1/5
3	1/5	0	1/5	2/5
$f_X(x)$	2/5	1/5	2/5	

There are five pairs with positive probability; for these, the products of the deviations from the respective means ($\mu_X = 4$ and $\mu_Y = 2$) are $-2, 2, 0, 2$, and -2. With equal weights ($\frac{1}{5}$ each), their average is zero. So from (1), $\sigma_{X,Y} = 0$, and the variables X and Y are uncorrelated. However, X and Y are *not* independent, as is evident from the 0's in the table of joint probabilities. ■

> Variables X and Y are said to be **uncorrelated** when $\sigma_{X,Y} = 0$. Independent variables are always uncorrelated, but $\sigma_{X,Y} = 0$ does *not* imply independence.

Linear Transformation   The deviations in (1) do not depend on the location of the reference origin; they do not change with a translation of origin. So if we apply linear transformations to the variables X and Y, the covariance is changed only by multiplication by the scale factors:

$$\text{cov}(a + bX, c + dX) = E[(bX - b\mu_X)(dY - d\mu_Y)] = bd \cdot \text{cov}(X, Y). \tag{4}$$

The linear transformation

$$\begin{cases} U = a + bX \\ V = c + dY \end{cases} \tag{5}$$

changes the covariance to

$$\sigma_{U,V} = bd\sigma_{X,Y}. \tag{6}$$

Equation (6) shows that covariance changes when the units of measurement are changed. For example, suppose X is measured in pounds and Y in inches. The same quantities measured in kilograms and centimeters are

$$U = 2.20X \quad \text{and} \quad V = .394Y.$$

From (6), we have

$$\text{cov}(U, V) = 2.20 \times .394 \,\text{cov}(X, Y) = .87\, \sigma_{X,Y} \,(\text{kg-cm}).$$

The *correlation coefficient* is the unitless measure obtained from the covariance by dividing it by the product of the standard deviations:

Coefficient of correlation:

$$\rho_{X,Y} = \frac{\sigma_{X,Y}}{\sigma_X \sigma_Y}. \tag{7}$$

The correlation coefficient does *not* change when the units of measurement are changed by the linear transformation (5), provided $bd > 0$. Using (6) above and (8) of Section 3.3, we have

$$\rho_{U,V} = \frac{\sigma_{U,V}}{\sigma_U \sigma_V} = \frac{bd\sigma_{X,Y}}{|b|\sigma_X|d|\sigma_Y} = \rho_{X,Y} \quad \text{if} \quad bd > 0.$$

Example 3.4c   Let (X, Y) have the joint distribution defined by the following table of probabilities:

x \ y	1	2	3	4	$f_X(x)$
1	0	0	0	.2	.2
2	0	.1	.2	0	.3
3	0	.2	.1	0	.3
4	.2	0	0	0	.2
$f_Y(y)$	.2	.3	.3	.2	1

In this distribution, all of the probability is on or near the line $x + y = 5$. From the symmetry, it follows that the mean values of X and Y are equal: $EX = EY = 2.5$. The variances are also equal:

$$\text{var } X = \text{var } Y = E(X^2) - (2.5)^2 = 7.3 - 6.25 = 1.05.$$

The expected product is

$$E(XY) = .2(4 + 6 + 6 + 4) + .1(4 + 9) = 5.3.$$

From (7), the correlation coefficient is

$$\rho_{X,Y} = \frac{5.3 - (2.5)^2}{\sqrt{1.05}\sqrt{1.05}} = -.905.$$

The correlation is negative because of the "inverse" nature of the relationship—X tends to be large when Y is small, and vice versa. The closeness to -1 reflects the near collinearity of the joint distribution.   ∎

Correlation Bounded

A correlation coefficient cannot exceed 1 in magnitude. This is a consequence of the Schwarz inequality, which is derived as follows. Given random variables X and Y, let

$$U = X - \mu_X \qquad \text{and} \qquad V = Y - \mu_Y.$$

For arbitrary real z, we have

$$0 \le E[(V - zU)^2] = E(V^2) - 2zE(UV) + z^2E(U^2)$$
$$= az^2 + bz + c. \tag{8}$$

Since this quadratic is nonnegative for all real z, its discriminant $(b^2 - 4ac)$ must be nonpositive, for if it were positive, there would be two real zeros. In that case the quadratic would be negative for some values of z. Thus,

$$b^2 - 4ac = 4[E(UV)]^2 - 4E(U^2)E(V^2) \le 0, \tag{9}$$

or

$$[\text{cov}(X, Y)]^2 \le (\text{var } X)(\text{var } Y). \tag{10}$$

This is (one form of) the Schwarz inequality. Thus, from (7),

$$\rho_{X,Y}^2 \le 1. \tag{11}$$

When $\rho^2 = 1$, the inequality (9) is an equality, so the discriminant $b^2 - 4ac$ vanishes. The quadratic function in (8) then has double zero at some point $z = z_0$, and

$$E[(V - z_0U)^2] = 0. \tag{12}$$

The average of a nonnegative quantity can only vanish if that quantity is zero with probability 1. So (12) can hold only if $V = z_0U$ with probability 1—all the probability in the joint distribution of U and V must be on this line. If z_0 (the slope of the line) is negative, then

$$\text{cov}(U, V) = \text{cov}(U, z_0 U) = z_0 \text{ var } U < 0,$$

and $\rho = -1$. Similarly, if the line has a positive slope, $\rho = +1$. If the line has zero slope, then var $Y = 0$, and ρ is undefined.

We'll encounter the correlation coefficient again—for continuous distributions in Section 5.7 and for samples in Section 7.6.

3.5   Sums of Random Variables

In many applications, statisticians add or average the observations in a sample, so we need to study the *sum* of several random variables. We know from Property (iii) of Section 3.1 that the expected value of a sum of two random variables is the sum of their expected values:

$$E(X + Y) = E(X) + E(Y),$$

or, in other notation,

$$\mu_{X+Y} = \mu_X + \mu_Y. \tag{1}$$

From this, it follows by induction that additivity holds for any finite set of random variables:

$$E(X_1 + \cdots + X_n) = E(X_1) + \cdots + E(X_n). \tag{2}$$

Thus, taking expected values is an *additive* operation.

What about the standard deviation—is it additive? Not in general; neither is the variance. However, in some important special cases, the variance is additive, as we show next. The variance of a sum $X + Y$ is the average squared deviation about its mean:

$$\text{var}(X + Y) = E[(X + Y - \mu_{X+Y})^2].$$

Using (1), we have

$$\text{var}(X + Y) = E([(X - \mu_X) + (Y - \mu_Y)]^2).$$

Now expand the square of the binomial in parentheses:

$$[(X - \mu_X) + (Y - \mu_Y)]^2 = (X - \mu_X)^2 + (Y - \mu_Y)^2 + 2(X - \mu_X)(Y - \mu_Y).$$

Averaging term by term, we obtain

$$\text{var}(X + Y) = E[(X - \mu_X)^2] + E[(Y - \mu_Y)^2] + 2E[(X - \mu_X)(Y - \mu_Y)],$$

or

$$\text{var}(X + Y) = \text{var } X + \text{var } Y + 2 \text{ cov}(X, Y). \tag{3}$$

Example 3.5a   Two chips are drawn from a bowl with four chips, numbered 1, 2, 3, and 4. Let X and Y denote, respectively, the larger and smaller of the numbers drawn. The joint distribution is given in the following table:

y x	1	2	3	$f_X(x)$
2	1/6	0	0	1/6
3	1/6	1/6	0	1/3
4	1/6	1/6	1/6	1/2
$f_Y(y)$	1/2	1/3	1/6	1

The means are $\mu_X = \frac{5}{3}$ and $\mu_Y = \frac{10}{3}$, and the variances are equal: var $X =$ var $Y = \frac{5}{9}$. (You should check these.) The average product is the sum of six terms (three cells have probability zero):

$$E(XY) = (2 + 3 + 4 + 6 + 8 + 12)\frac{1}{6} = \frac{35}{6}.$$

So

$$\text{cov}(X, Y) = \frac{35}{6} - \left(\frac{5}{3}\right)\left(\frac{10}{3}\right) = \frac{5}{18}.$$

From (3), the variance of the sum is then

$$\text{var}(X + Y) = \frac{5}{9} + \frac{5}{9} + 2\left(\frac{5}{18}\right) = \frac{5}{3}. \tag{4}$$

We can also find var$(X + Y)$ by first finding the p.f. of the sum $Z = X + Y$:

$$f(z) = \begin{cases} 1/6, & z = 3, 4, 6, 7, \\ 2/6, & z = 5. \end{cases}$$

So $E(Z) = 5$, which of course is $E(X) + E(Y)$, and the variance is

$$\text{var } Z = \frac{1}{6}(4 + 1 + 0 + 1 + 4) = \frac{5}{3}.$$

This agrees with (4).   ■

In general, induction based on (3) shows that the variance of the sum of any finite number of random variables can be expressed in terms of their individual variances and their covariances as follows:

Variance of a sum of random variables:

$$\text{var}\left(\sum_{i=1}^{n} X_i\right) = \sum_{i=1}^{n} \text{var } X_i + 2 \sum_{i>j} \text{cov}(X_i, X_j), \tag{5}$$

the latter sum extending over the $\binom{n}{2}$ pairs (i, j) in which $i > j$.

Formula (5) reduces to (3) when $n = 2$.

In many instances in the chapters to follow, the variables we add will be *independent*. Independent variables are uncorrelated (see Section 3.4), so the covariance terms in (5) vanish.

If $X_1, X_2, \ldots, X_n$ are **independent**, then

$$\text{var}(X_1 + \cdots + X_n) = \text{var } X_1 + \cdots + \text{var } X_n. \tag{6}$$

and

$$\sigma_{\Sigma X} = \sqrt{\sigma_{X_1}^2 + \cdots + \sigma_{X_n}^2}. \tag{7}$$

Example 3.5b **Two Dice**

The total number of points thrown with two dice is a random variable equal to $X + Y$, where X is the number of points on one die and Y the number on the other. In Example 3.3b, we found that the variance of the number of points on a single die is $\frac{35}{12}$. Assuming independence of X and Y, we have

$$\text{var}(X + Y) = \text{var } X + \text{var } Y = \frac{35}{12} + \frac{35}{12} = \frac{35}{6}.$$

The standard deviation of $X + Y$ is the square root, or about 2.4. (Notice that this is not equal to the sum of the standard deviations.) ■

Sampling with Replacement

The additivity of the variance (6) applies in particular to sampling with replacement from a population of finite size N. The successive observations are then independent and all have the same variance—the variance σ^2 of a single selection—so

$$\sigma_{\Sigma X}^2 = n\sigma^2. \tag{8}$$

Sampling without Replacement

However, in sampling *without* replacement (see Sections 1.5 and 2.3), this is not correct. We have seen in Section 2.6 that in this case, successive observations $(X_1, \ldots, X_n)$ are *exchangeable*, and we can exploit this property to derive a formula for the variance of their sum. Exchangeability implies that the X_i all have the same distribution, with the same mean μ and the same variance σ^2. Also, the joint distributions of any pair of X's are the same, so for all i and j,

$$\text{cov}(X_i, X_j) = \text{cov}(X_1, X_2).$$

Call this common value C. Since there are $n(n-1)$ covariances with $i \neq j$, (5) becomes

$$\text{var}\left(\sum_{i=1}^{n} X_i\right) = \sum_{i=1}^{n} \text{var } X_i + \sum_{i \neq j} \text{cov}(X_i, X_j) = n\sigma^2 + n(n-1)C. \tag{9}$$

When $n = N$, the sum ΣX is just the population sum; this is constant, so its variance is 0. With $n = N$ in (9), we obtain

$$0 = N\sigma^2 + N(N-1)C, \qquad \text{or} \qquad C = \frac{-\sigma^2}{N-1}.$$

Substituting this for C in Equation (9) yields the desired result:

Let $(X_1, \ldots, X_n)$ be a sample drawn at random without replacement from a population of size N, mean μ, and variance σ^2. Then

$$E\left(\sum_{i=1}^{n} X_i\right) = n\mu \tag{10}$$

and

$$\text{var}\left(\sum_{i=1}^{n} X_i\right) = n\sigma^2\left(\frac{N-n}{N-1}\right). \tag{11}$$

The factor $(N-n)/(N-1)$ in (10) is termed the **finite population correction factor**. Observe that as N becomes infinite (with n fixed), this factor approaches 1, and (11) approaches (8).

3.6   Probability Generating Functions

Generating functions are tools with many important applications in statistical theory. The first one we take up generates probabilities.

Many of the random variables in the first four chapters take values in the set of nonnegative integers: $0, 1, 2, \ldots$. For such a random variable X, we temporarily adopt the subscript notation usually used for sequences:

$$f(k) = P(X = k) = p_k, \qquad k = 0, 1, 2, \ldots.$$

Consider the power series formed with the elements of the sequence $\{p_k\}$ as coefficients of powers of t:

$$\eta(t) = p_0 t^0 + p_1 t^1 + p_2 t^2 + \cdots = \sum_{0}^{\infty} p_k t^k. \tag{1}$$

This is the expected value of t^X [see (1) in Section 3.2, page 100]:

$$\eta(t) = E(t^X). \tag{2}$$

It is called the **probability generating function** (p.g.f.) of X or of the probability sequence $\{p_n\}$.

Example 3.6a   **A Fair Die**

The model for the throw of a fair die assigns equal probabilities to the six sides: $p_k = \frac{1}{6}$, for $k = 1, 2, 3, 4, 5, 6$. The p.g.f. is then a finite power series:

$$\eta(t) = \frac{1}{6}t + \frac{1}{6}t^2 + \frac{1}{6}t^3 + \frac{1}{6}t^4 + \frac{1}{6}t^5 + \frac{1}{6}t^6.$$

We reduce this using the formula for the sum of a finite geometric series:

$$\eta(t) = \frac{1 - t^7}{6(1 - t)}.$$

You can check this by multiplying $t + t^2 + \cdots + t^6$ by $1 - t$.   ■

Probability generating function (p.g.f.):

$$\eta_X(t) = E(t^X) = \sum_0^\infty p_k t^k, \tag{3}$$

where $p_k = P(X = k)$.

Probability generating functions are useful in calculations involving sums of independent random variables because of this familiar law of exponents: $t^{a+b} = t^a t^b$. Thus, suppose X and Y are independent. The p.g.f. of $Z = X + Y$ is

$$\eta_Z(t) = E(t^Z) = E(t^{X+Y}) = E(t^X t^Y). \tag{4}$$

According to (5) of Section 3.2, the expected product of functions of independent variables is the product of the expected values, so

$$\eta_Z(t) = E(t^X)E(t^Y) = \eta_X(t)\eta_Y(t).$$

This property extends to the sum of any finite number of independent random variables.

The probability generating function (p.g.f.) of the sum of independent random variables is the product of their p.g.f.'s:

$$\eta_{\Sigma X_i}(t) = \prod \eta_{X_i}(t). \tag{5}$$

Example **3.6b**   **A Tetrahedral Die**

A four-sided die is a tetrahedron whose four sides are numbered 1, 2, 3, and 4. Let X denote the number of points showing when we throw the die. Figure 3.4 shows such a die, with the markings that indicate the number of points thrown. Taking the four sides to be equally likely, we have

$$p_k = \begin{cases} 1/4, & k = 1, 2, 3, 4, \\ 0, & \text{otherwise.} \end{cases} \tag{6}$$

The p.g.f. is then

$$\eta_X(t) = \frac{1}{4}t + \frac{1}{4}t^2 + \frac{1}{4}t^3 + \frac{1}{4}t^4.$$

Now suppose we throw *two* such dice. The total number of points thrown is the sum of the two independent variables X_1 and X_2, each with the distribution (6). Let $Y = X_1 + X_2$. Then

$$\eta_Y(t) = \eta_{X_1}(t)\eta_{X_2}(t) = [\eta_X(t)]^2 = \left[(t + t^2 + t^3 + t^4)\cdot\frac{1}{4}\right]^2$$

$$= (t^2 + 2t^3 + 3t^4 + 4t^5 + 3t^6 + 2t^7 + t^8)\frac{1}{16}.$$

We recognize this as the p.g.f. of a discrete random variable whose integer values k have probabilities given by the coefficient of t^k:

k	2	3	4	5	6	7	8
$f_Y(k)$	$\dfrac{1}{16}$	$\dfrac{2}{16}$	$\dfrac{3}{16}$	$\dfrac{4}{16}$	$\dfrac{3}{16}$	$\dfrac{2}{16}$	$\dfrac{1}{16}$

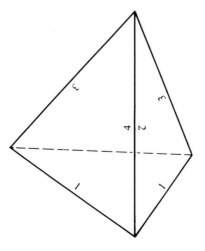

Figure 3.4

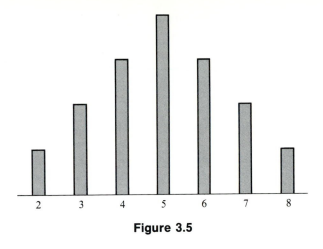

Figure 3.5

Figure 3.5 shows this triangular distribution. A similar distribution results when adding any two identically distributed, uniform random variables. (Compare with Example 3.1f.) ■

A closed form expression for a p.g.f. can help in calculating probabilities. The next example is a case in point.

Example 3.6c   **Razzle Dazzle**

Example 2.5d described a carnival game in which the player throws eight dice at each move and advances according to the total number of points thrown. A chart shows how far to advance for each possible sum: 8, 9, ..., 48. The operator announces that these 41 possibilities are equally likely! We'll calculate the correct probabilities assuming independence of the dice.[2]

In principle, we could calculate the probabilities by enumerating all configurations with a given total and dividing by 6^8, the total number of possible configurations. This is quite involved. (Try it!) It is much easier to use generating functions. The p.g.f. of the number of points on a single die (from Example 3.6a) is

$$\eta(t) = (t + t^2 + t^3 + t^4 + t^5 + t^6) \cdot \frac{1}{6} = \frac{t}{6} \cdot \frac{1 - t^6}{1 - t}.$$

The p.g.f. for Y, the total number of points on the eight dice, is then the eighth power of $\eta(t)$:

$$\eta_Y(t) = \left(\frac{t}{6} \cdot \frac{1 - t^6}{1 - t}\right)^8 = \left(\frac{t}{6}\right)^8 \cdot \sum_0^8 \binom{8}{i}(-t^6)^i \cdot \sum_0^\infty \binom{-8}{j}(-t)^j. \tag{7}$$

The first sum on the right is an ordinary binomial expansion, and the second is a power series expansion of the binomial $(1 - t)^{-8}$. In the latter expansion, an infinite

[2] See D. A. Berry and R. R. Regal, "Probabilities of winning a certain carnival game," *American Statistician 32* (1978), 126–129.

series, the coefficients are calculated formally in the same way as binomial coefficients—a product of j factors starting with -8 and decreasing in steps of 1, divided by $j!$:

$$\binom{-8}{j} = \frac{(-8)(-9)\cdots(-8-j+1)}{j!}.$$

Moving the first factor on the r.h.s. of (7) inside the second summation, we obtain

$$\eta_Y(t) = \sum_0^\infty \sum_0^8 \binom{8}{i}\binom{-8}{j}(-1)^{i+j}t^{8+6i+j}\left(\frac{1}{6}\right)^8.$$

Each term of this double sum includes a power of t, and collecting the terms involving t^k, we can find the coefficient that gives the probability p_k.

As an example, we calculate the probability that $Y = 13$. This is the coefficient of t^{13}, and the only term with this power of t is the one in which $i = 0$, $j = 5$:

$$P(Y = 13) = (-1)^5\binom{8}{0}\binom{-8}{5}\left(\frac{1}{6}\right)^8$$

$$= (-1)\cdot\frac{(-8)(-9)(-10)(-11)(-12)}{5!\cdot 6^8} \doteq .00047.$$

This is a far cry from the $\frac{1}{41}$ announced by the operator of the game.

Some sums require a bit more work. For instance, the probability of a total of 20 is the sum of terms for $i = 0$, $j = 12$; $i = 1$, $j = 6$; and $i = 2$, $j = 0$:

$$P(Y = 20) = \left\{\binom{8}{0}\binom{-8}{12} - \binom{8}{1}\binom{-8}{6} + \binom{8}{2}\binom{-8}{0}\right\}\cdot\frac{1}{6^8} = .02184.$$

The complete probability function of Y is given in Table 3.1 and shown graphically in Figure 3.6. ■

Table 3.1

y	$f(y)$	y	$f(y)$
8	5.95×10^{-7}	19	.01517
9	4.76×10^{-6}	20	.02184
10	2.14×10^{-5}	21	.02994
11	7.14×10^{-5}	22	.03918
12	1.96×10^{-4}	23	.04905
13	4.72×10^{-4}	24	.05883
14	.00102	25	.06769
15	.00200	26	.07477
16	.00366	27	.07935
17	.00624	28	.08094
18	.01001	$28 + k$	Same as for $28 - k$

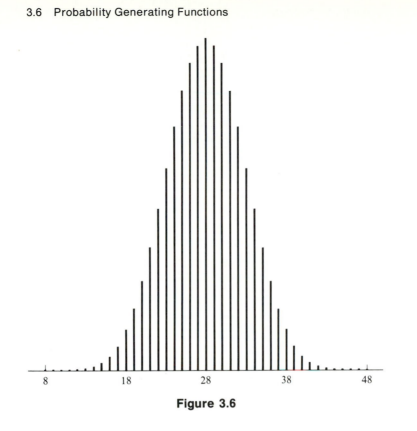

Figure 3.6

The p.g.f. has been defined here for random variables with nonnegative integer values. However, we can define the p.g.f. for any random variable X by the same formula: $\eta_X(t) = E(t^X)$, provided this expectation exists in the neighborhood of $t = 1$. We'll return to this idea in Section 5.10.

Chapter Perspective

Various kinds of averages are used to describe a distribution. The most commonly used averages are the mean, a kind of middle or average location, and the standard deviation, a "typical" deviation from the mean. In this chapter, we have defined averages for discrete random variables. In Chapter 5, we'll extend the definitions of this chapter to the case of continuous probability distributions. In Chapter 7, we'll apply the notion of averaging to sample distributions.

The notion of population average is important in applications. For instance, in Chapters 10, 12, and 14, we assess the effects of treatments on the basis of the average changes they produce in a population. In Chapter 15, we discuss the relationship between two variables in terms of conditional means. Averaging also plays an important role in statistical theory. In estimation (Chapter 9), an estimate of an unknown quantity is judged in terms of how well it does on the average. In Chapters 12 and 16, we use averages in developing procedures for testing hypotheses and for making decisions.

We have introduced the probability generating function as a transform of the p.f.—another application of averaging. This and related transforms are useful tools for studying the distributions that are important in statistical inference. As with any tool, you may not be comfortable with generating functions at first, but with practice you will gain in understanding and skill.

The next chapter applies the ideas of the first three chapters to some particular families of discrete models that are commonly assumed in applications of statistics.

Solved Problems

Sections 3.1–3.2

A. A town council consists of four liberals and four conservatives. Three council members are selected at random to serve on a committee; let X denote the number of liberals selected. Give the probability function for X in table form, and find its mean value.

Solution:

"Selected at random" means that all of the $\binom{8}{3} = 56$ distinct selections of three are equally likely. Of these, the numbers with 0, 1, 2, 3 liberals are

$$\#(0) = \binom{4}{0}\binom{4}{3} = 4, \qquad \#(1) = \binom{4}{1}\binom{4}{2} = 24,$$

$$\#(2) = \binom{4}{2}\binom{4}{1} = 24, \qquad \#(3) = \binom{4}{3}\binom{4}{0} = 4.$$

x	$f(x)$	$xf(x)$
0	$\dfrac{4}{56}$	0
1	$\dfrac{24}{56}$	$\dfrac{24}{56}$
2	$\dfrac{24}{56}$	$\dfrac{48}{56}$
3	$\dfrac{4}{56}$	$\dfrac{12}{56}$
	1	$\dfrac{3}{2}$

The column headed $f(x)$ gives the probability function for X. The mean value is the sum of the products $xf(x)$: $E(X) = 1.5$. Note the *symmetry* of this distribution: $3 - X$ (the number of conservatives) has the same distribution as X. The mean is the center of symmetry—the balance point.

B. Three numbers are selected at random without replacement from $1, 2, \ldots, 9$. Let X denote the largest of the three numbers. Find $E(X)$.

Solution:

The largest is k if and only if k is selected and the two smaller ones are chosen from $1, 2, \ldots, k-1$. Thus, the probability function for X is

$$f(k) = \binom{k-1}{2}/\binom{9}{3}, \text{ for } k = 3, 4, \ldots, 9.$$

The numerators are 1, 3, 6, 10, 15, 21, 28, which do sum to 84, or $\binom{9}{3}$. The mean value is

$$E(X) = \sum_{3}^{9} k\binom{k-1}{2}/\binom{9}{3}$$

$$= \frac{3 + 12 + 30 + 60 + 105 + 168 + 252}{84} = 7.5.$$

Another way to arrive at the p.f. is to find the probability that $X \le k$ and subtract the probability that $X \le k - 1$:

$$P(X \le k) = P(\text{all three come from } 1, \ldots, k) = \binom{k}{3}/\binom{9}{3},$$

and

$$P(X \le k) - P(X \le k-1) = \binom{k}{3}/\binom{9}{3} - \binom{k-1}{3}/\binom{9}{3} = \binom{k-1}{2}/\binom{9}{3}.$$

(You should check the last step by expressing things in terms of factorials.)

C. A children's game includes a spinner with six equal sectors marked $-5, 1, 2, 3, 4, 5$. After each spin, the player's token advances a corresponding number of spaces on the game board.
   (a) Find the mean number of spaces advanced in one spin.
   (b) Find the mean number of spaces advanced in six spins.

Solution:
   (a) Since $f(x) = \frac{1}{6}$ for each possible value x, the mean is

$$E(X) = \sum xf(x) = \frac{1}{6} \cdot (-5 + 1 + 2 + 3 + 4 + 5) \cdot \frac{1}{6} = \frac{10}{6}.$$

   (b) The number advanced in six spins is the sum of the numbers advanced in the individual spins. The average of this sum is the sum of the averages, and at each turn the average is $\frac{10}{6}$. So the average total advance is $6 \cdot \frac{10}{6} = 10$ spaces.

D. Two digits are selected at random and independently. In Problem B of Chapter 2, we found the probability function for the sum S of two random digits:

$$f(k) = \frac{10 - |k - 9|}{100}, \quad \text{for } k = 0, 1, \ldots, 18.$$

Find $E(S)$.

Solution:

There are at least three ways to do this.

(1)   The long way is to add the products $kf(k)$:

$$E(S) = 0(.01) + 1(.02) + 2(.03) + \cdots + 18(.01) = 9.$$

(2)   A shorter way is to observe that the sum S can be thought of as the sum of two random variables: $S = X + Y$, where X is the first digit and Y the second. Each of the variables X and Y is uniform on the integers $0, 1, \ldots, 9$, with mean 4.5. Hence, $E(S) = E(X) + E(Y) = 4.5 + 4.5 = 9$.

(3)   Easier still, we could simply note that the probability function of S is symmetric about 9:

$$f(9 - k) = f(9 + k).$$

The mean must therefore be 9. [See (4) in Section 3.1.]

E.   On the sample space of pairs of digits in Problem D, define Z to be the first digit minus the second. Find $E(Z)$.

Solution:

Again, there is more than one way to proceed.

(1)   In the notation of Problem D, $Z = X - Y$, so

$$E(Z) = EX - EY = 0.$$

(2)   The p.f. of Z (from Problem B of Chapter 2) is symmetric about 0, which is then the mean value.

(3)   Should you fail to notice symmetry or realize that you can use the linearity of expectations, you can calculate $E(Z)$ the long way, finding the probability function of Z and then calculating $\Sigma \, zf(z)$.

Section 3.3

F.   Suppose in a certain population the number of phones per household has the p.d.f. f shown in the table. Find the mean and standard deviation of the number of phones per household.

x	$f(x)$	$xf(x)$	$xf^2(x)$
1	.35	.35	.35
2	.45	.90	1.80
3	.15	.45	1.35
4	.04	.16	.64
5	.01	.05	.25
Sums:	1.00	1.91	4.39

Solution:

We append the column of products $xf(x)$, shown in the table, and add these to obtain the mean: $\mu = \Sigma \, xf(x) = 1.91$. Adding the column of products $x^2f(x)$, we obtain

$E(X^2) = 4.39$. The variance is then the average square minus the square of the average:

$$\sigma^2 = E(X^2) - \mu^2 = 4.39 - (1.91)^2 = .742.$$

The s.d. is the square root of this: $\sigma \doteq .86$. This should be checked against the distribution to see that it is reasonable as a typical deviation from the balance point.

G. Find the standard deviation of the random variable X in Problem A above.

x	$f(x)$	$x^2 f(x)$
0	1/14	0
1	6/14	6/14
2	6/14	24/14
3	1/14	9/14
Sums:	1	39/14

Solution:

We append a column of weighted squares to the table in Problem A, as shown here. The sum of this column is the mean square: $E(X^2) = \frac{39}{14}$. We found the mean in Problem A: $\mu = \frac{3}{2}$, and the variance is the mean square minus the square of the mean: $\sigma = \frac{78}{28} - (\frac{3}{2})^2 = \frac{15}{28}$, so $\sigma = .732$. (Again, this appears to be about right as a typical deviation from 1.5.)

H. Find the mean deviation of the random variable X in Problem G.

x	$f(x)$	$\lvert x - 1.5 \rvert f(x)$
0	1/14	3/28
1	6/14	6/28
2	6/14	6/28
3	1/14	3/28
Sum:		18/28

Solution:

From Problem A, we know that $\mu = 1.5$. The last column gives the weighted absolute deviations from the mean; its sum is the mean deviation: $E|X - \mu| = \frac{18}{28} = .643$. (In the preceding problem, we found $\sigma = .732$; as is true generally, σ is a bit larger than the mean deviation.)

Sections 3.4–3.5

I. Let X and Y have the joint distribution defined by the probability table shown. Find $\text{cov}(X, Y)$ and the correlation coefficient ρ.

y x	1	2	3	
1	0	0	1/2	1/2
2	0	1/3	0	1/3
3	1/6	0	0	1/6
	1/6	1/3	1/2	1

Solution:

The means (obtained from the marginal distributions in the usual way) are $\mu_X = \frac{5}{3}$ and $\mu_Y = \frac{7}{3}$. The variances are equal, since the pattern for Y is the same as the pattern for X, except reversed:

$$\sigma_X^2 = \frac{1}{2} + \frac{4}{3} + \frac{9}{6} - \left(\frac{5}{3}\right)^2 = \frac{5}{9} = \sigma_Y^2.$$

The covariance is

$$\sigma_{X,Y} = E(XY) - \mu_X\mu_Y = \left(\frac{3}{6} + \frac{4}{3} + \frac{3}{2}\right) - \frac{5}{3} \cdot \frac{7}{3} = \frac{-5}{9},$$

so

$$\rho = \frac{\sigma_{X,Y}}{\sigma_X\sigma_Y} = \frac{-5/9}{\sqrt{5/9} \cdot \sqrt{5/9}} = -1.$$

We could have predicted this. All of the probability lies on a line, so the correlation has to be $+1$ or -1; it's -1 because the line is $X + Y = 4$, which has a negative slope.

J. Two chips are picked at random from five chips numbered 1, 2, 3, 4, and 5. Let $X =$ the larger and $Y =$ the smaller number on the chips selected.

   (a) Find EX, EY, var X, var Y, $E(XY)$, and $\text{cov}(X, Y)$.

   (b) Give the probability table for $Z = X + Y$ and find the mean and variance of Z from this table. Show that you get the same answers by calculating the right-hand sides of (1) and (3) of Section 3.5.

Solution:

(a) The joint p.f. for (X, Y) is shown in the following table.

x y	1	2	3	4	$f_X(x)$	$xf_X(x)$	$x^2f_X(x)$
2	.1	0	0	0	.1	.2	.4
3	.1	.1	0	0	.2	.6	1.8
4	.1	.1	.1	0	.3	1.2	4.8
5	.1	.1	.1	.1	.4	2.0	10.0
	.4	.3	.2	.1	1	4.0	17.0

From this, we see that $E(X) = 4$ and var $X = 17 - 4^2 = 1$. The distribution for Y is the same as for $6 - X$. So $E(Y) = 2$, and var $Y =$ var $X = 1$. To find $E(XY)$, we multiply x and y in each pair and multiply by the probability, either 0 or .1. The tables of products and weighted products are as follows:

x y	1	2	3	4
		xy:		
2	2	4	6	8
3	3	6	9	12
4	4	8	12	16
5	5	10	15	20

x y	1	2	3	4
		xyf(x, y):		
2	.2	0	0	0
3	.3	.6	0	0
4	.4	.8	1.2	0
5	.5	1.0	1.5	2.0

So the average product is

$$.2 + .3 + .6 + .4 + .8 + 1.2 + .5 + 1.0 + 1.5 + 2.0 = 8.5,$$

and $\text{cov}(X, Y) = E(XY) - E(X)E(Y) = 8.5 - 2 \times 4 = .5$.

(b) The probability table for Z, with columns for finding the mean and mean square, is as follows:

z	$f_Z(z)$	$zf_Z(z)$	$z^2f_Z(z)$
3	.1	.3	.9
4	.1	.4	1.6
5	.2	1.0	5.0
6	.2	1.2	7.2
7	.2	1.4	9.8
8	.1	.8	6.4
9	.1	.9	8.1
Sums:	1	6.0	39.0

So $E(Z) = 6 = 2 + 4 = E(X) + E(Y)$, in agreement with (1) of Section 3.5; and

$$\text{var } Z = 39 - 6^2 = 3 = 1 + 1 + 2 \times .5$$
$$= \text{var } X + \text{var } Y + 2 \text{ cov}(X, Y),$$

in agreement with (3) of Section 3.5.

K. Find the correlation coefficient of X and Y in Problem J above.

Solution:
We saw in Problem J that the covariance is .5 and that both variances equal 1. So the s.d.'s are both 1, and the correlation coefficient [from (7) of Section 3.4] is just the covariance, .5.

L. Find the variance of the sum of two independent random digits. (See Problem D.)

Solution:
By symmetry, the mean for one digit X is 4.5. The mean square is

$$E(X^2) = (1^2 + 2^2 + \cdots + 9^2) \cdot \frac{1}{10} = 28.5.$$

So $\text{var } X = 28.5 - (4.5^2) = 8.25$, and $\sigma_X = 2.87$. Since the digits are independent, $\text{var}(X_1 + X_2) = 8.25 + 8.25 = 16.5$.

M. In Problem 34 of Chapter 2, we described a system of bidding points often used in the game of bridge. For a single card drawn at random, the mean number of assigned points is $\frac{10}{13}$, and the variance is $\frac{290}{169}$. (See Problems 4 and 17 in this chapter.) Find the mean and variance of the total number of points in a hand of 13 cards selected at random from the 52 in the deck.

Solution:

We represent the total number of the points as $X_1 + \cdots + X_{13}$, where X_i is the number of points for the ith card dealt. The X's are exchangeable, so all have the same mean, $\frac{10}{13}$. The mean of the sum is then 13 times the mean for a single card, or 10. This result can also be deduced by symmetry: The four hands must all have the same mean number of points, and the total number of points in the four hands must be 40 (the total number in the deck). So each hand gets 10 points on the average. The variance of the sum is 13 times the variance for a single card, multiplied by the "finite population correction factor" [given in (7) of Section 3.5]:

$$\text{var}(\Sigma X) = 13 \cdot \frac{290}{169} \cdot \frac{52 - 13}{52 - 1} = \frac{290}{17}.$$

Section 3.6

N. The system of bridge bidding referred to in Problem M (and described in Problem 34 of Chapter 2) assigns points as follows: $X(\text{ace}) = 4$, $X(\text{king}) = 3$, $X(\text{queen}) = 2$, $X(\text{jack}) = 1$, and $X(\omega) = 0$ for any other card ω. Find the p.g.f. of X and use this to obtain its mean value.

Solution:

The p.g.f. is $E(t^X) = (9t^0 + t^1 + t^2 + t^3 + t^4) \cdot \frac{1}{13}$. Differentiating, we get $(1 + 2t + 3t^2 + 4t^3) \cdot \frac{1}{13}$. Evaluating this at $t = 1$ gives us the mean: $(1 + 2 + 3 + 4) \cdot \frac{1}{3} = \frac{10}{13}$. (See Problem 4 of this chapter.)

O. You toss a coin and receive $\$Y$, where $Y = 2$ if it lands heads and $Y = 0$ if it lands tails. Find the p.g.f. for Y. Use this to find the probability that after four plays of this game, you have received a total of $6.

Solution:

For a single play, the p.g.f. is

$$E(t^Y) = \Sigma t^k f(k) = t^2 \cdot \frac{1}{2} + t^0 \cdot \frac{1}{2} = (t^2 + 1) \cdot \frac{1}{2}.$$

Let $S = Y_1 + Y_2 + Y_3 + Y_4$ be the total received in four independent plays. The p.g.f. of S is the fourth power of the p.g.f. for one play:

$$E(t^S) = (1 + t^2)^4 \cdot \frac{1}{16}.$$

$$= (1 + 4t^2 + 6t^4 + 4t^6 + t^8) \cdot \frac{1}{16}.$$

The probability that $S = 6$ is the coefficient of t^6:

$$P(\Sigma Y = 6) = \frac{4}{16} = \frac{1}{4}.$$

(The fact that all odd powers of t have coefficient 0 is consistent with the obvious fact that S cannot be odd.)

P. Find the probability of throwing a total of 7 with four ordinary six-sided dice.

Solution:

We could calculate this directly by enumerating the ways of getting a total of 7 and dividing this count by 6^4. Using the p.g.f. is somewhat less tedious (and perhaps safer). The p.g.f. for the roll of a single die is

$$(t + t^2 + t^3 + t^4 + t^5 + t^6) \cdot \frac{1}{6} = \frac{t(1 - t^6)}{6(1 - t)}.$$

The p.g.f. of the sum for four dice is the fourth power:

$$\frac{t^4(1 - t^6)^4}{1{,}296(1 - t)^4}.$$

To work with this, we expand the numerator:

$$t^4(1 - t^6)^4 = t^4(1 - 4t^6 + 6t^{12} - \cdots), \tag{1}$$

and write a power series for $(1 - t)^{-4}$:

$$= (1 - t)^{-4} = 1 + 4t + 10t^2 + 20t^3 + \cdots. \tag{2}$$

The product of (1) and (2) is a power series that starts out

$$t^4 + 4t^5 + 10t^6 + 20t^7 + \cdots.$$

Dividing by 1,296, we obtain the p.g.f. of the total number of points; the probability of a sum of 7 is the coefficient of t^7, or $\frac{20}{1296} \doteq .015$.

Problems

For problems marked with an asterisk, answers are provided in Appendix 2.

Sections 3.1–3.2

*1. Random variables X and Y are distributed as shown in the following table:

k	0	1	2	3	4
$f_X(k)$	.6	.3	.1	0	0
$f_Y(k)$	.1	.3	.3	.1	.2

Find $E(X)$ and $E(Y)$.

*2. Find the mean value of each of the variables S, D, and U in Problem 4 of Chapter 2 (page 85).

*3. A carton of 12 eggs includes two that are rotten. In Problem 3 of Chapter 2, you obtained the probability function for X = the number of rotten eggs in a random selection from the 12:

$$f(0) = \frac{12}{22}, \qquad f(1) = \frac{9}{22}, \qquad f(2) = \frac{1}{22}.$$

(a) Find $E(X)$, the mean number of rotten eggs among those selected.
(b) Find $E(X^2)$.

4. Let X denote the number of bidding points assigned to a card selected at random from a deck of cards in bridge (see Problem 34 in Chapter 2). Find $E(X)$ using the p.f.: $f(x) = \frac{1}{13}$, for $x = 1, 2, 3, 4$, and $f(0) = \frac{9}{13}$.

*5. Three fair coins are tossed independently. Let $X(\omega)$ denote the number of heads in the sample sequence ω, and let $Y(\omega)$ denote the indicator function of the event $X = 3$.
  (a)   List the eight ω's in the sample space Ω and give the values of $X(\omega)$ and $Y(\omega)$ for each.
  (b)   Find $E(X)$ and $E(Y)$.

6. Ten tickets are sold in a lottery for \$1 each. Three tickets are to be drawn without replacement. The first gets a \$5 prize, the second a \$2 prize, and the third a \$2 prize.
  (a)   Find the expected worth of a single ticket that you've bought.
  (b)   Find the total expected worth of four tickets.
  (c)   Is this lottery fair?

*7. Three digits are picked at random without replacement from the list $1, 2, 3, \ldots, 8$. Let Y denote the largest of the three selected.
  (a)   Find the probability function, $f(k) = P(Y = k)$.
  (b)   Find $P(Y \geq 5)$.
  (c)   Find $E(Y)$.

8. The Octopus Car Wash once ran a promotion, a "happy hour" from 4 to 6 in which each customer was asked to roll a die. If the die showed 1, 2, or 3, the customer got \$1 off the regular price of \$7.50; if the die showed 4 or 5, the customer got \$2 off; and if it showed 6, the wash was half price. Find the average cost of a wash during that period.

9. In Problem N of Chapter 1, we considered a party of four men who hung up their coats and later picked them up at random. Here we ask how many matches there are—how many men get the right coats? Define indicator variables

$$Y_i = \begin{cases} 1 & \text{if Mr. } i \text{ gets his own coat,} \\ 0 & \text{otherwise.} \end{cases}$$

  *(a)   Find $P(Y_1 = 1)$.
  *(b)   Find $E(Y_1)$.
  *(c)   Note that the Y_i are exchangeable, and use this fact to find the expected number of men who get their own coats.
  (d)   Show that if there are n men and n coats, the expected number who get the right coat is the same for all n.

10. In Examples 3.1a and 3.2a, a contestant was offered a random prospect with probabilities $\frac{1}{3}$ for each of the possible rewards \$2,000, \$5,000, and \$20,000, in exchange for his current fortune of \$9,000.
  (a)   Suppose the contestant owes a murderous loan shark \$9,000, due the next day. Then \$2,000 and \$5,000 are virtually useless to him, and the \$20,000 is really no better than \$9,000. Devise a suitable utility function and find the mean utility. Should he do the exchange?
  (b)   Find the expected utility (and whether the contestant should trade) for each of the following utility functions:
    (i) $u(M) = M$,   (ii) $u(M) = M^2$,   (iii) $u(M) = \sqrt{M}$.

11. Show that $P(X + Y = k \mid Y = j) = P(X = k - j \mid Y = j)$.

12. In Problem E of Chapter 2, we gave the joint distribution of random variables R (number of red socks) and G (number of green socks) in this table of joint probabilities:

R G	0	1	2	3	f_R
0	1/30	6/30	6/30	1/30	14/30
1	3/30	8/30	3/30	0	14/30
2	1/30	1/30	0	0	2/30
f_G	5/30	15/30	9/30	1/30	1

(a) Verify (2) of Section 3.2 (with $X = R$ and $Y = G$). (See page 101.)

(b) Verify the formula for iterated expectations, (3) of Section 3.2 (page 101), applying it to find $E(G)$ from $E(G \mid R = r)$.

Section 3.3

*13. Given $E(X^2) = 65$ and $E(X) = 7$, find the s.d. of X.

14. Given $E(X) = 3$ and $E[X(X - 1)] = 6$, find var X.

*15. Find the standard deviation of the number of heads in three independent tosses of a fair coin by using the fact that the number of heads can be expressed as a sum of three independent variables. (Do not use the distribution of the number of heads.)

16. Find the mean and standard deviation of the total number of points in a throw of eight dice. Assume independence and use the fact that the number of points is a sum of independent variables.

*17. In Problem 4, you found the mean number of points assigned to a card for bidding in bridge to be $\frac{10}{13}$. Find the standard deviation.

18. A census questionnaire once asked for the number of flush toilets per household. Assume these results:

Number	0	1	2	3	4
Proportion	.05	.55	.30	.08	.02

Find the mean and standard deviation of the number per household.

*19. Find the standard deviation of the random variable X in Problem 3.

20. I have four similar keys, just one of which opens my office door. I try them one at a time, selecting keys at random and without replacement. Let X denote the number of keys I put in the lock before the door will open (including the key that works).

(a) Give the probability table for X.

(b) Find the mean and standard deviation of X.

(c) Find the probability function for X if the setting is modified by making the selection at random but *with* replacement.

21. Scores on a certain hole on a golf course have been observed over a period years for professional golfers and for ordinary club members. Relative frequencies (which may be interpreted as probabilities) are as follows:

Score	Pro	Member
2 (eagle)	.02	
3 (birdie)	.16	.03
4 (par)	.68	.22
5 (bogie)	.13	.27
6	.01	.27
7		.13
8		.05
9		.02
10		.01

Find the mean and standard deviation of the score for each type of golfer.

*22. A bowl contains three white and six black chips. Let Y be the number of white chips in a random selection of four (without replacement). The probability function of Y is

$$f(y) = \frac{\binom{3}{y}\binom{6}{4-y}}{\binom{9}{4}}, \qquad y = 0, 1, 2, 3.$$

Find the mean and variance of Y.

23. Show that the mean deviation (1) is never larger than the standard deviation (7). [*Hint:* Use (3) of Section 3.3, applied to the variable $|X - \mu|$, and the fact that a variance is nonnegative.]

*24. Find the mean deviation of Y, the number of heads in three independent tosses of a fair coin. The p.f. of Y is given as follows: $f(0) = f(3) = \frac{1}{8}$, $f(1) = f(2) = \frac{3}{8}$. (Your answer will be smaller than the answer to Problem 15.)

Sections 3.4–3.5

*25. Given the discrete bivariate distribution in the accompanying table, find

(a) $\text{cov}(X, Y)$.   (c) $\text{var}(X + Y)$.

(b) $P(X = 1 \mid X + Y = 3)$.   (d) $\text{var}(Y \mid X = 1)$.

Y \ X	1	2
0	.2	0
1	.1	.2
2	.3	.2

*26. Given var $X =$ var $Y = \text{cov}(X, Y) = 1$, find

(a) $\text{var}(3 - X)$.   (d) $\text{cov}(X, X)$.

(b) $\text{var}(2X + 4)$.   (e) $\text{cov}(X, X + Y)$.

(c) $\text{var}(X - Y)$.   (f) $\text{var}(4X + Y - 7)$.

27. Calculate the covariance in Example 3.4a using the defining formula (1) of Section 3.4.

28. Given that X and Y are independent, with $\sigma_X = \sigma_Y = 1$, find
    (a) $\text{var}(2X + Y)$.   (c) $\rho_{X,Y}$.
    (b) $\text{cov}(2X + Y, X - Y)$.   (d) $\rho_{U,V}$, where $U = 2X + Y$, $V = X - Y$.

*29. Find the table of joint probabilities for Y_1 and Y_2 in Problem 9 (random distribution of coats). (To check that you are on the right track, verify that the marginal probabilities agree with those found in Problem 9.) Also,
    (a) Find the covariance of Y_1 and Y_2.
    (b) Find the variance of X, the total number of matches. (*Hint:* Observe that $X = Y_1 + Y_2 + Y_3 + Y_4$ and use the fact that the Y's are exchangeable.)

*30. Find the correlation coefficient $\rho_{X,Y}$ for the variables in Problem 25 above.

31. A city has streets laid out in a grid, with North-South streets perpendicular to East-West streets. You start at an arbitrary intersection and walk in a sequence of one-block moves. At each move, you choose a direction at random from N, E, W, S. Let $X_i =$ the number of blocks you move North at the ith move (-1 means one block South) and $Y_i =$ the number of blocks East. Thus, the p.f. of (X_i, Y_i) is as follows:

$$f(1, 0) = f(-1, 0) = f(0, 1) = f(0, -1) = \frac{1}{4}.$$

    (a) Find the variance of X_1.
    (b) Are X_1 and Y_1 independent? (Explain.)
    (c) Let U_n be the number of blocks you are North of the starting point after n moves and V_n the number of blocks East. Thus,

$$U_n = \sum_{i=1}^{n} X_i, \quad \text{and} \quad V_n = \sum_{i=1}^{n} Y_i.$$

    Find $E(U_n)$, $E(V_n)$, var U_n, and var V_n.
    (d) Find the distance from the starting point after n steps:

$$\sqrt{E(U_n^2 + V_n^2)}.$$

Section 3.6

*32. Let X be the indicator variable for an event A (equal to 1 if A occurs and 0 otherwise). Obtain the probability generating function for X in terms of $p = P(A)$.

*33. Let X_1, X_2, and X_3 denote the results of three independent trials of the experiment in the preceding problem. The sum $Y = X_1 + X_2 + X_3$ is then the number of times A occurs in the three trials. Find the probability generating function for Y and use it to find $P(Y = 2)$.

34. Let S denote the sum of the points thrown with two ordinary six-sided dice. Use the result of Example 3.6a to find the p.g.f. of S, and find $P(S = 5)$ from the p.g.f.

*35. In Example 1.7c we gave probabilities for $X =$ the number of tosses of a coin required to obtain heads for the first time.

$$P(\text{first heads at } k\text{th toss}) = \left(\frac{1}{2}\right)^k.$$

(a) Find the p.g.f. of X as a power series, and find the sum of the series as an elementary function.

(b) The number of tosses to obtain two heads can be thought of as the sum of X_1, the number required to get the first, and X_2, the number of additional tosses required to get the next heads. Both X_1 and X_2 have the p.g.f. found in (a), and they are independent. Find the p.g.f. of the sum $Y = X_1 + X_2$ and use it to find $P(Y = 3)$.

36. Use the p.g.f. in Problem 35(a) to find the mean number of tosses required to obtain the first heads. Also use the p.g.f. in Problem 35(b) to find the mean number of tosses to obtain two heads.

CHAPTER 4

Bernoulli and Related Variables

Categorical variables with just two categories occur frequently in statistical practice: A person selected at random smokes or does not; a product is good or defective; a mission succeeds or fails; blood pressure is lowered or not; an ICBM silo survives an enemy's preemptive nuclear strike or not; a TV viewer is or is not watching a particular program. In each of these examples, outcomes are of just one of two types. We code these types 0 and 1 and call them *failure* and *success*. In the case of the TV viewer, for example, *success* means that the viewer is watching the program in question.

A **Bernoulli population**[1] is one whose members are 0's and 1's. An observation from such a population is a **Bernoulli random variable**. In this chapter, we investigate various distributions associated with Bernoulli populations. We'll be interested in the number of successes in a sample of given size from a Bernoulli population, under sampling with and without replacement, and in the sample size required to obtain a specified number of successes. Extending the process of sampling success–failure populations in different ways will lead us to the Poisson and multinomial distributions. The generating function introduced in Section 3.6 will be a useful tool.

4.1   Sampling Bernoulli Populations

The distribution of a Bernoulli variable X is completely defined by the probability that $X = 1$, or the proportion of 1's in a Bernoulli population:

$$p = P(X = 1) = P(\text{success}).$$

---

[1] Named after Jakob Bernoulli (1654–1705).

The probability that $X = 0$ is then

$$q = P(X = 0) = P(\text{failure}) = 1 - P(\text{success}) = 1 - p.$$

Thus, in table form, the p.f. of X is quite simple:

x	$f(x)$
1	p
0	q

Because the p.f. involves the parameter p, we may exhibit this dependence with the augmented notation $f(x\,|\,p)$.[2] It is often useful to express $f(x\,|\,p)$ in terms of an equivalent algebraic formula.

Probability function for a Bernoulli variable:

$$f(x\,|\,p) = p^x(1-p)^{1-x}, \qquad x = 0 \text{ or } 1. \tag{1}$$

When X has this p.f., we write $X \sim \text{Ber}(p)$.

Moments of a Bernoulli distribution are easy to calculate. The mean value is

$$\mu = E(X) = 1 \cdot p + 0 \cdot q = p.$$

The second moment is the same, because $X^2 = X$ when $X = 0$ or 1 (the only values X can have):

$$E(X^2) = E(X) = p.$$

Hence,

$$\sigma^2 = E(X^2) - [E(X)]^2 = p - p^2 = p(1-p) = pq.$$

Mean and variance of a Bernoulli random variable:

$$\mu = p, \qquad \sigma^2 = pq. \tag{2}$$

---

[2] We've used the vertical bar previously in a notation for conditional probability. Here we use it to indicate dependence of a distribution on the parameter that follows the bar. In Chapter 16, where we treat parameters as random variables, the two uses are consistent.

Example 4.1a   Suppose a population is 20% black and 80% white. When an individual is selected at random, define

$$X = \begin{cases} 1 & \text{if the person selected is black,} \\ 0 & \text{if that person is white.} \end{cases}$$

Then $X \sim \text{Ber}(.20)$. From (2),

$$\mu = p = .2,$$
$$\sigma = \sqrt{pq} = \sqrt{.2 \text{x} .8} = .4.$$ ∎

As in the examples that opened this chapter, the population proportion p is usually not known. To learn about p, we must sample the population. The individual random variables in such a sample are called trials. We may sample with replacement or without replacement (see Section 1.5).

Sampling with Replacement   Sampling with replacement (and with thorough mixing between selections) yields a sequence of *independent* observations. A sequence of independent observations from a Bernoulli population is a **Bernoulli process**. The joint probability function for any finite set $X_1, \ldots, X_n$ of such independent Bernoulli trials is the product of the p.f.'s of the individual trials [from (1) above]:

$$f(x_1, \ldots, x_n \mid p) = \prod_{i=1}^{n} f(x_i \mid p)$$
$$= \prod_{i=1}^{n} p^{x_i} q^{1-x_i} = p^{\Sigma x_i} q^{n-\Sigma x_i}, \qquad x_i = 0, 1. \tag{3}$$

Because each x_i is a 0 or a 1, the sum Σx is just the number of 1's in the sequence of x's. [For example, for the sample sequence (0, 1, 0, 0, 1), the sum Σx is $0 + 1 + 0 + 0 + 1 = 2$.] Similarly, the complementary count $\Sigma(1 - x) = n - \Sigma x$ is the number of 0's.

Two characteristics of a Bernoulli process are important:

(a)   Nonoverlapping sequences of trials are independent.
(b)   The process has no memory: At any point in the sequence of trials, the process starts afresh—independent of the results up to that point.

Sampling Without Replacement   Consider next the case of sampling *without* replacement from a finite Bernoulli population. This method of sampling is usually more practical and gives more information about p than sampling with replacement. The sample is again a sequence $X_1, \ldots, X_n$, where $X \sim \text{Ber}(p)$ for each i, but now the X's are not independent. The *conditional* distribution of X_2 given X_1, for instance, depends on whether X_1 is 1 or 0. However, the X_i's are exchangeable, as we explained in Section 2.6, so all X_i's have the same distribution.

Example 4.1b   Consider a sample $(X_1, \ldots, X_5)$ drawn at random without replacement from a population of size $N = 20$ in which there are eight 1's and twelve 0's. Then $p = \frac{8}{20}$. To illustrate a general scheme, we'll find the probability of a particular sequence of five observations: 0, 0, 1, 0, 1.

We apply the extended multiplication rule for dependent events:

$$P(0, 0, 1, 0, 1) = \frac{12}{20} \cdot \frac{11}{19} \cdot \frac{8}{18} \cdot \frac{10}{17} \cdot \frac{7}{16} = \frac{\binom{12}{3}\binom{8}{2}}{\binom{20}{5}\binom{5}{3}}.$$

The 3 that appears in the last expression is the sum of the observations, and we can write the joint p.f. of the five observations, more generally, as

$$f(x_1, \ldots, x_5) = \frac{\binom{12}{\Sigma x}\binom{20-8}{5-\Sigma x}}{\binom{20}{5}\binom{5}{\Sigma x}}, \qquad x_i = 0, 1.$$

This is clearly a symmetric function of the x_i's, so the X's are exchangeable. This implies that the joint p.f. for (X_3, X_5), for example, is the same as the joint p.f. for (X_1, X_2). Thus, for example,

$$f_{X_3, X_5}(1, 1) = P(X_1 = 1, X_2 = 1)$$

$$= P(X_2 = 1 \mid X_1 = 1) \cdot P(X_1 = 1) = \frac{8}{20} \cdot \frac{7}{19} = \frac{14}{95}.$$

The probability table for any pair (X_i, X_j) is then as follows:

		X_i		
		1	0	
X_j	1	$\frac{14}{95}$	$\frac{24}{95}$	$\frac{8}{20}$
	0	$\frac{24}{95}$	$\frac{33}{95}$	$\frac{12}{20}$
		$\frac{8}{20}$	$\frac{12}{20}$	1

Each (univariate) marginal distribution is just the Bernoulli distribution of the population: Ber($\frac{8}{20}$).   ∎

Consider a population of size $N < \infty$ consisting of M 1's and $N - M$ 2's, and a sequence of observations drawn without replacement. Following the line of reasoning in Example 4.1b, we find their joint probability function:

$$f(x_1, \ldots, x_n) = \frac{\binom{M}{\Sigma x}\binom{N-M}{5-\Sigma x}}{\binom{N}{n}\binom{n}{\Sigma x}}, \qquad x_i = 0, 1. \tag{4}$$

Again, the observations in a sample sequence are exchangeable. All marginal distributions of given dimension are therefore identical.

An important function of the observations $X_i, \ldots, X_n$ in a Bernoulli process is ΣX_i, the number of successes in n trials. The next two sections deal with the distribution of ΣX_i, first for sampling with replacement and then for sampling without replacement.

4.2	**The Binomial Distribution**

We focus now on the number of successes in a specified number of independent Bernoulli trials. This is a random variable, whose distribution is called **binomial**. In the next example, we illustrate the calculation of binomial probabilities.

Example 4.2a

Broiled Burgers Better Bets?
A TV commercial for Burger King (1982) included this statement: "Our survey shows that three-fourths of the people prefer their hamburgers broiled." The speaker (dressed as a Burger King employee) then went on to say, "I'll make you a bet. Call four of your friends; at least three of them will say they prefer them broiled." What are appropriate odds for this bet?

A sample of three of "your friends" is not apt to be a random selection from "the people." For purposes of argument, suppose it is, and suppose further that the number of "the people" is infinite and that three-fourths of them prefer broiled: $p = \frac{3}{4}$. The following sample sequences, with "broiled" coded 1, would satisfy the condition "at least three say broiled":

Friend no.				*Probability*
1	2	3	4	
1	1	1	1	p^4
1	1	1	0	$p^3(1-p)$
1	1	0	1	$p^3(1-p)$
1	0	1	1	$p^3(1-p)$
0	1	1	1	$p^3(1-p)$

The probability of at least three 1's is the sum of the probabilities of these sequences:

$$P(3 \text{ or } 4 \text{ 1's}) = p^4 + 4p^3(1-p).$$

When $p = \frac{3}{4}$, this is .738, so the odds are about 14 to 5. The Burger King worker had a good chance of winning her bet—if the assumptions are correct.

Calculations similar to the above give probabilities for the other possible numbers correct: For two 1's, there are six sequences with two 1's and two 0's, each with probability p^2q^2, where $q = 1 - p$, and so on:

k	0	1	2	3	4
$f(k\mid p)$	q^4	$4pq^3$	$6p^2q^2$	$4p^3q$	p^3

∎

More generally, let Y denote the number of successes in n independent trials of a Bernoulli experiment. We can write $Y = \Sigma X_i$, where X_i is the indicator variable for success, scoring a 1 for success and 0 for failure at the ith trial. In view of (3) of Section 4.1, the probability of any particular sequence in which $Y = k$ is p^kq^{n-k}. Since there are $\binom{n}{k}$ such sequences, the probability of $Y = k$ is this number of sequences times the probability of a particular sequence:

Probability function for a binomial Y:

$$f_Y(k \mid p) = \binom{n}{k} p^k q^{n-k}, \qquad \text{for } k = 0, 1, \ldots, n, \tag{1}$$

where p is the probability of success at each of n independent trials and $q = 1 - p$. We write: $Y \sim \text{Bin}(n, p)$.

Binomial probabilities are given in Table Ia of Appendix 1 for $n = 5, 6, \ldots, 12$, and for selected values of p from .01 to .50. Table Ib gives *cumulative* binomial probabilities—probabilities of events of the form $Y \le k$. For $n > 12$, the methods of Sections 4.5 and 4.6 will provide approximations for binomial probabilities.

Example 4.2b   **Monitoring Quality**

A machine produces a stream of parts, each one of which is either good or defective. Quality is monitored by taking a sample of parts from the production line. How many defective parts will there be in a sample of 12 if there is one chance in ten that a particular part is defective?

Assume that producing 12 parts is a sequence of 12 independent Bernoulli trials with $p = .1$, where *success* means that the part is defective. Then Y, the number of defective parts among the 12, is Bin(12, .1). Its p.f. (1) is

$$f(k \mid .1) = \binom{12}{k} (.1)^k (.9)^{12-k} \qquad \text{for } k = 0, 1, \ldots, 12.$$

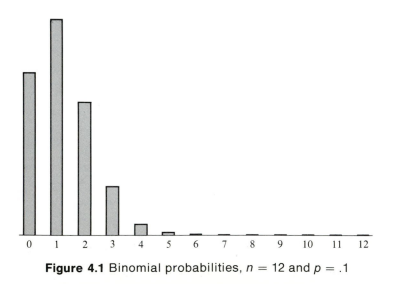

Figure 4.1 Binomial probabilities, $n = 12$ and $p = .1$

These probabilities are given in Table I of Appendix 1. Figure 4.1 shows f as a bar graph. The distribution is skewed to the right: Small values of Y are most likely, and probabilities tail off to the right. ∎

Although Table I in Appendix 1 gives binomial probabilities only for $p \le .5$, we can also use it for $p > .5$ by simply interchanging the roles of 1 and 0. If the number of successes is $Y \sim \text{Bin}(n, p)$, then the number of failures is $n - Y \sim \text{Bin}(n, 1 - p)$. The next example illustrates this use of Table I.

Example 4.2c Suppose we want the probability of at least seven successes in nine independent trials when $p = \frac{2}{3}$: $P(Y \ge 7 \,|\, p = \frac{2}{3})$. This is the same as the probability of at most two failures:

$$P(Y \ge 7) = \sum_{7}^{9} f\left(k \,\middle|\, \frac{2}{3}\right) = P(9 - Y \le 2) = \sum_{0}^{2} f\left(k \,\middle|\, \frac{1}{3}\right),$$

or

$$\binom{9}{9}\left(\frac{2}{3}\right)^9 + \binom{9}{8}\left(\frac{2}{3}\right)^8\left(\frac{1}{3}\right)^1 + \binom{9}{7}\left(\frac{2}{3}\right)^7\left(\frac{1}{3}\right)^2$$

$$= \binom{9}{0}\left(\frac{2}{3}\right)^9 + \binom{9}{1}\left(\frac{1}{3}\right)^1\left(\frac{2}{3}\right)^8 + \binom{9}{2}\left(\frac{1}{3}\right)^2\left(\frac{2}{3}\right)^7.$$

So the desired probability can be found in Table Ia with $n = 9$ and "p" $= \frac{1}{3}$: $.0260 + .1171 + .2341 = .3772$. It can also be found from a single entry in Table Ib:

$$P(9 - Y \le 2) = 1 - P(9 - Y \ge 3) = 1 - .6228 = .3772. \qquad ∎$$

Binomial probabilities get their name from the binomial theorem [(1) of Section 1.6]. They are terms in a binomial expansion:

$$(p + q)^n = \sum_{k=0}^{n} \binom{n}{k} p^k q^{n-k} = \sum_{k=0}^{n} f(k \,|\, p). \tag{2}$$

Since $p + q = 1$, (2) shows that the binomial probabilities $f(k \,|\, p)$ add up to 1 for any n and p, as of course they must.

In the special case $p = q = \frac{1}{2}$, the product $p^k q^{n-k}$ equals $(\frac{1}{2})^n$ for all k. The binomial probabilities (1) are then proportional to the combination counts $\binom{n}{k}$:

$$f\left(k \,\middle|\, \frac{1}{2}\right) = \binom{n}{k}\left(\frac{1}{2}\right)^n.$$

Because $\binom{n}{k} = \binom{n}{n-k}$, the binomial distribution with $p = \frac{1}{2}$ is *symmetric* about $\frac{n}{2}$. However, when $p \ne \frac{1}{2}$, the distribution is skewed—to the left when $p > \frac{1}{2}$, and to the right when $p < \frac{1}{2}$ (as in Figure 4.1).

Binomial Mean and Variance

To find the mean and variance of a binomial variable Y, we simply write Y as the sum of the 0's and 1's in the sequence of results of the n Bernoulli trials:

$$Y = X_1 + \cdots + X_n,$$

where $X_i \sim \text{Ber}(p)$ for each i. Since averaging is additive [Property (iii) of Section 3.1], and $E(X_i) = p$ for each i [(2) of Section 4.1], it follows that

$$E(Y) = E\left(\sum X_i\right) = \sum E(X_i) = np.$$

Moreover, because Y is the sum of n *independent* variables, its variance is also additive [(6) of Section 3.5]:

$$\text{var } Y = \text{var}\left(\sum X_i\right) = \sum \text{var } X_i.$$

But the variance of each Bernoulli X_i is pq [again by (2) of Section 4.1], so var $Y = npq$.

When $Y \sim \text{Bin}(n, p)$,

$$E(Y) = np, \tag{3}$$

and

$$\text{var } Y = npq, \tag{4}$$

where $q = 1 - p$.

The formula np for the mean is intuitively appealing and therefore easy to remember. For example, in 100 tosses of a coin you "expect" to get 50 heads: $np = 100 \times .5$. In throwing a die, you expect two "sixes" in twelve tries: $np = 12 \times 1/6 = 2$.

Example 4.2d   **Monitoring Quality (continued)**
In Example 4.2b, we considered the number of defectives in a sample of 12 parts taken from a production line, assuming that defectives occurred at the rate of one in ten: $p = .1$. The mean number of defectives among the 12 is then

$$E(Y) = np = 12 \times .1 = 1.2.$$

The standard deviation [see (4)] is:

$$\sigma = \sqrt{npq} = \sqrt{12 \times .1 \times .9} = 1.04.$$

The graph of the probability function in Figure 4.1 suggests that this mean and s.d. are reasonable. (Routine visual checks help to build your intuition as well as guard against mistakes in arithmetic.)   ∎

Reproductive Property   The binomial distribution has a reproductive property: The sum of independent binomial variables with the same p is again binomial. This follows from the definition of a binomial variable as the sum of independent Bernoulli variables. Thus, if $Y_1 \sim \text{Bin}(n_1, p)$, $Y_2 \sim \text{Bin}(n_2, p)$, and Y_1 and Y_2 are independent, then $Y_1 + Y_2$ is the sum of $n_1 + n_2$ independent Bernoulli variables with parameter p. Therefore, $Y_1 + Y_2 \sim \text{Bin}(n_1 + n_2, p)$.

The binomial distribution applies to sampling *with* replacement from a finite Bernoulli population, since successive selections are independent. When sampling *without* replacement, however, the number of successes is no longer binomial. Its distribution is the topic of the next section.

4.3

Hypergeometric Distributions

Consider sampling at random and *without* replacement from a finite Bernoulli population. Successive observations are not independent, because the pool of individuals available at any selection depends on preceding observations. Let N denote the population size (before any selections) and M the number of successes in the population, so that $p = M/N$. Let n denote the sample size.

The p.f. of the sample sequence in this case is given by (4) of Section 4.1. Multiplying this by $\binom{n}{\Sigma x}$, just as in the binomial derivation, we obtain the probability function for the sum—the number of 1's in the sample:

$$P(\sum X = k) = \frac{\binom{M}{k}\binom{N-M}{n-k}}{\binom{N}{n}}. \tag{1}$$

This formula applies provided $0 \le k \le M$ and $0 \le n - k \le N - M$. If one of these four inequalities fails, we'd have something like $\binom{1}{2}$, and of course we cannot pick two things from one. Defining the binomial coefficient to have the value 0 in such cases makes (1) correct for any integer k.

Example 4.3a

Sampling Oranges

A bag of 12 oranges is to be checked by inspecting two drawn at random without replacement. An orange is either bad or good. Let Y denote the number of bad ones among the two drawn. This number is a random variable, whose possible values are included in the list $\{0, 1, 2\}$. The probabilities of these values depend on M, the number of bad oranges among the 12 in the bag. They are as follows:

$$f(0 \mid M) = \frac{\binom{M}{0}\binom{12-M}{2}}{\binom{12}{2}}, \qquad f(1 \mid M) = \frac{\binom{M}{1}\binom{12-M}{1}}{\binom{12}{2}}, \qquad f(2 \mid M) = \frac{\binom{M}{2}\binom{12-M}{0}}{\binom{12}{2}}.$$

These define a different distribution of Y for each value of M from 0 to 12. Some of the 13 possible distributions are given in the columns of the following table:

			M					
			0	1	2	3	4	5	6
	0	1	$\frac{55}{66}$	$\frac{45}{66}$	$\frac{36}{66}$	$\frac{28}{66}$	$\frac{21}{66}$	$\frac{15}{66}$
Y	*1*	0	$\frac{11}{66}$	$\frac{20}{66}$	$\frac{27}{66}$	$\frac{32}{66}$	$\frac{35}{66}$	$\frac{36}{66}$
	2	0	0	$\frac{1}{66}$	$\frac{3}{66}$	$\frac{6}{66}$	$\frac{10}{66}$	$\frac{15}{66}$

Completing the table for $M = 7, \ldots, 12$ isn't necessary, since

$$f(k \mid M) = f(2 - k \mid 12 - M).$$

For instance, the column for $M = 9$ is the column for $M = 3$, reversed:

	9
0	$\dfrac{3}{66}$
1	$\dfrac{27}{66}$
2	$\dfrac{36}{66}$

■

The distributions defined by (1) and illustrated in the preceding example are called **hypergeometric**.

Hypergeometric Probability Function

When n items are selected at random without replacement from a population of size N, M of which are of type 1, the p.f. of the number of 1's among the n is

$$f(k \mid M) = \frac{\binom{M}{k}\binom{N-M}{n-k}}{\binom{N}{n}}, \tag{2}$$

where $\binom{a}{b} = 0$ when $b > a$ or $b < 0$.

There is no table of hypergeometric probabilities in Appendix 1. This is in part because the p.f. depends on two parameters (M and N) in addition to n. Such a table requires one more dimension than the binomial and would take too much space. Also, as we'll see shortly, binomial probabilities sometimes serve as good approximations to hypergeometric probabilities. This makes hypergeometric tables less important than otherwise.

Before giving general formulas for the mean and variance of a hypergeometric distribution, we calculate these moments for the distribution in Example 4.3a.

Example 4.3b   **Sampling Oranges (continued)**
In Example 4.3a, we considered the random variable Y, the number of bad oranges in a random selection of two from a bag of 12 oranges. We can calculate $E(Y \mid M)$ using the p.f. from Example 4.3a and the usual formula for a mean. Thus, when $M = 3$,

$$\mu = E(Y \mid 3) = \sum k \cdot f(k \mid 3) = 0 \cdot \frac{36}{66} + 1 \cdot \frac{27}{66} + 2 \cdot \frac{3}{66} = \frac{1}{2}.$$

So with one-fourth of the bag bad, we expect one-fourth of the two in the sample to be bad. This is just $np = 2 \times \frac{1}{4}$, the same as the mean number of bad oranges among two selected *with* replacement. (We'll see that this is true in general.)

Next consider the variance of Y when $M = 3$. We first calculate the expected square of Y:

$$E(Y^2 \mid 3) = \sum k^2 f(k \mid 3) = 0 \cdot \frac{33}{66} + 1 \cdot \frac{27}{66} + 4 \cdot \frac{3}{66} = \frac{13}{22}.$$

Subtracting the square of the mean, we obtain

$$\text{var}(Y \mid 3) = E(Y^2 \mid 3) - \mu^2 = \frac{13}{22} - \frac{1}{4} = \frac{15}{44}.$$

The binomial variance, appropriate when sampling with replacement, is $npq = \frac{3}{8} = .375$, slightly larger than $\frac{15}{44} = .341$. It seems reasonable that the variance should be smaller without replacement than with. Indeed, if $n = N$, a sample taken without replacement is the whole population, and there is *no* variability in Y. ∎

The setting of this section is a special case of Section 3.5 for sampling without replacement from a population with mean $\mu = p$ and variance $\sigma^2 = pq$. So (10) and (11) of Section 3.5 apply to give the mean and variance of a hypergeometric variable.

Mean and variance of a hypergeometric distribution:

$$E(Y) = np, \qquad \text{var } Y = npq \cdot \frac{N - n}{N - 1}, \tag{3}$$

where Y is the number of successes in a sample of n drawn without replacement from a population of N, and $p = M/N$ is the population proportion of successes.

Binomial Approximations

Suppose N is very large in comparison with the sample size n, so that the finite population correction factor in (3) is nearly equal to 1. The hypergeometric variance is then close to the binomial variance npq [(4) of Section 4.2]. Indeed, in this case the hypergeometric probabilities themselves are very close to the corresponding binomial probabilities. When the sample size is small compared with the population size, the proportion of successes in the population changes only slightly at each selection.

The larger N is, the better the binomial approximates the hypergeometric. Whether you should use this approximation depends on how accurate you want to be. A rule of thumb is that N should be at least ten times as large as n. If $N = 10n$, the finite population correction factor is about .9, and the s.d. of the hypergeometric distribution is about 95% as large as the binomial s.d. The following example compares binomial and hypergeometric probabilities in a particular case.

Example 4.3c   Consider samples of size $n = 2$ from a Bernoulli population of size N in which $p = 1/4$. The following table gives hypergeometric probabilities for selected values of N along with the limiting binomial probabilities ($N = \infty$).

N $\quad$ k	4	8	16	32	64	128	∞
0	.500	.536	.550	.556	.560	.5610	.5625
1	.500	.429	.400	.387	.381	.3780	.3750
2	0	.036	.050	.056	.060	.0610	.0625

The mathematical basis of the binomial approximation is that as M and N become infinite with $p = M/N$ fixed, hypergeometric probabilities converge to binomial probabilities. Thus, for each possible value k in $\{0, 1, \ldots, n\}$,

$$\frac{\binom{M}{k}\binom{N-M}{n-k}}{\binom{N}{n}} \to \binom{n}{k}p^k q^{n-k} \qquad \text{as } N \to \infty \text{ with } p = \frac{M}{N}. \tag{4}$$

This follows upon writing out the combination quantities on the left in terms of factorials, suitably pairing factors in numerator and denominator, and taking limits.

If the p.f.'s $\{f_N\}$ of a sequence of random variables $\{Y_N\}$ converge to a limiting p.f. f, we can approximate $f_N(y)$ by $f(y)$ when N is large. We'll say that Y_N is approximately distributed according to f, and use the symbol $\approx$: The hypergeometric $Y \approx \text{Bin}(n, M/N)$.

When $N \gg n$ but n itself is large, the binomial approximation to hypergeometric probabilities may be awkward to calculate, as in the next example.

Example 4.3d   **Sampling Voters**
Suppose 40% of the 100,000 voters in a certain city are Republican. Five hundred voters are selected at random without replacement. What is the probability that the sample proportion of Republicans is at least 38%?

Let Y denote the number of Republicans in the sample. This number has a hypergeometric distribution $[N = 100,000, n = 500, M = 40,000, p = .4]$. Since N is much larger than n, Y is approximately binomial: $Y \approx \text{Bin}(500, .4)$. The desired probability is

$$P(Y \geq 190) = \sum_{190}^{500} \frac{\binom{40,000}{k}\binom{60,000}{500-k}}{\binom{100,000}{500}} \doteq \sum_{190}^{500} \binom{500}{k}(.40)^k(.60)^{500-k}.$$

Calculating this binomial approximation is not easy (although possible if you have a table of logarithms of factorials). In Section 4.5, we'll give a further approximation that greatly simplifies the calculation.

The proportion of Republicans in a sample of size 500 would have the same approximate binomial distribution as above if the city involved included 5,000,000 voters. In sampling from large populations, it is essentially the sample size n that determines accuracy, not the ratio of sample size to population size.

| **4.4** | **Inverse Sampling** |

Example 4.4a **Monopoly**

In the game of Monopoly, the players throw a pair of dice at each turn. When a player is in "jail," it takes a "double" (a matched pair) to get out of jail without paying a fine. How many turns will it take to get out of jail if the player is allowed to keep throwing the dice indefinitely? (This requires relaxing the rules.) The number of turns required (call it Z) is a random variable. We'll derive its probability distribution.

The probability of a double on any one throw of the dice is $p = \frac{1}{6}$. [See Problem 31, Chapter 2.] Thus, $P(Z = 1) = \frac{1}{6}$. A double does *not* occur on the first throw with probability $q = \frac{5}{6}$. With the assumption of independent trials, the probability of failing on the first throw and then succeeding on the second ($Z = 2$) is $qp = \frac{5}{6} \times \frac{1}{6}$. Similarly, the probability that a double appears for the first time on the third throw of the dice ($Z = 3$) is $q \cdot q \cdot p = \frac{5}{6} \times \frac{5}{6} \times \frac{1}{6}$. And so on. In general,

$$f_Z(k) = P(Z = k) = P(\text{first double on } k\text{th toss})$$

$$= \left(\frac{5}{6}\right)^{k-1}\left(\frac{1}{6}\right), \qquad k = 1, 2, 3, \ldots . \qquad \blacksquare$$

Geometric Variables

The *waiting time* distribution in this example is a *geometric* distribution. The derivation of the probability function for general p follows the reasoning of the preceding example exactly:

$$f_Z(k \mid p) = q^{k-1}p, \qquad k = 1, 2, \ldots ,$$

where, as before, $q = 1 - p$. This is the p.f. of Z, the number of independent Bernoulli trials required to obtain a success. We write $Z \sim \text{Geo}(p)$. The name *geometric* stems from the fact that the probabilities $f(k \mid p)$ are terms in a **geometric series**. This series sums to 1:

$$\sum_{k=0}^{\infty} f(k \mid p) = p + pq + pq^2 + pq^3 + \cdots$$

$$= p(1 + q + q^2 + \cdots) = \frac{p}{1 - q} = 1. \qquad (1)$$

The mean of a geometric distribution is very intuitive. As in the preceding example, suppose $p = 1/6$. You might guess that it takes an average of six ($= 1/p$) trials to get out of jail. This is correct:

$$E(Z) = \sum_{k=1}^{\infty} kf(k \mid p) = 1 \cdot p + 2 \cdot pq + 3 \cdot pq^2 + 4 \cdot pq^3 + \cdots$$

$$= p + pq + pq^2 + pq^3 + \cdots$$

$$+ pq + pq^2 + pq^3 + \cdots$$

$$+ pq^2 + pq^3 + \cdots$$

$$+ \cdots$$

$$E(Z) = p(1 + q + q^2 + \cdots)$$
$$+ pq(1 + q + q^2 + \cdots) + \cdots$$
$$+ pq^2(1 + q + q^2 + \cdots) + \cdots$$
$$= (1 + q + q^2 + \cdots)(p + pq + pq^2 + \cdots) = \frac{1}{1-q} \cdot 1 = \frac{1}{p},$$

in view of (1).

The variance of the geometric distribution is q/p^2; we'll show this in Section 4.9 using generating functions.

When $Z \sim \text{Geo}(p)$, its p.f. is

$$f(k \mid p) = q^{k-1}p \qquad \text{for } k = 1, 2, 3, \ldots \tag{2}$$

and

$$E(Z) = \frac{1}{p}, \qquad \text{var } Z = \frac{q}{p^2}. \tag{3}$$

Still in the context of independent Bernoulli trials, suppose we want to know how many trials are required to obtain r successes. This random variable W is the sum of r waiting times: the time to the first success plus the time from the first success to the second, ..., plus the time from the $(r-1)$th to the rth. Let Z_i denote the waiting time from the $(i-1)$th to the ith success. Then

$$W = Z_1 + Z_2 + \cdots + Z_r,$$

Negative Binomial Distribution

where each $Z_i \sim \text{Geo}(p)$. The probability distribution of W is called **negative binomial**: $W \sim \text{Negbin}(r, p)$. We'll develop its p.f. in a particular case in the next example. However, we can find its mean and variance directly from the mean and variance of the Z's. Since $E(Z_i) = 1/p$ for each i,

$$E(W) = \frac{1}{p} + \frac{1}{p} + \cdots + \frac{1}{p} = \frac{r}{p}. \tag{4}$$

The Z_i's are independent, so the variance of W is the sum of the variances of the Z_i's. So, from (3),

$$\text{var } W = \text{var} \sum Z_i = \frac{q}{p^2} + \frac{q}{p^2} + \cdots + \frac{q}{p^2} = \frac{rq}{p^2}. \tag{5}$$

Example 4.4b   A couple wants three girls. How many children must they have to fulfill this desire?

We can answer this in probabilistic terms, provided (as usual!) we make enough assumptions. Suppose there is a constant probability p that a baby will be a girl, and that sexes are independent from one child to another. The proportion of female births in the population at large is about .48, but it could be argued that this proportion is not the probability of a girl for a given couple, that the p for this couple may actually change with time, and that the trials are not independent. We put these arguments aside and proceed as though we were dealing with independent Bernoulli trials, with $p = .48$.

The variable W, the (minimum) number of trials required, is an integer that must be at least 3: $W = 3, 4, 5, \ldots$. Using the assumed independence, we find

$$P(W = 3) = P(GGG \,|\, p) = p^3,$$
$$P(W = 4) = P(2\ G \text{ in 1st 3 trials, } G \text{ on } 4th \,|\, p)$$
$$= P(2\ G \text{ in 1st 3 trials} \,|\, p) \cdot P(G \text{ on 4th} \,|\, p) = \binom{3}{2} p^2 q \cdot p,$$

and so on. For $W = k$,

$$f(k \,|\, p) = P(W = k \,|\, p) = P(G \text{ on } k\text{th, 2 } G\text{'s in first } k - 1 \text{ trials} \,|\, p)$$
$$= \binom{k-1}{2} p^2 q^{k-3} \cdot p = \frac{(k-1)(k-2)}{2} \cdot p^3 q^{k-3}, \qquad k = 3, 4, \ldots .$$

With $p = .48$, these probabilities are as follows:

k	3	4	5	6	7	8	$\cdots$
$f(k)$	.111	.173	.179	.156	.121	.088	$\cdots$

(About 46% of the couples who have the goal of three girls would require five or fewer children.) The mean number of children required, from (5), is

$$E(W) = \frac{r}{p} = \frac{3}{.48} \doteq 6.25.$$

In making these calculations, we have assumed a truly "ideal" couple. Few actual couples would be willing to have more than 20 children in attempting to have three girls (and none we know would be able to have 100!). To be more realistic, then, we might put a limit of 20 on W and recalculate the mean with this constraint; this would be a little less than 6.25 (actually, about 6.245). ∎

To find the p.f. in the general case, the reasoning is exactly like that of the last example, with p in place of .48.

Negative binomial probability function:

$$f(k) = \binom{k-1}{r-1} p^r q^{k-r} \qquad \text{for } k = r, r + 1, r + 2, \ldots . \tag{6}$$

When $W \sim \text{Negbin}(r, p)$,

$$E(W) = \frac{r}{p}, \qquad \text{var } W = \frac{rq}{p^2}.$$

A common way to sample is to take a specified number of observations—say n. However, in some contexts it is natural to specify a number of successes and to stop sampling when this number is reached. This is called **inverse sampling**. With inverse sampling, the number of trials is random, whereas in direct sampling the number of successes is random. The next example gives a setting in which inverse sampling is preferable.

Inverse Sampling

Example 4.4c   **Inverse Sampling to Control Sample Size**

A serum being tested on dogs may induce a fatal reaction. To estimate the probability of a fatal reaction, one might administer the serum to 20 dogs and see how many survive. If 19 out of the 20 die, much information has been obtained, but at great cost. Instead, one might try the serum on one dog at a time and stop the experiment when two dogs have died. With such a rule, the number of fatalities is kept under control. (In practice, it may be advisable to stop after the 20th dog if there are fewer than two fatalities.)   ∎

Sampling Finite Populations

We consider next inverse sampling when the sampling is without replacement from a finite Bernoulli population. Suppose we sample until a specified number of successes are obtained. Denote by W the number of observations required. Successive observations are not independent, so W is not a binomial variable. To derive the p.f. of W, we take the same approach as in sampling with replacement but use instead the multiplication rule for *dependent* events [(4) in Section 2.3].

Specifically, let W denote the number of observations required to obtain r successes. The event $W = k$ can be expressed as EF, where

$E = \{r\text{th success occurs on the }k\text{th trial}\}$,

$F = \{r - 1 \text{ successes in the first } k - 1 \text{ trials}\}$.

Then, since

$$P(F) = \frac{\binom{M}{r-1}\binom{N-M}{k-r}}{\binom{N}{k-1}} \qquad \text{and} \qquad P(E \mid F) = \frac{M - r + 1}{N - k + 1},$$

it follows from the multiplication rule that

$$f_W(k) = P(EF) = P(F)P(E \mid F) = \frac{\binom{M}{r-1}\binom{N-M}{k-r}}{\binom{N}{k-1}} \cdot \frac{M - r + 1}{N - k + 1}. \tag{7}$$

Another way of arriving at (7) is to imagine that all N population items are arranged in a random sequence. There are $\binom{N}{M}$ ways of assigning the label *success*. The rth success occurs in the kth position when there are $r - 1$ successes in the preceding $k - 1$ positions and $M - r$ successes in the preceding $k - 1$ positions and $M - r$ successes in the remaining $N - k$ positions:

Probability function for a **hypergeometric waiting time**:

$$f_W(k) = \frac{\binom{k-1}{r-1}\binom{N-k}{M-r}}{\binom{N}{M}}. \tag{8}$$

The equivalence of (7) and (8) is easy to check by expressing each combination symbol in terms of factorials. [By analogy with the negative binomial, the distribution defined by (8) is sometimes called *negative hypergeometric*.]

The mean value turns out to be $r(N + 1)/(M + 1)$, close but not quite equal to $r/p = rN/M$, the mean of the corresponding negative binomial. (The formula for the variance is quite complicated, and we omit it.)

4.5   Approximating Binomial Probabilities

In this section, we introduce two different approximations for binomial probabilities. Both approximations are for large samples. Binomial probabilities are cumbersome to deal with when n is large. Many hand calculators have a factorial key, but they will generally overflow at 70! (the first factorial greater than 10^{100}). Tables of binomial probabilities with values of n up to 100 exist but are not readily available and are necessarily limited to selected values of p.

Complicated calculations and formulas for finite n sometimes become simpler in the limit as n becomes infinite. The simpler limiting formulas provide approximations for situations in which n is finite but large.

The first approximation we take up applies when n is large and the parameter p is neither close to 1 nor close to 0. The second, the *Poisson* approximation, is also for large n but for values of p that are close either to 1 or to 0.

The Normal Approximation   A formula for approximating binomial probabilities was found by the mathematicians DeMoivre (in 1733 for the case $p = \frac{1}{2}$) and Laplace (in 1812 for general p). They showed that the cumulative binomial probability $P(Y \le k)$ can be quite well approximated using a smooth function that is now readily available in tables and computer software:

> If Y is binomial with parameters n and p such that npq is moderately large (say, $npq > 5$), then
>
> $$P(Y \le k) \doteq \Phi\left(\frac{k + \frac{1}{2} - np}{\sqrt{npq}}\right), \tag{1}$$
>
> where values of $\Phi(z)$ are given in Table II of Appendix 1.

The function $\Phi(z)$ is actually the area to the left of the value z under the "normal curve"—the graph of an exponential function with a quadratic exponent: $(1/\sqrt{2\pi})\exp[-z^2/2]$. We'll return to this function in Chapter 6; for the present we only need the table of areas.

[The approximating curve is centered at 0, and $Y - np$ has mean 0, so one may wonder about the "$\frac{1}{2}$" in (1). It usually improves the approximation and is called a *continuity correction*.]

Example **4.5a**   **Sampling a Large Electorate**

An opinion poll uses a random selection of 400 individuals from the adult population of a very large city. Assume that the city is half male and half female. What is the probability that among those selected there are no more than 190 women?

Let Y denote the number of women among the 400 selected. For a very large city, sampling 400 without replacement is practically equivalent to sampling with replacement. So we assume that $Y \approx \text{Bin}(400, .5)$. [See (4) in Section 4.3.] In this case, $npq = 100$, so we can expect the approximation (1) to do quite well. With $np = 200$ and $\sqrt{npq} = 10$, we have

$$P(Y \le 190) \doteq \Phi\left(\frac{190.5 - 200}{10}\right) = \Phi(-.95).$$

To find $\Phi(-.95)$ in Table II, look opposite $-.9$ in the left-hand column and under the 5 in the top row—the top row gives the second decimal place. There we find: $\Phi(-.95) = .1711.$ ■

Example **4.5b**   **A Check on Accuracy**

To check the accuracy of the approximation (1), suppose $n = 8$ and $p = \frac{1}{2}$. Figure 4.2 shows the binomial probabilities with $n = 8$ and $p = \frac{1}{2}$ as a bar diagram. The area of each bar is the probability of the corresponding value of Y (from Table I).

The probability that $Y \le 3$ is the shaded area in Figure 4.2. From Table Ib, $P(Y \le 3) = .3632$. The figure also shows a normal curve with mean $np = 4$ and variance $npq = 2$. The area under that smooth curve to the left of $Y = k + \frac{1}{2} = 3.5$ is quite close to the shaded area. Following (1), we calculate

$$z = \frac{k + .5 - np}{\sqrt{npq}} = \frac{3.5 - 4}{\sqrt{8 \times .5 \times .5}} \doteq -.35.$$

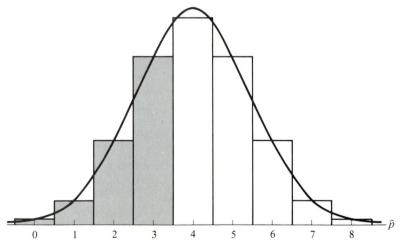

Figure 4.2 Normal approximation to $\text{Bin}(8, \frac{1}{2})$

From Table II, $\Phi(-.35) = .3632$. The approximation (1) is accurate to four decimal places, even though n is only 8, and even though $npq = 2$ (less than 5). However, approximation (1) is not equally successful for all values of k, as the following table shows:

	$Y \leq 0$	$Y \leq 1$	$Y \leq 2$	$Y \leq 3$	$Y \leq 4$
Exact probability	.0039	.0351	.1445	.3632	.6366
Normal approximation	.0067	.0386	.1444	.3632	.6368

[We have omitted $k > 5$ from the table because Y is symmetric about 4: $P(Y \leq k) = 1 - P(Y \leq 8 - k)$, for $k = 0, 1, \ldots, 8$. The same symmetry applies to the normal approximation $\Phi(-z) = 1 - \Phi(z)$.] ∎

This example shows that the normal approximation can be quite good for n as small as 8. How large should n be in general? For any given n, the accuracy of the approximation depends on the value of p. The approximating normal curve is symmetric about 0. When $p = 1/2$, the binomial distribution is symmetric, so the normal approximation is at its best. However, if p is close to 0 or 1, the binomial distribution is quite skewed unless n is very large, and the approximation is not as good. The rule of thumb, $npq > 5$, says that if p or q is close to 0, n has to be very large to compensate for the small size of pq.

Example 4.5c

A Further Check—Small p
Consider again the case $n = 8$, this time with $p = .1$. So $npq = .72$. Table 4.1 gives cumulative binomial probabilities and the corresponding normal approximations. The approximations are clearly not as accurate as in the previous example. ∎

Table 4.1

	Binomial probability	*Normal approximation*
$k \leq 0$	.4305	.3619
$k \leq 1$	.8131	.8807
$k \leq 2$	.9619	.9774
$k \leq 3$	.9950	.9992
$k \leq 4$	.9996	1.0000
$k \leq 5$	1.0000	1.0000
$\vdots$	$\vdots$	$\vdots$

The Poisson Approximation

When p is near 0 or near 1 and $npq < 5$, the normal approximation is usually not very good. However, there is a good approximation to binomial probabilities when n is large and p is small—for example, for $n = 100$ and $p = .03$. This approximation was proposed by Poisson in about 1837:

Poisson approximation:

If $Y \sim \text{Bin}(n, p)$, with large n and small p,

$$P(Y = k) \doteq \frac{(np)^k}{k!} e^{-np}. \tag{2}$$

The approximating quantities in (2) are called *Poisson probabilities*. Table IV in Appendix 1 gives Poisson probabilities for some values of $m = np$. The table entries are cumulative, so they are useful in approximating binomial probabilities of the form

$$P(Y \le c) = \sum_{k=0}^{c} \binom{n}{k} p^k q^{n-k}.$$

If we need the probability of a single value, we subtract successive entries from Table IV:

$$P(Y = c) = P(Y \le c) - P(Y \le c - 1).$$

However, individual approximations in (2) are easy to calculate with a hand calculator.

Table IV stops at $m = 15$ because if $np > 15$ and p is small, npq is usually large enough for the normal approximation to work well. This is not to say that the Poisson approximation would not work, but when p is small and np large, the Poisson probabilities are themselves well approximated by areas under the normal curve.

Example 4.5d   **An Uncommon Blood Type**

People with type AB blood constitute about 4% of the U.S. population. What is the probability of finding more than ten such individuals in a random selection of 100 people from the U.S.?

Because the population is very large compared with the sample size, we can use binomial probabilities in place of the more exact hypergeometric probabilities. Let Y be the number of individuals who have type AB blood in the sample of 100. The desired probability is

$$P(Y > 10) = 1 - P(Y \le 10) \doteq 1 - \sum_{0}^{10} \binom{100}{k} (.04)^k (.96)^{100-k}.$$

Using (2) and entering Table IV with $m = np = 100 \times .4 = 4$, we find

$$P(Y > 10) \doteq 1 - \sum_{0}^{10} \frac{4^k}{k!} e^{-4} = 1 - .997 = .003.$$

(The exact probability is .00224.)   ■

To verify (2), let $np = m$ be fixed (so that $p \to 0$ as $n \to \infty$). Then

$$P(Y = k) = \binom{n}{k} p^k (1 - p)^{n-k} = \frac{n(n-1) \cdots (n-k+1)}{k!} \left(\frac{m}{n}\right)^k (1 - p)^{n-k}$$

$$= \frac{n}{n} \cdot \frac{n-1}{n} \cdots \frac{n-k+1}{n} \cdot \frac{m^k}{k!} \cdot (1-p)^{-k}(1-p)^n.$$

As n becomes infinite, each of the first k factors tends to 1: the factor $m^k/k!$ doesn't change, since it doesn't depend on n or p); and the factor $(1 - p)^{-k}$ tends to 1. The last factor is

$$(1 - p)^n = ([1 - p]^{-1/p})^{-m} = [(1 + x)^{1/x}]^{-m},$$

where $x = -p$. As $n \to \infty$, x tends to 0, and the limiting value of $(1 + x)^{1/x}$ defines the constant $e = 2.71828 \ldots$. So the last factor approaches e^{-m}, and we see that (2) holds.

Both the Poisson and normal approximations lead to probability models that are important in their own right. The normal curve defines an important continuous model that will be studied in detail in Chapter 6. The Poisson formula defines a discrete model that is the topic of the next section.

4.6   Poisson Distributions

The Poisson formula (2) of Section 4.5 serves to define an important family of discrete distributions, indexed by the parameter $m = np$. We first verify that the quantities

$$f(k \mid m) = \frac{m^k}{k!} e^{-m}, \qquad k = 0, 1, 2, \ldots, \tag{1}$$

are actually probabilities. They are nonnegative, so we need only show that they sum to 1. To do this, we apply the Maclaurin series expansion for e^m:

$$\sum_{k=0}^{\infty} f(k \mid m) = e^{-m} + m e^{-m} + \frac{m^2}{2!} e^{-m} + \frac{m^3}{3!} e^{-m} + \cdots$$

$$= e^{-m}\left(1 + m + \frac{m^2}{2!} + \frac{m^3}{3!} + \cdots\right) = e^{-m} e^m = 1.$$

Therefore, (1) is a p.f. When X has this p.f., we write $X \sim \text{Poi}(m)$.

Suppose the random variable Y has the p.f. (1). The mean value $E(Y)$ is easy to calculate:

$$E(Y) = \sum_{k=0}^{\infty} k f(k \mid m) = 1 \cdot m e^{-m} + 2 \cdot \frac{m^2}{2!} e^{-m} + 3 \cdot \frac{m^3}{3!} e^{-m} + \cdots$$

$$= m e^{-m}\left(1 + m + \frac{m^2}{2!} + \cdots\right) = m e^{-m} e^m = m.$$

This is not surprising: In finding the Poisson p.f. as the limit of the binomial p.f. as n becomes infinite, we held $np = m$ (the binomial mean) fixed.

It takes only a little more effort to calculate the variance, although we can anticipate the result, since the binomial variance $np(1 - p)$ approaches m, as $n \to \infty$ and $p \to 0$, with $np = m$. It is convenient to find first the expected value of $Y^2 - Y$:

$$E[Y(Y - 1)] = \sum_{k=0}^{\infty} k(k - 1)f(k \mid m) = \sum_{k=2}^{\infty} k(k - 1)\frac{m^k}{k!}e^{-m}$$

$$= e^{-m}m^2\left(1 + m + \frac{m^2}{2!}\cdots\right) = m^2.$$

So then

$$\text{var } Y = E(Y^2) - m^2 = E(Y^2 - Y) + E(Y) - m^2$$

$$= m^2 + m - m^2 = m.$$

Poisson probability function:

$$f(k \mid m) = \frac{m^k}{k!}e^{-m}, \qquad k = 0, 1, 2, \ldots. \tag{1}$$

When $Y \sim \text{Poi}(m)$,

$$E(Y) = m, \qquad \text{var } Y = m. \tag{2}$$

Bernoulli Process

The Poisson distribution arises in the context of a continuous-time process that generalizes the Bernoulli process (see Section 4.1). In observing a Bernoulli process, we are concerned with the occurrence of some event or phenomenon, referred to generically as *success* and coded "1." When the probability of a success is small, successes occur infrequently overall, but can come close together or far apart.

For example, consider the random digits in Table XIV in Appendix 1. Let the 1's stand and replace the other digits $(0, 2, 3, \ldots, 9)$ with 0's. The result is a sequence of independent Bernoulli trials with $p = .1$. One row of Table XIV gives this sequence:

00100 00100 10000 11000 00001 10000 00000 00000 00000 00100.

The 1's come at unpredictable points in the sequence. This particular sequence of length 50 has eight 1's (16%), but over longer sequences the 1's will make up close to 10% of the total—no matter where the sequence starts.

A Bernoulli process has some characteristics that we want to preserve in going over to continuous time:

(a) The probability of a given number of 1's in a sequence of given length doesn't depend on where the sequence starts.
(b) Events having to do with sequences that don't overlap are independent.
(c) If p is small, 1's rarely occur right next to each other.

(d)   If p is small, the probability of a 1 in a short sequence is approximately proportional to the length of the sequence.

The independence we've assumed in defining a Bernoulli process implies (a) and (b). As to (c), the probability of two 1's in two trials is p^2, an order of magnitude smaller than the probability of one 1 in two trials $[2p(1 - p) \doteq 2p]$ when p is small. To illustrate (d): When $p = .05$,

$$P(\text{one 1 in 2 trials}) = .095, \qquad P(\text{one 1 in 4 trials}) \doteq .17.$$

The second interval is twice as long as the first, and the second probability is nearly twice the first.

We consider a continuous-time process with characteristics analogous to (a)–(d) above. A fairly infrequent phenomenon or event occurs at "random times"—with irregular spacing, occasionally close together, sometimes far apart, and with no apparent periodicities. Over any two long periods of time of equal length, there are about the same number of occurrences, so the *rate* of occurrence is constant over time. Some possible examples: fatal automobile accidents at a busy intersection, nuclear reactor meltdowns, goals in a hockey game, emissions of α-particles from a radioactive substance, meteors striking the earth, and arrivals of customers at a service facility.

The Poisson distribution is also applicable in certain situations in which linear or spatial dimensions take the place of time. Some examples: bacterial colonies on a Petri dish, raisins in cookie batter, flaws in a wire or on a sheet of photographic paper, flat tires in driving (as a function of miles covered, rather than time), and fish in a lake.

In each of these settings, it may be reasonable to assume properties like (a)–(d) of a Bernoulli sequence. Specifically, using the term *event* for the phenomenon or occurrence we're watching for, we may assume:

The Poisson Process

(a)   The probability of a given number of events in any region of size t doesn't depend on its location.
(b)   Events in regions that don't overlap are independent.
(c)   The possibility of more than one event in a small region can be neglected.
(d)   In a small region, the probability of occurrence of an event is approximately proportional to the size of the region.

To understand assumptions (a)–(d), it helps to see what kinds of processes they exclude. Consider fatal accidents at an intersection: If accidents are more likely to occur in daylight than at night, assumption (a) is violated, although it may apply as an approximation. Consider flaws in a wire: A flaw might be very rare overall; but when one occurs, this may suggest that the production process has developed a malfunction. In such a case, it is more likely that the next ten feet of wire will have a flaw than had the first flaw not occurred, so assumption (b) is violated. Consider fish in a lake: For species of fish that move in schools, both (b) and (c) do not hold.

In deriving the Poisson p.f. from assumptions (a)–(d), we'll reason from the Bernoulli process, with trials occurring in discrete time, to a continuous-time process. Assumption (c) implies that in a time interval of length h, there is either no event or exactly one event (as an approximation). Thus, the random variable

$$X = \text{number of events in } (t, t + h) \tag{3}$$

is approximately Ber(p), where

$$p = P(X = 1) \doteq \lambda h. \tag{4}$$

As in the case of a Bernoulli process, there are several random variables that might be of concern in the analogous continuous-time process. In Chapter 6 we'll study the (continuous) distribution of the waiting time from any point in time until the next occurrence of an event. Here we consider the discrete random variable

$$Y_t = \text{Number of events in an interval of length } t. \tag{5}$$

To obtain the distribution of Y_t for an arbitrary t, we take the interval as starting at time 0 and divide the interval $(0, t)$ into n subintervals of length $h = t/n$. A Bernoulli variable X_i is defined (approximately) by (3) for each subinterval. These variables $(X_1, \ldots, X_n)$ are independent, according to (b). Since $Y_t = \Sigma X_i$, it is (approximately) Bin(n, λh). Its p.f. [(2) of Section 4.2] is

$$P(Y_t = k) = P(Y_t = k) \doteq \binom{n}{k}(\lambda h)^k (1 - \lambda h)^{n-k}.$$

We showed in the preceding section that as n becomes infinite with $np = n\lambda h = \lambda t$ fixed, this converges to the Poisson probability function (1) with $m = \lambda t$:

$$P(Y_t = k) = \frac{(\lambda t)^k}{k!} e^{-\lambda t}. \tag{6}$$

The above informal discussion suggests the following formal definition, for those applications in which the indexing parameter is time or length.

A family of random variables Y_t ($t > 0$) is a **Poisson process** with rate λ when

(i)   For each $s > 0$,

$$Y_{t+s} - Y_t \sim \text{Poi}(\lambda s);$$

(ii)   Given a collection of nonoverlapping intervals $(t, t + s)$, the corresponding increments $Y_{t+s} - Y_t$ are independent.

In view of (2), the mean number of events in an interval of width t is λt. Setting $t = 1$, we have an interpretation of the rate parameter λ:

In a Poisson process, the rate parameter λ is the *mean* number of events per *unit* interval.

Example **4.6a** **Customer Arrivals**

Suppose customers arrive at a service counter according to a Poisson process at the rate of five per hour: $\lambda = 5$. The number of arrivals in a given one-hour period is Y_1. To find the probability of exactly k arrivals in that time interval, set $t = 1$ and $\lambda = 5$ in (6):

$$f(k \mid 5) = P(Y_1 = k) = \frac{e^{-5}5^k}{k!}.$$

The probability of *at least* two arrivals in one hour, for example, is

$$P(Y_1 \geq 2) = 1 - P(Y_1 \leq 1)$$
$$= 1 - f(0 \mid 5) - f(1 \mid 5) = 1 - e^{-5}(1 + 5) \doteq 0.96.$$

Alternatively, we can find this using Table IV with $m = 5$ and $c = 1$:

$$P(Y_1 \geq 2) = 1 - P(Y_1 \leq 1) = 1 - .040.$$

We can also calculate probabilities for intervals of other lengths. For instance, Y_2 denotes the number of arrivals in a two-hour period. The average per hour is 5, so the average in a two-hour period is twice that, or 10. For example, the probability of at most five arrivals in a two-hour period is a Poisson probability with $m = \lambda t = 10$:

$$P(Y_2 \leq 5) = \sum_0^5 \frac{2^k}{k!} e^{-2} \doteq .067,$$

according to Table IV.

For these calculations, we chose an hour as the unit of time. We could have chosen any time period as one unit. For example, had we chosen a minute as the unit of time, λ would be $\frac{1}{12}$. The answers would not change, since the number of arrivals in an hour is now Y_{60}, with $\lambda = \frac{1}{12}$ and $t = 60$: $\lambda t = 5$, as before. ■

In this example, the Poisson process unfolds in time. As indicated earlier, the Poisson distribution may also describe such things as the locations of defects along a wire, where the time dimension is replaced by distance, or the distribution of flaws in sheets of metal and particles in a volume of air.

Example **4.6b** **Flaws in Cloth**

Suppose flaws in bolts of a certain cloth are distributed according to a Poisson law, with an average of one flaw per 20 square yards. Taking a square yard as the unit of area means $\lambda = \frac{1}{20}$. The expected number of flaws in a piece of 10 square yards is $10\lambda = .5$. Then, for example,

$$P(\text{no flaws in 10 sq. yds.}) = P(Y_{10} = 0) = e^{-.5} \doteq .607,$$

from (6). To find the probability of two or more flaws in a piece containing 50 square yards, first find the expected number: $50/20 = 2.5$. Table IV (with $m = 2.5$ and $c = 1$) gives the probability of $Y_{10} \leq 1$ as .287, so $P(Y_{10} \geq 2) = .713$. We can also calculate this using (5):

$$P(Y_{10} \geq 2) = 1 - P(Y_{10} = 0 \text{ or } 1)$$
$$= 1 - e^{-2.5}(1 + 2.5) = .713. \qquad \blacksquare$$

Reproductive Property

Adding any two independent Poisson variables produces another Poisson variable. The reasoning is the same as we used in the case of two independent binomial variables (see Section 4.2). An alternative demonstration using generating functions will be given in Example 4.9b. The extension to any finite number of independent Poisson variables follows by induction:

> If $X_1, \ldots, X_k$ are independent Poisson variables with corresponding parameters $m_1, \ldots, m_k$, then their sum $X_1 + \cdots + X_k$ has a Poisson distribution with parameter $m = m_1 + \cdots + m_k$.

Example 4.6c   **Combining Different Customer Types**

Three different types of customers arrive according to independent Poisson processes with averages 3 per hour, 5 per hour, and 10 per hour, respectively. The total number of customers per hour is the *sum* of the numbers of the three types, so it is a Poisson process with $\lambda = 3 + 5 + 10 = 18$. The mean number of customers in a five-minute period is $m = \lambda t = 18 \cdot 5/60 = 3/2$, and the number of customers in that period is therefore $\text{Poi}(\frac{3}{2})$. For instance, the probability of no arrivals of any type during the five-minute period is

$$P(Y_{1/12} = 0) = e^{-3/2} = .223. \qquad \blacksquare$$

4.7   **Sample Proportions and the Law of Large Numbers**

Consider any repeatable experiment. For a particular event A in its sample space, let $p = P(A)$. Let $X = X_A$ be the indicator variable for A (see Section 2.5):

$$X = \begin{cases} 1 & \text{if } A \text{ occurs,} \\ 0 & \text{if } A \text{ fails to occur.} \end{cases}$$

Then $X \sim \text{Ber}(p)$. To learn about p, we perform the experiment n times in independent trials. The results $X_1, \ldots, X_n$ constitute a random sample. The sum $Y = \Sigma X_i$ is the number of times A occurs in the n trials, a binomial variable with mean np and variance npq (Section 4.2).

The *relative frequency* or *sample proportion* of A's among the n trials is denoted by $\hat{p}$ ("p-hat"), or by $\hat{p}_n$ when we want to stress its dependence on n:

$$\hat{p} = \hat{p}_n = \frac{Y}{n}.$$

By Property (ii) of Section 3.1,

$$E(\hat{p}) = E\left(\frac{Y}{n}\right) = \frac{E(Y)}{n} = \frac{np}{n} = p.$$

Also, from (8) of Section 3.3:

$$\text{var}(\hat{p}) = \text{var}\left(\frac{Y}{n}\right) = \frac{\text{var } Y}{n^2} = \frac{npq}{n^2} = \frac{pq}{n}.$$

The relative frequency $\hat{p}$ of an event A in n independent trials has mean and standard deviation

$$E(\hat{p}) = p, \qquad \sigma_{\hat{p}} = \sqrt{pq/n}, \tag{1}$$

where $p = P(A)$ at a single trial.

From (1), we see that the s.d. of $\hat{p}$ goes to 0 as $n \to \infty$. So the distribution of $\hat{p}$ becomes more and more tightly concentrated around its mean p. In this sense, the random variable $\hat{p}$ tends to the constant p. This tendency is called the *law of large numbers* or the *law of averages*. There are formal mathematical theorems expressing this law; we'll give a proof of one such theorem in Section 8.8. For the present, we give an informal statement.

Law of Averages (Law of Large Numbers)

In a sequence of independent Bernoulli trials, the proportion of successes converges to the probability of success as the number of trials becomes infinite.

Example 4.7a To see the law of large numbers in action, toss a coin repeatedly and keep track of $\hat{p}_n$, the sample proportion of heads. It may take a lot of perseverance to convince yourself that $\hat{p}$ really tends to $\frac{1}{2}$. You could also program a computer to mimic tossing a coin. We have done this and have plotted $\hat{p}_n$ as a function of n in Figure 4.3. The decreasing variability in the sequence reflects the fact that the variance of $\hat{p}_n$ tends to 0 and suggests that $\hat{p}_n$ has a limiting tendency. However, it is never clear that in any finite sequence the limiting value will be exactly $p = \frac{1}{2}$. ∎

Some people think that the law of averages says that a success is very likely after a string of failures. This is a misconception. Things do even out "in the long run," and an imbalance would indeed be corrected *in the limit*. However, any finite string of observations (no matter how long!) has no effect on the limiting value of $\hat{p}_n$. The law of averages assumes independent trials, so the result of any one trial is *not* affected by the result of any other trial. In particular, the mechanism producing the

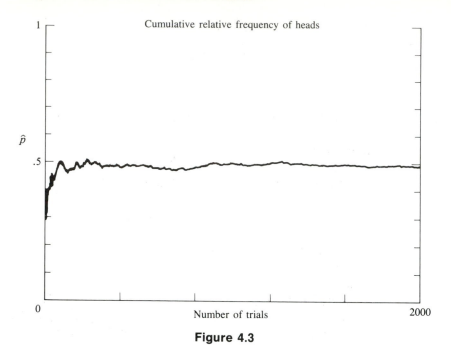

Figure 4.3

results does not "remember" previous results, so it cannot adapt itself to make up for any short-run imbalance in them. It does so in "the long run" because any imbalance is finite, and the long run is infinite.

4.8 Multinomial Distributions

In this section, we extend the notion of a Bernoulli variable, which has two possible values, to categorical variables that have any finite number of values. With this generality, a numerical coding is not helpful—the categories may not even be ordered. Let the values of X be denoted by A_i and the corresponding probabilities by p_i (where $p_i \leq 0$ and $\Sigma p_i = 1$):

Value of X	A_1	A_2	$\cdots$	A_k
Probability	p_1	p_2	$\cdots$	p_k

$$(1)$$

Consider a sequence of n independent trials of this experiment. Generalizing the Bernoulli case, we are interested in Y_j, the number of times A_j occurs in the n trials, $j = 1, \ldots, k$. The frequencies Y_j are random variables, and we want their joint distribution. The following example will lead to the general formula.

Example 4.8a **Roulette**

In roulette, the ball lands on one of three colors: red, black, and green. The wheel has 38 slots, which are presumed to be equally likely landing spots for the ball. Of these,

18 are red, 18 are black, and 2 are green (in most casinos). So the color resulting from a spin of the wheel is a random variable with three categories ($k = 3$) and the following probabilities:

Color (X)	A_1: red	A_2: black	A_3: green
Probability	$\dfrac{18}{38}$	$\dfrac{18}{38}$	$\dfrac{2}{38}$

Suppose we spin the wheel five times independently. What is the probability that two spins result in red, two in black, and one in green?

We proceed just as in the Bernoulli case. The probability for a particular sequence of results (say, $A_1 A_1 A_2 A_2 A_3$) is the product of the individual probabilities, because of the assumed independence:

$$P(A_1 A_1 A_2 A_2 A_3) = p_1 p_1 p_2 p_2 p_3 = p_1^2 p_2^2 p_3 = \left(\frac{18}{38}\right)^2 \left(\frac{18}{38}\right)^2 \left(\frac{2}{38}\right)^1. \tag{2}$$

This is also the probability of any other sequence in which there are two reds, two blacks, and a green. Multiplying by the number of such sequences, we obtain $P(Y_1 = 2, Y_2 = 2, Y_3 = 1)$. According to (3) in Section 1.4, the number of ways of arranging two reds, two blacks, and a green in a sequence is

$$\binom{5}{2, 2, 1} = \frac{5!}{2!2!1!} = 30. \tag{3}$$

So

$$P(Y_1 = 2, Y_2 = 2, Y_3 = 1) = \binom{5}{2, 2, 1}\left(\frac{18}{38}\right)^2 \left(\frac{18}{38}\right)^2 \left(\frac{2}{38}\right)^1 = .0795.$$

The joint distribution of Y_1, Y_2, Y_3 is called **trinomial**, and (3) is a **trinomial coefficient**. [See (2) of Section 1.6.] ■

The **multinomial formula** for the general case is derived exactly as for the special case in the preceding example:

Multinomial p.f.

$$P(Y_1 = y_1, \ldots, Y_k = y_k) = \binom{n}{y_1, y_2, \ldots, y_k} p_1^{y_1} p_2^{y_2} \cdots p_k^{y_k}, \tag{4}$$

where $p_i = P(A_i)$, and Y_i is the frequency of A_i in n independent trials of the experiment (1).

Since the frequencies Y_j must add up to n, any one of them is uniquely determined by the other $k - 1$. Thus, a trinomial distribution is determined by the distribution of (Y_1, Y_2), since $Y_3 = n - Y_1 - Y_2$, so it is really bivariate—just as a binomial distribution is univariate.

Now focus on *one* of the frequencies (say, Y_1) and regard the outcomes that are not A_1 as lumped together in a second category:

$$A_1^c = A_2 \cup \cdots \cup A_k.$$

Then $P(A_1) = p_1$, and $P(A_1^c) = 1 - p_1$. Thus, and by the same token, $Y_i \sim \text{Bin}(n, p_i)$. The same kind of reasoning shows that the marginal distribution of any subset of the category frequencies in a multinomial distribution is again multinomial (with fewer categories).

Example 4.8b   Suppose a question is asked in a sample survey with possible answers "For," "Against," and "Neutral." The population proportions of these categories are their probabilities: p_F, p_A, and p_N. For a random sample of size n, the frequencies (Y_F, Y_A, Y_N) have a trinomial distribution. If we are interested only in the number or proportion of "yes" responses, we can lump the others into a single class. Thus, $Y_F \sim \text{Bin}(n, p_F)$. For example, if $n = 600$, and the population proportions are 60% "For," 30% "Against," and 10% "Neutral," then the number of "For" responses in the sample is Bin(600, .6).   ∎

4.9   **Applying the p.g.f.**

We saw in Section 3.6 that the probability generating function of a sum of *independent* random variables is the product of their p.g.f.'s. We now use this property to derive or rederive some properties of the various distributions we've introduced.

Example 4.9a   The p.g.f. of a Bernoulli variable X is

$$\eta(t) = E(t^X) = t^1 P(X = 1) + t^0 P(X = 0) = pt + q. \tag{1}$$

A sequence of n independent trials $(X_1, \ldots, X_n)$ is a sequence of 1's and 0's. In Section 4.2, we found the distribution of the sum: $Y = \Sigma X_i \sim \text{Bin}(n, p)$. We now rederive this result using p.g.f.'s.

The p.g.f. of Y, the sum of n independent Bernoulli variables, is the product of n factors (1):

$$\eta_{\Sigma X}(t) = \prod_1^n (p + q) = (pt + q)^n. \tag{2}$$

When we apply the binomial theorem [(2) of Section 4.2] to expand $(pt + q)^n$, (2) becomes

$$\eta_Y(t) = \sum_{k=0}^n \binom{n}{k} (pt)^k q^{n-k} = \sum_{k=0}^n \left\{ \binom{n}{k} p^k q^{n-k} \right\} t^k. \tag{3}$$

According to (1) of Section 3.6, the coefficient of t^k in (3) is the probability that $Y = k$. This coefficient is precisely the binomial probability given as (1) of Section 4.2.   ∎

Example 4.9b   In Section 4.6, we claimed that the sum of independent Poisson variables is a Poisson variable, appealing to the corresponding result for binomials. We now give a proof using p.g.f.'s.

Let $Y \sim \text{Poi}(m)$. The p.g.f. of Y [see (1) of Section 3.6] is

$$\eta_Y(t) = E(t^Y) = \sum_{k=0}^{\infty} t^k f_Y(k \mid m) = \sum_{k=0}^{\infty} t^k \cdot \frac{m^k}{k!} e^{-m}$$

$$= e^{-m} \sum_{k=0}^{\infty} \frac{(mt)^k}{k!} = e^{-m} e^{mt} = e^{-(1-t)m}. \tag{4}$$

Suppose $Y_i \sim \text{Poi}(m_i)$, for $i = 1, 2, \ldots, n$. If $Y_1, \ldots, Y_n$ are independent, the p.g.f. of their sum [by (5) of Section 3.6] is the product of their p.g.f.'s:

$$\eta_{\Sigma Y}(t) = E(t^{\Sigma Y}) = \Pi \, e^{-(1-t)m_i} = e^{-(1-t)\Sigma m_i}.$$

In view of (4), this is the p.g.f. of a Poisson variable with parameter Σm_i. So $\Sigma Y_i \sim \text{Poi}(\Sigma m_i)$. In particular, if the Y_i all have the *same* parameter m, the sum is $\text{Poi}(nm)$. ∎

Factorial Moments

At the end of Section 3.6, we showed how to find the mean of a random variable with values $0, 1, 2, \ldots$ as the derivative of the p.g.f. at $t = 1$. The method applies generally to any random variable X for which the p.g.f. exists. We differentiate

$$\eta_X(t) = E(t^X) = \sum_x t^x f_X(x) \tag{5}$$

with respect to t:

$$\eta'(t) = \frac{d}{dt} \sum_x t^x f_X(x) = \sum_x \frac{d}{dt} (t^x) f_X(x) = \sum_x x t^{x-1} f_X(x). \tag{6}$$

Substituting $t = 1$ in (6) yields the mean:

$$\eta'(1) = \sum x f_X(x) = E(X). \tag{7}$$

Differentiating once more, we find

$$\eta''(t) = \frac{d^2}{dt^2} (E(t^X)) = E[X(X-1)t^{X-2}].$$

From this, we obtain

$$\eta''(1) = E[X(X-1)].$$

This is called the second **factorial moment** of X. Differentiating $\eta(t)$ k times and substituting $t = 1$, we obtain

$$\psi^{(k)}(1) = E[X(X-1)(X-2) \cdots (X-k+1)] = E[X_{(k)}] \tag{8}$$

(in the notation of Section 1.4), called the kth factorial moment of X.

The p.g.f. generates probabilities as power series coefficients only in the special case of a variable with values that are nonnegative integers. Because (8) applies more generally, the p.g.f. is sometimes called the *factorial moment generating function* (f.m.g.f.).

Example **4.9c**   We illustrate the use of the p.g.f. to generate factorial moments in the case of a Poisson distribution. If Y is Poi(m), its p.g.f. is given by (4) in Example 4.9b:

$$\eta(t) = e^{-(1-t)m}.$$

The kth derivative is

$$\eta^{(k)}(t) = m^k e^{m(t-1)}.$$

So

$$E(Y_{(k)}) = n^{(k)}(1) = m^k.$$

Thus, $E(Y) = m$, and $E[Y(Y-1)] = m^2$. In Section 4.6, we used these two factorial moments to calculate the variance, as follows:

$$\text{var } Y = E[Y^2 - (EY)^2]$$
$$= E[Y(Y-1)] + E(Y) - (EY)^2 = m^2 + m - m^2 = m.$$

Higher-order central moments can be calculated from the factorial moments in similar fashion. ∎

Table 4.2 gives the probability generating functions for some of the distributions discussed in this chapter. We have derived some of these in this section; the problems call for derivations of the others.

Table 4.2

Model	Notation	p.f.	f.m.g.f.
Bernoulli	Ber (p)	$p^x q^{1-x}, \quad x = 0, 1$	$pt + q$
Binomial	Bin(n, p)	$\binom{n}{x} p^x q^{n-x}, \quad x = 0, \ldots, n$	$(pt + q)^n$
Geometric	Geo(p)	$pq^{x-1}, \quad x = 1, 2, \ldots$	$\dfrac{tp}{1 - tq}$
Negative binomial	Negbin(r, p)	$\binom{x-1}{r-1} p^r q^{x-r}, \quad x = r, r+1, \ldots$	$\left(\dfrac{tp}{1 - qt}\right)^r$
Poisson	Poi(m)	$\dfrac{m^x}{x!} e^{-m}, \quad x = 0, 1, \ldots$	$e^{-(1-t)m}$

Chapter Perspective

Bernoulli trials are simple, but they are fundamental in many applications of statistics. Different sampling schemes for Bernoulli trials lead to different distributions. When the trials are independent, the number of successes in a specified number of trials has a binomial distribution; the geometric and negative binomial distributions apply for the number of trials needed to obtain a specified number of successes. The analogous distributions for sampling without replacement (in which case the trials are dependent) are the hypergeometric and the hypergeometric waiting time distributions. These distributions have applications in statistical inference for p, the population proportion of successes.

This chapter introduced the normal curve for approximating binomial probabilities. This is merely one application of this important continuous distribution, which we'll treat in detail in Chapter 6.

The Poisson distribution provides a convenient approximation for some binomial probabilities but also serves as a useful model in its own right. In Chapter 6, we'll return to the Poisson process and take up some continuous distributions for the waiting times.

The multinomial distribution is useful for describing category frequencies for discrete phenomena with a finite number of categories. We'll see in Chapter 13 that the distribution also applies for sample data from continuous variables when the data are summarized in class intervals, as is commonly the case.

We've exploited the p.g.f. in finding distributions of sums of independent random variables and in finding moments of certain discrete distributions. Chapter 5 will introduce the moment generating function, which is closely related to the p.g.f. Both functions characterize a distribution in the sense that there is a one-to-one correspondence between distributions and generating functions.

Solved Problems

Sections 4.1–4.2

A. Find the probability that there is at most one 3 in four tosses of a fair die.

Solution:
The probability of a 3 in a single toss is $\frac{1}{6}$. Let Y denote the number of 3's in four tosses. This is binomial with $n = 4$ and $p = \frac{1}{6}$. We want $P(Y = 1 \text{ or } 0)$. Using the binomial formula, we have

$$P(Y = 0 \text{ or } Y = 1) = \binom{4}{1}\left(\frac{1}{6}\right)^1\left(\frac{5}{6}\right)^3 + \left(\frac{5}{6}\right)^4 \doteq \frac{125}{144} = .868.$$

B. Each of twelve people tastes two colas. One glass is marked Q and the other M. Each person is asked to pick the Pepsi. If both glasses contain Pepsi, what is the probability that 9 or more pick glass M (assuming that they must pick one or the other and that the label has no effect on their choice)?

[Pepsi conducted such an experiment in 1977 to show that in a test result advertised by Coke, people really were expressing preference for the letter M over the letter Q.]

Solution:
With identical drinks, the probability of picking glass M is $\frac{1}{2}$. Assuming independent trials seems reasonable, in which case the number choosing glass M is Bin(12, $\frac{1}{2}$). Then

(with the aid of Table Ib in Appendix 1),

$$P(9 \text{ or more}) = \sum_{k=9}^{12} f(k) = \sum_{k=9}^{12} \binom{12}{k}\left(\frac{1}{2}\right)^k\left(\frac{1}{2}\right)^{12-k} = .0730.$$

C. In Problems 32 and 33 of Chapter 2, we regarded a baseball player's successive times at bat (AB's) as independent trials of the same experiment. With this assumption, find the probability that a "300 hitter" gets at least one hit in four AB's. (By "300 hitter," we mean one whose probability of a hit in a single time at bat is .300.)

Solution:

Let $Y = $ number of hits in four AB's. We assume $Y \sim \text{Bin}(4, .3)$. Then

$$P(Y \geq 1) = 4(.3)(.7)^3 + 6(.3)^2(.7)^2 + 4(.3)^3(.7) + (.3)^3$$

$$= .4116 + .2646 + .0756 + .0081 = .7599.$$

Alternatively,

$$P(Y \geq 1) = 1 - P(Y = 0) = 1 - (.7)^4 = .7599.$$

D. Suppose that the batter in the preceding problem has four at-bats in each of ten games. Assuming that all 40 at-bats are part of a Bernoulli process, find the probability that the batter gets at least one hit in each of the ten games.

Solution:

From the solution to Problem C, we know that the probability of at least one hit in a game is .7599. So the probability that this happens in each of the ten games is $(.7599)^{10} \doteq .0642.$

E. Find the probability that the batter in Problem C gets *exactly* one hit in a particular game, given that he gets at least one hit in that game.

Solution:

"Exactly 1" is included in the event "at least one," so

$$P(\text{exactly } 1 \mid \text{at least } 1) = \frac{P(\text{exactly } 1)}{P(\text{at least } 1)}.$$

In Problem D, we found $P(\text{at least } 1) = .7599$, and $P(1 \text{ hit}) = .4116$. So

$$P(\text{exactly } 1 \mid \text{at least } 1) = \frac{.4116}{.7599} \doteq .542.$$

F. Find the value of the following sum:

$$\sum_{k=0}^{15} k^2 \binom{15}{k}(.4)^k(.6)^{15-k}.$$

Solution:

The expression $\binom{15}{k}(.4)^k(1.6)^{15-k}$ is the probability that $Y = k$ when $Y \sim \text{Bin}(15, .4)$. The sum in the problem is therefore $E(Y^2) = \Sigma k^2 f(k)$. We can calculate this using (6) of

Section 3.3 as the sum of the variance and the square of the mean. These in turn are given by (3) of Section 4.2:

$$\sum_{k=0}^{15} k^2 \binom{15}{k}(.4)^k(.6)^{15-k} = E(Y^2) = \text{var } Y + [E(Y)]^2$$

$$= npq + (np)^2 = 15 \times .4 \times .6 + (15 \times .4)^2 = 39.6.$$

Sections 4.3–4.4

G. An instructor assigned seven possible essay topics, from which three would be selected for the final exam. One student studied five of the seven topics thoroughly but ignored the other two. Assuming a random selection on the part of the instructor, find the probability distribution of the number of questions on the exam for which the student was prepared.

Solution:

The distribution is hypergeometric ($N = 7$, $M = 5$, $n = 3$). [See (2) and (3) in Section 4.3.] Thus,

$$f(x) = \frac{\binom{5}{x}\binom{2}{3-x}}{\binom{7}{3}}, \qquad \text{for } x = 0, 1, 2, 3.$$

In table form:

x	0	1	2	3
$f(x)$	0	$\frac{1}{7}$	$\frac{4}{7}$	$\frac{2}{7}$

The mean is $np = n(M/N) = 3(5/7)$. To verify this formula, we can find the mean from the table as $\Sigma\, xf(x) = \frac{1}{7} + \frac{8}{7} + \frac{6}{7}$. The variance is

$$npq \cdot \frac{N-n}{N-1} = 3 \cdot \frac{5}{7} \cdot \frac{2}{7} \cdot \frac{7-3}{7-1} \doteq \frac{20}{49}.$$

(Again, the formula can be verified by calculating var X from the probability table— you should do this.)

H. A group of twelve patients to be used in evaluating a drug includes six men and six women. Six patients are to be selected at random to receive the drug; the others are to be used as controls and will receive an inert placebo. Find the probability that the control group is made up of three men and three women.

Solution:

There are $\binom{12}{6} = 924$ ways to choose the control group without regard to sex. To get three men for the control group, there are $\binom{6}{3} = 20$ ways; and by the same calculation, 20 ways to select three women. So there are $20 \times 20 = 400$ ways of selecting the control group with three of each sex. The desired probability is $\frac{400}{924}$. This is a hypergeometric probability with $N = 12$, $M = 6$, and $n = 6$:

$$f(3) = \frac{\binom{6}{3}\binom{6}{3}}{\binom{12}{6}} \doteq .433.$$

I. A bag of 50 jelly beans includes exactly eight that are black. Suppose we select beans one at a time until we get one that is black. Find the probability that we select exactly four beans.

Solution:

We'll give two methods:

(1) For the first black to appear on the fourth selection, the sequence of selections must be N, N, N, B (where N = not black). The probability of this sequence [using (6) from Section 2.3] is

$$\frac{42}{50} \cdot \frac{41}{49} \cdot \frac{40}{48} \cdot \frac{8}{47} \doteq .0997.$$

(2) This is a special case of the negative hypergeometric with $r = 1$, given by (6) or (7) of Section 4.4:

$$\frac{\binom{42}{3}}{\binom{50}{3}} \cdot \frac{8}{47} = \frac{\binom{3}{0}\binom{46}{7}}{\binom{50}{8}} = .0997.$$

(The first black one appears on the fourth selection if and only if there are no black ones among the first three *and* the fourth one is black.)

J. Compare this hypergeometric probability with its binomial approximation:

$$\frac{\binom{500}{4}\binom{500}{4}}{\binom{1000}{8}}.$$

Solution:

This is a hypergeometric $f(k)$ with $N = 1000$, $M = 500$, $n = 8$, and $k = 4$. The exact value is

$$\frac{\left(\dfrac{500 \cdot 499 \cdot 498 \cdot 497}{4!}\right)^2}{\left(\dfrac{1000 \cdot 999 \cdot 998 \cdot 997 \cdot 996 \cdot 995 \cdot 994 \cdot 993}{8!}\right)} \doteq .2745.$$

The sample size (8) is much smaller than the population size (500), so a binomial approximation should work well. With $n = 8$ and $p = M/N = 500/1000 = .5$, the binomial probability is

$$f(4) = \binom{8}{4}(.5)^4(.5)^4 = .2734.$$

K. The makers of M&M candies claimed[3] that 10% of plain M&M's are tan. We take handfuls of ten M&M's from a bowl until we get a tan one. Let X denote the number of handfuls required to produce at least one tan M&M. Assuming that the maker's claim is correct and that we're sampling randomly from an infinite population,

(a) find $P(X = 3)$.

(b) find $E(X)$.

---

[3] Since this information was published, they have begun including red M&M's, so present proportions may be different.

Solution:

The probability that a handful includes at least one tan M&M is

$$1 - P(\text{no tan}) = 1 - (.9)^{10} \doteq .6513.$$

The number of handfuls required to produce a tan M&M is geometric with $p = .6513$. Its p.f. is given by (2) of Section 4.4. So

$$P(X = 3) = f(3) = .6513(.3487)^2 = .0792.$$

From (3) of Section 4.4,

$$E(X) = \frac{1}{p} = \frac{1}{.6513} \doteq 1.54.$$

L. In screening donors for someone with type O blood,
   (a) find the mean number needed, if donors are tested one at a time.
   (b) what is the probability that at most twelve need to be tested to locate four with that type?
   Assume that 45% of the population have type O blood.

Solution:

   (a) The number required is negative binomial with mean [given by (3) of Section 4.4] $r/p = 4/.45 \doteq 8.9$.
   (b) We'll give two approaches to finding the desired probability:
   (1) What's wanted is the probability that at least four individuals of twelve selected from the population will be type O. The complement of this event is the event that fewer than four are type O, so

   $$P(\text{at least 4 type O among 12}) = 1 - P(\text{at most 3 among 12})$$
   $$= 1 - .9579 = .0421,$$

   where .9579 is read from Table Ib ($n = 12$, $p = .45$, $c = 3$).
   (2) Another way to calculate this probability is to express it in terms of the negative binomial distribution. The probability that *exactly* k individuals must be tested before finding the fourth one with type O blood is

   $$\binom{k-1}{3}(.55)^{k-4}(.45)^4, \qquad \text{for } k \geq 4.$$

   Summing these probabilities from $k = 4$ to $k = 12$ would yield the desired probability (with a good bit of work).

M. Find the mean number of rolls of an ordinary die required so that each face turns up at least once.

Solution:

The total number of rolls is 1 plus the number it takes to get something different from the first roll, plus the number it takes to get something still different, etc., until the sixth face is seen. After the first roll of the die, no matter how it turns out, it takes on average $(\frac{5}{6})^{-1}$ rolls to get something other than what we got first. [See (3) of Section 4.4.] After that, it takes on average $(\frac{4}{6})^{-1}$ to turn up one of the remaining four sides, and so on. The mean number of rolls required is the sum of these mean numbers required for seeing a new face:

$$1 + \frac{6}{5} + \frac{6}{4} + \frac{6}{3} + \frac{6}{2} + \frac{6}{1} = 14.7.$$

Section 4.5

N. Suppose 35% of the TV viewers in a metropolitan area are watching a certain special program. Find the probability that in a random sample of 1,000 TV viewers, at most 330 are found to be watching the program.

Solution:

The number in the sample that are watching the program is hypergeometric, but there is no way to calculate exact hypergeometric probabilities when we don't know the size of the population. So we use a binomial approximation:

$$P(\text{at most } 330) = \sum_{k=0}^{330} \binom{1000}{k} (.35)^k (.65)^{1000-k}.$$

Calculating this is impractical, so we'll appeal to the normal approximation [(1) of Section 4.5]. For this, we need the mean and s.d.:

$$np = 1000 \times .35, \qquad \sqrt{npq} = \sqrt{1000 \times .35 \times .65} = 15.08.$$

Then

$$P(\text{at most } 330) \doteq \Phi\left(\frac{330.5 - 350}{15.08}\right) = \Phi(-1.293) = .0980.$$

O. Find the probability of exactly 32 heads in 60 tosses of a fair coin.

Solution:

Let Y denote the number of heads in 60 tosses; this is Bin(60, .5). Thus,

$$P(Y = 32) = f_Y(32) = \binom{60}{32}(.5)^{32}(.5)^{28} = \frac{60!}{32!28!}(.5)^{60} = .08996.$$

A normal approximation uses the mean $np = 30$ and s.d. $= \sqrt{npq} = \sqrt{15} = 3.873$. Normal probabilities are *cumulative*, so to use the normal table to find the probability of exactly 32, we take a difference:

$$P(Y = 32) = P(Y \le 32) - P(Y \le 31)$$

and approximate each probability on the right with (1) of Section 4.5:

$$P(Y \le 32) - P(Y \le 31) \doteq \Phi\left(\frac{32.5 - 30}{3.873}\right) - \Phi\left(\frac{31.5 - 30}{3.873}\right)$$

$$\doteq .7407 - .6507 = .0900.$$

P. In Problem K, we assumed that 10% of plain M&M candies are tan. Find the probability that there are exactly two tan M&M's in a random selection of 20. Compare the binomial probability with its Poisson approximation.

Solution:

When the population is infinite, the number X of tan candies is binomial:

$$P(X = 2) = f(2) = \binom{20}{2}(.1)^2(.9)^{18} = .285.$$

In a Poisson approximation, we take $m = np = 20 \times .1 = 2$:

$$f(2) \doteq \frac{2^2}{2!}e^{-2} = .271.$$

[We can also find this probability in Table IV by subtracting $f(1)$ from $f(2)$ in the column for $m = 2$: $.677 - .406 = .271$.]

Q. Early in a presidential election year, only 2% of the voters in a certain state favor candidate A for president, whereas 30% favor candidate B. In a sample of 400 voters selected at random from this state, find

   (a) the mean and standard deviation of the number who favor B.
   (b) the probability that more than 100 favor B.
   (c) the probability that none is in favor of A.
   (d) the probability that at most six favor A.

Solution:

We're not given the population size, so we can't calculate exact hypergeometric probabilities. We use binomial approximations: $n = 400$, $p = .02$ and $.30$. Let $X =$ number favoring A, and $Y =$ number favoring B in the sample of 400.

   (a) $E(Y) = np = 400 \times .3 = 120$, and var $Y = npq = 400 \times .3 \times .7 = 84$. The standard deviation is $\sqrt{84} \doteq 9.165$.

   (b) With $n = 400$ and $npq = 84$, we use the normal approximation to the distribution of Y:

$$P(Y > 100) = 1 - P(Y \le 100)$$

$$\doteq 1 - \Phi\left(\frac{100.5 - 120}{9.165}\right) = \Phi(2.13) \doteq .983.$$

   (c) We also use a binomial distribution for X, with $n = 400$ and $p = .02$. The exact binomial probability that none favors A is

$$P(X = 0) = f(0) = \binom{400}{0}(.02)^0(.98)^{400} = .000309.$$

Since n is quite large and p is small, we can approximate this using a Poisson probability, with $m = np = 400 \times .02 = 8$: $f(0) = e^{-8} = .000335$. Although n is large, the small value of p means that the distribution is very skewed and npq is only 7.84, so a normal approximation is not as successful:

$$P(X = 0) = P(X \le 0) \doteq \Phi\left(\frac{0 + .5 - 8}{2.8}\right) = .0037.$$

(Probabilities in the tail of a distribution are especially hard to approximate well.)

(d) The exact binomial probability is

$$P(X \leq 6) = \sum_{k=0}^{6} \binom{400}{k} (.02)^k (.98)^{400-k} = .311.$$

This is quite tedious to calculate, and we don't have a cumulative binomial table for $n = 400$. A Poisson approximation works well and is readily available in Table IV (with $m = 8$ and $c = 6$): $P(X \leq 6) \doteq .313$. On the other hand, the normal approximation with continuity correction may work better than it did in (c), since $X = 6$ is not so far out in the tail:

$$P(X \leq 6) \doteq \Phi\left(\frac{6.5 - 8}{2.8}\right) = \Phi(-.536) \doteq .296.$$

This is not bad, but we recommend always using the Poisson approximation when p is small; if you fall outside Table IV, then you have to use the normal.

Section 4.6

R. Flaws in one manufacturer's carpet occur on the average of one in 50 square yards. Assuming a Poisson distribution, find the probability of
(a) at most one flaw in a 10-square-yard piece.
(b) exactly one flaw in a 10-square-yard piece, given that it has at least one flaw.

Solution:
Let X be the number of flaws in a 10-square-yard piece of material. With one flaw in 50 square yards, the average in 10 square yards is one-fifth of this: $m = .2$.
(a) $P(X = 0 \text{ or } 1) = f(0) + f(1) = e^{-.2}(1 + .2) = .982$. (This is given in Table IV under $m = .2$ opposite $c = 1$.)
(b) Here we use the definition of conditional probability. Also, since $\{X = 1\} \subset \{X \leq 1\}$, their intersection is $\{X = 1\}$:

$$P(X = 1 \mid X \leq 1) = \frac{P(X = 1 \text{ and } X \leq 1)}{P(X \leq 1)} = \frac{P(X = 1)}{P(X \leq 1)} = \frac{.163}{.982} = .166.$$

S. An arrival process is assumed to be Poisson with parameter $\lambda = 24/\text{hr}$.
(a) Find the probability of no more than 10 arrivals in a half-hour period.
(b) Find the probability that there is at most one arrival in a particular one-minute period.
(c) Find the probability that the time to the second arrival exceeds one minute.

Solution:
(a) The expected number in a half-hour period is $\lambda t = 12$, so $P(\text{at most } 10) = .347$ (from Table IV with $m = 12$, $c = 10$).
(b) The mean number of arrivals in one minute is $24 \times 1/60 = .4$. From the column $m = .4$ in Table VI, we find the probability .938 opposite $c = 1$. Or we can easily calculate this without a table:

$$P(0 \text{ or } 1 \text{ arrival}) = e^{-.4}(1 + .4) \doteq .938.$$

(c) The time to the second arrival will exceed one minute if and only if there is at most one arrival in one minute. So the event in question is the same as the event in (b); the answer is again .938.

T.   Given that X has the probability function

$$\frac{4^x}{x!}e^{-4}, \qquad x = 0, 1, 2, \ldots,$$

find $E(X^2)$.

Solution:
The given distribution is Poisson with $m = 4$. So $E(X) = 4$ and var $X = 4$. Hence, $E(X^2) = \text{var } X + [E(X)]^2 = 4 + 16 = 20$.

U.   Suppose X is the number of arrivals in a given unit of time in a Poisson arrival process, and Y is the number of arrivals in the next unit of time. Find the conditional distribution of X given that $X + Y = c$.

Solution:
We know from Poisson assumption (b) that X and Y are independent. Using this and the definition of conditional probability, we find

$$P(X = x \mid X + Y = c) = \frac{P(X = x \text{ and } X + Y = c)}{P(X + Y = c)} = \frac{P(X = x) \cdot P(Y = c - x)}{P(X + Y = c)}$$

$$= \frac{\left(\dfrac{\lambda^x}{x!}e^{-\lambda}\right)\left(\dfrac{\lambda^{c-x}}{(c-x)!}e^{-\lambda}\right)}{\dfrac{(2\lambda)^c}{c!}e^{-2\lambda}} = \frac{c!}{x!(c-x)!2^c}$$

for $x = 0, 1, \ldots, c$. This is the p.f. of Bin$(c, \frac{1}{2})$. So if there are c arrivals in the interval of length 2, the number in the first half of the interval is binomially distributed with $n = c$ and $p = \frac{1}{2}$.

Section 4.8–4.9

V.   Before red M&M's were reintroduced, plain M&M's came in five colors: brown, orange, yellow, tan, and green. The manufacturer claimed they were mixed in the proportions .4, .2, .2, .1, .1, respectively. Let $B = $ number of brown ones, $O = $ the number of orange, etc., in a random sample of 20.
(a)   Find the probability of getting four of each color.
(b)   Given that there are no greens and two tans among the 20, what is the distribution of (B, O, Y)?

Solution:
(a)   The joint distribution of the five frequencies is multinomial:

$$P(4 \text{ of each}) = \binom{20}{4, 4, 4, 4, 4}(.4)^4(.2)^4(.2)^4(.1)^4(.1)^4 \doteq .020.$$

(b) If there are two tan and no green M&M's among the 20, the others 18 are drawn from the three colors brown, orange, and yellow. These three are produced in the ratio 4:2:2, so their conditional probabilities are $\frac{4}{8}$, $\frac{2}{8}$, $\frac{2}{8}$. Thus, the required distribution is trinomial with $n = 18$ and probabilities $(.5, .25, .25)$. In case this reasoning is not convincing, we calculate one value of the conditional joint p.f.:

$$P(B = 10, O = 5, Y = 3 \mid T = 2, G = 0) = \frac{f(10, 5, 3, 2, 0)}{f_{T,G}(2, 0)}.$$

The numerator is a multinomial p.f. (five categories):

$$f(10, 5, 3, 2, 0) = \frac{20!}{10!5!3!2!0!}(.4)^{10}(.2)^5(.2)^3(.1)^2(.1)^0.$$

The denominator is the marginal (trinomial) p.f. of (T, G):

$$f_{T,G}(2, 0) = \frac{20!}{2!0!18!}(.1)^2(.1)^0(.8)^{18}.$$

The quotient is

$$P(B = 10, O = 5, Y = 3 \mid T = 2, G = 0) = \frac{18!}{10!5!3}\left(\frac{4}{8}\right)^{12}\left(\frac{2}{8}\right)^5\left(\frac{2}{8}\right)^3.$$

This is a trinomial probability with $n = 18$ and cell probabilities $(.5, .25, .25)$.

W. In a population of 500 people, the distribution according to blood group is as follows: 225 are type O, 200 are type A, 50 are type B, and 25 are type AB. We draw a random sample of size eight without replacement. Let

$X = $ number of type A in the sample,

$Y = $ number of type B in the sample,

$Z = $ number of type AB in the sample.

(a) Find an exact expression for the probability that $X = 4$.
(b) Approximate the probability in (a) using the fact that the sample size is small compared to 500.
(c) Find the exact probability that $X = 4$, $Y = 2$, and $Z = 1$.
(d) Approximate the probability in (c) using the multinomial distribution.
(e) Find $E(Y)$ and var Y.
(f) Use the Poisson p.f. to approximate $P(Z > 2)$.

Solution:
(a) The probability is hypergeometric $(N = 500, M = 200, n = 8)$:

$$P(X = 4) = \frac{\binom{200}{4}\binom{300}{4}}{\binom{500}{8}} \doteq .234.$$

(b) Since $N/n > 50$, $X \approx \text{Bin}(8, p)$, where $p = \frac{200}{500} = .40$:

$$\binom{8}{4}(.4)^4(.6)^4 = .232.$$

(c)   This probability is to the hypergeometric as the multinomial is to the binomial. We reason much as we do in obtaining a hypergeometric probability: The possible combinations of 8 from 500 are equally likely, so we simply count the combinations with four type A from 200, two type B from 50, and one type AB from 25 (and then of course one type 0 from 225):

$$P(4 \text{ A's, 2 B's, 1 AB}) = \frac{\binom{200}{4}\binom{50}{2}\binom{25}{1}\binom{225}{1}}{\binom{500}{8}} \doteq .00487.$$

(d)   Assuming sampling without replacement as an approximation, we have a multinomial distribution with category probabilities .45, .40, .10, and .05 for O, A, B, and AB, respectively. Apply (3) of Section 4.8:

$$\binom{8}{4, 2, 1, 1}(.45)^1(.40)^4(.10)^2(.05)^1 \doteq .00487.$$

(e)   Like X, the variable Y is hypergeometric, and

$$E(Y) = np = 8 \times .10 = .8.$$

From (3) of Section 4.3,

$$\text{var } Y = npq \cdot \frac{N - n}{N - 1} = 8 \times .1 \times .9 \times \frac{492}{499} \doteq .710.$$

(f)   With $n = 8$ and $p = M/N = .05$, the mean is $m = np = .4$. Opposite $c = 2$ in Table IV, we find $P(Z \leq 2) \doteq .992$. The desired probability is $P(Z < 2) = .008$.

X.   Consider the random variable X, the number of bidding points assigned a card in the system of counting for bridge bidding described in Problem 34, Chapter 2. In Problem N of Chapter 3, we used the p.g.f. of X to find the mean, $E(X) = \frac{10}{13}$. Find the variance of X using the p.g.f.

Solution:

The p.g.f. (from the solution to Problem N, Chapter 3) is

$$\eta(t) = (9 + t + t^2 + t^3 + t^4) \cdot \frac{1}{13}.$$

The first two derivatives are

$$\eta'(t) = (1 + 2t + 3t^2 + 4t^3) \cdot \frac{1}{13},$$

$$\eta''(t) = (2 + 6t + 12t^2) \cdot \frac{1}{13}.$$

Evaluating the second derivative at $t = 1$, we obtain

$$\eta''(1) = (2 + 6 + 12) \cdot \frac{1}{13} = \frac{20}{13} = E(X^2) - E(X).$$

So $\text{var } X = [E(X^2) - E(X)] + E(X) - [(E(X)]^2$

$$= \frac{20}{13} + \frac{10}{13} - \left(\frac{10}{13}\right)^2 = \frac{290}{169}.$$

Y.   The random variable X has this p.g.f.:

$$\eta(t) = .4t + .3t^2 + .2t^3 + .1t^4.$$

(a)   Use η to find the mean and variance of X.
(b)   Find the probability function of X.

Solution:
(a)   Differentiating, we have

$$\eta'(t) = .4 + .6t + .6t^2 + .4t^3,$$

and

$$\eta''(t) = .6 + 1.2t + 1.2t^2.$$

Substituting $t = 1$ yields

$$E(X) = \eta'(1) = .4 + .6 + .6 + .4 = 2$$

and

$$E[X(X - 1)] = \eta''(t) = .6 + 1.2 + 1.2 = 3.$$

So

$$\text{var } X = E(X^2) - [E(X)]^2$$
$$= E[X(X - 1)] + E(X) - [E(X)]^2 = 3 + 2 - 2^2 = 1.$$

(b)   Since $P(X = k)$ is the coefficient of t^k,

$$f(1) = .4, \ f(2) = .3, \ f(3) = .2, \text{ and } f(4) = .1.$$

Z.   Let $Y = X_1 + X_2$, where X_1 and X_2 are independent observations on the variable X of the preceding problem. Find the probability function of Y.

Solution:
There are various ways to do this. The first we give is in the spirit of Section 4.9, and the second uses first principles.
(1)   The p.g.f. of Y is the square of the p.g.f. for X:

$$\eta_Y = [\eta_X]^2 = (.4t + .3t^2 + .2t^3 + .1t^4)^2$$
$$= .16t^2 + .24t^3 + .25t^4 + .20t^5 + .10t^6 + .04t^7 + .01t^8.$$

Then $f_Y(k)$ is the coefficient of t^k:

k	2	3	4	5	6	7	8
$f_Y(k)$	.16	.24	.25	.20	.10	.04	.01

(2)   The accompanying table gives the joint probability function, obtained by multiplying the marginal probabilities. From this, we find $P(Y = k)$ as the sum of the joint probabilities along diagonals on which $x_1 + x_2 = k$. The result is as given in (1). For instance, $f_Y(5) = .20$, the sum of the circled quantities.

x_2 \\ x_1	1	2	3	4	$f_2(x_2)$
1	.16	.12	.08	.04	.4
2	.12	.09	.06	.03	.3
3	.08	.06	.04	.02	.2
4	.04	.03	.02	.01	.1
$f_1(x_1)$	.4	.3	.2	.1	

Problems

For problems marked with an asterisk, answers are provided in Appendix 2.

Sections 4.1–4.2

*1. Suppose each of ten subjects is to be assigned to treatment A or to treatment B according to the toss of a coin: heads to A, tails to B. Find
   (a) the probability that exactly 5 get each treatment.
   (b) the probability that at most 3 get treatment A.
   (c) the mean and s.d. of the number who get treatment A.

*2. A basketball player's free throws are successful 90% of the time. Assume that free throws are independent trials, each with probability $p = .90$ of success. For eight free throws, find
   (a) the probability of exactly one miss.
   (b) the probability of at most one miss.
   (c) the probability of at least one miss.
   (d) the expected number of misses.

3. To test a new type of golf ball, 20 golfers are paired in such a way that both golfers in a pair have the same handicap. One golfer in each of the ten pairs is assigned the new ball and the other a standard ball by the toss of a coin. The pairs play a round of golf (including a tie-breaker if necessary). Let X denote the number of pairs in which the player using the new type of ball wins. Suppose the new ball has precisely the same performance characteristics as the old, so that which ball wins is like the toss of a coin. Find the following:
   (a) The distribution of X.   (c) $P(X \leq 3)$.
   (b) $P(X = 10)$.   (d) $P(X \geq 5)$.

*4. A multiple-choice quiz consists of 20 questions, each with four choices. A student who guesses has one chance in four of being correct for each question. Let U denote the student's score when each question is worth five points. Find
   (a) the expected number of correct answers and the expected score.
   (b) the s.d. of U.
   (c) the probability that the student scores exactly 50.
   (d) the probability that the student scores at most 20.

5. A random sample of size 500 is drawn from the population of Washington, D.C.

Assume 55% are blacks. Let X denote the number of blacks included in the sample. Find the mean and standard deviation of X.

*6. Suppose P (male birth) $= P$ (female birth) $= 1/2$ and that the sexes of successive children in a family are independent. Among 160 families with four children each, how many would be expected to have 0, 1, 2, 3, and 4 boys?

7. (a) Calculate the covariance of X_1 and X_2 in Example 4.1b.
    (b) Obtain a general formula for cov(X_1, X_2), for sampling a population of size N in which M are "successes."

*8. An archer hits the bull's-eye one time in ten on the average. She will give even odds that she will hit the bull's-eye at least once in n tries. How large must n be for this bet to be favorable for her? ("Even odds" means odds of $1 against $1.)

9. Find the value:

$$\sum_{k=0}^{20} k \binom{20}{k} (.05)^k (.95)^{20-k}.$$

Sections 4.3–4.4

*10. A coffee taster tries to identify 20 cups of coffee. The taster knows only that 15 of the cups were brewed from regular coffee and is to pick out the five brewed from instant coffee. Let X be the number of cups among the five selected that actually are instant. Suppose the taster really cannot tell which is which. Find
    (a) $E(X)$.   (b) var X.   (c) $P(X = 3)$.   (d) $P(X \geq 3)$.

11. A bridge hand is a random selection of 13 cards from a standard deck of 52 playing cards.
    (a) Find the probability that a bridge hand contains at most one ace. (The deck contains four aces.)
    (b) Find $E(X)$, where $X =$ the number of red cards in a hand. (Half of the 52 cards in the deck are red.)
    (c) Find var X, for the X in (b).

*12. A carton contains twelve articles, four of which are defective and eight are good. Three are selected at random (without replacement). Let X denote the number of defectives in the sample of three. Find
    (a) $P(X \leq 1)$.
    (b) $E(X)$.
    (c) The probability that the first two selected are good and the third is defective.

13. Suppose that of the 10,000 individuals at a sporting event, 5,500 are male and 4,500 female. Five ticket stubs are drawn at random (without replacement) to receive prizes.
    (a) Find the exact probability that three males and two females receive the prizes.
    (b) Approximate the probability in (a) using the binomial formula.

*14. In the Massachusetts "Megabucks" lottery, you pay $1 and pick any six of the numbers $1, 2, \ldots, 36$ with no duplication. Order is immaterial. The six winning numbers are selected randomly. When the winning selection is announced, you win the grand prize of several million dollars if you have them all correct, $400 if you have any five correct, and $40 if you have four correct.

(a) Find the probability of winning the grand prize.

(b) Find the probability of winning $400.

(c) Find the probability of winning $40.

(d) We know someone who said he was very unlucky because he had no matches at all. Find the probability of this event.

(e) Calculate the expected value of your ticket if the grand prize is $10,000,000, and no one else has your combination so that you don't have to share the prize. (The Commonwealth skims off about half of the total money paid in, so a $1 ticket would be worth about 50 cents except for the fact that money not won in previous weeks is added to the next week's prize.)

*15. Suppose the archer of Problem 8, who hits the bull's-eye once in ten tries on average, keeps shooting until she hits it.

(a) Find the probability that she hits it for the first time on the 15th try.

(b) Find the expected number of tries to get the first hit.

(c) Find the expected number of tries to get the third hit.

16. Referring to Problem 12 (the carton with four defective items among twelve), find the probability that, in selecting one at a time, the first defective item is selected on the third try, if they are drawn at random

(a) with replacement.

(b) without replacement.

*17. The game of "Russian Roulette" was defined in Problem 38 of Chapter 2. Find the average length of the game (as a number of attempts) for one who plays it repeatedly,

(a) if the cylinder is spun between attempts.

(b) if the cylinder is not spun after each attempt but just advances to the next position.

18. To estimate the number of fish in a pond (call it N), 20 fish are caught, tagged, and replaced. A few days later, 10 fish are caught from the pond. Assuming this to be a random selection without replacement,

(a) find the distribution of the number of tagged fish in the second catch (size 10), assuming there were originally 100 fish in the pond.

(b) give an expression for the expected number of tagged fish in the catch of ten (in terms of N).

(c) if you found four tagged fish in your catch of ten, what would you estimate N to be?

*19. Find the value or an approximate value of each of the following without using a calculator:

(a) $\displaystyle\sum_{k=0}^{8} k^2 \frac{\binom{10}{k}\binom{10}{8-k}}{\binom{20}{8}}$ (b) $\displaystyle\frac{\binom{500}{4}\binom{500}{4}}{\binom{1000}{8}}$ (c) $\displaystyle\sum_{12}^{\infty} n\binom{n-1}{11}(.5)^n.$

20. Find the conditional distribution of X, the number of successes in the first m of a sequence of independent Bernoulli trials, *given* that there are c successes in the first $m + n$ trials.

*21. Until recently, peanut M&M's came in four colors. Assuming these to be mixed by the manufacturer in equal proportions, find the expected number of random selections of one M&M (from an inexhaustible supply) that would be needed to get at least one of each color.

22. Verify the asserted equality of (7) and (8) in Section 4.4.

Section 4.5

*23. For 25 tosses of a fair coin, approximate the probability of more than 15 heads.

24. Check the accuracy of the normal approximation to the binomial probability function in the case $n = 5$ and $p = .5$ by comparing exact and approximate values of $P(X \le k)$, $k = 0, 1, \ldots, 5$.

*25. One-third of the voters in a large population favor a certain proposition. Approximate the probability that fewer than 18 favor it in a random selection of 72 voters.

26. In a certain population 40% have type O blood. Use a normal approximation to find the probability that there are more than 25 people with type O blood in a random selection of size 50 from the population.

*27. Use the normal table to approximate the binomial probability

$$\binom{100}{48}(.5)^{100}.$$

28. A certain type of inoculation has been found to be about 99% effective. Find the probability that it is effective for at least 399 among a group of 400 individuals who are inoculated.

*29. Of the resistors made by a certain company, .5% are defective. Find the probability that in a shipment of 300,
   (a) none are defective.
   (b) exactly one is defective.
   (c) at most three are defective.

30. The company in Problem 29 must decide how many resistors to package in a box. Their policy is to refund the purchase price of an entire box if any resistor in it is defective. On the other hand, they'd like a reasonably large number of resistors in a box. Find the maximum number they can put in each box if they want to give refunds for no more than 10% of the boxes sold.

*31. A company will accept a shipment of 1,000 items if no more than one defective is found in a random selection of 20 items.
   (a) Give an exact expression for the probability that the shipment is accepted when it actually contains 50 defectives.
   (b) Give a binomial approximation for the probability in (a).
   (c) Give a Poisson approximation for the probability in (a).
   (d) For comparison with (c), find a normal approximation for the probability in (a). (The Poisson is preferred.)

32. Use the Poisson table to approximate the value of

$$\binom{100}{8}(.94)^{92}(.06)^{8}.$$

Section 4.6

*33. Phone calls arrive at an exchange at a rate of eight per minute. Assuming a Poisson arrival process, find the probability of

(a) at most eight calls in a one-minute period.

(b) no calls in a fifteen-second period.

(c) more than ten calls in a thirty-second period.

(d) at least two calls in a ten-second period.

34. A large metropolitan bus company experiences bus breakdowns at the rate of two per day. Assuming a Poisson distribution, find the probability of

(a) no breakdowns in a particular day.

(b) no breakdowns in two consecutive days.

(c) no breakdowns in a particular day, given that there were three the previous day.

(d) more than two breakdowns in a particular day.

(e) more than five breakdowns in a particular five-day week.

∗35. A radioactive substance emits particles according to a Poisson process. Assuming that the probability of no emissions in a one-second interval is .165, find

(a) the expected number of emissions per second.

(b) the probability of no emissions in a two-second interval.

(c) the probability of at most two emissions in a four-second interval.

(d) the probability of more than five emissions in five seconds.

36. Flaws appear at random on sheets of a certain type of photographic paper. The average number of flaws is $\frac{1}{2}$ per sheet. Assuming a Poisson distribution, find the probability

(a) that a sheet has no flaws.

(b) of no flaws in a batch of five sheets.

(c) that there is at least one flawless sheet in a batch of five sheets.

(d) of at most two sheets with more than one flaw, in a box of 100 sheets.

∗37. Find a numerical value for this sum:

$$\sum_{k=0}^{\infty} \frac{k^2 2^k}{k!}.$$

38. In a Poisson arrival process, let X and Y denote the numbers of arrivals in nonoverlapping intervals of lengths s and t, respectively. Find the conditional distribution of X given that $X + Y = c$.

39. Consider a Poisson process with parameter λ. Let $p(k)$ denote the probability function of the number of events in an increment of time Δt. Show the following:

(a) $\dfrac{p(1)}{\Delta t} \to \lambda$ as $\Delta t \to 0$. [Assumption (d) of Section 4.6.]

(b) $\dfrac{[1 - p(0) - p(1)]}{\Delta t} \to 0$ as $\Delta t \to 0$. [Assumption (c) of Section 4.6.]

Sections 4.8–4.9

∗40. The 5,000 people in a certain town are classified according to blood group as shown. In a random selection of ten (without replacement), let

$X = $ number of type A in the sample,

$Y = $ number of type B in the sample,

$Z = $ number of type AB in the sample.

Type	Proportion
O	.45
A	.40
B	.10
AB	.05

(a) Find the probability that $X = 4$, $Y = 3$, and $Z = 0$.
(b) Find the probability that $X = 4$ and $Y = 3$.
(c) Find the probability that $X = 4$.
(d) Find the probability that $Z \geq 2$.
(e) Approximate (d) using Poisson.

41. In a throw of eight dice, let X_i denote the number of dice that result in side i. Find
    (a) $E(X_i)$, $i = 1, 2, \ldots, 6$.
    (b) var X_i.
    (c) the probability that $X_1 = X_2 = X_3 = X_4 = 2$.
    (d) the distribution of X_1, given $X_2 = 3$.
    (e) the joint distribution of X_1 and X_2, given $X_3 = 0$.

*42. An ordinary six-sided die is painted red on two sides, white on two sides, and blue on two sides. Find the probability that in 24 tosses, we get
    (a) an equal number of reds, whites, and blues.
    (b) ten reds and eight whites.
    (c) ten reds.

43. In a small packet of 20 peanut M&M's (see Problem 21), find the distribution of the number of brown ones, given that there are six each of the green and yellow ones.

*44. Given that (X_1, X_2) is trinomial with parameters $(n; p_1, p_2)$,
    (a) what is the distribution of X_1?
    (b) what is the distribution of $X_1 + X_2$?
    (c) find the distribution of X_1 given $X_2 = k$.
    (d) find the distribution of X_1 given $X_1 + X_2 = k$.

45. Find the factorial moment generating function (in closed form) for the geometric distribution: $f(x \mid p) = p(1 - p)^x$, for $x = 0, 1, 2, \ldots$. [You will need the formula for the sum of a geometric series: $1 + a + a^2 + \cdots = (1 - a)^{-1}$.]

46. Suppose $Y_1 \sim \mathrm{Bin}(n_1, p)$ and $Y_2 \sim \mathrm{Bin}(n_2, p)$, and suppose further that Y_1 and Y_2 are independent. Use generating functions to show that $Y_1 + Y_2$ is binomial.

*47. Let $X = $ the number of failures before the rth success in a Bernoulli process. This is the sum of r independent, geometric variables. Its distribution is negative binomial, with probability function

$$f(j) = \binom{j + r - 1}{r - 1} p^r q^j, \qquad j = 0, 1, 2, \ldots.$$

Use the result of Problem 45 to find the f.m.g.f. of X. [Note that $X = W - r$, where W is the minimum number of trials required, as in (6) of Section 4.4.]

48. Use the f.m.g.f. of the distribution of the number of points obtained in the toss of an ordinary fair die to find its mean and variance.

49. Calculate the mean and variance of the geometric distribution in Problem 45 using the f.m.g.f.

50. Calculate the mean and variance of the negative binomial distribution in Problem 47 using the f.m.g.f.

51. Given the f.m.g.f. $\eta(t) = 1 - p + p(t + 1/t)/2$,
   (a) interpret this as an expected value of t^X and so deduce the probability function.
   (b) calculate the mean and variance both from the f.m.g.f. and from the probability function.

CHAPTER 5

Continuous Random Variables

In Chapter 3, we introduced continuous random variables and gave several examples. The events of interest for continuous variables are intervals rather than isolated values. Defining a model by listing possible values and their probabilities, as in the discrete case, is not feasible in the continuous case. The numbers in an interval are too numerous to list. However, we want to preserve the basic properties of a probability model [(1)–(3) of Section 1.7], so we need new building blocks and new tools. With these and a new definition for expected value, we can define moments, generating functions, and independence much as in the discrete case.

Example 5a **Waiting Times**

Buses go past a campus bus stop every ten minutes throughout the day. We set out to take the bus without knowing the precise schedule. How long will we have to wait for the next bus?

The waiting time T (in minutes) is a *continuous* random variable: It can't be predicted with certainty (so it's random), and it can take on any value on the continuous time-scale from 0 to 10 minutes (so it's continuous). In its decimal representation, a value of T is an infinite sequence of digits. Sequences starting with 0 (such as 0.2471. . .) represent waiting times of less than one minute; sequences starting with 1 (such as 1.902. . .) represent waiting times of between one and two minutes; and so on.

In representing our ignorance of the exact arrival times, we might feel that the ten one-minute intervals 0 to 1, 1 to 2, and so on, are equally likely. Further, we might feel the ten .1-minute intervals within any particular one-minute interval are equally likely, and so on. These assumptions imply that the digits in the sequence defining a T-value are independent random digits (Section 1.5). On this basis, we can calculate the probability of *any* interval of T-values. For example, the probability that T starts out with the three digits 3, 6, and 4 (in that order) is

$$P(T = 3.64. \ldots) = P(\text{1st digit} = 3, \text{2nd digit} = 6, \text{3rd digit} = 4)$$
$$= P(\text{1st digit} = 3) \cdot P(\text{2nd digit} = 6) \cdot P(\text{3rd digit} = 4)$$
$$= (.1)^3.$$

On the other hand, the probability of any *particular value* of T is 0. For example, the probability that $T = \frac{1}{3} = .33\overline{3}$ (the bar indicates infinite repetition of the 3's) is

$$P(T = \tfrac{1}{3}) = \lim_{n \to \infty} P(T \text{ has 3 in its first } n \text{ places})$$

$$= \lim_{n \to \infty} (.1)^n = 0.$$

Suppose we want the probability that the wait is at least two minutes but not as long as four minutes. The decimal representation of a T in this interval is a 2 or a 3 followed by any sequence of digits after the decimal point. So

$$P(2 \le T < 4) = P(T \text{ starts out 2. or 3.}) = \frac{1}{10} + \frac{1}{10} = .2.$$

Because single values have probability 0, the intervals $2 < T < 4$, $2 < T \le 4$, and $2 \le T \le 4$ all have this same probability, .2. Similarly, the probability that T is between 1.42 and 3.61 is the probability that the sequence of the first three digits of T is one of the 219 sequences 142, 143, . . . , 360:

$$P(1.42 < T < 3.61) = 219 \times (.1)^3 = .219 = \frac{3.61 - 1.42}{10}.$$

The same kind of calculation shows that the probability of the interval $a < T < b$ is equal to $b - a/10$, for any a and b between 0 and 10.

Thus, we have defined a probability model for the continuous variable T as a limit of discrete models with equally likely outcomes. ∎

In this example, the number of possible values is uncountably infinite, and each value has probability zero. This is true of continuous models generally. If single values have probability zero, then any set of at most countably many values has probability zero. On the other hand, *intervals* consist of uncountably many values and can have positive probability in a continuous model.

Specifying a distribution for a continuous variable is like specifying the distribution of mass along a line, such as the mass in a wire. Physicists adopt the convention that although a section of wire of positive length has mass, there is no mass *at* any point. The mass in a length of wire can be found by accumulating or integrating a function that gives the density of mass at each point. So the analogy of probability and mass, which we exploited in modeling discrete variables, carries over to the continuous case. We'll discuss "density" for probability distributions in Section 5.2, but first we define continuous distributions in terms of intervals.

| 5.1 | **The Distribution Function** |

The probability of an interval is fundamental to a continuous probability model on the real line. The probability of semi-infinite intervals of the form $X \le x$, where x is any real number, determine the probabilities of any other interval or any countable union of intervals. For example, finite intervals can be expressed as the "difference" between two semi-infinite intervals. Thus, if $b > a$,

$$\{X \leq b\} = \{a < X \leq b\} \cup \{X \leq a\},$$

a union of disjoint events. It follows that

$$P(X \leq b) = P(a < X \leq b) + P(X \leq a),$$

or

$$P(a < X \leq b) = P(X \leq b) - P(X \leq a). \tag{1}$$

So if we know the probability $P(X \leq x)$ for all x, we can find the probability of any interval.

The quantity $P(X \leq x)$, as a function of x, is called the **distribution function** of the random variable X. It is also called the **cumulative distribution function** and commonly referred to as the **c.d.f.**

Distribution function (c.d.f.) of the random variable X:

$$F(x) = P(X \leq x). \tag{2}$$

The probability of an interval (a, b) is obtained from the c.d.f. as a difference:

$$P(a < X \leq b) = F(b) - F(a). \tag{3}$$

[(3) is simply a restatement of (1) using (2).]

In a continuous distribution, the probability of a single point is 0, so the intervals $a < x \leq b$, $a < x < b$, and $a \leq x \leq b$ all have the same probability. We've written (3) so that it's true in general, applicable to discrete distributions as well.

Example 5.1a   **Waiting Times (continued)**

In Example 5.a, we considered the random variable T, the time we have to wait until the next bus comes along. We assumed that the next bus arrives at a time which is "equally likely" to be anywhere in the next ten minutes. We found the probability that the bus arrives in any subinterval (a, b) of the interval $(0, 10)$ to be proportional to its length:

$$P(a < T < b) = \frac{b - a}{10} \qquad \text{for } 0 < a < b < 10.$$

So for any t in the interval $(0, 10)$,

$$P(T \leq t) = P(-\infty < T \leq 0) + P(0 < T \leq t) = 0 + \frac{t}{10} = \frac{t}{10}.$$

Of course, if t is negative, the probability of $T \leq t$ is zero; and when t is larger than 10, the event $T \leq t$ happens with probability 1. The complete specification of the c.d.f. is as follows:

$$F(t) = P(T \leq t) = \begin{cases} 0, & t < 0, \\ \dfrac{t}{10}, & 0 \leq t \leq 10, \\ 1, & t > 10. \end{cases}$$

Figure 5.1 shows the graph of this "ramp" function.

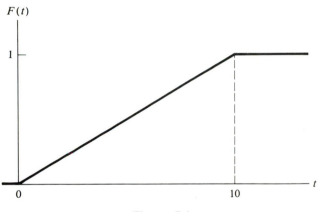

Figure 5.1

Uniform
Distribution

The distribution in this example is quite special, in that there are no preferred portions of the set of possible values: The probability of a subinterval is proportional to its length. In such cases, we say the distribution is **uniform**. When T has a uniform distribution over the interval (a, b), we'll write $T \sim U(a, b)$.

Some properties of the distribution function in Figure 5.1 apply to distribution functions generally. First, $F(x)$ is a probability. Its values are between 0 and 1; and its values at $-\infty$ and $+\infty$ are

$$F(-\infty) = \lim_{x \to -\infty} F(x) = \lim_{x \to -\infty} P(X \leq x) = 0,$$

$$F(\infty) = \lim_{x \to \infty} F(x) = \lim_{x \to \infty} P(X \leq x) = 1.$$

Second, $F(x)$ is nondecreasing: Since the interval $(-\infty, x)$ widens as x increases, its probability $F(x)$ cannot get smaller. [See Property (vii) in Section 1.7.] Third, the c.d.f. of a continuous variable is continuous. This amounts to saying that as x moves from left to right, probability is not added in discrete lumps. For technical reasons (beyond our scope) we'll assume that F is differentiable except possibly at a finite number of points.

Properties of $F(x)$, the c.d.f. of a **continuous** random variable:

  (i) $F(-\infty) = 0$ and $F(\infty) = 1$.
  (ii) $F(a) \leq F(b)$ whenever $a < b$.
  (iii) $F(x)$ is continuous for all x, and $F'(x)$ exists except possibly for finitely many x.

If a function $F(x)$ has Properties (i)–(iii), defining $P(X \leq x)$ as $F(x)$ determines a distribution on the real line that satisfies the probability axioms [(1)–(3) in Section 1.7].

The analogy between probability and mass is especially helpful in understanding a distribution function. Imagine that probability is represented by a unit measure of dust or snow distributed along a line (give the line width if you like). Then picture starting at the extreme left (at $-\infty$, if necessary) and sweeping up the dust or shoveling the snow as you move from left to right along the line. The amount collected up to any point is $F(x)$. Since you can only *add* dust as you sweep, the collected amount $F(x)$ cannot decrease.

Analogy with Mass

Starting from the *left*, and thus using $P(X \leq x)$ to describe a distribution, was an arbitrary choice. We could just as well start from the right and use the function

$$R(x) = P(X > x) = 1 - F(x). \tag{4}$$

This complementary function is sometimes more natural to consider (see Section 6.7). As we proceed from left to right along the x-axis, the value of $R(x)$, the *un*collected amount of probability, can only decrease or remain constant—$R(x)$ is a nonincreasing function. In view of (4), it is clear that knowing $R(x)$ is equivalent to knowing $F(x)$.

Example 5.1b   **Waiting Times in a Poisson Process**

Suppose arrivals at a service center follow a Poisson law (Section 4.6) with an average of six arrivals per minute: $\lambda = 6$. The number of arrivals in an interval of t minutes has a Poisson distribution, given by (1) of Section 4.6, with mean $m = 6t$:

$$P(k \text{ arrivals in time } t) = \frac{(6t)^k}{k!} e^{-6t}, \qquad k = 0, 1, 2, \ldots . \tag{5}$$

A variable of interest is T, the elapsed time from a given point t_0 to the next arrival. [The initial value t_0 is immaterial in view of (5) of Section 4.6.] The variable T is continuous. The function $R(t)$ defined by (4) is easy to find by expressing the condition $T > t$ in terms of numbers of arrivals:

$$R(t) = P(T > t)$$
$$= P(0 \text{ arrivals in time } t) = e^{-6t}.$$

The c.d.f. is then

$$F(t) = 1 - R(t) = 1 - e^{-6t}, \qquad t > 0.$$

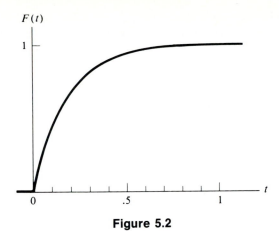

Figure 5.2

Clearly, $F(t)$ is 0 when $t < 0$. Figure 5.2 shows this c.d.f. The distribution is one of a family of distributions to be studied in more detail in Section 6.2.   ∎

Transformation of Variables

In Chapter 3 (Sections 3.2 and 3.3), we studied transformations of a discrete random variable: $Y = g(X)$. When X is continuous, the distribution of $Y = g(X)$ is readily derived in terms of c.d.f.'s. Suppose $g(x)$ is strictly increasing, so that the inverse function is single-valued: $x = g^{-1}(y)$. Then the event $g(X) \leq y$ is equivalent to $X \leq g^{-1}(y)$. So the c.d.f. for Y is

$$F_Y(y) = P(Y \leq y) = P[g(X) \leq y] = P[X \leq g^{-1}(y)]. \tag{6}$$

But this is just the c.d.f. of X evaluated at $g^{-1}(y)$. Thus,

$$F_Y(y) = F_X(g^{-1}(y)). \tag{7}$$

When g is a strictly *decreasing* function, the event $Y \leq y$ is equivalent to $X \geq g^{-1}(y)$. Since there is only one x corresponding to each y, the probability that $g(X) = y$ is zero. Thus, instead of (6), we have

$$\begin{aligned} F_Y(y) = P(Y \leq y) &= P[g(X) - y] \\ &= P[X > g^{-1}(y)] = 1 - F[g^{-1}(y)]. \end{aligned} \tag{8}$$

Example 5.1c   **Linear Transformation of a Uniform Variable**

Suppose X is uniformly distributed on the interval $(0, 1)$:

$$F_X(x) = \begin{cases} 0, & x < 0 \\ x, & 0 < x < 1, \\ 1, & x > 1. \end{cases} \tag{9}$$

Consider the *linear* transformation $Y = g(X)$, where

$$g(x) = 6 + 4x.$$

The point $x = 0$ corresponds to $y = 6$, and $x = 1$ corresponds to $y = 10$. To find the inverse function g^{-1}, we solve for x in terms of y:

$$x = \frac{y - 6}{4} = g^{-1}(y). \tag{10}$$

Using (10) in (7), we obtain

$$F_Y(y) = F_X\left(\frac{y - 6}{4}\right) = \begin{cases} 0, & y < 6, \\ \dfrac{y - 6}{4}, & 6 < y < 10, \\ 1, & y > 10. \end{cases} \tag{11}$$

Like the uniform c.d.f. for X in Figure 5.1, the c.d.f. defined by (11) is a ramp function, increasing linearly from 0 to 1 over the interval (6, 10). Thus, $Y \sim U(6, 10)$.

More generally, consider $g(X) = c + (d - c)X$, where $c < d$. In place of (11), we have

$$F_Y(y) = F_X\left(\frac{y - c}{d - c}\right) = \begin{cases} 0, & y < c, \\ \dfrac{y - c}{d - c}, & c < y < d, \\ 1, & y > d. \end{cases} \tag{12}$$

This c.d.f. is again a ramp function, and $Y \sim U(c, d)$. ∎

When g does not have a *unique* inverse g^{-1}, the same kind of approach works, but solving $g(X) \leq y$ for X takes some care. Each inverse must be taken into account, as we illustrate in the following example.

Example 5.1d   **A Double-Valued Inverse**
Suppose X is uniformly distributed on the interval $(-1, 1)$. That is, $c = -1$ and $d = 1$ in (12). Then

$$F_X(x) = \frac{x + 1}{2}, \quad -1 < x < 1.$$

Define $U = X^2$. For $0 < u < 1$, the c.d.f. of U is

$$F_U(u) = P(U \leq u) = P(X^2 \leq u) = P(-\sqrt{u} \leq X \leq \sqrt{u})$$

$$= F_X(\sqrt{u}) - F_X(-\sqrt{u}) = \frac{1}{2}(\sqrt{u} + 1) - \frac{1}{2}(-\sqrt{u} + 1) = \sqrt{u}.$$

Of course, $F_U(u)$ is 0 when $u < 0$ and 1 when $u > 1$. Figure 5.3 shows the transformation $u = x^2$ and the region $U \leq u$ corresponding to the region $-\sqrt{u} \leq X \leq \sqrt{u}$. Figure 5.4 shows $F_U(u)$.

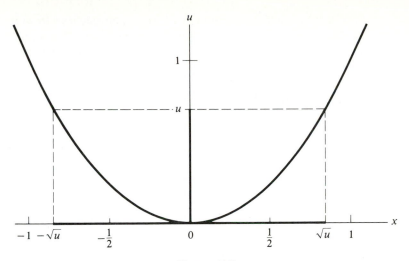

Figure 5.3

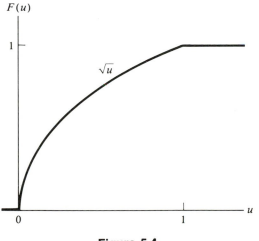

Figure 5.4

The definition of c.d.f. given by (2) need not be restricted to the continuous case. Indeed, (2) defines the c.d.f. for *any* distribution, continuous, discrete, or neither. In (3), we were careful to write $P(a < X \leq b)$ as the difference between $F(b)$ and $F(a)$, so that the equation is correct when the single values a and b have positive probability. When a single value k does have positive probability, F will suddenly increase or jump by that amount as x passes through the value k:

c.d.f's for Noncontinuous Cases

$$P(X = k) = \lim_{n \to \infty} P\left(k - \frac{1}{n} < X \leq k\right) = \lim_{n \to \infty} \left[F(k) - F\left(k - \frac{1}{n}\right)\right]$$
$$= F(k) - F(k-),$$

where $F(k-)$ denotes the limit of $F(x)$ as x approaches k from the left. Thus, the size of the jump in $F(x)$ at k is the probability that $X = k$. The value of F at a jump point is the limit from the right because $F(x)$ is $P(X \leq k)$, which includes the probability at k.

Example **5.1e**
The Binomial c.d.f.
Suppose X is the number of successes in four independent trials of a Bernoulli experiment with $p = \frac{1}{3}$: $X \sim \text{Bin}(4, \frac{1}{3})$. (See Section 4.2.) The probabilities of 0, 1, 2, 3, 4 follow from (1) of Section 4.2: $\frac{16}{81}, \frac{32}{81}, \frac{24}{81}, \frac{8}{81}, \frac{1}{81}$, respectively. The c.d.f. defined by this distribution is a step function, rising from the value 0 (to the left of $x = 0$) in jumps of $\frac{16}{81}$ at $x = 0$, $\frac{32}{82}$ at $x = 1$, and so on, to a final height of 1 at $x = 5$. The graph is shown in Figure 5.5.

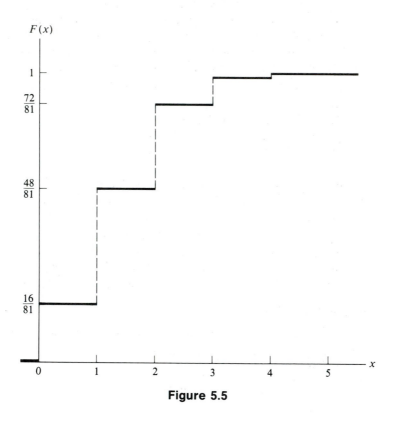

Figure 5.5

The c.d.f. of a *continuous* distribution is a continuous function, with no jumps. The c.d.f. of a discrete distribution increases *only* in jumps. Occasionally, one encounters the need for a distribution which is a mixture of these types.

Example **5.1f**
A Distribution of Mixed Type
Let T denote the time to failure of an appliance—say, a TV set. We may think of this as a continuous random variable, except that there may be a positive probability

$F(t)$

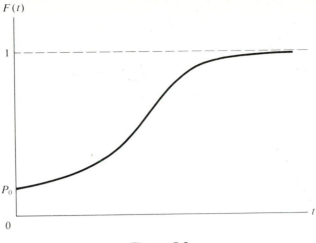

Figure 5.6

that the set would not work when first plugged in:

$$P_0 = P(T = 0) > 0.$$

The c.d.f. might look like the graph in Figure 5.6, which describes a distribution that is neither continuous nor discrete. It allots a positive probability to the single value $T = 0$ but distributes the remaining probability continuously over the positive real line. ∎

In the next section, we take up another way to describe a continuous distribution.

| 5.2 | **Density and the Probability Element** |

A c.d.f. describes the distribution of probability for a numerical random variable X by giving the fraction of the probability that lies at or to the left of each x. In the discrete case, an alternative description gives the list of possible values and corresponding probabilities. When X is a continuous variable, an alternative description of its distribution gives the *density* or concentration of probability at each point x.

The probability that a continuous random variable X falls in the interval from x to $x + \Delta x$ is given in terms of the c.d.f. by (3) of Section 5.1:

$$P(x < X < x + \Delta x) = F(x + \Delta x) - F(x) = \Delta F(x). \tag{1}$$

When $F(x)$ is differentiable at x, we can approximate this increment ΔF by the *differential*:

$$dF(x) = F'(x)\,\Delta x. \tag{2}$$

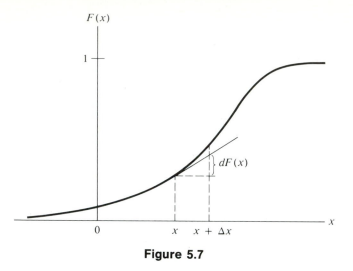

Figure 5.7

(See Figure 5.7.) This is termed the **probability element** at x. It approximates the value of ΔF, under the assumption that F is nearly linear over the small interval $(x, x + \Delta x)$, with slope $F'(x)$. This derivative is the rate of increase in F—the *rate* at which probability is added at x as we move from left to right along the x-axis. Where the c.d.f. is steep, the derivative is large—probability is added at a high rate. When it is nearly flat, the derivative is small—probability is added at a low rate.

The ratio of the amount of probability in the interval $(x, x + \Delta x)$ to the length of the interval Δx is the **average density** of probability in that interval:

$$\text{Average density} = \frac{P(x < X < x + \Delta x)}{\Delta x} = \frac{\Delta F(x)}{\Delta x}. \tag{3}$$

The limit of the average density (3), as Δx tends to 0, defines the probability density *at* the point x:

$$\text{Density at } x = \lim_{\Delta x \to 0} \frac{\Delta F(x)}{\Delta x} = F'(x).$$

This derivative is called the **density function** of X (or of its distribution) and is commonly denoted by $f(x)$. It is convenient to refer to the density function by the initials p.d.f. for *probability density function*. Like the c.d.f., a p.d.f. may have a subscript to indicate the random variable whose distribution it describes.

Probability density function (p.d.f.) for a continuous random variable X:

$$f(x) = F'(x). \tag{4}$$

The probability element at x is

$$f(x)\,dx \doteq P(x < X < x + dx). \tag{5}$$

Example **5.2a**   **A Uniform Density**

Example 5.1a gave the c.d.f. of waiting time at a bus stop as $\frac{t}{10}$ for $0 < t < 10$. The derivative is constant on that interval and 0 outside it:

$$f_T(t) = \begin{cases} \dfrac{1}{10}, & 0 < t < 10, \\ 0, & \text{elsewhere.} \end{cases}$$

This rectangular p.d.f. (shown in Figure 5.8) and the distribution it defines are *uniform* on the interval (0, 10): $T \sim U(0, 10)$.

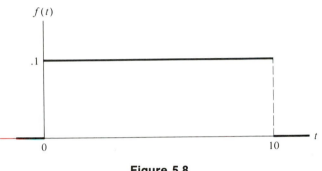

Figure 5.8

Example **5.2b**   **Waiting Time in a Poisson Process**

In the Poisson process of Example 5.1b, we found the c.d.f. of the time to the next arrival (after any point in time) to be

$$F(t) = 1 - e^{-6t}, \qquad t > 0.$$

The corresponding p.d.f. is the derivative,

$$f(t) = F'(t) = \begin{cases} 6e^{-6t}, & t > 0, \\ 0, & t < 0, \end{cases}$$

shown in Figure 5.9. The probability element at $t = .30$, for example, is

$$f(.30)\, dt = 6e^{-1.80}\, dt \doteq .992\, dt.$$

The probability between $t = .30$ and $t = .34$, for instance, is then approximately equal to $0.4 \times .992 = .0397$. Using (1), we find

$$P(.30 < X < .34) = (1 - e^{-2.04}) - (1 - e^{-1.80}) = .0353.$$

(To be continued.)

Many of our examples will involve p.d.f.'s that are 0 outside some region. **Support of a Distribution** The interval where $f(x) > 0$ is called the **support** of the distribution; in defining a distribution, we'll often give a formula for the p.d.f. that applies only for points in the support of the distribution, with the understanding that the density is zero outside

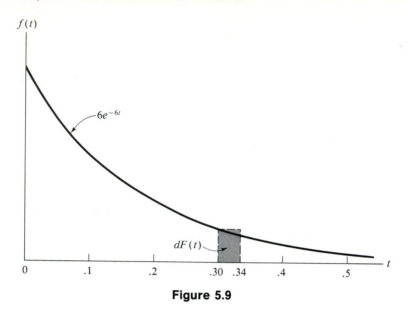

Figure 5.9

that support. Thus, in the case of Example 5.2b, we'd write

$$f(t) = 6e^{-6t}, \qquad t > 0,$$

with the understanding that $f(t) = 0$ when $t \leq 0$. [Indeed, since the integral of this $f(t)$ over $t > 0$ is 1, there can be no probability in $t \leq 0$, so it is almost redundant to say that $f(t) = 0$ when $t < 0$.]

Not all continuous, increasing functions are differentiable at enough points to be useful as a c.d.f. The restriction in (iii) of Section 5.1 (namely, that F' exists except possibly at a finite number of points) guarantees that the c.d.f. can be recovered by integrating its derivative, the p.d.f.:

$$F(x) = \int_{\infty}^{x} F'(u)\,du = \int_{\infty}^{x} f(u)\,du. \tag{6}$$

(The c.d.f. in Example 5.2b is such a function; its derivative is continuous except at one point, $t = 0$.) The integrand function in a Riemann integral can be undefined or arbitrarily defined at finitely many points without affecting the value of the integral.

Figure 5.10 shows a typical c.d.f. and its derivative function, $f(x)$. The area under $f(x)$ to the left of x is a number (of square units) equal to the height (in linear units) of the c.d.f. at x. The height of the density function at x is the slope of the c.d.f. at that point.

Condition (6) will always be satisfied when we construct a model by specifying a density function $f(x)$ as a nonnegative, integrable function such that the area under its graph is equal to 1. We then use (6) to *define* the c.d.f. The integral in (6) is a continuous, increasing function of the upper limit x, and

Model Defined by a p.d.f.

$$\int_{-\infty}^{\infty} f(x)\,dx = F(\infty) = 1. \tag{7}$$

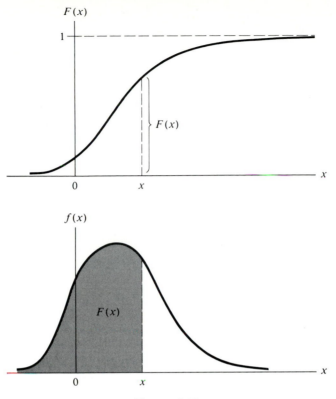

Figure 5.10

Thus, the function $F(x)$ defined from the specified p.d.f. by means of (6) satisfies Properties (i)–(iii) of a c.d.f. in Section 5.1. Moreover, if we differentiate it to recover the p.d.f, we end up where we started, according to the *fundamental theorem of integral calculus*:

$$F'(x) = \frac{d}{dx} \int_{-\infty}^{x} f(u)\, du = f(x)$$

at all continuity points of $f(x)$.

Any integrable nonnegative function $f(x)$ that satisfies the condition

$$\int_{-\infty}^{\infty} f(x)\, dx = 1 \qquad (8)$$

defines a distribution for X, with c.d.f.

$$F(x) = \int_{-\infty}^{x} f(u)\, du. \qquad (9)$$

Example 5.2c   **A Cauchy Density**

The function $(1 + x^2)^{-1}$ is nonnegative, and the area under its graph is finite:

$$\int_{-\infty}^{\infty} \frac{dx}{1 + x^2} = \text{Arctan}(\infty) - \text{Arctan}(-\infty) = \pi,$$

where Arctan means the principal value: $-\pi/2 < \text{Arctan } \theta < \pi/2$. Dividing by π produces a function,

$$f(x) = \frac{1/\pi}{1 + x^2}, \tag{10}$$

whose integral over $(-\infty, \infty)$ is 1. This **Cauchy p.d.f.** has c.d.f.

$$F(x) = \int_{-\infty}^{x} \frac{1/\pi}{1 + u^2}\, du = \frac{1}{\pi}\text{Arctan } x + \frac{1}{2}. \tag{11}$$

The c.d.f. (11) and the p.d.f. (10) are shown in Figure 5.11.

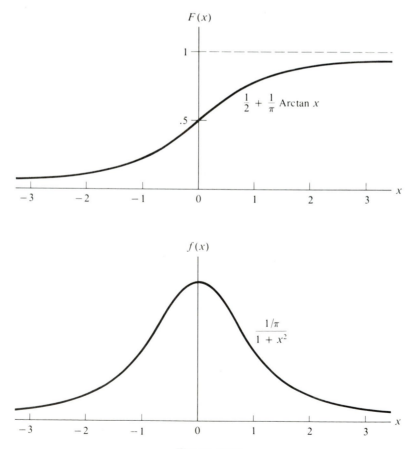

Figure 5.11

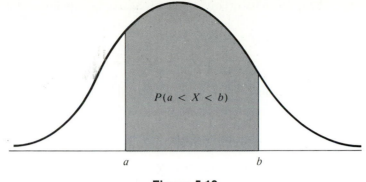

Figure 5.12

Probability of an Interval

In Section 5.1, we found that the probability of an interval is given by the amount of increase in the c.d.f. over that interval:

$$P(a < X \le b) = F(b) - F(a).$$

This probability can now be expressed in terms of the p.d.f., since both $F(b)$ and $F(a)$ are integrals of the p.d.f.:

$$P(a < X < b) = \int_{-\infty}^{b} f(x)\,dx - \int_{-\infty}^{a} f(x)\,dx = \int_{a}^{b} f(x)\,dx. \tag{12}$$

This is the area under the graph of $y = f(x)$ between $x = a$ and $x = b$, as shown for a typical density in Figure 5.12. We may also think of this integral or area as approximately equal to the sum of probability elements corresponding to subdivisions of the interval from a to b:

$$P(a < X < b) \doteq \sum f(x_i)\Delta x_i.$$

The integral (12) is the limit of such approximating sums as the number of subdivisions increases and $\Delta x_i \to 0$.

When X is continuous with p.d.f. $f(x)$,

$$P(a < X < b) = \int_{a}^{b} f(x)\,dx = F(b) - F(a). \tag{13}$$

Example 5.2d **Waiting Time in a Poisson Process (continued)**

In Example 5.2b, we considered the p.d.f. $6e^{-6x}$ for $x > 0$. If we start with this nonnegative function as a p.d.f. and define a c.d.f. as its integral:

$$F(x) = \int_{0}^{x} 6e^{-6t}\,dt = e^{-6t}\big|_{0}^{x} = -e^{6x} + 1,$$

we obtain a function whose derivative is the p.d.f. we started with.

The probability of any interval is also a definite integral; for example,

$$P(.1 < T < .3) = \int_{.1}^{.3} 6e^{-6t}\,dt$$

$$= -e^{-6t}\big|_{.1}^{.3} = e^{-.6} - e^{-1.8} \doteq .384.$$

Alternatively, this can be calculated directly from F:

$$P(.1 < T < .3) = F(.3) - F(.1)$$

$$= (1 - e^{-1.8}) - (1 - e^{-.6}) \doteq .384. \qquad \blacksquare$$

The Differential Method

Since the p.d.f. is the coefficient of dx (or Δx) in a probability element, we can sometimes derive a p.d.f. using a differential approach. We first find the probability of each infinitesimal interval of length Δx. The coefficient of Δx is then the desired p.d.f. An example will help make this clear.

Example 5.2e

Waiting Time p.d.f. by the Differential Method

Let X denote the time to the rth arrival in a Poisson process with parameter λ. For any given $x > 0$, X will fall between x and $x + dx$ if and only if (a) there are exactly $r - 1$ arrivals before time x, and (b) there is exactly one arrival between x and $x + dx$. Since these two events involve nonoverlapping intervals, the probability of their intersection factors [by Property (b), page 157] is:

$$P(x < X < x + dx) = P[r - 1 \text{ arrivals in } (0, x) \text{ and } 1 \text{ in } (x, x + dx)]$$

$$= P[r - 1 \text{ arrivals in } (0, x)] \cdot P[1 \text{ arrival in } (x, x + dx)].$$

The first factor on the right is a Poisson probability with $m = \lambda x$, and the second factor is $\lambda\,dx$ [according to Poisson Property (d), page 157]. Therefore,

$$P(x < X < x + dx) \doteq \left(\frac{(\lambda x)^{r-1}}{(r-1)!}e^{-\lambda x}\right) \cdot (\lambda\,dx) = \frac{\lambda^r}{(r-1)!}x^{r-1}e^{-\lambda x}\,dx. \qquad (14)$$

This approximation becomes exact as dx tends to 0. The coefficient of dx in the probability element (14) is the p.d.f.:

$$P(x < X < x + dx) \doteq f_X(x)\,dx. \qquad (15)$$

So, comparing (14) and (15), we see that

$$f_X(x) = \frac{\lambda^r}{(r-1)!}x^{r-1}e^{-\lambda x}, \qquad x > 0. \qquad (16)$$

Example 5.2b treated the special case in which X is the time to the first arrival, with $\lambda = 6$. When $r = 1$, (16) reduces to the p.d.f. of that example. $\qquad \blacksquare$

Transforming the p.d.f.

In Section 5.1, we showed how to obtain the c.d.f. of a function of a random variable $Y = g(X)$ from the c.d.f. of X. We could then find the p.d.f. of Y by first finding its c.d.f. and differentiating. The differential method of finding a p.d.f., illustrated in Example 5.2e, is a little more direct. We'll now use it to obtain a general formula for the p.d.f. of a transformed variable.

Suppose first that $g(x)$ is an increasing function of x with a single-valued inverse function, $x = g^{-1}(y)$, and consider an increment in y, from y to $y + \Delta y$, produced by an increment Δx in x, from x to $x + \Delta x$. The probability assigned to the Δy is just the probability in the corresponding increment Δx:

$$P(y < Y < y + \Delta y) = P(x < X < x + \Delta x) \doteq f_X(x)\,\Delta x.$$

The probability element for $Y = g(X)$ corresponding to the increment Δy is

$$f_Y(y)\,dy \doteq P(y < Y < y + \Delta y) = f_X(x)\,dx.$$

Since $dx = \Delta x$,

$$f_Y(y) \doteq f_X(x) \cdot \frac{dx}{dy} = f_X[g^{-1}(y)] \cdot \frac{dg^{-1}(y)}{dy}.$$

If we replace "increasing" by "decreasing" in this derivation, the relation between probability elements must be in terms of $|\Delta x|$ and $|\Delta y|$. The following formula covers both cases:

If $Y = g(X)$ is one-to-one, with inverse function $g^{-1}(y)$, the p.d.f. of Y is obtained from the p.d.f. of X as

$$f_Y(y) = f[g^{-1}(y)] \cdot \left| \frac{dg^{-1}(y)}{dy} \right| \tag{17}$$

In particular, if $Y = a + bX$, $\qquad b > 0$,

$$f_Y(y) = f\left(\frac{y - a}{b} \right) \cdot \frac{1}{b}. \tag{18}$$

Example 5.2f

Let X have the p.d.f. $2x$ for $0 < x < 1$ and consider the variable $Y = X^2$. On the interval of support, the function x^2 has a single-valued inverse: $x = \sqrt{y}$, with derivative $1/(2\sqrt{y})$. So for any y such that $0 < y < 1$, we apply (17) to obtain

$$f_Y(y) = f_X[g^{-1}(y)] \cdot \frac{dx}{dy} = f(\sqrt{y}) \cdot \frac{1}{2\sqrt{y}} = 1.$$

Thus, Y is uniform on $(0, 1)$.

With density $2x$ on $(0, 1)$, the variable X is quite likely to be close to 1. On the interval $(0, 1)$, however, X^2 is *smaller* than X; squaring pushes the probability toward 0 just enough to make the density uniform. ∎

Multiple Inverses

The differential method for finding the p.d.f. of $Y = g(X)$ is also adaptable to the more general case in which the inverse function g^{-1} is multiple-valued. In such cases, we trace the probability in an increment Δy back to each of the inverses that contribute to it and sum their various contributions. The next example illustrates the method.

Example 5.2g   Suppose $X \sim U(-1, 1)$. The transformation $Y = X^2$ is *not* one-to-one, since X can be negative. The inverse function is double-valued: $X = \pm\sqrt{Y}$. The probability increment at y, for any y on $(0, 1)$, comes from corresponding increments at both $-\sqrt{y}$ and $+\sqrt{y}$ (see Figure 5.13):

$$P(y < Y < y + dy) \doteq f_X(-\sqrt{y}) \cdot \frac{\Delta y}{2\sqrt{y}} + f_X(\sqrt{y}) \cdot \frac{\Delta y}{2\sqrt{y}} = \frac{1}{2\sqrt{y}} \cdot \Delta y.$$

The density of Y is the coefficient of Δy:

$$f_Y(y) = \frac{1}{2\sqrt{y}}, \qquad 0 < y < 1.$$

This same distribution of Y was obtained using c.d.f.'s in Example 5.1d.

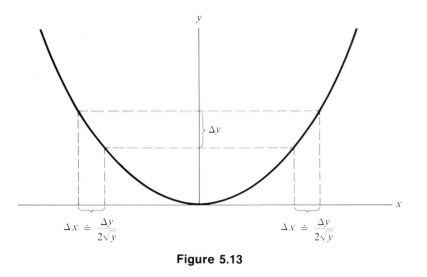

Figure 5.13

5.3   The Median and Other Percentiles

The median of a continuous distribution is a number such that half the *probability* is to its left and half to its right. We can always find such a number, since the c.d.f. is continuous and so takes on every value between 0 and 1. The median of X is any value x at which $F_X(x) = \frac{1}{2}$.

> **A median** of a continuous distribution with c.d.f. $F(x)$ is any number m such that
>
> $$F(m) = \frac{1}{2}. \qquad\qquad (1)$$

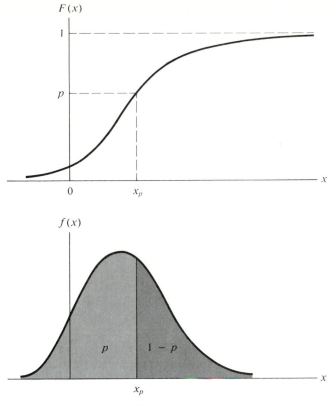

Figure 5.14

There may be more than one number m that satisfies (1), so the median may not be unique. For instance, suppose a random variable has half of its probability on the interval from 0 to 1 and half on the interval from 2 to 3. Then $F(x) = \frac{1}{2}$ for all x between 1 and 2, and any such number is a median.

Percentiles
　　　　　Since $\frac{1}{2} = 50\%$, we also refer to a median as a *50th percentile*. More generally, a $100p$th percentile of the continuous random variable X is a number x_p (not necessarily unique) such that the proportion of probability to its left is p:

$$F(x_p) = P(X \le x_p) = p \tag{2}$$

(see Figure 5.14). We also refer to the 25th percentile as the first *quartile*, and the 75th percentile as the third quartile.

Example 5.3a
Median Waiting Time
In the Poisson process of Example 5.1b, the time T to the next arrival after a specified point in time has the c.d.f.

$$F(t) = 1 - e^{-6t}, \qquad t > 0.$$

We find the median by setting this equal to $\frac{1}{2}$ and solving for t:

$$1 - e^{-6t} = \frac{1}{2}, \qquad e^{-6t} = \frac{1}{2}, \qquad 6t = -\log .5 \doteq .693.$$

So the median is $\log 2/6$, or about .116.[1] Similarly, we can find the 90th percentile, by setting $F(t)$ equal to .9 and solving for t:

$$1 - e^{-6t} = .9, \qquad \text{or} \qquad t = \frac{-\log(.1)}{6} \doteq .384,$$

and the first quartile is a value of t such that

$$1 - e^{-6t} = .25, \qquad \text{or } t = \frac{-(\log .75)}{6} = .048. \qquad \blacksquare$$

The median, quartiles, and other percentiles have interpretations in terms of areas under the graph of the density function. Writing $F(x)$ in terms of the p.d.f., we see from (2) that

$$F(x_p) = \int_{-\infty}^{x_p} f(u)\, du = p. \tag{3}$$

Thus, x_p is a value of X that divides the area under $f(x)$ into two parts, the proportion p to its left and the proportion $1 - p$ to its right. (Again, see Figure 5.14.) In particular, a median (50th percentile) divides the total area into two parts of equal area.

Example 5.3b   **Finding Quartiles**

Suppose X has the triangular distribution defined by the p.d.f. $f(x) = 2x$ for $0 < x < 1$. The first quartile, $x_{.25}$, is the value of X such that 25% of the area under the p.d.f. is to the left:

$$.25 = P(X < x_{.25}) = \int_{-\infty}^{x_{.25}} f(x)\, dx = \int_0^{x_{.25}} 2x\, dx = (x_{.25})^2.$$

Taking the square root, we find $x_{(.25)} = \sqrt{.25} = .5$. [It is clear from Figure 5.15 that the triangle over the interval $(0, 1)$ is four times as large as the triangle over $(0, .5)$.] $\qquad \blacksquare$

The median is especially easy to calculate for a **symmetric** distribution. A distribution is symmetric when there is a point, the **center of symmetry**, such that the pattern of probability on one side is a reflection of the pattern on the other side.

---

[1] Here and in what follows, "log" means logarithm to the base e.

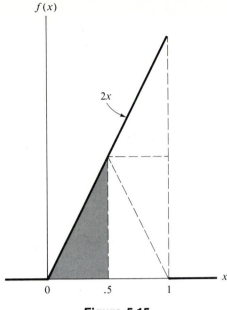

Figure 5.15

More precisely:

A distribution is **symmetric** about $x = c$ when, for every x,

$$F(c - x) = 1 - F(c + x),$$ (4)

or, in terms of densities,

$$f(c - x) = f(c + x).$$ (5)

If X has a symmetric distribution, the center of symmetry is a median. To see this, set $x = 0$ in (4) to obtain $F(c) = \frac{1}{2}$. Moreover, when (4) holds, the first and third quartiles are equidistant from the median, as are the $100p$th and the $100(1 - p)$th percentiles generally.

If a distribution is **symmetric** about $\tilde{x}$, then $\tilde{x}$ is a median of the distribution. Moreover, for any symmetric distribution,

$$\frac{x_p + x_{1-p}}{2} = \tilde{x}, \qquad 0 < p < 1.$$ (6)

Example 5.3c   **Quartiles of a Cauchy Distribution**
Generalizing the p.d.f. of Example 5.2c, we consider the p.d.f.

$$f(x) = \frac{1/\pi}{1 + (x - \theta)^2}. \tag{7}$$

This is symmetric about $x = \theta$, since $f(\theta + x) = f(\theta - x)$. Thus, θ is the median of the distribution. We can also see this by referring to the c.d.f.:

$$F(x) = \int_{-\infty}^{x} \frac{(1/\pi)\,du}{1 + (u - \theta)^2} = \frac{1}{\pi}[\text{Arctan}\,(x - \theta) - \text{Arctan}(-\infty)]$$

$$= \frac{1}{2} + \frac{\text{Arctan}(x - \theta)}{\pi}. \tag{8}$$

Setting $F(x)$ equal to $\frac{1}{2}$ yields the median value, $x = \theta$. Since F is strictly increasing, the median is unique.

The first quartile of the distribution, $Q_1 = x_{.25}$, is found by setting $F(x)$ equal to $\frac{1}{4}$:

$$\text{Arctan}(Q_1 - \theta) = -\pi/4.$$

Then

$$Q_1 - \theta = -1, \quad \text{or} \quad Q_1 = \theta - 1.$$

From symmetry, we see by (6) that the *third* quartile is $Q_3 = x_{.75} = \theta + 1$.   ■

5.4   **Expected Value**

We define the mean value or expected value of a continuous random variable much as we defined it in the discrete case [(1) of Section 3.1], as a "sum" of possible values weighted according to their probabilities. Thus, we multiply the value x by $f(x)\,dx$, the probability element at x, and "sum" the resulting products—the way we sum over a continuous index, by *integration*.

Example 5.4a   **An Approximate Integration**
In practice, integrations must often be done by numerical approximations. To illustrate, suppose X has the triangular density $f(x) = 2x$ for $0 < x < 1$. A discrete distribution approximating the distribution of X is obtained by partitioning $(0, 1)$ into (for example) ten equal class intervals and rounding the values in an interval to its midpoint. Let X^* denote this rounded variable, which is discrete. We can calculate its mean value according to the formulas of Section 3.1.

The ten class intervals have width $\Delta x = .1$. The midpoints are $.05, .15, \ldots, .95$. For each of these values x, the class probability is the probability element,

$$P(X^* = x) \doteq f(x)\,\Delta x = 2x\,\Delta x.$$

Table 5.1 gives values x, probabilities $f(x)\,\Delta x$, and products $xf(x)\,\Delta x$ for this discretized version of X. The mean of X^* is the sum .6650. This is an approximating

Table 5.1

x	$f(x)\Delta x$	$x \cdot f(x)\Delta x$
.05	.01	.0005
.15	.03	.0045
.25	.05	.0125
.35	.07	.0245
.45	.09	.0405
.55	.11	.0605
.65	.13	.0845
.75	.15	.1125
.85	.17	.1445
.95	.19	.1805
	1.00	.6650

sum for the definite integral

$$\int_0^1 xf(x)\,dx = \int_0^1 2x^2\,dx = \frac{2}{3}. \tag{1}$$

The approximation of the given continuous distribution by the discrete distribution of X^* improves as we let Δx tend to 0. The approximating sums converge to the integral (1), so we take the integral (1) as the mean of the continuous variable X. ∎

Mean Value

As in the example, we define the mean value of a continuous random variable as a definite integral. In calculus, the integral over a finite range is defined as the limit of approximating sums like the one in the above example. So when X has a distribution whose support is the finite interval (a, b), we define the mean or expected value of X as

$$E(X) = \int_a^b xf(x)\,dx. \tag{2}$$

When the support is not finite, the integral corresponding to (2) is improper:

$$E(X) = \int_{-\infty}^{\infty} xf(x)\,dx. \tag{3}$$

The value of the integral (3) is defined as the limit of the integral (2) as a and b independently tend to $-\infty$ and $+\infty$, respectively:

$$E(X) = \lim_{\substack{b\to\infty \\ a\to-\infty}} \int_a^b xf(x)\,dx = \lim_{a\to-\infty} \int_a^0 xf(x)\,dx + \lim_{b\to\infty} \int_0^b xf(x)\,dx. \tag{4}$$

If each limit on the right is finite, the integral (3) is said to *converge*, and the mean is (4). If the sum (4) is of the form $-\infty + C$ or $C + \infty$ for some finite C, we say that the mean value is infinite. If it is of the form $-\infty + \infty$, the integral is said not to exist, and the mean value is undefined—the distribution has no mean.

As in the case of discrete variables, the terms *mean, average, expectation,* and *expected value* are synonymous; the notations $E(X)$, μ, and μ_X will be used interchangeably.

The mean of a continuous random variable with p.d.f. $f(x)$ is

$$E(X) = \int_{-\infty}^{\infty} xf(x)\,dx, \tag{3}$$

when this integral exists.

Example 5.4b   **Mean Waiting Time**

Suppose X has the p.d.f. $6e^{-6x}$ for $x > 0$, as in Example 5.2b. The mean value, from (3), is

$$E(X) = \int_{-\infty}^{\infty} xf(x)\,dx = \int_{-\infty}^{0} 0 \cdot dx + \int_{0}^{\infty} x \cdot 6e^{-6x}\,dx$$

$$= \lim_{b \to \infty}\left\{-\frac{1}{6}(6x + 1)e^{-6x}\Big|_{0}^{b}\right\} = \frac{1}{6},$$

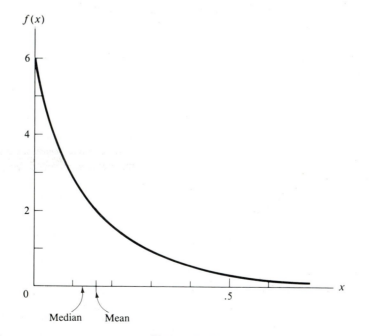

Figure 5.16

since at the upper limit be^{-6b} tends to 0 as $b \to \infty$. Figure 5.16 shows the p.d.f. and its mean, as well as the median: $m = \pm.116$ (see Example 5.2b). As is true generally for distributions skewed to the right, the mean is larger than the median. ∎

Exploiting Symmetry

When $f(x)$ is a mass density, (3) is just the formula for the center of gravity of the mass. In particular, if the distribution is symmetric [(4) and (5) of Section 5.3], the center of symmetry is the balance point. To see this, assume (without loss of generality) that the point of symmetry is 0. [If it is something else (say $x = k$), then $X - k$ is symmetric about 0, and $E(X - k) = 0$ implies $E(X) = k$.] Breaking the range $(-\infty, \infty)$ into the two pieces $(-\infty, 0)$ and $(0, \infty)$, we have

$$E(X) = \int_{-\infty}^{\infty} xf(x)\,dx = \int_{-\infty}^{0} xf(x)\,dx + \int_{0}^{\infty} xf(x)\,dx.$$

Change the variable to $-x$ in the first term on the right:

$$E(X) = \int_{+\infty}^{0} (-x)f(-x)(-dx) + \int_{0}^{\infty} xf(x)\,dx.$$

Use one minus sign to reverse the direction of the first integral:

$$E(X) = \int_{0}^{\infty} (-x)f(-x)\,dx + \int_{0}^{\infty} xf(x)\,dx = \int_{0}^{\infty} x[-f(-x) + f(x)]\,dx.$$

From symmetry about 0 $[f(x) = f(-x)]$, it follows that

$$E(X) = \int_{0}^{\infty} x[-f(x) + f(x)]\,dx = 0.$$

If the distribution of X is symmetric about x^*, then either $E(X) = x^*$ or the mean does not exist.

Example 5.4c   **A Symmetric Triangular Distribution**

Suppose X has a triangular distribution on the interval $(0, 2)$:

$$f(x) = \begin{cases} x, & 0 < x < 1, \\ 2 - x, & 1 < x < 2. \end{cases}$$

This is symmetric about $x = 1$. The mean exists, since the support of X is a finite interval $(0 < x < 2)$. So $E(X) = 1$.

Although an integration is avoided when we exploit symmetry, the integral formula (3) could be used to the same end:

$$E(X) = \int_{-\infty}^{\infty} xf(x)\,dx = \int_{0}^{1} x \cdot x\,dx + \int_{1}^{2} x \cdot (2 - x)\,dx = \frac{1}{3} + \frac{2}{3} = 1.$$ ∎

Example 5.4d   **Cauchy Distributions Have No Means**

In Example 5.3c, we considered the Cauchy p.d.f.:

$$f(x\,|\,\theta) = \frac{1/\pi}{1 + (x - \theta)^2}.$$

The median of the distribution is θ, the point of symmetry. Symmetry also implies that the mean value is θ if it exists. However, consider the integral expression (4) for $E(X)$:

$$\lim_{a \to -\infty} \int_a^0 \frac{x/\pi}{1 + (x - \theta)^2}\, dx + \lim_{b \to +\infty} \int_0^b \frac{x/\pi}{1 + (x - \theta)^2}\, dx. \tag{5}$$

The first term in (5) is $-\infty$, since the logarithm of the denominator is infinite at $\pm\infty$. Similarly, the second term in (5) is $+\infty$. Since (5) is of the form $-\infty + \infty$, the mean does not exist.   ∎

Expected Value
of $g(X)$

A function of a random variable is itself a random variable. When X is continuous, calculation of the mean value of $g(X)$ parallels the calculation in the discrete case, the probability element $f(x)\,\Delta x$ now playing the role of the probability of x. Thus, we multiply the value of $g(x)$ by the probability element at x, for each x, and "sum":

When X is continuous with p.d.f. f,

$$E[g(X)] = \int_{-\infty}^{\infty} g(x)f(x)\,dx. \tag{6}$$

An alternative way of calculating $E[g(X)]$ is to obtain the p.d.f. $f_Y(y)$ of the random variable $Y = g(X)$ and apply (3), with f_Y in place of f_X. The advantage of (6) is that it avoids the extra step of finding f_Y. When g is monotonic (nondecreasing or nonincreasing), the equivalence of these two methods amounts to a change of variable in the integral (6). A rigorous proof of the equivalence generally is beyond our scope.

Example 5.4e   Suppose $X \sim U(0, 1)$. Let $Y = X^3 + 1$. From (6),

$$E(X^3 + 1) = \int_0^1 (x^3 + 1) \cdot 1 \cdot dx = \frac{1}{4} + 1 = \frac{5}{4}.$$

Alternatively, we could find the p.d.f. of $Y = X^3 + 1$, using (17) of Section 5.2:

$$f_Y(y) = \frac{1}{3(y - 1)^{2/3}} \qquad 1 < y < 2.$$

Then from (3),

$$E(Y) = \int_0^1 y \cdot \frac{dy}{3(y - 1)^{2/3}} = \frac{5}{4}.$$ ∎

Applying (6) to the function $g(x) = c$ (a constant), yields $E(c) = c$, as in the discrete case. Further, since the integral is *linear* (additive and homogeneous), the averaging operation is linear, as in the discrete case:

For functions g and h of a random variable X and any constants a, b, and c,

$$E[ag(X) + bh(X) + c] = aE[g(X)] + bE[h(X)] + c, \tag{7}$$

provided the expected values exist.

The property that the expected value of a sum is the sum of expected values is often useful.

5.5   Average Deviations

As in the discrete case (Section 3.3), we describe variability in terms of deviations from the mean. Variance and standard deviation are defined for continuous variables by using (6) of Section 5.4, exactly as in the discrete case [(2) and (7) of Section 3.3]. The difference is that here $E(\cdot)$ means a weighted *integral* instead of a weighted sum.

Variance of a continuous random variable X:

$$\sigma^2 = \text{var } X = \int_{-\infty}^{\infty} (x - \mu)^2 f(x)\,dx. \tag{1}$$

Alternatively,

$$\sigma^2 = E(X^2) - \mu^2 = \int_{-\infty}^{\infty} x^2 f(x)\,dx - \mu^2. \tag{2}$$

The equality of (1) and (2) is a special case of the parallel axis theorem:

$$E[(X - c)^2] = E[(X - \mu)^2] + (\mu - c)^2. \tag{3}$$

The proof of (3) is the same as in the discrete case and, as in that case, (3) shows that the second moment is smallest when taken about the point $c = \mu$.

The (positive) square root of the variance is again called the *standard deviation*—s.d. for short. A somewhat more natural (though not as traditional) measure is the average absolute deviation or *mean deviation*:

$$\text{m.a.d.} = E|X - \mu|.$$

(As in the discrete case, the absolute moment $E|X - c|$ is smallest when c is the *median*,[2] and the m.a.d. could also be defined as the mean absolute deviation about the median.)

Standard deviation of X:

$$\sigma = \sqrt{\text{var } X}. \tag{4}$$

Mean deviation of X:

$$\text{m.a.d.} = \int_{-\infty}^{\infty} |x - \mu| f(x) \, dx. \tag{5}$$

As in the discrete case, the mean deviation is never larger than the standard deviation (see Problem 23 of Chapter 3). This follows from the fact that the variance of any random variable is nonnegative:

$$\text{var}|Y| = E(Y^2) - [E|Y|]^2 \geq 0, \quad \text{or} \quad E(Y^2) \geq [E|Y|]^2.$$

With $Y = X - \mu$, this implies

$$\sigma_X^2 \geq [E|X - \mu|]^2, \quad \text{or} \quad \sigma_X \geq E|X - \mu|.$$

All absolute deviations are nonnegative and are no larger than the largest possible deviation. So both the mean deviation and the standard deviation have this property; both are "average" or "typical" deviations.

It is usually a good idea to guess the s.d. even before you do any integrating. This provides a check on your calculations. It is generally easier to guess the value of the mean deviation. You can do this by covering either half of the p.d.f. (say to the left of μ) and estimating the mean of the uncovered portion. The distance from this "mean" to μ is an estimate of the m.a.d. Increasing it slightly gives a ballpark estimate of the standard deviation. If your calculation of the s.d. results in a something far from your guess, recheck the work. If it's larger than the largest deviation or smaller than the smallest, then you know you're wrong.

Example 5.5a   **The Variance of $U(0, 1)$**

Suppose $X \sim U(0, 1)$. To use (2), we need the average square:

$$E(X^2) = \int_0^1 x^2 (1) \, dx = \frac{1}{3}.$$

---

[2] For a proof of this in general, see Cramér [5], pp. 178–179. (Numbers in brackets refer to entries in the list of "References and Further Readings" that follows Chapter 16.)

The distribution is symmetric about $\frac{1}{2}$, so $\mu = \frac{1}{2}$ (see Section 5.4, p. 211):

$$\sigma^2 = E(X^2) - \mu^2 = \frac{1}{3} - \left(\frac{1}{2}\right)^2 = \frac{1}{12}.$$

The s.d. is the square root: $\sigma = 1/\sqrt{12} \doteq .289$.

The mean deviation is given by (5):

$$\text{m.a.d.} = \int_0^1 \left| x - \frac{1}{2} \right| dx = \int_0^{1/2} \left(\frac{1}{2} - x \right) dx + \int_{1/2}^1 \left(x - \frac{1}{2} \right) dx = .25.$$

As we claimed, this is slightly less than the standard deviation. Figure 5.17 shows the p.d.f., the mean deviation, and the standard deviation.

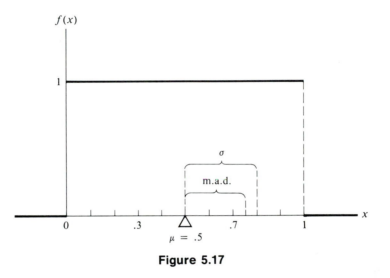

Figure 5.17   ■

Changing Units

Sometimes it is necessary to change the units of measurement. Such a change is usually a *linear* transformation, one of the form $Y = a + bX$, such as we considered in Example 5.1c. A linear transformation shifts the origin and changes the size of the unit. With expected value now defined as an integral, the argument we used in connection with discrete variables (Section 3.3) applies in the continuous case as well:

If $Y = a + bX$, then

$\mu_Y = a + b\mu_X,$ (6)

and

$\sigma_Y = |b|\sigma_X.$ (7)

Example 5.5b   **Linear Transformation of a Uniform Variable**
Suppose $X \sim U(0, 1)$, and consider

$$Y = c + (d - c)X.$$

According to Example 5.1c, $Y \sim U(c, d)$. From symmetry, we have $E(X) = .5$, so from (6),

$$E(Y) = c + (d - c) \cdot \frac{1}{2} = \frac{c + d}{2},$$

(Alternatively, this is the center of symmetry of Y.)

In Example 5.5a, we found that $\sigma_X = 1/\sqrt{12}$. So from (7),

$$\sigma_Y = (d - c)\sigma_X = \frac{d - c}{\sqrt{12}}.$$ ■

Standard Scores

There is a particular transformed scale that is especially useful, for any random variable X. The origin or reference point on this new scale is the mean of X, and the unit for the new scale is the standard deviation of X. The variable obtained by this linear transformation is a **standard score** called the **Z-score**:

$$Z = \frac{X - \mu_X}{\sigma_X}. \tag{8}$$

The interpretation of this Z-score is that X lies Z standard deviations to the right of its mean (negative Z means "lies to the left"):

$$X = \mu_X + Z \cdot \sigma_X. \tag{9}$$

This follows from (8) upon solving for X. For a given Z-score, (9) gives the corresponding value of X on the original scale.

A Z-score has mean 0 and s.d. 1:

$$E(X - \mu_X) = E(X) - \mu_X = 0,$$

so $E(Z) = 0$. From (7),

$$\sigma_Z = \frac{\sigma_X}{\sigma_X} = 1.$$

Example 5.5c   **Standardizing GRE Scores**
In any given year, hundreds of thousands of college students take the Graduate Record Exam (GRE), and the collection of their scores can be approximated by a continuous distribution. Suppose the mean score is 521 and the s.d. is 123 (as was the case for the "analytical ability" scores in 1977–1978). The Z-score for a student whose raw score is 562 is then

$$Z = \frac{562 - 521}{123} = \frac{1}{3}.$$

This means that 562 is one-third of a standard deviation above average. The Z-score (8) corresponding to a raw score of $X = 320$ is

$$Z = \frac{320 - 521}{123} = \frac{-67}{41} \doteq -1.63.$$

So 320 is 1.63 standard deviations below average.

Going the other way, we can find the GRE corresponding to a given Z-score using (9). For instance, when $Z = 1.7$, the X-score is

$$X = 521 + 1.7 \times 123 = 730.$$

(See Figure 5.18.)

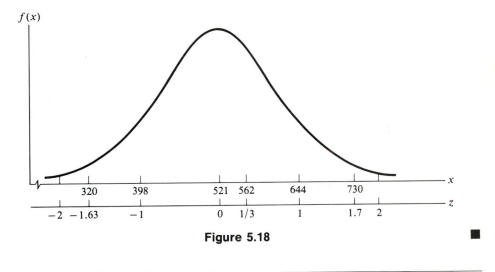

Figure 5.18

5.6 Several Variables

In Section 2.2, we considered discrete bivariate probability distributions. Here we take up continuous bivariate probability distributions.

A bivariate p.d.f. $f(x, y)$ gives the *density* of probability, or probability per unit area, at the point (x, y). The probability that (X, Y) lies in a small rectangle of dimensions Δx and Δy at (x, y) is approximately the product of the density at that point and the area of the rectangle:

$$P(x < X < x + \Delta x, y < Y < y + \Delta y) = f(x, y)\,\Delta x\,\Delta y. \tag{1}$$

This is the **probability element** at the point (x, y). Its graphical representation is the volume of the infinitesimal cylinder whose base in the xy-plane is a rectangle with sides Δx and Δy and whose top is the surface $z = f(x, y)$. (See Figure 5.19.) The area of the cylinder's base is $\Delta x\,\Delta y$, and its height is approximately $f(x, y)$.

Calculating Probabilities

The probability that (X, Y) lies in a region A of the plane is the "sum" of the probability elements over A. One can make this notion precise in two dimensions, as in the case of one dimension, by a limiting process in which Δx and Δy tend to zero.

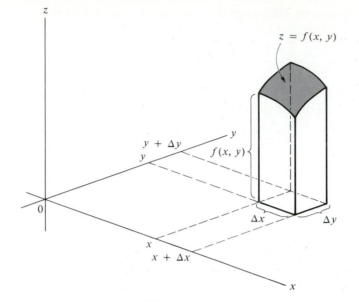

Figure 5.19

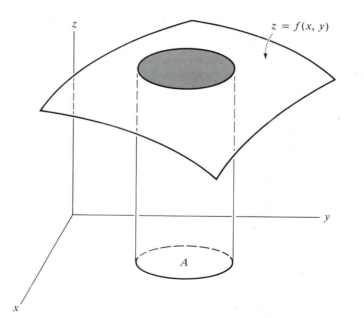

Figure 5.20

The result is a *double integral* over A:

$$P[(X, Y) \text{ in } A] = \int_A \int f(x, y)\, dx\, dy. \tag{2}$$

Interpreted graphically, the integral (2) is approximated by the sum of the volumes of all rectangular cylinders corresponding to probability elements in A. Its value is thus the volume under the surface $f(x, y)$, above the xy-plane, and within a cylinder with base A. (See Figure 5.20.)

Any nonnegative function of two variables can serve as the basis of a joint p.d.f. provided the volume under its graph is finite: Dividing the function by that volume produces a function whose double integral (volume under its graph) is equal to 1.

A function $f(x, y)$ is a joint p.d.f. if

(i) $f(x, y) \geq 0,$

(ii) $\displaystyle\iint_{xy\text{-plane}} f(x, y)\, dx\, dy = 1.$

Example 5.6a   **A Uniform Bivariate Density**

Suppose (X, Y) is uniformly distributed in the unit square:

$$f(x, y) = \begin{cases} 1, & 0 < x < 1 \text{ and } 0 < y < 1, \\ 0, & \text{elsewhere.} \end{cases}$$

The graph of f is a flat surface above the square, like a square table top. The volume under the surface is 1, the volume of a unit cube. The probability of a subregion A of the unit square, represented as the volume of a cylinder of base A and constant height 1, is simply the area of A. Thus, for instance, the probability that $X^2 + Y^2 < 1$ is equal to the area of a quarter of a circle of radius 1:

$$P(X^2 + Y^2 < 1) = \frac{\pi}{4}.$$

(See Figure 5.21, on page 220.)   ∎

When the joint p.d.f. is constant, as in Example 5.6a, calculating volumes amounts to calculating areas in the xy-plane. Calculating a double integral whose integrand is not constant is seldom that easy. When the joint p.d.f. is given by a sufficiently simple formula, the value of the double integral can be found by a process of repeated single integrations, as in the following example.

Example 5.6b   Let X and Y have the joint p.d.f. given by the formula e^{-x-y} for $x > 0$ and $y > 0$. The graph over the first quadrant is a surface that meets the coordinate planes in the decaying exponential curves e^{-x} and e^{-y}. (See Figure 5.22.)

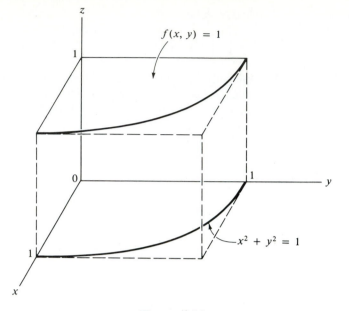

$f(x, y) = 1$

$x^2 + y^2 = 1$

Figure 5.21

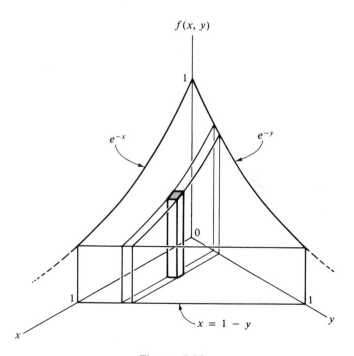

$f(x, y)$

e^{-x}

e^{-y}

$x = 1 - y$

Figure 5.22

The probability of the event $X + Y < 1$, for example, is the volume under the p.d.f. above the region where this inequality holds. We write this as a double integral over this region:

$$P(X + Y < 1) = \iint\limits_{x+y<1} e^{-x-y} dy\, dx.$$

We find the volume by adding the probability elements in A systematically. Adding them first in the x-direction at a fixed y, we find the volume in a slab of width dy; we then add these slab volumes in the y-direction to find the desired volume (again see Figure 5.22). The first summation is a single integration with respect to x with y held fixed; this summation extends from $x = 0$ to the value at the boundary line, $x = 1 - y$. The second summation is also a single integration, this time with respect to y, from $y = 0$ to $y = 1$:

$$P(X + Y < 1) = \int_0^1 \left\{ \int_0^{1-y} e^{-x-y} dx \right\} dy = \int_0^1 e^{-y} \left\{ \int_0^{1-y} e^{-x} dx \right\} dy$$

$$= \int_0^1 e^{-y}(1 - e^{-(1-x)})\, dy = 1 - 2e^{-1}. \qquad \blacksquare$$

Sometimes, as in the case of a single integral, the only way to evaluate a double integral is by numerical methods.

Example 5.6c   Consider a joint p.d.f. proportional to $\exp[-(x^2 + y^2)/2]$, and the region defined by the inequality $x + y < 1$. The probability of this region is the integral of the p.d.f. over the region:

$$P(X + Y < 1) = \frac{1}{K} \int_{-\infty}^{\infty} \int_{-\infty}^{1-y} e^{-(x^2+y^2)/2}\, dx\, dy,$$

where

$$K = \int_{-\infty}^{\infty} \int_{-\infty}^{\infty} e^{-(x^2+y^2)/2}\, dx\, dy.$$

The indefinite integral of $\exp(-x^2/2)$ is not an elementary function—the standard formulas of calculus fail. In Section 6.4 we show that $K = 2\pi$, but numerical methods are required to calculate $P(X + Y < 1)$. $\qquad \blacksquare$

Suppose we are interested in only one of the two variables X and Y whose joint distribution is defined by a given p.d.f., $f(x, y)$. The distribution of X alone is called its *marginal* distribution, as in the discrete case. When the distribution of (X, Y) is **Marginal p.d.f.'s**   discrete, we find the **marginal probability function** $f_X(x)$ by summing the joint probabilities in the row corresponding to $X = x$—that is, by "summing out" the y from the joint p.f. So too in the continuous case, we find the probability element $f_X(x)\Delta x$ at a given value x by summing the probability elements $f(x, y)\Delta x\,\Delta y$ with respect to y in the strip from x to $x + \Delta x$:

$$f_X(x)\Delta x = P(x < X < x + \Delta x) \doteq \sum_y f(x, y)\Delta y\,\Delta x.$$

The marginal p.d.f. of X is then the coefficient of Δx:

$$f_X(x) \doteq \sum_y f(x, y) \Delta y. \tag{3}$$

We thus define the limit of the right-hand side of (3), the ordinary single integral of the p.d.f. with respect to y, to be the marginal density of X at the point x. As a function of x, it is the marginal p.d.f. of X. The marginal p.d.f. of Y is defined in similar fashion.

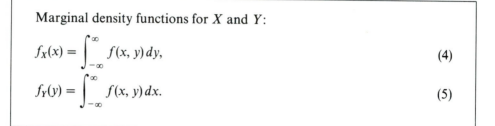

Marginal density functions for X and Y:

$$f_X(x) = \int_{-\infty}^{\infty} f(x, y) \, dy, \tag{4}$$

$$f_Y(y) = \int_{-\infty}^{\infty} f(x, y) \, dx. \tag{5}$$

Interpreted geometrically, $f_X(a)$, the marginal density of X at $x = a$, is the *area* under the graph of $f(a, y)$. This graph is a cross-section of the curve of intersection of the surface representing $f(x, y)$ with the plane $x = a$.

Example 5.6d   **Uniform Distribution on a Triangle**

Suppose the joint distribution of (X, Y) is uniform on the triangle bounded by the coordinate axes and the line $x + y = 1$. The joint p.d.f. is

$$f(x, y) = \begin{cases} 2, & x + y < 1, x > 0, y > 0, \\ 0, & \text{elsewhere.} \end{cases}$$

Its graph is a triangular portion of a plane parallel to the xy-plane and 2 units above it, shown in Figure 5.23. Using (4), we find the marginal density of X at the point x by integrating over y. For $0 < x < 1$,

$$f_X(x) = \int_0^{1-x} 2 \cdot dy = 2(1 - x). \tag{6}$$

In this simple case, where the integrand is constant, the integral in (6) is just the area of the rectangular cross-section of the solid region under the p.d.f., indicated by shading in Figure 5.23. (Imagine that the solid under $f(x, y)$ is a triangular wedge of cheese, and that you see the cross-section by slicing through the wedge at x.) The height of the rectangle is 2, for every x, and the length of the base is $1 - x$. Its area is therefore $2(1 - x)$.

Thus, the (univariate) distribution of X is triangular on the unit interval. It is clear from symmetry that the marginal distribution of Y is the same as that of X: $f_Y(y) = f_X(y)$.

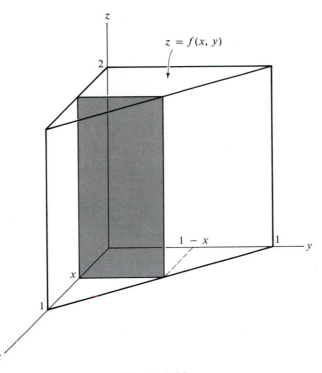

$z = f(x, y)$

Figure 5.23 ■

Example 5.6e   Suppose the random pair (X, Y) has the joint p.d.f.

$$f(x, y) = \begin{cases} k(x - y), & 0 < y < x < 1, \\ 0, & \text{elsewhere.} \end{cases}$$

The constant k is determined by the condition that the integral of a p.d.f. is 1. We obtain the marginal density for Y by integrating the joint p.d.f. with respect to x, holding y fixed. For $0 < y < 1$, the p.d.f. is zero except when $y < x < 1$. Thus, from (5), for $0 < y < 1$,

$$f_Y(y) = \int_y^1 k(x - y)\,dx = k\frac{x^2}{2} - kxy\Big|_y^1 = \frac{k}{2}(1 - y)^2.$$

Integrating this from 0 to 1 yields $k/6$. So k must equal 6 if $f_Y(y)$ is to be a density.

The two single integrations, first on x and then on y, complete a calculation of the double integral of the joint p.d.f.:

$$\int_{-\infty}^{\infty} f_Y(y)\,dy = \int_{-\infty}^{\infty} \left\{ \int_{-\infty}^{\infty} f(x, y)\,dx \right\} dy = \iint_{xy\text{-plane}} f(x, y)\,dx\,dy = \frac{k}{6} = 1.$$

Integrating over x to obtain the marginal p.d.f. of Y gives the area of a cross-section at y. Integrating with respect to y sums the volume elements defined by the increments Δy to give the volume under the surface. (See Figure 5.24.)

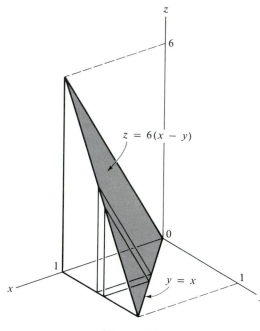

Figure 5.24   ■

Paralleling the definition in the discrete case, we define the expected value of a random variable $W = g(X, Y)$ to be the "sum" of the products of its value at (x, y) and the probability element at that point. This "sum" is again a double integral:

Expected value of a function of (X, Y):

$$E[g(X, Y)] = \iint\limits_{xy\text{-plane}} g(x, y) f(x, y)\, dx\, dy. \tag{7}$$

Example 5.6f   We calculate $E(XY)$ for the distribution of the preceding example, where the p.d.f. is $6(x - y)$ for $0 < y < x < 1$:

$$E(XY) = \int_0^1 \int_0^x xy \cdot 6(x - y)\, dy\, dx$$

$$= 6 \int_0^1 x \left\{ \int_0^x (xy - y^2)\, dy \right\} dx = \int_0^1 x^4\, dx = \frac{1}{5}.$$

■

Many Variables

In the case of more than two variables, the concepts and formulas are quite analogous to those for two variables. Consider random variables $X_1, \ldots, X_n$. Their joint p.d.f. is a function f of n arguments which is nonnegative and whose integral over the whole n-space is 1:

The joint p.d.f. of $(X_1, \ldots, X_n)$ is a function f satisfying the conditions

(i) $f(x_1, \ldots, x_n) \geq 0,$

(ii) $\displaystyle\int_{-\infty}^{\infty} \cdots \int_{-\infty}^{\infty} f(x_1, \ldots, x_n)\, dx_1 \cdots dx_n = 1.$

It will be convenient to omit limits in integrals such as (ii) when the integral is taken over the whole of n-space. Also, we'll use boldface $\mathbf{x}$ in place of $x_1, \ldots, x_n$ and $d\mathbf{x}$ in place of $dx_1 \cdots dx_n$.

By analogy with the two-dimensional case, we define a *probability element* as the probability of an infinitesimal rectangular parallelepiped:

$$P(x_1 < X_1 < x_1 + dx_1, \ldots, x_n < X_n < x_n + dx_n) \doteq f(\mathbf{x})\, d\mathbf{x}. \tag{8}$$

For a region A in n-space, we define

$$P(A) = \int \cdots \int_A f(\mathbf{x})\, d\mathbf{x}. \tag{9}$$

For a function g, we define

$$E[g(\mathbf{X})] = \int \cdots \int g(\mathbf{x}) f(\mathbf{x})\, d\mathbf{x}. \tag{10}$$

Applying (10) when g is linear in its arguments, we see that linearity of expectations [given by (iv) of Section 3.1 and (2) of Section 3.5 in the discrete case] extends to the continuous case:

Linearity of Expectation: For any random variables $X_1, \ldots, X_n$,

$$E\left(\sum_{i=1}^{n} a_i X_i \right) = \sum_{i=1}^{n} a_i E(X_i). \tag{11}$$

Exchangeable Variables

When the joint p.d.f. of $(X_1, \ldots, X_n)$ is a symmetric function of its arguments (as in Example 5.6c above), then the joint p.d.f. for any rearrangement of the variables is the same. We extend the notion of exchangeable variables (Section 2.6) to the continuous case:

Random variables $X_1, \ldots, X_n$ are **exchangeable** when

$$f_{\mathbf{X}}(x_1, \ldots, x_n) = f_{\mathbf{X}}(x_{i_1}, \ldots, x_{i_n}), \tag{12}$$

for every permutation $(i_1, \ldots, i_n)$ of $(1, 2, \ldots, n)$.

Example 5.6g   **Uniform Distribution in a Tetrahedron**

Suppose the p.d.f. of (X, Y, Z) is constant within the tetrahedron defined by the plane $x + y + z = 1$ and the coordinate planes:

$$f(x, y, z) = \begin{cases} 6, & x + y + z < 1, x > 0, y > 0, z > 0, \\ 0, & \text{elsewhere.} \end{cases}$$

This is a symmetric function—permuting (x, y, z) does not change f—so X, Y, and Z are exchangeable. In particular, they are identically distributed, with p.d.f.

$$f_X(u) = f_Y(u) = f_Z(u) = \int_0^{1-u} \int_0^{1-u-y} 6 \, dz \, dy = 3(1 - u)^2, \qquad 0 < u < 1.$$

Also, the joint distribution of any pair has the same p.d.f. as that of any other pair. So, for example,

$$E(XY) = E(XZ) = E(YZ) = \int_0^1 \int_0^{1-u} \int_0^{1-u-v} (6uv) \, dw \, dv \, du = \frac{1}{20}. \qquad \blacksquare$$

5.7   **Covariance and Correlation**

Covariance and correlation are defined for a continuous bivariate pair (X, Y) exactly as in the discrete case:

Covariance and Correlation of X and Y:

$$\text{cov}(X, Y) = \sigma_{X,Y} = E[(X - \mu_X)(Y - \mu_Y)] \tag{1}$$

$$\rho_{X,Y} = \frac{\sigma_{X,Y}}{\sigma_X \sigma_Y}. \tag{2}$$

The various relations we developed in Section 3.4 continue to hold in the continuous case:

$$\sigma_{X,Y} = E(XY) - \mu_X\mu_Y. \tag{3}$$
$$\text{cov}(X, Y) = \text{cov}(Y, X), \tag{4}$$
$$\text{cov}(aX, bY) = ab \cdot \text{cov}(X, Y), \tag{5}$$
$$\text{cov}(X + Y, Z) = \text{cov}(X, Z) + \text{cov}(Y, Z), \tag{6}$$
$$\text{var}\left\{\sum X_i\right\} = \sum \text{var } X_i + \sum_{i \neq j} \text{cov}(X_i, X_j). \tag{7}$$
$$\text{cov}\left(\sum_i a_i X_i, \sum_j b_j Y_j\right) = \sum_i \sum_j a_i b_j \text{cov}(X_i, Y_j). \tag{8}$$

[Property (8) includes (5), (6), and (7) as special cases.]

Example 5.7a   Consider again the joint p.d.f. from Example 5.6e:

$$f(x, y) = 6(x - y), \qquad 0 < y < x < 1.$$

We found the marginal p.d.f. for Y to be

$$f_Y(y) = 3(1 - y)^2, \qquad 0 < y < 1.$$

The marginal p.d.f. of X is

$$f_X(x) = \int_0^x 6(x - y)\,dy = 3x^2, \qquad 0 < x < 1.$$

From these, one easily finds the means and variances:

$$\mu_X = \frac{3}{4}, \qquad \mu_Y = \frac{1}{4}, \qquad \sigma_X^2 = \sigma_Y^2 = \frac{3}{20}.$$

In Example 5.6f, we found $E(XY) = \frac{1}{5}$, so from (3),

$$\sigma_{X,Y} = \frac{1}{5} - \frac{3}{4} \cdot \frac{1}{4} = \frac{1}{80},$$

and

$$\rho_{X,Y} = \frac{1/80}{3/20} = \frac{1}{12}. \qquad\blacksquare$$

Example 5.7b   Consider this transformation from (X, Y) to (U, V):

$$\begin{cases} U = X + Y \\ V = X - Y. \end{cases}$$

From (6),

$$\text{cov}(U, V) = \text{cov}(X, X) - \text{cov}(X, Y) + \text{cov}(X, Y) - \text{cov}(Y, Y)$$
$$= \text{var } X - \text{var } Y.$$

In particular, if var $X = $ var Y, then U and V are uncorrelated.   ∎

Schwarz's Inequality

In the discrete case (Section 3.4) we showed that the correlation coefficient (2) is bounded in magnitude: $\rho^2 \leq 1$. This followed from *Schwarz's inequality*:

$$[E(XY)]^2 \leq E(X^2)E(Y^2),$$

which holds in the continuous case as well. (The proof is exactly the same). As before, when $\rho^2 = 1$, there is a constant c such that with probability 1,

$$Y - EY = c(X - EX). \tag{11}$$

Thus, when the correlation coefficient is ± 1, there is a straight line that carries all the probability—one variable is a linear function of the other with probability 1. [The slope c in (11) is positive or negative, according as the correlation is $+1$ or -1.] In such a case, one can predict Y without error given $X = x$: Just substitute $X = x$ in (11) to find Y.

The correlation coefficient is expressible as the average product of the Z-scores formed from X and Y:

$$\rho_{X,Y} = E\left(\frac{X - EX}{\sigma_X} \cdot \frac{Y - EY}{\sigma_Y}\right).$$

A Z-score is unitless and unchanged in magnitude by a linear change of variable, although its sign may change. Thus, if $U = a + bX$,

$$\frac{U - EU}{\sigma_U} = \frac{b}{|b|} \cdot \frac{X - EX}{\sigma_X}.$$

The implication for ρ is this:

The linear transformation

$$\begin{cases} U = a + bX \\ V = c + dY \end{cases}$$

preserves $|\rho|$:

$$\rho_{U,V} = \begin{cases} \rho_{X,Y} & \text{if } bd > 0, \\ -\rho_{X,Y} & \text{if } bd < 0. \end{cases}$$

5.8   **Independence**

We define **independence** of two continuous random variables by analogy with the discrete case:

> Continuous variables X and Y are **independent** whenever their joint p.d.f. is the product of the marginal p.d.f.'s:
>
> $$f(x, y) = f_X(x)f_Y(y), \tag{1}$$
>
> for all pairs (x, y).

In practice, criterion (1) is most often used for *constructing* a joint distribution for variables that are assumed to be independent. In this section, we'll use it mainly to check independence.

Example **5.8a**   A sociologist is interested in whether the lengths of life of the husband and wife in a married couple are independent. Let f denote the joint p.d.f. and f_H and f_W denote the (marginal) p.d.f.'s of the husband's and wife's lengths of life. If they are independent, then by (1) their joint p.d.f. is

$$f(x, y) = f_H(x)f_W(y) \qquad \text{for all } x, y.$$

The statistical problem for the sociologist is to determine whether this proposed structure is supported or contradicted by available data. (We'll consider such questions in Chapters 13 and 15.)   ■

Consider a set A of x-values and a set B of y-values. As an event in the sample space for (X, Y), the set A defines a strip or cylinder set in the plane, a strip parallel to the y-axis. Similarly, B defines a strip or cylinder parallel to set parallel to the x-axis.
Product   The intersection C of these two cylinder sets is a **product set**, and we write $C =$
Sets $A \times B$.
Figure 5.25 shows how intervals A and B define such strips. The intersection C of these two strips, the product set, is a rectangle with sides parallel to the coordinate axes. If either interval is infinite, the product rectangle may be infinite; for example, the product of the sets defined by the conditions $X > 0$ and $Y > 0$ is the first quadrant.
Suppose X and Y are independent, having a joint p.d.f. satisfying (1). Consider a set A of x-values and a set B of y-values. The probability that X is in A *and* Y is in B is the double integral of their joint p.d.f. over the product set, $C = A \times B$. Evaluating the double integral over C as a succession of single integrals, first over A and then

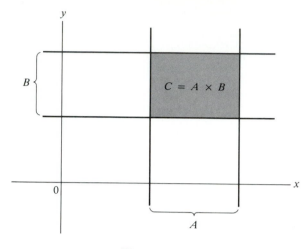

Figure 5.25

over B, we find [using (1)] that

$$P[(X, Y) \text{ in } C] = \int_C \int f(x, y)\, dx\, dy = \int_C \int f_X(x) f_Y(y)\, dx\, dy$$

$$= \int_B \int_A f_X(x) f_Y(y)\, dx\, dy = \int_B f_Y(y) \left\{ \int_A f_X(x)\, dx \right\} dy$$

$$= \int_A f_X(x)\, dx \cdot \int_B f_Y(y)\, dy = P(X \text{ in } A) \cdot P(Y \text{ in } B).$$

That is, when X and Y are independent, the probability of a product set is the product of the probabilities of the marginal sets that define it. Conversely, taking $A = (-\infty, x)$ and $B = (-\infty, y)$, this factorization of $A \times B$ implies that the joint c.d.f. factors into the product of the marginal c.d.f.'s, which in turn implies independence of X and Y.

If and only if X and Y are independent,

$$P[(X, Y) \text{ in } A \times B] = P(X \text{ in } A) \cdot P(Y \text{ in } B) \tag{2}$$

for any sets of real numbers A and B.

In the discrete case (Section 2.5), we pointed out that a joint probability cannot be zero at a point where the corresponding marginal probabilities are both positive. The same sort of thing holds in the continuous case. This means in particular that the support of independent variables is the product of the support of X and the support of Y. Therefore, if the support of a joint distribution is *not* a product set, the variables cannot be independent!

Example 5.8b Suppose the random pair (X, Y) has this joint p.d.f.:

$$f(x, y) = \begin{cases} 24xy, & x + y < 1, x > 0, \text{ and } y > 0, \\ 0, & \text{otherwise.} \end{cases}$$

The support of f is a triangle, and a triangle is not a product set. So X and Y are *not* independent. (Incidentally, since f is symmetric in its arguments, X and Y are exchangeable. Again, we see that exchangeable variables need not be independent.)

We can also see dependence by calculating the marginal p.d.f.'s. You should verify that

$$f_Y(u) = f_X(u) = 12u(1 - u)^2, \qquad 0 < u < 1.$$

The product of f_X and f_Y is not equal to f, as would be the case if X and Y were independent. ∎

Independence and Correlation

In the discrete case, we showed that two variables that are independent are uncorrelated, because then $E(XY) = E(X)E(Y)$. This factorization holds in the continuous case with essentially the same proof except that sums are replaced by integrals. However, as in the discrete case, uncorrelated variables can be dependent, as the next example illustrates.

Example 5.8c **Dependent Variables with 0 Correlation**
Suppose (X, Y) is uniform on the unit disc: $x^2 + y^2 < 1$. Thus, X and Y are exchangeable, and their distribution is symmetric about the origin. So $E(X) = E(Y) = 0$. Moreover, $E(XY) = 0$, since for every positive product xy there is a corresponding negative product in an adjacent quadrant that cancels it. Then

$$\text{cov}(X, Y) = E(XY) - E(X)E(Y) = 0.$$

However, the variables are not independent, since their joint support is not a product set. ∎

When X and Y are independent,

$$\sigma_{X,Y} = E(XY) - E(X)E(Y) = 0. \tag{3}$$

However, dependent variables can be uncorrelated.

Functions of Independent Variables

Suppose now that X and Y are independent, and consider variables $U = g(X)$ and $V = h(Y)$ and a product set $C \times D$ in the uv-space. Then

$$P[(U, V) \text{ in } C \times D] = P[(X, Y) \text{ in } A \times B],$$

where A and B are the inverse images of C and D, respectively. However, by independence of X and Y,

$$P[(X, Y) \text{ in } A \times B] = P(X \text{ in } A) \cdot P(Y \text{ in } B)$$
$$= P(U \text{ in } C) \cdot P(C \text{ in } D).$$

Since C and D were arbitrary, this shows independence of U and V. [See (2).]

When X and Y are independent, so are $g(X)$ and $h(Y)$.

These ideas are readily extended to the case of several continuous random variables. The criterion for independence of several variables is that their joint p.d.f. factors into the product of the individual (marginal) p.d.f.'s.

Random variables $X_1, \ldots, X_n$ with joint p.d.f. $f(x_1, \ldots, x_n)$ are **independent** if and only if for all $(x_1, \ldots, x_n)$,

$$f(x_1, \ldots, x_n) = f_{X_1}(x_1) \cdots f_{X_n}(x_n). \tag{4}$$

If $X_1, \ldots, X_n$ are independent, so are $g_1(X_1), \ldots, g_n(X_n)$.

When random variables $X_1, \ldots, X_n$ are independent as defined by (4), the variables in any subset of them are independent. For example, integrating (4) over $(x_3, \ldots, x_n)$, we obtain the marginal density for X_1 and X_2 as the product of the densities of X_1 and X_2. In particular, independent variables are pairwise independent and so are uncorrelated.

In studying sums of random variables in the chapters to follow, we'll exploit the fact that when the variables are independent, the variance of their sum is the sum of the variances. This follows from (7) of Section 5.7.

If $X_1, \ldots, X_n$ are independent, then

$$\text{var}(X_1 + \cdots + X_n) = \text{var } X_1 + \cdots + \text{var } X_n. \tag{5}$$

An important special case is that in which $X_1, \ldots, X_n$ are independent *and* have a common distribution. In terms of the mean μ and standard deviation σ of the common distribution, the sum ΣX_i has mean value

$$E(X_1 + \cdots + X_n) = E(X_1) + \cdots + E(X_n) = n\mu, \tag{6}$$

and variance

$$\text{var}(X_1 + \cdots + X_n) = \text{var}(X_1) + \cdots + \text{var}(X_n) = n\sigma^2. \tag{7}$$

Example 5.8d   **Elevator Passengers**

Suppose a population of potential elevator riders has mean $\mu = 150$ lb and standard deviation $\sigma = 18$ lb. We select 20 passengers at random from this population and assume that their weights are independent random variables. Consider now the random variable W, the total weight of all 20 passengers. Applying (6) and (7) yields

$$E(W) = 20\mu = 3000 \text{ lb}$$

and

$$\operatorname{var} W = 20 \cdot 18^2, \qquad \text{or} \qquad \sigma_W = \sqrt{20} \times 18 = 80.5.$$

So the standard deviation of the total weight is only $\sqrt{20}$ times the s.d. of a single weight, *not* 20 times as large.   ∎

5.9   **Conditional Distributions**

We considered conditional probability distributions for the discrete case in Section 2.3. The conditional probability function for Y given the value of X uses the basic definition of conditional probability:

$$f(y \mid x) = P(Y = y \mid X = x) = \frac{P(X = x \text{ and } Y = y)}{P(X = x)} = \frac{f(x, y)}{f_X(x)}, \tag{1}$$

provided $f_X(x) > 0$. In the continuous case, however, the probability of any single value of X or Y is zero. So we work instead with probability elements, or with p.d.f.'s. We define the conditional p.d.f. for $Y \mid x$ by analogy with (1):

$$f(y \mid x) = \frac{f(x, y)}{f_X(x)} \qquad \text{if } f_X(x) > 0. \tag{2}$$

The function defined by (2) is indeed a p.d.f.—it is nonnegative, and the area under its graph is 1:

$$\int_{-\infty}^{\infty} f(y \mid x)\, dy = \int_{-\infty}^{\infty} \frac{f(x, y)}{f_X(x)}\, dy = \frac{1}{f_X(x)} \int_{-\infty}^{\infty} f(x, y)\, dy = \frac{f_X(x)}{f_X(x)} = 1.$$

Moreover, averaging with respect to X yields the p.d.f. of Y:

$$E[f(y \mid X)] = \int_{-\infty}^{\infty} f(y \mid x) f_X(x)\, dx = \int_{-\infty}^{\infty} f(x, y)\, dx = f_Y(y), \tag{3}$$

as in the discrete case (see Example 3.2b).

The function $f(y \mid x)$ is different from the function $f(x \mid y)$. This stretches conventional mathematical notation, but a more consistent notation such as $f_{Y \mid X = x}(y)$ seems too cumbersome.

When X and Y are continuous random variables, the conditional p.d.f. of the random variable Y given $X = x$ is

$$f(y \mid x) = \frac{f(x, y)}{f_X(x)} \tag{2}$$

for any x such that $f_X(x) > 0$.

The shape of $f(y \mid x)$ as a function of y is that of the cross-section of the surface $z = f(x, y)$ in the plane defined by the condition $X = x$. So in particular, if the joint distribution is uniform, $Y \mid x$ is uniform on its support, for each x.

Example 5.9a   **Uniform Distribution on a Disc**

Suppose (X, Y) is uniformly distributed on the unit disc centered at the origin. (See Example 5.8c.) Because the area of this disc is π,

$$f(x, y) = \frac{1}{\pi} \qquad \text{for } x^2 + y^2 < 1.$$

If it becomes known that $x = \frac{1}{2}$, the only possible points are those along the chord of the circle at $x = \frac{1}{2}$. So the conditional distribution is restricted to that chord. The conditional p.d.f. is proportional to $f(\frac{1}{2}, y)$:

$$f\left(y \mid \frac{1}{2}\right) = \frac{f(\frac{1}{2}, y)}{f_X(\frac{1}{2})} = \frac{1/\pi}{\sqrt{3}/\pi} = \sqrt{3}, \qquad |y| < \frac{\sqrt{3}}{2}.$$

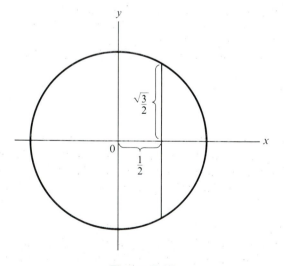

Figure 5.26

The denominator $f_X(\frac{1}{2})$ is just the area of the cross-section at $x = \frac{1}{2}$. Dividing by this area results in a function of y whose shape is that of the cross-section but whose area is 1. (See Figure 5.26.) So the conditional distribution of Y given $X = \frac{1}{2}$ is $U(-\sqrt{3}/2, \sqrt{3}/2)$. ∎

Regression Function

The conditional distribution of Y given X is a distribution. The word *conditional* refers only to the fact that it is obtained from a joint distribution by conditioning on the value of one variable. The conditional mean of Y given $X = x$ is defined by the usual formulas for a mean, in which the p.d.f. happens to be conditional:

$$E(Y \mid X = x) = \int_{-\infty}^{\infty} y f(y \mid x)\, dy. \qquad (4)$$

This is a function of x, called the **regression function of Y on X**. (The use of the term *regression* is explained in Chapter 15.)

Equation (3), which gives the marginal p.d.f. as the expected value of a conditional p.d.f., has a counterpart for expectations:

Iterated Expectations

$$E(Y) = \iint y f(x, y)\, dx\, dy = \iint y f(y \mid x) f_X(x)\, dy\, dx$$
$$= \int \left\{ \int y f(y \mid x)\, dy \right\} f_X(x)\, dx = E[E(Y \mid X)].$$

This iterated expectation result parallels (3) of Section 3.2 for discrete distributions.

One might expect, by analogy, that the expected value of a conditional variance is the unconditional variance. This is not the case. To obtain the correct relationship, we first write the parallel axis theorem:

$$\text{var}(Y \mid x) = E(Y^2 \mid x) - [E(Y \mid x)]^2.$$

Replacing x by X and taking expectations, we obtain

$$E[\text{var}(Y \mid x)] = E[E(Y^2 \mid X)] - \{\text{var}[E(Y \mid X)] + [EE(Y \mid X)]^2\}$$
$$= E(Y^2) - \text{var}[E(Y \mid X)] + (EY)^2 = \text{var } Y - \text{var}[E(Y \mid X)].$$

Averaging conditional p.d.f.'s and moments:

$$E[f(y \mid X)] = f_Y(y), \qquad (5)$$
$$E[E(Y \mid X)] = E(Y), \qquad (6)$$
$$E[\text{var}(Y \mid X)] = \text{var } Y - \text{var}[E(Y \mid X)]. \qquad (7)$$

Example 5.9b   Let (X, Y) have the joint p.d.f. introduced in Example 5.6e:

$$f(x, y) = 6(x - y), \qquad 0 < y < x < 1.$$

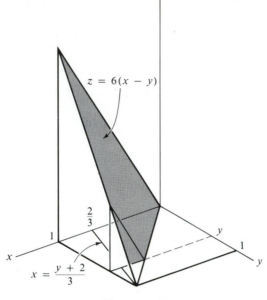

Figure 5.27

The marginal p.d.f.'s (see Example 5.6e) are

$$f_Y(y) = 3(1 - y)^2, \qquad 0 < y < 1, \qquad \text{and } f_X(x) = 3x^2, \qquad 0 < x < 1.$$

It's easy to calculate the means: $E(X) = \frac{3}{4}$ and $E(Y) = \frac{1}{4}$.

According to (2), the conditional p.d.f. of X given $Y = y$ (where $0 < y < 1$) is proportional to the joint density, for fixed y:

$$f(x \mid y) = \frac{6(x - y)}{3(1 - y)^2}, \qquad y < x < 1.$$

From (4), with the roles of X and Y reversed, the regression function of X on Y is

$$E(X \mid y) = \int_y^1 2x \frac{(x - y)}{(1 - y)^2} \, dx = \frac{y + 2}{3}.$$

This is shown in Figure 5.27. To illustrate (6), we find the unconditional mean as the expected value of the conditional mean:

$$E(X) = E[E(X \mid Y)] = E\left(\frac{Y + 2}{3}\right) = \frac{E(Y)}{3} + \frac{2}{3} = \frac{3}{4},$$

which agrees with the value found directly from f_X.   ■

5.10   Moment Generating Functions

In Sections 3.6 and 4.9, we used the probability generating function (p.g.f.) in dealing with sums of independent, discrete random variables. Its definition applies also to a continuous random variable X:

$$\eta_X(t) = E(t^X).$$

A closely related function, the **moment generating function** (m.g.f.), handles sums in the same way but is more convenient for the particular distributions introduced in this chapter:

Moment generating function of X:

$$\psi_X(t) = E(e^{tX}).$$ (1)

(As with p.d.f.'s, we may drop the subscript on ψ when there is no danger of confusion.) The relation between the p.g.f. and the m.g.f. is this:

$$\psi(t) = \eta(e^t).$$ (2)

The variable X in (1) can be discrete or continuous. When it is continuous with p.d.f. f, the expectation in (1) is an integral:

$$\psi(t) = \int_{-\infty}^{\infty} e^{tx} f(x)\,dx.$$ (3)

This expression transforms a function f into a function ψ. Depending on the value of t and the function f, the integral in (3) may or may not be defined. It is always defined for $t = 0$: $\psi(0) = 1$. We'll be interested in $\psi(t)$ only for values of t near 0.

Moment-generating functions are mysterious for many students. Don't look for physical interpretations for ψ. Rather, view it as a powerful mathematical device useful in finding distributions of functions of random variables. A typical scheme is to transform p.d.f.'s to m.g.f.'s, do some mathematics, and then come back to p.d.f.'s.

Example 5.10a   **Waiting Time Distribution**
The waiting time distribution of Example 5.2b has density $6e^{-6x}$, for $x > 0$. From (3),

$$\psi(t) = \int_0^{\infty} e^{tx} 6e^{-6x}\,dx = 6 \int_0^{\infty} e^{-x(6-t)}\,dx = \frac{6}{6-t} = \frac{1}{1 - t/6}.$$

This calculation is valid only for $t < 6$, for only then is the integral convergent. Figure 5.28 shows the graph of ψ.   ∎

Example 5.10b   **Binomial m.g.f.**
Suppose $Y \sim \text{Bin}(n, p)$. In Section 4.9, we calculated the p.g.f.:

$$\eta(t) = (pt + q)^n.$$

Applying (2), we see that the m.g.f. is

$$\psi(t) = \eta(e^t) = (pe^t + q)^n.$$

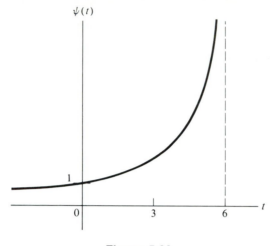

Figure 5.28

A direct calculation yields the same result:

$$\psi(t) = E(e^{tY}) = \sum_{0}^{n} e^{tk} \binom{n}{k} p^k q^{n-k} = (pe^t + q)^n. \qquad \blacksquare$$

The effect of a linear transformation of a random variable on its m.g.f. is easy to see. Suppose $Y = a + bX$. Then, from the laws of exponents,

$$\psi_Y(t) = E(e^{tY}) = E[e^{t(a+bX)}] = E[e^{at}e^{(bt)X}].$$

Since e^{at} is constant with respect to X, we can move it out:

The m.g.f. of $a + bX$ is

$$\psi_{a+bX}(t) = e^{at}\psi_X(bt). \qquad (4)$$

Example 5.10c Consider a random variable Z with p.d.f. $f(z) = K \exp(-z^2/2)$, where K is a constant needed to make the area under the p.d.f. equal to 1:

$$\frac{1}{K} = \int_{-\infty}^{\infty} e^{-z^2/2} dz.$$

The m.g.f. of Z is

$$\psi_Z(t) = E(e^{tZ}) = K \int_{-\infty}^{\infty} e^{tz - z^2/2} dz. \qquad (5)$$

We can rewrite the exponent in (5) by completing the square:

$$tz - \frac{z^2}{2} = -\frac{1}{2}[z^2 - 2tz + t^2] + \frac{t^2}{2} = -\frac{1}{2}[z - t]^2 + \frac{t^2}{2}.$$

So (5) becomes

$$\psi_Z(t) = e^{t^2/2} \int_{-\infty}^{\infty} Ke^{-(z-t)^2/2}\, dz = e^{t^2/2}, \tag{6}$$

since the area under $\exp[-(z - t)^2/2]$ is the same as the area under $\exp(-z^2/2)$, namely, $1/K$.

Suppose now we make a linear transformation, moving the center of the distribution to μ and taking the unit to be σ:

$$Y = \mu + \sigma Z.$$

Applying (4) to (6), with $a = \mu$ and $b = \sigma$, we obtain the m.g.f. of Y:

$$\psi_Y(t) = e^{\mu t}\psi_Z(\sigma t) = \exp\left[\mu t + \frac{\sigma^2 t^2}{2}\right]. \tag{7}$$

∎

Sums of Random Variables

Like the p.g.f., the m.g.f. is important in statistical theory for studying sums of independent variables. In particular, it provides a convenient way to find the distribution of a sum. To see that there may be difficulties with a more direct derivation, consider first a sample of two independent observations X and Y from a population with p.d.f. $f(x)$. To find the c.d.f. of $U = X + Y$, we can use the joint distribution of the pair (X, Y):

$$F_U(u) = P(X + Y \le u) = \int_A \int f(x, y)\, dx\, dy,$$

where A is the region in which $x + y \le u$, shown in Figure 5.29. Replacing the joint p.d.f. with the product of the marginal p.d.f.'s (using independence) and calculating

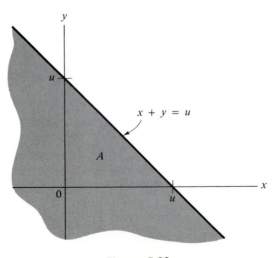

Figure 5.29

the double integral as a repeated integral, first with respect to x, we have

$$F_U(u) = \int_{-\infty}^{\infty} \left\{ \int_{-\infty}^{U-y} f(x)\,dx \right\} f(y)\,dy.$$

The p.d.f. of the sum U is then the derivative of this c.d.f. To find it, we use Leibniz's rule from calculus:

$$f_U(u) = F'_U(u) = \int_{-\infty}^{\infty} f(u - y)f(y)\,dy. \tag{8}$$

The integral in (8) is a *convolution* of f with itself.

Following this path to the distribution of the sum of the n independent observations in a random sample, we'd need an n-fold convolution. This is apt to be difficult or even impossible. The m.g.f. provides a way around the difficulties of such calculations.

Consider first two random variables, X and Y. The m.g.f. of their sum is

$$\psi_{X+Y}(t) = E(e^{t(X+Y)}) = E(e^{tX}e^{tY}).$$

When X and Y are *independent*, e^{tX} and e^{tY} are independent, so the expected product factors [according to (3) of Section 5.8]:

$$\psi_{X+Y}(t) = E(e^{tX})E(e^{tY}) = \psi_X(t)\psi_Y(t).$$

The analogous result for any finite number of independent summands follows from the discussion of that case in Section 5.8.

If $X_1, X_2, \ldots, X_n$ are independent, the m.g.f. of their sum is the product of their m.g.f.'s:

$$\psi_{\Sigma X}(t) = \psi_{X_1}(t) \cdots \psi_{X_n}(t).$$

If, in addition, the X_i's all have the same distribution, with common m.g.f. $\psi(t)$, then

$$\psi_{\Sigma X}(t) = [\psi(t)]^n. \tag{9}$$

Example 5.10d   Consider n independent observations on X, where X has the waiting time distribution of Example 5.10a. The m.g.f. of a single observation on X with p.d.f. $f(x \mid \lambda) = \lambda e^{-\lambda x}$, $x > 0$, is

$$\psi(t) = \left(1 - \frac{t}{\lambda}\right)^{-1}.$$

The m.g.f. of $Y = \Sigma X_i$ is the nth power:

$$\psi_Y(t) = [\psi(t)]^n = \left(1 - \frac{t}{\lambda}\right)^{-n}. \qquad \blacksquare$$

Suppose we find a distribution whose m.g.f. is the same as ψ_Y in this example. Is this necessarily the distribution of Y? It is, if a moment generating function uniquely determines a distribution. We mentioned that this one-to-one correspondence between generating functions and c.d.f.'s holds for discrete variables. A similar uniqueness property holds for m.g.f.'s[3]

As its name suggests, the m.g.f. generates moments. When we differentiate the m.g.f., and move the differentiation operation inside the expected value, we obtain

Moments from $\psi(t)$

$$\psi'(t) = \frac{d}{dt}\psi(t) = E\left(\frac{d}{dt}e^{tX}\right) = E(Xe^{tX}).$$

(Interchanging derivative and expected value is not always legitimate, but it is in the cases we'll encounter.) Substituting $t = 0$, we obtain the first moment:

$$\psi'(0) = E(Xe^0) = E(X).$$

A second differentiation yields

$$\psi''(t) = \frac{d^2}{dt^2}\psi(t) = \frac{d}{dt}E(Xe^{tX}) = E\left(\frac{d}{dt}(Xe^{tX})\right) = E(X^2 e^{tX}),$$

and at $t = 0$:

$$\psi''(0) = E(X^2).$$

The pattern is clear:

$$E(X^n) = \psi^{(n)}(0). \tag{10}$$

Another way of viewing this result is through the Maclaurin expansion of e^{tX} in a power series:

$$e^{tX} = \sum_0^\infty \frac{(tX)^n}{n!}.$$

Taking the expected value of this sum as the sum of the expected values, we obtain

$$\psi(t) = E(e^{tX}) = \sum_{n=0}^\infty E(X^n)\frac{t^n}{n!}.$$

So, if we happen to know the power series formula for ψ, we can read out the moments as the coefficients of $t^n/n!$.

The nth moment of X can be found from its m.g.f. either as

$$E(X^n) = \psi^{(n)}(0) \tag{11}$$

or as the coefficient of $\dfrac{t^n}{n!}$ in the power series expansion of ψ.

---

[3] See Feller [7], p. 430.

Example 5.10e　**Moments of $U(0, 1)$**

Suppose $X \sim U(0, 1)$. Then

$$\psi(t) = \int_0^1 e^{tx}\,dx = \frac{e^t - 1}{t} = \frac{1 + t + t^2/2 + \cdots - 1}{t}$$

$$= 1 + \frac{t}{2} + \frac{t^2}{6} + \frac{t^3}{24} + \cdots.$$

The mean value is the coefficient of t: $E(X) = \frac{1}{2}$. The mean square is the coefficient of $t^2/2$: $E(X^2) = \frac{1}{3}$. And so on.

Alternatively, we can find the mean from (11) as the value of $\psi'(0)$, where

$$\psi'(t) = \frac{t(e^t - 1) - e^t}{t^2}.$$

In this case, it happens that the expression for $\psi'(t)$ becomes indeterminate at $t = 0$. It can be evaluated using L'Hospital's rule to obtain $E(X) = \frac{1}{2}$, as before. ∎

For the m.g.f. to exist for $t \neq 0$, it is necessary that all moments exist. The Cauchy distribution, for example, has no moments of any positive integer order. In such a case, the integral (3) is not convergent. A more generally useful tool is the **characteristic function**:[4]

$$\varphi(t) = E(e^{itX}) = E(\cos tX) + iE(\sin tX) \tag{12}$$

where i is the imaginary unit, $\sqrt{-1}$. This exists for every distribution, since the sine and cosine functions are bounded. When the m.g.f. exists, there is a simple relationship:

$$\varphi(t) = \psi(it).$$

We'll use the m.g.f. because it serves our purposes and is easier to deal with mathematically.

Chapter Perspective

Continuous distributions provide simple representations for variables observed in a wide variety of phenomena. The simplicity is provided by methods of calculus, with integration playing the role of summation.

In continuous distributions, single values have probability zero, so we've described such distributions by giving either the probabilities of intervals or the density of probability at a point. Expected values are defined in terms of integration with respect to the density as a weighting function, and the concepts and formulas for discrete variables in terms of expected value are then carried over to the continuous case.

---

[4] In mathematics, the m.g.f. is known as a (bilateral) Laplace transform. The characteristic function is a Fourier transform. Using the characteristic function involves complex analysis. (See Feller [7], Chapter 15.)

Because sums of independent variables are so commonly encountered in statistical applications, the generating function is an important theoretical tool that we'll use often in the development to follow.

The distribution theory we need for addressing problems of inference is nearly complete. In the next chapter, we take up a number of families of continuous distributions that serve to represent real phenomena, at least approximately. We'll also take up a number of distribution families that arise in studying the sampling variation of a statistic.

Solved Problems

Section 5.1

A. The c.d.f. of Y is $F(y) = \frac{y}{2}$, for $0 < y < 2$. Find the following:
   (a) the value of $F(y)$ for $y > 2$.
   (b) $P(\frac{1}{2} < Y < \frac{3}{2})$.
   (c) $P(Y < \frac{1}{2} \mid Y < 1)$.

Solution:
   (a) Since $F(2) = 1$, and a c.d.f. is increasing but cannot exceed 1, $F(y) = 1$ for $y > 2$.
   (b) The probability of the interval is the increase in the c.d.f. over that interval [(3) of Section 5.1]:

$$F\left(\frac{3}{2}\right) - F\left(\frac{1}{2}\right) = \frac{3}{4} - \frac{1}{4} = \frac{1}{2}.$$

   (c) From the definition of conditional probability,

$$P\left(Y < \frac{1}{2} \,\middle|\, Y < 1\right) = \frac{P(Y < \frac{1}{2} \text{ and } Y < 1)}{P(Y < 1)} = \frac{P(Y < \frac{1}{2})}{P(Y < 1)} = \frac{F(\frac{1}{2})}{F(1)} = \frac{1}{2}.$$

B. Verify that the function $F(x) = 1/2 + (1/\pi)\text{Arctan } x$ is a c.d.f. ("Arctan" means the principal value of the inverse tangent, the value on the range $-\pi/2$ to $\pi/2$.)

Solution:
We need to verify that the function is nondecreasing, is 0 at $-\infty$ and 1 at $+\infty$. It is well known from trigonometry that the tangent function and its inverse are increasing:

$$\frac{d(\text{Arctan } x)}{dx} = \frac{1}{1 + x^2} > 0.$$

Also, $\text{Arctan}(-\infty) = -\pi/2$, so $F(-\infty) = \frac{1}{2} - \frac{1}{2} = 0$, and $\text{Arctan}(+\infty) = \pi/2$, so $F(+\infty) = \frac{1}{2} + \frac{1}{2} = 1$.

C. Given that X has c.d.f. $\sin x$ for $0 < x < \pi/2$, find the c.d.f. of the random variables
   (a) $Y = \sin X$.
   (b) $Z = \sin^2 X$.

Solution:

(a) We find the c.d.f. by turning to its definition:

$$F_Y(y) = P(Y \le y) = P(\sin X \le y).$$

If $0 < y < 1$, the event $\sin X \le y$ is equivalent to the event $X \le \text{Arcsin } y$. (The principal value of the inverse sine function is increasing.) So

$$F_Y(y) = P(X \le \text{Arcsin } y) = F_X(\text{Arcsin } y)$$
$$= \sin(\text{Arcsin } y) = y, \qquad 0 < y < 1.$$

Thus, Y is *uniform* on $(0, 1)$.

(b) Since the support of Y is $(0, 1)$, the transformation $Z = Y^2$ is one-to-one. So the support of Z is also $(0, 1)$, where then

$$P(Z \le z) = P(Y^2 \le Z) = P(Y \le \sqrt{z}) = F_Y(\sqrt{z}) = \sqrt{z},$$

from the result in (a).

D. Suppose X and Y have a distribution defined by the following:

$$P(X \le x \mid Y = 0) = x \qquad \text{for } 0 < x < 1,$$

$$P(X \le x \mid Y = 1) = \begin{cases} 0, & x < 1/2, \\ 1, & x \ge 1/2, \end{cases}$$

and

$$P(Y = 1) = 1 - P(Y = 0) = p.$$

(a) Find the (unconditional) c.d.f. of X and $P(X = \frac{1}{2})$.
(b) Find $P(Y = 0 \mid X = \frac{1}{2})$.
(c) Find $P(Y = 1 \mid .49 < X < .51)$.

Solution:

(a) According to the law of total probability (page 29),

$$P(X \le x) = P(X \le x \mid Y = 0) \cdot P(Y = 0) + P(X \le x \mid Y = 1) \cdot P(Y = 1).$$

When $x < \frac{1}{2}$, this is $x(1 - p) + 0 \cdot p$, and when $x \ge \frac{1}{2}$, it is $x(1 - p) + 1 \cdot p$. So,

$$F(x) = \begin{cases} (1 - p)x, & 0 \le x < 1/2, \\ p + (1 - p)x, & 1/2 \le x \le 1. \end{cases}$$

The graph jumps an amount p at $x = \frac{1}{2}$, as shown in Figure 5.30 for the case $p = \frac{1}{2}$, so $P(X = \frac{1}{2}) = p$.

(b) According to Bayes's theorem [(2) of Section 2.4],

$$P\left(Y = 0 \mid X = \frac{1}{2}\right) = \frac{P(X = \frac{1}{2} \mid Y = 0)P(Y = 0)}{P(X = \frac{1}{2})} = \frac{0 \cdot (1 - p)}{p} = 0.$$

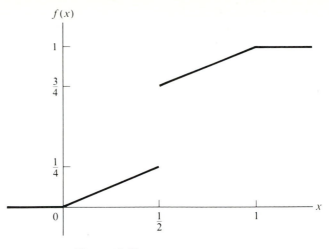

Figure 5.30

So then $P(Y = 1 | X = \frac{1}{2}) = 1$, which should agree with your conviction that $Y = 1$ if you know X to be exactly $\frac{1}{2}$.

(c) $P(.49 < X < .51) = .02(1 - p) + p = .02 + .98p.$ So, again from Bayes's theorem, the desired probability is

$$\frac{P(.49 < X < .51 | Y = 1)P(Y = 1)}{P(.49 < X < .51)} = \frac{1 \cdot p}{.02 + .98p},$$

or $\frac{50}{51}$ when $p = \frac{1}{2}$—still heavy odds in favor of $Y = 1$.

Section 5.2

E. Suppose the random variable Θ has the c.d.f. $F(\theta) = 1 - \cos \theta$, for $0 < \theta < \pi/2$. Find
   (a) $P(\pi/6 < \Theta < \pi/3)$.
   (b) the p.d.f. of Θ.

Solution:
   (a) $P(\pi/6 < \Theta < \pi/3) = F(\pi/3) - F(\pi/6) = (1 - \cos \pi/3) - (1 - \cos \pi/6) = .366.$
   (b) The p.d.f. is the derivative:

$$f(\theta) = \frac{d}{d\theta}(1 - \cos \theta) = \sin \theta, \qquad \text{for } 0 < \theta < \frac{\pi}{2}.$$

F. Given the density function

$$f(x) = \begin{cases} \dfrac{1}{3}, & 0 < x < 1, \\ \dfrac{2}{3}, & 1 < x < 2, \end{cases}$$

find the following:
   (a) $P(X > \frac{1}{2})$.

(b) $P(|X - 1| < \frac{1}{2})$.

(c) $F(x)$.

Solution:

It's always a good idea to sketch the p.d.f., as in Figure 5.31.

(a) The area to the right of $\frac{1}{2}$ is clearly $\frac{5}{6}$ of the total, so $P(X > \frac{1}{2}) = \frac{5}{6}$.

(b) $|X - 1| < \frac{1}{2}$ means $\frac{-1}{2} < X - 1 < \frac{1}{2}$, or $\frac{1}{2} < X < \frac{3}{2}$. So $P(|X - 1| < \frac{1}{2})$ is the area above the interval $(\frac{1}{2}, \frac{3}{2})$ or $\frac{3}{6}$.

(c) Moving from left to right, we accumulate area at a constant rate (slope) of $\frac{1}{3}$ from $x = 0$ to $x = 1$; the rate then doubles to $\frac{2}{3}$, until at $x = 2$ we have all the area. Figure 5.32 shows $F(x)$, the amount accumulated up to x.

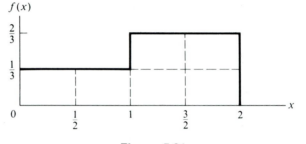

Figure 5.31

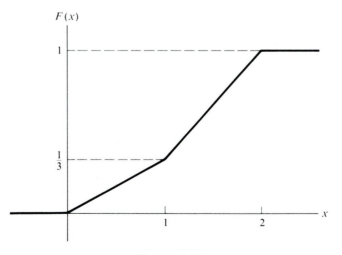

Figure 5.32

The formula has to be given in several pieces:

$$F(x) = \begin{cases} 0, & \text{for } x < 0, \\[2mm] \dfrac{x}{3}, & \text{for } 0 < x < 1, \\[2mm] \dfrac{1}{2} + \dfrac{2}{3}(x - 1), & \text{for } 1 < x < 2, \\[2mm] 1, & \text{for } x > 2. \end{cases}$$

If you prefer, a formal integration yields the same result. For instance, when x is on the interval $(1, 2)$,

$$P(X \leq x) = \frac{1}{2} + \int_1^x \frac{2}{3}\,dx = \frac{1}{2} + \frac{2}{3}(x - 1).$$

G. Let X have the p.d.f. $f(x) = e^{-x}$, for $x > 0$. Find

(a) $P(X > 1)$. (b) $P(X > 2 \mid X > 1)$. (c) $P(X < 2 \mid X < 1)$

Solution:

With so many events to consider, it pays to find the c.d.f.:

$$F(x) = P(X \leq x) = P(X < x) = \int_0^x e^{-u}\,du = 1 - e^{-x} \qquad \text{for } x > 0$$

and/or the tail probability: $R(x) = 1 - F(x) = e^{-x}, \qquad x > 0.$

(a) $P(X > 1) = 1 - F(1) = R(1) = e^{-1}.$

(b) $P(X > 2 \mid X > 1) = \dfrac{P(X > 2 \text{ and } X > 1)}{P(X > 1)} = \dfrac{R(2)}{R(1)} = \dfrac{e^{-2}}{e^{-1}} = \dfrac{1}{e}.$

(c) The answer is 1, since $X < 2$ implies $X < 1$. Formally:

$$P(X < 2 \mid X < 1) = \frac{P(X < 2 \text{ and } X < 1)}{P(X > 1)} = \frac{P(X < 1)}{P(X < 1)} = 1.$$

H. Let $f(x) = \dfrac{1}{2}e^{-|x|}$ (see Figure 5.33). Find

(a) $P(|X| > 1)$. (b) $P(X > 1 \mid |X| > 1)$.

Solution:

(a) The complement of $|X| > 1$ is $-1 < X < 1$:

$$P(|X| < 1) = \int_{-1}^1 \frac{1}{2}e^{-|x|}\,dx = \int_{-1}^0 \frac{1}{2}e^{-|x|}\,dx + \int_0^1 e^{-|x|}\,dx.$$

By breaking up the integral in this way, we can get rid of the absolute values. For $x < 0, |x| = -x$. For $x > 0, |x| = x$.

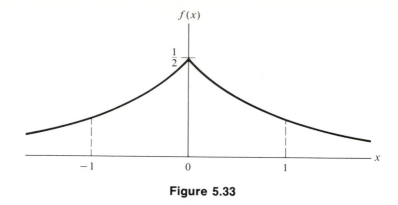

Figure 5.33

$$P(|X| < 1) = \int_{-1}^{0} \frac{1}{2} e^x dx + \int_{0}^{1} \frac{1}{2} e^{-x} dx = 1 - e^{-1}.$$

Alternatively, $|X|$ has the distribution of X in Problem G, so this answer is the same as for (a) of Problem G.

(b)   To find $P(X > 1 \,|\, |X| > 1)$, observe that the p.d.f. and the set $|X| > 1$ are both symmetric about 0. So half the probability of $|X| > 1$ is to the right of 1 and half to the left of -1: $P(X > 1 \,|\, |X| > 1) = \frac{1}{2}$.

I. Let T denote the time to the third arrival in a Poisson arrival process with mean $\lambda = 1$. Find

(a)   the c.d.f. of T

(b)   the p.d.f. of T.

(c)   Use the probability element $[dF(t)]$ to find an approximate probability that $1.00 < T < 1.02$. Check the accuracy by finding the exact probability using the c.d.f.

Solution:

(a)   Observe that $T \le t$ is equivalent to the condition that there are at least three arrivals by time t. Thus,

$$F_T(t) = P(T \le t) = P(\text{at least 3 by time } t)$$
$$= 1 - P(\text{at most 2 by time } t) = 1 - P(0, 1, \text{ or } 2)$$
$$= 1 - e^{-t}\left(1 + \frac{t}{1!} + \frac{t^2}{2!}\right) \qquad [\text{Poisson formula, } \lambda = 1].$$

(b) $f_T(t) = F'_T(t) = -e^{-t}(1 + t) + e^{-t}\left(1 + \frac{t}{1!} + \frac{t^2}{2!}\right)$

$$= \frac{t^2}{2!} e^{-t}.$$

(c)   With $dt = .02$ and $t = 1.00$,

$$f(t)\, dt = \frac{1}{2} e^{-1}(.02) = .003679.$$

The actual increment is $F(1.02) - F(1) = .003715$, so the approximation is accurate to two significant digits.

J.   Let Θ have the p.d.f. $f(\theta) = \sin \theta$ for $0 < \theta < \pi/2$, as in Problem E above.
   (a)   Find the p.d.f. of $U = \cos \Theta$.
   (b)   Find the p.d.f. of $V = \sin \Theta$.

Solution:
Each transformation is monotonic (the first one decreasing and the second increasing), so we can apply (17) of Section 5.2.

(a) $f_U(u) = \sin(\cos^{-1} u) \left| \dfrac{d}{du} (\cos^{-1} u) \right| = 1, \qquad 0 < u < 1.$

(b) $f_V(v) = \sin(\sin^{-1} v) \left| \dfrac{d}{dv} (\sin^{-1} v) \right| = \dfrac{v}{\sqrt{1 - v^2}}, \qquad 0 < v < 1.$

K.   Suppose we have ten observations from a population with p.d.f. $12x^2(1 - x)$, for $0 < x < 1$. Divide this interval into five equal subintervals, $(0, .2)$, and so on. Let Y_i denote the number of observations that fall in the ith subinterval. Give the joint distribution of $Y_1, \ldots, Y_5$.

Solution:
The probability that any single observation falls in a subinterval is the area over that subinterval under the p.d.f. Thus,

$$p_1 = \int_0^{.2} 12x^2(1 - x)\, dx = (4x^3 - 3x^4)\big|_0^{.2} = .0272.$$

Similarly,

$$p_2 = P(.2 < X < .4) = .1520, \qquad p_3 = P(.4 < X < .6) = .2960,$$
$$p_4 = P(.6 < X < .8) = .3440, \qquad p_5 = P(.8 < X < 1) = .1808.$$

The distribution of the $Y_1, \ldots, Y_4$ is then *multinomial* [see (4) of Section 4.8], defined by $n = 10$ and $p_1, \ldots, p_4$.

$$f_{Y_1, Y_2, Y_3, Y_4}(a, b, c, d) = \binom{10}{a, b, c, d, e} p_1^a p_2^b p_3^c p_4^d p_5^e, \text{ where } e = 10 - a - b - c - d.$$

Of course, $Y_5 = 1 - (Y_1 + \cdots + Y_4)$, and $p_5 = 1 - (p_1 + \cdots + p_4)$.

Sections 5.3–5.4

L.   Given the c.d.f. $F(x) = x^4$ for $0 < x < 1$, find
   (a)   the median of X.
   (b)   the mean of X.
   (c)   the 25th percentile of X.

Solution:
(a)   Set $F(m)$ equal to .5 and solve for m: $m^4 = .5$, or $m \doteq .841$.
(b)   To find the mean, we need the p.d.f.: $f(x) = F'(x) = 4x^3$, for $0 < x < 1$. Then

$$E(X) = \int_{-\infty}^{\infty} xf(x)\,dx = \int_{0}^{1} x \cdot 4x^3\,dx = \frac{4}{5} = .80.$$

(c)   Set $F(x)$ equal to .25 and solve: $x^4 = .25$, or $x \doteq .707$.

M.   Let X have the p.d.f. $f(x) = 6x(1 - x)$ for $0 < x < 1$, shown in Figure 5.34.
(a)   Find the $E(X)$.
(b)   Find the c.d.f. of X.
(c)   Find the median of X.
(d)   Find the probability that X falls between its 10th and 30th percentiles.

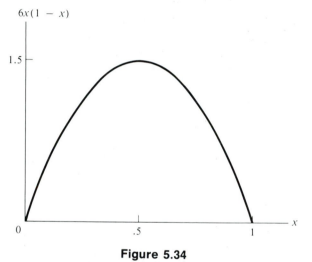

Figure 5.34

Solution:
(a)   The p.d.f. is symmetrical about $x = .5$, so .5 is both the mean and median. Alternatively,

$$E(X) = \int_{0}^{1} x \cdot 6x(1 - x)\,dx = \int_{0}^{1}(6x^2 - 6x^3)\,dx = \left(2x^3 - \frac{3x^4}{2}\right)\Bigg|_{0}^{1} = \frac{1}{2}.$$

(b)   The c.d.f. is the integral of the p.d.f. For $0 \le x \le 1$,

$$F(x) = \int_{0}^{x} 6u(1 - u)\,du = (3u^2 - 2u^3)\Bigg|_{0}^{x} = 3x^2 - 2x^3.$$

[Therefore, $F = 0$ for $x < 0$, and $F = 1$ for $x > 1$.]

(c)   The median m is $\frac{1}{2}$, by symmetry; or we can set $F(m)$ equal to $\frac{1}{2}$ and solve: $3m^2 - 2m^3 = \frac{1}{2}$, or $m = \frac{1}{2}$. [The other roots are outside the interval $(0, 1)$.]
(d)   This is $.30 - .10 = .20$ for any continuous random variable.

N. Let X have p.d.f. $f(x) = \frac{1}{2}e^{-|x|}$. (See Figure 5.33, page 248.) Find

(a) the median and quartiles.

(b) $E(X)$.

(c) $E(|X|)$ and $E(X^2)$.

Solution:

(a) The graph of $f(x)$ is symmetric about $x = 0$ $[f(-x) = f(x)]$, so the median is 0. (Alternatively, we could use the c.d.f. F found in Problem H above, setting it equal to $\frac{1}{2}$.) The first and third quartiles are symmetrically located about the median. The first is a value Q such that

$$.25 = F(Q) = \int_{-\infty}^{Q} \frac{1}{2}e^{-|u|}\,du = \int_{-\infty}^{Q} \frac{1}{2}e^{u}\,du = \frac{e^{Q}}{2},$$

or $Q = \log(.5) = -.693$. (When $x < 0$, $|x| = -x$.) The third quartile is then $.693$.

(b) The expected value is 0 because of the symmetry.

(c) The functions $|x|$ and x^2 are both symmetric about 0, so their expected values [(6) of Section 5.4] can be calculated by integrating over $(0, \infty)$, where $|x| = x$, and multiplying by 2:

$$E(|X|) = \int_{-\infty}^{\infty} |x| \cdot \frac{1}{2}e^{-|x|}\,dx = 2\int_{0}^{\infty} x \cdot \frac{1}{2}e^{-x}\,dx = 1,$$

$$E(X^2) = \int_{-\infty}^{\infty} x^2 \frac{1}{2}e^{-|x|}\,dx = 2\int_{0}^{\infty} x^2 \frac{1}{2}e^{-x}\,dx = 2.$$

O. Let X have the p.d.f.

$$\frac{x^k e^{-x}}{k!}, \qquad \text{for } x > 0,$$

where k is a nonnegative integer. Find $E(X^2 - X)$.

Solution:

Because the given function is a p.d.f., its integral is 1. This implies an integration formula for any nonnegative integer n:

$$\int_{0}^{\infty} x^n e^{-x}\,dx = n!.$$

Using this formula, we find (with $n = k + 1$)

$$E(X) = \int_{0}^{\infty} x \cdot \frac{x^k e^{-x}}{k!}\,dx = \frac{1}{k!}\int_{0}^{\infty} x^{k+1}e^{-x}\,dx = \frac{(k+1)!}{k!} = k + 1,$$

and (with $n = k + 2$)

$$E(X^2) = \int_{0}^{\infty} x^2 \cdot \frac{x^k e^{-x}}{k!}\,dx = \frac{1}{k!}\int_{0}^{\infty} x^{k+2}e^{-x}\,dx$$

$$= \frac{(k+2)!}{k!} = (k+2)(k+1).$$

From (7) of Section 5.4, we have $E(X^2 - X) = E(X^2) - E(X)$. So

$$E(X^2 - X) = (k + 2)(k + 1) - (k + 1) = (k + 1)^2.$$

Section 5.5

P. Let X have the p.d.f. $f(x) = 6x(1 - x)$ for $0 < x < 1$ (as in Problem M above). Find

(a) the standard deviation of X.

(b) the mean deviation of X.

(c) $\text{var}(1 - X)$.

(d) the Z-score for $X = \frac{3}{4}$.

Solution:

(a) In Problem M, we found $E(X) = \frac{1}{2}$. For the variance of X, we first find its average square:

$$E(X^2) = \int_0^1 x^2 \cdot 6x(1 - x)\,dx = 6\left(\frac{x^4}{4} - \frac{x^5}{5}\right)\Big|_0^1 = \frac{6}{20}.$$

Then $\text{var } X = \sigma^2 = E(X^2) - (EX)^2 = \frac{6}{20} - (\frac{1}{2})^2 = \frac{1}{20}$, and $\sigma_X = \sqrt{\frac{1}{20}} \doteq .224$.

(b) To find the mean deviation, we use (6) of Section 5.4:

$$E\left|X - \frac{1}{2}\right| = \int_0^1 \left|x - \frac{1}{2}\right| \cdot 6x(1 - x)\,dx.$$

To evaluate this integral, we write it as the sum of the integral over $(0, \frac{1}{2})$, where $|x - \frac{1}{2}| = \frac{1}{2} - x$, and the integral over $(\frac{1}{2}, 1)$, where $|x - \frac{1}{2}| = x - \frac{1}{2}$. Or, since both $|x - \frac{1}{2}|$ and the p.d.f. are symmetric about $x = \frac{1}{2}$, we can evaluate either half and multiply by 2:

$$E\left|X - \frac{1}{2}\right| = 2\int_0^{1/2} \left(\frac{1}{2} - x\right) \cdot 6x(1 - x)\,dx = \frac{3}{16} = .1875.$$

This m.a.d. is less than the s.d. (.224), as is generally the case.

(c) From (7) of Section 5.5, we know that $\text{var}(1 - X) = \text{var } X$, and this is $\frac{1}{20}$, from (a). [Interestingly, in this problem not only the variance but also the entire distribution of $1 - X$ is the same as that of X. This follows from the symmetry of the distribution about $x = \frac{1}{2}$.]

(d) The Z-score is the distance from X to the center ($\mu = \frac{1}{2}$) expressed as a multiple of the s.d. ($\sigma = .224$):

$$Z = \frac{\frac{3}{4} - \frac{1}{2}}{.224} \doteq 1.12.$$

So $x = \frac{3}{4}$ is about 1.12 s.d.'s to the right of the mean.

Q. As in Problems E and J above, let X have the p.d.f. $f(x) = \sin x$, for $0 < x < \pi/2$. Find the standard deviation.

Solution:

Using a table of integrals, we find

$$E(X) = \int_0^{\pi/2} x \sin x \, dx = \sin x - x \cos x \Big|_0^{\pi/2} = 1,$$

$$E(X^2) = \int_0^{\pi/2} x^2 \sin x \, dx = -x^2 \cos x + 2x \sin x + 2 \cos x \Big|_0^{\pi/2} = \pi - 2.$$

The variance is then $\sigma^2 = \pi - 2 - 1 = \pi - 3$, so $\sigma \doteq .376$.

Sections 5.6–5.7

R. Suppose the random pair (X, Y) has a uniform density within the circle $x^2 + y^2 = 1$.
   (a) Write the joint p.d.f. of (X, Y).
   (b) Find the marginal p.d.f.'s.
   (c) Find the covariance of X and Y.
   (d) Find the c.d.f. of the variable $Z = X^2 + Y^2$.
   (e) Find $E(X^2)$.

Solution:

(a) The density is constant within the given unit circle, and the constant is such that its product with the area of the circle is 1. Thus,

$$f(x, y) = \begin{cases} \dfrac{1}{\pi}, & x^2 + y^2 \le 1, \\ 0, & \text{otherwise.} \end{cases}$$

(b) The density f is symmetric in x and y (interchanging them does not change the density), so the variables X and Y are exchangeable. In particular, X and Y have the same marginal distribution. So we only need calculate f_X, by integrating over y. From Figure 5.35, it is clear that we have only to integrate between $\pm(1 - x^2)^{1/2}$. By symmetry, we can integrate over half this range and double the value:

$$f_X(x) = 2 \int_0^{\sqrt{1 - x^2}} \left(\frac{1}{\pi} \right) dy = \frac{2}{\pi} \sqrt{1 - x^2} \qquad \text{for } 0 < x < 1.$$

Because $f(x, y)$ is constant, what we have just found is the area of a rectangular cross-section of the solid under its graph—the product of the height $(1/\pi)$ and the base length (the chord of the circle at x, as shown in Figure 5.35).

(c) Since $E(X) = 0$, $\text{cov}(X, Y) = E(XY)$. Thus,

$$\text{cov}(X, Y) = \iint\limits_{x^2 + y^2 \le 1} xy \cdot \frac{1}{\pi} \, dx \, dy.$$

The product $xy \, dx \, dy$ in the first quadrant will exactly cancel the same product at the symmetric point $(-x, y)$ in the second quadrant. Similarly, the probability elements in the third and fourth quadrants will exactly cancel. The net result is $E(XY) = 0$.

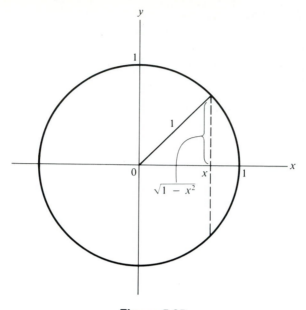

Figure 5.35

(d)   The c.d.f. of Z is the probability inside a circle of radius $\sqrt{z}$. Since f is constant, probability is proportional to area:

$$F(z) = \pi z/\pi = z \qquad \text{for} \qquad 0 < z < 1.$$

[So $Z \sim U(0, 1)$.]

(e)   From (d), we see that $E(Z) = \frac{1}{2}$. Because X and Y have the same distribution,

$$E(Z) = E(X^2) + E(Y^2) = 2E(X^2) = \frac{1}{2},$$

so $E(X^2) = \frac{1}{4}$. [Of course, we could have found $E(X^2)$ by integrating x^2 with respect to $f_X(x)$.]

S.   Given $\mu_X = 1$, $\mu_Y = 2$, $\text{cov}(X, Y) = -1$, $\text{var } X = 4$, and $\text{var } Y = 1$, find

(a) $E(2X - Y + 4)$.   **(c)** $\text{cov}(X + Y, X - Y)$.

(b) $\text{var}(2X - Y + 4)$.   **(d)** $E[(X + Y)^2]$.

Solution:

(a)   Taking expected values is a *linear* operation [(11) in Section 5.7], so

$$E(2X - Y + 4) = 2E(X) - E(Y) + 4 = 2 - 2 + 4 = 4.$$

(b)   Applying (7) of Section 5.5 and (5) and (7) of Section 5.6, we have

$$\begin{aligned}
\text{var}(2X - Y + 4) &= \text{var}(2X - Y) \\
&= \text{var}(2X) + \text{var}(-Y) + 2\,\text{cov}(2X, -Y) \\
&= 4\,\text{var } X + \text{var } Y - 2\,\text{cov}(X, Y) \\
&= 16 + 1 + 2 = 19.
\end{aligned}$$

(c) Apply (6) of Section 5.7:

$$\operatorname{cov}(X + Y, X - Y) = \operatorname{cov}(X, X) + \operatorname{cov}(X, -Y) + \operatorname{cov}(Y, X) + \operatorname{cov}(Y, -Y)$$
$$= \operatorname{var} X - \operatorname{cov}(X, Y) + \operatorname{cov}(X, Y) - \operatorname{var} Y = 3.$$

(d) Apply (7) of Section 5.7:

$$\operatorname{var}(X + Y) = \operatorname{var} X + \operatorname{var} Y + 2 \operatorname{cov}(X, Y) = 4 + 1 - 2 = 3,$$

to find

$$E[(X + Y)^2] = \operatorname{var}(X + Y) + [E(X + Y)]^2 = 3 + 9 = 12.$$

T. Let $f(x, y) = 24xy$, for (x, y) in the triangle bounded by the coordinate axes and $x + y = 1$.

(a) Find the marginal p.d.f.'s f_X and f_Y, means, and variances.
(b) Calculate the probability that $X + Y < \frac{1}{2}$.
(c) Calculate $\sigma_{X,Y}$ and $\rho_{X,Y}$.
(d) Find $\operatorname{var}(X + Y) = \operatorname{var} X + \operatorname{var} Y + 2 \operatorname{cov}(X, Y)$
    (i) by calculating each term on the right, and
    (ii) from the distribution of $X + Y$.

Solution:

(a) To find f_X, we integrate out the y from $f(x, y)$, over the range where $f > 0$. For fixed x, y ranges from 0 to $1 - x$:

$$f_X(x) = \int_0^{1-x} 24xy \, dy = 12x \int_0^{1-x} 2y \, dy = 12x(1 - x)^2, \qquad 0 < x < 1.$$

Because X and Y are exchangeable (the p.d.f. is unchanged when x and y are interchanged), the p.d.f. for Y is the same: $f_Y(y) = f_X(y) = 12y(1 - y)^2$, for $0 < y < 1$. The means and variances of X and Y are equal because f_X and f_Y are the same.

$$\mu = E(X) = E(Y) = \int_0^1 y \cdot 12y(1 - y)^2 \, dy = \frac{2}{5},$$

and

$$\operatorname{var} X = \operatorname{var} Y = \int_0^1 x^2 \cdot 12x(1 - x)^2 \, dx = \mu^2 = \frac{1}{5} - \frac{4}{25} = \frac{1}{25}.$$

(b) The probability that $X + Y < \frac{1}{2}$ is the double integral of the joint p.d.f. over that portion of the support where $x + y < \frac{1}{2}$:

$$P\left(X + Y < \frac{1}{2}\right) = \int_0^{1/2} \int_0^{1/2-x} (24xy) \, dy \, dx$$
$$= \int_0^{1/2} 12x\left(\frac{1}{2} - x\right)^2 dx = \frac{1}{16}.$$

(c) For the covariance, we need the expected product:

$$E(XY) = \int_0^1 \left(\int_0^{1-x} xy \cdot 24xy \, dy \right) dx = \int_0^1 8x^2(1 - x)^3 \, dx = \frac{2}{15}.$$

Then $\text{cov}(X, Y) = E(XY) - E(X)E(Y) = \frac{2}{15} - (\frac{2}{5})^2 = -2/75$. (That this is negative is consistent with the shape of the region of support: When x is large, y tends to be small.) The correlation coefficient is

$$\rho = \frac{-2/75}{1/25} = \frac{2}{3}.$$

(d) $\text{var } X + \text{var } Y + 2\,\text{cov}(X, Y) = \frac{1}{25} + \frac{1}{25} + 2(-2/75) = \frac{2}{75}.$

For method (ii), we first find the c.d.f. of $Z = X + Y$, following the pattern of (b), with a general z in place of $\frac{1}{2}$:

$$F(z) = \int_0^z \int_0^{z-x} 24xy\,dx\,dy = \int_0^z 12x(z-x)^2\,dx = z^4, \qquad 0 < z < 1.$$

So $f_Z(z) = F'(z) = 4z^3$ for $0 < z < 1$, and

$$\text{var } Z = \int_0^1 z^2 4z^3\,dz - (EZ)^2 = \frac{2}{3} - \left(\frac{4}{5}\right)^2 = \frac{2}{75}.$$

Sections 5.8–5.9

U. Suppose X is uniform on $(0, 1)$ and Y has p.d.f. $2y$ for $0 < y < 1$. Find the p.d.f. of the pair (X, Y), when X and Y are independent.

Solution:

Because X and Y are independent, we multiply their marginal p.d.f.'s to obtain the joint p.d.f.:

$$f(x, y) = 1 \cdot 2y = 2y, \qquad 0 < x < 1, \quad 0 < y < 1.$$

V. Suppose X and Y are independent, each uniform on the unit interval $(0, 1)$. Find the p.d.f. of their sum.

Solution:

A standard approach in finding a p.d.f. is to obtain the c.d.f. and differentiate. The joint p.d.f. of X and Y is the product of the marginal p.d.f.'s, which is 1 on the unit square, the product set defined by the two unit intervals that support the marginal distributions. The c.d.f. of the sum $Z = X + Y$ is

$$F_Z(z) = P(Z \le z) = P(X + Y \le z).$$

We could evaluate the last expression as a double integral of the joint p.d.f. over the region in the unit square where $x + y \le z$. However, because the density is constant, this is just the area of the region of integration. When $z < 1$, this is the area of a right triangle with base $=$ height $= z$, or $z^2/2$. When $z > \frac{1}{2}$, the region is the complement of a right triangle with legs equal to $2 - z$, so its area is $1 - (2 - z)^2/2$ (see Figure 5.36):

$$F_Z(z) = \begin{cases} \dfrac{z^2}{2}, & 0 < z < 1, \\[2mm] 1 - \dfrac{(2-z)^2}{2}, & 1 < z < 2. \end{cases}$$

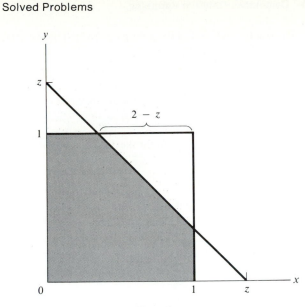

Figure 5.36

The p.d.f. is then the derivative:

$$f_Z(z) = \begin{cases} z, & 0 < z < 1, \\ 2 - z, & 1 < z < 2. \end{cases}$$

This is a triangular density on (0, 2).

W. Suppose the random pair (X, Y) has a uniform density within the circle $x^2 + y^2 = 1$. Are X and Y independent? (Explain.)

Solution:
The support is not a product set, so they are dependent.

X. Suppose $X \mid Y = y$ is uniform on $(0, y)$, and that Y has p.d.f. $f_Y(y) = 2y, 0 < y < 1$.
(a) Find the marginal p.d.f. of X.
(b) Find the conditional p.d.f. of Y given $X = x$.

Solution:
(a) The joint p.d.f. is the product of the conditional p.d.f. of X given y and the marginal p.d.f. of Y:

$$f(x, y) = \frac{1}{y} \cdot 2y = 2, \qquad 0 < x < y < 1.$$

We find the marginal p.d.f. for X by integrating out y:

$$f_X(x) = \int_x^1 2 \, dy = 2(1 - x), \qquad 0 < x < 1.$$

(b) The conditional p.d.f. is the joint p.d.f. divided by the marginal p.d.f. of X:

$$f(y|x) = \frac{f(x, y)}{f_X(x)} = \frac{1}{1 - x}, \qquad x < y < 1.$$

Thus, $Y|x$ is uniform on $(x, 1)$, as is also clear from the fact that (X, Y) is uniform on $0 < x < y < 1$.

Y. Find the two regression functions for the distribution whose density is constant on the triangle $0 < x < y < 1$.

Solution:

The conditional distributions are also uniform, being proportional to the joint p.d.f. along the line defined by the given value. Thus, given $X = x$, the distribution of Y is uniform on the interval from $y = x$ to $y = 1$; the mean is $(1 + x)/2$, which is the regression function of Y on X. Given $Y = y$, the distribution of X is uniform on the interval from $x = 0$ to $x = y$; the mean is $y/2$, which is the regression function of X on Y.

Section 5.10

Z. Find the m.g.f. of a random variable with p.d.f. $\frac{1}{2}e^{-|x|}$.

Solution:

We evaluate (3) in Section 5.10, integrating over the positive and negative axes separately. The absolute value $|x|$ is $-x$ on $(-\infty, 0)$ and $+x$ on $(0, \infty)$:

$$E(e^{tX}) = \int_{-\infty}^{\infty} e^{tx} \cdot \frac{1}{2}e^{-|x|}\,dx = \frac{1}{2}\int_{-\infty}^{0} e^{x+tx}\,dx + \frac{1}{2}\int_{\infty}^{\infty} e^{-x+tx}\,dx$$

$$= \frac{\frac{1}{2}}{1 + t} + \frac{\frac{1}{2}}{1 - t} = (1 - t^2)^{-1}.$$

AA. Find the first few moments of the distribution with m.g.f.

$$\psi(t) = (1 - t^2)^{-1/2}.$$

Solution:

We expand $\psi(t)$ by the extended binomial theorem:

$$[1 + (-t^2)]^{-1/2} = 1^{-1/2} + \left(-\frac{1}{2}\right)1^{-3/2}(-t^2)^1 + \frac{(-1/2)(-3/2)}{2!}1^{-5/2}(-t^2)^2 + \cdots$$

$$= 1 + \frac{t^2}{2} + \frac{3t^4}{8} + \cdots.$$

Since there are only even powers of t, all odd-order moments are 0. The second moment is 1 (the coefficient of $t^2/2$), and the fourth moment is 9 (the coefficient of $t^4/4!$). Incidentally, since

$$(1 - t^2)^{-1/2} = (1 - t)^{-1/2}(1 + t)^{-1/2} = h(t)h(-t),$$

it follows that the distribution is that of the difference $U - V$, where U and V are independent and identically distributed—if indeed $h(t)$ is an m.g.f. (It is. See Section 6.3 and Solved Problem DD in Chapter 6.)

BB.   For a random variable X with the triangular p.d.f.

$$f(x) = 1 - |x| \qquad \text{for } -1 < x < 1,$$

(a)   find the moment generating function.
(b)   find var X from (a).

Solution:

(a)   To obtain this by direct integration involves integrating such things as $(1 + x)e^{tx}$, perhaps using integration by parts. However, in Problem V above, we encountered essentially this same triangular p.d.f., except that its support was the interval $(0, 2)$. That is, the distribution of this problem is that of $U + V - 1$, where U and V are independent and $U(0, 1)$:

$$E(e^{t(U+V-1)}) = e^{-t}E(e^{tU})E(e^{tV}) = e^{-t}[\psi(t)]^2,$$

where $\psi(t)$ is the m.g.f. of $U(0, 1)$:

$$\psi(t) = \frac{(e^t - 1)}{t}.$$

Then

$$\psi_X(t) = e^{-t}(e^t - 1)^2 = \frac{(e^t - 2 + e^{-t})}{t^2} = 1 + \frac{t}{12} + \cdots$$

(To see this, put in the power series for the two exponentials.) Then $E(X) = 0$, the coefficient of t; and $E(X^2) = \frac{1}{6}$, the coefficient of $\frac{t^2}{2}$. So var $X = \frac{1}{6}$.

CC.   Find the characteristic function of the distribution defined by the p.d.f.

$$f(x) = \frac{1/\pi}{1 + x^2}.$$

Solution:

From the definition [(12) of Section 5.10], we have

$$\varphi(t) = \frac{1}{\pi} \int_{-\infty}^{\infty} \frac{\cos tx}{1 + x^2} dx + \frac{i}{\pi} \int_{-\infty}^{\infty} \frac{\sin tx}{1 + x^2} dx$$

The second integral vanishes, since the integrand is an odd function. The first integral is found in a table of integrals:

$$\varphi(t) = e^{-|t|}.$$

(This function does not have derivatives at $t = 0$, which corresponds to the fact that there are no moments of any integral order.)

DD.   Given that U is uniformly distributed on the interval $(0, 1)$, find the distribution of $-\log U$, using m.g.f.'s.

Solution:

By definition, the m.g.f. of $V = -\log U$ is

$$\psi_V(t) = E(e^{tV}) = E(e^{t(-\log U)}) = E(e^{\log(U^{-t})})$$

$$= E(U^{-t}) = \int_0^1 u^{-t} du = \frac{1}{1-t}.$$

We recognize this from Example 5.10d as the m.g.f. of a random variable with p.d.f. e^{-x}, for $x > 0$. By the uniqueness theorem, this must be the p.d.f. of V.

EE.   Given that X has the m.g.f.

$$\psi(t) = \frac{(e^t + e^{-t})}{2},$$

find the probability function of X.

Solution:

The definition of the m.g.f. for a discrete variable is

$$\psi(t) = \sum_{x_i} e^{tx} f(x) = f(x_1)e^{tx_1} + f(x_2)e^{tx_2} + \cdots.$$

The given m.g.f. is a series of this form with just two terms, one with $x_1 = 1$ and one with $x_2 = -1$. We see then that the given m.g.f. is precisely the m.g.f. of a discrete variable with two possible values, 1 and -1, each with probability $\frac{1}{2}$. From the uniqueness theorem for m.g.f.'s, we know that X must have this distribution, so $f_X(1) = f_X(-1) = \frac{1}{2}$.

Problems

For problems marked with an asterisk, answers are provided in Appendix 2.

Section 5.1

*1.   Suppose X is $U(-1, 1)$—that is, uniformly distributed on the interval $(-1, 1)$: $F(x) = x + 1/2$, $-1 < x < 1$. (See Example 5.1c.) Find
   (a) $P(X > \frac{3}{4})$.
   (b) $P(X > \frac{3}{4} \mid X > \frac{1}{2})$.
   (c) $P(X^2 \le \frac{1}{4})$.
   (d)   the 75th percentile (3rd quartile) of X.

2.   Let $F(x) = x^2$ for $0 < x < 1$. Find the following:
   (a) $P(X < \frac{1}{2})$.
   (b) $P(X < \frac{1}{4} \mid X < \frac{1}{2})$.
   (c) $P(\sqrt{X} < \frac{1}{2})$.
   (d) $P(\sqrt{X} < y)$. (This is the c.d.f. of the variable $Y = \sqrt{X}$.)

*3.   Given $F_V(v) = 1 - \cos v$ for $0 < v < \pi/2$,
  (a)   deduce the value of $F_V(v)$ for $v \le 0$ and for $v \ge \pi/2$.
  (b)   find $P(V > \pi/4)$.

4.   Verify that the following is a c.d.f.:

$$F(x) = \begin{cases} 0, & x < 0, \\ \dfrac{1 - e^{-4x}}{1 - e^{-4}}, & 0 \le x \le 1, \\ 1, & x > 1. \end{cases}$$

*5.   A point Q is chosen at random within a circle of radius 1 foot (imagine throwing a dart, for example).
  (a)   Find the c.d.f. of X, the distance from Q to the center of the circle in feet.
  (b)   Find the c.d.f. of $Y = 12X$ (the distance in inches).

*6.   Let X have a uniform distribution on the interval $(0, 1)$. Find
  (a) $P(|Y - \frac{1}{2}| < \frac{1}{4})$.
  (b)   the c.d.f. of the variable $Y = 2X - 1$.
  (c)   the c.d.f. of the variable $Z = \sqrt{X}$.

7.   Suppose X has the c.d.f. $F(x) = x^4$, for $0 < x < 1$.
  (a) $P(X > .5)$.
  (b) $P(X^2 < .5)$.
  (c)   Find the c.d.f. of the variable $Y = X^2$.
  (d)   Find the c.d.f. of the variable $Z = X^4$.

*8.   Let X have the c.d.f. defined (in part) as follows:

$$F(x) = \begin{cases} \dfrac{x}{3}, & 0 < x < 1, \\ \dfrac{x}{3} + \dfrac{1}{3}, & 1 < x < 2. \end{cases}$$

Find the following [a sketch of the graph of $F(x)$ may help]:
  (a) $P(X < \frac{1}{2})$.   **(c)** $P(X \ge 1)$.   **(e)** $P(\frac{1}{2} < X < 3)$.
  (b) $P(X = 1)$.   **(d)** $P(X > 1 \,|\, X \ge 1)$.   **(f)** $P(X = \frac{3}{2})$.

9.   Suppose that X has a strictly increasing c.d.f., $F(x)$. [This means that F has a unique inverse function, F^{-1}, and that the inequalities $F(x) \le u$ and $x \le F^{-1}(u)$ are equivalent.] Show that the random variable $U = F(X)$ is $U(0, 1)$.

*10.   A point is selected at random on the unit interval, dividing it into two pieces with total length 1. Find the probability that the ratio of the length of the shorter piece to the length of the longer piece is less than $\frac{1}{4}$.

Section 5.2

*11.   The random variable X has the c.d.f. $F(x) = x/2$, for $0 < x < 2$. Find the p.d.f., and sketch both the c.d.f. and the p.d.f.

12. Let X have this c.d.f.:

$$F(x) = \begin{cases} 2x^2, & 0 < x < \dfrac{1}{2}, \\[2mm] 1 - 2(1-x)^2, & \dfrac{1}{2} < x < 1, \end{cases}$$

with appropriate values outside these intervals. Find and sketch the density function of X.

*13. Let X have the c.d.f. shown in Figure 5.37.

    (a) Find and sketch the p.d.f.

    (b) Find $P(X > 2 \mid X > 1)$ by considering appropriate areas under the graph of the p.d.f.

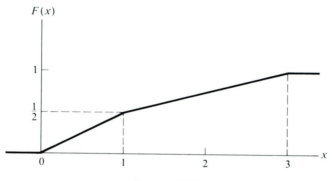

Figure 5.37

14. Let X have the p.d.f. shown in Figure 5.38.

    (a) Find $P(X < -1)$.

    (b) Find $P(X < -1 \mid X < 1)$.

    (c) Find $P(|X| < \frac{1}{2})$.

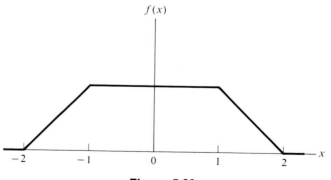

Figure 5.38

*15. Let X have p.d.f. $f(x) = c(1 - |x|)$, for $|x| < 1$, where c is a positive constant. Find
    (a)   the value of c.
    (b) $P(|X| > \frac{1}{2})$.
    (c)   the median and quartiles of X.
    (d)   the c.d.f. of X.

16. Let X have the density $30x^2(1 - x)^2$, for $0 < x < 1$. Find
    (a)   the c.d.f. of X.
    (b) $P(\frac{1}{4} < X < \frac{3}{4})$.
    (c) $P(X < \frac{1}{4} | X < \frac{1}{2})$.

*17. Given the density of X:

$$f(x) = \begin{cases} k(2 - |x|), & 1 < |x| < 2, \\ k, & |x| < 1, \end{cases}$$

    (a)   find the value of k.
    (b)   find the density function of $Y = |X|$.

18. Let T denote the time to the second arrival in a Poisson arrival process with mean λ.
    (a)   Find the c.d.f. of T.
    (b)   Find the p.d.f. of T.

*19. Use the probability element to find the approximate probability that $0 < X < .1$, where X is a random variable with the p.d.f.

$$f(x) = \frac{1/\pi}{1 + x^2}.$$

20. Let X be uniformly distributed on $(0, 1)$. Find the distribution of $Y = \theta X$, where θ is an arbitrary positive constant.

*21. Let X have a constant density on the interval $(-1, 1)$.
    (a)   Find the p.d.f. of X.
    (b)   Find the p.d.f. of $Y = |X|$.
    (c)   Find the p.d.f. of $Z = X^2$.

22. Let X have p.d.f. $1 - |x|$, for $-1 < x < 1$, and consider n independent observations on X. Classify these according to the four equal class intervals, $(-1, -.5), (-.5, 0)$, and so on, and let Y_i denote the number of observations in the ith class interval. Give the joint distribution of Y_1, Y_2, Y_3.

Sections 5.3–5.4

*23. Suppose Y has the p.d.f. $f(y) = y/2$ for $0 < y < 2$. Find
    (a)   the median of Y.
    (b) $E(Y)$.
    (c)   the probability that Y exceeds its tenth percentile.

24. Let X have the density

$$\frac{3}{4}(1 - x^2) \qquad \text{for } |x| < 1.$$

(a) Find $E(X)$.

(b) Find the median of the distribution.

***25.** Let Y have the triangular distribution with this p.d.f.: $f(y) = 1 - |y|$ for $|y| < 1$. Find
(a) $E(Y^2)$.   (b) $E|Y|$.

26. Given that X has this p.d.f.:

$$f(x) = \begin{cases} \dfrac{1}{2}, & 0 < x < 1, \\[2mm] \dfrac{1}{4}, & 1 < x < 3, \end{cases}$$

(a) find $P(\frac{1}{2} < X < 2)$.

(b) find $E(X)$.

***27.** Let X have the c.d.f. $F(x) = x^2$, for $0 < x < 1$. Find
(a) the median of X.

(b) $E(X^2 - X)$.

(c) the probability that X falls between its 20th and 65th percentiles.

28. Given that U is uniform on the unit interval $(0, 1)$, find
(a) $E(4U - 1)$.

(b) $E[(U - \frac{1}{2})^2]$.

(c) $E(e^{2U})$.

***29.** Given that X has p.d.f. $\sin x$ for $0 < x < \pi/2$. Find
(a) $E(\cos X)$.

(b) $E(X)$. (Use a table of integrals or integrate by parts.)

30. Consider the distribution defined by the c.d.f.

$$F(x) = \frac{1}{1 + e^{-\pi x/\sqrt{2}}}.$$

(a) Show that the distribution is symmetric about $x = 0$.

(b) Find the median and quartiles of this "logistic" distribution.

***31.** The Maxwell distribution, which arises in mathematical physics, is defined by the p.d.f.

$$\sqrt{\frac{2}{\pi}} y^2 e^{-y^2/2}, \qquad y > 0.$$

Find the mean of the distribution. (Integrate by parts.)

32. Let X have the "beta" density

$$f(x) = K x^{\alpha - 1}(1 - x)^{\beta - 1}, \qquad 0 < x < 1,$$

where α and β are positive integers. Given the integration formula

$$\int_0^1 x^{r - 1}(1 - x)^{s - 1}\, dx = \frac{(r - 1)!(s - 1)!}{(r + s - 1)!},$$

(a) find K.

(b) find the mean value of X.

Section 5.5

*33. Find the standard deviation of X, defined in Problem 24, where

$$f(x) = \frac{3}{4}(1 - x^2), \qquad \text{for } |x| < 1.$$

34. Find the standard deviation of X, a random variable with c.d.f. $F(x) = x^4$, for $0 \le x \le 1$, $F(x) = 0$ for $x < 0$, $F(x) = 1$ for $x > 1$, as in Problem 7.

*35. Find the mean deviation of X defined in Problem 33.

36. Problem 25 dealt with the variable Y with p.d.f. $f(y) = 1 - |y|$ for $|y| < 1$. Find
    (a) the standard deviation of Y.
    (b) the mean deviation of Y.

*37. Problem 27 called for $E(X^2 - X)$ for a random variable X with c.d.f. x^2, $0 < x < 1$. Find the variance of X.

*38. Suppose ACT (American College Testing) scores of a population of students have mean 26 and standard deviation 2.4. Find the Z-score of a student whose ACT score is
    (a) 29.   (b) 22.2.

*39. Let X have density $f(x) = 20x^3(1 - x)$, $0 < x < 1$. Find
    (a) the c.d.f., F.
    (b) $E(X)$.
    (c) var X.
    (d) $P(X < \frac{1}{4} | X < \frac{1}{2})$.

40. Find the variance of a random variable X whose p.d.f. is given by the formula $Kx^3(1 - x)^5$ for $0 < x < 1$. (*Hint:* Use the integration formula given in Problem 32.)

41. The function $(1 + |x|^3)^{-1}$ has a finite integral over $(-\infty, \infty)$. With an appropriate constant multiplier, it can be used as a p.d.f. Because of its symmetry about 0, the mean value is 0. Show that the mean deviation exists but the variance does not.

Sections 5.6–5.7

*42. Consider the random pair (X, Y), with joint p.d.f. $f(x, y) = 2$ in the portion of the first quadrant bounded by $x + y < 1$. Find
    (a) $P(X < \frac{1}{2})$.
    (b) $P(X < \frac{1}{2} \text{ and } Y < \frac{1}{2})$.
    (c) the marginal p.d.f.'s for X and Y.
    (d) $P(X + Y < \frac{1}{2})$.

*43. The joint distribution of (X, Y) concentrates probability uniformly on the line segment $x = y$, $0 < x < 1$, $0 < y < 1$.
    (a) Determine the marginal distributions of X and Y.
    (b) Find the covariance of X and Y.

44. Let (X, Y) have the joint density $4xy$ for $0 < x < 1$, $0 < y < 1$. Find the covariance of X and Y.

*45. Let (X, Y) have the joint density $x + y$ for $0 < x < 1$, $0 < y < 1$. Find the covariance and the correlation coefficient of X and Y.

46. Find the correlation coefficient of X and Y, given that their joint density is e^{-y} for $0 < x < y$. (Given: $\int_0^\infty y^n e^{-y} dy = n!$)

*47. Problem 42 deals with a joint p.d.f. that is constant on the triangle bounded by the coordinate axes and $x + y = 1$. For this distribution,
    (a) find $P(X + Y < z)$ for $0 < z < 1$ and from this, the p.d.f. of the random variable $Z = X + Y$.
    (b) verify the following relation:

    $$\text{var}(X + Y) = \text{var } X + \text{var } Y + 2 \text{ cov}(X, Y),$$

    by calculating each term.

48. Show that $\text{cov}(X + Y, Z) = \text{cov}(X, Y) + \text{cov}(Y, Z)$, with a similar expression for $\text{cov}(X, Y + Z)$.

49. Show that $\text{cov}(aX, bY) = ab \cdot \text{cov}(X, Y)$.

50. Show that if X and Y undergo a linear transformation, the correlation coefficient is unchanged.

*51. Given $\sigma_X = \sigma_Y = \sigma$, find $\rho_{X,Y-X}$ in terms of $\rho_{X,Y}$.

Section 5.8

*52. Let (X, Y) be uniformly distributed on the region consisting of two unit squares, one in the first quadrant $(0 < x < 1,$ and $0 < y < 1)$ and the other symmetrically located in the third quadrant $(-1 < x < 0, -1 < y < 0)$.
    (a) Are X and Y independent? (Explain.)
    (b) Find $P(X + Y > 1)$.
    (c) Find the marginal p.d.f.'s of X and Y.
    (d) Find the covariance of X and Y.

53. Given that (X, Y) has the p.d.f. $f(x, y) = 4xy$ for $0 < x < 1$ and $0 < y < 1$, determine
    (a) the marginal p.d.f.'s of X and Y.
    (b) whether X and Y are independent. (Justify your answer.)
    (c) $P(X < \frac{1}{2}$ and $Y < \frac{1}{2})$.

*54. Given that X and Y are independent, each uniform on $(0, 1)$, find and sketch the p.d.f. of each of the following. [For (b)–(f), use intuition rather than calculations.]
    (a) $X + Y$ [Find its c.d.f. as an area in the xy-plane.]
    (b) $X + Y - 1$
    (c) $1 - Y$
    (d) $X - Y$ [Hint: Write $X - Y = X + (1 - Y) - 1$.]
    (e) $2X - 1$
    (f) $2|X + Y - 1|$

*55. Let (X, Y) have a p.d.f. that is constant in the region $|x| + |y| < 1$. (This is a square with vertices on the axes.)
    (a) Find the marginal p.d.f.'s of X and Y.
    (b) Are the X and Y independent?
    (c) Find the covariance of X and Y. (Think about symmetries before calculating.)

     (d) Find $E(|X|)$. (This is the mean deviation of X.)

     (e) Find the c.d.f. of $X + Y$. (*Hint*: This is easy to do geometrically without a formal integration.)

56. The variables X and Y in Problem 42 are not independent. Give the joint p.d.f. f^* for *independent* variables X^* and Y^* such that $f_X = f_{X^*}$ and $f_Y = f_{Y^*}$.

*57. Find the joint p.d.f. of n independent, identically distributed variables with common p.d.f. e^{-x} for $x > 0$.

58. Find the joint p.d.f. of a sequence of n independent, identically distributed variables with common p.d.f. $f(x) = 2x$, for $0 < x < 1$.

*59. Suppose X_1 and X_2 are independent, with common p.d.f. $f(x) = 4x^3$ for $0 < x < 1$. Find the following:

     (a) $E(X_1 + X_2)$.   **(c)** $\text{var}(X_1 + X_2)$.

     (b) $E(X_1 - X_2)$.   **(d)** $\text{var}(X_1 - X_2)$.

60. Given that the pair (X, Y) has the joint p.d.f.

$$f(x, y) = 2\exp(-2x - y), \qquad x > 0, y > 0,$$

     (a) are X and Y independent? (Explain.)

     (b) find the marginal p.d.f.'s.

     (c) find $E(X + Y)$.

     (d) find $\text{var}(X + Y)$.

Section 5.9

*61. Suppose X and Y are independent, each with mean 0 and s.d. 1. Find the correlation coefficient of X and $X + Y$.

*62. Find $E(X \mid y)$ for the distribution in Problem 44, and verify that $E[E(X \mid Y)] = E(X)$ by calculating both sides.

63. For the distribution of Problem 46,

     (a) find the conditional density of Y given $X = x$.

     (b) from (a), obtain $E(Y \mid x)$ and verify that $E[E(Y \mid X)] = E(Y)$ by calculating both sides.

     (c) verify (7) of Section 5.9: $E[\text{var}(Y \mid X)] = \text{var } Y - \text{var}[E(Y \mid X)]$.

*64. Find the conditional density of Y given $X = x$ for the distribution of Problem 47, whose density is constant on the triangle bounded by the coordinate axes and $x + y = 1$. From this, find the regression function of Y on X.

65. Find the regression function of X on Y for the distribution of Problem 45, whose p.d.f. is $x + y$ for $0 < x < 1, 0 < y < 1$.

Section 5.10

*66. Find the moment generating function of the uniform distribution on the interval $(0, 1)$ and use it to find a formula for the kth moment.

*67. Given that X has m.g.f. $\psi(t) = (1 - t^2)^{-1}$,

     (a) obtain a formula for $E(X^k)$.

     (b) find var X.

68. Let X have its distribution concentrated at the single value $x = b$. That is, $P(X = b) = 1$. Find the m.g.f. of X and use it to find a formula for the kth moment, $E(X^k)$.

*69. Obtain a formula for the m.g.f. of $Y = a + bX$ in terms of the m.g.f. of X. Apply this to $Y = X - \mu_X$ where X is $U(0, 1)$ (see Problem 66), and find the central moments of X as moments of Y.

70. Use the result of Problem 66 to find the m.g.f. of the uniform distribution on the interval $(0, \theta)$.

*71. Find the m.g.f. of the distribution with p.d.f. $2x$ on the interval $(0, 1)$. Use it
(a) to find the mean.
(b) to find the variance.
[You may need this integration formula:
$$\int x \cdot e^{ax}\, dx = \frac{ax - 1}{a^2} \exp(ax) + C.]$$

72. Let X be symmetric about 0: $f(x) = f(-x)$. Show that in this case $\psi(t) = \psi(-t)$.

*73. Is $\psi(t) = (1 + t^2)^{-1}$ a moment generating function? (*Hint*: Expand ψ in a power series.)

*74. Find the m.g.f. of a Poisson variable with mean m.

75. Let $\psi(t) = 1 - p + p(e^t + e^{-t})$. Find the distribution of a random variable X with this m.g.f. and determine its mean and variance. (*Hint*: X is discrete.)

CHAPTER 6

Families of Continuous Distributions

In developing tools for continuous cases in Chapter 5, we introduced a number of particular continuous distributions as examples. Many of these belong to parametric *families* of distributions—families whose individual members are identified by the value of a parameter. Here we study these distributions further and introduce some new, useful families of continuous distributions.

6.1 Normal Distributions

One of the most important families of distributions in statistics and probability goes by the names **normal** and **Gaussian**.[1] (The latter is used mainly in the physical sciences.) In Section 4.5, we encountered a particular member of this family, the *standard normal* distribution, in approximating binomial probabilities.

The distributions in the general normal family have the same basic shape as the standard normal distribution: Any member of the family can be transformed into any other by translating the origin and changing the scale unit (that is, by a linear transformation). There are at least two reasons for the importance of this class of distributions. (1) Many populations encountered in practice are well approximated by a normal distribution. (2) Many of the commonly used statistics have distributions that are approximately normal in the important case of large samples.

---

[1] After the German mathematician Karl Friedrich Gauss (1777–1855), who used the distribution extensively in his theory of errors.

In Section 4.5, we gave the p.d.f. of the *standard* normal distribution, and we repeat it here, along with its c.d.f.:

Standard Normal p.d.f.

$$f(z) = \frac{1}{\sqrt{2\pi}} e^{-z^2/2}. \tag{1}$$

The c.d.f. is

$$\Phi(z) = \int_{-\infty}^{z} \frac{1}{\sqrt{2\pi}} e^{-u^2/2} \, du. \tag{2}$$

The constant $1/\sqrt{2\pi}$ in (1) makes the area under the p.d.f. 1, as we'll show in Section 6.4. The graph of the standard normal p.d.f. is shown in Figure 6.1. Figure 6.2

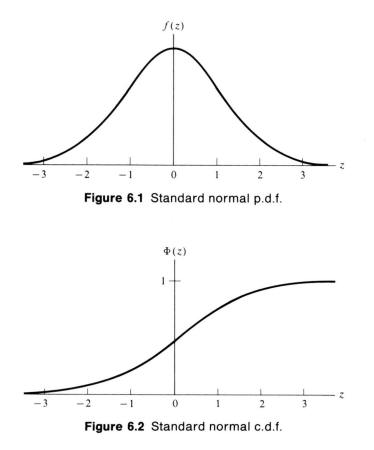

Figure 6.1 Standard normal p.d.f.

Figure 6.2 Standard normal c.d.f.

shows the graph of the c.d.f. (2). The c.d.f. $\Phi(z)$ cannot be expressed in terms of elementary functions, so a special table is required. Table II of Appendix 1 gives values of $\Phi(z)$ for selected z-values.

The standard normal distribution is clearly symmetric about $z = 0$: $f(-z) = f(z)$. So both the mean and the median are 0. (The mean value does exist.)

We can find the higher-order moments of the standard normal distribution using the moment generating function found in Example 5.8b:

$$\psi(t) = e^{t^2/2}. \tag{3}$$

The power series expansion of (3) is

$$\psi(t) = e^{t^2/2} = 1 + \frac{t^2}{2} + \frac{(t^2/2)^2}{2!} + \frac{(t^2/2)^3}{3!} + \cdots. \tag{4}$$

The kth moment is the coefficient of $t^k/k!$, so we rewrite (4) to exhibit these coefficients:

$$\psi(t) = 1 + 0t + 1\left(\frac{t^2}{2}\right) + 0\left(\frac{t^3}{3!}\right) + 3\left(\frac{t^4}{4!}\right) + \cdots.$$

Thus,

$$E(Z) = 0, \qquad E(Z^2) = 1, \qquad E(Z^3) = 0, \qquad E(Z^4) = 3, \ldots.$$

All moments of odd order vanish—the power series has no odd powers of t. (This is also clear from the symmetry of the p.d.f. about 0.)

Mean and s.d. of the standard normal variable Z:

$$E(Z) = 0, \qquad \operatorname{var} Z = E(Z^2) = 1. \tag{5}$$

More generally,

$$E(Z^n) = \begin{cases} (n-1)(n-3)\cdots 5 \cdot 3 \cdot 1 & \text{for } n \text{ even,} \\ 0 & \text{for } n \text{ odd.} \end{cases} \tag{6}$$

We turn now to the *general* normal distribution, which is useful in modeling heights, weights, examination grades, IQ, blood pressures, and the like. These distributions are bell-shaped, like the standard normal, but their means are not 0, nor are the s.d.'s likely to be exactly 1. Yet it is often possible to fit such distributions to a standard normal curve by making a linear transformation of the scale. When this is possible, we'll call the distribution *normal*.

General Normal Distributions

A variable X with mean μ and s.d. σ is said to have a normal distribution when the standard score

$$Z = \frac{X - \mu}{\sigma} \tag{7}$$

has a *standard* normal distribution. An equivalent way of saying this is that X is normal if

$$X = \mu + \sigma Z, \tag{8}$$

where Z is standard normal. Since $E(Z) = 0$ and var $Z = 1$, it follows that

$$E(X) = \mu \qquad \text{and} \qquad \text{var } X = \sigma^2.$$

[This is why we used the notation μ and σ in (7)].

When X has a normal distribution with mean μ and variance σ^2, we write $X \sim N(\mu, \sigma^2)$.

Using (8), we can express the c.d.f. of a general normal random variable in terms of the *standard* normal c.d.f.:

$$P(X \leq x) = P(\mu + \sigma Z \leq x) = P\left(Z < \frac{x - \mu}{\sigma}\right) = \Phi\left(\frac{x - \mu}{\sigma}\right),$$

where $\Phi(z)$ is given by (2). So to find $P(X \leq x)$, we simply enter Table II at the Z-score corresponding to x.

When $X \sim N(\mu, \sigma^2)$, its c.d.f. is

$$F(x) = P(X \leq x) = \Phi\left(\frac{x - \mu}{\sigma}\right), \tag{9}$$

where $\Phi(z)$ is the standard normal c.d.f., given in Table II.

Example 6.1a   Blood pressures vary from individual to individual, as well as within the same individual at different times. The blood pressure of an individual selected at random from some population may be considered to be a random variable. Its distribution is the distribution of blood pressures in the population. Suppose systolic blood pressures X in a particular population are normally distributed with mean $\mu = 117$ and s.d. $\sigma = 12$ (in mm Hg). Figure 6.3 shows the p.d.f. of X. The proportion of individuals in the population with pressures between, say, 100 and 135, is the probability that an individual picked at random has a blood pressure in this interval. We find this probability using (9) to give us F_X in terms of Φ:

$$P(100 < X < 135) = F_X(135) - F_X(100)$$

$$= \Phi\left(\frac{135 - 117}{12}\right) - \Phi\left(\frac{100 - 117}{12}\right)$$

$$\doteq \Phi(1.50) - \Phi(-1.42)$$

$$\doteq .9332 - .0778 \doteq .855.$$

Figure 6.3 shows this probability as a shaded area under the graph of the p.d.f. (Had we found the Z-scores to three decimal places and interpolated between table

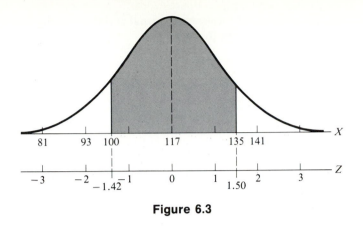

Figure 6.3

entries, the result would be .8549. Such accuracy is seldom warranted.   ■

We can find the p.d.f. of the general normal distribution by applying the transformation (8) to the standard normal p.d.f., or alternatively, by differentiating the c.d.f. as given by (9).

> When $X \sim N(\mu, \sigma^2)$, its p.d.f. is
>
> $$f(x \mid \mu, \sigma^2) = \frac{1}{\sigma\sqrt{2\pi}} \exp\left\{ -\frac{(x-\mu)^2}{2\sigma^2} \right\}. \tag{10}$$

To find a particular percentile of a general normal distribution, we first use Table II (or IIa) to find that percentile of the standard normal distribution and then transform to the more general scale. The following example illustrates this process.

Example 6.1b   Finding a Percentile

In Example 5.5c, we considered scores on the Graduate Record Exam (GRE). The "analytical ability" score X is nearly normally distributed. In 1977–1978, $X \approx N(521, 123)$. If we assume that distribution, what is the score of a student at the 63rd percentile?

The closest entry to .6300 in Table II is .6293, corresponding to $Z = .33$. (This percentile is not one of those listed in Table IIa.) Interpolating between that entry and the next larger, we take the 63rd percentile of Z to be $Z = .332$. The corresponding GRE score, from (8), is then

$$X = \mu + Z\sigma = 521 + (.332)(123) \doteq 562.$$ ■

Central moments of the general normal variable X follow from the corresponding moments of Z given by (6), since

$$E[(X - \mu)^n] = E[(\sigma Z)^n] = \sigma^n E(Z^n).$$

When $X \sim N(\mu, \sigma^2)$,

$$E[(X - \mu)^n] = \begin{cases} (n - 1)(n - 3) \cdots 5 \cdot 3 \cdot \sigma^n & \text{for } n \text{ even,} \\ 0 & \text{for } n \text{ odd.} \end{cases} \tag{11}$$

Linear Function of a Normal X

A linear function of a normal variable is normal. To see this, we note that if $Y = a + bX$, where $X \sim N(\mu, \sigma^2)$ and $b > 0$, the Z-score for Y is

$$\frac{Y - \mu_Y}{\sigma_Y} = \frac{a + bX - (a + b\mu)}{b\sigma} = \frac{X - \mu}{\sigma},$$

which is standard normal. Therefore (by definition), Y is normal. [The same is true when $b < 0$, because $-Z \sim N(0, 1)$.]

The moment generating function is useful in studying the distribution of the sum of independent normal variables. In Example 5.10c, we found the m.g.f. of a general normal X:

$$\psi_X(t) = e^{t\mu} \psi_Z(t\sigma) = \exp\left(\mu t + \frac{\sigma^2 t^2}{2} \right). \tag{12}$$

Sums of Independent Normals

Now suppose X and Y are independent normal variables: $X \sim N(\mu, \sigma^2)$, and $Y \sim N(v, \tau^2)$. We know [from (7) of Section 5.4 and (5) of Section 5.8] that $X + Y$ has mean $\mu + v$ and variance $\sigma^2 + \tau^2$. Since X and Y are independent, the m.g.f. of their sum is the product of their m.g.f.'s:

$$\psi_{X+Y}(t) = \psi_X(t) \cdot \psi_Y(t) = \exp\left(\mu t + \frac{1}{2}\sigma^2 t^2 \right) \cdot \exp\left(vt + \frac{1}{2}\tau^2 t^2 \right)$$

$$= \exp\left[(\mu + v)t + \frac{1}{2}(\sigma^2 + \tau^2)t^2 \right]. \tag{13}$$

We recognize this from (12) as the m.g.f. of $N(\mu + v, \sigma^2 + \tau^2)$. Since there is only one distribution with a given m.g.f., the distribution of the sum of two independent normal variables is normal.

In comparing populations in Chapter 12, we'll need to know about the *difference* between two independent normal variables. However, $X - Y = X + (-Y)$, and $-Y$ is a linear transformation of Y. So with X and Y defined as above, $-Y \sim N(-v, \tau^2)$, and $X + (-Y) \sim N(\mu - v, \sigma^2 + \tau^2)$.

If $X \sim N(\mu, \sigma^2)$ and $Y \sim N(v, \tau^2)$, and if X and Y are independent, then $X \pm Y \sim N(\mu \pm v, \sigma^2 + \tau^2)$.

Derivation (13) generalizes to any finite number of independent normal variables. The following property will be used often in later chapters and warrants special emphasis:

If $X_1, X_2, \ldots,$ are independent, and $X_i \sim N(\mu_i, \sigma_i^2)$, then

$$\sum X_i \sim N(\sum \mu_i, \sum \sigma_i^2).$$

6.2   Exponential Distributions

In Example 5.2d, we derived the density function of the distribution of the waiting time T for a particular Poisson process (one with parameter $\lambda = 6$):

$$f_T(x) = 6e^{-6x}, \qquad x > 0.$$

More generally, consider a Poisson process with rate parameter λ. Following the same reasoning as in Example 5.2d, we find that the distance along the x-axis from any given point to the next "event" (arrival, or failure, or particle emission, or the like) has p.d.f.

$$f(x \mid \lambda) = \lambda e^{-\lambda x}, \qquad x > 0.$$

The parameter λ indexes this family of distributions. They are called **exponential** (or negative exponential) because of the form of the p.d.f. We write $X \sim \text{Exp}(\lambda)$.

The c.d.f. corresponding to the exponential p.d.f. (see Example 5.1b) is

$$P(T \leq x) = \int_0^x \lambda e^{-\lambda u} du = 1 - e^{-\lambda x}, \qquad x > 0.$$

An **exponential** distribution is one with p.d.f.

$$f(x \mid \lambda) = \lambda e^{-\lambda x}, \qquad x > 0, \tag{1}$$

and c.d.f.

$$F(x \mid \lambda) = 1 - e^{-\lambda x}, \qquad x > 0. \tag{2}$$

(Graphs of f and F were shown in Figures 5.2 and 5.11 for $\lambda = 6$.) Because the exponential function is conveniently available on inexpensive hand calculators, there is no need to have a table of the cumulative probabilities given by (2).

The moments of an exponential distribution are easy to find by integration. When $X \sim \text{Exp}(\lambda)$,

$$E(X) = \int_0^\infty x\lambda e^{-\lambda x}\,dx = \frac{1}{\lambda}. \tag{3}$$

Similarly,

$$E(X^2) = \int_0^\infty x^2\lambda e^{-\lambda x}\,dx = \frac{2}{\lambda^2}.$$

The variance is the mean square minus the square of the mean:

$$\text{var } X = E(X^2) - [E(X)]^2 = \frac{1}{\lambda^2}. \tag{4}$$

The first two moments are produced along with all higher moments by a single integration when we calculate the moment generating function. As in Example 5.10a, we have

$$\psi(t) = \int_0^\infty e^{tx}\lambda e^{-\lambda x}\,dx = \lambda\int_0^\infty e^{-x(\lambda - t)}\,dx = \frac{1}{1 - t/\lambda}. \tag{5}$$

This calculation is valid for any value of t less than λ, since the integral is then convergent. In particular, since $\lambda > 0$, it converges in an interval about $t = 0$. The power series expansion is as follows:

$$\frac{1}{1 - t/\lambda} = 1 + \frac{t}{\lambda} + \left(\frac{t}{\lambda}\right)^2 + \left(\frac{t}{\lambda}\right)^3 + \cdots$$

$$= 1 + \left(\frac{1}{\lambda}\right)t + \left(\frac{2}{\lambda^2}\right)\left(\frac{t^2}{2}\right) + \left(\frac{3!}{\lambda^3}\right)\left(\frac{t^3}{3!}\right) + \cdots.$$

Reading out the coefficients of $t^k/k!$, we obtain the moments:

$$E(X) = \frac{1}{\lambda}, \qquad E(X^2) = \frac{2!}{\lambda^2}, \qquad E(X^3) = \frac{3!}{\lambda^3}, \ldots.$$

The median of the exponential distribution is a number m such that

$$F(m) = 1 - e^{-\lambda m} = \frac{1}{2},$$

or

$$m = \frac{\log 2}{\lambda} \doteq \frac{.693}{\lambda}. \tag{6}$$

(As in this case, the mean is typically larger than the median when the distribution is skewed to the right.)

When $X \sim \mathrm{Exp}(\lambda)$, its median, mean, standard deviation, and m.g.f. are

$$m = \frac{\log 2}{\lambda},$$

$$\mu = \frac{1}{\lambda}, \qquad \sigma = \frac{1}{\lambda},$$

$$\psi(t) = \frac{1}{1 - \frac{t}{\lambda}}.$$

Example 6.2a   **Particle Emissions**

Alpha-particles emitted by carbon-14 are recorded using a Geiger counter. Suppose emissions occur according to a Poisson process at the rate $\lambda = 16$ per second. The time X to the next emission from any given point in time is exponential, with mean $\frac{1}{\lambda} = \frac{1}{16}$ sec. Its m.g.f. is

$$\psi(t) = \left(1 - \frac{t}{16}\right)^{-1}.$$

The median of X is $.693/\lambda = .0433$ sec. The mean and standard deviation of X are both $\frac{1}{\lambda} = \frac{1}{16}$ sec.   ∎

Lack of Memory

The exponential distribution is called *memoryless*, for the following reason. Suppose $X \sim \mathrm{Exp}(\lambda)$; suppose also we have been waiting for time c and that the event in question has not occurred. How much longer do we have to wait? The probability we have to wait an additional time y is

$$P(X > c + y \mid X > c) = \frac{P(X > c + y)}{P(X > c)} = \frac{1 - F_X(c + y)}{1 - F_X(c)} = \frac{e^{-\lambda(c+y)}}{e^{-\lambda c}} = e^{-\lambda y}.$$

Since this is $1 - F_Y(y)$, where $Y \sim \mathrm{Exp}(\lambda)$, we see that the additional waiting time $Y = X - c$ is $\mathrm{Exp}(\lambda)$. The remarkable thing about this answer is that it does not depend on c! Thus, the distribution of further waiting time is exponential, with the same mean as before, no matter how long we've been waiting.

The memoryless feature of the exponential distribution derives from the Poisson process, which is a continuous-time version of a Bernoulli process. The discrete-time analogue of the exponential distribution is the geometric distribution, which also has the memoryless property. (Compare Section 4.1.) In fact, the geometric is the only discrete distribution with this property, and the exponential is the only continuous distribution with this property.

Independence of Successive Waiting Times

Consider a Poisson arrival process with rate λ, and let $T_1, \ldots, T_r$ be successive interarrival times. Then $T_i \sim \mathrm{Exp}(\lambda)$. We'll show now that these are independent as a result of the lack of memory in the process. We'll argue this using the differential method (Section 5.2) to obtain the joint p.d.f. of $T_1, \ldots, T_r$. Consider an infinitesimal increment dt_i in T_i. We'll find the probability of the intersection of the events $\{t_i < T_i < t_i + dt_i\}$ for $i = 1, \ldots, r$. These r conditions are satisfied if there is exactly one arrival in each incremental interval $(t_i, t_i + dt_i)$ and no other arrivals between 0 and $\Sigma\, t_i$. (See Figure 6.4.)

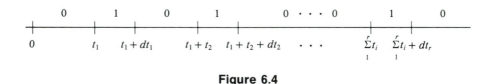

Figure 6.4

The probability of an arrival in any interval of length dt_i is $\lambda\, dt_i$, and the probability of no arrival in an interval of approximate length t_i is $\exp(-\lambda t_i)$. The numbers of arrivals in these various intervals are independent because the intervals do not overlap, so

$$P(t_1 < T_1 < t_1 + dt_1, \ldots, t_r < T_r < t_r + dt_r)$$
$$\doteq \prod_{i=1}^{r} (\lambda\, dt_i) \cdot \prod_{i=1}^{r} \exp(-\lambda t_i) = (\lambda^r \exp(-\lambda \Sigma\, t_i))\, dt_1\, dt_2 \cdots dt_r.$$

The coefficient of the volume element $\Pi\, dt_i$ is the joint p.d.f. of the successive interarrival times:

$$f(t_1, \ldots, t_r) = \lambda^r \exp(-\lambda \Sigma\, t_i) = \prod_{i=1}^{r} (\lambda \exp(-\lambda t_i)) = \prod_{i=1}^{r} f_{T_i}(t_i).$$

Because this joint p.d.f. is the product of the marginal p.d.f.'s, it follows that the successive interarrival times $T_1, \ldots, T_r$ are *independent* random variables.

6.3   Gamma Distributions

Example 6.3a   **Time to the *r*th Arrival**
Consider a Poisson arrival process with rate parameter λ and the random variable Y, the waiting time to the rth arrival. We've just seen that successive interarrival times X_i in a Poisson process are *independent* exponential variables. The time to the rth arrival is the *sum* of r successive waiting times,

$$Y = X_1 + \cdots + X_r.$$

We apply the differential method to find the p.d.f.:

$$P(t < Y < t + dt) = P[r - 1 \text{ arrivals in } (0, t) \text{ and } 1 \text{ in } (t, t + dt)]$$
$$= P[r - 1 \text{ arrivals in } (0, t)]P[1 \text{ arrival in } (t, t + dt)]$$
$$= \frac{(\lambda t)^{r-1}}{(r - 1)!} e^{-\lambda t} \cdot (\lambda\, dt).$$

The p.d.f. of Y is the coefficient of dt:

$$f_Y(y) = \frac{\lambda^r}{(r-1)!} y^{r-1} e^{-\lambda y}, \qquad y > 0. \tag{1}$$

To find the c.d.f., we refer back to the underlying Poisson process and use the equivalence of the following events:

$$\{Y > y\} \qquad \text{and} \qquad \{\text{at most } r \text{ arrivals in time } y\}.$$

Since these have the same probability,

$$F_Y(y) = 1 - P(Y > y) = 1 - P(\text{at most } r - 1 \text{ arrivals in time } y)$$

or

$$F_Y(y) = 1 - \sum_{k=0}^{r-1} \frac{(\lambda y)^k}{k!} e^{-\lambda y} \tag{2} \blacksquare$$

The p.d.f. (1) assumed a positive integer r, but now we replace r with a positive real number α and consider the function $t^\alpha e^{-\lambda t}$ as a potential p.d.f. For this we need integrals of the form

$$\Gamma(\alpha) = \int_0^\infty x^{\alpha-1} e^{-x} dx. \tag{3}$$

Gamma Functions

This is a convergent improper integral provided $\alpha > 0$, a function of α called the **gamma function**. Its values for arbitrary $\alpha > 0$ can be calculated recursively from its values for $0 < \alpha \le 1$ by means of the identity,

$$\Gamma(\alpha + 1) = \alpha \Gamma(\alpha). \tag{4}$$

This follows upon integrating by parts:

$$\alpha \Gamma(\alpha) = \int_0^\infty e^{-x}(\alpha x^{\alpha-1} dx) = -x^\alpha e^{-x}\Big|_0^\infty + \int_0^\infty x^\alpha e^{-x} dx = \Gamma(\alpha+1). \tag{5}$$

Since $\Gamma(1) = 1$, it follows that $\Gamma(2) = 1$, $\Gamma(3) = 2\Gamma(2) = 2 \cdot 1$, $\Gamma(4) = 3\Gamma(3) = 3 \cdot 2 \cdot 1$, and so on. Thus, when α is an integer, $\Gamma(\alpha) = (\alpha - 1)!$. So $\Gamma(\alpha)$ is a generalized factorial function. Its graph is shown in Figure 6.5.

The change of variable $x = \lambda y$ and $x = z^2/2$ yields useful alternative expressions for $\Gamma(\alpha)$:

$$\Gamma(\alpha) = \lambda^\alpha \int_0^\infty y^{\alpha-1} e^{-\lambda y} dy, \tag{6}$$

$$\Gamma(\alpha) = \frac{1}{2^{\alpha-1}} \int_0^\infty z^{2\alpha-1} e^{-z^2/2}. \tag{7}$$

In view of (6), the following is a p.d.f.:

$$f(y \mid \alpha, \lambda) = \frac{\lambda^\alpha}{\Gamma(\alpha)} y^{\alpha-1} e^{-\lambda y}, \qquad y > 0, \alpha > 0. \tag{8}$$

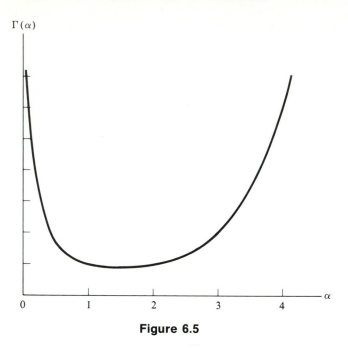

Figure 6.5

Gamma
Distributions

A random variable Y with this p.d.f. is said to have a **gamma distribution**: $Y \sim \text{Gam}(\alpha, \lambda)$. When α is not an integer, the c.d.f. of Y does not have a closed form expression [like (2), when $\alpha = r$, an integer]. The c.d.f. is called an *incomplete gamma function*:

$$F(y \mid \alpha, \lambda) = \frac{\lambda^{\alpha}}{\Gamma(\alpha)} \int_{0}^{y} u^{\alpha-1} e^{-\lambda u} \, du.$$

The m.g.f. of the gamma distribution is

$$\psi_{Y}(t) = \int_{0}^{\infty} e^{ty} \cdot \frac{\lambda^{\alpha}}{\Gamma(\alpha)} y^{\alpha-1} e^{-\lambda y} \, dy = \frac{\lambda^{\alpha}}{\Gamma(\alpha)} \int_{0}^{\infty} y^{\alpha-1} e^{-(\lambda-t)y} \, dy.$$

Making the change of variable $u = (\lambda - t)y$, we obtain

$$\psi(t) = \frac{\lambda^{\alpha}}{(\lambda - t)^{\alpha}} \cdot \frac{1}{\Gamma(\alpha)} \int_{0}^{\infty} u^{\alpha-1} e^{-u} \, du.$$

The last integral is just $\Gamma(\alpha)$ [from (3)], so

$$\psi_{Y}(t) = \left(\frac{1}{1 - t/\lambda} \right)^{\alpha} = \left(1 - \frac{t}{\lambda} \right)^{-\alpha}. \tag{9}$$

We can obtain the power series for ψ_{Y} by finding the values of its derivatives at $t = 0$. However, the same series results when we follow the scheme used in expanding a positive integer power of a binomial (compare Example 3.6c):

$$\psi_Y(t) = \left(1 - \frac{t}{\lambda}\right)^{-\alpha}$$

$$= 1^{-\alpha} + (-\alpha)1^{-\alpha-1}\left(\frac{-t}{\lambda}\right) + \frac{(-\alpha)(-\alpha-1)}{2!}1^{-\alpha-2}\left(\frac{-t}{\lambda}\right)^2 + \cdots$$

$$= 1 + \left(\frac{\alpha}{\lambda}\right)t + \left[\frac{\alpha(\alpha+1)}{\lambda^2}\right]\left(\frac{t^2}{2}\right) + \cdots.$$

So

$$E(Y) = \frac{\alpha}{\lambda}, \qquad E(Y^2) = \frac{\alpha(\alpha+1)}{\lambda^2},$$

and

$$\sigma^2 = E(Y^2) - [E(Y)]^2 = \frac{\alpha(\alpha+1)}{\lambda^2} - \left(\frac{\alpha}{\lambda}\right)^2 = \frac{\alpha}{\lambda^2}.$$

(In Problem 27, we ask you to obtain these moments more directly when α is an integer.)

If $Y \sim \text{Gam}(\alpha, \lambda)$,

$$f_Y(y \mid \alpha, \lambda) = \frac{\lambda^\alpha}{\Gamma(\alpha)}y^{\alpha-1}e^{-\lambda y}, \qquad y > 0, \tag{8}$$

$$\psi_Y(t) = \left(1 - \frac{t}{\lambda}\right)^{-\alpha}, \tag{9}$$

$$E(Y) = \frac{\alpha}{\lambda}, \qquad \sigma_Y = \frac{\sqrt{\alpha}}{\lambda}. \tag{10}$$

The normal family of distributions is related to the gamma distribution with $\alpha = \frac{1}{2}$. In particular, applying (7) with $\alpha = \frac{1}{2}$, we find

$$\Gamma\left(\frac{1}{2}\right) = \sqrt{2}\int_0^\infty e^{-z^2/2}\,dz = \frac{1}{\sqrt{2}}\int_{-\infty}^\infty e^{-z^2/2}\,dz = \frac{\sqrt{2\pi}}{\sqrt{2}} = \sqrt{\pi}. \tag{11}$$

The value $\sqrt{2\pi}$ for the integral comes from the fact that the area under the normal p.d.f. is 1 [see (1) of Section 6.1].

Having found the value of the gamma function at $\alpha = \frac{1}{2}$, we can easily find its value at any odd multiple of $\frac{1}{2}$ by means of the recursion formula (4). For instance,

$$\Gamma\left(\frac{7}{2}\right) = \frac{5}{2}\Gamma\left(\frac{5}{2}\right) = \frac{5}{2}\cdot\frac{3}{2}\cdot\frac{1}{2}\Gamma\left(\frac{1}{2}\right) = \frac{15}{8}\sqrt{\pi}.$$

These values also follow from (6) of Section 6.1, the formula for the even-order moments of the standard normal distribution (compare Problem 25).

The connection with the normal distribution is developed further in Section 6.5.

6.4   Beta Distributions

Closely related to gamma functions and gamma distributions are **beta functions** and beta distributions. For $r > 0$ and $s > 0$, the beta function $B(r, s)$ is defined as

$$B(r, s) = \frac{\Gamma(r)\Gamma(s)}{\Gamma(r + s)}. \tag{1}$$

Various integral formulas for $B(r, s)$ can be given:

$$B(r, s) = \int_0^1 x^{r-1}(1 - x)^{s-1}\, dx, \tag{2}$$

$$B(r, s) = \int_0^\infty \frac{w^{r-1}\, dw}{(1 + w)^{r+s}}, \tag{3}$$

$$B(r, s) = \int_0^{\pi/2} 2\cos^{2r-1}\theta \sin^{2s-1}\theta \, d\theta. \tag{4}$$

Formula (3) follows from (2) upon substituting $x = w/(1 + w)$, and (4) follows from (2) upon substituting $x = \cos^2\theta$. The derivation of (1) as the value of the integral (2) is outlined in Problem 26.

When $r = s = \frac{1}{2}$, the integral in (4) is trivially equal to π. So from (1),

$$\pi = B\left(\frac{1}{2}, \frac{1}{2}\right) = \frac{\Gamma(\frac{1}{2})\Gamma(\frac{1}{2})}{\Gamma(1)} = \left[\Gamma\left(\frac{1}{2}\right)\right]^2. \tag{5}$$

This shows that $\Gamma(\frac{1}{2}) = \sqrt{\pi}$. We obtained this result in (11) of Section 6.3, assuming that (1) of Section 6.1 is a p.d.f.; thus (5) validates that assumption.

Each of the integrands in (2)–(4) is a positive function on an interval of the real line and so serves to define a p.d.f. Thus, (2) defines a p.d.f. on the unit interval:

$$f(x \mid r, s) = \frac{1}{B(r, s)} x^{r-1}(1 - x)^{s-1}, \qquad 0 < x < 1, \tag{6}$$

called a **beta density**. When X has this density, we write $X \sim \text{Beta}(r, s)$.

The beta family of densities is fairly rich, including a wide variety of shapes. Some of these are shown in Figure 6.6. The mean value of $\text{Beta}(r, s)$ involves an integral that is again a beta function as given by (2):

$$E(X) = \frac{1}{B(r, s)} \int_0^1 x \cdot x^{r-1}(1 - x)^{s-1}\, dx = \frac{B(r + 1, s)}{B(r, s)}$$

$$= \frac{\Gamma(r + 1)\Gamma(s)}{\Gamma(r + s + 1)\Gamma(r)} \cdot \frac{\Gamma(r + s)}{\Gamma(s)} = \frac{r}{r + s}. \tag{7}$$

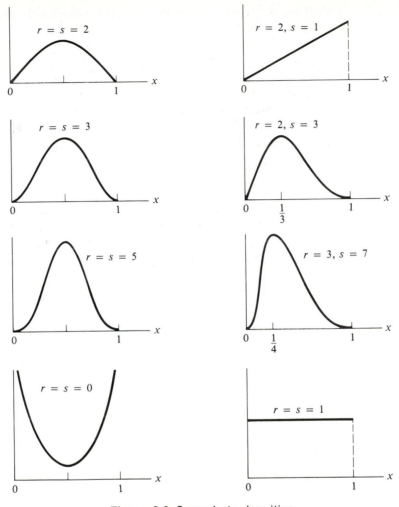

Figure 6.6 Some beta densities

Higher-order moments can be found similarly (Problem 29).

If $X \sim \text{Beta}(r, s)$,

$$f(x \mid r, s) = \frac{1}{B(r, s)} x^{r-1}(1 - x)^{s-1}, \qquad 0 < x < 1, \tag{6}$$

$$E(X) = \frac{r}{r + s}, \qquad \text{var } X = \frac{rs}{(r + s)^2(r + s + 1)}. \tag{8}$$

Distributions closely related to the beta distribution are implied by (3) and (4); the p.d.f.'s are as follows:

$$f(w \mid r, s) = \frac{1}{B(r, s)} w^{r-1} (1 + w)^{-(r+s)}, \qquad w > 0, \tag{8}$$

and

$$f(\theta \mid r, s) = \frac{2}{B(r, s)} \cos^{2r-1} \theta \, \sin^{2s-1} \theta, \qquad 0 < \theta < \frac{\pi}{2}. \tag{9}$$

The moments of a distribution defined by (8) are easily evaluated using (3). (See Solved Problem P.)

6.5   Chi-Square Distributions

The **chi-square** family of distributions is basic to the study of sample variances and, more generally, to the analysis of variance (see Chapter 14).

The simplest chi-square distribution is that of Z^2, where Z is standard normal. According to (1) of Section 5.10, the m.g.f. of Z^2 is

$$\psi_{Z^2}(t) = E(e^{tZ^2}) = \frac{1}{\sqrt{2\pi}} \int_{-\infty}^{\infty} e^{tz^2} e^{-z^2/2} \, dz = \frac{1}{\sqrt{2\pi}} \int_{-\infty}^{\infty} e^{-(1-2t)z^2/2} \, dz.$$

Making the change of variable $u = (1 - 2t)^{1/2} z$, we obtain

$$\psi_{Z^2}(t) = \frac{1}{\sqrt{1 - 2t}} \int_{-\infty}^{\infty} \frac{1}{\sqrt{2\pi}} e^{-u^2/2} \, du = \frac{1}{\sqrt{1 - 2t}}. \tag{1}$$

Comparing (1) with (9) of Section 6.3, we see thất $Z^2 \sim \text{Gam}(\frac{1}{2}, \frac{1}{2})$.

The **chi-square distribution with k degrees of freedom** is defined as the distribution of the sum of the squares of k independent, standard normal variables: $Z_1, \ldots, Z_k$. Let

$$\chi^2 = Z_1^2 + Z_2^2 + \cdots + Z_k^2. \tag{2}$$

The m.g.f. of this sum is the kth power of the m.g.f. (1), which is common to the summands:

$$\psi_{\chi^2}(t) = [(1 - 2t)^{-1/2}]^k = (1 - 2t)^{-k/2}. \tag{3}$$

To find the density of χ^2, we look for a p.d.f. with this m.g.f. Since (3) is a special case of (9) in Section 6.3, with $\alpha = \frac{k}{2}$ and $\lambda = \frac{1}{2}$, it follows that $\chi^2 \sim \text{Gam}(\frac{k}{2}, \frac{1}{2})$. Its p.d.f. is

$$f_{\chi^2}(x) = \frac{x^{k/2-1} e^{-x/2}}{2^{k/2} \Gamma(\frac{k}{2})}, \qquad x > 0. \tag{4}$$

Graphs of chi-square p.d.f.'s for several values of k are shown in Figure 6.7. When Y has a chi-square distribution with k degrees of freedom, we'll write $Y \sim \chi^2(k)$.

function $h(x)$ is the rate of failure relative to the number surviving at time x:

$$h(x)\,dx = P(x < L < x + dx \,|\, L > x)$$
$$= \frac{P(x < L < x + dx)}{P(L > x)} = \frac{f(x)\,dx}{R(x)}.$$

Example 6.7a   **Constant Hazard**

The reliability function for the exponential distribution is $e^{-\lambda x}$ for $x > 0$. So the hazard is *constant*:

$$h(x) = \frac{f(x)}{R(x)} = \frac{\lambda e^{-\lambda x}}{e^{-\lambda x}} = \lambda.$$

The hazard determines R, by (4), and R defines the distribution. So the exponential lifetime distribution is the *only* one with constant hazard for all $x > 0$. Having constant hazard is equivalent to having lack of memory (see Section 6.2). This is intuitively clear: Lack of memory in a distribution is the same as not aging—not wearing out with time. ■

A hazard function that increases with time may be appropriate in cases where failure is the result of wear, as in the case of the life of a manufactured article. Suppose the hazard function is a power of x:

$$h(x) = \frac{\alpha x^{\alpha - 1}}{\beta^{\alpha}}, \qquad x > 0, \tag{5}$$

where α and β are positive parameters. Using (3), we have

$$-\ln R(x) = \int_0^x \frac{\alpha u^{\alpha - 1}}{\beta^{\alpha}}\,du = \left(\frac{x}{\beta}\right)^{\alpha}.$$

So

$$R(x) = e^{-(x/\beta)^{\alpha}},$$

and

$$F(x) = 1 - e^{-(x/\beta)^{\alpha}}.$$

This c.d.f. defines the **Weibull distribution**; its derivative is the Weibull p.d.f.:

$$f(x \,|\, \alpha, \beta) = \frac{\alpha x^{\alpha - 1}}{\beta^{\alpha}} e^{-(x/\beta)^{\alpha}}, \qquad x > 0.$$

(When $\alpha = 1$, this is an exponential p.d.f., so the exponential distribution is a particular Weibull distribution.) The Weibull distribution is often used in modeling system reliability.

Another setting in which hazard increases with time is that of lifetimes in human populations.

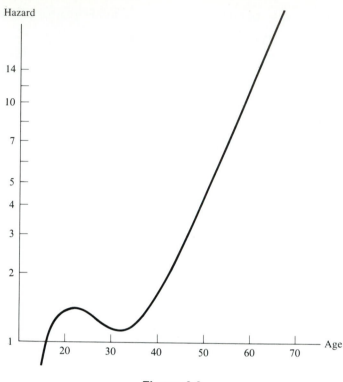

Figure 6.9

Example 6.7b Let L denote a person's age at death. Insurance companies want to know the chance that a 40-year-old man, for example, will die in the next year, *given that he has reached the age of 40*. This is the hazard at $t = 40$. (Actuaries call hazard the *force of mortality*.) It is estimated as the ratio of the number of deaths among 40-year-old males to the number of 40-year-old males "at risk." Figure 6.9 shows the approximate hazard function for males of ages 15–70.[3] The bump in the hazard function in the late teens corresponds to the time that people start to drive automobiles!

Past age 40, the hazard curve is nearly linear, when plotted (as in Figure 6.9) on a logarithmic scale. This suggests that the hazard is exponential:

$$h(x) = ae^{bx}, \qquad x > 0,$$

for some constants a and b. Integrating, we find

$$R(x) = \exp\left[-\frac{a}{b}(e^{bx} - 1) \right].$$

---

[3] The graph is plotted from tables in the 1982 *Transactions* of the Society of Actuaries, which give the figures for ages 15–100.

So

$$f(x) = h(x)R(x) = a \exp\left(bx - \frac{a}{b}e^{bx} + \frac{a}{b}\right), \qquad x > 0.$$

This is the p.d.f. of the *Gompertz distribution*, which is important in actuarial science.

∎

Reliability of Combinations

The reliability of a system consisting of several units is derivable from reliabilities of the individual units when we can assume that their times to failure are independent. Particularly easy to analyze are series and parallel combinations; components interconnected so that the system fails when any unit fails are said to be *in series*. The time to failure of a series system is the *minimum* of the times to failure of its components. If those times are *independent*, the system reliability is

$$P(\text{system life} > x) = P(\text{all units survive to } x)$$
$$= \prod P(\text{unit } i \text{ survives to } x).$$

That is,

$$R(x) = \prod R_i(x), \tag{6}$$

where R_i is the reliability function for unit i.

Components interconnected so that the system fails only when all component units fail are said to be *in parallel*. The time to failure of a parallel system is the *maximum* of the times to failure of its components. If those times are independent, the reliability function for a parallel system is

$$P(\text{system life} > t) = 1 - P(\text{all units fail before } t)$$
$$= 1 - \prod P(\text{unit } i \text{ fails before } t).$$

That is,

$$R(t) = 1 - \prod [1 - R_i(t)]. \tag{7}$$

Figure 6.10 shows schematic diagrams of series and parallel combinations of

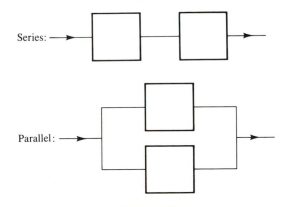

Series:

Parallel:

Figure 6.10

two component units. The next example considers a system that is neither series nor parallel.

Example **6.7c** Suppose three units are connected according to the schematic diagram of Figure 6.11. Assume times to failure of the three units are independent. Components 2 and 3 are connected in series, so the reliability function of the lower branch is $R_2 R_3$. The reliability function of the parallel combination of the upper and lower branches is

$$R = 1 - (1 - R_1)(1 - R_2 R_3).$$

Suppose the hazards are constant: 2 for unit 1, 4 for unit 2, and 5 for unit 3. Then

$$R(t) = 1 - (1 - e^{-2t})(1 - e^{-4t}e^{-5t}) = e^{-2t} + e^{-9t} - e^{-11t}.$$

From (2) the mean time to failure is the integral of $R(t)$:

$$\frac{1}{2} + \frac{1}{9} - \frac{1}{11} \doteq .52.$$

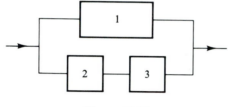

Figure 6.11　　　　　　　　　　　　■

6.8　　**Bivariate Normal Distributions**

In predicting a random variable Y from a related random variable X (Chapter 15), we'll assume that these variables have a bivariate normal distribution.

> The random pair (X, Y) is said to have a **bivariate normal distribution** if and only if every linear combination $aX + bY$ has a univariate normal distribution.

This definition has some immediate and important consequences. The linear combination in which $a = 1$ and $b = 0$, which is X itself, is univariate normal. Similarly, Y is univariate normal.

> The marginal distributions of a bivariate normal distribution are uni-variate normal.

Next, consider the linear transformation

$$\begin{cases} U = aX + bY \\ V = cX + dY. \end{cases}$$

A linear combination of U and V is also a linear combination of X and Y:

$$\alpha U + \beta V = \alpha(aX + bY) + \beta(cX + dY)$$
$$= (\alpha a + \beta c)X + (\alpha b + \beta d)\, Y.$$

So (U, V) is bivariate normal.

> If (U, V) is obtained from a bivariate normal pair (X, Y) by a linear transformation, then (U, V) is bivariate normal.

Bivariate m.g.f.'s

A convenient tool for studying bivariate normal distributions is the bivariate moment generating function. The m.g.f. of the random pair (X, Y) is

$$\psi(s, t) = E(e^{sX + tY}). \tag{1}$$

This function of two variables yields the univariate marginal m.g.f.'s upon setting one or the other argument equal to 0:

$$\psi_X(s) = E(e^{sX}) = \psi(s, 0), \qquad \psi_Y(t) = E(e^{tY}) = \psi(0, t).$$

When X and Y are *independent*, their joint m.g.f. factors into the product of the marginal m.g.f.'s:

$$\psi(s, t) = E(e^{sX + tY}) = E(e^{sX})E(e^{tY}) = \psi_X(s)\psi_Y(t).$$

Moreover, if $\psi(s, t)$ factors into the product of a function of s alone and a function of t alone, then these functions (with constant factors suitably allocated) are the m.g.f.'s of the marginal variables. By the uniqueness property for the m.g.f., which extends to bivariate m.g.f.'s, the marginal variables are independent. (See Solved Problem BB.)

The m.g.f. of bivariate normal pair (X, Y) is easy to find. According to our definition of bivariate normal, the variable $U = sX + tY$ is univariate normal. Hence,

$$\psi_{X,Y}(s, t) = E(e^{sX + tY}) = E(e^{U}) = \psi_U(1) = \exp\left(\mu_U + \frac{\sigma_U^2}{2}\right). \tag{2}$$

The mean and variance of U are

$$\mu_U = E(sX + tY) = s\mu_X + t\mu_Y,$$

and

$$\sigma_U^2 = s^2\sigma_X^2 + t^2\sigma_Y^2 + 2st\sigma_{X,Y}.$$

Substituting these in (2), we obtain

$$\psi_{X,Y}(s, t) = \exp\left(s\mu_X + t\mu_Y + \frac{1}{2}[s^2\sigma_X^2 + t^2\sigma_Y^2 + 2st\sigma_{X,Y}]\right). \tag{3}$$

From (3), it is clear that if X and Y are *uncorrelated*, the m.g.f. factors into the product of a function of s and a function of t. So we have the following important result:

> When X and Y are bivariate normal, they are independent if and only if they are uncorrelated.

Example 6.8a   Suppose (X, Y) has a bivariate normal distribution with $\mu_X = \mu_Y = 0$, $\sigma_X^2 = \sigma_Y^2 = 2$, and $\sigma_{X,Y} = -1$. Define $U = X + Y$ and $V = X - Y$. This pair is bivariate normal, with

$$\text{cov}(U, V) = \text{cov}(X + Y, X - Y) = \text{cov}(X, X) - \text{cov}(Y, Y) = 2 - 2 = 0.$$

Since U and V are uncorrelated, they are independent. ∎

Whenever $\sigma_X = \sigma_Y$, the covariance of $X + Y$ and $X - Y$ will vanish, as in the preceding example, because $\text{cov}(X, Y) = \text{cov}(Y, X)$. This has an important implication. Consider a random sample (X_1, X_2) from a normal population. Since the observations are identically distributed, their standard deviations are equal. This means that the variables $X_1 + X_2$ and $X_1 - X_2$ are independent. (This independence is a special case of a general, useful result: The mean and variance of a random sample from a normal population are independent. See Section 8.10.)

One can always find a linear transformation of a bivariate normal pair (X, Y) such that the new variables are uncorrelated, and hence independent (as in Example 6.8a). Although it is not the only way, a rotation transformation

$$\begin{cases} U = (\cos \theta)X - (\sin \theta)Y \\ V = (\sin \theta)X + (\cos \theta)Y \end{cases} \tag{4}$$

produces uncorrelated variables if θ satisfies the condition

$$\cot 2\theta = \frac{\text{var } Y - \text{var } X}{2 \text{ cov}(X, Y)}.$$

In particular, if var $Y = $ var X, a rotation through 45 degrees will produce independent variables. This is the same transformation (except for an irrelevant constant) as in Example 6.8a.

If we can always find a linear transformation that transforms dependent normals into independent normals, then we can find the form of the most general bivariate normal p.d.f. by starting with independent normal variables and considering a general linear transformation. The joint density of a pair of independent standard normal variables (U, V) is the product of univariate standard normal densities:

$$f_{U,V}(u, v) = \frac{1}{2\pi} \exp\left[-\frac{u^2}{2} - \frac{v^2}{2}\right].$$

Suppose we now define (X, Y) by linear transformation:

$$\begin{cases} X = aU + bV + h \\ Y = cU + dV + k. \end{cases} \tag{5}$$

Expressions for the means, variances, and covariance of X and Y are called for in Problem 52 in a special case. The joint p.d.f. of (X, Y), except for a constant factor, is obtained by substituting expressions for u and v in terms of x and y, defined by (5), into $f_{U,V}(u, v)$. Since there are linear, it is clear that the new density is an exponential with a quadratic expression in u and v in the exponent. The result (after quite a bit of algebra) is as follows:

$$f(x, y) = \frac{\exp(-\frac{1}{2}Q)}{2\pi\sigma_X\sigma_Y\sqrt{1 - p^2}}, \tag{6}$$

where

$$Q = \frac{1}{1 - \rho^2}\left\{\left(\frac{x - \mu_X}{\sigma_X}\right)^2 - 2\rho\left(\frac{x - \mu_X}{\sigma_X}\right)\left(\frac{y - \mu_Y}{\sigma_Y}\right) + \left(\frac{y - \mu_Y}{\sigma_Y}\right)^2\right\}. \tag{7}$$

The graph of this density is a bell-shaped surface with elliptical level curves. Figure 6.12 shows a bivariate normal density surface. (If $\rho = 0$, the cross-product

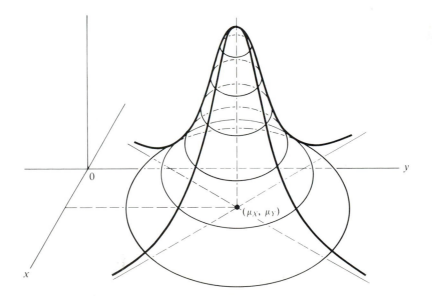

Figure 6.12

term in (7) vanishes. In this case, the p.d.f. *factors* into the product of the marginal p.d.f.'s. This means that the variables X and Y are *independent*, as we saw earlier using m.g.f.'s.)

Conditional Densities

Because the exponent in the joint density of a bivariate normal distribution is quadratic in each variable, the *conditional* density of X given the value of Y, for example, is an exponential function with a quadratic function of x in the exponent. That is, it's a *normal* p.d.f. Completing the square shows that the conditional mean is linear in y and the conditional variance is constant. This process is illustrated in the next example.

Example 6.8b Suppose the joint density of (X, Y) is an exponential function with a quadratic exponent:

$$f(x, y) \propto \exp[-x^2 + xy - 2y^2 + 5x - 6y].$$

Given that $Y = y_0$, the conditional p.d.f. of X (Section 5.9) is proportional to $f(x, y_0)$:

$$f(x \mid y_0) \propto \exp[-\{x^2 - (y_0 + 5)x\}].$$

(The constant of proportionality may depend on y_0.) As a function of x, this is a *normal* p.d.f. To find the mean and variance of X given $Y = y_0$, we complete the square in x in the exponent:

$$x^2 - (y_0 + 5)x = \left[x - \frac{y_0 + 5}{2} \right]^2 + \text{term in } y_0$$

and rewrite the exponent as follows:

$$-[x^2 - (y_0 + 5)x] = -\frac{1}{2(\frac{1}{2})}\left(x - \frac{(y_0 + 5)}{2} \right)^2 + \text{term in } y_0.$$

We see then that the conditional mean is $\frac{(y_0 + 5)}{2}$. In the terminology of Section 5.9, $(y + 5)/2$ is the regression function of X on y. The conditional variance is $\frac{1}{2}$. ∎

The technique of this last example yields the conditional distribution for a general bivariate normal distribution:

If X and Y are bivariate normal, the conditional distribution of Y given $X = x$ is normal. The regression function of Y on x is linear in x:

$$E(Y \mid x) = \mu_Y + \rho\frac{\sigma_Y}{\sigma_X}(x - \mu_X). \tag{8}$$

Its variance is constant:

$$\text{var}(Y \mid x) = \sigma_Y^2(1 - \rho^2). \tag{9}$$

Chapter Perspective

In Chapters 2 through 6, we have studied population models—both discrete and continuous—in some detail. We've given various ways of characterizing and describing distributions and discussed important families of distributions. In Chapter 7, we'll define a sample distribution as a special case of a discrete distribution and give various ways of summarizing the information in a sample. We then proceed to show how to utilize the information in a random sample for the purpose of drawing inferences about the population from which it was drawn.

Solved Problems

Section 6.1

A. Let X denote the net weight in ounces of a "half-pound" bag of potato chips. Given that X is normal with mean 8 ounces and s.d. $\frac{1}{4}$ ounce, find

    (a) $P(X > 7.5)$.

    (b) $E(X^2)$.

    (c) the value of c such that 90% of the bags weigh between $8 - c$ and $8 + c$ ounces.

Solution:

(a) The Z-score for $X = 7.5$ is

$$Z = \frac{X - \mu}{\sigma} = \frac{7.5 - 8}{.25} = -2,$$

so

$$P(X > 7.5) = 1 - P(X < 7.5) = 1 - P(Z < -2) = \Phi(2) = .9772.$$

(b) $E(X^2) = \sigma^2 + \mu^2 = \dfrac{1}{16} + 64 = 64.0625.$

(c) The given interval is symmetric about 8. Since the p.d.f. of X is symmetric about $x = 8$, there must be 5% of the probability to the left of $8 - c$ (and 5% to the right of $8 + c$). The fifth percentile of Z is -1.645, so for x it is

$$8 - c = 8 - 1.645 \times \frac{1}{4} = 8 - .411,$$

and $c = .411$.

B. Given that X has the p.d.f. $K \cdot \exp[-4x^2 + 6x]$, find $E(X)$, var X, and the value of K.

Solution:

The exponent is quadratic in x, so the distribution is normal. To find the mean and variance, we complete the square in the exponent:

$$-4x^2 + 6x = -4\left[x^2 - \left(\frac{3}{2}\right)x + \left(\frac{3}{4}\right)^2\right] + \frac{9}{4} = -4\left(x - \frac{3}{2}\right)^2 + \frac{9}{4}.$$

Comparing this with the exponent of a general normal p.d.f. [(10) in Section 6.1], we see that

$$\mu = \frac{3}{2}, \quad \text{and} \quad \frac{-1}{2\sigma^2} = -4, \quad \text{or} \quad \sigma^2 = \frac{1}{8}.$$

The constant $K \cdot e^{9/4}$ must be $1/\sqrt{2\pi}\sigma$, so

$$K = \frac{2e^{-9/4}}{\sqrt{\pi}}.$$

C. Given that $X \sim N(\mu, \sigma^2)$, find $E(X^4)$.

Solution:

Since in (11) of Section 6.1 we have a formula for $E[(X - \mu)^n]$, we proceed by adding and subtracting μ from X and expanding the fourth power by the binomial theorem:

$$X^4 = (X - \mu + \mu)^4 = [(X - \mu) + \mu]^4$$
$$= (X - \mu)^4 + 4\mu(X - \mu)^3 + 6\mu^2(X - \mu)^2 + 4\mu^3(X - \mu) + \mu^4.$$

We now obtain $E(X^4)$ by adding the expected values of the five terms on the right. The first has mean $3\sigma^4$, the second and fourth have mean 0, the middle term has mean $6\mu^2\sigma^2$, and of course the mean of the last term is μ^4. So $E(X^4) = 3\sigma^4 + 6\mu^2\sigma^2 + \mu^4$.

D. Given that the 10th percentile of a normal variable X is 15 and the 60th percentile is 25, find $\mu = E(X)$ and $\sigma = \sqrt{\text{var } X}$.

Solution:

The corresponding Z-percentiles are -1.2816 and $.2533$ (from Table IIb in Appendix 1). Using $X = \mu + Z\sigma$, we have

$$\begin{cases} 15 = \mu - 1.2816\sigma, \\ 25 = \mu + .2533\sigma. \end{cases}$$

To solve simultaneously, we subtract to obtain $10 = 1.5349\sigma$, or $\sigma = 6.515$. Substituting this in either of the above equations, we find $\mu = 23.25$.

E. Suppose $X \sim N(10, 3)$ and $Y \sim N(6, 9)$. Given that they are independent, find the distribution of the following:
(a) $X + Y$.   (b) $X - Y$.   (c) $2X + 5Y$.

Solution:

In each case, the given combination is normal. The means and variances are
(a) $\mu = 10 + 6 = 16$, $\sigma^2 = 3 + 9 = 12$.
(b) $\mu = 10 - 6 = 4$, $\sigma^2 = 3 + 9 = 12$.
(c) $\mu = 2 \cdot 10 + 3 \cdot 5 = 35$, $\sigma^2 = 4 \cdot 3 + 25 \cdot 9 = 237$.

F. Suppose the lengths of 1800-foot audio tapes of a certain brand are normally distributed, with mean 1800 ft and s.d. 5 ft. Find the probability that the lengths of two tapes selected independently and at random differ by more than 10 ft.

Solution:

The difference D between the two lengths is normal, with mean $0 (= 1800 - 1800)$ and variance $50 \text{ ft}^2 (= 5^2 + 5^2)$. So

$$P(|D| > 10) = 1 - P(-10 < D < 10)$$

$$= 1 - \left[\Phi\left(\frac{10}{\sqrt{50}}\right) - \Phi\left(\frac{-10}{\sqrt{50}}\right) \right] = 2(.0787) = .1574.$$

G. Imagine throwing a dart at a target on a wall. Assume that the vertical and horizontal components of the point hit by the dart are independent, each with a normal distribution centered at the target center. Given that the standard deviations of the two components are equal, find the p.d.f. of the distance R from the landing point to the center of the target.

Solution:

Let σ denote the common s.d. of X and Y, the horizontal and vertical components, respectively. The joint p.d.f. of these components is the product of their univariate p.d.f.'s:

$$f(x, y) = \frac{1}{2\pi\sigma^2} e^{-(x^2 + y^2)/2}.$$

Then $P(R \leq r)$ is the volume under this surface of revolution within a circle of radius r:

$$P(R \leq r) = \frac{1}{2\pi\sigma^2} \int_0^r e^{-x^2/(2\sigma^2)}(2\pi x)\, dx.$$

Differentiating to find the p.d.f., we evaluate the integrand at the upper limit:

$$f_R(r) = \frac{r}{\sigma^2} e^{-r^2/(2\sigma^2)}, \qquad \text{for } r \geq 0.$$

[Physicists call this a *Rayleigh density*. It gives the distribution of the distance that a molecule confined to a plane travels in a fixed time when it is bombarded by other molecules in random fashion but with no particular drift.]

Sections 6.2–6.4

H. Phone calls arrive at an exchange at the rate of eight per minute, according to a Poisson law. Let X denote the time from any particular time t_0 to the next call. Find
 (a) the probability that X exceeds 10 seconds.
 (b) the mean and standard deviation of the time between calls.
 (c) the mean time between the most recent call and the next call.

Solution:

The time T from any point to the next incoming call is exponential, with mean $\frac{1}{8}$ minute $= 7.5$ seconds.
 (a) The probability that this time exceeds $\frac{1}{6}$ minute is

$$1 - F_T\left(\frac{1}{6}\right) = e^{-8/6} = .2636.$$

(b)   The standard deviation T is equal to the mean, or $\frac{1}{8}$ minute.

(c)   If you're like most people, you'd say 7.5 seconds. That's wrong! A Poisson process looks the same whether we look ahead or look back in time. So the mean time since the last call is 7.5 seconds—the same as the mean time to the next call. The correct answer is the sum of these two times: 15 seconds. [It sounds as though we're giving two answers to finding the mean time from one call to the next, depending on our point of view. This is true; at the time of one call, the mean time to the next is 7.5 seconds. For the present question, we enter the picture at a randomly selected reference point on the time line. However, spaces are selected in proportion to their sizes, and we are more likely to end up in a large space than in a small one. The distribution of the width of the space chosen is given by (1) of Section 6.3 ($r = 2$), and the mean is twice $\frac{1}{\lambda}$.]

I.   For the arrival process in the preceding problem, let U denote the time to the tenth call after t_0. Find

(a) $E(U)$.

(b) $P(U > 1$ minute$)$.

Solution:

(a)   The mean time to the tenth call is ten times the mean to the first call: $\frac{10}{8}$ minutes $= 75$ seconds.

(b) $P(U > 1) = 1 - F_U(1)$, where F_U is the c.d.f. given by (2) of Section 6.3. To find this, we enter the Poisson table with $m = 8$ and $c = 10 - 1 = 9$ to obtain .717.

J.   Three different types of customer arrive at a service facility, each according to a Poisson process. Type A customers arrive with mean 10 per hour, type B with mean 15 per hour, and type C with mean 5 per hour. Assuming independence of the three processes, find the mean time to the arrival of the tenth customer (without regard to type).

Solution:

The number of arrivals in an hour is the sum of the number of type A, the number of type B, and the number of type C. With independence assumed, this sum is Poisson with mean $10 + 15 + 5$ or 30 per hour. The mean time to an arrival is then $\frac{1}{30}$ hr or 2 minutes, and to ten arrivals, 10 times 2 minutes or 20 minutes.

K.   In the setting of Problem J, where type A customers arrive at the rate of 10 per hour and type B customers arrive at the rate of 15 per hour, find the probability that a customer of type A arrives before one of type B.

Solution:

The simplest approach is to use the fact that over a long period of time, $\frac{2}{5}$ of the customers are of type A. Selecting a time at random is equivalent to picking the next customer at random, so the probability that the next one is of type A is $\frac{2}{5}$. If you should miss seeing such simple reasoning, you can always follow the long road to the same answer: Let U and V denote the times to a type A and a type B arrival, respectively. Their joint p.d.f. is the product of their individual p.d.f.'s:

$$f_{U,V}(u, v) = 10e^{-10u} \cdot 15e^{-15v}, \qquad u > 0, v > 0.$$

The probability of $U < V$ is the integral of this p.d.f. over the region where $u < v$:

$$P(U < V) = \iint\limits_{0 < u < v} 15e^{-10u-15v}\,du\,dv = 15 \int_0^\infty \left(\int_0^v 10e^{-10u}\,du \right) e^{-15v}\,dv$$

$$= 15 \int_0^\infty (1 - e^{-10v})e^{-15v}\,dv = 1 - \frac{15}{25} \int_0^\infty e^{-25v}\,d(25v) = \frac{2}{5}.$$

L. Find the value of the following integrals:

(a) $\displaystyle\int_0^\infty x^{13}e^{-3x}\,dx.$ (b) $\displaystyle\int_0^\infty x^{9/2}e^{-x}\,dx.$

Solution:

(a) We recognize this, except for a constant factor, as a gamma function [see (3) of Section 6.3]:

$$\frac{\Gamma(14)}{3^{14}} = \frac{13!}{3^{14}} \doteq 1301.9.$$

(b) Using (8) and (16) of Section 6.3, we find

$$\Gamma\left(\frac{11}{2}\right) = \frac{9}{2} \cdot \frac{7}{2} \cdot \frac{5}{2} \cdot \frac{3}{2} \cdot \frac{1}{2} \Gamma\left(\frac{1}{2}\right) = 29.53\sqrt{\pi}.$$

M. When $Z \sim N(0, 1)$, $E(Z^4) = 3$. Use this fact to find the value of $\Gamma(\frac{5}{2})$. [See (11) in Section 6.3.]

Solution:

From (7) of Section 6.3,

$$\Gamma\left(\frac{5}{2}\right) = 2^{-3/2} \int_0^\infty z^4 e^{-z^2/2}\,dz = 2^{-3/2} \int_{-\infty}^\infty z^4 e^{-z^2/2}\frac{dz}{2}$$

$$= 2^{-5/2}\sqrt{2\pi}\,E(Z^4) = \frac{3\sqrt{\pi}}{4}.$$

N. Find the value of each of the following:

(a) $B(\frac{5}{2}, \frac{7}{2}).$ (b) $\displaystyle\int_0^1 x^9(1 - x)^6\,dx.$ (c) $\displaystyle\int_0^\infty \frac{x^6\,dx}{(1 + 2x)^8}.$

Solution:

(a) We use (1) of Section 6.4, to write this beta function in terms of gamma functions, and then the recurrence relation for gamma functions:

$$B\left(\frac{5}{2}, \frac{7}{2}\right) = \frac{\Gamma(\frac{5}{2})\Gamma(\frac{7}{2})}{\Gamma(6)}$$

$$= \frac{\frac{3}{2} \cdot \frac{1}{2}\Gamma(\frac{1}{2}) \cdot \frac{5}{2} \cdot \frac{3}{2} \cdot \frac{1}{2}\Gamma(\frac{1}{2})}{5!} = \frac{45\pi}{32 \cdot 120},$$

or about .0368.

(b)   This is $B(10, 7) = 9!6!/16! = 1.25 \times 10^{-5}$.

(c)   Except for a factor 2^7 (needed with the transformation $y = 2x$), this is $B(r, s)$, where $r - 1 = 6$ and $r + s = 8$, or $r = 7$ and $s = 1$ [see (3) of Section 6.4]. Since $B(7, 1) = 7!1!/9! = \frac{1}{72}$, the value of the integral is $[72 \times 2^7]^{-1} = 1.085 \times 10^{-4}$.

Sections 6.5–6.6

O.   Find the mean of the beta-type distribution on $(0, \infty)$ given by the p.d.f. (9) in Section 6.4.

Solution:

The fact that the density integrates to 1 gives us the integration formula (3) of Section 6.4:

$$\int_0^\infty \frac{w^{r-1}\,dw}{(1+w)^{r+s}} = B(r, s).$$

The mean is an integral of the same type, with r replaced by $r + 1$:

$$\int_0^\infty w\,\frac{1}{B(r, s)}\frac{w^{r-1}\,dw}{(1+w)^{r+s}} = \frac{1}{B(r, s)}\int_0^\infty \frac{w^r\,dw}{(1+w)^{r+s}} = \frac{B(r+1, s-1)}{B(r, s)}.$$

Evaluating the beta functions in terms of gamma functions and using the recurrence relation for gamma functions, we find the mean value to be $r/(s - 1)$.

P.   Let $Y = 2\lambda X$, where $X \sim \text{Exp}(\lambda)$.
   (a)   Show that $Y \sim \chi^2(2)$ or $\text{Exp}(\frac{1}{2})$.
   (b)   Use the result in (a) and the chi-square table to find the probability that $X < 5$ when $\lambda = .4$.
   (c)   Check the probability in (b) using the exponential c.d.f.

Solution:

   (a)   The m.g.f. of X is $[(1 - t)/\lambda]^{-1}$, and the m.g.f. of Y is then

$$\psi_Y(t) = E(e^{Yt}) = E(e^{2\lambda Xt}) = \psi_X(2\lambda t)$$

$$= \left(1 - \frac{2\lambda t}{\lambda}\right)^{-1} = (1 - 2t)^{-1}.$$

This is the m.g.f. of a chi-square variable with 2 d.f. and also the m.g.f. of an exponential variable with $\lambda = \frac{1}{2}$.

   (b) $P(X < 5) = P(2\lambda X < 10\lambda) = F_Y(4) = .135$.
   (c)   When X is $\text{Exp}(.4)$, $P(X < 5) = F_X(5) = 1 - e^{-.4 \times 5} = .135$.

Q.   Find the distribution of $1/F$, where $F \sim F(r, s)$.

Solution:

The reciprocal of F has the same structure as F: the ratio of independent chi-square variables each divided by d.f. However, now the variable with s d.f. is on top. Hence, $1/F \sim F(s, r)$.

R. Show that $Y = \sqrt{2\chi^2} - \sqrt{2k-1}$ is approximately $N(0, 1)$ for large values of k, given that χ^2 is approximately normal.

Solution:

Since $E(\chi^2) = k$ and var $\chi^2 = 2k$, $\dfrac{\chi^2 - k}{\sqrt{2k}} \approx N(0, 1)$. The c.d.f. of Y is

$$P(Y \le y) = P(\sqrt{2\chi^2} \le \sqrt{2k-1} + y)$$

$$= P\left(\chi^2 \le \frac{2k-1}{2} + y\sqrt{2k-1} + \frac{y^2}{2}\right)$$

As $k \to \infty$, we can neglect terms not involving k:

$$P(Y \le y) \doteq P(\chi^2 \le k + \sqrt{2k}\, y) = P\left(\frac{\chi^2 - k}{\sqrt{2k}} \le y\right) \doteq \Phi(y).$$

S. Approximate the 90th percentile of a chi-square distribution with 25 degrees of freedom,

    (a) using the fact that $\chi^2(k) \approx N(k, 2k)$.

    (b) using the fact that $\sqrt{2\chi^2} - \sqrt{2k-1} \approx N(0, 1)$.

Solution:

    (a) The 90th percentile of Z is 1.28, so the approximate 90th percentile of the chi-square is $k + 1.28\sqrt{2k} = 25 + 1.28\sqrt{50}$ or 34.05.

    (b) The 90th percentile of χ^2 satisfies

$$\sqrt{2\chi^2} - \sqrt{49} \doteq 1.28, \quad \text{or} \quad \chi^2 \doteq \frac{1}{2}(1.28 + 7)^2 = 34.28.$$

The exact value to one decimal place, from Table Vb, is 34.4, so the approximation in (b) is better than that in (a). (This is generally the case.)

T. Given that $X \sim \chi^2(10)$, find $E(1/X)$.

Solution:

Using the integral formula for $E[g(X)]$, (5) of Section 6.4, we obtain

$$E\left(\frac{1}{X}\right) = \int_{-\infty}^{\infty} \frac{1}{x} f(x)\, dx = \int_0^\infty \frac{1}{x} \cdot \frac{1}{2^5 4!} x^4 e^{-x/2}\, dx = \frac{1}{2^5 4!} \int_0^\infty x^3 e^{-x/2}\, dx$$

$$= \frac{2^4 3!}{2^5 4!} = \frac{1}{8}.$$

[More generally, for $\chi^2(k)$, the result is $1/(k-2)$. Note that here $E(\frac{1}{X}) \ge 1/E(X)$. It can be shown that this is always the case provided only that $X \ge 0$.]

U. In Section 6.6, we saw that when $U \sim F(k, m)$, its p.d.f. is

$$f_U(u) = \frac{C \cdot u^{k/2 - 1}}{(1 + ku/m)^{(k+m)/2}}, \quad u > 0.$$

Find the constant C.

Solution:

Suppose we let $V = kU/m$. Then

$$f_V(v) = f_U\left(\frac{mv}{k}\right) \cdot \frac{m}{k} = \frac{C \cdot (mv/k)^{k/2-1} m/k}{(1+v)^{(k+m)/2}} = C \cdot \left(\frac{m}{k}\right)^{k/2} \cdot \frac{v^{k/2-1}}{(1+v)^{(k+m)/2}}.$$

From (7) of Section 6.4, we see that $C(m/k)^{k/2} = 1/B(k/2, m/2)$. So

$$C = \frac{\left(\frac{k}{m}\right)^{k/2}}{B\left(\frac{k}{2}, \frac{m}{2}\right)}.$$

V.   Find the mean of the variable V with p.d.f. (7) of Section 6.6, given in Problem U.

Solution:

With $K = 1/B(k/2, m/2)$,

$$E(V) = K \int_0^\infty u \cdot \frac{u^{k/2-1}}{(1+u)^{(k+m)/2}}\, du = K \int_0^\infty \frac{u^{(k+2)/2-1}}{(1+u)^{(k+m)/2}}\, du.$$

The integral on the right is just like the integral evaluated in the preceding problem, except that k is now $k + 2$, and m is $m - 2$. Hence,

$$E(U) = \frac{B(k/2 + 1, m/2 - 1)}{B(k/2, m/2)} = \frac{k}{m-2},$$

where we have simply replaced the beta functions by appropriate combinations of gamma functions [(1) in Section 6.4] and used the recursion relation (5) of Section 6.3.

Section 6.7

W.   A certain type of unit has an exponential time to failure, with mean two hours. Find the probability

(a)   that one unit lasts at least 4 hours.

(b)   that at least one of a pair of units operating independently lasts four hours—that is, that $Y \geq 4$, where Y is the longer of two lifetimes.

(c) $E(Y)$, with Y as in (b).

(d) $E(W)$, where W is the first failure in the pair in (b).

Solution:

(a)   Let T denote the time to failure. What's asked for is

$$R(4) = P(T > 4) = \int_4^\infty \frac{1}{2} e^{-t/2}\, dt = e^{-4/2} \doteq .135.$$

(b)   We want the reliability, at 4 hours, of a *parallel* system of two units, each with reliability function $e^{-t/2}$. Denoting the time to failure of unit i by T_i and using (7) of Section 6.5, we have

$$R_{\text{sys}}(4) = P[\max(T_1, T_2) > 4] = 1 - (1 - e^{-2})^2 \doteq .252.$$

(c)   From (b), the c.d.f. of $Y = \max(T_1, T_2)$ is

$$F_Y(t) = 1 - R_{\text{sys}}(t) = (1 - e^{-t/2})^2 = 1 - 2e^{-t/2} + e^{-t}.$$

The p.d.f. is the derivative, $e^{-t/2} - e^{-t}$, and the mean is

$$E(Y) = \int_{-\infty}^{\infty} t f_Y(t)\,dt = \int_0^{\infty} t(e^{-t/2} - e^{-t})\,dt = 4 - 1 = 3 \text{ hours.}$$

A more intuitive approach, also correct, is to pick one unit and wait until it fails. Then go to the other; by symmetry, it is alive with probability $\frac{1}{2}$. If it is alive, its mean time to failure is two hours. So

$$E(Y) = E(T_1) + \frac{1}{2}E(T_2) = 2 + \frac{1}{2} \cdot 2 = 3 \text{ hours.}$$

(d) This is the mean life of a *series* combination of the two units: $W = \min(T_1, T_2)$. The reliability function of the combination [by (6) of Section 6.7] is

$$P[W > t] = [R(t)]^2 = (e^{-t/2})^2 = e^{-t},$$

an exponential reliability with $\lambda = 1$. So $E(W) = 1/\lambda = 1$ hour. There is also an intuitive way of finding this, similar to that in (c), but since we have the answer to (c), there is an even simpler way. Clearly, $Y + W = T_1 + T_2$, so

$$E(W) = E(T_1) + E(T_2) - E(Y) = 2 + 2 - 3 = 1.$$

X. Three units are linked in a system as shown in Figure 6.13. (The system fails if either units 1 and 2 both fail, or units 2 and 3 both fail.) Find the reliability function of the system in terms of the unit reliability functions, R_i for system i.

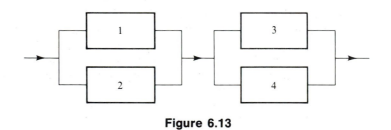

Figure 6.13

Solution:

For the system consisting of the first two blocks in parallel, the reliability function is $1 - (1 - R_1)(1 - R_2)$. A similar expression applies for the second two blocks. Putting these in series, we find the system reliability function to be

$$R = \{1 - [1 - R_1][1 - R_2]\}\{1 - [1 - R_3][1 - R_4]\}.$$

Y. The Weibull distribution with parameters $\alpha = 2$ and $\beta = 1$ [see (5) in Section 6.7] has the reliability function $R(t) = \exp(-t^2)$. Given that T has this distribution, find
(a) $P(T < 1)$.   **(b)** $E(T)$.   **(c)** var T.

Solution:

(a) This is the c.d.f. at $t = 1$: $F(1) = 1 - R(1) = 1 - 1/e = .632$.

(b)   The mean [from (2) of Section 6.7] is integral of $R(t)$:

$$\int_0^\infty e^{-t^2}\,dt = \frac{\sqrt{\pi}}{2}.$$

(Except for the factor $1/(\sigma\sqrt{\pi})$, the integral is half the area under a normal curve with mean zero and variance $\frac{1}{2}$.)

(c)   The p.d.f. is $-R'(t) = 2t[\exp(-t^2)]$, so the expected square is

$$\int_0^\infty t^2 \cdot 2te^{-t^2}\,dt = \int_0^\infty t^2 e^{-t^2}\,d(t^2) = \int_0^\infty ue^{-u}\,du = 1.$$

Subtracting the square of the mean yields var $T = 1 - \pi/4$.

Z.   Given that X has an exponential distribution with mean $1/\lambda$, define $Y = be^{\lambda X/a}$, for positive parameters a and b.
   (a)   Find the p.d.f. of Y.
   (b)   Find $E(Y)$.
   (c)   Find the median value of Y.
   [The variable Y is said to have a *Pareto* distribution; it has been found useful in describing income distributions.]

Solution:
(a)   The transformation $g(x) = be^{\lambda x/a}$ takes the interval $(0,\,\infty)$ into $(b,\,\infty)$. The inverse transformation and the p.d.f. of X are

$$x = \frac{a}{\lambda}\log\frac{y}{b} = g^{-1}(y), \quad\text{and}\quad f_X(x) = \lambda e^{-\lambda x} \quad\text{for } x > 0.$$

Thus,

$$f_Y(y) = f_X[g^{-1}(y)]\frac{dx}{dy} = \lambda \cdot \frac{a}{\lambda y}\exp\left(-\lambda \cdot \frac{a}{\lambda}\log\frac{y}{b}\right) = \frac{ab^a}{y^{a+1}}, \qquad y > b.$$

The c.d.f. is

$$F_Y(y) = \int_b^y \frac{ab^a}{u^{a+1}}\,du = 1 - \left(\frac{b}{y}\right)^a, \qquad y > b.$$

(b)   If $a > 1$, the mean value of Y is

$$E(Y) = b^a a \int_b^\infty y \cdot y^{-a-1}\,dy = \frac{ab}{a-1}.$$

(If $a \le 1$, the mean does not exist.)
(c)   To find the median of Y, we set $F(y) = \frac{1}{2}$: $y = b2^{1/a}$.

Section 6.8

AA.   Consider the joint p.d.f. $f(x, y) \propto \exp[-10x^2 + 8xy - 2y^2]$.
   (a)   Find the marginal distributions.
   (b)   Find the conditional distributions.
   (c)   Find the correlation coefficient and the constant of proportionality in the p.d.f.

Solution:

a) To integrate out y, we first complete the square in y:

$$-2[y^2 - 4xy] - 10x^2 = -2(y - 2x)^2 - 10x^2 + 8x^2.$$

Performing the integration produces a constant times e^{-2x^2}. This means that the density of X is normal, with mean 0 and

$$\frac{1}{2\sigma_X^2} = 2, \quad \text{or} \quad \sigma_X = \frac{1}{2}.$$

Similarly, integrating out x with the exponent written as

$$-10\left(x^2 - \frac{8}{10}xy + \frac{16}{100}y^2\right) - 2y^2 + \frac{16}{10}y^2$$

leaves a constant times $e^{-4y^2/10}$. So Y is normal with mean 0 and $2\sigma_Y^2 = \frac{10}{4}$, or $\sigma_Y = \sqrt{5}/2$.

(b) The first completion of the square in (a) shows that when x is held fixed, the p.d.f. varies in y as a normal p.d.f. with mean $2x$ and variance $\frac{1}{4}$. The second completion shows that when y is held fixed, the p.d.f. varies in x as a normal p.d.f. with mean $4y/10$ and variance $\frac{1}{20}$. So the conditional distributions are normal, with these respective parameters. [The conditional means $2x$ and $2y/5$ are the regression functions of Y on X and of X on Y, respectively.]

(c) We can find the correlation coefficient by comparing the coefficients of x^2 in the general normal p.d.f. [(7) of Section 6.8] and the given p.d.f.:

$$10 = \frac{1}{2(1 - \rho^2)\sigma_X^2} \quad \text{or} \quad \rho^2 = \frac{4}{5}.$$

Since the coefficient of xy is positive, $\rho = 2/\sqrt{5}$. The proportionality constant is

$$\frac{1}{2\pi\sigma_X\sigma_Y\sqrt{1 - \rho^2}} = \frac{2}{\pi}.$$

BB. Explain in detail why (as claimed in Section 6.8) factorization of a bivariate m.g.f. $\psi(s, t)$ into the product of a function of s and a function of t implies independence of the marginal variables.

Solution:

Suppose $\psi(s, t) = g(s)h(t)$. The marginal m.g.f.'s are then

$$\psi_X(s) = \psi(s, 0) = g(s)h(0),$$
$$\psi_Y(t) = \psi(0, t) = g(0)h(t).$$

Since $\psi(0, 0) = g(0)h(0) = 1$,

$$\psi_X(s)\psi_Y(t) = g(s)h(0)g(0)h(t) = g(s)h(t) = \psi(s, t).$$

This is precisely how we obtain the m.g.f. of X and Y when they are independent. So by the uniqueness of m.g.f.'s, $\psi(s, t)$ is the m.g.f. of independent variables.

CC. Given that X and Y are bivariate normal with

$$\mu_X = 1, \qquad \mu_Y = 0, \qquad \sigma_X = 2, \qquad \sigma_Y = 5, \qquad \rho = .8,$$

(a) give the m.g.f. and p.d.f. of X and Y.
(b) find the joint distribution of $U = X - 2Y$ and $V = 2X + Y + 2$.

Solution:
(a) The covariance is $\sigma_X \sigma_Y \cdot \rho = 8$. Substitute this and the given parameters in (3) and (7):

$$\psi(s, t) = \exp\left\{ s + 2s^2 + 8st + \frac{25t^2}{2} \right\}$$

and

$$f(x, y) = \frac{1}{12\pi} \exp\left\{ \frac{-25(x - 1)^2}{72} + \frac{2y(x - 1)}{9} - \frac{y^2}{18} \right\}.$$

(b) The joint distribution of (U, V) is bivariate normal, in view of the box on page 295. Since averaging is linear,

$$\mu_U = 1 - 0 = 1, \qquad \mu_V = 2 + 0 - 2 = 0.$$

Applying (7) and (8) of Section 5.7 produces the second moments:

$$\operatorname{var} U = \operatorname{var} X + 4 \operatorname{var} Y - 4 \operatorname{cov}(X, Y) = 4 + 100 - 104 = 72,$$
$$\operatorname{var} V = 4 \operatorname{var} X + \operatorname{var} Y + 4 \operatorname{cov}(X, Y) = 16 + 25 + 32 = 73,$$
$$\operatorname{cov}(U, V) = 2 \operatorname{var} X - 3 \operatorname{cov}(X, Y) - 2 \operatorname{var} Y = -68.$$

DD. Given that X and Y are independent, standard normal variables, find the m.g.f. of their product XY.

Solution:
Here's a neat trick! The product can be expressed as

$$XY = \frac{1}{4}[(X + Y)^2 - (X - Y)^2].$$

Define $U = X + Y, V = X - Y, Z = U^2/2$, and $W = V^2/2$. Then

$$XY = \frac{U^2 - V^2}{4} = \frac{Z - W}{2}.$$

We know that U and V are independent (see Example 6.8a and the discussion following it), and each is normal with mean 0 and variance 2. So Z and W are independent chi-square variables, each with d.f. $= 1$. Hence,

$$\psi_{XY}(t) = E(e^{tXY}) = E(e^{t(Z - W)/2}) = E(e^{tZ/2})E(e^{-tW/2})$$

$$= \psi_Z\left(\frac{t}{2}\right)\psi_W\left(\frac{t}{2}\right) = (1 - t)^{-1/2}(1 + t)^{-1/2} = (1 - t^2)^{-1/2}.$$

Problems

For problems marked with an asterisk, answers are provided in Appendix 2.

Section 6.1

*1. Find the standard scores (Z-scores) for
(a) $X = 14.2$, where $\mu_X = 12.6$ and $\sigma_X = 0.4$.
(b) $Y = 135$, where $\mu_Y = 150$ and $\sigma_Y^2 = 100$.

*2. The height of female students at the University of Baroda follows approximately a normal distribution, with mean 60 inches and s.d. 2 inches. Find the probability that a female student selected at random is
(a) shorter than 58 inches.
(b) between 58 and 62 inches tall.

3. A candy bar wrapper says "Net weight, 1.4 oz." The bars actually vary in weight. To be reasonably confident that most bars weigh at least 1.4 oz, the manufacturer adjusts the production so that the mean weight is 1.5 oz. Assume the distribution of weights to be approximately normal with standard deviation .05 oz and find the proportion of bars that weigh less than 1.4 oz.

*4. For ACT scores for a large group of entering freshmen in one particular year, the mean is 22.2 and s.d. 4.6. Quartiles were reported to be $Q_1 = 19$, $Q_2 = 23$, $Q_3 = 26$. Suppose such scores are normally distributed with mean 22.2 and s.d. 4.6.
(a) What proportion of scores are less than 18?
(b) Are the reported quartiles consistent with the assumption that the distribution of ACT scores is normal? We are not yet prepared for this inferential question, but as a crude check, find the quartiles of a normal distribution with mean 22.2 and s.d. 4.6.
(c) What fraction of the scores are between the quartiles found in (b)?
(d) Assuming the given normal distribution, what fraction of test scores would be between the reported quartiles?

5. Graduate record exam (GRE) scores of two students are reported as follows:

Student #1: 760 (96th percentile)

Student #2: 520 (49th percentile).

Find the mean and standard deviation of all GRE scores, assuming them to be normally distributed.

*6. The frequency distribution of increases in human heart rate (beats per minute) after taking a particular drug is given approximately by the p.d.f.

$$f(x) \propto \exp\left[-\frac{1}{128}(x + 6)^2 \right].$$

(a) Find the mean and variance.
(b) Find the m.g.f. of this distribution.

7. Given the p.d.f. $f(x) \propto \exp[-x^2 - 5x]$, find the mean and variance of the distribution.

*8. Given that X is normal with mean 2 and variance 1, find $E(X^3)$
     (a) using the m.g.f.
     (b) using the formula for the central moments of a normal variable and the identity

$$X^3 = (X - 2 + 2)^3 = (X - 2)^3 + 6(X - 2)^2 + 12(X - 2) + 8.$$

9. (a) Given that Z is standard normal, find $E(e^Z)$.
     (b) Given that the m.g.f. of X is $\exp(2t + t^2)$, find EX and var X.

*10. The random variable R has the p.d.f. $(r/\theta \exp[r^2/(2\theta)]$ for $r > 0$. (See Solved Problem G.) Find
     (a) the mean value of R.
     (b) the c.d.f. of R.

11. Suppose $X \sim N(4, 1)$, $Y \sim N(5, 4)$, $Z \sim N(2, 2)$, and that the three variables are independent.
     *(a) Find the distribution of
         (i) $X + Y + Z$.   (ii) $X - Y$.   (iii) $2X - Y - Z$.
     (b) Find $P(X + Y) > 10$.
     (c) Find $P(X < Y)$.

*12. The heights of men students at the University of Baroda are approximately normal, with mean 64 in. and s.d. 2.5 in. A male student and a female student (see Problem 2) are selected at random. Find the probability that the male student is taller than the female student.

*13. Given that $X_1, \ldots, X_n$ are independent and $X_i \sim N(\mu_i, \sigma_i^2)$, obtain the distribution of the linear combination $\Sigma a_i X_i$ for given constants a_i.

14. Show that if X has a normal distribution, so does $Y = a + bX$ for any constants a and b,
     (a) using the basic definition of a normal distribution.
     (b) using moment generating functions.

Sections 6.2–6.4

*15. Fuses in an electric circuit fail when overloaded. Assume that in a particular circuit, overloads follow a Poisson process with a mean time between occurrences of six months. Find the probability
     (a) that a fuse will last at least 12 months.
     (b) that a fuse will fail before 3 months.
     (c) that there will be more than two failures in a one-year period, assuming that fuses are replaced immediately after they fail.

16. During the busy periods of the day, customer arrivals at an airport car-rental counter are Poisson, averaging one in two minutes.
     (a) Find the probability that no customers arrive in the next minute.
     (b) Find the mean time for six more customers to arrive.
     (c) Find the probability that fewer than four customers arrive in the next minute.

*17. A computer is subject to major breakdowns and minor breakdowns. Suppose these two types of breakdown follow independent Poisson distributions. The average time to failure is ten days for major breakdowns and two days for minor breakdowns. Find
     (a) the average time to the first breakdown of either type.

(b) the probability that the next breakdown is minor.

18. For the fuses of Problem 15, find
    (a) the probability that the time to the third failure exceeds one year.
    (b) the mean time to the third failure.

*19. Find the joint p.d.f. of a sequence of n independent variables, each of which is exponential with mean $1/\lambda$.

*20. Calculate:
    (a) $\Gamma(\frac{15}{2})$.   (b) $B(6, 9)$.   (c) $B(\frac{1}{2}, \frac{5}{2})$.

21. Calculate:

   *(a) $\displaystyle\int_0^\infty x^9 e^{-2x}\,dx$.   (b) $\displaystyle\int_0^\infty x^{5/2} e^{-x/2}\,dx$.   (c) $\displaystyle\int_0^1 x^8(1-x)^{12}\,dx$.

22. Calculate:

   *(a) $\displaystyle\int_0^\infty \frac{x^5\,dx}{(1+x)^9}$.   (b) $\displaystyle\int_0^{\pi/2} \cos^7\theta \sin^{10}\theta\,d\theta$.

*23. Find (to the nearest hundredth) the median of a gamma distribution with $\alpha = 2$ and $\lambda = 1$. (The p.d.f. is xe^{-x} for $x > 0$.)
    (*Hint*: Use integration by parts and solve the resulting equation iteratively or by trial and error.)

*24. Suppose X_1 and X_2 are independent, where $X_i \sim \text{Gam}(\alpha_i, \lambda)$, $i = 1, 2$. Find the distribution of $X_1 + X_2$.

25. Write the integral that defines the variance of a standard normal distribution. Use the symmetry of the normal p.d.f. to express the variance as an integral over the positive axis. Identify this as an integral of the form (7) of Section 6.3, and deduce the value of $\Gamma(\frac{3}{2})$.

*26. Use the method of the preceding problem to obtain a formula for $\Gamma(n + \frac{1}{2})$, where n is a nonnegative integer.

27. Obtain the mean and variance of a gamma distribution for $\alpha = r$, a nonnegative integer [see (10) of Section 6.3], using the structure of a gamma variable in that case as a sum of exponential variables.

28. Express $\Gamma(r)$ and $\Gamma(s)$ as integrals of the form (7) in Section 6.3, using different dummy variables. From their product, write this as a repeated integral and then a double integral; make a change of variables to polar coordinates and recognize that the result involves $\Gamma(r + s)$ and the formula for $B(r, s)$ given by (4) of Section 6.4. This will establish (1) of Section 6.4.

*29. Derive the formula for the variance of the beta distribution (see page 283).

Sections 6.5–6.6

30. Given that $Z_1, Z_2, \ldots$ are independent standard normal variables, use the chi-square tables (Table V of Appendix 1) to find

   *(a) $P\left(\displaystyle\sum_1^4 Z_i^2 > 10\right)$.   (b) $P\left(\displaystyle\sum_1^{18} Z_i^2 < 26\right)$.

31. Suppose $(X_1, \ldots, X_n)$ is a random sample from $\text{Exp}(\lambda)$. Show that the variable $V = (2\lambda)\,\Sigma\,X_i \sim \chi^2(2n)$. [Compare Problem R.]

*32.   Suppose Y is the sum of nine independent variables, each Exp(5). Use the result of Problem 31 and Table V to find $P(Y < 3)$. Check your answer using (2) of Section 6.3 (for the c.d.f. of the sum) and the Poisson table.

33.   Approximate the probability in Problem 30(b)
   (a)   using the normal table and the fact that for large k, $\chi^2(k)$ is approximately $N(k, 2k)$.
   (b)   using the approximation suggested in Solved Problem R and the footnote to Table Vb. (This is generally better.)

34.   Prove that the sum of independent chi-square variables is chi-square, using moment generating functions.

*35.   Suppose $U \sim \chi^2(4)$ and $V \sim \chi^2(6)$. Assuming further that U and V are independent, find $E(U/V)$.
   (*Hint*: Use $U/V = U \cdot (1/V)$ and Solved Problem T.)

36.   Suppose the joint p.d.f. of (X, Y, Z) is

$$f(x, y, z) \propto \exp\left\{ -\frac{1}{2}(x^2 + y^2 + z^2) \right\}.$$

Find the distribution of $X^2 + Y^2 + Z^2$, the squared distance from the origin to the point (X, Y, Z).

*37.   If $T \sim t(m)$, its p.d.f. is $f_T(t) = C \cdot (1 + t^2/m)^{-(m+1)/2}$. Find C.
   [*Hint*: The p.d.f. is $f_T(t) = t \cdot f_U(t^2)$, where U is $F(1, m)$. See Solved Problem U.]

38.   Use the result of Problem V to find
   (a)   the mean of $F(k, m)$.
   (b)   the variance of $t(m)$.

Section 6.7

*39.   Let L denote the length of time (in minutes) an automobile battery randomly selected from a certain production line will continue to crank an engine. Assume that $L \sim N(10, 4)$.
   (a)   Find the probability that the battery will crank the engine longer than $10 + x$ minutes given that it is still cranking after 10 minutes, as an expression depending on x and involving the standard normal c.d.f. Φ.
   (b)   Evaluate the probability in (a) for $x = 2$ and for $x = 6$.
   (c)   Explain why an exponential distribution for L is not likely to be appropriate.

*40.   A system consists of two components connected in series. The operating life of each component is exponential with mean one hour, and their times to failure are independent. Find the p.d.f. and mean value of T, the time to breakdown of the system.

41.   Repeat Problem 40 with this change: The two components are connected in parallel.

*42.   Buses in a certain fleet have independent exponential times to breakdown, with mean 3 months. Suppose there are n buses in the fleet. Find the mean time to the first breakdown.

43.   A system has the following p.d.f. for its effective lifetime T:

$$f(t) = \frac{1}{(t + 1)^2}, \qquad t > 0.$$

(a) Find the reliability function $R(t)$.

(b) Find $E(T)$.

(c) Calculate $P(T > t_0 + t \mid T > t_0)$ for t and $t_0 > 0$.

(d) Find the hazard function. [Observe whether it increases or decreases and see how this bears on the answers to (a) and (c).]

*44. Three units are connected to form a system according to the block diagram in Figure 6.14. Find the system reliability function R in terms of the reliability functions R_i for unit i.

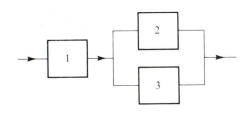

Figure 6.14

45. When exposed to high temperatures, a system fails at time T, with reliability function $R(t) = 8/(2 + t)^3, t > 0$.

(a) Find the c.d.f. and p.d.f. of T.

(b) Find the mean and variance of T.

*46. Suppose X has c.d.f. $1 - 1/x^2$ for $x > 1$.

(a) Find the mean and median of X.

(b) Show that $Y = \log X$ has an exponential distribution.

Section 6.8

*47. Find the marginal distributions of X and Y, given that their joint p.d.f. is proportional to

$$\exp\left[-\frac{x^2}{4} - \frac{y^2}{8} + x - \frac{y}{2} \right].$$

48. Given the joint p.d.f.

$$f(x, y) \propto \exp\left(-\frac{1}{4}(x - 2y + 1)^2 - \frac{1}{2}y^2 \right),$$

find

(a) the marginal density of Y.

(b) the conditional mean, $E(X \mid y)$.

(c) the conditional variance, $\operatorname{var}(X \mid y)$.

*49. Write the p.d.f. of a bivariate normal pair (X, Y) with $EX = 0$, $EY = 4$, $\operatorname{var} X = 1$, $\operatorname{var} Y = 9$, and $\operatorname{cov}(X, Y) = 2$.

50. Find $E(X \mid y)$ and $\operatorname{var}(X \mid y)$ for (X, Y) defined in Problem 49.

∗51. For the distribution with p.d.f. proportional to

$$\exp\left[-\frac{x^2}{2} + xy - y^2 - x + 2y\right],$$

(a) find the regression function of Y on X, and the regression function of X on Y.

(b) find the means, variances, covariance, and correlation.

∗52. Given that U and V are independent, standard normal variables, find the joint distribution of $X = U + 2V$ and $Y = 3U - V$.

∗53. Given that the m.g.f. of (X, Y) is

$$\psi_{X,Y}(s, t) = \exp[4s^2 - 4st + 9t^2 - 8s + 6t],$$

(a) find the means and variances and the correlation ρ.

(b) find the regression function of X on Y.

54. For (X, Y) defined in Problem 49, find the joint distribution of $U = X - 2Y$ and $V = 3X + Y$.

CHAPTER 7

Organizing and Describing Data

In the first six chapters, we have developed the notion of probability for describing (or modeling) an experiment of chance. Probability models describe population distributions. They represent the experiment of selecting a population member at random. When an experiment does not involve sampling a tangible population, it is still useful to think of the experiment as sampling from a conceptual population.

In practice, populations are seldom known or understood. The fundamental problem of statistical inference is to learn about a population from *sample* data. In this chapter, we take up ways of organizing and describing data using tables, graphs, and summary statistics. We first treat data from discrete populations, univariate and bivariate. We then take up methods for displaying and summarizing data from continuous populations.

7.1 Frequency Distributions

In Chapter 4, we considered sampling from categorical populations. We summarized sample results by giving the numbers of times the various categories occurred. These numbers are the category **frequencies**; the ratios of frequencies to sample size are **relative frequencies**. The list of categories and corresponding frequencies or relative frequencies is a **frequency distribution**.

Example **7.1a** **Statistics Students' Statistics**

Table 7.1 gives data collected from the male students in a statistics class in October, 1987. Some of the variables are numerical; others are categorical. The number of siblings is discrete, with values 0, 1, 2, ... Although one's true height is continuous, the recorded height is discrete, being necessarily rounded off (in this case) to the nearest inch.

The classification according to use of marijuana has two categories. The questionnaire defined the category "yes" to mean "more than just experimented once or twice." We obtain the category frequencies by simply counting Y's and N's in

Table 7.1 Survey of males in a statistics class

No.	Height	Class	No. of siblings	Marijuana use	Behavior motive	No.	Height	Class	No. of siblings	Marijuana use	Behavior motive
1	76	3	5	Y	B	25	68	4	1	Y	B
2	70	4	7	N	B	26	69	4	3	N	B
3	72	3	1	Y	A	27	72	3	2	N	B
4	73	4	3	N	B	28	73	3	2	Y	B
5	74	3	2	N	M	29	66	4	3	Y	B
6	68	3	5	Y	B	30	62	4	2	N	M
7	71	4	2	Y	B	31	73	3	10	Y	B
8	68	3	4	N	B	32	74	3	3	N	M
9	69	2	7	Y	M	33	68	3	2	N	M
10	72	4	2	Y	B	34	77	3	6	Y	B
11	68	3	1	Y	B	35	70	3	8	N	M
12	70	2	1	Y	M	36	70	5	2	N	B
13	70	5	2	Y	M	37	74	4	4	Y	B
14	70	3	8	Y	A	38	71	5	1	N	B
15	70	3	2	Y	B	39	71	5	5	N	M
16	74	2	3	Y	M	40	70	5	4	Y	B
17	69	3	3	N	M	41	70	4	3	Y	M
18	70	2	2	N	B	42	75	5	1	N	B
19	73	4	2	N	M	43	68	3	3	Y	B
20	72	3	2	N	M	44	69	2	5	Y	M
21	70	2	1	N	B	45	70	4	1	Y	M
22	74	3	4	N	M	46	73	3	2	N	M
23	68	3	1	N	M	47	66	2	6	N	M
24	76	3	3	Y	A	48	70	5	3	Y	B

Class: 1 = Freshman
       2 = Sophomore
       3 = Junior
       4 = Senior
       5 = Adult Special

Marijuana use:
Y = yes (I use)
N = no (I don't use)

Behavior motive:
M = moral code
A = fear of AIDS
B = both

the "Marijuana" column of Table 7.1:

Use marijuana?	Frequency	Relative frequency
Yes	25	$\frac{25}{48} = .521$
No	23	$\frac{23}{48} = .479$
Total	48	

The question on behavior motivation was worded this way: "Would you describe your behavior with respect to socializing with the opposite sex as motivated or guided mostly by moral considerations (M), or fear of AIDS (A), or some of both

(B)?" A count of M's, A's, and B's yields the following frequency distribution of students' answers:

Motivation	Frequency	Relative frequency
Moral code (M)	20	$\dfrac{20}{48} = .417$
Fear of AIDS (A)	3	$\dfrac{3}{48} = .063$
Both (B)	25	$\dfrac{25}{48} = .521$
Total	40	1

In such tables, the sum of the frequencies always equals the sample size. Equivalently, the sum of the relative frequencies is always 1. ∎

We saw in Section 2.2 that studying relationships between two random variables requires their *joint* probability distribution. The sample analogue is a joint *frequency* distribution. In the case of categorical variables, joint frequencies can be given two-way arrays called **contingency tables**. In a contingency table, each combination of a category of one variable with a category of the other defines a **cell**. The number we put in a particular cell is the frequency of this combination in the sample.

Contingency Tables

The row sums and the column sums in a contingency table are frequencies of categories of the corresponding variables considered separately. These two univariate distributions are *marginal* distributions, as we see in the following example.

Example 7.1b

We'll cross-classify the 48 males listed in Table 7.1 on Marijuana Use and Behavior Motive. There are two categories for Marijuana Use and three for Behavior Motive, making six cells. When all students on the sample list are tallied, the number of tally marks in each cell is the frequency of that cell. The contingency table for Marijuana Use and Behavior Motive is as follows:

		Marijuana use		
		Y	N	
	M	7	13	20
Behavior motive	A	3	0	3
	B	15	10	25
		25	23	48

The right-hand margin gives the frequency distribution for Behavior Motive and the lower margin gives the frequency distribution for Marijuana Use. These marginals agree with the distributions given in Example 7.1a.

Consider next a three-way cross-classification, including Class as a third variable. In this we show four two-way classifications, one for each class:

2 (Soph.)	Y	N	
M	4	1	5
A	0	0	0
B	0	2	2
	3	3	7

3 (Jr.)	Y	N	
M	0	9	9
A	3	0	3
B	8	2	10
	11	11	22

4 (Sr.)	Y	N	
M	2	2	5
A	0	0	0
B	5	3	8
	7	5	12

5 (A.S.)	Y	N	
M	1	1	2
A	0	0	0
B	2	3	5
	3	4	7

These two-way tables taken together constitute a *three-way* contingency table. Imagine that the four tables are stacked on top of each other, and think of each as a *layer* in a three-dimensional array of frequencies.

We could also layer the three-way table according to either of the other variables. For example, one of the three behavior-motive layers is as follows:

		Class				
Behavior motive = M		2	3	4	5	
Marijuana use	Yes	4	0	2	1	7
	No	1	9	2	1	13
		5	9	4	2	20

■

Numerical Data

Sometimes the categories of a discrete random variable are *numbers*. There are ways of describing numerical data that are not possible with nonnumerical variables. The natural ordering of numbers serves to order the categories, so numerical values can be shown graphically on a number scale. (Some nonnumerical categories are also naturally ordered.)

Example 7.1c   **Siblings**

The number of siblings an individual has is a discrete variable; the only possible values are integers. Counting the number of students having 0, 1, 2, . . . siblings in Table 7.1, we obtain the following frequency distribution:

Number of Siblings	1	2	3	4	5	6	7	8	10
Frequency	9	14	10	4	4	2	2	2	1

Figure 7.1 shows this distribution as a bar graph in which the height of each bar is proportional to the number of siblings.   ■

Bivariate numerical data consist of pairs of numbers (X, Y). Any relationship between X and Y is exhibited by plotting each data pair as a point in the xy-plane. When there are repeated pairs in the data, we can represent the frequency of a particular pair as a bar (in a direction perpendicular to the XY-plane).

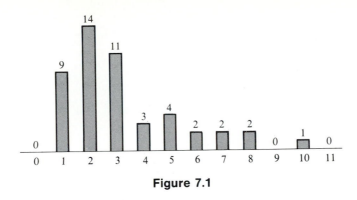

Figure 7.1

Example **7.1d**

Class Versus Siblings

It would be surprising to find much of a relationship between the variables Class and Number of Siblings, but any such relationship would show up in a cross-classification of the 48 students on these two variables. Many of the possible pairs in Table 7.1 occur more than once, so we summarize the data in a contingency table, as follows:

				Number of Siblings						
Class	1	2	3	4	5	6	7	8	10	
2	2	1	1	0	1	1	1	0	0	7
3	3	7	4	2	2	1	0	2	1	22
4	2	4	4	1	0	0	1	0	0	12
5	2	2	1	1	1	0	0	0	0	7
	9	14	10	4	4	2	2	2	1	48

Figure 7.2 is a bivariate bar diagram for these data. Pictures like this can be awkward because some bars may hide others. Computer software is available that can overcome such problems by drawing the diagrams from various perspectives. ∎

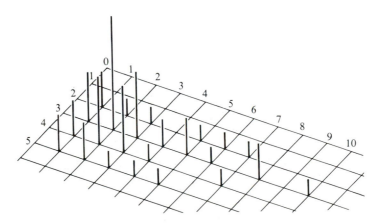

Figure 7.2

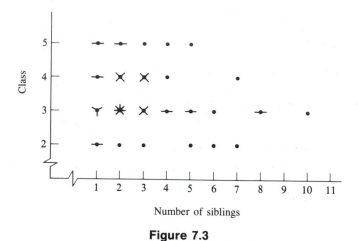

Figure 7.3

A different way of showing bivariate frequencies is a **star plot** (or **sunflower plot**[1]): At each point (X, Y) that occurs once in the data, we plot a point; at points that occur more than once, we plot a point and draw a short ray from the point to represent each occurrence of that pair. A frequency of 2 has two rays, in opposite directions; a frequency of 3, three rays in equally spaced directions; and so on. A star plot for Class versus Number of Siblings in Table 7.1 is shown in Figure 7.3.

7.2   Data on Continuous Variables

Variables such as time, temperature, and velocity are thought of as continuous. However, we measure and record observations on such a variable on a finite, discrete scale. So some of the devices described in Section 7.1 for categorical variables—bar charts and star plots—are useful for continuous variables as well.

When there is not much data, we can convey the information they contain by simply listing the observations, but graphical representations are useful no matter how small the data set. For example, simply marking an "x" or a dot on a number scale for each observation produces a plot that can be very revealing. We'll refer to

Dot
Diagrams

such a plot as a **dot diagram**. Such diagrams are awkward when the marks begin piling up on repeated values. The remainder of this section is concerned with methods specifically adapted to displaying larger data sets.

A first step in organizing numerical data is to put them in numerical order. Ordering a large mass of numbers is not easy without a system. One such is the **stem-leaf diagram**, which not only orders the data numerically but also provides a visual display. We explain this easy-to-learn device with an example.

Stem-Leaf
Diagrams

---

[1] See W. S. Cleveland, *The Elements of Graphing Data* (Pacific Grove, Calif.: Wadsworth Advanced Books and Software, 1985).

Example **7.2a** Strength of spot welds made by a particular welding tool and operator vary. The following are weld strengths (in psi) in a sample of 50 welds:

400	395	398	421	445	389	372	408	398	401
399	386	423	364	394	414	390	412	398	363
388	431	392	438	411	399	399	408	390	420
400	389	430	426	388	406	431	411	404	424
450	416	397	404	388	405	392	405	379	419

To make a stem-leaf diagram, take the first two digits as "stems" and record measurements according to the last digits, as "leaves" on the appropriate stems. Thus, 400 is a 0-leaf on the 40-stem, 395 is a 5-leaf on the 39-stem, and so on. The result is shown in Table 7.2. The table includes a second stem-leaf diagram with the leaves on each stem in numerical order. Having the leaves in order is convenient for a number of purposes, as we shall soon see.

Table 7.2

Stem	Leaves	Stem	Ordered leaves
36	43	36	34
37	29	37	29
38	968988	38	688899
39	5889408299072	39	0022457888999
40	0818064455	40	0014455688
41	421169	41	112469
42	13064	42	01346
43	1801	43	0118
44	5	44	5
45	0	45	0

Not all data sets lend themselves usefully to display in a stem-leaf diagram. For example, the student heights in Table 7.1 are given as two-digit numbers, and the first digit is either a 6 or a 7; there would be thus only two stems. At the other extreme, data such as the populations of the 50 states differ in not just the last digit or two but in the last five to seven digits. There are ways of adapting stem-leaf diagrams to such awkward cases.[2]

The stems of a stem-leaf diagram constitute categories of nearby observations. Counting the leaves on each stem results in a frequency distribution. Stem frequencies are proportional to stem length if one uses a typewriter (for example), so that all leaves take the same amount of space. The outline formed by the leaves shows the frequency distribution as a bar diagram in which the stems define categories. Figure 7.4 outlines the stem-leaf diagram from Table 7.2.

In a stem-leaf diagram, the choice of stems is restricted by the decimal system, but there are other ways of grouping values. For any particular grouping, partition

Regrouping
Observations

---

[2] See J. Tukey, *Exploratory Data Analysis* (Reading, MA: Addison-Wesley, 1977).

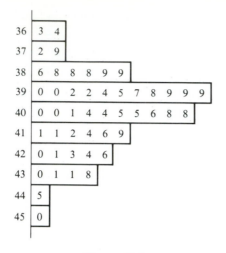

36	3	4								
37	2	9								
38	6	8	8	8	9	9				
39	0	0	2	2	4	5	7	8	9	9	9
40	0	0	1	4	4	5	5	6	8	8
41	1	1	2	4	6	9				
42	0	1	3	4	6					
43	0	1	1	8						
44	5									
45	0									

Figure 7.4

the range of values in the data set into subintervals, usually of equal sizes, called **class intervals**. Then tally the data in these class intervals and count the observations in each subinterval. These counts define a *frequency distribution.*

A frequency distribution loses some of the sample detail, whereas a stem-leaf diagram preserves all the data. However, the details are often unimportant. The usual reason for sampling is to find out about a population, and a second sample from the same population would be different in detail. Nevertheless, the two sample distributions will be roughly similar in broad outline—especially when both sample sizes are large.

A frequency distribution depends on the (arbitrary) choice of class intervals. Some guidelines are useful: The intervals should be wide enough that most of them get more than a few observations, but there should be a reasonable number of class intervals (perhaps between 5 and 20).

It is common practice to mark each class interval with its midpoint, its **class mark**. When we calculate sample moments in the next section, the class mark will represent all the values in the class interval. This amounts to rounding sample values to class marks.

Histograms

A frequency distribution is often represented graphically by a **histogram**—a bar diagram in which the categories are class intervals. The height of each bar is such that the area of the bar is proportional to the frequency of the corresponding class interval. Since continuous data have no gaps between class intervals, the bars for adjacent intervals should touch.

Example **7.2b**

The spot welds in Example 7.2a range from the 360's to the 450's. Suppose we divide this range into nine equal subintervals, each about eleven units wide. We arbitrarily take the first interval to be 361–371 (inclusive). Its class mark is the midpoint, 366. The class mark for the next interval, 372–382, is 377; and so on. Table 7.3 shows the tallies and frequencies.

Table 7.3 Frequency distribution of spot-weld strengths (psi)

Class mark	Class interval	Tally	Frequency
366	361–371	||	2
377	372–382	||	2
388	383–393	‖‖‖ ‖‖‖	10
399	394–404	‖‖‖ ‖‖‖ ‖‖‖	15
410	405–415	‖‖‖ |||	8
421	416–426	‖‖‖ ||	7
432	427–437	|||	3
443	438–448	||	2
454	449–459	|	1

Measurements are rounded to the nearest psi, but strength is inherently continuous. So the class interval 361–371, for example, actually covers the range from 360.5 to 371.5: All numbers in that interval should be rounded off to integers from 361 to 371, and the class mark is 366. The second bar covers 371.5–382.5—the dividing line between the first and second class intervals is 371.5. The dividing line between the second and third intervals is 382.5, and so on. Thus, no observation falls on a dividing line between class intervals. Figure 7.5 gives the histogram for this frequency distribution. Because the class interval widths are equal, bar heights are proportional to bar areas, which in turn are proportional to class frequencies.

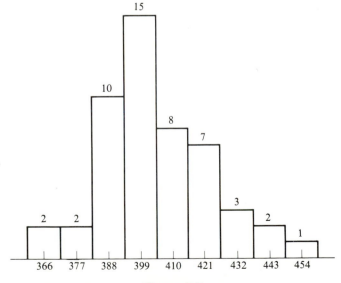

Figure 7.5

Unequal
Class
Intervals

In the above example, we used class intervals of equal size. We recommend doing so generally, but frequency distributions are sometimes constructed using unequal class intervals. For a histogram with class intervals of differing widths, it is necessary to adjust the heights of the bars, to avoid giving a wrong impression. The eye perceives *area*, so we want one unit of area to represent one observation, no matter which interval contains it. The next example will illustrate this point.

Example **7.2c**

The following frequency table summarizes plasma clearance of a certain drug (in ml/min/kg) for 44 smokers:[3]

Clearance	Frequency
4.5–8.5	6
8.6–12.6	18
12.7–16.7	14
16.8–20.8	4
20.9–24.9	1
25.0–29.0	1

Suppose we were to combine the last three class intervals, making one interval (16.8–29.0) with frequency 6. Figure 7.6 shows an incorrectly drawn histogram. The equal heights representing the two frequencies of 6 suggest that there are three times as many observations in the wider interval as in the narrower one. The histogram is drawn correctly in Figure 7.7. Superimposed is the histogram for the original frequency distribution, which shows that the height 2 of the new wide bar is the average of the heights 4, 1, and 1 of the three bars for the original class intervals.

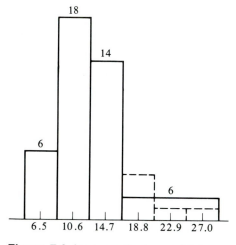

Figure 7.6 Incorrectly drawn histogram

[3] From the consulting practice of one of the authors.

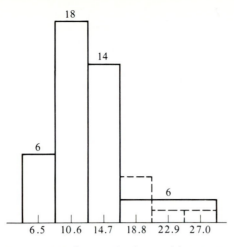

Figure 7.7 Correctly drawn histogram   ■

A **histogram** is a bar chart for continuous data in which the areas of the bars are proportional to the corresponding class frequencies.

Scatter Diagrams

To organize and display *bivariate* data, we interpret the two variables as *x*- and *y*-coordinates in the plane. For data sets of moderate size, marking a dot or other symbol at each point provides an excellent visual representation of the data. Such a plot is called a **scatter diagram**. These are helpful in studying relationships between two variables. (If data points occur more than once in a data set, one can mark each point with its frequency or any other covenient code such as that used in the star plot described in Section 7.1.)

Example 7.2d   **LSAT Versus GPA**
Table 7.4 gives average LSAT scores (X) and average GPA's (Y) for entering students in each of 15 law schools in 1973.[4] (The LSAT is a national aptitude test taken by those who want to enter a law school. GPA is undergraduate grade-point average.) Figure 7.8 shows these fifteen (x, y)-pairs in a scatter diagram. As one might expect, high LSAT scores tend to be associated with high GPA's. This is only a tendency, however. For instance, the first pair in the data set doesn't fit this description very well.   ■

A third variable that is categorical can be included in a bivariate display by using its category labels, or any other system of symbols (such as triangles for one

---

[4] Taken from B. Efron, "Computers and the theory of statistics...," *SIAM Review 21*, 460–480.

Table 7.4

x	y	x	y	x	y
576	3.39	580	3.07	653	3.12
635	3.30	555	3.00	575	2.74
558	2.81	661	3.43	545	2.76
578	3.03	651	3.36	572	2.88
666	3.44	605	3.13	594	2.96

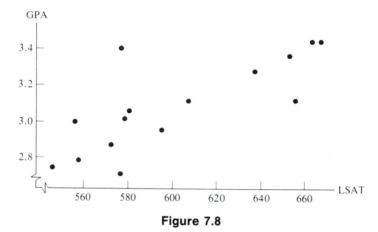

Figure 7.8

category and circles for another) to mark the location of an (x, y)-pair. Or a scatter diagram can be made for each category.

Bivariate Histograms

Two-dimensional frequency distributions for large data sets employ a system of class intervals on the x-axis and another on the y-axis. Each interval defines a strip (perpendicular to the axis), and the two systems of strips intersect to form rectangular cells (like graph paper). The cells play the role played by the class intervals in a univariate distribution. To summarize a set of data, we can construct a two-way table of cell frequencies. A *histogram* for such a frequency distribution consists of a column above each cell, with volume proportional to the cell frequency. As in the case of discrete data (Example 7.1c), such pictures are awkward without computer graphics.

Frequency distributions and their graphical representations convey the sense of a data set and also give a partial picture of the population from which they are drawn. The following example shows two sample histograms that by themselves offer convincing evidence that the corresponding populations sampled are different.

Example 7.2e

Platelet Count

Figure 7.9 gives two histograms, one representing blood platelet count data for 35 healthy males and the other, data for a group of 153 male cancer patients.[5] No one

---

[5] S. Silvis et al., "Thrombocytosis in patients with lung cancer," *J. Amer. Med. Assn. 211* (1970), p. 1852.

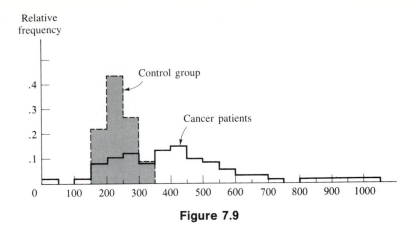

Figure 7.9

looking at these graphs could fail to see a difference between the two kinds of individuals. ■

Conclusions drawn from histograms are not always this obvious. Histograms are visually appealing but difficult to work with analytically. It is usually easier to work with a number, a sample characteristic especially relevant to the question at hand.

A **statistic** is a number calculated from the observations in a sample.

The remaining sections of this chapter introduce some commonly used statistics.

7.3   Order Statistics

In statistical applications, sample observations are often regarded as exchangeable: The order in which they were collected is not important (see Section 8.3). However, many important statistics depend on the *numerical order* of the sample observations.

Given a sample $(X_1, \ldots, X_n)$ from a continuous population, we denote the smallest observation by $X_{(1)}$, the second smallest by $X_{(2)}$, and so on. The nth smallest is the largest. Each $X_{(i)}$ is a statistic—a number calculated from the sample. The vector

$$[X_{(1)}, X_{(2)}, \ldots, X_{(n)}]$$

is called **the order statistic**. Each component $X_{(i)}$ is called *an* order statistic.

The Median   The most commonly used order statistic is the number in the middle of the ordered observations: the **sample median**, $\tilde{X}$. If n is odd, there is a middle order statistic. For instance, if $n = 15$, then $\tilde{X} = X_{(8)}$. This is the eighth smallest observation and also the eighth largest. If n is an even number, there are two numbers in the middle; somewhat arbitrarily, we take their average (half their sum) as the median. For example, when $n = 24$, $X_{(12)}$ and $X_{(13)}$ are in the middle; the median is halfway between these: $(X_{(12)} + X_{(13)})/2$. There are twelve observations on either side of $\tilde{X}$.

$$
\tilde{X} = \begin{cases} X_{([n+1]/2)} & \text{for odd } n, \\ \frac{1}{2}(X_{(n/2)} + X_{(n/2+1)}) & \text{for even } n. \end{cases} \tag{1}
$$

Sample median:

Ordering the leaves on the individual stems of a stem-leaf diagram orders all the data. This makes it easier to find the sample median.

Example 7.3a   **Median of Spot-Weld Data**

The stem-leaf diagram for the spot-weld data of earlier examples (7.2a, 7.2b) is repeated as Table 7.5. We've added a column called *depth*. The depth is a cumulative count of the observations from top to bottom toward the middle. The depth for the 400-stem is special. In cumulating from the bottom of the diagram (the larger numbers), adding in the 10 leaves on the 400-stem gives a count of 27; from the top of the diagram, the cumulative count for that stem is 33. Both of these are greater than $25 = n/2$, so instead of cumulating, we give the number of leaves on the 400-stem in the depth column and isolate it between horizontal lines.

Since more than half the data are at least 400 and more than half are less than 410, the median is on the 400-stem. According to (1), the median is the average of the 25th and 26th from either end. There are 23 leaves above the 400-stem; the 24th,

Table 7.5

Depth	Stem	Leaves
2	36	34
4	37	29
10	38	688899
23	39	0022457888999
10	40	0014455688
17	41	112469
11	42	01346
6	43	0118
2	44	5
1	45	0

$n = 50$

25th, and 26th smallest numbers are thus the first three leaves on the 400-stem: 400, 400, and 401. So the median is $(400 + 401)/2 = 400.5$.

(When using a stem-leaf diagram in which the leaves have not been ordered, remember to enter a stem at the smallest number when you count from the low end of the scale and at the largest number when you count from the high end of the scale.) ∎

Quartiles and Percentiles

The median divides a set of data in half: Half the observations are smaller than the median and half are larger. When n is large, it is sometimes convenient to divide the data set further and identify *quartiles* and *percentiles*. The three **quartiles** of a set of numerical data are numbers that divide the ordered data into *four* segments of about equal size. When n is a multiple of 4, the first quartile Q_1 has one-quarter of the observations to its left and three-quarters of them to its right. The second quartile is the median $\tilde{X}$, since $\frac{2}{4} = \frac{1}{2}$. The third quartile Q_3 has three-quarters of the observations to its left and one-quarter to its right. When n is not a multiple of 4, such division can only be approximate. A convenient convention is to take the first and third quartiles as the medians of the left and right halves, respectively, of the data. (When n is odd, include the median in both halves.) Sometimes the first and third quartiles are referred to as *hinges*.

Example 7.3b

There are 50 spot-weld strengths in the stem-leaf diagram of Table 7.5. In Example 7.3a, we found $\tilde{X} = 400.5$. The first quartile is the median of the 25 numbers smaller than 400.5—the 13th smallest: $Q_1 = X_{(13)} = 392$. There are 12 observations smaller than 392 and 37 larger—not quite a $\frac{1}{4}$ and $\frac{3}{4}$ division. Similarly, $Q_3 = X_{(38)} = 416$. ∎

When there are hundreds of observations, **percentiles** may be useful. The 37th percentile, for example, is a number such that 37% of the observations are smaller and 63% larger.

Sample Range

Applied to a data set, the word *range* could refer to the interval from the smallest to the largest. In statistics, the word has a technical meaning. The statistic **sample range** is defined as the *width* of the interval from the smallest to the largest observation:

$$R = X_{(n)} - X_{(1)}. \tag{2}$$

It is a crude measure of the dispersion or amount of spread in a data set.

The range is obviously greatly affected by the most extreme observations, $X_{(n)}$ and $X_{(1)}$. Sometimes an extreme value is extreme because of an error in recording or some other aberration and should not be included as a sample observation. For example, if the decimal point is wrong on any observation, that observation is apt to end up as either the smallest or largest. We say that the range is not very "robust" with respect to unusually large or small observations. A more robust measure of dispersion is the **interquartile range**:

$$\text{IQR} = Q_3 - Q_1. \tag{3}$$

A statistic that competes with the median as an indicator of the middle of a data set is the **midrange**:

$$\text{Midrange} = \frac{X_{(n)} + X_{(1)}}{2}. \tag{4}$$

This is midway on the number scale from the smallest to the largest observation, whereas the median is halfway from smallest to largest as counted in the set of ordered values.

Another competitor of the median in measuring "middle" is the **midhinge**. This is the number halfway between the quartiles on the number scale:

$$\text{Midhinge} = \frac{Q_1 + Q_3}{2}. \tag{5}$$

Five numbers that describe a data set well are the smallest observation, the three quartiles, and the largest observation. We refer these as a *five-number summary*. A **box plot** shows these five numbers graphically. It is constructed along a number scale as a box whose ends are at the first and third quartiles. "Whiskers" extend out from both ends of the box to the smallest and largest observations. A line through the box indicates the median. The various measures (2)–(5) are easily obtained from the five-number summary or the box plot.

Box
Plots

Example 7.3c   **Spot-Weld Data**
Examples 7.3a and 7.3b give the three quartiles of the data in Table 7.5, and the extreme values $X_{(1)}$ and X_n are easy to read off the table, so the five-number summary is as follows:

$$X_{(1)} = 363$$
$$Q_1 = 392$$
$$\tilde{X} = 400.5$$
$$Q_3 = 416$$
$$X_{(n)} = 450$$

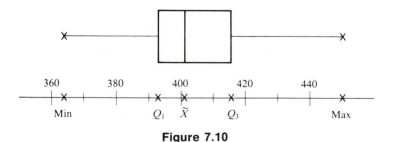

Figure 7.10

From these we obtain:

$$\text{Range} = X_{(n)} - X_{(1)} = 450 - 363 = 87,$$
$$\text{Interquartile range} = Q_3 - Q_1 = 416 - 392 = 24,$$
$$\text{Midrange} = \frac{1}{2}(X_{(n)} + X_{(1)}) = \frac{450 + 363}{2} = 406.5,$$
$$\text{Midhinge} = \frac{1}{2}(Q_1 + Q_3) = \frac{392 + 416}{2} = 404.$$

The box plot representing the five-number summary is shown in Figure 7.10.   ■

In the next section, we take up the most commonly used measure of the middle of a set of numerical data.

7.4   The Sample Mean

The mean of a sample $(X_1, \ldots, X_n)$ is defined in the same way as the mean or expected value of a random variable with equally likely values. Think of the sample as a finite population of size n from which one member is selected at random. Each sample value then has probability $1/n$. The mean value of this "selection" is the **sample mean**:

$$\bar{X} = \sum_{i=1}^{n} X_i \cdot \left(\frac{1}{n}\right) = \frac{1}{n} \sum_{i=1}^{n} X_i. \tag{1}$$

This is the ordinary arithmetic average of the n observations.

Because $\bar{X}$ is an expected value, it follows from (3) of Section 3.1 that the average deviation about the mean is 0. Hence,

$$\sum (X_i - \bar{X}) = 0. \tag{2}$$

Mean from a Frequency Distribution

Samples sometimes include repeated values and are summarized in a frequency distribution. Formula (1) counts an X-value once each time it occurs in the sample. So a value that occurs f times has total weight f/n in (1). For a sample with k *distinct* values $x_1, x_2, \ldots, x_k$, the sample mean is

$$\bar{X} = \frac{1}{n} \sum_{j=1}^{k} x_j f_j. \tag{3}$$

(Lowercase x's denote the distinct values in a data set and capital X's the individual observations.)

Sample mean for a data set $X_1, \ldots, X_n$:

$$\bar{X} = \frac{1}{n} \sum_{i=1}^{n} X_i. \tag{1}$$

For data given in a frequency distribution with distinct values $x_1, \ldots, x_k$ and corresponding frequencies $f_1, \ldots, f_k$,

$$\bar{X} = \frac{1}{n} \sum_{j=1}^{k} x_j f_j. \tag{3}$$

Example 7.4a   A student counted the number of licorice "snaps" in each of 25 boxes. His results are shown in the following frequency table:

x_j	f_j	$x_j f_j$
11	3	33
12	3	36
13	10	130
14	8	112
15	1	15
Sums:	25	326

The column of products $x_j f_j$ aids in calculating the sample sum. Using (3) to find $\bar{X}$, we divide the sample sum by n: $\bar{X} = \frac{326}{25} = 13.04$. This is reasonable as a center of gravity of the bar chart shown in Figure 7.11.

The sample sum 326 has a simple interpretation in this example: It is the total number of snaps in the 25 boxes! Dividing them equally among the 25 boxes would mean $\bar{X} = 13.04$ snaps per box.

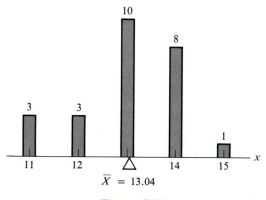

$\bar{X} = 13.04$

Figure 7.11

Means for Regrouped Data

Consider data on a continuous variable that are grouped into class intervals and presented as a frequency distribution. We can calculate $\bar{X}$ using (1) only if the individual observations are also available. If only the grouped data are available, we can approximate $\bar{X}$ by assuming that every observation in a class interval is located at the interval's midpoint, the class mark. This is the same as using (3) and taking the x_j's to be the class marks and the f_j's to be the class frequencies; k is the number of class intervals.

This procedure is only approximate, but the mean calculated using (3) can't be off by more than half the interval width. In practice, some of the observations will be smaller and some larger than the class mark. So the errors involved in assuming they are all equal to the class mark tend to cancel. The approximation using (3) is usually remarkably accurate.

Example 7.4b   **Spot-Weld Data**

We use formula (1) to find the mean of the spot-weld data of Example 7.2b:

$$\bar{X} = \frac{20,210}{50} = 404.2.$$

If only the frequency distribution given in Table 7.3 were available, we could approximate $\bar{X}$ using (3). The class marks (x) and class frequencies (f) are repeated here in Table 7.6, along with their products.

Table 7.6

x_j	f_j	$x_j f_j$
366	2	732
377	2	754
388	10	3880
399	15	5985
410	8	3280
421	7	2947
432	3	1296
443	2	886
454	1	454
Sums:	50	20214

From (3),

$$\bar{X} = \frac{20,214}{50} = 404.28.$$

This is very close to the exact value, 404.2. (The sum of the actual observations is 20,210 and the sum approximated from the frequency table is 20,214. The errors in replacing actual values by class marks did not completely cancel.) ∎

Mean or
Median?

Both the median and mean are measures of the middle of a distribution. Which one is better? The answer depends in part on what it's going to be used for (see Section 8.3) and also in part on the population. Both the mean and median are "typical" observations. The median is relatively unaffected by extreme observations. Thus, for example, whether the most expensive house in a city has a market value of $1 million or $10 million, the median market value of a house in the city is the same. Not so the mean, because the total market value in the city is $9 million larger in the second case. Of great practical importance is the possibility of gross recording or measuring errors. The mean is influenced by such errors, whereas the median is much less apt to be affected. However, when a population does not contain "extreme" values, the mean has the advantage of using all the data and using them efficiently.

Trimmed
Means

A procedure that may well become more commonplace, because it is a compromise between the mean and the median without having the disadvantages of either, is to delete a specified number of the largest and the same number of the smallest observations in a data set and then calculate the mean of the rest. (This is common practice in some athletic competitions in calculating a score based on the ratings of several judges.) The result is a **trimmed mean**. In the special case in which the largest 25% and the smallest 25% of the observations are trimmed, the trimmed mean is called the **midmean**—the mean of the middle half of the data. [Compare this with the midhinge, defined as (5) of Section 7.3.]

Example **7.4c**

A student with a part-time job collected data on the time for an extruding machine to process polyethylene tubing. The machine was rated at 45 minutes for 6,000 feet. The following times were recorded:

45, 93, 45, 50, 51, 49, 45, 48, 40, 45, 45, 65, 55, 40.

The order statistic is

40, 40, 45, 45, 45, 45, 45, 48, 49, 50, 51, 55, 65, 93.

The mean of these 14 times is $\bar{X} = \frac{716}{14} = 51.1$. Dropping the smallest and largest observations, we find the mean of the remaining 12 to be $\frac{583}{12} = 48.6$. This trimmed mean may be more reliable than the mean. The 93 seems out of line—an "outlier." Indeed, a check revealed that this time included a change of the screen pack during the run. ∎

One reason for calculating the mean of a sample is to estimate the mean of the population that was sampled. How well the sample mean estimates the population mean depends on the sample size and on the amount of variability in the population. (This will be explained in more detail in Chapter 9.) One way to learn about population variability is to examine the variability in a sample. The next section takes up sample variability.

7.5   **Some Measures of Dispersion**

In Section 7.3, we defined the range (R) and interquartile range (IQR) as measures of the width or spread of a sample distribution. Here we take up measures of variability based on deviations from the mean, measures that make better use of the data and are more widely used.

The traditional way of averaging deviations about the sample mean, as in the case of populations (Section 5.5), is to take the square root of their average square.

Sample Variance

We tentatively define the **variance** of a sample as we defined the mean, by relating it to a corresponding calculation for populations. As in Section 7.4, think of selecting at random from the n observations in the sample. This assigns weight $1/n$ to each observation, or in the case of a frequency distribution, weights f_j/n to each of the distinct values x_j. The variance of the one selected depends on the sample and so is a statistic:

$$V = \frac{1}{n} \sum_{i=1}^{n} (X_i - \bar{X})^2. \tag{1}$$

For grouped data, with x_j occurring f_j times or with class marks x_j and class frequencies f_j,

$$V = \frac{1}{n} \sum_{i=1}^{k} (x_j - \bar{X})^2 f_j. \tag{2}$$

Because V is a special case of the variance defined in Section 3.3, applied here to the "population" of sample values, the general properties of variance apply to V. In particular, these formulas are alternatives to (1) and (2) [see (3) in Section 3.3]:

$$V = \frac{1}{n} \sum_{i=1}^{n} X_i^2 - (\bar{X})^2. \tag{3}$$

$$V = \frac{1}{n} \sum_{i=1}^{k} x_j^2 f_j - (\bar{X})^2. \tag{4}$$

Formula (3) is a special case of the parallel axis theorem [(6) of Section 3.3], which reads as follows in the present context:

$$\frac{1}{n} \sum (X_i - c)^2 = \frac{1}{n} \sum (X_i - \bar{X})^2 + (\bar{X} - c)^2. \tag{5}$$

As a function of c, this is clearly smallest when $c = \bar{X}$: The average squared deviation is smallest about $\bar{X}$.

We described (1) as a "tentative" definition of variance. The usual statistical convention is to replace n in the denominator with $n - 1$:

Sample variance:

$$S^2 = \frac{1}{n-1}\sum(X_i - \bar{X})^2. \tag{6}$$

Comparing (1) and (6), we see that

$$S^2 = \frac{n}{n-1}V. \tag{7}$$

So the difference between using S^2 and using V is slight, especially for large n. Although V seems more natural and simple, we'll conform to the standard in statistical practice and computer software and use S^2.

(The "$n - 1$" is called the "degrees of freedom" in S^2. This refers to the fact that the n deviations whose squares are summed are not completely free but satisfy the one constraint that they sum to zero. In other estimates of variance, to be encountered in Chapters 14 and 15, the degrees of freedom used as a divisor may be something other than $n - 1$. Although there is no theoretical basis[6] for preferring division by degrees of freedom, the practice is convenient for dealing with sampling distributions and providing tables of probabilities.)

With sample variance defined by (5), we define the **sample standard deviation** as its square root, given by various equivalent formulas:

Standard deviation:

For a sample $(X_1, \ldots, X_n)$:

$$S = \sqrt{\frac{1}{n-1}\sum(X_i - \bar{X})^2} = \sqrt{\frac{\sum X^2 - (\sum X)^2/n}{n-1}}. \tag{8}$$

For a frequency distribution with distinct values or class marks x_j and frequencies f_j:

$$S = \sqrt{\frac{1}{n-1}\sum(x - \bar{X})^2 f} = \sqrt{\frac{\sum x^2 f - (\sum xf)^2/n}{n-1}}. \tag{9}$$

The second formula in each of (8) and (9) is usually easier to use for calculations. To make them less cluttered, we have suppressed the subscripts i and j. When we deal

---

[6] B. W. Lindgren, "To bias or not to bias," *Amer. Statistician 37* (1983), 254.

with the sample observations X_i, the summations are over the sample, from $i = 1$ to $i = n$; when working with a frequency table with k classes and class marks x_j, the summations are over the class marks, from $j = 1$ to $j = k$.

The mean deviation (1) of Section 3.3, applied to the "population" of sample observations each assigned probability $1/n$, defines the sample mean deviation:

Mean Deviation

$$\text{m.a.d.} = \frac{1}{n} \sum_{i=1}^{n} |X_i - \bar{X}|. \tag{10}$$

Sometimes defined as the mean absolute deviation from the median, the mean deviation is at present not much used in statistical practice.

Example 7.5a

6-MP and Leukemia
The drug 6-MP (mercaptopurine) has been used in therapy to extend remission in leukemia patients. The lengths of remission (in weeks) for 21 patients receiving this treatment were observed[7] as follows:

10, 7, 32, 21, 22, 6, 16, 34, 32, 25,
11, 20, 19, 6, 17, 35, 6, 13, 9, 6, 10.

The mean length of remission is $\frac{357}{21} = 17$ weeks. A patient about to undergo therapy would be interested in deviations from the average:

$-7, -10, 15, 4, 5, -11, -1, 17, 15, 8,$
$-6, 3, 2, -11, 0, 18, -11, -4, -8, -11, -7.$

The mean of these 21 deviations is 0, as we know it must be from (3) of Section 3.1. The mean of the *absolute* deviations is m.a.d. $= \frac{174}{21} = 8.29$. The plot in Figure 7.12 suggests that this is about right as an average distance from 17.

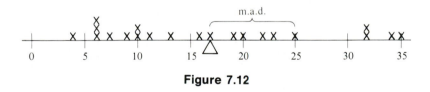

Figure 7.12   ∎

To find the standard deviation using the first formula in (7), we square the deviations and find their sum to be 1,980. Then

$$S = \sqrt{\frac{1,980}{20}} = 9.95.$$

---

[7] E. J. Freireich et al., "The effect of 6-Mercaptopurine on the duration of steroid-induced remissions in acute leukemia...," *Blood 21* (1963), 699. [We have altered one observation slightly (the 21 was really 23) to make the mean an integer, so we could avoid decimals.]

Alternatively, to use the second formula in (9), we first find $\Sigma X = 357$ and $\Sigma X^2 = 8,049$:

$$S = \sqrt{\frac{1}{20}\left(8,049 - \frac{357^2}{21}\right)} = 9.95.$$

The standard deviation is a little larger than the mean deviation (8.29). This is true generally. (See Solved Problem I.)

Example 7.5b   **Spot-Weld Data**

The histogram for the frequency distribution of the spot-weld data in Examples 7.2a and 7.2b is shown in Figure 7.13. Before calculating S, try to guess its value from this figure. It may help to think of it as just a little larger than the average distance from $\bar{X} = 404.3$. The frequency table is repeated below in Table 7.7, with additional columns of products xf and x^2f. We'll calculate S^2 two ways, first using (9) and the frequency distribution, and then using (8) with the original observations from Table 7.2.

Substituting the sums from Table 7.7 in (9), we find

$$S^2 = \frac{1}{49}\left(8,189,598 - \frac{20,214^2}{50}\right) = 356.8.$$

The standard deviation is then the square root:

$$S = \sqrt{356.8} = 18.89.$$

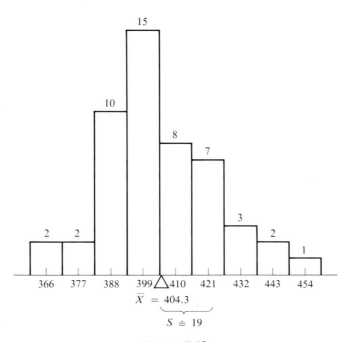

Figure 7.13

Table 7.7

x_j	f_j	$x_j f_j$	$x_j^2 f_j$
366	2	732	267,912
377	2	754	284,258
388	10	3,880	1,505,440
399	15	5,985	2,388,015
410	8	3,280	1,344,800
421	7	2,947	1,240,687
432	3	1,296	559,872
443	2	886	392,498
454	1	454	206,116
Sums:	50	20,214	8,189,598

How does this compare with your guess? If it is not close, your intuition may need adjusting. Making a guess as to the value of S before calculating it can serve as a crude check on your calculations. It will also help teach you the meaning of a standard deviation.

To use (8), we first find $\Sigma X = 20,210$ and $\Sigma X^2 = 8,186,032$:

$$S = \sqrt{\frac{1}{49}\left(8,186,032 - \frac{20,210^2}{50}\right)} = 18.71.$$

[We kept many significant digits in the calculation. Rounding off can present a real problem when using the second formulas in (8) and (9), owing to the subtraction. For instance, if we round the sample sum to 20,200 and the sum of squares to 8,190,000, we obtain

$$S^2 = \frac{1}{49}\left(8,190,000 - \frac{20,200^2}{50}\right) = 595.9.$$

The square root is 24.4, quite far from 18.9.   ■

Calculating a standard deviation is easy for hand calculators and computers. Yet just *entering* data on a calculator or computer keyboard involves risks of error related to the complexity of the data. Sometimes, performing a simple linear transformation first will substantially eliminate this source of error. For this and other reasons, we present next a method involving linear transformations of data.

Linear Transformations   We saw in Section 3.3 that when a random variable undergoes the linear transformation

$$Y = a + bX, \tag{11}$$

its mean value undergoes the same transformation, and its standard deviation is multiplied by the scale factor $|b|$. The same is true for the *sample* mean and standard deviation:

When observations X are changed to Y by the linear transformation $Y = a + bX$,

$$\bar{Y} = a + b\bar{X}, \tag{12}$$

$$S_Y = |b|S_X. \tag{13}$$

These facts can be used to advantage in choosing a scale of values that will simplify calculations of the mean and variance, especially when data are grouped into equal class intervals.

Example 7.5c   To show how a transformation of the measurement scale can simplify calculations, we recalculate the mean and standard deviation for the preceding example. Choose any class mark near the center of the data (say, 399) as a new reference point; call this 0 on the new scale. Then take the spacing between class marks as a unit on the new scale. This entails the following transformation:

$$Y = \frac{X - 399}{11}, \quad \text{or} \quad X = 399 + 11Y.$$

In (11), $a = \frac{1}{11}$ and $b = -\frac{399}{11}$. So Y is the number of class marks to the right of 399—positive to the right and negative to the left. Table 7.8 gives the frequencies, the Y-values, and products for calculating mean and standard

Table 7.8

x_j	f_j	y_j	$y_j f_j$	$y_i^2 f_j$
366	2	-3	-6	18
377	2	-2	-4	8
388	10	-1	-10	10
399	15	0	0	0
410	8	1	8	8
421	7	2	14	28
432	3	3	9	27
443	2	4	8	32
454	1	5	5	25
Sums:	50	—	24	156

deviation. From the sums, we calculate the mean and s.d.'s of the Y's:

$$\bar{Y} = \frac{24}{50} = .48, \qquad S_Y = \sqrt{\frac{1}{49}\left[156 - \frac{(24)^2}{50}\right]} = 1.717.$$

So from (12) and (13),

$$\bar{X} = 399 + 11 \times .48 = 404.28, \qquad \text{and} \qquad S_X = 11(1.717) = 18.89.$$

These agree with the values obtained in Examples 7.4b and 7.5b. An advantage of transforming to Y is that we can do most of the calculations by hand.   ∎

7.6   Correlation

In Section 7.2, we described how a set of bivariate data can be displayed graphically in a scatter diagram. Such pictures show the extent and nature of a relationship between two variables in qualitative terms. However, for purposes of quantitative inference and for succinct summaries, it is useful to have a numerical measure of the relationship. The most commonly used measure is the "product moment correlation coefficient," or "coefficient of linear çorrelation," commonly referred to as simply the **correlation coefficient**.

We defined a correlation coefficient for populations in Section 3.4. To apply this definition to a sample (as usual), we imagine a distribution that assigns weight $1/n$ to each sample pair. With this assignment, μ_X and μ_Y become $\bar{X}$ and $\bar{Y}$; σ_X^2 and σ_Y^2 become V_X and V_Y; and the covariance $\sigma_{X,Y}$ [see (1) of Section 3.4] becomes

$$C_{X,Y} = \frac{1}{n}\sum(X_i - \bar{X})(Y_i - \bar{Y}) = \frac{1}{n}\sum X_i Y_i - \bar{X} \cdot \bar{Y}. \tag{1}$$

The correlation coefficient r or $r_{x,y}$ is the covariance divided by the square root of the product of the V's:[8]

$$r_{X,Y} = \frac{C_{X,Y}}{\sqrt{V_X V_Y}} = \frac{\sum(X_i - \bar{X})(Y_i - \bar{Y})}{\sqrt{\sum(X_i - \bar{X})^2}\sqrt{\sum(Y_i - \bar{Y})^2}}. \tag{2}$$

---

[8] In terms of the standard deviations with divisors $n - 1$, we can write

$$r = \frac{S_{X,Y}}{S_X S_Y}$$

if we define the covariance using the divisor $n - 1$:

$$S_{X,Y} = \frac{1}{n-1}\sum(X_i - \bar{X})(Y_i - \bar{Y}).$$

(We've canceled $1/n$'s in numerator and denominator to get the last expression.) The following equivalent formula is useful for calculating r:

Sample correlation coefficient, given data pairs $(X_1, Y_1), \ldots, (X_n, Y_n)$:

$$r = \frac{\sum XY - \dfrac{(\sum X)(\sum Y)}{n}}{\sqrt{\left(\sum X^2 - \dfrac{(\sum X)^2}{n}\right)\left(\sum Y^2 - \dfrac{(\sum Y)^2}{n}\right)}}. \qquad (3)$$

Example 7.6a   In Example 7.2d (page 327), we gave average LSAT scores (X) and average GPA's (Y) for entering students in each of 15 law schools in 1973. We repeat the data in Table 7.9 and include columns of the products and squares needed for calculating the correlation coefficient.

Table 7.9

X	Y	XY	X^2	Y^2
576	3.39	1,952.64	331,776	11.4921
635	3.30	2,095.5	403,225	10.8900
558	2.81	1,567.98	311,364	7.8961
578	3.03	1,751.34	334,084	9.1809
666	3.44	2,291.04	443,556	11.8336
580	3.07	1,780.6	336,400	9.4249
555	3.00	1,665	308,025	9.0000
661	3.43	2,267.23	436,921	11.7649
651	3.36	2,187.36	423,801	11.2896
605	3.13	1,893.65	366,025	9.7969
653	3.12	2,037.36	426,409	9.7344
575	2.74	1,575.5	330,625	7.5076
545	2.76	1,504.2	297,025	7.6176
572	2.88	1,647.36	327,184	8.2944
594	2.96	1,758.24	352,836	8.7616
Sums: 9,004	46.42	27,975.00	5,429,256	144.4846

Substituting these sums in (3) yields $r \doteq .776$. Now look at Figure 7.8, which shows the scatter diagram for these data. There seems to be a tendency toward linearity, but data point #1 is rather far from what would otherwise seem to be the "trend" line. Indeed, if we recompute r without that point, we find $r = .893$—an appreciably higher degree of linear correlation. ∎

The next two examples illustrate important special cases—symmetry and collinearity.

Example **7.6b**   Consider the five points listed in the table below and shown in Figure 7.14. The average X and the average product are zero, so the numerator of (3) is zero—and therefore, so is r. The X's and Y's are *uncorrelated*. This is because the data points are symmetrically located about the y-axis: Each positive product of deviations in the numerator is canceled by a negative product.   ■

	X	Y	XY	X^2	Y^2
	-2	0	0	4	0
	-1	1	-1	1	1
	1	1	1	1	1
	0	-1	0	0	1
	2	0	0	4	0
Sums:	0	1	0	10	3

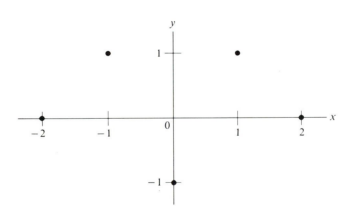

Figure 7.14

Example **7.6c**   Consider the five data points (X, Y) in the following table.

	X	Y	X^2	Y^2	XY
	0	-2	0	4	0
	2	-1	4	1	-2
	4	0	16	0	0
	6	1	36	1	6
	8	2	64	4	16
Sums:	20	0	120	10	20

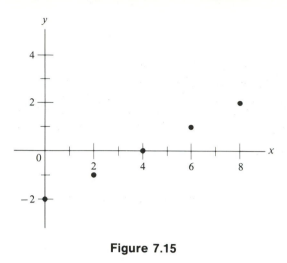

Figure 7.15

Substituting these sums in (3), we find:

$$\sum X^2 - \frac{(\sum X)^2}{n} = 120 - \frac{400}{5} = 40, \qquad \sum Y^2 - \frac{(\sum Y)^2}{n} = 10 - 0,$$

$$\sum XY - \frac{(\sum X)(\sum Y)}{n} = 20 - 0, \qquad r = \frac{20}{\sqrt{40 \times 10}} = 1.$$

The scatter plot in Figure 7.15 shows that the five data points lie on a straight line with positive slope. In Section 3.4, we showed that a population correlation coefficient equals 1 when and only when the distribution of (X, Y) is concentrated on a line with positive slope. This holds for the sample correlation coefficient as a special case.   ■

Linear Transformations

Calculating correlations by hand is tedious for large n. Some calculators are preprogrammed to calculate r. Even with this aid, however, simplifying the numbers to be entered can help to avoid mistakes. Both variances and covariances are based on *deviations*, and these do not change when a constant is added to or subtracted from all the X's or all the Y's. So r is unchanged if we add any constant to each X and another constant to each Y. Also, changing the scale of either variable ($U = bX$ or $V = dY$) introduces the same constant multiplier (b or d) in both numerator and denominator of (2). So linear transformations (such as changing units from degrees Fahrenheit to degrees Kelvin, or kilograms to pounds) have no effect on r. Problem 39 asks you to transform the law school data of Example 7.2d and verify this.

If bivariate data points (X_i, Y_i) are transformed by $U_i = a + bX_i$ and $V_i = c + dY_i$, with $b > 0$ and $d > 0$, the new correlation coefficient is the same as the old:

$$r_{U,V} = r_{X,Y}.$$

To acquire a feel for the meaning of a value of the correlation coefficient other than ± 1, it helps to see lots of examples. Figures 7.16–7.19 give four examples, each with 500 data points.

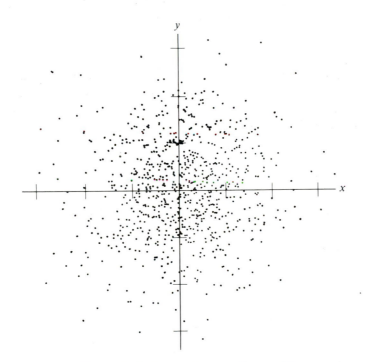

Figure 7.16 Data with correlation .05

Chapter Perspective

This chapter is different in many ways from those that precede it. It assumes a given data set—everything is known and nothing is random. For example, the word *probability* appears in this chapter only to relate the calculations of sample moments to those of population moments. Sample characteristics will be used in the following chapters as the basis for drawing conclusions about the populations from which the

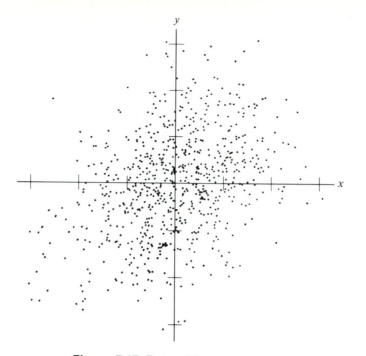

Figure 7.17 Data with correlation .3

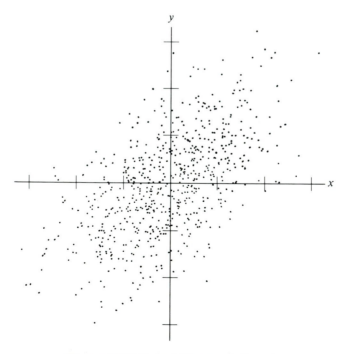

Figure 7.18 Data with correlation .6

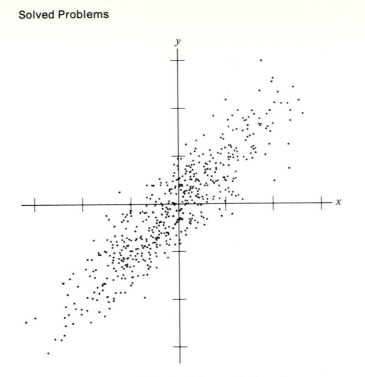

Figure 7.19 Data with correlation .9

samples are drawn. However, just as population parameters give us important features of a population, the descriptive methods developed in this chapter are also useful simply for understanding what a set of data is like. They are useful in *exploratory data analysis* (EDA)—in informal examination of data to get hints of what may be important and to suggest more formal experimentation.

Extrapolating from sample characteristics to unknown population characteristics is the fundamental problem of statistical inference, and it is this problem to which we turn next.

Solved Problems

Sections 7.1–7.2

A. The data[9] in Table 7.10, on 72 couples enrolled in an HMO, were taken at the time of birth of their child: the birthweight of the child (Wt), the education and the smoking status (E, Sm) of the father and of the mother, and the church-going habits of the family (C).

---

[9] Extracted from an extensive set of such data given in J. L. Hodges, D. Krech, and R. S. Crutchfield, *Stat Lab* (McGraw-Hill, 1975). The ages of the mothers represented in Table 7.10 were 31–33 years.

Table 7.10

Wt	Mother E	Mother Sm	Father E	Father Sm	C	Wt	Mother E	Mother Sm	Father E	Father Sm	C
10.9	2	N	2	Q	1	5.6	2	S	3	Q	1
6.1	2	S	2	S	1	6.6	4	N	4	S	6
5.6	4	N	4	N	1	7.8	1	N	2	Q	1
7.2	2	Q	4	Q	1	7.4	2	Q	4	Q	1
7.3	4	Q	2	Q	1	9.2	3	N	4	N	6
7.9	4	N	4	N	6	7.6	3	N	4	N	1
6.6	4	S	2	S	1	6.6	1	S	1	S	2
8.0	3	S	4	S	6	7.4	2	N	3	N	4
8.3	1	N	2	N	1	6.4	1	S	2	N	6
8.5	3	N	4	N	4	8.3	0	N	1	N	1
7.1	4	N	4	Q	6	5.4	1	S	2	Q	4
7.2	4	S	4	S	6	7.1	3	S	4	S	4
7.1	3	N	4	N	1	9.9	4	S	4	N	6
7.1	2	N	3	N	6	6.7	1	N	2	Q	3
6.0	4	S	4	N	6	6.8	4	Q	4	N	6
8.3	4	N	4	N	5	5.9	1	S	1	N	2
7.8	2	S	2	S	3	7.1	1	S	2	S	4
6.9	2	S	2	N	1	6.5	2	N	4	N	1
7.8	2	N	2	N	1	6.7	1	N	1	Q	1
7.9	3	N	2	S	1	6.8	2	Q	4	Q	1
5.4	3	Q	2	N	1	6.9	2	S	3	S	4
6.1	4	N	4	N	6	3.3	4	N	4	N	4
5.6	2	N	0	S	1	7.5	3	S	4	Q	5
7.8	3	N	4	Q	1	5.3	2	N	0	N	1
8.4	3	N	4	Q	6	6.6	2	N	2	N	1
7.4	4	N	4	Q	1	6.1	4	N	1	Q	1
8.6	4	N	4	Q	2	7.6	2	S	1	S	6
8.8	3	N	4	N	6	7.6	2	S	1	Q	4
9.2	4	N	4	S	1	7.3	4	N	4	Q	6
6.6	4	N	4	Q	1	5.6	2	N	4	N	2
6.6	1	N	2	N	1	7.1	2	N	2	S	4
6.1	4	N	4	Q	1	6.6	4	Q	4	N	6
6.6	2	S	4	N	6	6.9	4	N	4	N	4
6.9	3	Q	3	Q	6	7.8	2	N	3	Q	1
4.5	2	S	1	Q	1	8.6	0	N	3	S	1
7.8	4	N	4	Q	1	8.6	4	N	4	N	6

Education (E):

4 = College grad
3 = Some college
2 = HS grad
1 = 8–12 gr.
0 = < 8 gr.

Smoking (Sm):

S = Smokes
Q = Quit
N = Never

Church (C):

1 = All attend
2 = Mother and
    children
    attend
3 = Only children
    attend
4 = Sporadic
5 = Holy days
6 = None

(a) Make a cross-classification on the three variables church, father's smoking status, and mother's smoking status, using mother's smoking as the layering variable.

(b) Obtain a two-way classification of father's and mother's smoking status from (a).

(c) In what fraction of the couples in which the mothers smoke do the fathers also smoke?

(d) Construct the contingency table for education of the father and of the mother. Comment.

Solution

(a) The tables are as follows:

Mother never smoked:

		Church						
		1	2	3	4	5	6	
	N	10	1	0	4	1	6	22
Father's smoking:	Q	10	1	1	0	0	3	15
	S	4	0	0	1	0	1	6

Mother smoked but quit:

		Church						
		1	2	3	4	5	6	
	N	1	0	0	0	0	2	3
Father's smoking:	Q	4	0	0	0	0	1	5
	S	0	0	0	0	0	0	0

Mother smokes:

		Church						
		1	2	3	4	5	6	
	N	1	1	0	0	0	4	6
Father's smoking:	Q	2	0	0	2	1	0	5
	S	2	1	1	3	0	3	10

(b) The table is constructed from marginal totals shown in (a):

		Mother:		
		N	Q	S
	N	22	3	6
Father:	Q	15	5	5
	S	6	0	10

(c) Twenty-one $(6 + 5 + 10)$ mothers smoke; ten of their husbands also smoke: $\frac{10}{21}$.

(d)

		Mother's education:					
		0	1	2	3	4	
	0	0	0	2	0	0	2
	1	1	3	3	0	1	8
Father's education:	2	0	7	7	2	2	18
	3	1	0	5	1	0	7
	4	0	0	6	10	21	37
		2	10	23	13	24	72

[Our comment: The levels agree (main diagonal) in nearly half the couples; in all but twelve, they are within one level of each other (on the three middle diagonals).]

B. Make a stem-leaf diagram or other convenient display of the weight data in Problem A.

Solution

Most of the weights begin with 6 or 7; splitting the stems is in order. The class intervals in a stem-leaf diagram may not be the best choice. A frequency distribution depends on the choice of class interval and choice of initial endpoint. We show here a stem-leaf diagram and a distribution in which the interval width is .7, starting at 3.1.

Depth	Stem	Leaves (tenths)
1	3	1
1	3	
1	4	
2	4	5
5	5	443
10	5	66669
16	6	011114
33	6	56666666677889999
13	7	1111112233444
26	7	566688888899
14	8	033345
8	8	5668
4	9	22
1	9	9
1	10	
1	10	9

$n = 72$ (bracketing the rows with stems 6 | , 7 | , 7 | at depths 33, 13, 26)

Class	Frequency
3.1–3.7	1
3.8–4.4	0
4.5–5.1	1
5.2–5.8	7
5.9–6.5	8
6.6–7.2	24
7.3–7.9	17
8.0–8.6	9
8.7–9.3	3
9.4–10.0	1
10.1–10.7	0
10.8–11.4	1

Sections 7.3–7.4

C. Use the stem-leaf diagram in Problem B to find the five-number summary of the weight data and construct a box plot.

Solution

With 72 observations in all, the median is between the 36th and 37th smallest; the quartiles are the medians of the lower 36 and the upper 36. Thus, $\tilde{X} = 7.1$, the average of

the 3rd and 4th on the 7-stem; Q_1 = average of 18th and 19th = 6.6, the 2nd and 3rd on
the 6-stem; and Q_1 = 7.7, the average of the 18th and 19th from the largest. The box plot
is shown in the figure.

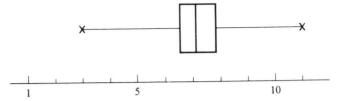

D. Tests of a process for removing nitrogen oxides from flue gases yielded the following
data[10] on the amount removed (in percent) from a coal-burning facility in Pittsburgh:

91	95	90	83	91	65	55	42	55
81	89	38	20	45	58	85	78	70

Construct a stem-leaf diagram and find the median, mean, quartiles, range, interquartile
range, and midrange.

Solution

With just two digits, the first is a stem and the second a leaf, and we need stems from
2 to 9:

Depth	Stem	Leaves
1	2	0
2	3	8
4	4	25
7	5	558
8	6	5
2	7	08
8	8	1359
4	9	0115

The median is on the 7-stem, which contains the 9th and 10th of the 18 observations:

$$M = \frac{70 + 78}{2} = 74.$$

The midrange is halfway from the smallest to the largest or m.r. = (20 + 95)/2 = 57.5.
The sum of the 18 numbers is 1,231, so the mean is $\bar{X} = \frac{1,231}{18} \doteq 68.4$. The range is the
width of the interval covered by the data, or the largest minus the smallest: $R = 95 - 20 = 75$. The first quartile is the median of the smallest nine ($= n/2$) observations,
or the fifth smallest: $Q_1 = 55$. The third quartile is the median of the largest nine:
$Q_3 = 89$. The interquartile range is IQR $= Q_3 - Q_1 = 89 - 55 = 34$.

---

[10] J. T. Yeh et al., *Environmental Progress* (1985), 223–228.

Section 7.5

E. Find the mean and s.d. of the birthweight data in Problem A from the frequency distribution shown in the accompanying table (columns 1 and 2).

x_i	f_i	y_i	$f_i y_i$	$f_i y_i^2$
3.4	1	−5	−5	25
4.8	1	−3	−3	9
5.5	7	−2	−14	28
6.2	8	−1	−8	8
6.9	24	0	0	0
7.6	17	1	17	17
8.3	9	2	18	36
9.0	3	3	9	27
9.7	1	4	4	16
10.4	0	5	0	0
11.1	1	6	6	36
$n = 72$			24	202

Solution

Arithmetic is simpler with the transformation

$$Y_i = \frac{X_i - 6.9}{.7}, \quad \text{or} \quad X_i - 6.9 + .7Y_i.$$

Then $\bar{Y} = \frac{24}{72} \doteq .333$, $\bar{X} = 6.9 + .7 \times .333 \doteq 7.13$, and

$$S_Y^2 = \frac{1}{71}(202 - 72 \times (.333)^2) \doteq 2.732, \quad \text{or} \quad S_Y = 1.653,$$

so [by (13) of Section 7.5] $S_X = .7 \times 1.653 \doteq 1.16$. (For the ungrouped data, $\bar{X} = 7.13$, $S_X = 1.22$.)

F. In the preceding problem, suppose we want to delete the entry 3.3 (perhaps it is a premature birth and we want only full-term weights). Find the mean and s.d. of the 71 remaining weights.

Solution

We first retrieve the sum and sum of squares from what's given:

$$\sum X = n\bar{X} = 72 \times 7.133 = 513.6,$$

and

$$\sum X^2 = (n - 1)S^2 + n\bar{X}^2 = 71 \times 1.157^2 + 72 \times 7.133^2 = 3{,}758.7.$$

Subtracting 3.3, we have the new sum of 510.3, and mean $\frac{510.3}{71} = 7.187$. Subtracting 3.3^2 from the sum of squares, we obtain 3,747.8; the new variance is then

$$\frac{3{,}747.8 - 71 \times 7.187^2}{70} = 1.149,$$

and the s.d. is the square root, or 1.07.

G. Try finding the s.d. of these numbers on a statistical hand calculator: 900,001; 900,002; 900,003. In view of (13) of Section 7.5, this should be the same as the s.d. of the numbers 1, 2, 3. Is it?

Solution

What you get depends on the particular calculator you use: most will give 0, but 1 is the correct answer. To see what's happening, if you just try squaring 900,001, the calculator rounds this to 810,001,800,000, whereas it should be 810,001,800,001.

H. Show that the average absolute deviation is smallest when taken about the sample median.

Solution

We'll find the minimum by examining the slope of $\Sigma|X_i - a|$ as a function of a. The derivative of the absolute value function is

$$\frac{d}{dy}|y| = \begin{cases} +1, & y > 0, \\ -1, & y < 0. \end{cases}$$

The derivative of $\Sigma|X_i - a|$ with respect to a is thus a sum of $+1$'s and -1's: For each term with $a > X_i$, the derivative is -1; for each with $a < X_i$, the derivative is $+1$. So the slope of $\Sigma|X_i - a|$ is $-n$ for $a < X_{(1)}$. As a increases, the slope increases by 2 as a passes each sample value. Eventually, for $a > X_{(n)}$, the slope becomes $+n$. If n is odd, the slope will jump from -1 to $+1$ as a passes the middle observation—the median; this produces the minimum. If n is even, the slope is 0 for any value of a between the middle two observations including the median. So in either case, $\Sigma|X_i - a|$ decreases to the left of the median and increases to the right; the median is a minimum point.

I. Show that the sample standard deviation is at least as large as the mean deviation.

Solution

Since this is true for discrete probability distributions (Problem 21 of Chapter 3), it holds for corresponding sample moments, which are defined for the sample distribution—the distribution that assigns probability $1/n$ to each sample observation. However, we should note that the sample standard deviation, defined with $n - 1$ in the denominator in place of n, is even larger than the standard deviation of the sample distribution.

Section 7.6

J. Mineral compositions were determined for 12 rocks taken from an area in England with many excellently preserved plant fossils.[11] Quartz and siderite concentrations (in

---

[11] C. R. Hill, D. T. Moore, J. T. Greensmith, and R. Williams, "Palaeobotany and petrology of a middle Jurassic ironstone bed at Wrack Hills, North Yorkshire," *Yorkshire Geological Society 45* (1985), 277—292.

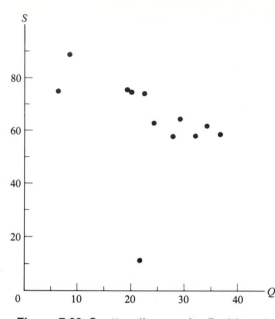

Figure 7.20 Scatter diagram for Problem J

percent) are as follows:

Rock	1	2	3	4	5	6	7	8	9	10	11	12
Quartz	34.3	37.5	22.3	18.7	27.9	36.7	19.8	21.6	7.9	31.5	7.0	24.4
Siderite	62.2	58.3	74.7	75.0	64.9	59.0	74.3	11.1	89.5	58.3	74.7	63.5

A plot (Figure 7.20) quickly reveals that #8 is very different from the rest. (One possibility is that there was an error in measuring or in recording it. Another is that the measurement is correct but that this rock was formed during a different geological period.) Calculate the correlation coefficient with and without observation #8, given these sums (for all 12 data points):

$$\sum X = 289.6, \qquad \sum Y = 765.5, \qquad \sum XY = 17{,}690.73,$$

$$\sum X^2 = 8{,}112.64, \qquad \sum Y^2 = 52{,}831.01.$$

Solution

We simply substitute the given sums and $n = 12$ into (3) of Section 7.6 to obtain $r = -.369$ $(n = 12)$. When we delete #8, $n = 11$, and $\Sigma X = 268.0$, $\Sigma Y = 754.4$, $\Sigma XY = 17{,}450.97$, $\Sigma X^2 = 7{,}647.08$, and $\Sigma Y^2 = 52{,}707.8$. Then $r = -.885$. Point #8 is clearly *influential*. We note that except for #8, the points are nearly on a line; the sum $x + y$ is roughly constant:

Rock #:	1	2	3	4	5	6	7	8	9	10	11	12
$x + y$:	96	96	97	94	93	96	94	34	97	90	82	88

This suggests that quartz and siderite together make up some nearly fixed proportion of the rocks—except for #8.

Problems

For problems marked with an asterisk, answers are provided in Appendix 2.

Sections 7.1–7.2

∗1. For the variables in Table 7.11:
  (a) Construct frequency tables for Class (Cla), Marijuana (Pot), and Number of siblings (Sib).
  (b) Construct a two-way frequency table for the joint distribution of the variables Class and Marijuana.

2. From the table in Problem 1(b), find
  (a) the proportion of junior girls who have not used marijuana.
  (b) the proportion of senior girls who have not used marijuana.
  (c) the proportion of girls who have not used marijuana.
  [Refer to the code in Table 7.1, page 318.]

∗3. **(a)** Construct a two-way table for the joint distribution of Class and Marijuana Use for the males in Table 7.1, page 318.
  (b) The table in (a), together with the table in Problem 2, constitutes a three-way cross-classification of the whole class. Construct another three-way table using Class as the layering variable.

Table 7.11 Female students—statistics class

#	Ht	Cla	Wt	Sib	Pot	Beh	Smo	#	Ht	Cla	Wt	Sib	Pot	Beh	Smo
1	67	3	122	4	N	M	N	24	64	4	170	1	N	M	N
2	72	3	130	2	N	B	N	25	64	2	115	7	Y	M	Q
3	64	2	120	3	N	M	S	26	63	3	100	7	Y	T	N
4	67	4	140	3	Y	B	S	27	67	1	125	3	Y	T	Q
5	66	3	125	3	N	B	S	28	72	1	130	5	N	M	N
6	66	3	127	1	Y	B	Q	29	59	3	110	5	N	M	N
7	67	3	130	3	Y	B	Q	30	69	4	130	3	Y	M	N
8	67	3	150	1	N	M	N	31	63	4	125	5	N	M	N
9	69	4	155	13	N	M	N	32	65	1	135	9	N	M	N
10	66	4	263	8	N	M	N	33	62	3	115	2	Y	M	Q
11	63	3	105	2	N	B	Q	34	62	2	115	0	Y	M	Q
12	64	3	115	2	N	M	N	35	68	3	135	3	N	M	N
13	66	4	135	1	Y	B	S	36	70	2	135	5	N	M	N
14	64	4	110	3	N	B	N	37	68	3	120	1	Y	M	N
15	65	3	115	2	N	M	N	38	64	2	130	3	Y	T	N
16	66	3	145	7	N	B	N	39	74	2	135	6	N	M	N
17	66	2	125	5	N	B	N	40	65	2	140	6	N	M	N
18	66	4	165	2	Y	B	Q	41	68	2	105	3	N	M	N
19	69	3	135	1	N	B	N	42	63	2	145	13	Y	H	N
20	66	3	135	2	N	B	N	43	62	5	135	1	N	M	N
21	64	2	120	1	N	M	N	44	69	2	130	2	N	M	N
22	63	5	200	1	N	B	N	45	64	3	140	1	Y	H	S
23	63	3	111	8	N	B	Q								

4. Obtain the two-way classification on Class and Sex from the three-way classification of Problem 3 as a marginal distribution.

*5. A group of 129 grade-school children was cross-classified[12] on race (black or white), sex (M or F), and whether their placement in reading groups was below, at, or above grade level (GL):

	Black		*White*	
	F	*M*	*F*	*M*
Below GL	6	4	15	11
At GL	14	12	9	16
Above GL	9	13	6	14

(a) Recast these data so that sex is the "layer variable." (That is, make two two-way tables, one each for males and females.)

(b) What fraction of those below grade level are black?

6. Regroup the spot-weld data of Table 7.2, (page 323) using eight class intervals as in Example 7.2b, but starting with the class interval 358–370. (Observe that the table you get is different from the one given in Example 7.2b. Are both correct?)

7. The slag (leftover solid waste) from a lead-smelting process was checked for concentration of lead (a toxic pollutant). These observations on the concentration X were obtained:[13]

2.83	209	.53	1300	3.86	141	543	21.9	125
17.9	4.88	31.8	774	493	409	1.58	21.2	16.6
35.3	15.7	59.1	2.26	291	4.18	380	39.4	340

(a) Consider ways of representing these data—dot diagram, stem-leaf diagram, frequency distribution, and so on. What problems do you see?

(b) Try a *transformation*: Let $Y = \log X$ (either base e or base 10). Transform each observation and represent the Y's in a dot diagram or in a frequency table.

8. Make a stem-leaf diagram for the heights of female students in Table 7.11,
   (a) splitting the stems in two [0–4 and 5–9].
   (b) splitting the stems in five [0–1, 2–3, 4–5, 6–7, 8–9].
   (c) Construct a histogram for the frequency distribution implied in the stem-leaf diagram of (b).

---

[12] Data from L. Grant, "Black females 'place' in desegregated classrooms," *Sociology of Education 57* (1984), 98–110.

[13] N. Woodley and J. V. Walters, "Hazardous waste characterization extraction procedures...," *Environmental Progress* (1986), 12–17.

9. Birthweights of 30 babies were recorded[14] in a maternity ward, as follows (given in ounces):

131	120	112	88	114	128	133	104	108	94
133	124	84	132	107	144	93	114	116	120
128	132	108	107	86	123	116	92	106	108

***(a)** Make a stem-leaf diagram for these data.

(b) Construct a histogram for these data.

10. The following are times to burnout of light bulbs of a certain type,[15] expressed in thousands of hours:

1.07	0.86	1.16	1.02	0.92	0.52	0.93	1.00	0.90	1.00
1.19	0.82	0.84	1.04	1.02	1.13	1.00	0.61	0.99	1.07
1.24	1.44	1.12	1.22	1.13	1.16	1.17	1.02	1.24	1.01
1.08	1.21	0.80	1.03	1.49	1.11	0.76	1.25	1.02	1.11

Construct a frequency table and a graphical representation.

11. Two judges of piano contestants turned in these ratings:

				Performer				
Judge	1	2	3	4	5	6	7	8
1	92	90	85	96	92	88	96	88
2	89	90	88	93	90	85	95	90

Make a scatter diagram.

12. The weight and height of each of ten students were reported as follows, in (lb, in):

(133, 65), (125, 68), (155, 71), (150, 70), (155, 68),
(155, 65), (125, 67), (105, 62), (170, 71), (175, 72).

Make a scatter diagram.

Sections 7.3–7.4

***13.** Plasma clearances of a certain drug (in ml/min/kg) for 32 nonsmokers[16] were recorded as follows:

3.1	3.7	4.2	4.4	4.8	4.8	5.4	5.7	5.8	6.0	6.0
6.4	7.0	7.2	7.2	7.3	7.4	7.5	7.8	8.0	8.2	8.6
9.0	9.4	10.2	10.3	11.2	13.2	13.6	14.5	15.5	26.0	

Find the mean, median, midmean, and midrange.

---

[14] From a student's class project.

[15] D. J. Davis, "An analysis of some failure data," *J. Amer. Stat. Assn. 47* (1952), 142.

[16] Data from a consulting project.

14. Find the mean time to burnout from the frequency table in Problem 10.

*15. For the birthweights in Problem 9,
(a) find the median birthweight.
(b) find the mean birthweight.

*16. Find the mean of the spot-weld data of Example 7.2b, from the table you obtained in Problem 6. (Notice that this is different from the mean obtained in Example 7.3b.)

17. Find the mean and median of the weights (column labeled "Wt") of the females in Table 7.11, page 357. (If you choose to use a frequency table summary, an approximation to the mean will suffice.)

*18. The population density per square mile in 1980 was reported by the Bureau of the Census for each of the 50 states, as follows (the order is alphabetical by states' names):

77	1	24	44	151	28	638	308	180	94
150	12	205	153	52	29	92	95	36	429
733	163	51	53	71	5	21	7	102	986
11	371	120	9	263	44	27	264	898	103
9	112	54	18	55	135	62	81	87	5

Find the median density.

19. The median age for each state in 1980, as reported by the Census Bureau, was as follows (same order as in Problem 18):

29.3	28.6	28.4	30.1	31.2	29.0	27.4	30.1	28.9	29.8
26.1	32.0	27.6	29.1	28.9	29.7	31.9	30.2	30.1	29.8
29.2	29.7	29.9	27.4	29.2	30.3	29.6	32.1	28.2	30.4
30.6	34.7	29.2	30.4	27.7	30.1	28.3	31.8	24.2	29.4
29.9	28.7	30.0	30.3	30.9	32.2	29.9	28.2	29.4	27.1

Use a stem-leaf diagram to find the *median* of these ages. Explain why this is *not* necessarily the median age in the United States.

*20. Construct a box plot
(a) for the data of Problem 13.
(b) for the birthweights of Problem 9.

21. Among the data collected from families participating in a California study[17] were children's birthweights. Birthweights of children of mothers in the age range 23–25 years are shown in the accompanying stem-leaf diagrams. Construct a box plot for these data.

---

[17] The family data are reported in Hodges, Krech, and Crutchfield, *Stat Lab* (New York: McGraw-Hill, 1975).

Stem (lb)		Leaves (tenths)
2		3
2		
3		
3		9
4		1
4		
5		3344
5		666899
6		111233344
6		5677788889999
7		01122333333444
7		566677899
8		33333
8		67
9		034
9		678
10		
10		5

Section 7.5

*22. Find the standard deviation of these numbers: 1, 1, 3, 3. [Note that all distances from the mean are 1. The mean deviation has to be 1, but using $n - 1$ in (7) of Section 7.5 gives an s.d. that is a bit larger. The divisor n in (2) of Section 7.5 gives an s.d. of 1.]

*23. For the plasma clearance data of Problem 13,
  (a) find the standard deviation.
  (b) find the interquartile range (third quartile minus the first quartile).

*24. Find the s.d. of the birthweights in Problem 9. (Compare Problem 15.)

*25. Find the s.d. of the spot-weld data of Example 7.2b from the table you obtained in Problem 6. (See Example 7.5b.)

26. Approximate the s.d. of the weights in Table 7.11 using the frequency distribution from Problem 17. (If you didn't construct one for that problem, do it now.)

*27. Given that the mean of 100 observations is 42.5 and that their s.d. is 6.313, find the sum of the squares of the observations.

28. For the birthweight data in Problem 9,
  (a) calculate the *mean deviation* about the mean and compare with the s.d. found in Problem 24. (Compare Problem 41.)
  (b) find the mean deviation about the median and compare with the result in (a). (See Solved Problem H.)

29. Find the s.d. of times to burnout in Problem 10 from the frequency table obtained there. (See also Problem 14.)

30. A sample of rivet head diameters is summarized in the accompanying frequency table. Using the coding

$$Y = \frac{X - 13.39}{.09},$$

find the mean and s.d. of the Y's. From these, find the mean and s.d. of the rivet head diameters.

Class interval	Frequency
13.17–13.25	1
13.26–13.34	7
13.35–13.43	10
13.44–13.52	5
13.53–13.61	6
13.62–13.70	1

*31. A sample of ten temperature readings in degrees Fahrenheit has mean 50 degrees and s.d. 9 degrees. Find the mean and s.d. in degrees Celsius [$C = \frac{9}{5}(F - 32)$].

32. The parallel axis theorem [(5) of Section 7.5] shows algebraically why the sample variance V is the smallest second moment. Show this minimum property using the method of calculus to minimize $\Sigma(X_i - c)^2$ with respect to c.

Section 7.6

*33. Find the correlation coefficient for the data in Problem 11.

34. Find the coefficient of linear correlation between height and weight for the ten students of Problem 12.

*35. Suppose, in Problem 12, the heights had been measured in centimeters and the weights in kilograms. Find the correlation coefficient between height and weight when they are measured in these units, from the answer to Problem 34.

36. Given $\bar{X} = 10$, $\bar{Y} = 5$, $S_X = 3$, $S_Y = 2$, and $r = .5$, find the corresponding statistics for (U, V), where $U = 32 + 9X/5$ and $V = 2.2Y$.

*37. Plot these points in a scatter diagram: $(1, 1), (1, 2), (2, 1), (2, 2)$, and $(4, 4)$. Calculate r, the correlation coefficient,
    (a) for all five points.
    (b) for the first four points [that is, omitting $(4, 4)$]. (You may be able to find r by inspection of the scatter diagram in this case.)
    (The point of this exercise is to show the tremendous influence one point can have on the correlation coefficient, but the possibility of such influential points is not restricted to small data sets.)

38. Ten states were selected at random (using a random number table). For these states, the per capita personal income (in dollars) and the divorce rate (per thousand) were as

follows (1980 census):

State	Income (X)	Divorce rate (Y)
Tennessee	7,720	6.8
California	10,938	5.8
Massachusetts	10,125	2.9
Wisconsin	9,348	3.7
Michigan	9,950	4.4
Nebraska	9,365	4.1
Mississippi	6,580	5.5
Colorado	10,025	6.4
Pennsylvania	9,434	3.0
North Dakota	8,747	3.3

Plot these data on a scatter diagram and calculate the correlation coefficient. How would the correlation change if both Tennessee and Mississippi were omitted? (The correlation for all 50 states, the population from which these ten were drawn, is about .02.)

39. In Example 7.2d, we gave data on 15 American law schools, consisting of pairs (X, Y), where $X =$ average LSAT score and $Y =$ average undergraduate GPA. The data can be simplified somewhat by subtracting 500 from each LSAT, and subtracting 3 from each GPA and multiplying by 100: $U = X - 500$, $V = 100(Y - 3)$. Do this, calculate r for the pairs (U, V), and so verify that r does not change under the linear transformation.

40. Defining the sample distribution by putting mass $\frac{1}{n}$ at each data pair (x_i, y_i), we interpret an expectation as an arithmetic mean. With this in mind, recast Schwarz's inequality (10) of Section 3.4, as an inequality involving sums of squares and products.

41. Show that the mean deviation is never larger than the standard deviation, when the latter is defined with divisor n (rather than with divisor $n - 1$),
    (a) using the inequality obtained in Problem 40. [*Hint*: Let $y_i = 1$ and replace x_i with $(x_i - \bar{x})$.]
    (b) as the corresponding fact was shown for probability distributions in Problem 23 of Chapter 3.

Samples, Statistics, and Sampling Distributions

A typical problem in disciplines that apply statistics is to learn something about a particular population. Scientists, market analysts, and others sample populations to learn about various population characteristics. A politician wants to know the opinion of a certain group of people; a dairy scientist is interested in the butterfat content of milk from cows that have been fed a certain diet; a physician needs to know how patients of a certain type respond to a treatment; and so on. It may be impractical or even impossible to examine the whole population, so one must rely on a sample from the population.

A sample is considered representative if it accurately reflects the population characteristics of interest. Unfortunately, no method of sampling short of taking the whole population *guarantees* a representative sample. Some methods are better than others at producing nearly representative samples, but many (including some in common use) usually produce samples that are quite unlike the population. For instance, samples are apt to be biased if they contain only individuals with similar characteristics, such as having the same religion, coming from the same home town, or being at a cocktail party together. Samples made up of those who volunteer to be included are notoriously misleading but are commonly used.

Example 8a **A Voluntary Response Sample**
In early 1988, "Dear Abby" asked her readers to write to her about whether they cheated on their spouses. She received 210,336 responses—149,786 females and 60,550 males. "The results were astonishing," she writes. "There are far more faithfully wed couples than I had surmised. . . . The marriage vow . . . is still honored by 85% of the females and 74% of the males who responded." She seems to assume that her voluntary response sample is representative of married folks generally, but it is almost surely not. ■

Random sampling (as defined in Chapter 1) has a good chance, roughly speaking, of producing a sample that is close to representative—especially if the sample is large. Although even random sampling cannot guarantee complete accuracy, it allows for assessing accuracy.

In most applications, the sampler is concerned with a particular population characteristic, rather than with every aspect of a population distribution. A population characteristic is called a **parameter**; the population mean is an example. Sample information about a parameter can often be summarized in some sample **statistic**. (A sample characteristic could be called a *sample parameter*, but calling it a *statistic* is standard terminology.) Good summary statistics are sometimes found by intuition; for instance, an obvious choice for estimating a given population parameter is the corresponding sample characteristic. The notions of likelihood and sufficiency (Sections 8.2 and 8.3) help us find appropriate statistics.

The value of any sample statistic depends, of course, on the particular sample one happens to obtain: *It varies from sample to sample.* Thus a statistic is a random variable. As such, it has a probability distribution, called its **sampling distribution**. Much of this chapter is devoted to obtaining and describing sampling distributions of commonly used statistics. The moment-generating function is useful in deriving sampling distributions for the sample mean.

8.1 Random Sampling

Chapter 1 introduced the idea of sampling at random from a finite population. In the lottery model for such a process, each ticket represents a population member. Imagine selecting one ticket by drawing blindly from the collection of all tickets. In random sampling *with replacement*, we replace each ticket and mix the tickets thoroughly before selecting the next one. In random sampling *without replacement*, tickets that are drawn are not replaced, so they are not available for subsequent selections.

> **Random sampling** from a finite population: Individuals are selected at random, one at a time from those available, until the prescribed sample size *n* is reached, either with replacement and mixing or without.

We saw in Chapter 1 that random sampling (with or without replacement) has the following properties:

(i) All possible samples are equally likely, and (as a result)
(ii) All population members have the same chance of being included.

The number of possible samples may be quite different under sampling with replacement and sampling without replacement.

In some cases, the population of interest is too unwieldy for random sampling in the sense defined above. For example, although one can imagine selecting an individual at random from the population of the United States, it is almost impossible actually to do so. Opinion polls, TV viewer surveys, and so on, employ methods that are harder to describe but easier to carry out than random sampling.

For instance, poll takers may divide the population into *strata*—say by locale, age, or income level—and sample at random in each stratum. Another technique is to *cluster* into convenient groups such as households or precincts, sample the clusters, and then sample some or all individuals from each cluster selected. These variations[1] are used partly out of practical necessity and partly to reduce sampling error.

We return to (strictly) random sampling. Suppose we are interested in a particular measurement X on the individuals in a population. The value of X for an individual selected at random from the population is a random variable, and we refer to its distribution as the *population distribution*. A random sample of size n is a sequence: $\mathbf{X} = (X_1, X_2, \ldots, X_n)$, where X_i is the X for the ith individual selected. The variables X_i are *exchangeable* (Sections 2.6 and 5.6), whether sampling is with or without replacement. In particular, the individual X_i's have a common distribution—the population distribution.

Example 8.1a   **Checking Lot Quality**

Consider sampling from a lot consisting of 20 articles. Suppose that two articles are defective and the rest are good. To determine whether or not an article is defective, it must be tested. Let

$$X = \begin{cases} 1 & \text{if the article selected is defective,} \\ 0 & \text{if it is good.} \end{cases}$$

The distribution of X is the population distribution, with p.f. $f(1) = \frac{2}{20}$ and $f(0) = \frac{18}{20}$.

A random sample of three articles is a sequence of 1's and 0's: X_1, X_2, X_3. In sampling with or without replacement, the X_i are exchangeable, and each has the distribution of X. With no replacement, the possible sample sequences are as follows (written without commas):

110, 101, 011, 001, 010, 100, 000.

(The sequence 111 is impossible in this case.) The sampling scheme implicitly defines the probability of each of these sample sequences. For instance, by the multiplication rule of Section 2.3,

$$P(001) = \frac{18}{20} \cdot \frac{17}{19} \cdot \frac{2}{18}.$$

Similar calculations yield the probabilities of the other possible samples shown in Table 8.1.

In sampling with replacement, there are eight possible sequences, since it is then possible to observe 111. In this case, the sequence elements are not only exchangeable but also independent. So each sequence probability is the product of three factors: a factor $\frac{18}{20}$ for each 0 and a factor $\frac{2}{20}$ for each 1. Table 8.1 shows these probabilities (to three decimal places).

---

[1] See W. G. Cochran, *Sampling Techniques*, 2nd ed. (New York: Wiley, 1963) for more precise descriptions of such variants of sampling and their advantages and disadvantages.

Table 8.1 Probabilities of possible samples sequences

Sequence	000	001	010	100	110	101	011	111
With Replacement	.729	.081	.081	.081	.009	.009	.009	.001
Without Replacement	.716	.089	.089	.089	.005	.005	.005	0

■

Sampling finite populations with replacement is not common. Indeed, in some settings it is impossible. Testing can even be destructive, as in testing automobile tires, bullets, or flashbulbs. As we saw for Bernoulli populations in Section 4.3, however, the simpler formulas based on sampling with replacement are usually adequate when the sample size is much smaller than the population size. Most of this chapter assumes random sampling with replacement.

We have seen that in sampling with replacement and mixing, the sample observations are independent and have a common distribution. When the "population" is only a conceptual one, it is often reasonable to assume (at least as a first approximation) that successive observations are independent and identically distributed. Examples of this sort are repeated measurements of time, mass, temperature, and the like—in general, the results of repeated trials of the same experiment. We refer to any sequence of independent variables with a common distribution as a *random sample*.[2] Whether or not there is an actual population, we refer to that common distribution as the *population distribution*.

Example 8.1b   **Standard Weights**

One of the standard weights used by the National Bureau of Standards is made of a chrome–steel alloy and has a nominal weight of 10 grams—about the weight of two nickels. Such standard weights are used in calibrating the various standard weights at the state and local levels. To assess the variability in its weighing apparatus, the Bureau weighs this 10-gram weight repeatedly (once a week!). Friedman, Pisani, and Purves[3] report 100 measurements made in the years 1962–1963. The first 20 of these are as follows, given as numbers of micrograms below the nominal weight of 10 grams:

409	400	406	399	402	406	401	403	401	403
398	403	407	402	401	399	400	401	405	402

Successive measurements are assumed to be the result of sampling at random from the "population" of all possible measurements.   ■

A *random sample* of size n is a sequence of independent observations $\mathbf{X} = (X_1, \ldots, X_n)$, each with the distribution of the population being sampled.

---

[2] Although our definition of the phrase *random sample* is one commonly used, a sample produced by random sampling is a "random sample" in this sense only if sampling is done with replacement.

[3] *Statistics* (New York: W. W. Norton, 1978).

The mathematical description of a random sample is especially simple: Because of the independence, the joint p.d.f. (continuous case) or p.f. (discrete case) of the sample observations is the product of their marginals. The marginal distribution of each observation in a random sample is the population distribution.

The joint p.d.f. or p.f. of the observations in a random sample is

$$f(x_1, \ldots, x_n) = \prod_1^n f_X(x_i), \tag{1}$$

where f_X is the population p.d.f. or p.f.

Example 8.1c   **Normal Population with Known Variance**

Consider a random sample $(X_1, \ldots, X_n)$ from $N(\mu, 1)$. The population p.d.f. [from (10) of Section 6.1] is

$$f_X(x) = \frac{1}{\sqrt{2\pi}} e^{-(x-\mu)^2/2}.$$

The joint p.d.f. of the sample observations is the product:

$$f(x_1, \ldots, x_n) = \prod_1^n \frac{1}{\sqrt{2\pi}} e^{-(x_i-\mu)^2/2}$$

$$= \frac{1}{(2\pi)^{n/2}} \exp\left\{ -\frac{1}{2} \sum_1^n (x_i - \mu)^2 \right\}. \tag{2}$$

(To be continued.)   ∎

8.2   Likelihood

The use of the term **likelihood** in statistics is adapted from its ordinary meaning in English.

Example 8.2a   **A Whodunnit**

Some cookies disappear from a plate in the kitchen of a family with two small boys. Which is the more likely explanation—is it the younger boy who, it is reckoned, would succumb to the temptation nine times out of ten, or is it his older brother, whose past experience leads him to be more cautious, giving in only five times out of ten? The probability $\frac{9}{10}$, in comparison with $\frac{5}{10}$, would point the finger at the younger boy first. He is the "more likely" culprit.

This does not mean that we take $\frac{9}{10}$ as the *probability* that the younger boy is guilty; $\frac{9}{10}$ and $\frac{5}{10}$ do not even add up to 1! [To find such a probability requires an application of Bayes' theorem (Section 2.3); we'll do this in Chapter 16.] With respect to the question of the younger versus the older boy, we call these probabilities *likelihoods*, although only their ratio 9:5 is relevant in inference.   ∎

We'll deal mainly with models (population distributions) that are identified by a parameter (say, θ) and write $f(x|\theta)$ for the population p.d.f. or p.f. A parameter θ indexes a family of models in the sense that each possible value of θ corresponds to a different model.

As in Example 8.2a, given a sample, we take the likelihood of any population model to be the probability of obtaining that sample, assuming that model. In the case of a continuous distribution, we take the likelihood of the model to be the joint p.d.f. of the sample observations, which is proportional to the probability of an infinitesimal interval at $\mathbf{x}$ [see (8) in Section 5.6].

In a ratio of likelihoods of two values of θ, any factor not involving θ cancels. For this reason, we modify the definition slightly and consider two likelihoods the same if they differ only by such a factor. So likelihood is defined only up to a multiplicative constant—an equivalence class of functions—but with this understanding, we may refer to the "the" likelihood function.

Given a sample $\mathbf{x} = (x_1, \ldots, x_n)$ with p.d.f. or p.f. $f(\mathbf{x}|\theta)$, the *likelihood function L* is any function of θ proportional to $f(\mathbf{x}|\theta)$:

$$L(\theta) \propto f(x_1, \ldots, x_n|\theta). \tag{1}$$

Example 8.2b

Bernoulli Trials

Consider a sequence $\mathbf{X} = (X_1, \ldots, X_n)$ of independent, Bernoulli trials with parameter p. If we observe six successes followed by four failures, what is the likelihood function $L(p)$?

The p.f. of a Bernoulli variable is

$$f(x|p) = p^x(1-p)^{1-x}, \qquad x = 0, 1.$$

The joint p.f. of the sequence $\mathbf{X}$ is the product of the individual Bernoulli p.f.'s, so

$$L(p) = f(x_1, \ldots, x_{10}|p) = \prod f(x_i|p) = p^6(1-p)^4, \qquad 0 \le p \le 1.$$

Another way of viewing the experimental result is that we made *one* observation on $Y \sim \text{Bin}(10, p)$ and found $Y = 6$. The binomial p.f. for $Y = 6$ is

$$f_b(6|p) = \binom{10}{6} p^6(1-p)^4, \qquad 0 \le p \le 1.$$

The likelihood function is the same as before, because we may omit any factor not involving p. The graph of $L(p)$ is shown in Figure 8.1. Values of p near $p = .6$ are the most likely.

Suppose that instead of conducting ten trials to learn about the Bernoulli parameter p, we had planned to carry out as many trials as needed to obtain six successes and found that it took ten trials. The probability function for the required number of trials is negative binomial [(6) of Section 4.4]:

$$f_N(10|p) = \binom{9}{5} p^6(1-p)^4, \qquad 0 \le p \le 1.$$

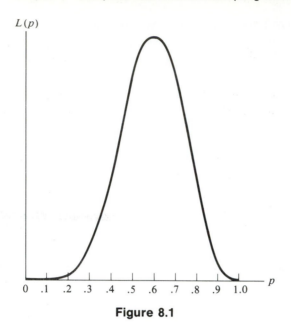

Figure 8.1

So again, the likelihood function is

$$L(p) = p^6(1 - p)^4, \qquad 0 \leq p \leq 1.$$

The sampling plan we used is irrelevant to the likelihood function. ■

Example 8.2c   **Normal X with Known σ (continued)**
Let $\mathbf{X} = (X_1, \ldots, X_n)$ denote a random sample of size n from $N(\mu, 1)$. We gave the joint p.d.f. of the sample observations in Example 8.1c:

$$f(\mathbf{x} \mid \mu) = \prod_{i=1}^{n} f(x_i \mid \mu) = (2\pi)^{-n/2} \exp\left(-\frac{1}{2} \sum (x_i - \mu)^2 \right).$$

This is the likelihood function, $L(\mu)$. We can see the dependence on μ more clearly if we expand the square, drop or insert factors independent of μ, and write $\Sigma x = n\bar{x}$:

$$L(\mu) = \exp\left(-\frac{n}{2}(\mu - \bar{x})^2 \right). \tag{2}$$

This shows that the likelihood function has the shape of a normal density centered at $\mu = \bar{x}$.

Figure 8.2 shows the likelihood function (2) for the case $n = 9$ and $\bar{x} = 2$. ■

In the definition of likelihood, there is no reason to restrict it to the case of a single-parameter family. In the next example, the population distribution depends on a multidimensional parameter.

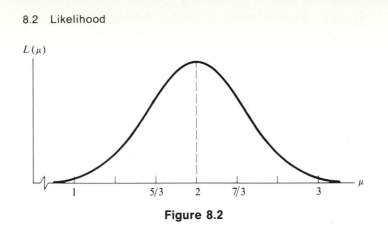

$L(\mu)$

$1 \qquad 5/3 \quad 2 \quad 7/3 \qquad 3 \qquad \mu$

Figure 8.2

Example **8.2d**   **Multinomial Sampling**

The biologist Gregor Mendel[4] crossed round yellow pea plants with wrinkled green pea plants, obtaining plants bearing peas in one of four categories. The numbers of plants in each category are shown in the accompanying table.

Category	Frequency	Probability
Round yellow	315	p_1
Round green	108	p_2
Wrinkled yellow	101	p_3
Wrinkled green	32	p_4
Sum:	556	1

The probability of this result, given the category probabilities p_i, is multinomial [(4) of Section 4.8]:

$$\binom{556}{315,\,108,\,101,\,32} p_1^{315} p_2^{108} p_3^{101} p_4^{32}.$$

We can take this as the likelihood function—or drop the constant multiplier. More generally, for a variable with k categories and sample frequencies f_i, the likelihood function is

$$L(\mathbf{p}) = p_1^{f_1} \cdots p_k^{f_k}. \tag{3}$$

∎

In later chapters, the notion of likelihood will play an important role. In the next section, we show how a consideration of the form of a likelihood function can lead to appropriate summary statistics.

[4] Gregor Mendel (1822–1884) is considered the father of modern genetics.

8.3   Sufficient Statistics

In Example 8.2c, we found the likelihood function for a random sample of size n from $N(\mu, 1)$:

$$L(\mu) = \exp\left[-\frac{n}{2}(\mu - \bar{X})^2 \right].$$

This depends on the sample only through the value of $\bar{X}$ (and n). So we could throw away the raw data, keeping only the value of $\bar{X}$ (and n), and still calculate L. This is at the heart of the notion of sufficiency: $\bar{X}$ is sufficient for μ in this case. However, the sample mean is not always sufficient for μ. Consider the next example.

Example 8.3a   **An Unusual Scale**

Imagine a scale with this rather peculiar characteristic: For a person whose weight (in lb) is θ, it gives a reading of $\theta + 1$ with probability $\frac{1}{2}$, and $\theta - 1$ with probability $\frac{1}{2}$. The mean weight is $EX = \theta$.

Consider two candidates for estimating θ, the sample mean and the sample midrange (the average of the smallest and largest observations). We weigh someone 25 times, independently, and get results $X_1, \ldots, X_{25}$. If all X_i are the same (an event with probability 6×10^{-8}), the mean and the midrange are equal and both in error by 1 unit. If not all X_i are the same, the midrange is exactly correct, but the sample mean is apt to be wrong. For instance, if the sample includes eleven 144.4's and fourteen 142.4's, the midrange is 143.4 (correct) and the mean is 143.28 (incorrect). So the midrange is clearly the better estimator.   ■

What's wrong with the sample mean in this example? One answer is that it's silly. Another is that it's not sufficient. We define *sufficiency* next.

Most statistical methods involve reducing the sample data to just a few numbers—to a statistic of low dimension. Such a reduction may lose important information. Roughly speaking, a statistic is **sufficient** if it contains all the useful information. However, if $\bar{X}$ (for example) contains this information, so does any one-to-one function of $\bar{X}$; if you know one, you can calculate the other. Sufficiency is essentially a property of a *partition*, which we define first.

Statistics and Partitions

A statistic Y is a function $Y(\mathbf{X})$ defined for each sample $\mathbf{X} = (X_1, \ldots, X_n)$. Each possible value y_0 of Y identifies a set of sample points, the set of all $\mathbf{X}$ such that $Y(\mathbf{X}) = y_0$. The collection of all such pre-image sets constitutes a **partition** of the sample space: The union of all these sets is the whole sample space, and the sets are disjoint [see Section 1.3, page 14].

Example 8.3b   **Lot Quality**

In Example 8.1a, we considered the eight sample sequences for three trials of a Bernoulli experiment:

          DDD, DDG, DGD, GDD, DGG, GDG, GGD, GGG.

Of interest for checking lot quality is the statistic Y = number of defectives (D's). The corresponding partition is

$$\{DDD\}, \{DDG, DGD, GDD\}, \{DGG, DGG, GGD\}, \{GGG\}.$$

These four partition sets correspond to the Y-values 3, 2, 1, and 0.   ∎

Example 8.3c   **$\bar{X}$-Partitions**
A particular sample of size n is a point in n-space: $\mathbf{x} = (x_1, \ldots, x_n)$. The mean $\bar{x}$ partitions the sample space into hyperplanes. Each hyperplane is the locus of points satisfying the condition $(x_1 + \cdots + x_n)/n = c$ for some c. (When $n = 2$, this is the equation of a line, and when $n = 3$, it defines a plane.) The hyperplanes that make up the partition are disjoint—in fact, parallel, and their union over all real c is the whole sample space.   ∎

Two statistics that define the same partition are in one-to-one correspondence and so are *equivalent* in their usefulness for inference. In the definition that follows, bear in mind that any statistic equivalent to a sufficient statistic is also sufficient.

A statistic $Y = Y(\mathbf{X})$ is **sufficient** for a family of distributions $f(\mathbf{x} \mid \theta)$ if and only if the likelihood function depends on $\mathbf{X}$ through the value of Y:

$$L(\theta) = g[Y(\mathbf{X}), \theta]. \tag{1}$$

Sufficiency is often defined in such a way that condition (1) is a theorem, called the *factorization criterion*. (See Solved Problem E.) When (1) holds, it is clear that knowing only the value of Y, one can calculate the likelihood of θ; Y is "sufficient" in this sense.

Example 8.3d   **Sufficiency of the Order Statistic**
Let $\mathbf{X}$ denote a sample obtained by random sampling, with or without replacement, from a population whose distribution has an arbitrary p.d.f. or p.f. $f(x \mid \theta)$. The joint p.d.f. or p.f. of the sample $\mathbf{X}$ is a symmetric function of its arguments: The X_i's are exchangeable random variables (see Section 2.6). So

$$L(\theta) = f_{\mathbf{X}}(X_1, \ldots, X_n \mid \theta) = f_{\mathbf{X}}(X_{(1)}, \ldots, X_{(n)} \mid \theta),$$

where $X_{(i)}$ is the ith order statistic. Since L is a function of θ and the order statistic, the order statistic is sufficient. The intuitive meaning of this is that with random sampling, ordering the observations does not lose any relevant information.   ∎

Going to the order statistic is only a slight reduction of raw data. Much greater reduction may be possible, depending on the form of f. In the next example, the data can be reduced to a single order statistic without sacrificing sufficiency.

Example 8.3e   Suppose $X \sim U(0, \theta)$:

$$f(x \mid \theta) = \begin{cases} \dfrac{1}{\theta}, & 0 < x < \theta, \\ 0, & \text{otherwise.} \end{cases}$$

The likelihood function is the joint p.d.f. of the observations in a random sample of size n. This is the product of the marginal p.d.f.'s, each of which is either 0 or $1/\theta$; so the product is either 0 or $(1/\theta)^n$:

$$L(\theta) = f(\mathbf{X} \mid \theta) = \begin{cases} \dfrac{1}{\theta^n}, & 0 < X_i < \theta \text{ for all } i, \\ 0, & \text{otherwise.} \end{cases}$$

All the X_i's are in the interval $(0, \theta)$ if and only if $X_{(n)} < \theta$. So

$$L(\theta) = \begin{cases} \dfrac{1}{\theta^n}, & \theta > X_{(n)}, \\ 0, & \text{otherwise.} \end{cases}$$

This is of the form $g(X_{(n)}, \theta)$. It follows that $X_{(n)}$ is sufficient for θ. In making inferences about θ, all we need to know from the sample is its size n and the largest observation. ∎

Example 8.3f   **The Exponential Family**

Consider a population whose p.d.f. or p.f. has this form:

$$f(x \mid \theta) = B(\theta)h(x)e^{Q(\theta)R(x)}. \tag{2}$$

This is called the **exponential family** and includes as special cases $\text{Bin}(n, \theta)$, $N(0, \theta)$, $\text{Poi}(\theta)$, $\text{Exp}(\theta)$, and many other special families of distributions.

The joint p.d.f. or p.f. of a random sample of size n from (3) is

$$f(\mathbf{x} \mid \theta) = \prod B(\theta)h(x_i)e^{Q(\theta)R(x_i)} = [B(\theta)]^n e^{Q(\theta)\Sigma R(x_i)} \prod h(x_i),$$

and the likelihood function is

$$L(\theta) = [B(\theta)]^n e^{Q(\theta)\Sigma R(X_i)} = g\left[\sum R(X_i), \theta\right].$$

So the statistic $Y = \Sigma R(X_i)$ is sufficient. ∎

Example 8.3g   **Sampling a Bernoulli Population**

Suppose $X \sim \text{Ber}(p)$. The p.f., for $x = 0$ or 1, is

$$f(x \mid p) = p^x(1 - p)^{1-x} = \left(\frac{p}{1 - p}\right)^x (1 - p)$$

$$= (1 - p)e^{x \log[p/(1 - p)]}.$$

This is a special case of (2), with $\theta = p$, $B(\theta) = 1 - p$, $Q(\theta) = \log[p/(1 - p)]$, and $R(x) = x$. So for a random sample of size n, $\Sigma R(X_i) = \Sigma X_i$ is sufficient for p. This is simply the number of successes in the n trials. ∎

Example 8.3h

Sampling $N(\mu, \sigma^2)$

The family of normal distributions is indexed by two parameters: $\theta = (\mu, \kappa)$, where $\kappa = \sigma^2$. Given a random sample of size n, the likelihood function is

$$L(\mu, \kappa) = \exp\left\{\frac{n}{2}\log\kappa - \frac{1}{2\kappa}\sum(X_i - \mu)^2\right\}. \tag{4}$$

According to the parallel axis theorem [(3) of Section 5.5],

$$\sum(X_i - \mu)^2 = nV + n(\bar{X} - \mu)^2,$$

where

$$V = \frac{1}{n}\sum(X_i - \bar{X})^2 = \frac{n-1}{n}S^2.$$

So we may write the likelihood function in this form:

$$L(\mu, \kappa) = \exp\left\{-\frac{n}{2}\log\kappa = \frac{1}{2\kappa}(nV + n(\bar{X} - \mu)^2)\right\}. \tag{5}$$

Since L depends on the data only through the pair $(\bar{X}, V)$, these statistics are sufficient for (μ, σ^2). ∎

In Example 8.2b, we saw that the likelihood function was the same whether we observed the sample data or the value of the statistic ΣX. This happens when the statistic is sufficient. We'll demonstrate this in the discrete case.

Consider a random sample $\mathbf{X}$ from a discrete population, and suppose $Y = Y(\mathbf{X})$ is sufficient for θ. When we observe $\mathbf{X} = \mathbf{x}$ such that $Y(\mathbf{x}) = y$, the likelihood function is

$$L(\theta) = g[Y(\mathbf{X}), \theta] = g(y, \theta). \tag{6}$$

On the other hand, suppose we observe $Y = y$ but not $\mathbf{X}$. Then

$$P(Y = y \mid \theta) = \sum_{Y(\mathbf{x})=y} f(\mathbf{x} \mid \theta) = \sum_{Y(\mathbf{x})=y} g(y(\mathbf{x}), \theta)h(\mathbf{x})$$

$$= g(y, \theta)\sum_{Y(\mathbf{x})=y} h(\mathbf{x}).$$

This is proportional to $L(\theta)$ in (6), so the likelihood function is the same whether we observe $\mathbf{X}$ or only the value of the sufficient statistic Y.

The converse is also true: If Y is not sufficient, then the likelihood function determined by $\mathbf{X}$ will be different from the likelihood function determined by Y.

8.4   **Sampling Distributions**

A statistic is a random variable. Its value varies from sample to sample. Its probability distribution, which describes this variation, is called its **sampling distribution**. The sampling distribution of a particular statistic depends on the statistic, on the population being sampled, and on the method of sampling.

When the population distribution is known, it may be possible to find a sampling distribution mathematically. For instance, suppose we sample a Bernoulli population at random. For a sample of given size, the sample *sum* is a statistic, one whose sampling distribution is binomial if sampling is with replacement and hypergeometric if sampling is without replacement. The next example compares these two cases. (See also Sections 4.2 and 4.3.)

Example 8.4a **Sampling Inspection**

As in Example 8.1a, consider a lot containing 20 articles, M of which are defective. To estimate M, we select three at random without replacement. The number Y of defective articles in the sample is sufficient for M (see Solved Problem C). Its sampling distribution is hypergeometric:

$$f(y \mid M) = P(y \text{ defectives}) = \frac{\binom{M}{y}\binom{20-M}{3-y}}{\binom{20}{3}}, \qquad y = 0, 1, 2, 3.$$

In practice, it may be pointless to put back an article once it has been tested, but for comparison, consider sampling with replacement. In this case, the number Y of defectives in the sample is again sufficient (see Example 8.3g). Its sampling distribution is $\text{Bin}(3, M/20)$:

$$f(y \mid M) = \binom{3}{y}\left(\frac{M}{20}\right)^y \left(1 - \frac{M}{20}\right)^{3-y}, \qquad y = 0, 1, 2, 3.$$

The two sampling distributions are shown in Table 8.2 for $M = 2$. (The probabilities in Table 8.2 can also be calculated by adding appropriate entries in Table 8.1.) The statistic Y is more variable under sampling with replacement; its standard deviation is $\sqrt{2(.1)(.9)} = .520$, compared with

$$.492 = .520 \sqrt{\frac{20-3}{20-1}}$$

under sampling without replacement.

Table 8.2

y	$f(y)$ Without replacement	$f(y)$ With replacement
0	$\frac{136}{190} \doteq .716$	$(.9)^3 = .729$
1	$\frac{51}{190} \doteq .268$	$3(.1)(.9)^2 = .243$
2	$\frac{3}{190} \doteq .016$	$3(.1)^2(.9) = .027$
3	0	$(.1)^3 = .001$

■

Example **8.4b**   **Sample Surveys**
Political pollsters periodically ask people what they think about the president's job performance. The proportion p in the population who think the president is doing a good job is unknown, and the purpose of the survey is to estimate this proportion.

A pollster might obtain a sample of 500 voters and assume that it is a random sample. Let Y denote the number in the sample who think the president is doing a good job. (According to Example 8.3g, Y is sufficient for p.) This statistic is approximately *binomial*:

$$f(y \mid p) = \binom{500}{y} p^y (1 - p)^{500-y}, \qquad y = 0, 1, 2, \ldots, 500.$$

Clearly, the sampling distribution of Y depends on the population parameter p; specifying the value of p defines a particular member of the binomial family. ■

As in the preceding examples, the sample sum is often a sufficient statistic. When it is, so is the sample mean—the sample sum divided by the sample size. The next example treats quite a different type of statistic.

Example **8.4c**   **The Sample Maximum**
A biologist gathers n individual organisms to learn about the maximum weight of such organisms. She assumes that they grow to their maximum linearly with time and then die. (All assumptions in science are subject to criticism and revision; this one may be more open to objection than most, but she has to assume something to get started.) She assumes further that the individual weights $X_1, \ldots, X_n$ constitute a random sample from a population whose distribution is $U(0, \theta)$, where $\theta > 0$. The population c.d.f. is then

$$F_X(x \mid \theta) = \frac{x}{\theta}, \qquad 0 < x < \theta, \tag{1}$$

and each X_i has this distribution.

According to Example 8.3e, the largest sample observation, $Y = X_{(n)}$, is sufficient for θ. The c.d.f. of its sampling distribution is

$$F_Y(y \mid \theta) = P(Y \le y \mid \theta) = P(\text{largest observation} \le y \mid \theta).$$

The largest observation is less than or equal to y if and only if every observation is less than or equal to y:

$$F_Y(y \mid \theta) = P(X_1 \le y, \ldots, X_n \le y \mid \theta).$$

Since the X's are independent, this probability factors:

$$F_Y(y \mid \theta) = \prod_{i=1}^n P(X_i \le y \mid \theta) = \prod_{i=1}^n F_{X_i}(y \mid \theta) = [F_X(y \mid \theta)]^n$$

In view of (1),

$$F_Y(y \mid \theta) = \left(\frac{y}{\theta}\right)^n, \qquad 0 < y < \theta.$$

Having found the c.d.f. of Y, we differentiate to find the p.d.f.:

$$f_Y(y \mid \theta) = \frac{d}{dy} F_Y(y \mid \theta) = \frac{ny^{n-1}}{\theta^n}, \qquad 0 < y < \theta.$$

The mean of the sampling distribution is

$$E(Y) = \int_{-\infty}^{\infty} y f_Y(y \mid \theta) \, dy = \int_0^{\theta} y \left(\frac{ny^{n-1}}{\theta^n} \right) dy = \frac{n}{n+1} \theta.$$

Not surprisingly, when n is large, this is close to θ, the largest possible population weight, and since $P(Y \le \theta) = 1$, the variance of Y must be small. You should verify that

$$\text{var } Y = \frac{n\theta^2}{(n+2)(n+1)^2}.$$

As anticipated, this tends to zero as n becomes infinite. The rate at which var Y goes to 0 is $1/n^2$, which, in view of some results to follow in Section 8.6, is extraordinarily fast. A practical implication is that the biologist can be satisfied that the largest observation in a sample of moderate size is a good approximation to θ. ■

8.5   Simulating Sample Distributions

Imagine drawing a large number of samples from some population and calculating the value of a particular statistic for each sample. The histogram of these values approximates the p.d.f. or p.f. of the sampling distribution of the statistic. Any sampling distribution can be approximated in this way using artificially generated samples—sampling a simulated population. This is particularly important when a sampling distribution cannot be derived mathematically.

Example 8.5a   **Distribution of the Sample Range**
The sample range is a statistic. In the next section, we show how to derive its distribution for any given population distribution. To illustrate the process of simulation described above, we drew 400 samples of size $n = 5$ from $U(0, 1)$. For each sample, we calculated R and tallied these 400 R-values using a system of 20 class intervals. Figure 8.3 shows the corresponding histogram, along with a smooth curve drawn in by eye. This curve is a rough estimate of the density of the sampling distribution. The exact p.d.f. is given by (7) in Section 8.6 as

$$f_R(r) = 20r^3(1 - r), \qquad 0 < r < 1.$$

The smooth curve in the figure is very close to this. ■

How did we get the 400 samples for the empirical distribution in the above example? One way of obtaining an observation from a $U(0, 1)$ population is to spin a pointer with a scale from 0 to 1 at its tip. Another is to use random digits.

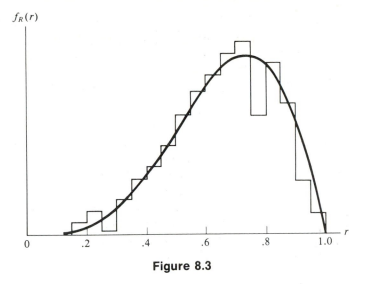

$f_R(r)$

Figure 8.3

In Section 1.5, we explained how to use sequences of random digits to obtain random integers. With a decimal point in front, a random integer becomes a rounded random observation from $U(0, 1)$. Table XIV in Appendix 1 is a table of random digits that one can use (starting at an arbitrary point) to produce uniform random numbers. Some hand calculators and most computer languages and statistical software packages have commands that produce sequences of uniform random numbers. For example, in BASIC, the command "RAND" generates numbers that behave like independent observations on a $U(0, 1)$ distribution.

Table XV of Appendix 1 gives a sequence of independent observations from a *standard normal* population. The statistical software package MINITAB has a routine, called "NRAND," which produces a sequence of random numbers from a normal population with mean and s.d. of your choosing. Usually, uniform and normal random numbers are the only ones available at a single command. However, as we explain next, it is possible to obtain a sequence of random observations from any specified distribution by suitably transforming random numbers from $U(0, 1)$.

Suppose we need a sequence of observations from a continuous population with c.d.f. F. The random variable $U = F(X)$ is uniformly distributed on $(0, 1)$, as shown in Problem 9 of Chapter 5. Conversely, if U is uniform on $(0, 1)$, then the transformed variable $X = F^{-1}(U)$ has the c.d.f. $F(x)$:

$$P(X \leq x) = P[F^{-1}(U) \leq x] = P[U \leq F(x)] = F(x).$$

This means that if $(U_1, \ldots, U_n)$ is a random sample from $U(0, 1)$, the vector of transformed quantities $[F^{-1}(U_1), \ldots, F^{-1}(U_n)]$ is a random sample from a population with c.d.f. F.

Example 8.5b   **Simulating Exponentials**
To simulate a random sample from $Exp(1/2)$, we first obtain a sequence of ten

(rounded) observations from $U(0, 1)$ from Table XIV in Appendix 1:

.4224   .0663   .9728   .8801   .4410,
.9582   .0402   .2316   .9240   .5073.

The population c.d.f. is $F(x) = 1 - e^{-x/2}$; to find $F^{-1}(u)$, we solve $u = F(x)$ for x in terms of u:

$$u = 1 - e^{-x/2}, \qquad 1 - u = e^{-x/2}, \qquad \log(1 - u) = \frac{-x}{2}.$$

Then

$$F^{-1}(u) = -2 \log(1 - u).$$

Applying this transformation to each of the above observations on U yields:

1.098   .137   7.209   4.242   1.163,
6.350   .082   .527   5.154   1.416.

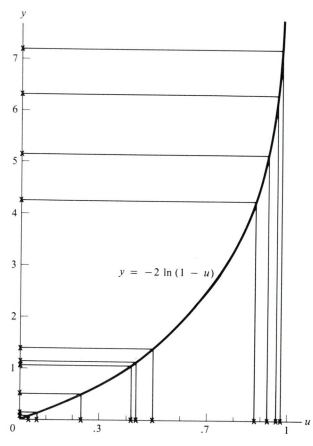

Figure 8.4

These ten values of X constitute a random sample from an exponential population with mean 2. (The sample mean is about 2.2, not far from the population mean.) Figure 8.4 shows the two sets of observations and their correspondence via the exponential c.d.f. ∎

One can always simulate a random sample from a population with given c.d.f. as in the preceding example, but sometimes there are more efficient methods. For example, suppose we want a sequence of normal observations but don't have access to a normal-number generator on a computer or calculator. Using the inverse normal c.d.f. requires a numerical approximation. A common method of generating **Box–Muller** normal random numbers is to use the Box-Muller transformation, which generates **Transformation** them in pairs. If U and V are independent and each is $U(0, 1)$, the variables

$$\begin{cases} X = \sigma(\sqrt{-2 \log U}) \cos(2\pi V) + \mu \\ Y = \sigma(\sqrt{-2 \log U}) \sin(2\pi/V) + \mu \end{cases} \tag{1}$$

are independent and each is $N(\mu, \sigma^2)$. So (1) can be used to transform $2n$ independent uniform variates into $2n$ independent normal random variables.

Example 8.5c **Generating Standard Normals**
To illustrate (1) with $\mu = 0$ and $\sigma = 1$, we take the ten uniform observations listed in Example 8.5b:

.4224	.0663	.9728	.8801	.4410
.9582	.0402	.2316	.9240	.5073.

Taking the first row as U's and the second as V's, we obtain

X:	1.2678	2.2557	.0271	.4489	-1.2783
Y:	$-.3409$	.5822	.2334	$-.2323$	.0587.

(You should reproduce some of these using a scientific calculator.) These ten numbers constitute a random sample from a standard normal population. ∎

8.6 Order Statistics

Example 8.6a **R-Charts**
A standard tool used for monitoring the variability in a production process is the R-chart. (It is used in conjunction with an $\bar{X}$-chart, but for this example we focus on the R-chart.) An R-chart is a record of the ranges of samples taken at regular intervals from the production line. The chart has lines to tell the operator when the process variability is "in control." The placement of these lines depends on the sampling distribution of the range:

$$R = X_{(n)} - X_{(1)}.$$

The distribution of R can be found approximately using a simulation (see Section 8.5) or derived mathematically from the joint distribution of $X_{(1)}$ and $X_{(n)}$. ∎

In this section, we show how to derive distributions of the order statistics and combinations of order statistics, such as R in the example, for random sampling from a continuous population.

For any single $X_{(k)}$, the p.d.f. could be found using the approach of Example 8.4c—deriving its c.d.f. and differentiating. However, the differential method introduced in Section 5.2 also yields the p.d.f. To apply this method, we'll find the probability that $X_{(k)}$ is in the infinitesimal interval from y to $y + dy$. The coefficient of dy in this differential or probability element is the p.d.f. of $X_{(k)}$.

Classify each observation X as lying in one of the intervals $(-\infty, y), (y, y + dy)$, and $(y + dy, \infty)$. The probability of the interval $(-\infty, y)$ defines the c.d.f. $F(y)$. The probability of the second interval is approximately the probability element of X corresponding to the increment dy:

$$P(y < X < y + dy) = \int_y^{y+dy} f(x)dx \doteq f(y)\,dy.$$

[See (5) in Section 5.2 and Figure 5.7.] For the interval $(y + dy, \infty)$,

$$P(X > y + dy) = 1 - F(y + dy) \doteq 1 - F(y).$$

To summarize:

Outcome	Approximate probability
$X < y$	$F(y)$
$y < X < y + dy$	$f(y)\,dy$
$X > y + dy$	$1 - F(y)$

The variable $X_{(k)}$ is in the interval $(y, y + dy)$ if and only if there are $k - 1$ X's to the left of y, exactly one X between y and $y + dy$, and the remaining $n - k$ X's are to the right of $y + dy$. (See Figure 8.5.) The probability of this event is *trinomial* [(4) of Section 4.8]:

$$P(y < X_{(k)} < y + dy) = \frac{n!}{(k-1)!1!(n-k)!}[F(y)]^{k-1}[f(y)\,dy]^1[1 - F(y)]^{n-k}.$$

The coefficient of dy in this probability element is the density function of $X_{(k)}$.

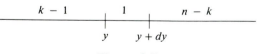

Figure 8.5

> The p.d.f. of the kth smallest observation in a random sample from a continuous population with c.d.f. $F(x)$ and p.d.f. $f(x)$ is
>
> $$f_{X_{(k)}}(y) = \frac{n!}{(k-1)!(n-k)!}[F(y)]^{k-1}[1 - F(y)]^{n-k}f(y). \tag{1}$$

Sample Maximum

In Example 8.4c, we found the p.d.f. of the largest observation in a random sample from a uniform distribution on $(0, \theta)$. Setting $k = n$ in (1) generalizes that result:

$$f_Y(y) = n[F(y)]^{n-1}f(y). \tag{2}$$

For the special case of a uniform density on $(0, \theta)$, set $f(y) = 1/\theta$ and $F(y) = y/\theta$, both for $0 < y < \theta$.

Sample Median

In Section 7.3, we defined the sample *median* X in terms of components of the order statistic. When n is odd, the median is $X_{(k)}$, where $k = (n + 1)/2$. (For instance, in a sample of size seven, the median is the fourth largest.) The p.d.f. of the median is then a special case of (1):

$$f_{\tilde{X}}(u) = \frac{n!}{\left(\dfrac{n-1}{2}\right)!\left(\dfrac{n-1}{2}\right)!}[F(u)]^{(n-1)/2}[1 - F(u)]^{(n-1)/2}f(u). \tag{3}$$

(When n is even, the median is the average of two order statistics, a complication we won't go into.)

Example 8.6b   **Order Statistics for $U(0, 1)$**

If $X \sim U(0, 1)$, the population c.d.f. is $F(x) = x$, and the p.d.f. is $f(x) = 1$, both for $0 < x < 1$. So

$$f_{X_{(1)}}(u) = n(1 - u)^{n-1}, \qquad 0 < u < 1.$$

The mean and variance (calculated in the usual way) are:

$$E(X_{(1)}) = \frac{1}{n+1}, \qquad \text{var } X_{(1)} = \frac{n}{(n+1)(n+2)^2}.$$

By symmetry, it follows that var $X_{(1)}$ = var $X_{(n)}$, and that

$$E[X_{(n)}] = 1 - E[X_{(1)}] = \frac{n}{n+1}.$$

Suppose $n = 15$. The p.d.f. of the median is given by (3):

$$f_{\tilde{X}}(y) = \frac{15!}{7!7!}y^7(1 - y)^7, \qquad 0 < y < 1.$$

This density is symmetric about $y = \frac{1}{2}$, so the mean and median of $\tilde{X}$ are both $\frac{1}{2}$. (Problem 28 asks you to show that whenever f_X is symmetric, so is $f_{\tilde{X}}$.)   ∎

Returning to the problem of the sample range, we now derive its sampling distribution from the joint distribution of the smallest and largest observations, $X_{(1)}$ and $X_{(n)}$. To obtain this, we use the differential approach. Suppose $n \geq 2$. Divide the set of possible X-values into the following classes, with probabilities as shown (for $u < v$):

Outcome	Approximate probability
$X < u$	$F(u)$
$u < X < u + du$	$f(u)\,du$
$u + du < X < v$	$F(v) - F(u)$
$v < X < v + dv$	$f(v)\,dv$
$X > v + dv$	$1 - F(v)$

Now $X_{(1)}$ is between u and $u + du$ and $X_{(n)}$ is between v and $v + dv$ if and only if the numbers of X's in the above class intervals are as shown in Figure 8.6. The joint distribution of the class frequencies is multinomial, so

$$P(u < X_{(1)} < u + du, v < X_{(n)} < v + dv)$$
$$= C \cdot [F(u)]^0 [f(u)\,du]^1 [F(v) - F(u)]^{n-2} [f(v)\,dv]^1 [1 - F(v)]^0, \qquad (4)$$

where C is the multinomial coefficient:

$$C = \binom{n}{0, 1, n-2, 1, 0} = \frac{n!}{(n-2)!} = n(n-1).$$

The desired joint density is the coefficient of $du\,dv$ in (4):

> Joint density of $X_{(1)}$ and $X_{(n)}$ in a random sample of size $n \geq 2$ from a continuous population with c.d.f. F and p.d.f. f:
>
> $$f_{X_{(1)}, X_{(n)}}(u, v) = n(n-1)[F(v) - F(u)]^{n-2} f(u) f(v), \qquad u < v. \qquad (5)$$

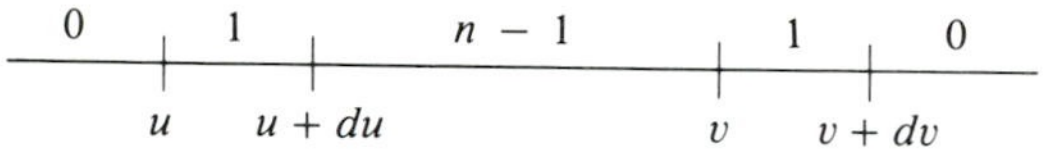

Figure 8.6

We can now use (5) to find the p.d.f. of any function of the smallest and largest observations, and in particular the **sample range R**. The p.d.f. of R is an integral involving F and f, which we'll derive next for a special case.

Example 8.6c

Sample Range for $U(0, 1)$

Suppose $X \sim U(0, 1)$. The p.d.f. is $f(x) = 1$, and the c.d.f. is $F(x) = x$, both for

$0 < x < 1$. In this case, (5) becomes

$$f_{X_{(1)}, X_{(n)}}(u, v) = n(n - 1)(v - u)^{n-2}, \qquad 0 < u < v < 1. \tag{6}$$

The c.d.f. of the sample range R is

$$F_R(r) = P(R \le r) = P[X_{(n)} - X_{(1)} \le r].$$

To find this, we integrate (6) over that subset of the support in which $v - u \le r$:
$A = \{(u, v): 0 < u < v < u + r < 1\}$:

$$P(R \le r) = \int_A \int n(n - 1)(v - u)^{n-2} \, du \, dv.$$

Evaluating this double integral and differentiating the result with respect to r yields

$$f_R(r) = n(n - 1)r^{n-2}(1 - r), \qquad 0 < r < 1. \tag{7}$$

For large n, the graph of this p.d.f. behaves like $1 - r$ near $r = 1$ and has a high order of tangency to the r-axis at $r = 0$, so most of the distribution is crowded near $r = 1$. Thus, the range of a large sample is very likely to be near the population range. ∎

We turn next to the joint p.d.f. of all components of the order statistic, $(X_{(1)}, \ldots, X_{(n)})$. We'll derive this p.d.f. also using the differential approach. Given $y_1 < \cdots < y_n$, the probability of finding exactly one sample observation in each interval $(y_i, y_i + dy_i)$ is multinomial:

$$n![f(y_1) \, dy_1] \cdots [f(y_n) \, dy_n].$$

The coefficient of $dy_1 \cdots dy_n$ is the desired p.d.f.:

The joint density of the ordered observations $X_{(1)}, \ldots, X_{(n)}$ in a random sample from a continuous population with density $f(x)$ is

$$f^*(y_1, \ldots, y_n) = n! \prod_{i=1}^{n} f(y_i), \tag{8}$$

for $y_1 < y_2 < \cdots < y_n$.

Example 8.6d   Consider a random sample of size $n \ge 2$ from $U(0, 1)$. The population p.d.f. is $f(x) = 1$ for $0 < x < 1$. From (8), the joint p.d.f. of the ordered observations in a random sample is

$$f^*(y_1, \ldots, y_n) = n!, \qquad 0 < y_1 < \cdots < y_n < 1.$$

The distribution is thus uniform in a simplex in n-space. Figure 8.7 shows this simplex when $n = 3$. ∎

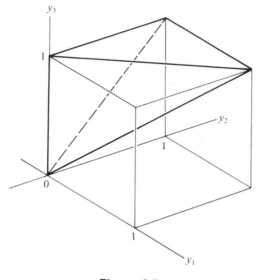

Figure 8.7

8.7   Moments of Sample Means and Proportions

A common problem in applications of statistics is estimating a population mean μ. The sample mean $\bar{X}$ is a natural estimator for μ. It is a sufficient statistic in the case of normal, exponential, Poisson, Bernoulli, and various other distributions. (See Section 8.3.) The sampling distribution of $\bar{X}$ depends on the population sampled and on the method of sampling. In Section 8.9, we'll show how to derive the distribution of the sample mean from the distribution of the population in the case of random samples. For the large sample inferences of the next few chapters, we'll need only its mean and variance.

When we sample at random, with or without replacement, the sample observations are identically distributed. In particular, all observations X_i have the same mean μ and the same variance σ^2. Hence, using the linearity of expected values [(11) of Section 5.6], we obtain

Expected Value of $\bar{X}$

$$E(\bar{X}) = E\left[\frac{1}{n}(X_1 + \cdots + X_n)\right] = \frac{1}{n}[E(X_1) + \cdots + E(X_n)]$$

$$= \frac{1}{n}(\mu + \cdots + \mu) = \mu.$$

This says that the center of the distribution of $\bar{X}$-values is the same as the center of the distribution of individual X-values.

On the other hand, the variance of the sample mean depends on whether the sampling is with or without replacement. In either case, we may write

$$\sigma_{\bar{X}} = \mathrm{var}(\bar{X}) = \mathrm{var}\left(\frac{1}{n}(X_1 + \cdots + X_n)\right)$$

and bring the factor $1/n$ out as $1/n^2$, following (7) of Section 5.5:

$$\text{var}(\bar{X}) = \frac{1}{n^2}\,\text{var}(X_1 + \cdots + X_n). \tag{1}$$

Consider first a random sample (sampling with replacement and mixing if the population is finite). The variance is additive [(5) of Section 5.8], and (1) becomes

**Variance
of $\bar{X}$**

$$\text{var}(\bar{X}) = \frac{1}{n^2}(\text{var } X_1 + \cdots + \text{var } X_n).$$

Since the variance of each X_i is the population variance σ^2,

$$\text{var}(\bar{X}) = \frac{1}{n^2}(\sigma^2 + \sigma^2 + \cdots + \sigma^2) = \frac{\sigma^2}{n}. \tag{2}$$

Because it is easier to interpret a standard deviation as a measure of variability, we usually focus on the square root of (2).

For a random sample from a population with mean μ and standard deviation σ,

$$E(\bar{X}) = \mu, \tag{3}$$

$$\sigma_{\bar{X}} = \frac{\sigma}{\sqrt{n}}. \tag{4}$$

Formula (4) shows that the sampling distribution of $\bar{X}$ narrows as n increases: *The larger the sample, the less the variability in the sample mean.*

Example 8.7a   **Mean of Three Dice Rolls**
The distribution of the number of points obtained at a single throw of a fair die is uniform on the integers 1, 2, 3, 4, 5, 6. In Examples 3.1e and 3.3b, we found that $\mu = 3.5$ and $\sigma^2 = \frac{35}{12}$. Consider now three successive throws. From (3),

$$E(\bar{X}) = \mu = 3.5.$$

Assuming independence of the three throws, we use (4) to find the standard deviation of $\bar{X}$:

$$\sigma_{\bar{X}} = \frac{\sigma}{\sqrt{3}} = \sqrt{\frac{35}{36}} \doteq .986.$$

This is only $1/\sqrt{3}$, (58%) as large as the standard deviation for a single throw.
    We can verify the above calculations using the p.f. for the sample sum: $\Sigma X = 3\bar{X}$. The sum of the points showing varies from 3 to 18, and we can find the

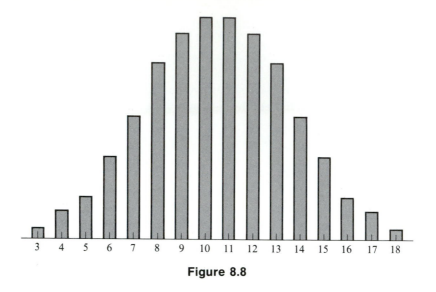

Figure 8.8

probability of any particular sum by counting.[5] For example, a sum of 7 can arise from (2, 2, 3) in three arrangements, from (1, 2, 4) in six arrangements, from (1, 1, 5) in three arrangements, and from (1, 3, 3) in three arrangements. Thus, $3 + 6 + 3 + 3 = 15$ of the 216 equally likely sample sequences add up to seven, so

$$P(\textstyle\sum X = 7) = \frac{15}{216}.$$

The complete distribution is shown in Figure 8.8. Its mean is $\frac{21}{2}$ and its variance is $\frac{35}{4}$. The mean of $\bar{X} = \sum X/3$ is therefore $\frac{21}{6}$ or $3.5 = \mu$, as we found above. The variance of $\bar{X}$ is $[\mathrm{var}(\sum X)]/9 = 35/36$, which also agrees with our earlier calculation.   ■

Example 8.7b   Consider 20 weighings of the standard weight described in Example 8.1b. Many weighings have shown that $\sigma = 6\ \mu g$. The sample mean $\bar{X}$ is again a random variable, with s.d. given by (4):

$$\sigma_{\bar{X}} = \frac{6}{\sqrt{20}} = 1.34\ (\mu g).$$

The distribution of $\bar{X}$-values is centered at the population mean but is only $1/\sqrt{20}$ (about 22%) as wide as the distribution for single weighings.   ■

Sample Proportions   Consider now the important special case of sampling from a Bernoulli population. The population mean is p, the population proportion of successes [see (2) of Section 4.1]. The sample mean, by a parallel calculation, is the sample

---

[5] We could also find these probabilities using the probability-generating function. (See Section 3.6 and Solved Problem P of Chapter 3.)

proportion of successes:

$$\bar{X} = \frac{1}{n}\sum X_i = \frac{1}{n}\cdot(\# \text{ of 1's}) = \hat{p}.$$

The population variance [again from (2) of Section 4.1] is $\sigma^2 = pq$, where $q = 1 - p$. Applying (3) and (4):

The mean and variance of $\hat{p}$, the proportion of successes in a random sample from a Bernoulli population with $P(\text{success}) = p$, are

$$E(\hat{p}) = p, \tag{6}$$

$$\sigma_{\hat{p}} = \sqrt{\frac{pq}{n}}. \tag{7}$$

We've seen closely related formulas in connection with the binomial distribution. The sample proportion $\hat{p}$ is just $1/n$ times the number of successes in the sample, and that number is $\text{Bin}(n, p)$. So (6) and (7) follow at once from (3) and (4) of Section 4.2.

Example 8.7c

Sample Survey

Suppose voter sentiment in a certain district is evenly divided on a proposal to increase the school levy. What kind of divisions might show up in a random sample of 25 voters?

"Evenly divided" means that $p = \frac{1}{2}$, where p is the proportion who favor the proposal. So [from (6)] $E(\hat{p}) = \frac{1}{2}$. The standard deviation of $\hat{p}$ [from (7)] is

$$\sigma_{\hat{p}} = \sqrt{\frac{pq}{n}} = \sqrt{\frac{.5 \times .5}{25}} = .10.$$

As is typical of sample statistics, sample proportions vary from sample to sample. In some samples, $\hat{p} < p$, and in others, $\hat{p} > p$. We interpret $\sigma_{\hat{p}} = .10$ to say that for samples of size 25, sample proportions typically deviate from $p = \frac{1}{2}$ by about ten percentage points when $p = \frac{1}{2}$. ∎

Sampling Without Replacement

When sampling is without replacement from a finite population, (3) continues to hold, since the expected value is additive even when the observations are not independent. However, (4) needs to be modified. In Section 3.5, we showed that

$$\text{var}(X_1 + \cdots + X_n) = n(\text{var } X) + n(n - 1)\,\text{cov}(X_1, X_2), \tag{8}$$

where

$$\text{cov}(X_1, X_2) = \frac{-\sigma^2}{N(N - 1)}. \tag{9}$$

[In the Bernoulli case, $\sigma^2 = p(1 - p)$.]. Combining (8) and (9), we find

$$\text{var}(\textstyle\sum X) = n\sigma^2 \cdot \frac{N - n}{N - 1}.$$

Since $\bar{X} = (\sum X)/n$, it follows that

$$\text{var }\bar{X} = \frac{\text{var}(\sum X)}{n^2}.$$

For random sampling without replacement from a population of size N with mean μ and s.d. σ,

$$E(\bar{X}) = \mu, \tag{10}$$

$$\sigma_{\bar{X}} = \frac{\sigma}{\sqrt{n}} \cdot \sqrt{\frac{N - n}{N - 1}}. \tag{11}$$

Comparing (10) and (11) with (3) and (4), we see that although the means are the same, the s.d. of $\bar{X}$ is smaller under sampling without replacement.

Example 8.7d

Sample Survey: A Small Population

In the preceding example, suppose the sample of 25 voters ($n = 25$) was drawn at random without replacement from the 100 voters of a small town ($N = 100$), instead of from the large population of a school district. Again assume an equal division on the issue of the levy. The expected value of the sample proportion is still $\frac{1}{2}$, but the standard deviation [from (11), with $\sigma = pq = .25$] is

$$\text{s.d.}(\hat{p}) = \frac{\sigma}{\sqrt{n}} \sqrt{\frac{N - n}{N - 1}} = .10 \sqrt{\frac{75}{99}} \doteq .10 \times .87 = .087.$$

The .10 we found in the preceding example for s.d. $(\hat{p})$ is reduced by about 13% by sampling without replacement. ∎

In the more detailed treatment of estimation in Chapter 9, we'll need the large sample distributions of sample means and proportions. We take these up next.

8.8   Sampling Distributions for Large Samples

Finding sampling distributions can be difficult. Yet, for many of the statistics we commonly use, the *normal* distribution turns out to give good approximations when sample sizes are large enough.

Sample Proportions

Consider first a Bernoulli population with parameter p, the population proportion of 1's. The sample proportion of 1's in a random sample of size n is $\hat{p} = Y/n$, where Y is the number of 1's in the sample. The DeMoivre–Laplace theorem of Section 4.5 gives the standard normal distribution as an approximation to the distribution of the standardized Y:

$$Z = \frac{Y - E(Y)}{\sigma_Y} = \frac{Y - np}{\sqrt{npq}} \approx N(0, 1). \tag{1}$$

[See (3) and (4) of Section 4.2.] Division of numerator and denominator by n shows that (1) is also the standardized value of $\hat{p}$:

$$Z = \frac{\frac{Y}{n} - p}{\sqrt{pq/n}} = \frac{\hat{p} - E(\hat{p})}{\sigma_{\hat{p}}}. \tag{2}$$

[See (6) and (2) of Section 8.7.]

   In view of (1) and (2), the sample proportion $\hat{p}$ is also approximately normal for "large" n. [In Section 4.5, we gave the rule of thumb $npq > 5$ for deciding if n is large enough.] So we can use the normal table to find approximate probabilities for intervals of $\hat{p}$-values.

For n sufficiently large ($npq > 5$),

$$\frac{\hat{p} - p}{\sqrt{pq/n}} \approx N(0, 1). \tag{3}$$

Example 8.8a   **TV Ratings**

Suppose that 30% of the nation's TV viewers are watching a particular presidential press conference. A Nielsen survey will sample about 1,000 viewers (a typical size). Suppose the sampling is random.[6] What is the probability that no more than 28% of the viewers in the sample are watching the press conference?

   The population proportion is $p = .30$. From (6) and (7) of Section 8.7,

$$E(\hat{p}) = p = .30, \quad \text{and} \quad \sigma_{\hat{p}} = \sqrt{\frac{pq}{n}} = \sqrt{\frac{.3 \times .7}{1000}} \doteq .0145.$$

Since $npq = 210 > 5$, we use (3) to conclude that $\hat{p}$ is approximately normal. Figure 8.9 shows the approximating normal density. We use Table II, in Appendix 1 to obtain

$$P(\hat{p} \le .28) \doteq \Phi\left(\frac{.28 - .30}{.0145}\right) = \Phi(-1.38) \doteq .084.$$

This probability is the shaded area in Figure 8.9.

---

[6] Although some elements of randomness are involved in their selection, Nielsen's samples are not random samples. Nielsen recognizes this but nonetheless bases its assessment of sampling errors on formulas for random sampling.

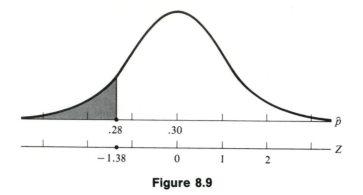

Figure 8.9

We could also calculate the desired probability using (1). The number Y who are watching is binomial (see Section 4.5) with $n = 1000$ and $p = .3$. So $E(Y) = np = 300$ and var $Y = npq = 210$, and

$$P(Y \leq 280) \doteq \Phi\left(\frac{280 - 300}{\sqrt{210}}\right) \doteq \Phi(-1.38),$$

the same as before.

In these calculations, we have not used the continuity correction given in Section 4.5 because it makes little difference here:

$$P(Y \leq 280) \doteq \Phi\left(\frac{280.5 - 300}{\sqrt{210}}\right) \doteq \Phi(-1.35) \doteq .091. \qquad \blacksquare$$

The Sample Mean

More generally, consider the distribution of $\bar{X}$, the mean of a random sample of size n. It turns out that, like $\hat{p}$, $\bar{X}$ has approximately a *normal* distribution when n is large. This is quite amazing, since it means that the shape of the limiting distribution is the same for any population distribution. This striking result is a generalization of the DeMoivre–Laplace approximation of Section 4.5:

Central limit theorem: Let $\bar{X}_n$ denote the mean of a sequence of n independent observations from a population with mean μ and finite variance σ^2. Then

$$\lim_{n \to \infty} P\left(\frac{\bar{X}_n - \mu}{\sigma/\sqrt{n}} \leq z\right) = \Phi(z), \qquad (4)$$

where $\Phi(z)$ is the standard normal c.d.f.

Various proofs of the central limit theorem have been given. In Section 8.9, we'll outline a proof using moment generating functions.

Example **8.8b** **Averaging $U(0, 1)$'s**
When $X \sim U(0, 1)$, with $E(X) = .5$ and var $X = \frac{1}{12}$, for a random sample of size n, the standardized mean is

$$Z_n = \frac{\bar{X} - .5}{\sqrt{1/12n}}$$

Figure 8.10 shows the p.d.f.'s of Z_n for $n = 1, 2, 4$, and 20. A graph of the standard normal density would be indistinguishable from the graph shown for $n = 20$, which in turn is not far from the graph for $n = 4$.

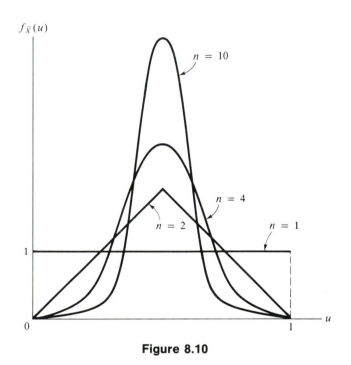

Figure 8.10 ■

    A convenient way of expressing the conclusion of the central limit theorem is to is to say that $\bar{X}$ is **asymptotically normal** with mean μ and variance σ^2/n, or that $\bar{X} \approx N(\mu, \sigma^2/n)$. The practical import of this is that when n is large, we may use the normal distribution to approximate the distribution of $\bar{X}$, *regardless of the shape of the underlying population*, provided only that the population variance is finite.

For random samples of size n where n is large, $\bar{X} \approx N(\mu, \sigma^2/n)$:

$$P(\bar{X} \le k) \doteq \Phi\left(\frac{k - \mu}{\sigma/\sqrt{n}}\right). \tag{5}$$

How Large is Large?

How large is "large"? Whether a sample size is large enough for (5) to provide a good approximation depends on k, on the desired accuracy, and on the population distribution. Roughly speaking, the closer the population is to being normal, the closer the distribution of the sample mean is to normal. In fact, if the population *is* normal, so is the sample mean. If the population is nearly symmetric and has light tails (its p.d.f. tending to zero rapidly near $\pm\infty$), a sample size as small as five or even less may give an approximation that is good enough for practical purposes. The uniform distribution of Example 8.8b is a case in point; it is symmetric and has no tails. However, if the population is markedly asymmetric or if there is a substantial amount of probability far from the mean (as measured in standard deviations), then a very large sample may be needed for the distribution of $\bar{X}$ to be well approximated by the normal.

Example 8.8c   **Elevator Load Limit**

We know an elevator with this sign on the wall:

> Maximum number of passengers:
>
> 16
>
> Maximum load:
>
> 2,500 lbs

Suppose the mean weight in the population of people who use the elevator is $\mu = 145$ lb, and the standard deviation is $\sigma = 35$ lb. The expected total weight of 16 people is $16 \times 145 = 2320$ lb. However, many people weigh more than $\frac{2500}{16}$ or 156.25 lb; if 16 such people were to get on the elevator together, their combined weight would exceed 2500 lb. Assuming that the 16 people are selected *randomly* from the population, what is the probability that their combined weight exceeds 2500 lbs?

Of course, we don't know the shape of the distribution of weights in the population. It could be close to normal or not. However, we don't need to know the shape in detail when applying the central limit theorem. (The population variance is surely finite, since the population range is finite.) In view of (5), the average weight of 16 individuals is approximately normally distributed. The Z-score corresponding to $\bar{X} = 156.25 = \frac{2500}{16}$ is

$$Z = \frac{156.25 - 145}{35/\sqrt{16}} = 1.286.$$

The probability of exceeding this value is $1 - \Phi(1.286)$, or about 10%.

(Although the precise population distribution is not known, the normal approximation for a sample of only 16 individuals is apt to be adequate. You might speculate about the shape of the population distribution on the basis of your experience. If the population of elevator riders includes both sexes, the distribution may even be bimodal; nevertheless, averaging 16 observations is apt to result in a unimodal distribution for $\bar{X}$, one that is close to normal.)   ▪

8.9   **The m.g.f. of the Sample Mean**

In Section 8.7, we saw how to find the first two moments (mean and s.d.) of the distribution of the sample mean. In Section 8.8, we saw that these two moments allow us to approximate probabilities of events such as $\bar{X} \leq k$ when n is large. However, the sample size may be such that these approximations are not adequate, and the complete distribution of the sample mean is then required. The m.g.f. is useful in finding it.

We first show how easy it is to get the m.g.f. of the mean from the population m.g.f. The observations in a random sample $(X_1, \ldots, X_n)$ are independent and identically distributed, each having the distribution of the population. Let X denote a generic variable with this distribution. The moment generating function of the sample mean $\bar{X}$ is

$$\psi_{\bar{X}}(t) = E(e^{t\bar{X}}) = E(e^{t \Sigma X / n}) = E(e^{(t/n) \Sigma X})$$

$$= \psi_{\Sigma X}\left(\frac{t}{n}\right) = \prod \psi_{X_i}\left(\frac{t}{n}\right) = \left[\psi_X\left(\frac{t}{n}\right)\right]^n.$$

The next-to-the-last step used the independence of the X's. The last step used the fact that all the X's have the same m.g.f.

> The m.g.f. of the mean of a random sample of size n is
>
> $$\psi_{\bar{X}}(t) = \left[\psi_X\left(\frac{t}{n}\right)\right]^n,$$
>
> where ψ_X is the population m.g.f.

(1)

To calculate the p.d.f. or p.f. of the sample mean, we carry out the steps of the following scheme:

$$f_X \rightarrow \psi_X \rightarrow \psi_{\bar{X}} \rightarrow f_{\bar{X}}.$$

We've seen how to carry out the first two steps—finding the population m.g.f. from the population p.d.f. and finding the m.g.f. of $\bar{X}$ from the population m.g.f. The last step, finding the p.d.f. of $\bar{X}$ from its m.g.f., is called *inverting* the transform that defines the m.g.f. There are integral expressions for the inverse, but it is often quite difficult to use such expressions. Alternatively, if the m.g.f. of $\bar{X}$ happens to be a function that we recognize as the m.g.f. of a familiar distribution, we're in business. The uniqueness theorem for m.g.f.'s mentioned in Section 5.10 says that (except for some pathological counterexamples) there is one and only one distribution having a given m.g.f. So if we know a distribution whose m.g.f. is the function $\psi_{\bar{X}}$ we've found, then we know that $\bar{X}$ has that distribution.

Example 8.9a   **Sample Mean for Exp(λ)**

In Example 5.10a, we found the m.g.f. of Exp(λ) to be

$$\psi_X(t) = \frac{\lambda}{\lambda - t}.$$

From (1), the m.g.f. of the mean of a random sample of size n is

$$\psi_{\bar{X}}(t) = \left[\psi_X\left(\frac{t}{n}\right) \right]^n = \left(\frac{\lambda}{\lambda - (t/n)} \right)^n. \tag{2}$$

In Example 5.10d, we found this to be $\psi_Y(t)$, where

$$f_Y(y) = \frac{\lambda^n}{(n-1)!} y^{n-1} e^{-\lambda y}, \qquad y > 0. \tag{3}$$

Since $\bar{X} = \frac{Y}{n}$, its p.d.f. is

$$f_{\bar{X}}(u) = n f_Y(nu) = \frac{(n\lambda)^n}{(n-1)!} u^{n-1} e^{-\lambda n u}, \qquad u > 0. \tag{4}$$

This is the p.d.f. of Gam($n, n\lambda$) [see (8) of Section 6.3]. ■

Example 8.9b   **Sample Mean from $N(\mu, \sigma^2)$**

When $X \sim N(\mu, \sigma^2)$, its m.g.f. [(12) of Section 6.1] is

$$\psi_X(t) = \exp(\mu t + \tfrac{1}{2}\sigma^2 t^2). \tag{5}$$

Using (2), we obtain the m.g.f. of the mean:

$$\psi_{\bar{X}}(t) = \left\{ \exp\left(\mu \frac{t}{n} + \frac{\sigma^2}{n}\left(\frac{t}{n}\right)^2 \right) \right\}^n = \exp\left(\mu t + \frac{\sigma^2}{n}\frac{t^2}{2} \right)$$

We recognize this, in view of (5), as the m.g.f. of $N(\mu, \frac{\sigma^2}{n})$. This is then the distribution of the sample mean. Thus, we have shown a property mentioned earlier: The approximation (5) of Section 8.8 actually is an *equality* for all n when the population sampled is normal. ■

Providing the Central Limit Theorem

We now outline a proof of the central limit theorem using moment generating functions, assuming that the population m.g.f. exists. (A similar but more general proof uses the characteristic function—see Section 5.10.)

Consider a sequence of independent, identically distributed random variables, $X_1, X_2, \ldots$. Let μ, σ^2, and ψ denote the mean, variance, and m.g.f. (respectively) of the common distribution $\psi_X(t)$, and assume $\sigma^2 < \infty$. The standardized mean of the first n observations is the same as the standardized sum:

$$Z_n = \frac{\bar{X}_{n-\mu}}{\sigma/\sqrt{n}} = \frac{X_1 + \cdots + X_n - n\mu}{\sqrt{n} \cdot \sigma} = U_1 + \cdots + U_n,$$

where

$$U_i = \frac{X_i - \mu}{\sqrt{n} \cdot \sigma}.$$

Since $E(U_i) = 0$ and var $U_i = \frac{1}{n}$, the series expansion for the m.g.f. of each U_i is

$$\psi_U(t) = 1 + \frac{t^2}{2n} + R,$$

where R is the remainder term. It can be shown that as $n \to \infty$, this remainder tends to 0 faster than $1/n^2$ and that therefore

$$\psi_{Z_n}(t) = [\psi_U(t)]^n \doteq \left[1 + \frac{t^2}{2n} \right]^n \to e^{t^2/2}. \tag{6}$$

This is the m.g.f. of a standard normal variable. So, by uniqueness, $Z_n \approx N(0, 1)$. This conclusion also depends on a "continuity theorem": If ψ_n is the m.g.f. of X_n, and ψ_n tends to ψ, the m.g.f. of X, then the c.d.f. of X_n tends to the c.d.f. of X at every continuity point of X.

8.10   Independence of Mean and Variance[7]

In showing the independence of mean and variance, the distribution of the sample variance is a useful by-product.

Let $X_1, \ldots, X_n$ be a random sample from a normal population with mean μ and variance σ^2, and let

$$S_n^2 = \frac{1}{n-1} \sum (X_i - \bar{X})^2.$$

Then if $n \geq 2$,

(a) $\bar{X} \sim N\left(\mu, \frac{\sigma^2}{n} \right)$.

(b) $(n-1)\frac{S_n^2}{\sigma^2} \sim \chi^2(n-1)$.

(c) $\bar{X}$ and S_n^2 are independent.

---

[7] The proof in this section depends on the material in Section 7.8.

In deriving the distribution of the one-sample t-statistic, we used the fact that for random samples from a normal population, the sample mean and sample variance are independent random variables. In Section 6.8, we showed that when $n = 2$, $X_1 + X_2$ and $X_1 - X_2$ are independent, which in turn implies that the sample mean and sample variance (functions of the sum and of the difference) are independent. In this section, we show this independence for arbitrary $n > 2$. The proof uses only bivariate distributions and an induction on n. Alternative proofs use linear algebra or multivariate generating functions.

Assertion (a) follows from the fact, given at the end of Section 6.1, that the sum of independent normal variables is normal. Also, we showed in Example 6.8a that (b) and (c) hold for $n = 2$.

The next step in a proof by induction is to show that if (b) and (c) are true for n, they must then hold for $n + 1$. We first write the mean of $n + 1$ observations in terms of the mean of the first n:

$$\bar{X}_{n+1} = \frac{n\bar{X}_n + X_{n+1}}{n + 1}. \tag{1}$$

From this we find the difference:

$$\bar{X}_n - \bar{X}_{n+1} = \bar{X}_n - \frac{n\bar{X}_n + X_{n+1}}{n + 1} = \frac{\bar{X}_n - X_{n+1}}{n + 1}. \tag{2}$$

Using (2), we can obtain an expression for the new sample variance in terms of the old:

$$nS_{n+1}^2 = \sum_1^{n+1}(X_i - \bar{X}_{n+1})^2 = \sum_1^{n+1}(X_i - \bar{X}_n + \bar{X}_n - \bar{X}_{n+1})^2$$

$$= \sum_1^n(X_i - \bar{X}_n)^2 + (X_{n+1} - \bar{X}_n)^2 + (n + 1)(\bar{X}_{n+1} - \bar{X}_n)^2$$

$$+ 2(\bar{X}_n - \bar{X}_{n+1})\sum_1^{n+1}(X_i - \bar{X}_n).$$

The first n terms $(X_i - \bar{X}_n)$ in the last sum add to 0. Thus

$$nS_{n+1}^2 = (n - 1)S_n^2 + (X_{n+1} - \bar{X}_n)^2\left(1 + \frac{1}{n+1} - \frac{2}{n+1}\right). \tag{3}$$

So

$$\frac{nS_{n+1}^2}{\sigma^2} = \frac{(n-1)S_n^2}{\sigma^2} + \frac{n}{n+1} \cdot \frac{(X_{n+1} - \bar{X}_n)^2}{\sigma^2}. \tag{4}$$

Because all the X_i's are independent, X_{n+1} is independent of S_n^2, which is a function of the first n X_i's. By the induction hypothesis, $\bar{X}_n$ is independent of S_n^2. This implies that $X_{n+1} - \bar{X}_n$ is independent of S_n^2, which in turn implies that the two terms on the right of (4) are independent.

By the induction hypothesis, the first term on the right of (4) is chi-square with $n - 1$ degrees of freedom. Since $X_{n+1} - \bar{X}_n$ is the difference between independent normal variables, it is also normal (see Section 6.1). Its mean is 0, and its variance is $(\sigma^2 n + 1)/n$. So the second term of (4) has a chi-square distribution with one d.f. Since it's independent of the first term, it follows that (b) holds for $n + 1$, and hence for all n.

To show that $\bar{X}_{n+1}$ is also independent of $X_{n+1} - \bar{X}_n$, we calculate the covariance:

$$\operatorname{cov}(\bar{X}_{n+1}, X_{n+1} - \bar{X}_n) = \operatorname{cov}\left(\frac{n\bar{X}_n + X_{n+1}}{n+1}, X_{n+1} - \bar{X}_n\right)$$

$$= 0 + \frac{\sigma^2}{n+1} - 0 - \frac{n}{n+1} \cdot \frac{\sigma^2}{n} = 0.$$

Because they are jointly normal and have covariance 0, the two variables are independent (see Section 6.8). Since $\bar{X}_{n+1}$ is independent of both terms on the right-hand side of (4), it is independent of S_{n+1}^2. So (c) follows by induction.

It is noteworthy that the independence of sample mean and variance in fact characterizes the normal distribution: The sample mean and variance are independent *only* when the population distribution is normal.[8]

Chapter Perspective

Statistical inference is the process of drawing conclusions about populations on the basis of sample information. In this chapter, we have seen how to extract information from a sample pertinent to a given population characteristic. We have also seen just how the reliability of sample information increases with the sample size.

In this chapter, we have addressed various mathematical issues dealing with distributions of sample statistics. In particular, we've studied how distributions of sample statistics depend on the various population characteristics. Later chapters will exploit this dependence in using samples to make inferences about population characteristics of interest.

Solved Problems

Sections 8.1–8.2

A. Write the joint density function of four observations in a random sample from a population with

  (a) uniform density on the interval $(0, 1)$.

  (b) $f(x \mid \alpha = \alpha x^{\alpha - 1}$ for $0 < x < 1$, $\alpha > 0$. [This includes (a) as the special case $\alpha = 1$.]

---

[8] See E. Lukacs and R. G. Laha, *Applications of Characteristic Functions* (New York: Hafner, 1964), p. 79.

Solution:

(a) The observations in a random sample are independent, so their joint p.d.f. is the product of the individual p.d.f.'s. The p.d.f. of one observation is

$$f(x) = \begin{cases} 1, & 0 < x < 1, \\ 0, & \text{elsewhere.} \end{cases}$$

The product of n such functions is a product of 1's and 0's; the result is 1 if and only if no factor is 0:

$$f(x_1, \ldots, x_n) = \begin{cases} 1, & 0 < x_i < 1 \text{ for } i = 1, \ldots, n, \\ 0, & \text{otherwise.} \end{cases}$$

The joint distribution is thus uniform on the unit n-cube.

(b) Again, the joint p.d.f. is a product of individual p.d.f.'s:

$$f(x_1, \ldots, x_n) = \prod_{i=1}^{n} f(x_i \mid \alpha) = \begin{cases} \alpha^n \prod_{i=1}^{n} x_i^{\alpha - 1}, & 0 < x_i < 1 \text{ for all } i, \\ 0, & \text{otherwise.} \end{cases}$$

B. Find the joint probability function of the observations in
   (a) a random sample from a Poisson population with mean m.
   (b) a sample of size n drawn without replacement from a Bernoulli population of size N, where $p = M/N$.

Solution:

(a) The p.f. for a single observation is

$$f(x \mid m) = \frac{m^x}{x!} e^{-m}.$$

The joint p.f. of the n observations is the product of the individual p.f.'s:

$$f(x_1, \ldots, x_n) = \prod_{i=1}^{n} f(x_i \mid m) = \prod_{i=1}^{n} \frac{m^{x_i}}{x_i!} e^{-m}$$

$$= e^{-nm} \left(\frac{m^{\Sigma x}}{x_1! \cdots x_n!} \right),$$

where each $x_i = 0, 1, 2, \ldots$, for $i = 1, \ldots, n$.

(b) The joint probability function of the observations $X_1, \ldots, X_n$ is the probability of $\{X_1 = x_1, \ldots, X_n = x_n\}$, where each x_i is 0 or 1. The exchangeability of the X's (Section 2.6) implies that this probability is the same for each possible sample sequence with a fixed sum (= number of 1's or successes). The probability for a given sum is hypergeometric. The probability of a specified number of 1's is therefore a hypergeometric probability divided by the number of sequences with that number of 1's:

$$f(x_1, \ldots, x_n) = \frac{\binom{M}{\Sigma x}\binom{N-M}{n-\Sigma x}}{\binom{n}{\Sigma x}\binom{N}{n}}, \qquad x_i = 0 \text{ or } 1 \text{ for } i = 1, \ldots, n.$$

Section 8.3

C.   Find a sufficient statistic for each case of the preceding problem.

Solution:

(a)   The likelihood function is the joint p.d.f.—or we can drop factors not involving m:
$L(m) = e^{-nm}m^{\Sigma x}$. Since this depends on the data through Σx, the statistic ΣX is
sufficient.

(b)   The likelihood function is the p.f., although we can omit the denominator since it
does not involve M:

$$L(M) = \binom{M}{\Sigma x}\binom{N - M}{n - \Sigma x}.$$

since this depends on the data through Σx, the statistic ΣX is sufficient.

D.   Consider the independent random variables $X_1, \ldots, X_n$, where $X_i \sim N(\alpha + \beta c_i, \theta)$, for
given constants c_i. Find three sample quantities that constitute a sufficient statistic.

Solution:

The likelihood function (except for a constant) is the product of the marginal p.d.f.'s:

$$L(\alpha, \beta; \theta) = \theta^{-n/2} \exp - \frac{1}{2}\Sigma(X_i - \alpha - \beta c_i)^2$$

$$\propto \theta^{-n/2} \exp\left(-\frac{1}{2}\Sigma X_i^2 + \Sigma c_i X_i + \alpha \Sigma X_i + \Sigma(\alpha + \beta c_i)^2\right)$$

This depends on the data through the value of the statistics ΣX_i^2, ΣX_i, and $\Sigma c_i X_i$, so
these are sufficient. [This is essentially a "regression" model—one that we'll take up in
Chapter 15.]

E.   Suppose X is geometric: $f(x \mid p) = p(1 - p)^{x-1}$, $x = 1, 2, \ldots$. Consider a random
sample from X: $(X_1, \ldots, X_n)$. Find the conditional distribution of $(X_1, \ldots, X_n)$ given
$\Sigma X_i = t$.

Solution:

The joint p.f. of the X_i's is

$$f(x_1, \ldots, x_n) = p^n(1 - p)^{\Sigma x - n}, \qquad x_i = 1, 2, \ldots.$$

The distribution of the statistic $T = \Sigma X_i$ is negative binomial:

$$f_T(t \mid p) = \binom{t - 1}{n - 1}p^n(1 - p)^{t-n}.$$

(See Section 4.4.) The desired conditional distribution is

$$P(\mathbf{X} = \mathbf{x} \mid \Sigma X_i = t) = \begin{cases} \dfrac{p^n(1 - p)^{t-n}}{\binom{t-n}{n-1}p^n(1 - p)^t} = \dfrac{1}{\binom{t-1}{n-1}}, & \text{if } \Sigma x = t, \\ 0, & \text{if } \Sigma x \neq t. \end{cases}$$

(That such a conditional probability does not involve the population parameter is sometimes taken as a definition of sufficiency; it implies the factorization [(1) of Section 8.3] used as the definition of sufficiency in this text.)

Sections 8.4–8.6

F. A bowl contains eight chips, numbered 1 through 8. Find the distribution of the median number on the chips in a random selection of three (no replacement).

Solution:

There are 56 equally likely combinations of three chips. The combinations with median equal to 2 are 123, 124, 125, 126, 127, and 128. Those with median 3 can start with either 1 or 2:

134, 135, 136, 137, 138, and 234, 235, 236, 237, 238.

Similarly, there are three groups of four with median 4, four groups of three with median 5, five groups of three with median 6, and six groups of three with median 7. The distribution is then as follows:

Median	2	3	4	5	6	7
Probability	$\frac{6}{56}$	$\frac{10}{56}$	$\frac{12}{56}$	$\frac{12}{56}$	$\frac{10}{56}$	$\frac{6}{56}$

(There are combinatorial methods that give these probabilities for arbitrary population and sample sizes but these are outside our scope.)

G. Find the probability that the smallest observation in a random sample of size 10 from a uniform distribution on (0, 1) is at least .15.

Solution:

The smallest observation is at least .15 if and only if every observation is at least .15:

$$P(X_{(1)} \geq .15) = P(X_i \geq .15, \quad i = 1, \ldots, 10) = (.85)^{10} = .197.$$

H. Use the p.d.f. of the range of a random sample from $U(0, 1)$ [(7) of Section 8.6] to derive the p.d.f. of the range of a random sample from $U(a, b)$, where $a < b$.

Solution:

If X is $U(a, b)$, the linear function

$$Y = \frac{X - a}{b - a}$$

is $U(0, 1)$. (See Example 5.1d.) Moreover, since the function is monotonic, the largest X is transformed into the largest Y and the smallest X into the smallest Y. So the range of the sample of X's is a multiple of the range of the corresponding Y's:

$$R_Y = Y_{(n)} - Y_{(1)} = \frac{X_{(n)} - a}{b - a} - \frac{X_{(1)} - a}{b - a} = \frac{R_X}{b - a}.$$

Hence, for $0 < r < b - a$,

$$f_{R_Y}(r) = f_{R_X}\left(\frac{r}{b - a}\right) \cdot \left(\frac{1}{b - a}\right) = \frac{n(n - 1)r^{n-2}(b - a - r)}{(b - a)^n}.$$

I. Find the covariance of the smallest and largest observations in a random sample from $U(0, 1)$.

Solution:

Let V and W, respectively, denote the smallest and largest. The joint p.d.f. [(5) of Section 8.6, with $F(x) = x$ and $f(x = 1)$] is

$$f_{V,W}(v, w) = n(n - 1)(w - v)^{n-2}, \qquad v < w.$$

We use this to find the expected product:

$$E(VW) = n(n - 1) \int_0^1 \int_0^w wv(w - v)^{n-2}\, dv\, dw.$$

The inner integral is

$$\int_0^w v(w - v)^{n-2}\, dv = \int_0^y w^{n-2}(1 - y)^{n-2} yw^2\, dy = w^n B(n - 1, 2) = \frac{w^n}{n(n - 1)},$$

so

$$E(VW) = \int_0^1 w^{n+1}\, dw = \frac{1}{n + 2}.$$

Then

$$\text{cov}(V, W) = E(VW) - E(V)E(W)$$

$$= \frac{1}{n + 2} - \frac{1}{n + 1} \cdot \frac{n}{n + 1} = \frac{1}{(n + 2)(n + 1)^2}.$$

J. Use the differential method (as we did in Section 8.6) to find the joint p.d.f. of the largest and second largest observations in a random sample. (The result will be in terms of the population c.d.f. F and p.d.f. f.)

Solution:

Denote by U and V the second largest and largest, respectively. For $u < v$, we want the probability of the event

$$\{u < U < u + du \text{ and } v < V < v + dv\}.$$

Class intervals and corresponding (approximate) probabilities defined by interval endpoints $u, u + du, v,$ and $v + dv$ are as follows, along with the frequencies (in a sample of size n) corresponding to the above event:

Interval	Probability	Frequency
$X < u$	$F(u)$	$n - 2$
$u < X < u + du$	$f(u)\,du$	1
$u + du < X < v$	$F(v) - F(u)$	0
$v < X < v + dv$	$f(v)\,dv$	1
$X > v + dv$	$1 - F(v)$	0

The desired probability is multinomial:

$$f_{U,V}(u, v)\,du\,dv = \frac{n!}{(n - 2)!1!0!1!0!}[F(u)]^{n-2}[f(u)\,du]^1[f(v)\,dv]^1,$$

and the joint density we seek is the coefficient of $du\,dv$:

$$f_{U,V}(u, v) = n(n - 1)[F(u)]^{n-2}f(u)f(v), \qquad u < v.$$

Sections 8.7–8.9

K. The average ACT score in Minnesota in 1986 was 19.4. Assuming a standard deviation of 2.0, find the probability that the mean of a random sample of 25 scores from Minnesota exceeds 20.0.

Solution:

According to (3) and (4) in Section 8.7, $E(\bar{X}) = \mu = 19.4$ and $\text{var}(\bar{X}) = (2.0)^2/25 = .16$. Using (5) of Section 8.8, we have

$$P(\bar{X} > 20) = 1 - F_{\bar{X}}(20) \doteq 1 - \Phi\left(\frac{20 - 19.4}{.4}\right) = \Phi(-1.5) = .0668.$$

L. A company packages paper clips in boxes labeled "100 clips." We've counted the clips in these boxes and found that the number in a box varies from as low as 94 to well over 100. Suppose that the number of clips in boxes produced by this company has mean 100 and standard deviation 3. Find the probability that a carton of 36 boxes contains fewer than 3,580 clips in all (more than 20 shy of 3,600).

Solution:

The central limit theorem makes it unnecessary to know the population shape. Given $\mu = 100$ and $\sigma = 4$, the mean number in 36 boxes is 3,600, and the s.d. is $\sqrt{36} \times \sigma = 12$. So

$$P(\text{total} < 3,580) \doteq \Phi\left(\frac{3,580 - 3,600}{12}\right) = \Phi\left(\frac{-20}{12}\right) \doteq .048.$$

M. Occasionally a machine screw is skipped or damaged in the threading process. Suppose .5% of the screws produced by a particular machine are defective. Find the probability that more than 1% of a carton of 200 screws are defective.

Solution:

If the probability of a defective were not so small, we could use a normal curve to find an approximate probability. As it is, with $n = 200$ and $p = .005$, the Poisson approximation is better. Since $np = 1$, $Y = n\hat{p} \approx \mathrm{Poi}(1)$:

$$P(\hat{p} > .01) = P(Y > 2) = 1 - P(Y = 0, 1, \text{ or } 2)$$
$$= 1 - e^{-1}(1 + 1 + 1.5) \doteq .0803.$$

N. Two polls independently obtain random samples of size 400 from the population of voters in a state. On a question with a "yes" or "no" answer, find the probability that the proportion of "yes" responses in the two polls differ by more than 2 percentage points.

Solution:

This natural and important kind of question anticipates Section 9.6, where we discuss the difference between sample proportions in more detail. To find the probability asked for, we need the distribution of the difference in sample proportions. Each sample proportion is approximately normal. Since they are independent, the difference is approximately normal. So we need the mean and variance of the difference. The mean and variance of $\hat{p}_i$, the sample proportion of "yes" answers in poll i, are p and pq/n, respectively. Thus,

$$E(\hat{p}_1 - \hat{p}_2) = 0,$$

and because the samples are independent,

$$\mathrm{var}(\hat{p}_1 - \hat{p}_2) = \mathrm{var}\,\hat{p}_1 + \mathrm{var}\,\hat{p}_2 = \frac{2pq}{400} = .005pq.$$

To go further, we need to know p; however, $p(1 - p)$ is largest when $p = \frac{1}{2}$, in which case $.005pq = .00125 = .0354^2$. Then

$$P(|\hat{p}_1 - \hat{p}_2| > .02) = 2P(\hat{p}_1 - \hat{p}_2 < -.02) \doteq 2\Phi\left(\frac{.02 - 0}{.0354}\right) \doteq .572.$$

The answer is not much different if p is near .5. For instance, it is .537 when $p = .3$.

O. Find the probability of getting a bridge hand with 26 or more points, assuming a random deal. (A bridge hand consists of a random selection of 13 cards from a standard deck. Each card is assigned points as described in Problem 34 of Chapter 2.)

Solution:

A card picked at random has a value X with $E(X) = \frac{10}{13}$ and $\mathrm{var}\,X = \frac{290}{269}$, according to Problems 4 and 17 of Chapter 3. The total number of points is the sum:

$$Y = X_1 + X_2 + \cdots + X_{13},$$

where X_i is the ith card dealt. The central limit theorem as we've stated it (in its simplest form) does not apply—even though the X's are identically distributed—because they are *not* independent. So this is not a fair question; you cannot answer it with what you've learned so far. However, there is a version of the central limit theorem that does apply—

it says that the standardized Y is approximately normal. Now, $E(Y) = 13E(X) = 10$, and

$$\text{var } Y = n(\text{var } X) \cdot \frac{N - n}{N - 1} = \frac{290}{13} \cdot \frac{52 - 13}{52 - 1} \doteq (4.13)^2.$$

Then

$$P(Y \geq 26) = 1 - \Phi\left(\frac{25.5 - 10}{4.13}\right) \doteq .0001.$$

(You might think of trying to find the exact probability by counting hands with $Y \geq 26$. Lots of luck!)

P. Consider a random sample $(X_1, \ldots, X_n)$ from an exponential population with mean θ. Find the limiting distribution of the sample mean using moment generating functions.

Solution:

The population m.g.f. is $\psi(t) = (1 - \theta t)^{-1}$. So the m.g.f. of $\bar{X}$ is

$$\psi_{\bar{X}}(t) = \left[\psi\left(\frac{t}{n}\right)\right]^n = \left(1 - \frac{\theta t}{n}\right)^{-n} = ((1 + w)^{1/w})^{\theta t},$$

where $w = -\theta t/n$, or $-n = \theta t/n$. Since the quantity in brackets tends to e as $n \to \infty$ (and $w \to 0$),

$$\psi_{\bar{X}}(t) \to e^{\theta t}.$$

The limit $e^{\theta t}$ is the m.g.f. of a singular distribution, one that puts probability 1 at $\bar{X} = \theta$. The random variable $\bar{X}$ "converges to the constant θ in distribution." This is true not only for this exponential example but also in general for any population for which the mean exists. (This is a version of the law of averages or law of large numbers stated in Section 4.7.)

Q. Find the characteristic function of the mean of a random sample from a population with p.d.f.

$$f(x) = \frac{1/\pi}{1 + x^2}.$$

Solution:

In Solved Problem CC of Chapter 5, we found the characteristic function of the Cauchy population with this p.d.f.:

$$\varphi_X(t) = e^{-|t|}.$$

The characteristic function of $\bar{X}$ is

$$\varphi_{\bar{X}}(t) = E(e^{it\bar{X}}) = E(e^{it\Sigma X/n}) = \prod_{j=1}^{n} E(e^{itX_j/n})$$

$$= \left[\varphi\left(\frac{t}{n}\right)\right]^n = (e^{-|t/n|})^n = e^{-|t|}.$$

This is the characteristic function of X, so the sample mean has the same distribution as the population!

Problems

For problems marked with an asterisk, answers are provided in Appendix 2.

Sections 8.1–8.2

1. Write the joint density function of four observations in a random sample from a population
   *(a) with a uniform distribution on the interval $(-1, 1)$.
   *(b) with the exponential distribution with p.d.f. $2e^{-2x}$, for $x > 0$.
   (c) with a normal distribution with mean 10 and s.d. 2.

2. Write an expression for the joint probability function of five independent observations on a random variable X whose distribution is
   *(a) binomial with $n = 3$ and $p = \frac{1}{4}$.
   (b) geometric with $p = \frac{1}{3}$.

*3. You obtain the following, in random sampling from $\mathrm{Exp}(\lambda)$: (2, 4, 7). Find and sketch the likelihood function.

4. Let $(X_1, \ldots, X_n)$ be a random sample. Write the likelihood function in the case of each of the following populations:
   (a) Normal with mean 0 and variance θ.
   (b) Normal with mean μ and variance θ.

*5. Suppose we are interviewing people one at a time, recording the variables $(X_1, X_2, \ldots)$ where X_1 is the number of interviews it takes to find a person with type B blood, X_2 is the additional number to find a second person with type B blood, and so on. Considering each person as a Bernoulli trial in which p is the population proportion with type B blood,
   (a) give the joint probability function of the $(X_1, \ldots X_n)$.
   (b) We find that it takes 30 interviews to locate three people who are type B. Give the likelihood function $L(p)$.

6. Write the likelihood function for a random sample of size 10 from a population whose distribution is
   (a) Cauchy, with location parameter θ:

   $$f(x \mid \theta) = \frac{1/\pi}{1 + (x - \theta)^2}.$$

   (b) gamma, with parameters (α, λ).
   (c) truncated exponential: $f(x \mid \theta) = \dfrac{e^{-x}}{1 - e^{-\theta}}$ for $0 \leq x \leq \theta$, and 0 elsewhere.

*7. Consider sampling one at a time without replacement from a carton of 12 items including an unknown number M of defective ones.
   (a) You decide to stop after drawing four and find two defective items among the four. Give the likelihood function.
   (b) You decide to stop after finding the second defective item and find that it takes four selections. Give the likelihood function.
   (c) Find the value of M for which the likelihood function in (a) has its maximum value.

8. Consider two persons: One plans to watch a Poisson process with parameter λ (counts per hour) until ten events have occurred; the other plans to watch for one hour and

count the number of events. Suppose the first person finds that it takes exactly one hour to the tenth event and the second person observes ten events in a one-hour period. Compare the likelihood functions.

Section 8.3

*9. Show that the exponential family of distributions [(2) of Section 8.3] includes the following and give a sufficient statistic in each case:
(a) The normal family with mean 0.
(b) Exponential distributions.
(c) Poisson distributions.

10. Show that the exponential family of distributions [(2) of Section 8.3] includes the following and give a sufficient statistic in each case:
(a) Geometric distributions.
(b) The gamma family with $\alpha = 2$.
(c) The family of distributions with p.d.f $f(x; \theta) = \theta x^{\theta-1}$ for $0 < x < 1$.

*11. Find a two-dimensional sufficient statistic for θ based on a random sample from the bivariate population with p.d.f. $\exp(-\theta x - (y/\theta))$ for $x > 0$ and $y > 0$, where θ is a positive parameter.

12. Consider a random sample $(X_1, \ldots, X_n)$ from a Poisson distribution with parameter m. Let $T = \Sigma X_i$. Find the conditional probability distribution of $(X_1, \ldots, X_n)$, given $T = t$.

*13. Find a sufficient statistic for (θ_1, θ_2), based on a random sample from a population that is uniform on the interval (θ_1, θ_2).

14. Consider a random sample from X whose distribution is Poi(λ). Let $Y = \Sigma X$. (This is sufficient for λ.) Find the likelihood function based on the sample of X's and compare it with the likelihood function based on the observed value of Y.

*15. Find a sufficient statistic for the parameter M in the discrete distribution with p.f. $f(x \mid M) = \frac{1}{M}, k = 1, 2, \ldots, M$, based on a random sample of size n.

16. Suppose $T = t(\mathbf{X})$ is sufficient for θ. Let $\varphi(\mathbf{x})$ be a function defined on the sample space and show that $E(\varphi(\mathbf{X}) \mid T = t)$ does not depend on θ. (*Hint:* See Solved Problem E.)

*17. Suppose X is Ber(p). Consider a random sample $\mathbf{X} = (X_1, \ldots, X_n)$ and find the conditional distribution of $\mathbf{X}$ given $\Sigma X = t$.

Sections 8.4–8.6

18. Suppose three chips are selected at random without replacement from a bowl containing chips numbered from 1 to 5. Define

Y = the *largest* number among the three chips in a sample.

S = the sum of the numbers on the three chips.

M = the smallest of the numbers on the three chips.

(a) List the $\binom{5}{3}$ equally likely samples.
(b) Obtain the sampling distributions of Y and of M.
*(c) Find the distribution of S.

19. Consider a sample of size $n = 3$ from the bowl in Problem 18 when sampling is *with* replacement.
    (a) Find the p.f. of $X_{(3)}$, the largest sample observation.
    (b) Find the sampling distribution of S, the sample sum. (Compare with the corresponding distributions in Problem 18.)

20. As a check on the answers to Problem 19, actually perform the experiment 50 times. (You could use five coins or buttons of equal size marked with numbers 1 to 5, taking care to mix thoroughly after each selection of three.) Record the values of S and M and construct a histogram for each.

21. Determine the sampling distribution of the sample sum for random samples of size n from each of the following populations, giving it by name and identifying parametric(s):
    *(a) $N(\mu, \theta)$.   *(b) $\text{Exp}(\lambda)$.   *(c) $\text{Geo}(p)$.
    (d) $\text{Poi}(m)$.   (e) $\text{Bin}(k, p)$.

*22. Find the mean value of the largest observation in a random sample of size n from a population with c.d.f. $F(x) = x^k, 0 < x < 1$.

23. Let $(X_1, \ldots, X_5)$ be a random sample from $U(0, 1)$.
    *(a) Find the sampling distribution of $\tilde{X}$, the sample median.
    *(b) Find the mean and variance of $\tilde{X}$.
    (c) Interpret each sequence of five digits in Table XV of Appendix 1 as the first five decimal digits of an observation from $U(0, 1)$. Each 5×5 block of digits in the table can be considered a random sample of size five. Find the median $\tilde{X}$ of each sample in 40 successive blocks, starting at an arbitrary point in the table. Find the mean and variance of the 40 $\tilde{X}$-values calculated from your 40 samples and compare them with your answers to (b). (They're apt not to be the same but should be close.)

*24. Find the probability that the largest observation in a random sample of size $n = 5$ from $N(10, 1)$ is less than 12.

25. Use (a) of Problem 23 to find the p.d.f. of the sampling distribution of the median of a random sample of size 5 from $U(a, b)$, where $a < b$.

*26. Use the differential method to find the joint p.d.f. of the second smallest and second largest observations in a random sample from a population with c.d.f. F and p.d.f. f.

*27. Use (5) of Section 8.6 and the transformation

$$Y = \frac{X - a}{b - a}$$

to obtain the joint p.d.f. of the smallest and largest observations in a random sample from $U(a, b)$.

28. Consider a continuous population whose p.d.f. is symmetric about its median. Show that the median of a random sample of size n is also symmetrically distributed when n is odd. (The result is also true for n even.) [Recall: f is symmetric about a means that for all x, $f(a - x) = f(a + x)$ or, equivalently, $F(a - x) = 1 - F(a + x)$.]

Sections 8.7–8.9

*29. Six families have incomes as follows: 22, 24, 24, 28, 32, 50 (in thousands of dollars). Suppose we select two of these families at random and calculate $\bar{X}$, the average income for this sample.
   (a)   Find the sampling distribution of $\bar{X}$.
   (b)   Using (a), calculate the mean and variance of $\bar{X}$.
   (c)   Check (b) using (10) and (11) of Section 8.7.

*30. A population of males has mean height $\mu = 70$ inches and standard deviation $\sigma = 3$ inches. Find the mean and s.d. of $\bar{X}$, the average height of the males in a random sample of size 400.

31. Find (approximately) the probability that $\bar{X}$ in Problem 30
   (a)   exceeds 70.3 inches.
   (b)   differs from the population mean (70 in.) by more than .4 inch.

*32. The mean and s.d. of GRE scores in a particular year were 521 and 123, respectively. Find the probability that the mean score in a random sample of 100 students taking the GRE is less than 500.

33. Boxes of detergent are filled by machine. The net weight is a random variable with mean 28 ounces and s.d. $\frac{1}{2}$ ounce. Find
   (a)   the mean and s.d. of the total weight of a carton of 24 boxes.
   (b)   the probability that the carton weight exceeds 42.3 lb. (There are 16 ounces in a pound.)

*34. In a carnival game (observed in Santa Barbara, California, in 1988), a player throws six darts at a board. The board has 396 squares (22 by 18), each with a number from 1 to 6, in no systematic order. We may assume that the number a dart hits has the following distribution:

x	1	2	3	4	5	6
$f(x)$	$\dfrac{57}{396}$	$\dfrac{29}{396}$	$\dfrac{142}{396}$	$\dfrac{114}{396}$	$\dfrac{27}{396}$	$\dfrac{27}{396}$

(Each probability is the fraction of squares with the corresponding number.) Assuming six independent tosses, let Y denote the total score—a number from 6 to 36. You win a "large prize" if $Y \geq 29$ and a "small prize" if $Y \leq 14$. Find, approximately,
   (a)   the probability of winning a large prize.
   (b)   the probability of winning a small prize.

35. Suppose 15% of the people in a certain population are left-handed. For a random sample of size 200,
   (a)   find the mean and s.d. of the proportion of left-handed people in the sample.
   (b)   approximate the probability that fewer than 10% in the sample are left-handed.

*36. Suppose 25% of all families with TVs are watching a particular program. Find the probability that more than 26% of a random sample of 1,000 TV viewers are watching it.

37. Suppose people with type AB blood constitute only 5% of a certain population. Find the probability that fewer than two in a random sample of 100 individuals have this type of blood.

*38. A poultry farmer has an order for 30 black-winged chicks. Suppose that one-fourth of the chicks that hatch are black-winged. How many chicks must hatch for the farmer to be 99.87% sure of having at least 30 black-winged chicks? (*Hint*: The equation $a + bn = c\sqrt{n}$ is quadratic in $\sqrt{n}$.)

39. Consider two independent random samples from the same normal population. Obtain the m.g.f. of the difference between the sample means (in terms of the parameters of the parent population) and deduce the distribution of the difference.

40. Find the m.g.f. of $\hat{p}$, the proportion of successes in a random sample from Ber(p). Show that as n becomes infinite, the m.g.f. converges to e^{pt}, the m.g.f. of a singular distribution at p. (*Hint*: Using L'Hospital's rule, show that the log of the m.g.f. tends to pt as n becomes infinite.)

Estimation

A random sample from a population contains information about the various aspects of the population. Estimating population means and estimating population proportions from sample information are important practical problems.

Example 9a

TV Ratings

Producers and advertisers are vitally interested in knowing what TV viewers watch. Surveying *all* viewers is impossible, but viewers can be *sampled*. Arbitron and Nielsen TV ratings are based on samples of TV owners. These ratings are estimates of population proportions based on sample information. Television executives make decisions with far-reaching consequences on the basis of such estimates.

A Nielsen rating of 22 for a particular TV program means that 22% of the sample were watching that program. The sample proportion .22 is an estimate of the proportion who watched the program in the population of all TV owners. This is a reasonable estimate if the sample is reasonably representative of that population. ∎

Example 9b

Auditing Accounts by Sampling

The book value of an account receivable of a large company may or may not agree with the actual value, owing to errors in the records. Auditors, faced with enormous numbers of accounts and transactions, may conduct an audit by taking a sample of accounts and examining the sampled accounts in detail. They use random sampling to estimate the error rate as well as the magnitudes of errors. ∎

An estimate based on a random sample is a statistic—a random variable. Different samples will generally lead to different estimates. An estimate calculated from a particular sample is apt to be in error.

Example 9c

Variability of Sample Means

In Example 8.1b, we referred to the very precise scales used by the National Bureau of Standards to calibrate standard weights. Despite the precision, different weighings of the same object give different results. The variation in these results can be modeled as a probability distribution. If it were possible to perform infinitely many weighings, their average would be the "true" weight—assuming that there is no systematic error caused by such factors as moisture on the scale. Bureau scales are periodically calibrated to eliminate such biases.

Suppose we want to calibrate a scale using a standard weight, a block of metal whose weight is rather precisely known. If we could weigh it an infinite number of times and average these weighings, this would tell us how to adjust the scale. We can't, so a finite number of weighings must suffice.

Suppose we weigh the block 20 times, observing these differences between the observed weight and the nominal weight (in micrograms):

8.82	7.39	18.86	2.65	12.92	2.07	4.26
−3.05	6.93	−15.40	5.95	15.97	−6.78	−3.80
4.52	5.54	13.45	−5.17	3.57	4.92	

The mean of the 20 differences is 4.181. Actually, we generated the data artificially, sampling from a population with mean 5.00 and standard deviation 6.00. So the estimate 4.181 is not correct. Whenever we base our estimate on a sample, the estimate is apt to be wrong. The question is, how large is the error? ∎

In this chapter, we consider a number of natural estimates of population parameters. For example, the largest sample value $X_{(n)}$ estimates the largest population value, and a sample proportion $\hat{p}$ estimates the corresponding population proportion p. When a parameter to be estimated is a function of population moments, an obvious candidate is the same function of the sample moments.

We begin by studying ways of assessing and reporting errors of estimation and then proceed to show how to determine the sample size required to keep the error within a specified level.

9.1 Errors in Estimation

We're going to be estimating various population parameters—means, proportions, variances, and so on. As in the preceding chapter, let θ denote a generic parameter. Let Y denote a statistic proposed for estimating θ. When used for this purpose, Y is called an **estimator**. We call the value of Y calculated from a particular sample an **estimate** of θ.

We saw in Example 9c that the value of an estimator varies from sample to sample, so it is too much to expect that Y *equals* θ. Yet we can hope that it is *near* θ in some sense. We'll consider Y to be a good estimator of θ if its sampling distribution is concentrated near θ.

Example 9.1a   **Estimating a Maximum**
Suppose $X \sim U(0, \theta)$, and consider estimating θ based on the observations in a random sample from X of size n. The largest sample observation $Y = X_{(n)}$ is sufficient for θ (see Example 8.3e) and is an obvious candidate for estimating the

largest population value θ. In Example 8.4c, we derived the c.d.f. of Y:

$$F_Y(y) = \left(\frac{y}{\theta}\right)^n, \qquad 0 < y < \theta.$$

When $n = 100$ (for example), the probability that Y is within 2% of θ is

$$P(.98\theta < Y < \theta) = F_Y(\theta) - F_Y(.98\theta) = 1 - (.98)^{100} \doteq .867.$$

This large a probability in such a small interval results from the enormous concentration of probability near θ. [The standard deviation of $X_{(100)}$ is only about $(.01)\theta$.] ∎

Estimation Error

The difference $Y - \theta$ is the **error** when we use Y to estimate θ. The error varies from sample to sample because Y does. The **absolute error** of estimation is $|Y - \theta|$. This is the distance between the estimated value Y and the actual value θ. An obvious overall measure of how poorly an estimator Y performs is the **mean absolute error** $E(|Y - \theta|)$, where the averaging is with respect to the sampling distribution of Y. However, a more commonly used criterion is the **mean squared error**:

Mean Squared Error

$$\text{m.s.e.}(Y) = E[(Y - \theta)^2]. \tag{1}$$

We use the parallel axis theorem [(6) in Section 3.3 and (3) in Section 5.5] to rewrite the m.s.e. as follows:

$$\text{m.s.e.}(Y) = E[(Y - \theta)^2] = \text{var } Y + [E(Y) - \theta]^2. \tag{2}$$

The quantity $E(Y) - \theta$ is termed the **bias** in Y as an estimator of θ. In words, (2) says that the mean squared error is the variance of Y plus the square of its bias.

The **mean squared error** using Y as an estimator of θ is

$$\text{m.s.e.}(Y) = E[(Y - \theta)^2] = \text{var } Y + [b_Y(\theta)]^2, \tag{3}$$

where $b_Y(\theta)$ is the bias:

$$b_Y(\theta) = E(Y) - \theta. \tag{4}$$

When $b_Y(t) = 0$, Y is **unbiased**.

Since both the variance and the squared bias are nonnegative, the m.s.e. will be small only when *both* the bias and the variance of Y are small. When they are both small, most of the sampling distribution of Y is concentrated near θ. However, unbiasedness is not in itself especially virtuous.

Example 9.1b **S^2 Unbiased for σ^2**

In Section 5.5, we defined this measure of sample variability:

$$V = \frac{1}{n} \sum_{i=1}^n (X_i - \bar{X})^2.$$

According to the parallel axis theorem [(4) of Section 5.5],

$$\frac{1}{n} \sum_{i=1}^{n} (X_i - \mu)^2 = V + (\bar{X} - \mu)^2.$$

Transposing and taking expected values, we obtain

$$E(V) = \frac{1}{n} \sum_{i=1}^{n} E(X_i - \mu)^2 - E[(\bar{X} - \mu)^2].$$ (5)

Each X_i has the population distribution, so

$$E[(X_i - \mu)^2] = \sigma^2 \qquad \text{for} \qquad i = 1, 2, \ldots, n.$$

The second term on the right of (5) is var $\bar{X} = \sigma^2/n$. Thus,

$$E(V) = \frac{1}{n} \cdot n\sigma^2 - \frac{\sigma^2}{n} = \frac{n-1}{n} \sigma^2.$$

It follows that V is biased in estimating σ^2. On the other hand, the sample variance defined in Section 7.5:

$$S^2 = \frac{n}{n-1} V = \frac{1}{n-1} \sum_{i=1}^{n} (X_i - \bar{X})^2,$$

is unbiased in estimating σ^2. However, we'll see in Section 9.7 that in the case of normal populations, it is V that has the smaller m.s.e. ∎

The units of m.s.e. are the units of Y^2. This makes m.s.e. hard to interpret. The square root of the m.s.e. has the same units as Y; we refer to this as the r.m.s. error of estimation.

> The **root-mean-squared error** of Y as an estimator of θ is
>
> $$\text{r.m.s.e.}(Y) = \sqrt{\text{m.s.e.}} = \sqrt{E[(Y - \theta)^2]}.$$ (6)
>
> When Y is unbiased, r.m.s.e.$(Y) = \sigma_Y$.

Estimating a Proportion

Estimating a population proportion p is a special case of estimating a mean. The corresponding sample proportion $\hat{p}$ is a natural estimator. If we assume random sampling, $\hat{p}$ is sufficient for p. According to (6) of Section 8.7, $E(\hat{p}) = p$, so $\hat{p}$ is unbiased. Its r.m.s.e. is therefore its standard deviation [(7) of Section 8.7]:

$$\text{r.m.s.e.}(\hat{p}) = \sqrt{\frac{p(1-p)}{n}}.$$ (7)

This depends on p, which is awkward in that we need to know the parameter we're estimating in order to assess its performance! However, we can approximate (7) when n is large, as we explain next.

For large n, the r.m.s error of $\hat{p}$ is small because n appears in the denominator of (7). Then $\hat{p}$ is likely to be close to p (a fact that is the essence of the law of large numbers, mentioned in Section 4.7). It follows that the r.m.s error can be well approximated by replacing p with $\hat{p}$ in (7); this approximate value is termed the **standard error**.

Standard error[1] of $\hat{p}$, as an estimate of a population proportion p:

$$\text{s.e.}(\hat{p}) = \sqrt{\frac{\hat{p}(1 - \hat{p})}{n}}. \qquad (8)$$

Example 9.1c

TV Ratings

A ratings company takes a random sample of 1,000 TV owners to estimate the proportion p who are watching a particular program. It finds that 290 of the viewers in the sample are watching the program. It then announces the program's rating as 29, taking the sample proportion $\hat{p} = .29$ as an estimate of the population proportion p. The standard error (8) is

$$\text{s.e.}(\hat{p}) = \sqrt{\frac{\hat{p}(1 - \hat{p})}{n}} = \sqrt{\frac{.29 \times .71}{1000}} \doteq .0143.$$

This may be thought of as a typical error involved in estimating p to be $\hat{p}$. The estimate is often reported in the form $.29 \pm .014$. ∎

Estimating a Mean

We turn next to estimating a population mean μ from a random sample $(X_1, \ldots, X_n)$. The sample mean $\bar{X}$ is an obvious estimator of μ. Its bias is zero: $E(\bar{X}) - \mu = 0$ [see (3) of Section 8.7], so the r.m.s. error of $\bar{X}$ is its standard deviation, given by (4) of Section 8.7:

$$\sigma_{\bar{X}} = \frac{\sigma}{\sqrt{n}}. \qquad (9)$$

Like (7), this depends on a population parameter—but not on the one being estimated. Still, σ is generally unknown, so we can only approximate the error s.d. using a sample statistic in place of σ. If the sample distribution approximates the population distribution, the sample standard deviation S will be close to σ when n is large. (We'll see in Section 9.7 that the variance of S tends to 0 as n becomes infinite.) We make the following definition:

Standard error of $\bar{X}$ as an estimate of a population mean μ:

$$\text{s.e.}(\bar{X}) = \frac{S}{\sqrt{n}}. \qquad (10)$$

---

[1] Some statisticians use the term *standard error* for the standard deviation of the error and call the approximation (8) the *estimated standard error*.

Example 9.1d   **Platelet Counts**

The report cited in Example 7.2e gives blood platelet counts for 35 healthy males. The mean count per microliter was $\bar{X} = 234$ and the s.d., $S = 45.05$. Assume random sampling. The standard error of $\bar{X}$ as an estimate of the population mean μ [from (10)] is

$$\text{s.e.}(\bar{X}) = \frac{S}{\sqrt{n}} = \frac{45.05}{\sqrt{35}} = 7.61.$$

The result could then be reported as 234 ± 7.6—the point estimate plus or minus the typical error. ■

Having established a measure of precision for estimation, we next see how specifying the desired precision can lead to a choice of sample size.

9.2   Determining Sample Size

In setting up a research study, one must decide how much data to collect. A factor in this decision is the specification of an allowable standard error. The standard error of an estimator is often a decreasing function of sample size: The larger the sample, the smaller the standard error. In such cases, we may achieve a specified level of precision by taking a sufficiently large sample.

Estimating a Mean

Consider estimating a population mean. To meet a requirement that s.d.$(\bar{X})$ does not exceed ε:

$$\sigma_X = \frac{\sigma}{\sqrt{n}} \le \varepsilon,$$

we need

$$\sqrt{n} \ge \frac{\sigma}{\varepsilon}, \quad \text{or} \quad n \ge \frac{\sigma^2}{\varepsilon^2}.$$

So we take n to be the next integer larger than $\dfrac{\sigma^2}{\varepsilon^2}$.

Example 9.2a   **Calibrating a Scale**

A standard weight is to be used in calibrating a scale. We require the standard deviation of $\bar{X}$ to be at most two micrograms. How many weighings will be needed?

From (1), $n = \sigma^2/4$. If we know that $\sigma = 6$ (as in Example 8.7b), we can find the required sample size: $n = \frac{36}{4} = 9$. ■

Using (1) to determine a sample size requires that we know σ. In most applications, σ is unknown. The only experimental settings we can imagine in which σ is known but μ is not known are those in which the error arises from using a measuring device whose variability is known from previous experiments. In such cases, the mean measurement μ will be unknown, since it depends on the object being measured. Even then, if we claimed to "know" $\sigma = 10$ but found $S = 100$ in a sample of size 50, we'd back down from our "knowledge."

What Estimate
for σ?

There are various ways of getting around the problem of unknown σ. The investigator may be able to estimate σ based on past experience; using this in (1) gives an approximate sample size. If the investigator can give only an interval of possible values for σ, taking σ in (1) to be the largest value in that interval leads to a sample size that is more than adequate for any other σ in the interval.

This approach has the obvious disadvantage that the n found using (1) is only as reliable as the estimate of variance on which it's based. Indeed, the current circumstances may turn out to be quite different from those that produced the estimate. A disadvantage of estimating n based on an interval of σ-values is that the n-interval can turn out to be enormous—for example, 10 to 1,000. The conservative 1,000 may cost more (in time and resources) than the investigator is willing to invest.

A second approach is more interesting. It is feasible only when data can be gathered in such a way that at least some intermediate results are available during the course of the experiment. The data available at any point provide an estimate of σ—namely, the current sample standard deviation S. Using S in place of σ in (1) yields an estimate of the required sample size. Such a calculation can be done periodically—even after every observation!

Example 9.2b

Sequential Adjustment of the Required n

A device for measuring the amount of stretch or "drawer" in the human knee is used to determine the extent of damage to an injured anterior cruciate ligament. For such measurements to be useful, the distribution of drawers of normal knees must be known, at least approximately. Although the entire distribution is of interest, we focus on the mean μ. Suppose this is to be estimated with a standard error no greater than .2 mm. How many knees should we measure to meet this specification of s.e.?

Table 9.1 lists the drawer measurements of the normal knees of 34 patients

Table 9.1

Subj. No.	Drawer (mm)	$\bar{X}$ (mm)	S (mm)	$\left(\dfrac{S}{.2}\right)^2$	Subj. No.	Drawer (mm)	$\bar{X}$ (mm)	S (mm)	$\left(\dfrac{S}{.2}\right)^2$
1	11	11.0	—	—	18	10	8.0	1.75	77
2	9	10.0	1.41	50	19	9	8.1	1.77	74
3	10	10.0	1.00	25	20	7	8.0	1.73	72
4	6	9.0	2.16	117	21	7	8.0	1.69	69
5	7	8.6	2.07	108	22	8	8.0	1.65	66
6	8	8.5	1.87	88	23	8	8.0	1.62	63
7	8	8.4	1.72	74	24	7	7.9	1.59	61
8	5	8.0	2.00	100	25	11	8.0	1.68	68
9	9	8.1	1.90	91	26	6	8.0	1.69	69
10	7	8.0	1.83	84	27	6	7.9	1.69	70
11	10	8.2	1.83	85	28	7	7.9	1.67	68
12	10	8.3	1.83	84	29	9	7.9	1.66	67
13	6	8.2	1.86	87	30	8	7.9	1.63	65
14	7	8.1	1.82	83	31	8	7.9	1.60	63
15	8	8.1	1.75	77	32	10	8.0	1.62	64
16	6	7.9	1.77	79	33	9	8.0	1.61	63
17	7	7.9	1.73	75	34	7	8.0	1.59	62

whose other knee was thought to have an injured ligament. Also shown, after each observation, is the current estimate of σ (based on subjects measured up to that point) along with the corresponding estimate of required sample size [from (1)]:

$$n = \frac{S^2}{(.2)^2}$$

(rounded up to an integer). Early estimates of sample size are unstable, but they eventually settle down.

The experiment could continue in this fashion beyond 34 patients to about 65; this would be the required sample size. The continual calculation of S could be abandoned, since the required sample size would have become pretty well stabilized. If feasible, one could check the value of the s.e. after 60 knees and continue sampling if it exceeds .2 mm. ∎

Pilot Samples

It may not be possible to incorporate the data into an estimate of σ after each observation. For example, it may be feasible only to collect all the data at once. However, something can be done if the data can be collected in even two batches: Take a pilot sample of m observations to estimate σ, and a second sample of size $n - m$ with n chosen to satisfy (1) using the estimated value of σ. However, σ may be badly underestimated, especially if n is small. (Overestimating σ does not present as much of a problem.) With σ underestimated, sampling might stop too soon. We recommend taking at least 20 observations in a pilot sample.

Estimating a Proportion

We consider next the estimation of a **population proportion** p using the sample proportion $\hat{p}$. Again in this case, the standard deviation of the estimator involves an unknown parameter, but this time the unknown parameter is the very one we're estimating:

$$\sigma_{\hat{p}} = \sqrt{\frac{p(1 - p)}{n}}. \tag{2}$$

Even so, the situation is actually somewhat better, because the s.d. (2) is *bounded* from above. The graph of $\sqrt{p(1 - p)}$ is a semicircle with center at $p = \frac{1}{2}$, shown in Figure 9.1. The maximum value of (2) therefore occurs at $p = \frac{1}{2}$; that is,

$$\sigma_{\hat{p}} \le \sqrt{\frac{.5 \times .5}{n}} = \frac{.5}{\sqrt{n}}.$$

If we choose n so that $.5/\sqrt{n} \le \varepsilon$, then $\sigma_{\hat{p}} \le \varepsilon$. Thus, to meet the specification that $\sigma_{\hat{p}} \le \varepsilon$, we take n to be the smallest integer such that

$$n \ge \frac{.25}{\varepsilon^2}. \tag{3}$$

[Compare this with (1).] With n determined by (3), the standard error of $\hat{p}$ will not exceed ε, whatever the true value of p. It actually will be substantially less than ε if p is near 0 or 1.

In a particular application, we may know that p is less than some $p_0 < .5$ or greater than some $p_0 > .5$. In either case, we can use p_0 in place of .5 as the "worst

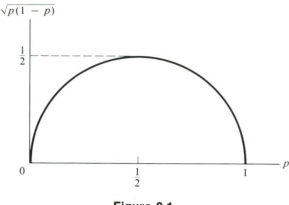

$$\sqrt{p(1-p)}$$

Figure 9.1

case" and get by with a smaller required sample size:

$$n \geq \frac{p_0(1 - p_0)}{\varepsilon^2}. \tag{4}$$

Example 9.2c   **A Political Poll**

Suppose we want to estimate the proportion of voters who favor a certain candidate with a standard error of at most .01. Setting $\varepsilon = .01$ in (3), we obtain $n = .25/(.01)^2 = 2500$. No matter what $\hat{p}$ turns out to be, its s.e. will not exceed .01. However, if $\hat{p} \neq \frac{1}{2}$, the s.e. will be smaller than the specified .01. For instance, if $\hat{p} = .2$, the standard error is only

$$\sqrt{\frac{.2 \times .8}{2500}} = .008.$$

Suppose we know that p cannot be close to $\frac{1}{2}$. For instance, suppose the candidate is running for president of the United States as a socialist, and we are quite sure that the population proportion who will vote for that candidate is less than 10%. Using $p_0 = .10$ in (4), we find

$$n \geq \frac{.1 \times .9}{.01^2} = 900. \qquad \blacksquare$$

When little is known about p in advance of sampling, it may be possible nonetheless to save on sampling costs. We can estimate $\sqrt{p(1 - p)}$ as we collect the data (just as we estimated σ in Example 9.2b) and continually revise the estimate of required sample size accordingly.

9.3   **Large-Sample Confidence Intervals**

In Example 9.1d, we reported an estimate of mean platelet count in the form 234 ± 7.6. Such a report suggests that the point estimate 234 could easily be in error by as much as 7.6. We can't say for sure that the population mean μ

lies between $234 - 7.6$ and $234 + 7.6$. Yet we are more confident that it is in the interval $(226.4, 241.6)$ than it is in, for example, the interval $(326.4, 341.6)$. A natural interpretation of "confidence" is in terms of probability. Can we calculate $P(226.4 < \mu < 241.6)$?

So long as we view μ as a constant, it does not make sense to ask for the probability that it lies in any specified interval. Chapter 16 presents an approach to statistical inference in which μ is treated as a random variable; probabilities such as $P(226.4 < \mu < 241.6)$ can then be calculated and have meaning. Not here, however. In the framework of this chapter—the framework of "classical" inference—we can answer only questions such as this: *Before* drawing the sample, what is the probability that the interval with endpoints

$$\bar{X} \pm \text{s.e.}(\bar{X}) \tag{1}$$

will contain the actual value of μ?

The interval (1) will include μ if and only if $\bar{X}$ is within one standard error of μ: $|\bar{X} - \mu| < S/\sqrt{n}$, or

$$\frac{|\bar{X} - \mu|}{S/\sqrt{n}} < 1. \tag{2}$$

As explained in Section 9.1, when n is large, S is apt to be close to σ. So by the central limit theorem,

$$\frac{|\bar{X} - \mu|}{S/\sqrt{n}} \doteq \frac{|\bar{X} - \mu|}{\sigma/\sqrt{n}} \approx N(0, 1). \tag{3}$$

The probability of (2) is then about 68% (from Table II in Appendix 1). In roughly two out of three large samples, $\bar{X}$ will be close enough to μ that the interval with endpoints (1) includes μ.

Confidence Intervals for μ

Standard practice is to calculate (1) from sample statistics and claim that the interval with these endpoints includes the population mean, with 68% confidence in this claim. (Whether the interval does include the actual value of μ is rarely verifiable.)

The level of confidence can be increased by using a wider interval, one that extends farther on both sides of the sample mean. For example,

$$P\left[|\bar{X} - \mu| < 2\left(\frac{S}{\sqrt{n}}\right)\right] \doteq P\left(\frac{|\bar{X} - \mu|}{\sigma/\sqrt{n}} < 2\right)$$
$$= \Phi(2) - \Phi(-2) = .9544,$$

from Table II. So in about 19 samples out of 20, the endpoints

$$\bar{X} \pm 2[\text{s.e.}(\bar{X})] \tag{4}$$

include the population mean. The interval

$$\bar{X} - 2[\text{s.e.}(\bar{X})] < \mu < \bar{X} + 2[\text{s.e.}(\bar{X})]$$

is a 95.44% **confidence interval** for μ.

Practitioners usually use a 95% confidence level; this requires changing the multiplier of s.e. slightly from 2 to 1.96, since

$$\Phi(1.96) - \Phi(-1.96) = .9750 - .0250 = .9500.$$

Approximate large-sample confidence limits for a population mean:

$$\bar{X} + k\frac{\sigma}{\sqrt{n}}, \qquad \sigma \text{ known}, \tag{5}$$

$$\bar{X} \pm k\frac{S}{\sqrt{n}}, \qquad \sigma \text{ unknown}. \tag{6}$$

The confidence level is determined by k; some common choices (from Table IIb):

Confidence level	k
68%	1.00
95%	1.96
99%	2.58

Example 9.3a   **Confidence Limits for a Mean**

In Example 9.1d, we gave the mean of 35 blood platelet data as $\bar{X} = 234$, with standard error $S/\sqrt{n} = 7.6$. An approximate 95% confidence interval for the population mean then has these limits:

$$\bar{X} \pm (1.96)\,\text{s.e.}(\bar{X}) = 234 \pm 1.96 \times 7.6 \doteq 234 \pm 14.9.$$

So the 95% confidence interval extends from 219.1 to 248.9. The population mean is somewhere between these limits *if* our sample happens to be one of the "19 out of 20" whose means are within 1.96 s.e.'s of μ.

In these calculations, we've assumed that $n = 35$ is large enough that the mean is approximately normal and that S is close to σ. In the next section, we address the question of whether this assumption is justified. ∎

Confidence intervals are frequently misinterpreted. Suppose $\bar{X} = 12$ and $\sigma/\sqrt{n} = 1$. The 68% confidence interval then extends from 11 to 13. Saying this is a 68% confidence interval sounds as though one is saying that the *probability* of $11 < \mu < 13$ is .68. It isn't. We explained at the outset that in this development, parameters such as μ are *not* random variables. So it doesn't make sense here to talk about the probability that μ is in a particular interval.

Large-sample confidence limits for a population *proportion* can be obtained using the approximate normality of sample proportions in random samples [see (1) in Section 4.5 and Section 8.8]. We proceed as in the case of μ: Take the point

Confidence Intervals for p

estimate $(\hat{p})$ plus and minus some number of standard errors. The s.e. [from (8) of Section 9.1] is

$$\text{s.e.}(\hat{p}) = \sqrt{\frac{\hat{p}(1 - \hat{p})}{n}}. \tag{7}$$

As before, Table IIb in Appendix 1 gives appropriate multipliers (in place of 1.96) for other confidence levels.

Large-sample confidence limits for a population proportion:

$$\hat{p} \pm k \sqrt{\frac{\hat{p}(1 - \hat{p})}{n}}, \tag{8}$$

with k given by Table IIb.

Example 9.3b

Nielsen Ratings

Nielsen ratings are based on samples of about 1,000 TV owners. The Nielsen company says that the sampling error for a single telecast with a rating of 22 is 1.3 percentage points (that is, .013). Although the actual sampling scheme is complicated, involving many stages and clusters, they still use (7) for s.e.:

$$\text{s.e.}(\hat{p}) = \sqrt{\frac{.22 \times .78}{1000}} \doteq .0131.$$

Then, taking $k = 1$ in (8), Nielsen then reports .23 $\pm$ .013 as 68% confidence limits.

A Nielsen official was quoted in *TV Guide* (Vol. 26, no. 25, 1978) as saying, "the reason we use the 68% level is that it makes it very convenient for the user." The "convenience" is not clear, but being narrower, a 68% confidence interval looks more accurate than a 95% confidence interval! A 95% interval for a rating of 22 is (19.4, 24.6), about twice as wide as the 68% interval. (In one survey, a show with a 19.4 rating was ranked 27th, while one with a 24.6 rating ranked 6th—quite a spread!) ■

9.4

When σ Is Unknown—Small n

In Section 9.3, we assumed sample sizes large enough that (i) the sample mean is approximately normal, and (ii) the sample s.d. S is close to the population s.d. σ. When the parent population is not too far from normal, the Z-score

$$Z = \frac{\bar{X} - \mu}{\sigma/\sqrt{n}} \tag{1}$$

may be close to standard normal even when n is as small as 5 or so. Indeed, if the population is exactly normal, so is $\bar{X}$. However, the approximate score

$$T = \frac{\bar{X} - \mu}{S/\sqrt{n}} \tag{2}$$

is more variable than Z because of uncertainty in S. When n is not large, a confidence interval based on T has to be wider to take this extra variability into account.

Using the t Distribution

We adjust the confidence limits by changing the multiplier of standard error in accordance with the null distribution of (2). This distribution depends on the distribution in the parent population. Later in this section, we show that when the population is normal, $T \sim t(n - 1)$. (See Section 6.6.) Table IIIa in Appendix 1 gives multipliers of the standard error for confidence intervals based on the t-distribution, for $n \leq 40$.

Like the large-sample multipliers based on the standard normal Z given in Table IIb, the entries in Table IIIa are percentiles of T. Thus, for example, to achieve a 95% level of confidence, we use the 2.5 and 97.5 percentiles of the t-distribution as multipliers, because the area between these is 95% of the total area. (You should check an entry for a particular sample size n in Table IIIa—say, for 95% confidence—and see that it is indeed the 97.5 percentile of the t-distribution with $n - 1$ d.f. shown in Table IIIc.)

When n is larger than 40, the t-distribution is close enough to the normal that the multipliers can be taken from Table II; these multipliers are also given in the last row of Table IIIa—the row labeled $n = \infty$. This is because S is likely to be close to σ when n is large, so T given by (2) is close to the Z-score we used in constructing large sample interval estimates. Indeed, as we saw in Section 6.6, the t-density tends to the standard normal density as the degrees of freedom tend to infinity.

Example 9.4a   **Adjusting Limits for Unknown σ**

In Example 9.3a, we gave confidence limits for mean blood platelet count. We assumed that $n = 35$ is large enough that the normal table can be used even though we replace the unknown σ with S. To use the t-distribution, we enter Table IIIa in the row with $n = 35$. We find 2.03 to be the proper multiplier of the standard error for 95% confidence, slightly larger than the 1.96 from the normal table. The adjusted confidence limits are

$$\bar{X} \pm 2.03 \frac{S}{\sqrt{n}} = 234 \pm 2.03 \times \frac{45.05}{\sqrt{35}},$$

or 218.5 and 249.5. Since $\frac{2.03}{1.96} \doteq 1.04$, the interval defined by these limits is about 4% wider than that found in Example 9.3a. The modification can be much greater when n is smaller. ∎

Suppose the population is *not* normal, σ is unknown, and n is not large. Although the t-distribution is not strictly correct, simulation studies show that it approximates the distribution of T quite well so long as the population is "not too far" from normal. The t-distribution does not give a good approximation if either or both tails of the distribution are heavy—that is, if there is too much probability far from the mean.

Approximate confidence limits for μ when the population is close to normal:

$$\bar{X} \pm k \frac{S}{\sqrt{n}}, \tag{3}$$

where k is found in Table IIIa of Appendix 1.

Normal or t?

When should one use Table II and when Table IIIa? The simplest rule is to use Table IIIa whenever σ is unknown and the population distribution is not especially heavy-tailed. If the sample size is beyond the range of sizes shown in Table IIIa, you will be led to the last row of the table ($n = \infty$), which gives the appropriate normal percentiles as multipliers. If σ is known, then T is never appropriate. Whether Z is appropriate in this case depends only on whether or not the sample size is large enough for $\bar{X}$ to be close enough to normal. If n is not sufficiently large, another approach is needed, but so far we've presented no alternatives to Z and T.

Small Sample Limits for p

We have seen how to modify large-sample confidence limits for a population mean when the sample size is small. Is there an analogous modification for a proportion? There is, but when n is small, the confidence interval turns out to be so wide that fine-tuning of the limits is not very useful. For example, with a sample of size $n = 25$ and a sample proportion of .32, the 95% limits given by (5) of Section 9.3 are .137 and .503. With an appropriate modification for small samples, they turn out to be .171 and .516. We'll explore this no further. [Lest there be confusion in this regard, you should *never* use the t-distribution (Tables IIIa and IIIb) in connection with sample proportions.]

We now show that T given by (2) has a t-distribution when the population is normal. We first rewrite it by dividing numerator and denominator by σ and rearranging the factors:

$$T = \frac{\sqrt{n}(\bar{X} - \mu)/\sigma}{S/\sqrt{n}} = \frac{U}{V}. \tag{4}$$

Proof that $T \sim t(n-1)$

When the population is normal, the mean is normal, so the numerator U is standard normal. The denominator V is proportional to $\chi(n-1)$, the square root of a chi-square variable with $n-1$ degrees of freedom (see Section 6.5). Moreover, being functions of independent variables, U and V are independent (Section 8.10).

To find the distribution of T in (4), recall the general formula for the p.d.f. of the ratio of two independent variables [(4) of Section 6.6] in which the denominator is nonnegative:

$$f_{U/V}(w) = \int_0^\infty v f_U(vw) f_V(v)\, dv. \tag{5}$$

Now, $U \sim N(0, 1)$, and from Section 8.10 we know that $(n-1)V^2 \sim \chi^2(n-1)$, so $\sqrt{(n-1)}\, V$ has the chi-distribution [see (7) in Section 6.5]. Thus,

$$f_U(u) \propto e^{-u^2/2}, \quad \text{and} \quad f_V(v) \propto v^{n-2} e^{-(n-1)v^2/2}.$$

Substituting in (5), we have

$$f_T(w) \propto \int_0^\infty v e^{-(vw)^2/2} v^{n-2} e^{-(n-1)v^2/2}\, dv$$

$$= \int_0^\infty v^{n-1} \exp\left\{-(n-1)v^2\left(1 + \frac{w^2}{n-1}\right)\right\} dv.$$

Using the fact that the integral is essentially a gamma function [(7) of Section 6.3], we find

$$f_T(w) \propto \frac{1}{\left(1 + \dfrac{w^2}{n-1}\right)^{n/2}}.$$

This is the p.d.f. of a t-distribution with $n-1$ degrees of freedom [(1) in Section 6.6].

9.5   Pivotal Quantities

In finding a confidence interval for the mean of a normal population, we used the fact that the distribution of

$$T = \frac{\bar{X} - \mu}{S/\sqrt{n}}$$

is $t(n-1)$, independent of the parameters μ and σ^2 of the population. Combinations of sample data and parameters whose distributions are independent of population parameters are called **pivotal quantities**, or more simply, **pivotals**. They are useful in constructing confidence intervals, as in Sections 9.3 and 9.4 and again in Section 9.6.

> A **pivotal** is a function of a statistic and population parameter(s) whose distribution is independent of population parameters.

Suppose there is a single parameter θ to be estimated, and suppose $g(\mathbf{X}, \theta)$ is pivotal. We can use the distribution of $g(\mathbf{X}, \theta)$ to find numbers $A(\mathbf{X})$ and $B(\mathbf{X})$ (depending on $\mathbf{X}$ but not on θ) such that

$$P[A(\mathbf{X}) < g(\mathbf{X}, \theta) < B(\mathbf{X})] = \gamma,$$

where γ is a specified confidence level. Solving the inequality for θ will often produce (two-sided) confidence limits.

Example 9.5a   Consider a random sample $(X_i, \ldots, X_n)$ from $\text{Exp}(\lambda)$, and let $W = \Sigma X_i$. We know from Section 6.3 that $W \sim \text{Gam}(n, \lambda)$. Making the change of variable $V = 2W\lambda$ in the p.d.f. (1) of Section 6.3, we obtain

$$f_V(v) = f_W\left(\frac{V}{2\lambda}\right) \cdot \frac{1}{2\lambda} = \frac{v^{n-1}}{2^n(n-1)!} e^{-v/2},$$

a chi-square p.d.f. with $2n$ d.f. So V is pivotal, and

$$P(\chi^2_{.025} < 2W\lambda < \chi^2_{.975}) = .95.$$

Solving for λ, we obtain 95% confidence limits for λ. For example, if $n = 10$,

$$P\left[\frac{9.59}{(2W)} < \lambda < \frac{34.2}{2W}\right] = .95.$$

The choice of 2.5 and 97.5 was of course arbitrary—we could use any two percentages differing by 95. ∎

To obtain a *one*-sided confidence interval (a one-sided *confidence bound*), we start with a one-sided inequality such as $A < Y$.

Example 9.5b   Continuing in the setting of Example 9.5a, suppose we want a one-sided 95 upper percent confidence bound for the mean, $\theta = 1/\lambda$. For this, we'd use the fifth percentile of $\chi^2(20)$, which is 10.9:

$$P(2W\lambda < 10.9) = .95,$$

or

$$P\left(\theta < \frac{2W}{10.9}\right) = .95.$$ ∎

9.6   Estimating a Mean Difference

Comparing two populations on the basis of a sample from each is a common and important problem in many fields. For example, treated experimental units may respond differently than those untreated. When populations are compared in terms of mean response, the *difference* $\delta = \mu_1 - \mu_2$ is a relevant parameter. A natural estimate of δ is the corresponding difference between sample means.

Example 9.6a   **Two Methods of Filtering Water**
A water purification plant was equipped with a new filtration device. To test its effectiveness, 20 test runs were carried out using one method of operation and 29 using a second method.[2] All runs lasted the same length of time, and the quantity measured was the total water filtered (cubic meters per square meter of filter).

   The sample means were 202.0 for the first method and 278.3 for the second. The difference between these sample means, 76.3, is an estimate of the population mean difference δ. Its reliability depends on the sampling distribution of the difference between the two sample means. (To be continued.) ∎

---

[2] Vosloo, P. B. V., P. G. Williams, and R. G. Rademan, "Pilot and full-scale investigations on the use of combined dissolved-air flotation and filtration for water treatment," *Water Pollution Control 85* (1986), 114–121.

Suppose population 1 has mean μ_1 and s.d. σ_1, and population 2 has mean μ_2 and s.d. σ_2. We take independent samples, one of size n_1 from population 1 with mean $\bar{X}_1$ and one of size n_2 from population 2 with mean $\bar{X}_2$. According to (7) of Section 5.4 and (5) of Section 5.8,

$$E(\bar{X}_1 - \bar{X}_2) = \mu_1 - \mu_2 = \delta, \tag{1}$$

$$\text{var}(\bar{X}_1 - \bar{X}_2) = \frac{\sigma_1^2}{n_1} + \frac{\sigma_2^2}{n_2}. \tag{2}$$

The population variances σ_i^2 are usually unknown and must be estimated from the samples to obtain a standard error—an approximate standard deviation of the error of estimation:

Standard error of the difference in means of independent random samples:

$$\text{s.e.}(\bar{X}_1 - \bar{X}_2) = \sqrt{\frac{S_1^2}{n_1} + \frac{S_2^2}{n_2}}. \tag{3}$$

Example 9.6b **Filtering Water (continued)**

Suppose the 20 and 29 test runs in Example 9.6a are independent random samples. The published report gives $S_1 = 74.35$ and $S_2 = 79.03$. The s.e. (3) is

$$\text{s.e}(\bar{X}_1 - \bar{X}_2) \doteq \sqrt{\frac{(74.35)^2}{20} + \frac{(79.03)^2}{29}} \doteq 22.18.$$

(To be continued.) ∎

Confidence Interval for δ

We can construct a confidence interval for the difference δ between two population means, just as we did for a single mean. First, we need the *distribution* of the difference between the sample means. We already know that when n_1 and n_2 are large, the two sample means $\bar{X}_1$ and $\bar{X}_2$ are approximately normal. We also know (from Section 6.1) that the difference of independent normal variables is again normal. So in the case of large samples, it is reasonable to expect—and it's true— that the distribution of $\bar{X}_1 - \bar{X}_2$ is approximately normal.

Confidence limits for $\delta = \mu_1 - \mu_2$, where n_1 and n_2 are large and the samples are independent:

$$\bar{X}_1 - \bar{X}_2 \pm k\sqrt{\frac{S_1^2}{n_1} + \frac{S_2^2}{n_2}}, \tag{4}$$

where k is given in Table IIb of Appendix 1.

Example 9.6c   **Filtering Water (continued)**
In the preceding examples, we found $\bar{X}_1 - \bar{X}_2 = 76.3$, with a standard error of 22.18. For the population mean difference δ, 95% confidence limits are given by (4) with $k = 1.96$: $76.3 \pm 1.96 \times 22.18$. So a 95% confidence interval is $44.1 < \delta < 98.5$. ■

Bernoulli Populations

The above development for a difference in means applies in particular to Bernoulli (0–1) populations. For 0–1 populations, μ_i is p_i (the proportion of 1's in population i), and σ_i^2 is $p_i(1 - p_i)$. Similarly, the sample mean $\bar{X}_i$ is $\hat{p}_i$ (the proportion of 1's in sample i), and the sample variance is $S_i^2 = \hat{p}_i(1 - \hat{p}_i)$. So the standard error (3) is

$$\text{s.e.}(\hat{p}_1 - \hat{p}_2) = \sqrt{\frac{\hat{p}_1(1 - \hat{p}_1)}{n_1} + \frac{\hat{p}_2(1 - \hat{p}_2)}{n_2}}. \tag{5}$$

Confidence limits for $p_1 - p_2$ based on sample proportions $\hat{p}_1$ and $\hat{p}_2$ from large, independent random samples of sizes n_1 and n_2:

$$\hat{p}_1 - \hat{p}_2 \pm k\sqrt{\frac{\hat{p}_1(1 - \hat{p}_1)}{n_1} + \frac{\hat{p}_2(1 - \hat{p}_2)}{n_2}}, \tag{6}$$

where k is found in Table IIb of Appendix 1.

Example 9.6d   **A Gallup Poll**
A 1987 Gallup survey (reported April 12, 1987) found "no evidence of growing public intolerance towards gays." The poll used a sample of 1,015 adults from "scientifically selected localities across the nation." One question was this: "Do you think homosexual relations between consenting adults should or should not be legal?" In 1986, the percentage who said "should not" was 54; in the 1987 poll, it was 55. Is this "no evidence" of growing intolerance?

The estimate of $p_1 - p_2$ is $\hat{p}_1 - \hat{p}_2 = .55 - .54 = .01$, or one percentage point. We know that $n_1 = 1015$; we'll assume that the 1986 sample size was $n_2 = 1,015$ and that the two samples are independent. From (5),

$$\text{s.e.}(\hat{p}_1 - \hat{p}_2) = \sqrt{\frac{.55 \times .45}{1015} + \frac{.54 \times .46}{1015}} = .022.$$

To find 95% confidence limits, we use $k = 1.96$ in (6):

$$\hat{p}_1 - \hat{p}_2 \pm 1.96 \times \text{s.e.}(\hat{p}_1 - \hat{p}_2) = .01 \pm 1.96 \times .022.$$

The 95% confidence interval is thus

$$-.033 < p_1 - p_2 < .053.$$

In particular, $p_1 - p_2 = 0$ is well within this interval. So the Gallup statement of "no evidence" seems reasonable. ■

9.7   Estimating Variability

We have needed estimates of population variability in judging the reliability of estimates of the population mean. For instance, in setting confidence limits for a population mean in Section 9.3, we used S as a substitute for the unknown σ. However, population variability is sometimes of interest for its own sake. For example, it may be necessary to control the variability of a production process to achieve product uniformity.

Example 9.7a   A circuit board in a large computer has many layers that must match up reasonably well, or else the board is defective. The board can be so bad that repairing it is impossible or prohibitively expensive. The process that matches layers can be absolutely perfect *on the average* yet be so variable that no single board is worth keeping. (Companies have experienced a rejection rate as high as 95% in the course of circuit board development.)   ∎

Obvious estimators of σ^2 and σ are the corresponding statistics, S^2 and S. We consider first the sample variance,

$$S^2 = \frac{1}{n-1} \sum_{i=1}^{n} (X_i - \bar{X})^2. \tag{1}$$

The sampling distribution of S^2 depends on the population being sampled.

Moments of S^2   However, in Example 9.1b, we found its mean value to be the population variance:

$$E(S^2) = \sigma^2. \tag{2}$$

Since S^2 is unbiased, the mean squared error in estimating σ^2 to be S^2 is the variance of S^2. However, calculating the variance of S^2 is quite tedious, and in general this variance is given by a complicated formula involving the first four population moments.[3] The formula shows that the variance tends to zero as n becomes infinite. So the distribution of S^2 becomes more and more concentrated near σ^2. The sample variance tends to the population variance in the same sense that the sample mean tends to the population mean: Its mean squared error tends to zero.

Another candidate for estimating σ^2 is the version of sample variance with denominator n:

$$V = \frac{1}{n} \sum_{i=1}^{n} (X_i - \bar{X})^2 = \frac{n-1}{n} S^2. \tag{3}$$

As we saw in Example 9.1b, the bias of V is

$$b_V(\sigma^2) = -\frac{\sigma^2}{n}.$$

---

[3] See Cramér [5], p. 348.

Its mean squared error is then

$$\text{m.s.e.}(V) = \text{var } V + b_V^2(\sigma^2) = \left(\frac{n-1}{n}\right)^2 \text{var } S^2 + \left(\frac{\sigma^2}{n}\right)^2.$$

Normal Populations

The *distribution* of the sample variance is known in the special case of a *normal* population; it was derived in Section 8.10:

> For a random sample $(X_1, \ldots, X_n)$ from a *normal* population, the distribution of
>
> $$\frac{(n-1)S^2}{\sigma^2} = \sum_{i=1}^{n} \left(\frac{X_i - \bar{X}}{\sigma}\right)^2. \tag{4}$$
>
> is $\chi^2(n-1)$, as defined in Section 6.5.

The formula for the m.s.e. of S^2 in this case follows from the variance of the chi-square distribution:

$$\text{m.s.e.}(S^2) = \text{var}(S^2) = \frac{2\sigma^4}{n-1}. \tag{5}$$

The mean squared error of the biased estimator (3) is then

$$\text{m.s.e.}(V) = \sigma^4 \left(\frac{2(n-1)}{n^2} + \frac{1}{n^2}\right) = \frac{(2n-1)\sigma^4}{n^2}. \tag{6}$$

This is always smaller than (5); the m.s.e. of the unbiased estimator S^2 is larger than the m.s.e. of the (slightly) biased V.

Since the distribution of $(n-1)S^2/\sigma^2$ does not depend on σ^2, this quantity is a pivotal. So, as outlined in Section 9.5, we can use it to construct a confidence interval for σ^2. This is illustrated in the next example.

Confidence Intervals

Example 9.7b

Blood Platelet Counts (continued)

For the blood platelet data of Example 9.1d, $S = 45.05$ and $n = 35$. If we assume that platelet counts are normally distributed and that the sampling is random, then (4) is $\chi^2(34)$. The 2.5 and 97.5 percentiles of this distribution, from Table V of Appendix 1, are 19.4 and 51.5, respectively. Thus,

$$P\left(19.4 < \frac{34S^2}{\sigma^2} < 51.5\right) = .95.$$

Taking reciprocals (remembering to reverse the direction of the inequalities) and multiplying through by $34S^2$ yields

$$P\left(\frac{34S^2}{19.4} > \sigma^2 > \frac{34S^2}{51.5}\right) = .95. \tag{7}$$

Thus $34S^2/51.5$ and $34S^2/19.4$ are 95% confidence limits for the population variance. There is a 95% chance that random sampling produces an S^2 close enough to σ^2 that these limits straddle σ^2. With $S = 45.05$, the limits are 1340 and 3557:

$$1340 < \sigma^2 < 3557.$$

A standard deviation is easier to interpret than a variance. Taking the square root of each member of the inequality in (7) does not change the probability, so by taking the square root of the confidence limits for σ^2, we get 95% confidence limits for σ: $36.6 < \sigma < 59.6$. ∎

9.8   The Method of Moments

In the preceding, we have often taken sample moments as obvious candidates for estimating corresponding population moments. This is an application of the **method of moments** for deriving an estimator. More generally, to estimate a parameter that is a function of population moments, the method of moments is to use the same function of corresponding sample moments. Also, to estimate k parameters, the method is first to express those parameters in terms of the first k population moments, and then to replace those population moments with corresponding sample moments. In particular, for estimating a single parameter $\theta = g(\mu)$, the estimator is $g(\bar{X})$.

Example 9.8a   Consider a random sample from a Exp(λ). The population mean is $\mu = 1/\lambda$. The method-of-moments estimate of $\lambda = 1/\mu$ is then $1/\bar{X}$. ∎

Example 9.8b   Consider a random sample from a gamma distribution: $X \sim \text{Gam}(\alpha, \lambda)$. The population mean and variance are

$$\mu = \frac{\alpha}{\lambda} \quad \text{and} \quad \text{var } X = E(X^2) - [EX]^2 = \frac{\alpha}{\lambda^2},$$

respectively [(10) of Section 6.3]. Solving, we find

$$\lambda = \frac{E(X)}{E(X^2) - [EX]^2}, \quad \alpha = \lambda\mu = \frac{[E(X)]^2}{E(X^2) - [EX]^2}.$$

Applying the method of moments, we replace EX with $\bar{X}$ and $E(X^2)$ with

$$\overline{X^2} = \frac{1}{n} \sum X_i^2.$$

The estimators are thus

$$\tilde{\lambda} = \frac{n\bar{X}}{(n-1)S^2}, \quad \tilde{\alpha} = \frac{n\bar{X}^2}{(n-1)S^2}. \quad ∎$$

In large samples, we expect a sample moment to be close to the corresponding population moment, so we'd naturally expect a method-of-moments estimator of a parameter to be close to that parameter. We'll make this idea more precise in Section 9.10.

9.9 Maximum Likelihood Estimation

We suggested in Section 8.2 that the plausibility of a model (or of a parameter value) should depend on its likelihood. Carrying this to an extreme, we now derive parameter estimates as values that maximize the likelihood function. Thus, a **maximum likelihood estimate** is a parameter value under which the data actually obtained have the highest probability.

Consider first the case where θ is one-dimensional, and suppose $L(\theta)$ is differentiable. We can maximize L by differentiating and setting the derivative equal to 0. If L has a maximum at an interior point $\hat{\theta}$ of the parameter space, then $L'(\hat{\theta}) = 0$. So we can locate maxima by checking the solutions of the equation $L'(\theta) = 0$ to see if any of them is a maximum point. Thus, if $L''(\hat{\theta}) < 0$, then $\hat{\theta}$ is a maximum point.

Since the logarithm function is increasing, maximizing $\log L(\theta)$ is equivalent to maximizing $L(\theta)$, and $\log L(\theta)$ is often easier to deal with when differentiating, as we'll see in examples to follow.

A **maximum likelihood estimate** of θ (m.l.e.) is any value $\hat{\theta}$ that maximizes the likelihood function $L(\theta)$. When the maximum occurs at an interior value of θ where L' exists,

$$L'(\hat{\theta}) = 0. \tag{1}$$

Example 9.9a

m.l.e. for Exp(λ)

Consider estimating the positive parameter λ in the exponential density

$$f(x \mid \lambda) = \lambda \exp(-\lambda x), \qquad x > 0,$$

using a random sample of size n. The likelihood function is (proportional to) the joint density of the observations:

$$L(\lambda) = \prod_{i=1}^{n} f(x_i \mid \lambda) = \lambda^n \exp\left(-\lambda \sum_{i=1}^{n} x_i\right) = \lambda^n \exp(-n\lambda\bar{x}), \qquad \lambda > 0.$$

So

$$\log L(\lambda) = n \log \lambda - n\lambda\bar{x}.$$

Then

$$\frac{d}{d\lambda} \log L = \frac{n}{\lambda} - n\bar{x}, \qquad \text{and} \qquad \frac{d^2}{d\lambda^2} \log L = -n/\lambda^2.$$

The first derivative vanishes only at $\hat{\lambda} = 1/\bar{x}$, and the second derivative is always negative. So L achieves a maximum at $1/\bar{x}$. This is the m.l.e. of λ. ∎

As this example clearly shows, the likelihood function and the m.l.e. depend on the particular sample on which they're based. As a function of the random sample $\mathbf{X} = (X_1, \ldots, X_n)$, an m.l.e. is a statistic—a random variable.

The derivative of L at a maximum point may or may not be 0. The next example, in which the range of parameter values is a finite interval, illustrates both possibilities.

Example 9.9b   **Estimating the Bernoulli Parameter**

We observe six successes in ten independent trials of a Bernoulli experiment with parameter p. The likelihood function is

$$L(p) = p^6(1 - p)^4, \qquad 0 \le p \le 1.$$

(See Example 8.2b.) The function $L(p)$ has a maximum in $[0, 1]$, being continuous and bounded above by 1. Since it is nonnegative, and $L(1) = L(0) = 0$, the maximum occurs at an interior point, so we can locate it by differentiation. The log-likelihood is

$$\log L(p) = 6 \log p + 4 \log(1 - p).$$

We differentiate this and set the result equal to zero:

$$\frac{d}{dp} \log L = \frac{6}{p} - \frac{4}{1 - p} = \frac{6 - 10p}{p(1 - p)} = 0.$$

The unique solution is $\hat{p} = \frac{6}{10}$. This is the m.l.e., since it is the only point between 0 and 1 where the derivative is zero.

More generally, consider Y successes in n independent trials. Suppose first that $0 < Y < n$. The log-likelihood is

$$\log L(p) = Y \log p - (n - Y) \log(1 - p).$$

Setting the derivative equal to zero, we find, for $0 < p < 1$,

$$\frac{Y}{p} - \frac{n - Y}{1 - p} = 0, \qquad \text{or} \qquad Y - np = 0.$$

The solution is the m.l.e., $\hat{p} = Y/n$.

If $Y = n$, the likelihood function is $L(p) = p^n$; this has maximum value on the range $0 \le p \le 1$ at $\hat{p} = Y/n = 1$. Similarly, if $Y = 0$, the m.l.e. is 0. So the m.l.e. is actually $\hat{p} = Y/n$ in all cases. ∎

Joint m.l.e.'s

When $\theta = (\theta_1, \ldots, \theta_k)$, values of the θ_j that maximize the likelihood function are said to be **joint** m.l.e.'s. When the maximum of a function of several variables occurs on the interior of the parameter space, each partial derivative must vanish there. So we find potential maximum points by setting each partial derivative equal to zero and solving this set of simultaneous equations. A solution is a critical point, but it may or may not be a maximizing point. It is, if the maximum of the likelihood function occurs in the interior of the parameter space and if there is just one critical point in the interior of the parameter space. This is the case in the next example.

Example 9.9c

Multinomial Sampling

In Example 9.9b, we saw that the m.l.e. of the Bernoulli parameter p is $\hat{p}$, the relative frequency of successes. More generally, consider a categorical population with a finite number of categories. Suppose there are k categories, and let p_j denote the probability of category j, where $\Sigma p_j = 1$.

Random sampling of such a population is called *multinomial sampling*. The category frequencies Y_i are clearly sufficient: The likelihood function (from Example 8.2d) is

$$L(p_1, \ldots, p_k) = p_1^{Y_1} \cdots p_k^{Y_k}, \qquad 0 \le p_i \le 1, \qquad i = 1, \ldots, k.$$

Suppose $Y_i > 0$ for all i. Substituting $p_k = 1 - p_1 - \cdots - p_{k-1}$ and taking logs, we have

$$\log L = \sum_{j=1}^{k-1} Y_j \log p_j + Y_k \log(1 - p_j - \cdots - p_{k-1}).$$

The derivative with respect to each p_j must vanish at a maximum:

$$\frac{Y_j}{p_j} - \frac{Y_k}{p_k} = 0 \qquad \text{for } i = 1, \ldots, k-1. \tag{2}$$

This says that the p's must be proportional to the Y's. Since they must add to 1, we divide the Y's by their total, $\Sigma Y_i = n$. So $\hat{p}_i = Y_i/n$ for $i = 1, \ldots, k$. These relative frequencies do indeed maximize L when no frequency is 0, for L is nonnegative, bounded above, and zero on the boundary of the $(k-1)$-parameter space. Thus, the maximum occurs on its interior—and there is only the one critical point.

Just as the relative frequencies are m.l.e.'s when one of them is 0 (see Example 9.9b), it can be shown here that the restriction $Y_i > 0$ can be dropped. Thus, in general, $\hat{p}_i = Y_i/n$. ∎

There is no unique way of parameterizing a family of distributions. Any one-to-one function of θ serves as well as θ.

Example 9.9d

Exp($1/\mu$)

In Example 9.9a, we considered the family Exp(λ) and found that $\hat{\lambda} = 1/\bar{X}$. This same family can be indexed equally well by the population mean, $\mu = 1/\lambda$:

$$f(x \mid \mu) = \frac{1}{\mu} \exp\left(-\frac{x}{\mu}\right), \qquad x > 0.$$

(The function $1/\lambda$ is a one-to-one function for $\lambda > 0$.) For a random sample of size n,

$$\log L(\mu) = -n \log \mu - \frac{n\bar{X}}{\mu}.$$

Setting the derivative equal to 0:

$$\frac{-n}{\mu} + \frac{n\bar{X}}{\mu^2} = 0,$$

we find the m.l.e. to be $\hat{\mu} = \bar{X}$. (The second derivative is

$$\frac{n(\mu - 2\bar{X})}{\mu^3},$$

which is negative at $\mu = \bar{X}$.) So $\hat{\mu} = 1/\hat{\lambda}$.   ∎

This example demonstrates a property that is true in general:

> If $g(\theta)$ is a one-to-one function and $\hat{\theta}$ is a maximum likelihood estimate of θ, then $g(\hat{\theta})$ is a m.l.e. of $g(\theta)$.

m.l.e.'s and Sufficiency

Maximum likelihood estimates are functions of sufficient statistics: If Y is sufficient for θ, then $L(\theta)$ is a function of only Y and θ. So maximizing L over θ results in a $\hat{\theta}$ that depends on the data only through Y. However, there is no guarantee that the m.l.e. itself is sufficient.

9.10   Consistency

The law of large numbers (Section 4.7), roughly speaking, says that a sample proportion tends to the population proportion as the sample size becomes infinite. An estimator that does not have this property is said to be *inconsistent*. Consistency is defined for a sequence of estimators:

$$Y_n = Y_n(X_1, \ldots, X_n) \qquad \text{for } i = 1, 2, \ldots, n.$$

For example, $\{\bar{X}_n\}$ is such a sequence. As in the case of $\bar{X}$, we sometimes drop the subscript and refer to the estimator simply as Y. Since each Y_n is a random variable, it is not clear what "Y_n tends to θ" means. We need a precise definition.

> A sequence of estimators $\{Y_n\}$ is said to be **consistent** as an estimate of θ when
>
> $$\lim_{n \to \infty} P(|Y_n - \theta| \le \varepsilon) = 1 \qquad (1)$$
>
> for every $\varepsilon > 0$.

Equation (1) says that any small interval centered at θ will eventually capture an arbitrarily large fraction of the distribution of Y_n.

Consistency can sometimes be shown using an inequality that relates probability near a point to the second moment about that point.

Chebyshev's inequality: If $E(X^2) < \infty$, then for any real θ and $\varepsilon > 0$,

$$P(|X - \theta| \geq \varepsilon) \leq \frac{1}{\varepsilon^2} E[(X - \theta)^2]. \tag{2}$$

To establish (2), we use (6) of Section 5.9:

$$E[h(X)] = E[h(X)|A] \cdot P(A) + E[h(X)|A^c] \cdot P(A^c). \tag{3}$$

Let $h(x) = (x - \theta)^2$, and take A to be the set where $h(x) \geq \varepsilon^2$. (See Figure 9.2.) The second term on the right of (3) is nonnegative, so we can drop it and preserve the inequality. Then, since $h(X) \geq \varepsilon^2$ on A,

$$E[h(X)] \geq E[h(X)|A] \cdot P(A) \geq \varepsilon^2 P(A).$$

Dividing through by ε^2, we obtain inequality (2). (This technique for establishing a probability inequality is powerful yet intuitive: We replaced one function h by a second, smaller function as shown in Figure 9.2, and then averaged.)

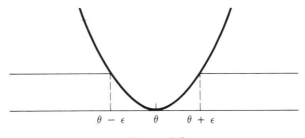

Figure 9.2

We now apply Chebyshev's inequality (2) with $X = Y_n$ and θ as the parameter to be estimated:

$$P(|Y_n - \theta| \geq \varepsilon) \leq \frac{1}{\varepsilon^2} E[(Y_n - \theta)^2] = \frac{1}{\varepsilon^2} \text{m.s.e.}(Y_n). \tag{4}$$

[See (1) of Section 9.1.] From this it follows that Y_n is consistent whenever its mean squared error tends to zero. From (3) of Section 9.1,

$$\text{m.s.e.} = E[(Y_n - \theta)^2] = \text{var } Y_n + b_{Y_n}^2(\theta). \tag{5}$$

Thus, Y_n is consistent whenever its bias and variance both tend to zero as n becomes infinite.

In particular, the m.s.e. of $\bar{X}$ under random sampling is $\frac{\sigma^2}{n}$, which tends to 0 when $\sigma < \infty$. So the sample mean is a consistent estimate of the population mean. [See (3) and (4) of Section 8.7.]

Example 9.10a

A pencil-written notebook compiled by Hall of Famer Lou Gehrig when he was a 17-year-old high school student was sold at auction in October 1988 for \$7,150.[4] It was compiled by Gehrig in 1920 for a statistics class at New York City's Commerce High School. After discussing rolling dice, he wrote, "if you want to find the average of any large number of items, take a small group (at random) and it will practically every time give you the average of the large group." He had rolled three dice, and gave the average of each successive group of 10 rolls:

10.7, 10.4, 10.9, 11.1, ,11.1, 10.5,

as well as the average of the 60: 10.78. (The expected sum of the points showing on three fair dice is 10.5.)

He went on to say, "To find the average height of boys in the H. S. of C., I would take 1 class from every term, measure each boy's height and I am almost sure that the average obtained would be the average height of the boys in the whole school." ∎

The Law of Averages

The consistency of the sample mean is a classical result in probability theory, known as the **law of large numbers** or the **law of averages**. [We quoted the special case of a sample proportion as (2) in Section 4.7.] Showing consistency of $\bar{X}$ by showing that the m.s.e. (4) tends to zero as n becomes infinite requires that the population variance be finite. However, the sample mean converges to the population mean in the sense of (1) even when the variance is infinite; all that is required is existence of the mean. This fact is **Khintchine's theorem**.[5] [Recall that a Cauchy p.d.f. (see Example 5.4d) has such heavy tails that the mean is not finite. In this case, the law of averages does not apply! Indeed, the distribution of the mean of a random sample turns out to be the same as the distribution of a single observation; the sample mean does not converge to a constant as the sample size increases.]

Applying the law of averages to sample moments other than the mean shows that these, too, are consistent estimators of corresponding population moments.

Example 9.10b

Second Moments

Consider estimating the population second moment $E(X^2)$ using the corresponding sample moment:

$$Y_n = \frac{1}{n} \sum X_i^2.$$

---

[4] *Sports Collectors' Digest* (Nov. 18, 1988), pp. 54–60.
[5] For a proof, see W. Feller [7], p. 235.

The mean of Y_n is $E(X^2)$, so Y_n is unbiased, and

$$\text{var } Y_n = \frac{1}{n^2} \sum \text{var}(X_i^2) = \frac{1}{n} \text{var}(X^2) = \frac{1}{n}(E(X^4) - [E(X^2)]^2).$$

This tends to zero as $n \to \infty$ provided $E(X^4) < \infty$, so Y_n is consistent. Indeed, according to Khintchine's theorem, Y_n is a consistent estimate of $E(X^2)$ provided only that this moment exists. ∎

Khintchine's theorem guarantees that sample moments about 0 are consistent estimators of the corresponding population moments. It can be shown that any function of sample moments, including any central moment, is therefore a consistent estimator of that function of the corresponding population moments (see, for example, Cramér [5], p. 367). Thus, the method of moments (Section 9.8) yields consistent estimators.

One of the nice properties of maximum likelihood estimators is that in "regular" cases, they too are consistent. Theorems on their consistency need to be carefully stated, because the likelihood can do some weird things in pathological cases.[6] In all of the examples we've considered, and in most practical situations, the m.l.e. is consistent.

9.11   Efficiency

When we use mean squared error as a criterion for accuracy of estimation, it is natural to choose between competing estimators in terms of m.s.e. In some cases, however, one estimator may have a smaller m.s.e. than another for some parameter values but not others. For instance, the m.s.e. estimator $Y \equiv 5$ of a population mean is $(5 - \mu)^2$. If $\mu = 5$, this is 0—smaller than the m.s.e. of the sample mean; but for μ far from 5, the inequality will go the other way.

Consider a distribution family indexed by a single parameter θ and estimators Y and Y'. When the *ratio* of their m.s.e.'s is independent of θ, that ratio is called the *efficiency* of Y' relative to Y:

Relative Efficiency

$$\text{eff}(Y', Y) = \frac{\text{m.s.e.}(Y)}{\text{m.s.e.}(Y')}.$$

When $\text{eff}(Y', Y) < 1$, then $\text{m.s.e.}(Y) < \text{m.s.e.}(Y')$; we say that Y makes more efficient use of the data than Y'.

When estimators Y and Y' are both unbiased, the relative efficiency of Y' with respect to Y is the ratio of their variances,

$$\text{eff}(Y', Y) = \frac{\text{var } Y}{\text{var } Y'}. \tag{1}$$

---

[6] See. E. Lehmann [13], pp. 410, 414.

Example 9.11a   **Mean of $U(0, \theta)$**

Consider a random sample of size n from $U(0, \theta)$. For this population, the mean is $\mu = \theta/2$ and the variance $\sigma^2 = \theta^2/12$. As always, the sample mean $\bar{X}$ is an unbiased estimate of the population mean. Its m.s.e. is

$$\text{var } \bar{X} = \frac{\sigma^2}{n} = \frac{\theta^2}{12n}.$$

However, $X_{(n)}$, the largest sample observation, has mean $\frac{n\theta}{(n+1)}$, so the estimator

$$Y = \frac{n+1}{2n} X_{(n)}$$

is also unbiased in estimating μ. From Example 8.6b,

$$\text{var } X_{(n)} = \frac{n\theta^2}{(n+2)(n+1)^2},$$

so

$$\text{var } Y = \left(\frac{n+1}{n}\right)^2 \frac{\text{var } X_{(n)}}{4} = \frac{\theta^2}{4n(n+2)}.$$

From (1),

$$\text{eff}(\bar{X}, Y) = \frac{\text{var } Y}{\text{var } \bar{X}} = \frac{3}{n+2}.$$

When $n > 1$, this is less than 1, meaning that Y is more efficient than $\bar{X}$. (Of course, when $n = 1$, the two statistics are the same.)   ■

If we could relate the mean squared error of Y to the smallest possible mean squared error, we'd have a measure of absolute efficiency. Such a measure is not possible in every case, but the following inequality provides a lower bound on the mean squared error of all *unbiased* estimators:

Cramér–Rao Inequality:

Let $Y = g(\mathbf{X})$ be an unbiased estimator of θ, where $\mathbf{X}$ is a random sample from $f(x \mid \theta)$ whose support is independent of θ. Under mild conditions of regularity,

$$\text{var } Y \geq \frac{1}{I_n(\theta)}, \tag{2}$$

where

$$I_n(\theta) = n \cdot E\left\{ \left[\frac{\partial}{\partial \theta} \log f(X \mid \theta) \right]^2 \right\}. \tag{3}$$

The quantity $I_n(\theta)$ is called the **information** in the sample regarding the parameter θ, and the inequality (2) is sometimes referred to as the *information inequality*.

A formal proof of the Cramér–Rao inequality (2) requires stating and using the "conditions of regularity." We give an informal derivation using the fact (from Sections 5.7 and 3.4) that a correlation coefficient cannot exceed 1 in magnitude. We'll assume the continuous case; the discrete case is similar. Our manipulations involve interchanges of differentiation that call for the regularity conditions. The only exceptional cases we'll encounter involve p.d.f.'s whose support depends on θ.

Consider the random variable W, defined as follows:

$$W = \frac{\partial}{\partial \theta} \log f(\mathbf{X} \mid \theta) = \frac{f'(\mathbf{X} \mid \theta)}{f(\mathbf{X} \mid \theta)}. \tag{4}$$

Now, W always has mean 0:

$$E(W) = \int \frac{f'(\mathbf{x} \mid \theta)}{f(\mathbf{x} \mid \theta)} f(\mathbf{x} \mid \theta) \, d\mathbf{x} = \frac{d}{d\theta} \int f(\mathbf{x} \mid \theta) \, d\mathbf{x} = \frac{d}{d\theta}(1) = 0. \tag{5}$$

(The integrals here are n-dimensional, and a prime means $\partial/\partial\theta$.) In view of (5),

$$\text{cov}(W, Y) = E(WY) = \int g(\mathbf{x}) \cdot \frac{f'(\mathbf{x} \mid \theta)}{f(\mathbf{x} \mid \theta)} f(\mathbf{x} \mid \theta) \, d\mathbf{x}$$

$$= \frac{d}{d\theta} \int g(\mathbf{x}) f(\mathbf{x} \mid \theta) \, d\mathbf{x} = \frac{d}{d\theta} E(Y) = \frac{d\theta}{d\theta} = 1. \tag{6}$$

Then

$$1 \geq \rho_{W,Y}^2 = \frac{\text{cov}^2(W, Y)}{(\text{var } Y)(\text{var } W)},$$

or

$$\text{var } Y \geq \frac{1}{\text{var } W}. \tag{7}$$

In a random sample, the observations X_i are independent, so

$$W = \frac{\partial}{\partial \theta} \log f(\mathbf{X} \mid \theta) = \frac{\partial}{\partial \theta} \log \prod f(X_i \mid \theta) = \sum \frac{\partial}{\partial \theta} \log f(X_i \mid \theta) = \sum W_i,$$

where

$$W_i = \frac{\partial}{\partial \theta} \log f(X_i \mid \theta).$$

The W_i are independent and identically distributed, with mean 0 and variance $I_1(\mu)$. Hence,

$$\text{var } W = \text{var}\left(\sum W_i\right) = nI_1(\mu) = n \cdot E\left\{\left[\frac{\partial}{\partial \theta} \log f_X(X \mid \theta)\right]^2\right\}.$$

Together with (7), this establishes (2).

Efficient
Estimators

According to (2), no unbiased estimator can have a variance smaller than $1/I_n(\theta)$. Can we find one whose variance is *equal* to $1/I_n(\theta)$? If so, such as estimator makes fullest use of the data, and we say it is **efficient**. We define the efficiency of an estimator as the ratio of the Cramér–Rao lower bound [in (2)] to the variance of the estimator.

The efficiency of an unbiased estimator Y_n in estimating θ is

$$\text{eff}(Y_n) = \frac{\frac{1}{I_n}(\theta)}{\text{var } Y_n}. \tag{8}$$

When $\text{eff}(Y_n) = 1$ [or var $Y_n = 1/I_n(\theta)$], Y_n is *efficient*.

Example 9.11b   Exp(1/μ)

Consider a random sample from a population with p.d.f.

$$f(x \mid \mu) = \frac{1}{\mu} \exp\left(-\frac{x}{\mu}\right), \qquad x > 0,$$

where μ is population mean. In Example 9.9d, we obtained

$$\log L(\mu) = -n \log \mu - \frac{n\bar{X}}{\mu}.$$

The partial derivative (with respect to μ) is

$$\frac{\partial}{\partial \mu} \log f(X \mid \mu) = -\frac{1}{\mu} + \frac{X}{\mu^2}. \tag{9}$$

This has mean zero, since $E(X) = \mu$. From (3),

$$I_n(\mu) = n \cdot E\left\{\left[\frac{\partial}{\partial \mu} \log f(X \mid \mu)\right]^2\right\}$$

$$= n \text{ var}\left(\frac{X}{\mu^2}\right) = \frac{n(\text{var } X)}{\mu^4} = \frac{n}{\sigma^2} = \frac{1}{\text{var } \bar{X}}.$$

Thus, since $\bar{X}$ is an unbiased estimate of μ, it is efficient.   ■

Example 9.11c   The Exponential Family

In Example 8.3f, we introduced the exponential family of distributions, those with p.d.f. of the form

$$f(x \mid \theta) = B(\theta)h(x) \exp[Q(\theta)R(x)].$$

If we assume a random sample,

$$W = \frac{\partial}{\partial \theta} \log \mathbf{f}(X \mid \theta) = n\frac{B'(\theta)}{B(\theta)} + Q'(\theta) \sum R(x_i).$$

So W is a linear function of the statistic $Y = \Sigma R(X_i)$. Thus, $\rho_{W,Y}^2 = 1$, so (2) is an equality: If Y is unbiased, it is efficient. ∎

Example 9.11d **Gam(2, $1/\theta$)**

Consider a random sample of size n from a population with p.d.f.

$$f(x \mid \theta) = \frac{x}{\theta^2} e^{-x/\theta}, \qquad x > 0, \qquad \theta > 0.$$

This gamma density is in the exponential family, with $R(x) = x$. The population mean [from (10) of Section 6.3] is $\mu = 2\theta$, so $\bar{X}/2$ is an unbiased estimator of θ. Moreover,

$$\frac{\bar{X}}{2} = \frac{\Sigma R(X_i)}{2n}$$

is a linear function of $\Sigma R(X_i)$, so it is efficient. ∎

Relation to Sufficiency

An efficient estimator is a sufficient statistic. If Y is efficient, the information inequality (2) is an equality. This means that the correlation between W and the statistic Y is ± 1. So (with probability 1) W is a linear function of Y, with coefficients that may involve θ:

$$W = a(\theta) + b(\theta)Y.$$

Since $\log f(\mathbf{X} \mid \theta)$ is an indefinite integral of W,

$$\log f(\mathbf{X} \mid \theta) = A(\theta) + B(\theta)Y + K(\mathbf{X}).$$

(The functions A and B are indefinite integrals of a and b, and K is a constant of integration—constant with respect to θ.) The likelihood function is then

$$L(\theta) = \exp[A(\theta) + B(\theta)Y],$$

which depends on the sample through the value of Y. By definition (Section 8.3), Y is *sufficient* for θ.

Example 9.11e In Example 9.9a, we found the m.l.e. of λ in Exp(λ) to be $\hat{\lambda} = 1/X$. Its mean value (Problem 53) is

$$E\left(\frac{1}{\bar{X}}\right) = nE\left(\frac{1}{\Sigma X}\right) = n \int_0^\infty \frac{1}{u} f_{\Sigma X}(u)\, du = \frac{\lambda n}{n-1},$$

so $Y = \dfrac{n-1}{\Sigma X}$ is unbiased for λ. Also (see Problem 53),

$$\mathrm{var}(Y) = \frac{\lambda^2}{n-2}.$$

No other unbiased estimate has smaller variance. The information in the sample is $I_n(\lambda) = n\lambda^2$, so the efficiency of Y is $1 - 2/n$. This is less than 1, and there is no efficient estimator. However, as n becomes infinite, the efficiency of y tends to 1. ∎

Asymptotic
Efficiency

When the efficiency of an estimator Y_n approaches 1 as the sample size tends to infinity, it is said to be **asymptotically efficient**. The notion of asymptotic efficiency can be given a more general definition, one that applies to sequences of biased estimators and estimators whose variance is not finite. In regular cases, maximum likelihood estimates can be shown to be asymptotically efficient in this extended sense.[7]

Example **9.11f**

The sample median and the sample mean are candidates for estimating the mean μ (which is also the median) of a normal population with known variance σ^2. The sample mean is efficient for any finite n, so it is asymptotically efficient. It can be shown that the sample median has asymptotic variance $\pi\sigma^2/(2n)$. The ratio of var $\bar{X} = \sigma^2/n$ to the variance of the median is asymptotically $2/\pi \doteq .637$. In large samples, using the sample mean requires only about 64% as many observations to estimate μ with specified precision as would be required using the sample median.

Don't read too much into this result. If the population is normal, the median is indeed less efficient than the mean. If it's not normal, the median can be much preferable to the mean. For instance, if the population is normal except for an occasional observation that has a larger variance (as when a response is recorded incorrectly), then the median may be much better than the mean. ∎

Chapter Perspective

We have studied the estimation of the population parameter θ, mainly for cases in which θ is a single real parameter and in terms of mean squared error as a criterion for accuracy. For unbiased (or nearly unbiased) estimators, we've assessed estimation accuracy in terms of the standard error—the estimated standard deviation of the estimator. Confidence intervals have provided a way of combining a point estimate with its standard error to give a range of parameter values in which, with a given degree of confidence, we think the true value lies. Confidence interval construction is closely related to the testing of hypotheses, as we'll see in Chapter 11.

For deriving estimates, we've emphasized only the most popular method, that of maximum likelihood. Maximizing the likelihood is intuitive and leads to estimates that are consistent and often efficient. It is this method that we'll use in problems of regression, categorical data, and analysis of variance (Chapters 13–15) and in constructing "likelihood ratio" tests (Chapter 11).

The maximum likelihood technique is applicable when several parameters are jointly estimated. The notions of m.s.e. and efficiency can be extended to these multiparameter situations. For efficiency, one needs an extended definition of information. (See, for example, Rao [16].)

Chapter 16 takes up Bayesian methods, based on a completely different approach to inference generally, including estimation.

---

[7] H. Cramér [5], p. 500.

Solved Problems

Sections 9.1–9.2

A. Measurements of tracheal compliance in six newborn lambs are given in a pediatric research study:[8]

.029, .043, .022, .012, .020, .034.

(a) Find the standard error of the mean.
(b) Suppose the investigators wanted to achieve an s.e. of .0020. Assuming S stays at .011, about how many lambs would be required?

Solution:
(a) The mean and s.d. are $\bar{X} = .02667$ and $S = .01102$. Then

$$\text{s.e.} = \frac{S}{\sqrt{n}} = \frac{.01102}{\sqrt{6}} = .0045.$$

(b) For s.e. $= .0020$, set $S/\sqrt{n} = .011/\sqrt{n} = .010$ and solve: $n = (.011/.0020)^2 \doteq 30$.

B. A newspaper poll reported that about 22% in a sample of U.S. voters (434 out of 1,549) said they were Independents. Assuming random sampling, find the standard error of the sample proportion as an estimate of the population proportion of Independents.

Solution:
With $n = 1549$, $\hat{p} = \frac{434}{1549}$ and $1 - \hat{p} = \frac{1115}{1549}$, we find the standard error to be

$$\sqrt{\frac{\hat{p}(1 - \hat{p})}{n}} = .0114.$$

C. Suppose we want to estimate the proportion of students on a campus who used alcohol in the last month.
(a) How large a random sample is required if we want the standard error of our estimate not to exceed .05?
(b) If we assume that the population proportion is at least .8, what would be the minimum sample size needed to meet the same condition as in (a)?

Solution:
(a) With no information of p, use (3) of Section 9.2: $n \geq .25/(.5)^2 = 100$.
(b) Use (4) of Section 9.2 with $p_0 = .8$: $n \geq (.8)(.2)/(.05)^2 = 64$.

D. When $X \sim U(0, \theta)$, the parameter θ is twice the mean: $\theta = 2\mu$. In Example 9.1a, we considered the largest sample observation as an estimate of θ. Suppose instead we use $2\bar{X}$ as an estimator for θ. Find its bias and the m.s.e.

---

[8] V. K. Bhutani, R. J. Koslo, and T. H. Shaffer, "The effect of tracheal smooth muscle tone on neonatal airway collapsability," *Pediatric Research 20* (1986), 492–495.

Solution:

The sample mean is always an unbiased estimate of the population mean ($E\bar{X} = \mu$). So the bias in $2\bar{X}$ as an estimate of $\theta = 2\mu$ is $E(2\bar{X}) - 2\mu = 0$. The m.s.e. is then just the variance:

$$\text{var } 2\bar{X} = 4 \text{ var } \bar{X} = \frac{4\sigma^2}{n} = \frac{\theta^2}{3n}.$$

Sections 9.3–9.5

E.   A newspaper headline proclaimed "Most U.S. Catholics want changes in sexual policies." The article was a report of a 1986 Gallup Poll based on telephone interviews with 264 Roman Catholics, and the proportion favoring change in its church's policies was 57%. The fine print at the end of the article said that "19 times out of 20, the error attributable to sampling and other random effects should not exceed 8 percentage points in either direction." Explain, and give 95% confidence limits for the proportion of all Catholics who want change. (The sampling was said to be conducted at "scientifically selected places nationwide," but the sampling scheme was not otherwise described.)

Solution:

We assume random sampling. The standard error of the estimate ($\hat{p} = .57$) is

$$\text{s.e.} = \sqrt{\frac{.57 \times .43}{264}} = .0305.$$

Confidence limits at 95% confidence are $.57 \pm 1.96 \times .0305$, or $.57 \pm .060$. So it's not clear where the news report's "8 percentage points" comes from, since Gallup usually uses two s.e.'s in reports of accuracy. (Perhaps "other random effects"?) If we assume 99% confidence, the limits are $.57 \pm .08$.

F.   Durations (in min) of 74 storms in the Tampa Bay area are reported in a 1984 study.[9] The mean is 75.7 and the standard deviation 60.7. Assuming random sampling, give 95% confidence limits for the mean duration of storms in the Tampa Bay area.

Solution:

The standard error is $60.7/\sqrt{74} = 7.056$. If we assume normality of the sample mean, the limits are

$$75.7 \pm 1.96 \times 7.056 \qquad \text{or} \qquad 75.7 \pm 13.8.$$

Since the durations are positive numbers, the size of S in relation to $\bar{X}$ suggests that the sample observations are quite skewed to the right. Nevertheless, for a sample of size 74, the sample mean should be very close to normal. (The mean duration has questionable usefulness. One might be more interested in the proportion of storms that last more than one hour, more than two hours, and so on.)

G.   A random sample of 15 cigarettes of a certain brand was tested for nicotine content. The

---

[9] "Lightning phenomenology in the Tampa Bay area," *J. Geophys. Res.* (1984), 11789–11805.

average content of these 15 cigarettes was found to be 20.3 mg, and the s.d., $S = 3.0$ m.g. Construct a 95% confidence interval for the mean content for all cigarettes of this brand. (Assume a random sample from a normal population.)

Solution:

The standard error is $3.0/\sqrt{15}$. The population s.d. is not given but estimated, so we turn to the *t*-distribution for the proper multiplier of s.e. Table IIIa in Appendix 1 gives $k = 2.18$:

$$20.3 \pm 2.18 \times \frac{3.0}{\sqrt{15}} = 20.3 \pm 1.69.$$

The 95% confidence interval is (18.6, 22.0).

H. Consider a random sample of size 10 from $U(0, \theta)$. From the distribution of $Y = \frac{X_{(n)}}{\theta}$, find a 95% "lower confidence bound"—a one-sided confidence interval for the parameter θ of the form $kX_{(10)} < \theta < \infty$.

Solution:

The p.d.f. of $X_{(n)}$ (from Example 9.1a) is nx^{n-1}/θ^n for $0 < x < \theta$. The p.d.f. and c.d.f. of $Y = \frac{X_{(n)}}{\theta}$ are

$$f_Y(y) = ny^{n-1} \quad \text{and} \quad F_Y(y) = y^n, \qquad 0 < y < 1.$$

So Y is pivotal. Then

$$P(kX_{(n)} < \theta) = P\left(\frac{X_{(n)}}{\theta} < \frac{1}{k}\right) = F_Y(k) = \frac{1}{k^n} = \frac{1}{k^{10}}.$$

This is .95 if k is the tenth root of $1/.95$, or about 1.005.

Sections 9.6–9.7

I. A research study[10] concerned with air pollutants in the Lincoln Tunnel under the Hudson River reported these summary statistics on concentrations in ppm (parts per million):

	n	Mean	s.d.
1970	28	65.8	14.9
1983	28	15.6	3.8

(a) The mean difference in concentration between these two years is $\bar{D} = 50.2$. Determine the standard error of this estimate.

(b) Find 95% confidence limits for the mean difference, assuming 28 is "large."

---

[10] "Non-methane organic composition in the Lincoln Tunnel," *Envir. Sci. & Technology 20* (1986), 789–805.

Solution:

(a) The s.e. of the difference in means is the square root of the sum of the squares of the s.e.'s:

$$\text{s.e.}(\bar{D}) = \sqrt{\frac{14.9^2}{28} + \frac{3.8^2}{28}} \doteq 2.9.$$

(b) For 95% confidence, we use the multiplier 1.96: $\bar{D} \pm (1.96)\,\text{s.e.}(\bar{D}) = 50.2 \pm (1.96)(2.9) = 50.2 \pm 5.7.$

J. Independent polls before ($n = 1200$) and after ($n = 950$) a particular crisis show a drop in approval of the president's performance from 65% before to 48% after. Find the standard error for the estimated 17% drop and construct 95% confidence limits for the population difference in proportions.

Solution:

We apply (5) in Section 9.6 with $\hat{p}_1 = .65$ and $\hat{p}_2 = .48$:

$$\text{s.e.} = \sqrt{.01377^2 + .01442^2} = \sqrt{\frac{.65 \times .35}{1200} + \frac{.48 \times .52}{1200}} = .020.$$

The 95% confidence limits are $.17 \pm 1.96 \times .020$, or about .13 to .21.

K. An experiment was conducted to check the variability in explosion times of detonators.[11] Actual times of the ignitions or explosions are variable, but if the variability is too great, the near simultaneity may not achieve the desired effect. Data for run #1 are as follows, expressed in milliseconds short of 2.7 seconds:

11, 23, 25, 9, 2, 6, −2, 2, −6, 8, 9, 19, 0, 2.

The sample standard deviation is 9.27, and the variance 85.9. Give 90% confidence limits for σ, the population s.d.

Solution:

A dot diagram of the data (Figure 9.3) suggests no obvious population nonnormality. We assume normality and follow the pattern of Example 9.7b. With $n - 1 = 13$ d.f., we find the 5th and 95th percentiles of $\chi^2\,(13)$ in Table Vb of Appendix 1 to be 5.89 and 22.4. The 90% limits for σ^2 are then $13 \cdot 85.9/22.4$ and $13 \cdot 85.9/5.89$. Taking square roots, we obtain limits for σ: $7.06 < \sigma < 13.77$.

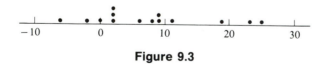

Figure 9.3

L. For the data of Problem K, construct a 90% confidence interval for σ, of the form $0 < \sigma < K$.

---

[11] Reported in G. L. Tietjen and M. E. Johnson, "Exact statistical tolerance limits for sample variances," *Technometrics 21* (1979), 107–110.

Solution:
From (4) of Section 9.7, we know that $13S^2/\sigma^2 \sim \chi^2(13)$. In the language of Section 9.5, this is a pivotal quantity. In Table Vb, we find the 10th percentile of $\chi^2(13)$ to be 7.04:

$$.90 = P\left(\frac{13S^2}{\sigma^2} > 7.04\right) = P\left(\sigma^2 < \frac{13S^2}{7.04}\right).$$

The confidence upper bound for σ^2 is $13S^2/7.04 = 158.6$, so $K = \sqrt{158.6} \doteq 12.6$.

M. The usual estimators for the variance of a normal population are multiples of the sum of squared deviations about the mean:

$$Y = k \cdot \sum_1^n (X_i - \bar{X})^2 = k(n-1)S^2.$$

Find the multiplier k that minimizes the mean squared error.

Solution:
The bias in Y is $b_Y(\sigma^2) = E(Y) - \sigma^2 = k(n-1)\sigma^2 - \sigma^2$, and its variance is

$$\text{var } Y = k^2(n-1)^2 \text{ var } S^2 = 2k^2(n-1)\sigma^4.$$

So the mean square error is

$$\text{var } Y + b_Y^2(\sigma^2) = 2k^2(n-1)\sigma^4 + [k(n-1) - 1]^2\sigma^4$$
$$= [(n-1)(n+1)k^2 - 2(n-1)k + 1]\sigma^4.$$

This quadratic function of k is smallest at $k = 1/(n+1)$. [The bias in Y with this value of k is $-2\sigma^2/(n+1)$.]

Sections 9.8–9.9

N. Find the m.l.e. of the parameter θ of a uniform distribution on the interval $(0, \theta)$, given a random sample of size n.

Solution:
The likelihood function (see Example 8.3e) is

$$L(\theta) = \begin{cases} \dfrac{1}{\theta^n}, & \theta > X_{(n)} \\ 0, & \theta < X_{(n)}. \end{cases}$$

Since $1/\theta^n$ is a decreasing function, $L(\theta)$ has a maximum at the largest observation: $\hat{\theta} = X_{(n)}$.

O. Given a random sample of size n from a normal population,
   (a) find the joint m.l.e.'s of the mean and variance.
   (b) show that $\hat{\sigma}$ is biased as an estimator of σ.

Solution:
   (a) Given that $X \sim N(\mu, \theta)$, the log of the likelihood function [from (4) in Example 8.3h] is

$$\log L(\mu, \theta) = -\frac{n}{2}\log \theta - \frac{n}{2\theta}[V + (\bar{X} - \mu)^2].$$

For any V, this is largest when μ is $\hat{\mu} = \bar{X}$. Then

$$\log L(\hat{\mu}, \theta) = -\frac{n}{2} \log \theta - \frac{nV}{2\theta},$$

We set the derivative with respect to θ equal to 0:

$$-\frac{n}{2\theta} + \frac{nV}{2\theta^2} = 0$$

Solving for θ, we find the m.l.e. to be $\hat{\theta} = V$. That is, the joint m.l.e.'s of μ and θ are, respectively, $\bar{X}$ and V (where V is the version of sample variance with divisor n).

(b) $\hat{\sigma}$ is the square root of V

$$\hat{\sigma} = \sqrt{\frac{1}{n} \sum (X_i - \bar{X})^2} = \sqrt{\frac{n-1}{n}} S^2 \,.$$

We saw in Section 9.4 that $\sqrt{n-1}\, S/\sigma$ has a chi distribution with $n-1$ degrees of freedom. So $E(\hat{\sigma}) = E(\chi/\sqrt{n})\sigma$, where χ is chi with $n-1$ d.f. Using the chi density [(7) of Section 6.5], we find

$$E(\chi) = \int_0^\infty \frac{y}{2^{k/2-1}\Gamma(k/2)}\, y^{k-1} e^{-y^2/2}\, dy = \frac{\sqrt{2}\,\Gamma[(k+1)/2]}{\Gamma(k/2)},$$

where $k = n - 1$. Division by $\sqrt{n}$ yields

$$E(\hat{\sigma}) = E\left(\frac{\chi\sigma}{\sqrt{n}}\right) = \sigma\sqrt{\frac{2}{n}} \cdot \frac{\Gamma(n/2)}{\Gamma[(n-1)/2]}$$

The multiplier of σ is not 1, so $\hat{\sigma}$ is biased, although it tends to 1 as $n \to \infty$. (When $n = 10$, it's about .923. Incidentally, S is also biased.)

P.   Given a random sample of size n from a population with c.d.f. $F(x\,|\,\theta) = x^\alpha$ for $0 < x < 1$, where $\alpha > 0$, obtain an estimator for α
(a)   using the method of moments.
(b)   using the method of maximum likelihood.

Solution:
(a)   The p.d.f. is $f = F' = \alpha x^{\alpha-1}$, so the likelihood function is

$$L(\alpha) = \prod f(x_i\,|\,\alpha) = \alpha^n \Pi x_i^{\alpha-1},$$

with logarithm

$$\log L(\alpha) = n \log \alpha + \sum (\alpha - 1) \log x_i.$$

Differentiating and setting the derivative equal to zero, we obtain

$$\hat{\alpha} = \frac{-n}{\sum \log x_i}.$$

(b)   The population mean is $\mu = \alpha/(\alpha + 1)$. Solving for α, we find $\alpha = \mu/(1 - \mu)$. We replace μ by $\bar{X}$: $\tilde{\alpha} = \bar{X}/(1 - \bar{X})$.

For example, suppose the sample observations are .4, .7, .9, .95. The mean is .7375, so $\tilde{\alpha} = 2.8095$. The m.l.e. is $\tilde{\alpha} = 2.7979$.

Sections 9.8–9.9

Q. Show that a sample proportion is consistent in estimating the corresponding population proportion p, assuming random sampling, by calculating the probability in an interval about p of width ε.

Solution:
The sample proportion $\hat{p}$ is asymptotically normal with mean p and variance $p(1-p)/n$ (see Section 8.8). Thus, for given $\varepsilon > 0$,

$$P(|\hat{p} - p| > \varepsilon) = 1 - P(p - \varepsilon < \hat{p} < p + \varepsilon)$$

$$\doteq 1 - \Phi\left(\frac{p + \varepsilon - p}{\sqrt{p(1-p)/n}}\right) + \Phi\left(\frac{p - \varepsilon - p}{\sqrt{p(1-p)/n}}\right)$$

$$\to 1 - \Phi(\infty) + \Phi(-\infty) = 0.$$

Then by definition [(1) of Section 9.10], $\hat{p}$ is consistent.

R. Consider a random sample from $U(\theta - \frac{1}{2}, \theta + \frac{1}{2})$ and let Y_n denote the midrange—the average of the largest and smallest sample observations. Show that Y_n is consistent as an estimate of θ.

Solution:
This is not easy. We could easily show that Y_n is unbiased, but the variance takes more work—it involves finding the covariance of the smallest and largest observations. The result:

$$\text{m.s.e.}(Y_n) = \text{var } Y_n = \frac{(n-1)/2}{(n+2)(n+1)^2}.$$

So Y_n is indeed consistent—the m.s.e. tends to 0. Alternatively, we can show that Y_n tends to θ in probability. Given: X is $U(\theta - \frac{1}{2}, \theta + \frac{1}{2})$. By the triangle inequality,

$$|X_{(1)} + X_{(n)} - 2\theta| = \left|X_{(1)} - \theta - \frac{1}{2} + X_{(n)} - \theta + \frac{1}{2}\right| \le \left|X_{(1)} - \theta - \frac{1}{2}\right| + \left|X_{(n)} - \theta + \frac{1}{2}\right|$$

We know that $X_{(1)}$ and $X_{(n)}$ are consistent estimates of, respectively, $\theta - \frac{1}{2}$ and $\theta + \frac{1}{2}$. If each term on the right is less than $\varepsilon/2$, for given $\varepsilon > 0$, the expression on the left is less than ε. Then

$$P(|X_{(1)} + X_{(n)} - 2\theta| > \varepsilon) \le P\left(\left|X_{(1)} - \theta - \frac{1}{2}\right| > \frac{\varepsilon}{2}\right) + P\left(\left|X_{(n)} - \theta + \frac{1}{2}\right| > \frac{\varepsilon}{2}\right).$$

Since both terms on the right tend to 0, so does the left-hand side, which means that Y_n is consistent.

S. If X, Y, and Z are independent and each $N(0, \theta)$, the distance U from the origin to (X, Y, Z) has a Maxwell distribution with p.d.f.

$$f(u \mid \theta) \propto \frac{u^2}{\theta^{3/2}} e^{-u^2/(2\theta)}, \qquad u > 0.$$

(See Problem 31 of Chapter 5.) Show that the m.l.e. of θ is efficient.

Solution:

The m.l.e. is $\hat{\theta} = \Sigma U^2/(3n)$. Since $E(U^2) = 3 \text{ var } X = 3\theta$, $E(\hat{\theta}) = \theta$. The m.l.e. is unbiased. We've shown efficiency for an unbiased one-to-one function of the sufficient statistic for any exponential family (Example 9.11c). Efficiency of $\hat{\theta}$ then follows as a special case once we see that the Maxwell distribution is a member of that family: Set

$$R(u) = u^2,$$

$$Q(\theta) = \frac{-1}{(2\theta)},$$

$$h(u) = u^2,$$

and

$$B(\theta) \propto \theta^{-3/2}.$$

Problems

For problems marked with an asterisk, answers are provided in Appendix 2.

Sections 9.1–9.2

∗1. A random survey of 1,500 individuals shows that 630 watched a televised presidential news conference. Give an estimate of the population proportion who watched it, together with its standard error.

2. A TV show gets a rating of 27 (i.e., 27%) based on a random sample of 1,000 individuals. Find the standard error.

∗3. A report on water pollution[12] included the following data on chemical oxygen demand (mg/l) of 18 water samples:

580	674	512	540	616	298	960	570	640
588	556	588	582	844	574	420	696	620.

Find the mean and the standard error of the mean.

∗4. The statistic $X_{(n)}$, the largest observation in a random sample of size n, is proposed in Example 9.1a as an estimator for the parameter θ of a uniform distribution on $(0, \theta)$. Its mean and variance are given in Example 8.4c, page 378.
  (a) Find the bias in $X_{(n)}$ as an estimator of θ.
  (b) Find its mean squared error.

5. Laplace's rule of succession for an unknown probability p of success estimates it to be $(Y + 1)/(n + 2)$, where Y is the number of successes in n independent trials.
  (a) Find the bias and the mean squared error for this estimator.
  (b) For what range of values of p is this estimate better than the sample proportion Y/n, in terms of m.s.e.?

---

[12] K. J. Shapland, "Industrial effluent treatability—A case study," *Water Pollution Control 85* (1986), 75–80.

*6. A newspaper poll reported that about 28% in a sample of U.S. voters (434 out of 1,549) said they were Independents. Assuming random sampling, find the standard error of the sample proportion as an estimate of the population proportion of Independents.

7. A newspaper report says that a sample proportion of .43 has a "margin of error" of five percentage points (in estimating the population proportion). Interpret .05 as two standard errors and find the approximate sample size, assuming random sampling.

*8. Consider estimating a population proportion p using a random sample of size n. The statistic X_1, the first observation in the sample, is an unbiased estimate, and the sample sum $Y = \Sigma X$ is sufficient for p. Find an unbiased estimator based on the sufficient statistic by conditioning X_1 on the value of Y and averaging.

*9. Previous experience suggests that the s.d. for a certain kind of measurement X is about 5.0. How large a sample should then be taken to estimate the population mean with
   (a) a standard error of 1.0?
   (b) a standard error of .50?

*10. To estimate the proportion of TV viewers watching a certain special, how large a random sample is required for a standard error of
   (a) at most .01?
   (b) at most .02?

11. In estimating the proportion p in a population having a disease, how large a random sample is needed to be sure the s.e. does not exceed .005,
   (a) if you have no idea as to the value of p?
   (b) if you assume that p does not exceed 5%?

12. Let $\bar{X}_1$ and $\bar{X}_2$ denote the means of independent random samples from populations with the same mean μ. Find the value of γ for which the mixture

$$\gamma\bar{X}_1 + (1 - \gamma)\bar{X}_2 \qquad (0 \le \gamma \le 1)$$

has the smallest m.s.e., when
   (a) the sample sizes are both n and the population variances are σ_1^2 and σ_2^2, respectively.
   (b) the population variances are the same, but the sample sizes are n_1 and n_2, respectively.

Sections 9.3–9.5

*13. Data[13] on time from origin to destination for 201 buses on a particular bus route in Chicago have mean $\bar{X} = 18.31$ and standard deviation $S = 2.111$. Find 95% confidence limits for the mean time μ.

14. In a survey[14] of college students, it was found that at least one parent of 69 of the 224 students surveyed had had heart disease. Give 95% confidence limits for the corresponding population proportion. (The only relevant population is that of all pairs

---

[13] A. Polus, "Modeling and measurements of bus service reliability," *Transportation Research 12* (1978), 253–256.

[14] R. N. Tamragouri, "Cardiovascular risk factors and health knowledge among freshman college students with a family history of cardiovascular disease," *J. Amer. College Health 34* (1986), 267–270.

of parents of students at that college; the incidence is apt to be different for adults without children in college and for different colleges.)

*15. The water samples referred to in Problem 3 were also checked as to biological oxygen demand. The mean and s.d. of the 18 measurements are $\bar{X} = 89.14$ and $S = 25.19$. Give 95% and 99% confidence limits for μ. (Assume normality and random sampling.)

16. Through careful examination of sound and film records, it is possible to measure the distance at which a bat first detects an insect. Given the following data (in cm):[15]

62, 52, 68, 23, 34, 45, 27, 42, 83, 56, 40,

construct a 90% confidence interval for the population mean distance (assuming normality and random sampling).

*17. A research report is to include 20 confidence intervals, each at the .95 level. Although one would expect some relationships, suppose the intervals are based on independent statistics.
   (a) About how many of intervals would you expect actually to include the true value of the parameter being estimated?
   (b) What is the probability that all 20 intervals will contain the true values of the parameters being estimated?

18. Find a 95% upper confidence interval for θ [that is, an interval of the form $X_{(n)} < \theta < kX_{(n)}$] based on the largest observation in a random sample of size 10 from $U(0, \theta)$. (See Example 9.1a and Solved Problem H.)

*19. A telephone switchboard records 25 incoming calls in a ten-minute period. Based on this, find 95% confidence limits for the rate parameter λ = expected number of calls per minute. [*Hint*: The number of calls Y is the sum of ten independent variables X_i, where $X_i \sim \text{Poi}(\lambda)$, so $(Y - 10\lambda)/\sqrt{10\lambda} \approx N(0, 1)$.]

20. Consider a random sample of size n from $\text{Gam}(2, 1/\theta)$:

$$f(x \mid \theta) \propto \frac{x}{\theta^2} e^{-x/\theta}, \qquad x > 0.$$

   (a) Show that $(2\Sigma X)/\theta$ is pivotal.
   (b) Construct 90% two-sided confidence limits for θ, given $n = 5$ and $\Sigma X = 32$.

Sections 9.6–9.7

*21. Blacks constitute 15% of the population in a certain city. Find the probability that the proportions of blacks in two independent random samples, each of size 500, will differ by more than two percentage points (that is, the numbers of blacks differ by more than 10).

*22. Aggregate with a low measure of thermal conductivity is desirable for paving. Two types

---

[15] Reported in Griffin et al., "The echolocation of flying insects by bats," *Animal Behavior* 8 (1960), 141–154.

of aggregate are tested for thermal conductivity, with 25 observations of each type:

X(low cost): mean = .485, s.d. = .180,

Y(higher cost): mean = .372, s.d. = .160.

(a) Find the s.e. of the difference in sample means as an estimate of the difference in population mean conductivities.

(b) Obtain 99% confidence limits for the difference between the mean conductivity of these two types of aggregate.

23. An experiment by Danish investigators[16] on the effectiveness of toothpaste involved 295 children who used fluoride toothpaste and 284 who used a nonfluoride toothpaste (colored and flavored to look the same as the fluoride version). The mean number of cavities per child, developed over a 30-month period, was 10.88 for the fluoride sample and 13.41 for the nonfluoride sample. The sample s.d.'s were 6.36 and 7.20, respectively. Give an estimate of the difference in mean number of cavities, together with the standard error of the estimate.

*24. The following[17] are summary statistics on SAT-M scores, the mathematical part of the Scholastic Aptitude Test, of children of seventh-grade age who were found in a Johns Hopkins talent search. (To be included, the children had to be in the top 3% of any standard achievement test in verbal, mathematical, or overall intellectual ability.)

	n	Mean	s.d.
Boys	19,883	416	87
Girls	19,937	386	74

Assuming that the participants can be regarded as constituting independent random samples from some population of interest, determine a 99% confidence interval for the population mean difference in boys' and girls' SAT-M scores.

25. In a large group of young couples, the s.d. of the husbands' ages is four years, and that of the wives' ages three years. Let D denote the age difference within a couple. Since $4^2 + 3^2 = 5^2$, you might expect to find the s.d. of D's in the group to be about five years. Instead you find it to be two years. One explanation is that the discrepancy is the result of random variability. Give another explanation.

*26. In a sample of 120 adults from the roster of a health maintenance organization in California, there were 39 smokers. In a sample of 100 adults from the roster of an HMO in Minnesota, there were 27 smokers. Find the standard error of the difference in sample proportions of smokers, as an estimate of the difference in population proportions.

27. An opinion poll surveyed a random sample of 1,549 adults prior to the 1984 election. The sample included 638 Democrats and 441 Independents. Of the Democrats, 43% felt

[16] I. J. Moller, J. J. Holst, and E. Sorensen, "Caries reducing effect of a sodium monofluorophosphate dentifrice," *Brit. Dental Jour. 124* (1968), 209–213.
[17] C. P. Benbow, and J. C. Stanley, "Sex differences in mathematical reasoning ability: More facts," *Science 213* (1983), 1029–1031.

that Walter Mondale would do a better job of maintaining prosperity than John Glenn; among the Independents, 28% felt that Mondale would do the better job. (Of course, many respondents said that Mondale and Glenn would do about the same, or declined to make the comparison.)

(a)   Give a 95% confidence interval for the proportion of Democrats in the population.

(b)   Assume that the 638 Democrats and the 441 Independents are random samples of Democrats and Independents. Give a 95% confidence interval for the difference in population proportions of Democrats and of Independents who felt that Mondale would do better.

*28.   Nine readings of an inlet oil temperature are as follows:

99, 93, 99, 97, 90, 96, 93, 88, 89.

Give 95% confidence limits for the population s.d., assuming random sampling and a normal population.

*29.   Consider a random sample of size n from a normal population with mean 0. Using the fact that $\frac{\Sigma X^2}{\sigma^2}$ is pivotal, obtain 95% confidence limits for σ^2.

30.   Cholesterol measurements of 100 males have mean 245.7 and s.d. 46.25. Assuming random sampling from a normal population, find 95% confidence limits for

(a)   the population mean.

(b)   the population standard deviation.

31.   For the data in Problem 15 ($\bar{X} = 89.14$, $S = 25.19$, $n = 18$), give 90% confidence limits for the population s.d.

*32.   Consider a random sample of size n from $N(\mu, \sigma^2)$. Because of (4) in Section 9.7, the distribution of S/σ depends only on n. Define $\alpha_n = E(S/\sigma)$ and $\beta_n^2 = \text{var}(\frac{S}{\sigma})$. It can be shown that

$$\alpha_n = \sqrt{\frac{2}{n}} \cdot \frac{\Gamma(\frac{n}{2})}{\Gamma(\frac{n-1}{2})}, \qquad \beta_n^2 = 1 - \alpha_n^2 - \frac{1}{n}.$$

In terms of these constants, n, and σ, find

(a)   the mean squared error of S as an estimator of σ.

(b)   the m.s.e. of $n\alpha_n S/(n-1)$ as an estimator of σ.

[We do have in hand the necessary facts for deriving the given formulas for α_n and β_n, in case you'd like to try doing it.]

33.   Calculate the m.s.e.'s in Problem 32 for $n = 10$.

34.   Verify the claim [see (5) in Section 9.7] that the m.s.e. of the biased estimator V is always less than the m.s.e. of S^2, in the case of a normal population.

Sections 9.8–9.9

*35.   Obtain estimators of the parameter θ based on a random sample from $N(0, 1/\theta)$, using

(a)   the method of moments.

(b)   the method of maximum likelihood.

*36.   Obtain the m.l.e. of σ^2 based on a random sample of size n from $N(0, \sigma^2)$,

(a)   by finding the likelihood function and maximizing.

(b)   by using (b) of the preceding problem.

37. Given a random sample from a geometric distribution with parameter p, obtain estimators for p using
   (a) the method of moments.
   (b) the method of maximum likelihood.

*38. Given a random sample of size n from a Poisson population, find the m.l.e. of the population mean.

*39. Find the m.l.e.'s of the parameters (θ_1, θ_2) of a population with a uniform distribution on the interval (θ_1, θ_2), based on a random sample of size n.

40. Referring to the preceding problem, find estimates of the parameters using the method of moments, given the sample (5, 7, 8, 12). (Compare with the m.l.e.'s.)

*41. Consider a random sample of pairs (X, Y) from a bivariate population with joint p.d.f.
   $f(x, y) = \exp(-\theta x - y/\theta)$ for $x > 0$ and $y > 0$, where θ is a positive parameter (see Problem 11 of Chapter 8).
   (a) Find $\hat{\theta}$ the m.l.e. of θ.
   (b) Show that $(\theta/\hat{\theta})^2$ is pivotal. [*Hint:* First show that $2\theta \Sigma X \sim \chi^2(2n)$.]
   (c) Use (b) to construct 90% confidence limits for θ when $n = 6$, $\Sigma X = 6.0$, and $\Sigma Y = .60$.

42. Find the m.l.e. of θ from a random sample of size n from a truncated exponential distribution with p.d.f.

   $$\frac{e^{-x}}{1 - e^{-\theta}}$$

   for $0 < x < \theta$.

*43. Find the m.l.e.'s of the parameters α, β, and σ^2, given that $Y_i \sim N(\alpha + \beta c_i, \sigma^2)$, for given constants c_i, assuming independence of the Y_i. (This is a regression model, to be taken up in Section 15.1.)

44. For large n, the mean squared error of S as an estimate of σ is approximately equal to $\frac{\sigma^2}{2n}$. The s.d. of platelet counts of 235 male cancer patients in the study cited in Example 9.1d is $S = 170$. Given that S is approximately normal for large n, construct an approximate 95% confidence interval for the population s.d.

Sections 9.10–9.11

*45. Consider the sample mean $\bar{X}$ as an estimate of μ, the mean of a population with finite variance. For given $\varepsilon > 0$, use the central limit theorem to approximate the probability outside the interval $(\mu - \varepsilon, \mu + \varepsilon)$ and show that it tends to 0 as the sample size becomes infinite.

46. Given a random sample from a population with c.d.f. F, let F_n denote the sample d.f. Show that for each x, $F_n(x)$ is a consistent estimator of $F(x)$. [Actually, something much stronger is true: Glivenko's theorem[18] says that $F_n(x)$ tends to $F(x)$ uniformly in x.]

47. Show that the largest observation in a random sample from $U(0, \theta)$ is a consistent

---

[18] See Rao, [16], p. 42.

estimator of θ

(a) by calculating $P(|X_{(n)} - \theta| > \varepsilon)$.

(b) by examining the limiting values of the mean and variance.

48. Consider a random sample from $N(0, \theta)$ and show that the m.l.e. $\hat{\theta}$ is consistent. (See Problem 36.)

49. Show that if Y_n is efficient for each n as an estimator of θ, it is also consistent.

50. For a random sample of size n, show that a sample proportion $\hat{p}$ is an efficient estimate of the corresponding population proportion

   *(a) by calculating the information $I(p)$.

   (b) using the fact that $Ber(p)$ is in the exponential family.

51. Show that the mean of a random sample from $Poi(\lambda)$ is efficient as an estimate of λ.

52. Consider a random sample of size n from $N(0, \theta)$. The likelihood function for a single observation is

$$L(\theta) = \theta^{-1/2} e^{-x^2/(2\theta)}.$$

Show that the m.l.e. $\hat{\theta}$ is efficient

   (a) by calculating $I(\hat{\theta})$.

   (b) using the fact that $N(0, \theta)$ is in the exponential family.

*53. Given that $V \sim Gam(n, \lambda)$, find $E(1/V)$ and $var(1/V)$. [For use in Example 9.11e.]

CHAPTER **10**

Significance Testing

Sample data provide evidence concerning hypotheses about the population from which they are drawn. The following examples give some settings in which hypotheses about populations need testing.

Example **10a** **Extrasensory Perception**

In one segment of the TV show "The Odd Couple," Felix claimed to have ESP. Naturally, Oscar was skeptical and suggested testing the claim. Oscar would draw a card at random from four large cards, each with a different geometric figure on it, and without showing it, ask Felix to identify the card. They repeated this basic experiment several times.

At each such trial, an individual without ESP has one chance in four of correctly identifying the card. In ten trials, Felix made six correct identifications. Although he didn't claim to be perfect, six is rather more than 2.5, the average number correct if he does *not* have ESP. What does this prove? Strictly speaking, it proves nothing, because it was possible for Felix to get six correct whether he used ESP or was only guessing. Nevertheless, his high score provides some evidence for ESP. (To be continued.) ∎

Example **10b** **Feeding Schedules and Blood Pressure**

Falk et al.[1] report on a study of the effect of intermittent feeding on the blood pressure of rats. They fed eight rats intermittently over a period of weeks, and at the end of that time measured the rats' blood pressures. The results were as follows, in mm of Hg:

170, 168, 115, 181, 162, 199, 207, 162.

---
[1] J. Falk, M. Tang, and S. Forman, "Schedule-induced chronic hypertension," in *Psychosomatic Medicine* 39 (1977), 252–263.

These numbers would mean little in the absence of data for comparison. So blood pressures of seven rats that had been fed normally were also measured, with these results:

169, 155, 134, 152, 133, 108, 145.

(There was an eighth rat in this group, but its result is missing because it died during induction of ether anesthesia.)

The data are plotted as dot diagrams in Figure 10.1. It is apparent that the measurements in the first group tend to be higher than those in the second group, although there is one rather low reading in the first group. (It is possible that this observation resulted from a misprint, a temporarily defective instrument, or a rat that was inadvertently exchanged with one in the second group; but without information of this nature, all we can do is proceed as though all measurements were equally valid.)

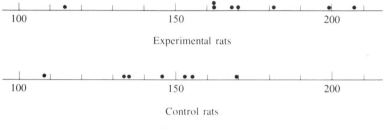

Figure 10.1

The question addressed in the study is whether intermittent feeding affects blood pressure. If it does, we might well see differences like those actually observed, but because of sampling variability, we would see differences even if the feeding schedule had *no* effect. The problem is to decide which explanation of the data is the more plausible—that feeding has an effect, or that it has no effect and that sampling variability is responsible for the discrepancy between the two samples. (To be continued in Example 12.4a.)   ■

Example **10c**   **Are Boys Better at Math?**

Do boys have a greater natural proclivity toward mathematics than girls?[2] Some people think so, but how could such a claim be tested? First, of course, we'd need a way of measuring "proclivity." This is not easy, but suppose that it can be done, using a battery of tests administered at an early age. Since one can't possibly test every child, we must be content with testing those in a sample. Naturally, since the basic question is one of comparison, both boys and girls must be tested.

---

[2] This question has been the subject of much research. For example, see L. H. Fox, L. Brody, and D. Tobin (Eds.), *Women and the Mathematical Mystique* (Baltimore: Johns Hopkins University Press, 1980).

The claim is not that *all* boys have a greater proclivity toward mathematics than all girls. The question is one of *tendency*, ordinarily measured in terms of an *average*. To draw conclusions about the difference between averages for boys and for girls generally, we could calculate the averages for the sample of boys and for the sample of girls. Especially when sample sizes are large, the sample means provide evidence as to whether there is a difference between population means. ■

10.1 Hypotheses

Examples 10a–10c have some common elements. In each, there is a theory or claim and (at least implicitly) a countertheory or counterclaim. These theories and countertheories are **hypotheses**. Each hypothesis can be expressed in terms of a probability model (or a set of probability models). The model describes a population, real or conceptual, and it is assumed that sample observations are selected at random from this population.

Oscar's claim in Example 10a that Felix does not have ESP corresponds to a model in which the probability p of correct identifications is one-fourth. If Felix has ESP, then $p > \frac{1}{4}$; this defines a set of probability models. In the feeding-schedule example, there are two populations—one of rats fed intermittently, the other of rats fed regularly. One hypothesis about these two populations is that their average blood pressures are the same—that how the rats are fed does not affect average blood pressure. A competing hypothesis, undoubtedly the one the investigators had in mind, is that the feeding schedule *is* a factor. In Example 10c, one hypothesis is that boys and girls have the same natural proclivity for mathematics, on average. An alternative hypothesis, suggested by the original question, is that boys have the greater proclivity.

The Null
Hypothesis

The hypothesis of "no difference," in some sense, is usually called the **null hypothesis** and denoted by H_0. In Example 10a, Oscar's claim that Felix has no special ability is the null hypothesis. In Example 10b, the null hypothesis is that intermittent feeding has no effect on average blood pressure—that the two populations have the same mean. In Example 10c, the null hypothesis is that on average there is no difference between the mathematical proclivities of boys and girls.

In each of the examples, the data are intended to provide evidence about a null hypothesis. Collecting data and using them as evidence about a null hypothesis is called **testing** the hypothesis.

The Alternative
Hypothesis

Suppose the null hypothesis is not true. What then is true? The **alternative** to H_0, denoted by H_A, is ordinarily either a particular probability distribution (for the population) or a family of distributions. For example, if the null hypothesis in Example 10a is $p = \frac{1}{4}$, the alternative expressing the existence of ESP is that $p > \frac{1}{4}$. When a researcher is conducting a test to establish that there *is* a difference or an effect (that is, that H_0 is not true), the alternative is sometimes called the *research hypothesis*.

Although it is common to have a rather well-defined set of probability distributions in mind for the alternative, there are settings in which the alternative is simply that H_0 is not true. An instance of this is the hypothesis that the stars are distributed at random; in this case, the alternative includes models that are not necessarily probability models.

In planning a test, it is important, before doing any sampling, to be clear as to the population of interest. If it is not possible to run an experiment on the population of interest, it may be possible to use another population and extrapolate the results. For example, one might extrapolate from animals to humans, or from moderately sick or healthy people to very sick people.

How do we get experimental subjects? In hypothesis testing, we assume that the sampling is random. (We have not considered some reasonable but more sophisticated methods such as stratified or clustered sampling schemes.) Obtaining random samples is usually very difficult and may be impossible. Often we must be content with a convenient sample; using this to carry out a statistical test assumes that it is something like a random sample, with a good chance of being "representative."

10.2   Assessing the Evidence

Samples vary from one sample to another. Because of this, the evidence in a particular sample can be misleading. There is no way to be absolutely sure we're not being misled by a sample, unless the sample is the entire population. However, properly designed tests using random samples have a good chance of leading to the right conclusion.

The essence of a statistical test of H_0 is a comparison of actual sample data with what might be expected when H_0 is true. Such comparisons are usually based on the value of a particularly relevant statistic, called the **test statistic**.

The sampling distribution of a statistic depends on what is assumed as the population distribution (see Section 8.4). The sampling distribution of a test statistic under H_0 is called its **null distribution**.

Example 10.2a   ESP (continued)

We return to "The Odd Couple" (Example 10a). Let p denote the probability that Felix correctly identifies a card. The null hypothesis H_0 is that Felix has no ESP and is only guessing or, in terms of the parameter p, that $p = \frac{1}{4}$. A natural test statistic is

Y = number of correct identifications.

(We know from Section 8.3 that Y is sufficient for p.) Suppose Oscar and Felix decided at the outset to conduct exactly ten trials, that the trials are independent, and that the probability p is the same at each trial. Then $Y \sim \text{Bin}(10, p)$. Under H_0 ($p = \frac{1}{4}$) the distribution of Y is given in Table Ia of Appendix 1 as follows:

y	0	1	2	3	4	5	6	7	8	9	10
$f(y)$	.056	.188	.282	.250	.146	.058	.016	.003	.000	.000	.000

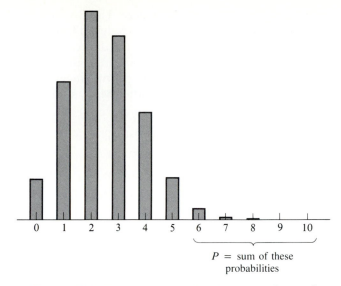

$P =$ sum of these
probabilities

Figure 10.2 Binomial probabilities: $n = 10$, $p = \frac{1}{4}$

Figure 10.2 gives a graph of this distribution. Its mean value is 2.5 ($= np = 10 \times \frac{1}{4}$, from Section 4.2).

Felix's score (six correct) is rather far from 2.5, out in a tail of the null distribution. In this sense, six correct is rather surprising when H_0 is true. A common way of indicating the discrepancy between what we expect and what we observe is to give the tail-probability in the null distribution: the probability of observing six or more correct when H_0 is true. From the table, we have

$$P(6 \text{ or more correct} \mid H_0) = .016 + .003 + .000 + .000 + .000 = .019.$$

That this probability is small tells us that $Y = 6$ is quite far from what we expect when $p = \frac{1}{4}$; the sample results are rather inconsistent with the null hypothesis.

On the other hand, six correct is *not* very surprising if Felix really has some degree of ESP. The evidence in this case appears to favor the conclusion that Felix has ESP. Of course, if you strongly disbelieve in the possibility of ESP, this evidence may not be convincing enough; you would attribute the evidence to chicanery, or perhaps to luck. (To be continued.) ∎

As in the above example, it is common practice to describe the location of the observed result in the null distribution of a test statistic by giving the tail-area or

P-Values

tail-probability beyond the observed value.[3] This probability is called the *P-value*. The smaller it is, the farther the observed value is from the expected value when H_0 is true, and the harder it is to accept the discrepancy as sampling variability. Thus, a very small *P*-value is evidence against the null hypothesis.

How small must a *P*-value be to be convincing evidence against a null hypothesis? This is purely a subjective matter. Many statisticians[4] take the following interpretations as benchmarks:

$P < .01$: Strong evidence against H_0;

$.01 < P < .05$: Moderate evidence against H_0;

$P > .10$: Little or no evidence against H_0.

Statistical Significance

There is a strong tradition that arbitrarily selects the values .01 and .05 as critical levels for *P*-values, using this language: When $P < .05$, the result is called *statistically significant*; and when $P < .01$, the result is called *highly statistically significant*. In many research journals, a statistically significant result is indicated by attaching an asterisk ($*$) to a *P*-value less than .05 and a highly statistically significant result, by attaching a double asterisk ($**$) to a *P*-value less than .01. There seems to be a common perception that the .01 and .05 critical levels have a theoretical basis, but they are in fact arbitrary.

Clearly, stating whether $P < .05$ or $P > .05$ is not as informative as giving the *P*-value itself. Thus, both $P = .049$ and $P = .011$ are described as statistically significant, but the latter is stronger evidence against H_0. A report of statistical significance is quite like the red light on a dashboard that tells you the water temperature has reached a certain arbitrary value; a thermometer that tells you the exact temperature is more useful.

Calculating a *P*-value from given data, or simply determining whether $P < .05$, is called a **significance test**. A *P*-value is sometimes called the *observed level of significance*.

One reason for the importance of significance tests is their championing by the great statistician and geneticist, Sir Ronald A. Fisher (1890–1962). Even Fisher noted: "in fact no scientific worker has a fixed level of significance at which from year to year, and in all circumstances, he rejects hypotheses; he rather gives his mind to each particular case in the light of his evidence and his ideas."[5]

Although testing should not be reduced to a stepwise procedure that is carried out without thinking, a checklist can help. We list the main ingredients and then

---

[3] Summarizing sample results in terms of tail-probabilities is controversial. Many people feel that the sample actually observed is all that is relevant, so that in Example 10.1 the probabilities of 7, 8, 9, and 10 successes in Example 10.1 are irrelevant. In this chapter, we are presenting the predominant view in applied statistics; another view will be presented in Chapter 16.

[4] For an example of dissent, see J. Berger and T. Sellke, "Testing a point null hypothesis: The irreconcilability of *P* values and evidence," *J. Amer. Stat. Assn.* 82 (1987), 112–122.

[5] R. A. Fisher, *Statistical Methods and Scientific Inference*, 3rd ed. (Edinburgh: Oliver & Boyd, 1956), p. 45. Fisher introduced the practice of using 5% and 1% levels, simply to overcome tabulation difficulties that no longer exist in these days of universal computer availability.

discuss them in greater detail:

To carry out a test of significance:

1. Identify the null hypothesis, H_0.
2. Identify H_A, the alternative to H_0.
3. Specify or construct a test statistic Y, one that discriminates between H_0 and H_A.
4. Specify values of Y that are "extreme" under H_0 and in the direction of H_A, suggesting that H_A provides a better explanation of the data than H_0.
5. Obtain data and calculate the corresponding P-value—the probability, assuming H_0, in the tail of the null distribution of Y at and beyond (more extreme than) the observed value of Y. The smaller the P-value, the stronger the evidence against H_0.

Step 1   The hypothesis H_0 is usually a very specific statement about the population or populations of interest. The null hypothesis in most applications is that there is no effect, no special ability, or no difference. Hypotheses are often expressed in terms of population parameters; for example, one might test the null hypothesis that the population mean is zero.

Step 2   The *alternative hypothesis* H_A is a theory or claim that is different from H_0. It may be a hypothesis that an investigator either has a special interest in establishing or would particularly like to guard against.

Step 3   The test statistic Y is especially designed to give evidence about H_0 and discriminate effectively between it and competing hypotheses. When H_0 is phrased in terms of a population parameter, the test statistic is based on a sample estimate of that parameter.

Step 4   Values of the test statistic Y that are in a tail of its null distribution are *extreme* when they tend to support H_A. Suppose H_0 specifies $\theta = \theta_0$, where θ is a parameter indexing the class of possible models. The alternatives $\theta < \theta_0$ and $\theta > \theta_0$ are *one-sided*. The alternative $\theta \neq \theta_0$ is *two-sided*. When values of the test statistic in *both* tails of its distribution are regarded as extreme, the *test* is two-sided; otherwise it is one-sided.

Step 5   The P-value corresponding to an observed result is calculated according to what is considered extreme in Step 4. If large values of Y are extreme and we observe $Y = y$, then the P-value is $P(Y \geq y)$, calculated assuming H_0. Figure 10.3(a) illustrates this in a typical continuous case. If small values are extreme, the P-value is $P(Y \leq y)$, as in Figure 10.3(b). If *both* large and small values of Y are extreme and we observe $Y = y$, we need to know more about the null distribution of Y. In many

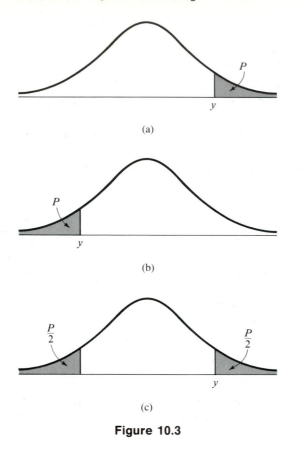

Figure 10.3

practical settings, this distribution is symmetric. When it is, we may give *twice* the tail area beyond $Y = y$ as a two-sided P-value. For example, if Y is in the right tail of its distribution, $P = 2 \cdot P(Y \geq y)$. (See Figure 10.3(c).)

Example **10.2b**   **ESP (conclusion)**

We return to the Odd Couple's experiment (Example 10.2a) and put our earlier analysis in the framework of the five steps given above.

1.   The null hypothesis is that Felix does not have ESP. In terms of the probability of a correct identification, the null hypothesis H_0 is that $p = \frac{1}{4}$.

2.   The alternative to H_0, corresponding to Felix's having ESP, is $H_A: p > \frac{1}{4}$. (A value of p *less* than $\frac{1}{4}$ is not an alternative to $p = \frac{1}{4}$, since Felix claims to do *better* than a random selection.)

3.   The test statistic Y is the number of correct identifications in the sequence of ten trials.

4.   Since H_A specifies large values of p, we take Y-values in the *right* tail of its null distribution as "extreme"—these are more typical under H_A than under H_0.

5. The *P*-value is the probability in the right tail at and beyond the observed number of successes: $Y = 6$. In Example 10.2a, we found this to be $P \doteq .019$, so six successes in ten trials are statistically significant against $p = \frac{1}{4}$. ■

When H_A is of the form $\theta > \theta_0$, the appropriate null hypothesis in some cases may be $\theta \leq \theta_0$, rather than $\theta = \theta_0$. Thus, the null hypothesis concerning a treatment intended to increase an average response is really that the average does *not* increase. (A treatment intended to increase a response would not be worthwhile if it turned out to decrease the average response!) We take the null hypothesis in such cases to be $\theta = \theta_0$ because we need a specific null distribution to calculate a *P*-value. If the data cast doubt on $\theta = \theta_0$, they would cast even more doubt on any θ less than θ_0. So testing $H_0: \theta = \theta_0$ tests $\theta \leq \theta_0$ as well.

One-Sided Versus Two-Sided P-Values

Whether one should use one-sided or two-sided *P*-values is a matter of controversy among some statisticians.[6] Some argue that *all P*-values should be two-sided. They reason that two-sided *P*-values are conservative, because they are larger and therefore indicate less significance. In our view, it would suffice to say what statistic was used and whether the reported *P*-value is one- or two-sided. The reader is then free to double or halve it, as may seem appropriate.

One reason for concern as to whether or not to double a *P*-value lies in the traditional language of significance testing. In that language, a *P*-value such as .04 indicates statistical significance, but the doubled *P*-value .08 does not. Clearly, whether or not *P* is doubled can affect your conclusions if you adhere to the standard benchmarks (.01 and .05).

A result that is not significant is sometimes taken as evidence that the null hypothesis is true. However, a result may be nonsignificant just because the sample was not large enough to reveal a difference that does exist — and one that may be important. (This has implications for the design of a study; sample sizes should be large enough that differences deemed important are likely to be detected. See Section 11.3.)

A common misinterpretation of "significance" is that a statistically significant result has great practical importance. However, a statistically significant deviation may have little or no practical significance! Moreover, a deviation that is significant in practice may not show up as statistically significant (perhaps the sample was not large enough).

Example 10.2c   **Statistical Versus Practical Significance**

A 1965 Alabama court case (*Swain* v. *Alabama*) involved a claim of discrimination against blacks in the selection of grand juries.[7] According to census data, the population proportion of blacks was about 25%, but among those called to appear for jury duty, the percentage of blacks was much smaller. Problem 12 calls for a test of $H_0: p = 25$; the result is a *P*-value of about 6×10^{-10} — so close to 0 as to be very

---

[6] For a more detailed discussion of this controversy see J. Gibbons and J. Pratt, "*P*-values: Interpretation and methodology," *Amer. Statistician 29* (1975), 20–25.
[7] Cited by D. Kaye in "Statistical evidence of discrimination," *J. Amer. Stat. Assn. 77* (1982), 773–783.

convincing evidence that the probability of a black's being called is less than 25%. However, the observed discrepancy did not strike the court as large enough for a *prima facie* case!

The author of the referenced article (a law professor) writes as follows: "That trivial differences can appear statistically significant only underscores the admonition that the *p*-value should not be considered in a vacuum. The courts are not likely to lose sight of the question of practical significance and to shut their eyes to the possibility that the degree of discrimination is itself *de minimis*." ■

A *P*-value is a "probability" *only* in a particular model, H_0, a model that may or may not be true (and is not true more often than true). It is better not to think of the *P*-value as a probability at all, since doing so only leads to confusion and misinterpretation. Rather, regard a *P*-value as a standard type of report of the degree of agreement between the null hypothesis model and the data, a smaller value indicating less compatibility.

Studies often report observations on many variables; some of the *P*-values are likely to be less than .05 even if all of the corresponding hypotheses are true. (See Problem 2.) Even if only two independent *P*-values are calculated, the probability that at least one is less than .05 is $1 - (.95)^2 = .0975$. More generally, the "overall error rate" for n independent tests is $1 - (.95)^n$. Investigators who search their data will assuredly find at least one result that is statistically significant.

How does one properly summarize the results of several experiments, drawing an *overall* conclusion? In making simultaneous inferences and determining overall error rates, it is possible to adjust "critical" levels for individual *P*-values so that the *overall* critical level is .05. For example, the probability that at least one of two *P*-values is less than .0253 is $1 - .9747^2 \doteq .0500$. Although this kind of adjustment is recommended by many statisticians, the cure seems worse than the ailment. For instance, in Example 10b, why should the mere fact that heart rates were measured, as in fact they were, have any bearing on the conclusions about blood pressure? This is a severe limitation of significance testing. Scientists have to decide subjectively how to weigh various experiments and their conclusions, and *P*-values are but one part of this process.

In the examples and problems to follow, we carry the analysis to the point of calculating a *P*-value, sometimes adding the words "strong evidence," or "some evidence," or "little or no evidence" against H_0, according to the benchmarks we've described above. We do not go farther and say which hypothesis you should regard as being true; this would take us into subjective interpretations, and in particular, into considerations of the particular field of application.

10.3   One-Sample Z-Tests

Many test statistics in common use have null distributions that are approximately *normal* when the sample size is large. If the null distribution of a test statistic Y is approximately normal, the null distribution of the standard score

$$Z = \frac{Y - E(Y \mid H_0)}{\text{s.d.}(Y \mid H_0)} \tag{1}$$

is approximately standard normal. The null hypothesis is ordinarily specific enough (for example, $\mu = \mu_0$) that the mean $E(Y \mid H_0)$ is a known constant; but the denominator can involve a parameter that may or may not be known.

If s.d.$(Y \mid H_0)$ is known, Z is a statistic, and we can use it in place of Y as a test statistic. But if s.d.$(Y \mid H_0)$ is not known, we have to approximate the s.d. in the denominator using sample data, substituting the standard error of Y for its standard deviation. In such cases, define

$$T = \frac{Y - E(Y \mid H_0)}{\text{s.e.}(Y \mid H_0)}. \tag{2}$$

When n is large, the null distribution of T is also approximately standard normal.

In both (1) and (2), the standard score measures the discrepancy between the observed Y and $E(Y \mid H_0)$, the Y we "expect" when H_0 is true. A Z- or T-score is unitless, being simply a number of s.d.'s [in (1)] or s.e.'s [in (2)]. [The units for the quantities in the numerator and denominator of (1) and (2) cancel.]

Now, suppose Y is approximately normal and it turns out that $Z = 1$. This means that the observed Y is about one standard deviation away from what is expected under H_0. Such a deviation is typical when H_0 is true and therefore does not provide much evidence against H_0. The one-sided P-value for $Z = 1$ (when H_A makes positive Z's likely) is

$$1 - \Phi(1) = .1587.$$

A P-value this large is not much evidence against H_0. On the other hand, if $Z = 3$, say, then the observed Y is three standard deviations from its mean under H_0. The one-sided P-value is .0013, the area to the right of $Z = 3$. So $Z = 3$ is usually considered as strong evidence against H_0.

Example 10.3a   **Average Speed Reduced?**

Over a long period of time in which the posted speed limit was 65 mph, the average speed along a certain stretch of highway was found to be 63.0 mph. After the speed limit dropped to 55 mph, the average speed of cars in a sample of size 100 was found to be 61.4 mph, and the standard deviation, $S = 4.6$ mph. Does this indicate a genuine reduction in mean speed, or could it simply be sampling variability? We follow the five-step outline of the preceding section:

1. The null hypothesis is H_0: $\mu = 63.0$.
2. The alternative hypothesis is $\mu < 63.0$, since setting the new limit was intended to reduce the average speed.
3. The test statistic is given by (2), calculated from the given data:

$$Z = \frac{61.4 - 63.0}{4.6 / \sqrt{100}} = -3.48.$$

4. Values of Z in the left tail of its null distribution are considered extreme, in view of H_A.
5. The P-value is the area in the tail of the null distribution of Z, the area under a standard normal curve to the left of -3.48:

$$P \doteq \Phi(-3.48) = .0002.$$

The evidence against H_0 is quite strong. The alternative explanation, that the average speed is lower, is more plausible. (Whether this lowering was *caused* by the new speed limit is another matter; the time periods were different, so there may be other factors involved.)

When such a result is reported as statistically significant, the media will often report that "a significant reduction" has been achieved. The language suggests that the mean speed has been lowered by an amount that has important consequences. Even if the 1.6 mph reduction is real, however, it may not be important. A statistically significant result signifies only that the sample result is far from what is expected under the null hypothesis.

Rather than simply claiming statistical significance, it may be better to give an interval estimate of the mean reduction. The observed mean 61.4 is an estimate of the new average speed, and $S/\sqrt{n} = .46$ is the standard error of estimate. For the new average speed, 95% confidence limits are $61.4 \pm 1.96 \times .46$, or $61.4 \pm .90$. (See Section 9.3.) This interval, from 60.5 to 62.3, excludes not only 63 but values near 63 as well. ∎

Significance test of H_0: $\mu = \mu_0$, based on the mean of a large random sample: Calculate

$$Z = \frac{\bar{X} - \mu_0}{\sigma/\sqrt{n}} \left(\text{or } Z \doteq \frac{\bar{X} - \mu_0}{S/\sqrt{n}} \right), \tag{3}$$

and find the P-value in Table II of Appendix 1:

$P = \Phi(Z)$ for H_A: $\mu < \mu_0$,

$P = 1 - \Phi(Z)$ for H_A: $\mu > \mu_0$,

$P = 2\Phi(-|Z|)$ for H_A: $\mu \neq \mu_0$.

Testing a Proportion

We know from Section 8.7 that a sample proportion is approximately normal when the sample size is large. So a Z-score based on a sample proportion provides a large-sample test for the sample test for the corresponding population proportion. The Z-score we use is a special case of (3)—the version with σ. A special feature of Bernoulli populations is that the population variance depends on the population proportion: $\sigma^2 = p(1 - p)$. So when we assume $p = p_0$, there is no need to approximate σ^2 from the sample; we simply set $\sigma^2 = p_0(1 - p_0)$.

Example 10.3b

Male Births and Timing of Conception

A long-standing theory holds that a greater proportion of male births occur when conception is late in the woman's fertile period. One investigation,[8] which studied thousands of births, reported 145 cases in which conception was reckoned as having occurred two days after ovulation. Among these there were 95 male births: $\frac{95}{145} = .655$. The proportion of male births among the rest was .520. Is this evidence

---

[8] S. Harlap, "Gender of infants conceived on different days of the menstrual cycle," *New Engl. Jour. Med*, 300 (1979), 1445–1448. Subjects of the study were Orthodox Jewish mothers, whose religious practices facilitated the documentation.

that the probability of a male child is higher when conception occurs late? Or is the sample ratio .655 attributable to sampling variations? To address this question, we'll test the hypothesis $H_0: p = .520$, where p is the proportion of male births among those who conceive two days after ovulation. (To be continued.) ■

The mean and variance of a sample proportion $\hat{p}$, according to (6) and (7) of Section 8.7, are

$$E(\hat{p}) = p \qquad \text{and} \qquad \text{var } \hat{p} = \frac{p(1-p)}{n}. \tag{4}$$

To test the hypothesis $p = p_0$, we use a Z-score of the form (1) based on the sample proportion [with $p = p_0$ in (4)]:

$$Z = \frac{\hat{p} - p_0}{\sqrt{p_0(1-p_0)/n}}. \tag{5}$$

[Compare (1).] The Z-score (5) is approximately standard normal when $p = p_0$, so once again we look to Table II in Appendix 1 for the corresponding P-value.

The Z-score (5) can also be found by standardizing $Y = n\hat{p}$, the *number* of successes in the sample. The variable Y is binomial with mean np and variance $np(1-p)$. The corresponding Z-score is

$$Z = \frac{Y - np_0}{\sqrt{np_0(1-p_0)}}. \tag{6}$$

This is algebraically identical with (5): Multiply both numerator and denominator of (5) by n to obtain (6).

Large-sample test for $H_0: p = p_0$: Calculate the Z-score for the observed $\hat{p}$ using (5). Find the P-value in Table II of Appendix 1:

$P = \Phi(Z)$ for $H_A: p < p_0$,

$P = 1 - \Phi(Z)$ for $H_A: p > p_0$,

$P = 2\Phi(-|Z|)$ for $H_A: p \neq p_0$.

Example 10.3c **Male Births (continued)**

We return to Example 10.3b to test the hypothesis $p = .520$. The proportion of boys in the sample of size $n = 145$ is $\hat{p} = .655$. The five steps of Section 10.2 are:

1. The null hypothesis is $H_0: p = .520$, where p is the probability of a male child when conception occurs two days after ovulation.
2. The alternative hypothesis is the long-standing theory that the probability of a male child is greater when conception is late in the fertile period, $H_A: p > .52$.
3. The test statistic is $Y = 95$, the number of male births in the sample, or equivalently, $\hat{p} = Y/n = .655$. The Z-score can be calculated from (5):

$$Z = \frac{.655 - .520}{\sqrt{.52 \times .48/145}} = 3.25.$$

or from (6):

$$Z = \frac{95 - 145 \times .520}{\sqrt{145 \times .52 \times .48}} = 3.25.$$

4. In view of H_A, large values of Z are considered extreme.
5. The one-sided P-value for the given data is the area in the tail of the normal distribution beyond $Z = 3.25$:

$$P = 1 - \Phi(3.25) = .0006.$$

The evidence strongly favors H_A. ∎

10.4   *t*-Tests

In a Z-test for $\mu = \mu_0$ (Section 10.3), the large sample size allowed us to assume (i) that the sample mean is approximately normal, and (ii) that the sample s.d. is a good approximation to the population s.d. As explained in Section 9.4, when the sample sizes are not large, (ii) is more worrisome than (i). For example, $\bar{X}$ can be fairly close to normal even when n is as small as 5, depending on the population distribution. (Indeed, $\bar{X}$ is exactly normal for *any* n if the population is normal.) However, the sample standard deviation S is quite unreliable as an estimate of σ when n is as small as 5.

In this section, we consider tests involving the population mean μ when σ is unknown and n is not large. To test $H_0: \mu = \mu_0$, we again measure the discrepancy between $\bar{X}$ and μ_0 in terms of standard errors, using the ratio

$$T = \frac{\bar{X} - \mu_0}{S/\sqrt{n}}. \tag{1}$$

When the sampling is random and the population is normal, the null distribution of T is $t(n-1)$, as we saw in Section 9.4. Table III in Appendix 1 gives percentiles and tail-areas of the t-distribution.

Although populations encountered in practice may be nearly normal, you can never be certain that one is *exactly* normal. The statistic T in (1) usually does not have a t-distribution when the population is nonnormal. However, probabilities given in Table III are nearly correct if the population is not far from normal—if it has a single hump, has tails that are not too heavy, and is not too skewed. So procedures based on T and the t-table work rather well when the population is only moderately different from normal. They are said to be *robust* with respect to the assumption of population normality.

Robustness
of *T*

Example 10.4a   Ventricular Premature Beats

To study the effectiveness of a drug[9] in reducing ventricular premature beats (VPB's), a clinician administered the drug intravenously to ten patients. The

---

[9] Private communication.

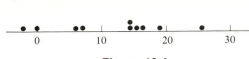

Figure 10.4

following are reductions (VPB's per minute) after a prescribed interval following administration of 2 mgs/kg:

0, 7, −2, 14, 15, 14, 6, 16, 19, 26.

A dot plot is shown in Figure 10.4. The mean and s.d. are $\bar{X} = 12.4$ and $S = 7.763$.

1.   The null hypothesis is that the mean reduction is $\mu = 0$.
2.   The alternative is one-sided: $\mu > \mu_0$.
3.   The test statistic is

$$T = \frac{\bar{X} - \mu_0}{S/\sqrt{n}} = \frac{12.4 - 0}{7.763/\sqrt{10}} = 5.05.$$

4.   In view of H_A, large values of T are extreme.
5.   Under H_0, $T \sim t(9)$, and from Table IIIb, $P < .001$.

The result is statistically significant: the observed mean reduction is not easy to account for as a phenomenon of random sampling. The evidence against H_0 is strong.   ■

Test of $H_0: \mu = \mu_0$ when σ^2 is unknown: Calculate

$$T = \frac{\bar{X} - \mu_0}{S/\sqrt{n}} \qquad\qquad (1)$$

and find the P-value in Table IIIb, entering at $n - 1$ degrees of freedom (d.f.). The population should be close to normal if $n < 40$.

   Is the population close to normal in Example 10.4a? Nonnormality is very hard to detect with a sample size as small as 10, unless there is a gross deviance from normality. For Example 10.4a, the plot in Figure 10.4 reveals neither skewness nor heavy tails that would bring the *t*-test into question. In Chapter 13, we'll consider some formal tests for normality. In Section 10.5, we'll take up a test that does not assume a normal population. The following example shows what nonnormality can do to a *t*-test.

Example 10.4b   The average normal heart rate for pigs is considered to be 114 beats/min, according to a 1984 research paper.[10] The paper gives the heart rates of four pigs under

---

[10] "Xylazine-Ketamine-Oxymorphone: An injectable anesthetic combination in swine," *J. Amer. Vet. Med. Assn.* (1984), 182–184.

anesthesia:

116   85   118   118.

Let's test the null hypothesis $\mu = 114$ against the alternative that $\mu \neq 114$. Large values of $|T|$, from (1) with $\mu_0 = 114$, are then extreme. For the given data, $\bar{X} = 109.25$ and $S = 16.194$, so

$$T = \frac{109.25 - 114}{16.19/\sqrt{4}} \doteq -.59,$$

with $4 - 1 = 3$ d.f. This is not even close to being statistically significant.

Suppose the second pig (the 85) had not been included in the sample. (Perhaps it did not qualify for some reason, or just had not been part of the sample.) This leaves us with only 116, 118, and 118. Now $\bar{X} = 117.33$, $S = 1.155$, and $T = 5.00$! (The mean is now actually closer to 114 and flipped to the other side. The reason $|T|$ is now so much larger is that S has dramatically decreased.) Leaving out one observation, we find very strong evidence *against* $\mu = 114$: $P \doteq .019$. This is peculiar, since the observation omitted should *add* to the evidence against H_0—it's the one farthest from 114!

Although $n = 4$ is really too small to draw definite conclusions, the observation 85 suggests that the population distribution may have a long left tail. This would be an important violation of the assumption of normality, and the t-test would not be appropriate. ∎

The result in the preceding example reflects the fact that the standard deviation S is much more dramatically affected by a large deviation than is $\bar{X}$. Consider H_0: $\mu = \mu_0$, and a particular sample of size n for which T is as large as you please, so that the P-value is small. Then take any single observation and let it get large. This increases $\bar{X}$. One might expect an even smaller P-value, but S also increases. The net result is that as the one observation grows, T eventually starts to *decrease* and in fact tends to 1! So the (one-sided) P-value grows to about .16.

The moral here is the same as in the example: Don't use the t-test when there are *outliers*—observations so unusual in size (either large or small) as to give them undue influence in the analysis. Outliers may be in error; if so, they should be deleted. If they are genuine, or if their legitimacy is in doubt, a t-test is not appropriate.

10.5   Some Nonparametric Tests

The tests considered so far have limited applicability. They have assumed either very large samples or, when sample sizes are only moderate and σ is unknown, populations that are not too different from normal populations. In this section, we take up some tests concerning location that require fewer assumptions.

Example **10.5a**   **Marijuana and Proficiency**

A study[11] on the effects of smoking marijuana included the following data. Each of nine subjects was given a "digit substitution test," before and again 15 minutes after a marijuana smoking session. Changes in the number correct from the baseline number (i.e., number correct after smoking minus number correct before) were reported as follows:

$$+5, -17, -7, -3, -7, -9, -6, +1, -3.$$

We can analyze these data using the t-test of Section 10.4. But suppose we don't want to assume normality? Then the t-test is not appropriate. Instead, we focus on whether the predominance of negative changes in the data set indicates a real effect or is simply random variation. (Even if marijuana has no effect, the inherent variability in testing will produce changes from before to after.) Under the hypothesis of no effect, the probability of an increase is $\frac{1}{2}$: we'd expect about as many negative changes ($-$'s) as positive changes ($+$'s). Let $p = P(+)$:

1.   The null hypothesis H_0 is that the marijuana has no effect, or $p = \frac{1}{2}$.
2.   The alternative H_A is that scores tend to get worse after marijuana smoking: $p < \frac{1}{2}$.
3.   The test statistic is Y, the number of $+$'s in the sequence of scores; with random sampling, its null distribution is binomial.
4.   Under H_A, we'd expect an unusually small number of $+$'s, so small values of Y are extreme.
5.   For the given data, $Y = 2$. The P-value is the probability of $Y \le 2$ under H_0:

$$P(Y \le 2 \mid p = .5) = \frac{\binom{9}{2} + \binom{9}{1} + \binom{9}{0}}{2^9} = \frac{46}{512} = .0898.$$

[Alternatively, entering Table Ib with $n = 9$ and $p = .5$, we find $P(Y \le 2) = 1 - P(Y \ge 3) = 1 - .9102 = .0898$.] This analysis provides little evidence against H_0. ∎

As in the preceding example, let $p = P(X > 0) = \frac{1}{2}$. When X is continuous, $P(X < 0) = \frac{1}{2}$, so testing $p = \frac{1}{2}$ is the same as testing that the population median is 0.

More generally, we can test the hypothesis that the population median is m_0 by using as a test statistic the number of sample observations greater than m_0. Consider a random sample $(X_1, \ldots, X_n)$ and denote by Y the number of X_i's greater than m_0—the number of "successes" in n independent trials, where "success" means $X_i > m_0$. Then $Y \sim \text{Bin}(n, p)$, where $p = P(X_i > m_0)$. If the median is indeed m_0, then $p = \frac{1}{2}$. Testing H_0: $p = \frac{1}{2}$ is called a *sign test* for the hypothesis that the population median is m_0. When a population is symmetric, its median is equal to its mean; in this case testing $m = m_0$ is equivalent to testing $\mu = m_0$.

When X is continuous, $P(X = m_0) = 0$, but in practice, zero differences can result from rounding. Such observations carry no information about the hypothesis

Sign Test
for $m = m_0$

---
[11] Reported in *Science 162*, 1234 (1968).

that positive and negative deviations are equally likely. So in applying the sign test, we ignore zeros and use the reduced sample consisting of the nonzero differences.

Sign test for H_0: $m = m_0$:

When there are n nonzero differences $X_i - m_0$ in a random sample, let Y denote the number of positive differences. Under H_0, $Y \sim \text{Bin}(n, \frac{1}{2})$. When $n \leq 12$, find P-values in Table I; when $n > 12$, calculate the standardized Y and find approximate P-values in Table II.

A useful property of the sign test is that the shape of the population distribution is immaterial, even when the sample size is small. The null distribution of the test depends only on n. In particular, it does not depend on the population distribution—it is **distribution-free**. Also, the distributions included under $H_0 \cup H_A$ are not a parametric family, so the sign test is a **nonparametric** test.

The sign test uses only the algebraic sign of the deviations from m_0, ignoring the numerical values of the deviations. Therefore, one might expect it to be less sensitive (in some sense) than a test that exploits the sizes of the deviations. For instance, suppose the observations are 24, 33, -3, -1, 29. Three of these are positive, so $Y = 3$, and the P-value for testing H_0: $m = 0$ against H_A: $m > 0$ is $P = \frac{5}{16}$—which is not significant. However, the sizes of these deviations from 0 suggest more of an imbalance than do their signs: those that are negative are small in magnitude, and those that are positive are quite large. (Using a t-test for H_0: $\mu = 0$ gives $T = 2.14$ and a one-sided P-value of .049.)

Signed-Rank Statistic

Wilcoxon's **signed-rank** test is a refinement of the sign test that takes the relative sizes of the deviations $X - m_0$ into account, but not their precise values. To carry out this test, we first order the deviations $X - m_0$ by their *magnitudes* and then assign *ranks*. To illustrate, consider again the observations 24, 33, -3, -1, 29. To test $m = 0$, these are the deviations $X - m_0$ arranged by magnitudes:

$$-1, \ -3, \ 24, \ 29, \ 33.$$

We then assign ranks 1, 2, 3, 4, 5 from smallest to largest. The two negative deviations have ranks 1 and 2, and the *sum* of their ranks is 3. This sum is the **signed-rank statistic**:

$$R_- = \text{sum of ranks of negative deviations.}$$

We could work equally well with R_+, the sum of the ranks of the positive deviations, since the sum of all the ranks is a constant—the sum of the first n integers:

$$R_- + R_+ = 1 + 2 + \cdots + n = \tfrac{1}{2}n(n + 1). \tag{1}$$

So $R_+ = \tfrac{1}{2}n(n + 1) - R_-$. We use whichever is convenient.

Signed-rank statistics for $H_0: m = m_0$:

$R_+ = $ sum of ranks of positive deviations $X - m_0$,

$R_- = $ sum of ranks of negative deviations $X - m_0$,

where the deviations have been ordered by magnitude and ranked from smallest to largest.

The statistic R_- will tend to be small when there are very few negative deviations or when the deviations that are negative are among the smallest in magnitude. The smallest possible value of R_- is 0 (all deviations positive), and it can be as large as $\frac{1}{2}n(n + 1)$ (all deviations negative).

Distribution of R_-

In order to calculate the P-value for the signed-rank test, we need to know the null distribution of R_-. The value of R_- depends only on the pattern of $+$'s and $-$'s in the sequence of observations arranged by magnitude. There are 2^n such patterns. To find their probabilities (under H_0), we require symmetry of the population—a much weaker assumption than normality. Symmetry implies that the observation smallest in magnitude is as likely to be $+$ as $-$. Given the smallest observation, the second smallest in magnitude is as likely to be $+$ as $-$, independent of whether the sign of the smallest is $+$ or $-$. Continuing in this fashion, we see that the 2^n sequences are equally likely. So the probability of any given value of R_- is the number of sign patterns in which R_- has that value, divided by 2^n.

When n is small, we simply list all 2^n sequences, find the value of R_- for each, and count. When n is large (say $n \geq 5$), it is practical to list only those sequences with the smallest values of R_- and count to find the probabilities for these small values. Table 10.1 gives the beginnings of such extreme sequences and the corresponding value of R_-. (Dots indicate that the omitted signs are all $+$.) From this table, it is easy to count the number of the 2^n sequences for each value of R_- up to 5. Dividing by 2^n, we obtain the probabilities in Table 10.2.

Table 10.1

Sequence	R_-	Sequence	R_-
$+ \ + \ + \ + \ + \ + \ \cdots$	0	$+ \ + \ + \ - \ + \ + \ \cdots$	4
$- \ + \ + \ + \ + \ + \ \cdots$	1	$- \ + \ - \ + \ + \ + \ \cdots$	4
$+ \ - \ + \ + \ + \ + \ \cdots$	2	$+ \ + \ + \ + \ - \ + \ \cdots$	5
$+ \ + \ - \ + \ + \ + \ \cdots$	3	$+ \ - \ - \ + \ + \ + \ \cdots$	5
$- \ - \ + \ + \ + \ + \ \cdots$	3	$- \ + \ + \ - \ + \ + \ \cdots$	5

Table 10.2

r	0	1	2	3	4	5
$P(R_- = r)$	$\dfrac{1}{2^n}$	$\dfrac{1}{2^n}$	$\dfrac{1}{2^n}$	$\dfrac{2}{2^n}$	$\dfrac{2}{2^n}$	$\dfrac{3}{2^n}$

By symmetry, the null distribution of R_+ is the same as that of R_-. This means [according to (1)] that we also know the probabilities in the right tail of the distribution for R_-:

$$P\left[R_- = \frac{n(n+1)}{2} - c\right] = P(R_+ = c) = P(R_- = c). \tag{2}$$

The null distribution of R_- is symmetric about its mean, so

$$E(R_-) = E(R_+) = \frac{n(n+1)}{4}. \tag{3}$$

Unfortunately, counting can be rather difficult for values of R_- that are not the most extreme. Table VI gives *cumulative* left-tail probabilities for samples of sizes up to $n = 15$. Right-tail probabilities are easily found using (2), but we can always use the left tail if we work with the smaller of the statistics R_- and R_+.

For larger samples, we exploit the fact (which we state without proof) that R_- is approximately normal, and use a Z-test based on R_-. To calculate the standard score Z, we need to know the mean and variance of R_-.

The mean is given by (3) and applies also to R_+. The formula for variance[12] is less obvious:

$$\text{var}(R_-) = \text{var}(R_+) = \frac{1}{24}n(n+1)(2n+1). \tag{4}$$

Signed-rank test for H_0: $m = m_0$ in a symmetric population:

For H_A: $m > m_0$, small values of R_- are extreme;

for H_A: $m < m_0$, small values of R_+ are extreme.

For $n \leq 15$, find P-values in Table VI. For $n > 15$, use Table II and the Z-score:

$$Z = \frac{R_- - \frac{1}{4}n(n+1)}{\sqrt{\frac{1}{24}n(n+1)(2n+1)}}. \tag{5}$$

The null distribution of the signed-rank statistic does not depend on the form of the population distribution, provided the latter is symmetric. Being distribution-free and not directed at parametric alternatives, the signed-rank test is *nonparametric*.

Example 10.5b   **Marijuana and Proficiency (continued)**

In Example 10.5a, the changes in test score from before to after a period of marijuana smoking were

$$+5, \ -17, \ -7, \ -3, \ -7, \ -9, \ -6, \ +1, \ -3.$$

---

[12] See Lindgren [15], p. 509.

We order these according to magnitude:

$$+1, -3, -3, +5, -6, -7, -7, -9, -17,$$

and assign ranks:

$$1, \quad 2, \quad 3, \quad 4, \quad 5, \quad 6, \quad 7, \quad 8, \quad 9.$$

The ranks of the two positive changes ($+1$ and $+5$) are 1 and 4. The test proceeds as follows:

1.   The null hypothesis is that marijuana smoking has no effect, or that the median change is 0: $m = 0$.
2.   The alternative hypothesis is that the median change is negative: $m < 0$.
3.   We take the test statistic to be R_+, the sum of the ranks of the positive scores: $R_+ = 1 + 4 = 5$. (We use R_+ because it's smaller than R_-.)
4.   Under H_A, we'd expect negative deviations from baseline scores, so small values of R_+ are extreme.
5.   The P-value corresponding to $R_+ = 5$ is $P = P(R_+ \leq 5)$, and in Table VI (with $n = 9$ and $c = 5$) we find $P = .020$. Alternatively, P is the sum of the probabilities shown in Table 10.2 with $n = 9$: $P(R_+ \leq 5) = \frac{10}{512} = .0195$.

For comparison with the exact P-value in Step 5, we'll approximate P using the normal table. To find the Z-score for $R_+ = 5$, we need the mean and variance of R_+ from (3) and (4):

$$E(R_+) = \frac{9 \cdot 10}{4} = 22.5, \qquad \text{var}(R_+) = \frac{9 \cdot 10 \cdot 19}{24} = 71.25.$$

Then

$$Z = \frac{5 - 22.5}{\sqrt{71.25}} = -2.07,$$

and $P \doteq \Phi(-2.07) = .0192$ (from Table II). (The normal approximation is usually not this accurate for such a small sample size.)

The P-value from the sign test is $P = .0898$ (Example 10.5a). That this is much larger than .020 reflects the greater sensitivity of the signed-rank test.

For yet another comparison, consider the t-test. (Although the observation -17 may seem a bit out of line, the population distribution could be close enough to normal.) The t-statistic is

$$T = \frac{\bar{X} - 0}{S/\sqrt{n}} = \frac{-5.111}{6.254/\sqrt{9}} = -2.45.$$

From Table IIIb (8 d.f.) we find $P = .020$, coincidentally the same P as for the signed-rank test.

In an experiment such as this, where subjects are given a test before and after a treatment, one might worry that there is learning—that scores would be better the second time around with no marijuana. Since the subjects tended to do *worse* on the second test, a conclusion against the null hypothesis seems conservative in this example.   ■

Sign Test
Versus
Signed-Rank
Test

The sign test, which employs a cruder summary of data than the signed-rank test, is the less sensitive. On the other hand, the sign test has advantages related to its simplicity. For one thing, calculations are easy. Moreover, it can be used when one can say whether a difference is positive or negative, but assigning specific numerical values is difficult. (For example, in judging the quality of fruit or wine, one may be able to say which of two is the better but not want to assign numerical ratings.) If X is survival time, we can stop observing at m_0, knowing that those surviving at m_0 will have positive deviations from m_0.

10.6 Testing a Population Variance

In Section 8.10 we gave the distribution of the sample variance S^2 for the case of normal populations. With this distribution, we can test the hypothesis that the population variance has a specified value. The need for such a test arises in settings such as process control, in which excessive product variability is undesirable (see Section 9.7).

Consider a random sample of size n from $N(\mu, \sigma^2)$ for testing the hypothesis

$$H_0: \sigma^2 = \sigma_0^2.$$

We'll compare the hypothesized value of σ^2 with S^2, using their ratio. To be precise, we use the statistic defined in (4) of Section 9.7:

$$Y = \frac{(n-1)S^2}{\sigma_0^2}. \tag{1}$$

We saw in Section 9.7 that the null distribution of (1) is $\chi^2(n-1)$. For the alternative $H_A: \sigma^2 > \sigma_0^2$, large values of Y are extreme. So the P-value is the area under the chi-square density to the right of the observed value of Y. For the alternative $H_A: \sigma^2 < \sigma_0^2$, the P-value is the area to the left of the observed Y. If the alternative is $H_A: \sigma^2 \neq \sigma_0^2$, a value of (1) in *either* tail is evidence against H_0, but the null distribution of (1) is not symmetric (except in the limit as $n \to \infty$). So there is some question as to what *extreme* should mean. One approach to a two-sided P-value is to double the area beyond the observed Y, just as though χ^2 were symmetric.

Example **10.6a**   The experiment referred to in Problem K of Chapter 9 was conducted to check the variability in explosion times of detonators. In each of five "shots," 14 detonators were fired. Times to detonation in microseconds for one run were as follows:

2.689	2.677	2.675	2.691	2.698	2.694	2.702
2.698	2.706	2.692	2.691	2.681	2.700	2.698

The sample standard deviation is .0092.

Suppose the specifications state that $\sigma \leq .007$. To see if this is met, we'll use the data to test $H_0: \sigma = .007$ against $H_A \; \sigma > .007$. The test statistic (1) is

$$\frac{13 \times (.0092)^2}{.007^2} = 22.8.$$

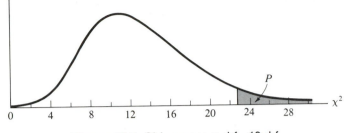

Figure 10.5 Chi-square p.d.f., 13 d.f.

If we assume that the population of detonation times is normal, the null distribution of (1) is $\chi^2(13)$, shown in Figure 10.5. In Table Va in Appendix 1 (at 13 d.f.) we find $P = .044$.   ∎

To test $H_0 : \sigma = \sigma_0$, calculate

$$Y = \frac{(n-1)S^2}{\sigma_0^2}.$$

Under H_0, $Y \sim \chi^2(n-1)$. For the alternative $\sigma > \sigma_0$, large values of Y are extreme. Against $\sigma \neq \sigma_0$, both large and small Y's are extreme.

Chapter Perspective

Significance tests are extensively used in every field that applies statistical ideas. Since they are easily misinterpreted, they are widely misunderstood, so it is important to understand what they are and what they are not.

P-values serve to standardize the reporting of statistical results on a scale from 0 to 1. The closer a P-value is to 0, the more evidence there is against the null hypothesis. However, the exact meaning of $P = .03$ (for example) is not clearly understood by users of statistics generally. A P-value is often incorrectly interpreted as the probability of H_0, although the framework of significance testing makes no provision for assigning probabilities to hypotheses. (In Chapter 16, we'll present an approach to statistical inference which does associate probabilities with parameters and with hypotheses.)

P-values are not always uniquely defined, for given data. The P-value corresponding to a given set of data can depend on what alternative an investigator has in mind (should it be doubled or not?), and on how he or she decided to stop sampling (see Problem 13).

A conclusion that H_0 is false or true, based on the information in a sample, is subject to error. This point is often overlooked in popular reporting of statistical analyses. The next chapter considers the errors one may make in drawing such conclusions.

The point of view of this chapter is that experimental results are published or otherwise placed on file to be assimilated along with previous or subsequent evidence. However, it is difficult or impossible to make coherent comparisons or combinations of P-values from different experiments. A simple numerical comparison of P-values does not tell the whole story. (See Problem 10.)

In the next chapter, we take up a type of inference related to significance testing. This approach treats statistical inference as a decision problem. Indeed, in some applications, data are used as an aid in making decisions. The decision approach also has relevance in purely scientific settings where there is not a clear-cut decision problem. With it, we can study the sensitivity of a test in detecting a treatment effect of specified size.

Solved Problems

Sections 10.1–10.3

A. Test the hypothesis that the mean number of arrivals in a Poisson arrival process is $m = 1$ against the alternative $m > 1$, given the following numbers of arrivals in ten successive unit time intervals:

2, 0, 1, 4, 0, 2, 2, 6, 3, 0.

Solution:

The null hypothesis is $H_0: m = 1$. The sample sum, the number of arrivals in ten time units, is sufficient for m [Problem 9(c) of Chapter 8]. Its distribution is Poi(10m). Since the one-sided alternative is $m > 1$, we consider large values of the sum to be extreme. The total number observed in ten periods is $\sum X = 20$. Under H_0, $\sum X \sim$ Poi(10), and we find the P-value from Table IV:

$$P\left(\sum X \geq 20\right) = 1 - P\left(\sum X \leq 19\right) = 1 - .997 = .003.$$

(The probability of $\sum X \leq 19$ is the entry in Table IV in the column for $m = 10$ opposite $c = 19$.) Since $P < .01$, the result is highly statistically significant.

B. Consider testing $H_0: \theta = 0$ against $H_A: \theta > 0$ in the Cauchy population with p.d.f.

$$f(x \mid \theta) = \frac{1/\pi}{1 + (x - \theta)^2}.$$

Find the P-value for each of the following:
(a) We take one observation and find $X = 4$.
(b) We take 100 independent observations and find $\bar{X} = 4$.
(c) We obtain three independent observations and find the median to be $\tilde{X} = 4$.

Solution:

(a) The c.d.f. is $F(x \mid \theta) = \frac{1}{2} + \frac{1}{\pi} \text{Arctan} (x - \theta)$. The P-value is the probability beyond $x = 4$ when $\theta = 0$:

$$P(X > 4 \mid \theta = 0) = 1 - F(4 \mid 0) = \frac{1}{2} - \frac{1}{\pi} \text{Arctan } 4 = .078.$$

(b)   The distribution of the mean of a random sample from a Cauchy population is the same as the distribution of a single observation [according to Solved Problem Q of Chapter 8]. So again $P = .078$.

(c)   When $n = 3$, the median is the second smallest observation:

$$\tilde{X} = X_{(2)}.$$

We find its c.d.f. in terms of the population c.d.f. as follows:

$$P(\tilde{X} \leq x) = F_{\tilde{X}}(x) = P(2 \text{ or } 3 \ X\text{'s} \leq x) = [F(x)]^3 + 3[F(x)]^2[1 - F(x)].$$

When X is Cauchy with $\theta = 0$, we have [from (a)]

$$F(4\,|\,0) = 1 - P(X \geq 4\,|\,\theta = 0) = 1 - .078 = .922.$$

So
$$P = 1 - F_{\tilde{X}}(4\,|\,0) = 1 - [(.922)^3 + 3(.922)^2(.078)] = .0173.$$

C.   A college administrator takes a random sample of 100 from more than 20,000 applicants. The sample average ACT score is 21.6 and the standard deviation is $S = 4.9$. The average score nationwide (1987) was 20.2. Is the mean score of all applicants to that college higher than the national average?

Solution:
We can't answer the question, but we can test the hypothesis $\mu = 20.2$ against the alternative $\mu > 20.2$. Large values of $\bar{X}$ are extreme. The Z-score of the sample mean is

$$Z = \frac{21.6 - 20.2}{4.9/\sqrt{100}} = \frac{1.4}{.49} \doteq 2.9.$$

The P-value is about .002. The observed sample mean is "highly statistically significant"—strong evidence against $\mu = 20.2$.

D.   Called to jury duty, one of the authors noticed that the panel of about 80 prospective jurors was all white. The population from which the panel was presumably chosen includes about 10% blacks. Is there evidence that there is bias in the selection process? (Assume random sampling. The selections are supposed to be made at random from various lists, although one who is selected may be excused from duty for valid reasons.)

Solution:
Interpreting this result is problematical. Every sample has some feature that will seem unusual. Merely finding an unusual feature is not evidence against random sampling. Suppose, prior to assembling with the rest of the panel, we had in mind to observe the number of blacks selected and use that number in a test of $H_0 : p = .1$ versus $H_A : p < .1$. The one-sided P-value for the result we found (0 out of 80) is

$$P = P(0 \text{ blacks}\,|\,p = .1) = (.9)^{80} \doteq .0002.$$

E.   The "normal deviates" in Table XV are intended to simulate successive, independent observations Z from $N(0, 1)$. If they do, then the probability of obtaining a negative Z is $p = \frac{1}{2}$. Count the negative entries among the 500 observations on the first page of the table and test the hypothesis $p = \frac{1}{2}$.

Solution:

We counted 264 minus signs: $\hat{p} = \frac{264}{500} = .528$. The Z-score is

$$Z = \frac{.528 - .500}{\sqrt{.5 \times .5/500}} \doteq 1.25.$$

The area beyond 1.25 is about .11, so the two-sided P-value is .22—not small enough to cast any doubt on H_0.

Section 10.4

F. Ten children were treated with a therapy[13] involving the drug ethosuximide, to see if it increases IQ scores. Increases in verbal IQ scores were as follows:

16, 7, -5, 24, 19, 11, 6, 2, 10, 30.

Do these data indicate an increase in mean score for treated children?

Solution:

The mean and s.d. are $\bar{X} = 12$ and $S = 10.48$. Our analysis proceeds as usual:

1. The null hypothesis is H_0: $\mu = 0$, where μ is the mean increase in IQ in the population of treated individuals.
2. The question asks whether the mean increase is positive, so H_A is one-sided: $\mu > 0$.
3. We use T, from (1) of Section 10.4, since σ is estimated by S:

$$T = \frac{12.0 - 0}{10.48/\sqrt{10}} = 3.62.$$

4. Large positive values of T are considered extreme.
5. Under H_0, T has a t-distribution with 9 degrees of freedom and the P-value (from Table IIIb) is about .003.

The evidence against H_0 is strong.

   Even if convinced by this evidence that $\mu > 0$, we cannot conclude that the drug therapy was responsible for the increase in IQ. In taking tests, subjects learn; they tend to do better on a second attempt. Any effect the therapy may have is confounded with this learning, which makes the study worthless for judging the effect of the therapy. One way to design a study to take learning into account is to have a second group of children who take the IQ test twice but are given a placebo or no treatment between tests.

G. We wanted to know if the sand in a "three-minute egg" timer actually takes 180 seconds to run out, so we timed one twice, obtaining this sample: 208, 198 (seconds). Test the hypothesis that the mean time is $\mu = 180$.

---

[13] W. L. Smith, "Facilitating verbal-symbolic functions in children with learning problems ...," in W. Smith (Ed.), *Drugs and Cerebral Function* (Thomas, 1970), p. 125.

Solution:
The sample mean is 203, and the standard deviation is $S = \sqrt{50}$. With these we obtain

$$T = \frac{203 - 180}{\sqrt{50}/\sqrt{2}} = 4.6.$$

Table IIIc shows this to be between the 90th and 95th percentiles of $t(1)$. Table IIIb doesn't have an entry for 1 d.f., but we can get an exact (one-sided) P-value using the fact that $t(1)$ is a Cauchy distribution [see (1) of Section 6.6]:

$$P(T > 4.6) = \int_{4.6}^{\infty} \frac{1/\eta}{1 + x^2}\, dx = \frac{1}{\pi}[\text{Arctan } \infty - \text{Arctan } 4.6] = .068.$$

The two-sided P-value is twice this, or .136.

Sections 10.5–10.6

H. In a study assessing the effectiveness of a certain drug in reducing VPB's (ventricular premature beats) the following decreases in VPB's were recorded about 20 minutes after administration of 2 mg/kg of the drug:[14]

1, 7, 17, 22, 5, 4, 5, 14, 9, 7, -4, 51.

Test the hypothesis that the mean decrease is 0 (i.e., no treatment effect).

Solution:
The summary statistics are $\bar{X} = 11.5$, $S = 14.29$, so

$$T = \frac{11.5}{14.29/\sqrt{12}} = 2.79.$$

Table IIIb (10 d.f.) gives $P = .009$, but is the t-test really appropriate? If we omit the most successful case (51), so that $\bar{X} = 7.82$, $S = 6.75$, we get $T = 3.84$—much more "significant." A dot diagram of the data suggests the reason for this strange state of affairs: The population is apparently not close to normal, and a t-test is inappropriate. A sign test (for H_0: median $= 0$) is quite convincing. The P-value for one negative observation in 12 is $(12 + 1)/2^{12} \doteq .003$. The signed-rank statistics is $R_- = 1$ (the one negative observation is the smallest in magnitude), and Table IV gives $P < .001$. (Another approach is to transform the data so that normality is more plausible and use a t-test on the transformed data. We'll return to this idea in Chapter 12.)

I. To test the hypothesis that $\mu = 0$ in a symmetric, continuous population, we take a random sample of size $n = 9$. The sequence of nine observations defines a sequence of signs ($+$ or $-$), and we can calculate the signed-rank statistic R_- for each sequence.
  (a) How many sign sequences are possible?
  (b) Find the one-sided P-value for $R_- = 5$ by listing and counting the sign sequences for which $R_- \le 5$ (and thus verify the entry in Table VI).
  (c) Find the (one-sided) P-value for $R_- = 40$.

---

[14] Reproduced in D. A. Berry, "Logarithmic transformations in ANOVA," *Biometrics 43* (1987), 439–456.

Solution:

(a) Since each sequence element can be $+$ or $-$, there are $2^9 = 512$ possible sequences.

(b) Here are the sequences with smallest values of R_-:

Sequence	R_-
$+\ +\ +\ +\ +\ +\ +\ +\ +$	0
$-\ +\ +\ +\ +\ +\ +\ +\ +$	1
$+\ -\ +\ +\ +\ +\ +\ +\ +$	2
$+\ +\ -\ +\ +\ +\ +\ +\ +$	3
$-\ -\ +\ +\ +\ +\ +\ +\ +$	3
$-\ +\ -\ +\ +\ +\ +\ +\ +$	4
$+\ +\ +\ -\ +\ +\ +\ +\ +$	4
$+\ -\ -\ +\ +\ +\ +\ +\ +$	5
$-\ +\ +\ -\ +\ +\ +\ +\ +$	5
$+\ +\ +\ +\ -\ +\ +\ +\ +$	5

Each of these ten sequences has probability $\frac{1}{512}$, so $P(R_- \le 5) = \frac{10}{512} \doteq .0195$.

(c) The sum of the integers from 1 to 9 is 45, so the symmetric point of 40 in the distribution of R_- is $45 - 40 = 5$. The right tail-probability for 40 is then the same as the left tail-probability for 5: $P = P(R_- \ge 40) = \frac{10}{512}$.

J. Suppose $X \sim N(0, \sigma^2)$. You are to test $H_0: \sigma^2 = 1$ against $H_A: \sigma^2 \ne 1$, using a random sample of size $n = 80$. What do you conclude if you find $\bar{X} = -.0320$ and $S = 1.07$?

Solution:

The m.l.e. of σ^2 is

$$Y = \frac{\sum X^2}{n} = \frac{(n-1)S^2}{n} + \bar{X}^2.$$

For the given data, $Y = 1.13$. According to the central limit theorem, Y is approximately normal, since it is an *average*—the average of the squares of the sample observations. The mean and variance of Y are σ^2 and $2\sigma^2/n$, respectively (see page 274). So under H_0, $Y \sim N(1, \frac{2}{80})$, and the Z-score for the observed value of Y is

$$Z = \frac{1.13 - 1}{\sqrt{\frac{2}{80}}} = .832.$$

The observed deviation of Y from its expected value when H_0 is true is less than what is typical under H_0, so there is little or no evidence against H_0.

Problems

For problems marked with an asterisk, answers are provided in Appendix 2.

In posing problems for your solution, we may ask questions that, although natural to ask, do not have pat answers. Our intention is that you should phrase the problem as one of hypothesis testing—giving the null and alternative hypotheses and finding a P-value. Give a one- or two-sided P-value as you think appropriate, but be sure to say which it is. In applying the various methods, you'll have to assume random sampling.

Sections 10.1–10.3

*1. Referring to Oscar Madison's testing of Felix Unger for ESP (Example 10a), suppose there were 25 successes in 80 trials. Find an approximate P-value for $H_0: p = \frac{1}{4}$.

*2. Again referring to Example 10a, suppose 60 subjects are given Oscar's test for ESP, ten trials for each subject. Suppose that *none* of the 60 subjects has ESP—all of them simply guess.

   (a) For a single subject, what is the probability of obtaining a result that is statistically significant?

   (b) How many of the 60 subjects would you expect to yield statistically significant results?

   (c) Find the probability that none of the results for the 60 subjects is statistically significant.

*3. An educational testing organization has prepared a new version of an aptitude test. The average score on the old version is 500, based on hundreds of thousands of individuals who took the test. The new version is tried out on a sample of 50 students selected at random from those taking the test during the course of a year; their average score is 480, with standard deviation $S = 80$. Does the evidence suggest that the new version is harder than the old?

4. Suppose the mean weight of a rancher's cattle (live weight at the time of slaughter) has been 400 lb. A feed salesman has induced him to try a new supplement on 60 cattle. The mean weight of these is 405 lb, with s.d. 20 lb. Is there evidence that the supplement is effective?

*5. For a number of years, the proportion of entering students who survived the first year in a certain engineering school was about 65%. The school then instituted various measures to improve the situation, including more diligent screening of admissions and more counseling help. In the first year after the change, there were 560 survivors in a class of 800, or 70%. Is this increase explainable as sampling variability?

6. In a TV commercial, the makers of the medium-priced car M reported that in a road test by 90 owners of a high-priced luxury car C, 57 preferred car M for overall ride. (Presumably, the makers of car M felt that if you saw that a substantial majority of luxury car owners preferred car M over their own cars, you'd at least want to try it out.) Is there evidence in the reported result that in the population sampled, a majority of drivers prefer the ride of car M?

*7. An observation X has a triangular distribution centered at $x = 0$:

$$f(x \mid \theta) = 1 - |x - \theta|, \qquad |x - \theta| < 1.$$

   Given that $X = .7$, find the P-value for testing the hypothesis $\theta = 0$ against the alternative $\theta \neq 0$.

8. Someone has advanced the theory that people tend to postpone their death dates until

after their birthdays.[15] If this is so, people are less likely to die in the month immediately preceding their birthdays than in any other month. The skeptic's hypothesis is that all twelve months are equally likely.

(a)   Data on 348 notables[16] showed that 16 died in the month preceding their birthdays. What evidence does this provide?

(b)   A study of death dates of 1,202 athletes[17] showed that 91 died in the month preceding their birthdays. Does this evidence support the theory?

*9.   Can the average consumer distinguish between cheddar cheese that has been aged 9 months and cheddar cheese that has been aged 18 months? To answer this, 45 clerical employees of a cheese manufacturer, with no previous experience in taste testing, are given three samples of cheese, two aged 9 months and one aged 18 months. They are told that one piece is different from the others and asked to single out that piece. The result is that 25 of the 45 picked the odd piece correctly. Can one conclude that consumers generally can tell 9-month-old from 18-month-old cheeses by taste?

*10.   Consider testing $H_0: \mu = 0$ against $H_A: \mu > 0$, where μ is the mean of a normal population with s.d. $\sigma = 1$.

(a)   Suppose you use a sample of size $n = 1$ and observe $X = 2$. Find the P-value.

(b)   Suppose you use a sample of size $n = 400$ and observe $\bar{X} = .1$. Find the P-value.

11.   Consider a sample of size 100 from $\exp(\lambda)$, a distribution with p.d.f. $\lambda \exp(-\lambda x)$, for $x > 0$. If $\bar{X} = 1.23$, find the P-value for a test of $\lambda = 1$ against $\lambda \neq 1$. (*Hint:* Use the approximate normality of the sample mean.)

*12.   Example 10.2c cited a 1965 Alabama court case (*Swain* v. *Alabama*) involving a claim of discrimination against blacks in the selection of grand juries. According to census data, the population proportion of blacks was about 25%, but there were only 177 blacks among 1,050 called to appear for jury duty. Test the hypothesis that the 1,050 constitute a random sample from a population in which the proportion of blacks is .25.

13.   Consider testing $H_0: p = \frac{1}{2}$ against $H_A: p < \frac{1}{2}$, where p is a Bernoulli parameter. Independent Bernoulli trials resulted in the following sequence: 0 0 1 0 0 0 0 0 0 1.

(a)   Suppose this sequence resulted from carrying out ten trials, where $n = 10$ was fixed. In view of H_A, a small number of successes would be evidence against H_0. Find the P-value.

(b)   Suppose the sequence resulted from sampling until two successes were obtained. In view of H_A, a large number of required trials would be evidence against H_0. Find the P-value. (*Hint:* The number of trials required is negative binomial, and it may be easier first to find the probability that it takes 9 or fewer.)

(c)   Find the likelihood function in each case, (a) and (b).

---

[15] See D. P. Phillips, "Deathday and birthday: An unexpected connection," in J. M. Tanur et al. (Eds.), *Statistics: A Guide to the Unknown* (San Francisco: Holden-Day, 1972), pp. 52–66.

[16] From R. B. Morris, (Ed.), *Four Hundred Notable Americans* (New York: Harper & Row, 1965).

[17] Data from R. Hickock, *Who Was Who in American Sports* (Hawthorne Books, 1971), reported in Larson & Stroup, *Statistics in the Real World* (New York: Macmillan, 1976).

Section 10.4

*14. Seven skulls found in a certain digging averaged 143.3 mm in width. The s.d. of the widths was $S = 5.62$. Use these statistics to test the hypothesis that these skulls belong to a race previously found nearby in which the average width is known to be 146 mm. (Assume a normal population.)

15. R. A. Fisher analyzed the results of testing two sleeping drugs on each of ten patients.[18] The additional hours of sleep obtained when using drug A instead of drug B are as follows:

    1.2, 2.4, 1.3, 1.3, 0.0, 1.0, 1.8, 0.8, 4.6, 1.3.

    Test the hypothesis of no average difference in hours of sleep with the two drugs (making necessary assumptions).

*16. Female killdeer usually lay four eggs each spring. A scientist finds the following differences in weights of the first-hatched and last-hatched birds in 8 broods:

    .02, −.10, .06, .23, .14, .14, −.04, .29

    (first-hatched minus last-hatched). Is there evidence of a difference between average weights of first- and last-hatched birds?

17. Ten subjects were tested to determine the increase in reaction time after being given a certain drug. The mean increase was 10 milliseconds and the s.d., $S = 10.6$ milliseconds. What does this say about the hypothesis of no average difference in before-drug and after-drug reaction times?

*18. Example c in the prologue of this book gave corneal thicknesses of the glaucomatous eye and the nonglaucomatous eye of eight patients. The differences in thickness are: −4, 0, 12, 18, −4, −2, 6, 16. Use a t-test for the hypothesis that the mean difference between the corneal thicknesses of glaucomatous and nonglaucomatous eyes is 0.

19. Six healthy young subjects were tested on a cycle ergometer.[19] The work rate was increased continuously and linearly, at a fast rate and at a slow rate. One of the variables measured was a difference in oxygen uptake at which the criterion for anaerobic threshold was met. These differences were (in ml-min):

    765, 650, 700, 550, 250, 20.

    Test the hypothesis that the mean change is zero.

Sections 10.5–10.6

*20. Redo Problem 16 using a signed-rank test.

---

[18] Quoted in P. Meier, "Statistical analysis of clinical trials," in S. Shapiro and T. Louis (Eds.), *Clinical Trials: Issues and Approaches* (New York: Marcel Dekker, 1983).

[19] R. L. Hughson and H. J. Green, "Blood acid-base and lactate relationships studied by ramp work tests," *Medicine and Science in Sports and Exercise 14* (1982), 297–301. (Curiously, two of the six subjects were female (nos. 4 and 5), but the researchers made no attempt to take this into account in the analysis.)

21. An alternative (but equivalent) version of the signed-rank test is based on the statistic

$$W = \sum(\text{signed-ranks}) = R_+ - R_- = \frac{n(n+1)}{2} - 2R_-.$$

(The "signed-rank" of an observation is its rank with a sign attached that is the sign of the original observation.)
   (a) Show that knowing n and any one of the three statistics W, R_+, R_-, one can calculate the other two.
   (b) Find the mean and variance of W.
   (c) Show that the Z-scores corresponding to W, R_+, R_- are the same.

*22. A sprinkler system is designed so that the mean activation time is 25 seconds. A series of tests resulted in the following times of first sprinkler activation:[20]

27, 41, 22, 27, 23, 35, 30, 33, 24, 27, 28, 22, 24.

Test the hypothesis that the mean time is 25 seconds against the alternative that it exceeds 25 seconds,
   (a) using the sign test.
   (b) using the signed-rank test: Find the P-value in Table VI and compare it with the approximate P-value found using a normal approximation.
   (c) using a t-test.

23. Referring to Problem 18, test the hypothesis that the median difference is 0 using
   (a) a sign test.
   (b) a signed-rank test.

*24. A random sample of size $n = 8$ is drawn from a symmetric population with mean 0. Consider the sequence of signs of the observations in a sample sequence and the signed-rank statistic R_-.
   (a) How many possible sign sequences are there?
   (b) Find $P(R_- \leq 4)$ by listing and counting the sign sequences for which $R_- \leq 4$ (and so verify the entry in Table VI).

25. The study referred to in Solved Problem H included these data, decreases in VPB's in the first 10 minutes after administration of the drug:

4, 5, 7, 3, 5, 12, 22, 7, 4, −20, 2, 15.

Plot these in a dot diagram (to see if you think the assumption of population normality is reasonable) and test $\mu = 0$ with an appropriate test.

*26. Consider again the data of Problem 15.
   (a) Make a dot plot. Is there reason to question the applicability of the t-test?
   (b) Use a sign test for the hypothesis that the median difference is zero.
   (c) Calculate the signed-rank statistic and test the hypothesis of no median difference.

27. Problem 28 of Chapter 9 gives nine readings of inlet oil temperature with $\bar{X} = 93.78$ and $S = 4.206$. Suppose specifications call for a population standard deviation not exceeding 3.00. Test the hypothesis that the specification is met.

---

[20] "Use of AFFF in sprinkler systems," *Fire Technology* (1976), 5.

Tests as Decision Rules

In Chapter 10, we considered issues of statistical inference in scientific investigations. We were concerned generally with studies that add to scientific knowledge but that may not require an immediate choice of a specific action. Some situations, especially in business and industry, medicine, law, and public policy-making, require more than a mere reporting of results. A decision must be made between two or more alternative courses of action.

Making the choice between two actions according to the results of a statistical analysis of data can be viewed as testing a null hypothesis H_0 against an alternative H_A. Thus, one action is preferred when H_0 is true and the other when H_A is true.

Example **11a** **Adjusting Lab Equipment**

Hospital laboratories use various instruments to determine white cell count, sedimentation rate, and the like. Many states require daily checks of these instruments. For such checks, a laboratory makes measurements using a small sample from a standard blood supply, one whose characteristics are known rather precisely.

Measurements of hemoglobin with the lab's test equipment on any given blood sample will vary from one determination to another, and we regard a measurement as a random variable. If the mean of this variable is equal to the hemoglobin level of the standard blood sample (call this H_0), the equipment is properly adjusted. In this case, the appropriate action is to proceed with the day's tests. If the mean is not equal to the hemoglobin level of the blood sample (this is H_A), the equipment is out of adjustment. In this case, the appropriate action is to make an adjustment.

A particular standard supply used for checking the instrumentation has a hemoglobin level of 15.1. Let Y denote the result of measuring the hemoglobin level of this standard supply on a particular day. The population of all possible such measurements has unknown mean μ_Y. If Y is much different from 15.1, there is evidence that $\mu_Y \neq 15.1$. The lab technician must have a rule for deciding whether an observed value of Y is far enough from 15.1 to warrant a readjustment—to reject $H_0: \mu = 15.1$.

The decision rule used in one hospital lab is this: If the measured hemoglobin level is between 14.3 and 15.9, proceed with the day's lab work; otherwise, check the equipment and make necessary adjustments. Following this rule can result in a wrong decision—checking the equipment when it's working properly or not checking when it's out of adjustment.   ■

The difference between this example and those in Chapter 10 is rather subtle. Here, a rule is established *before* observing the data, a rule that will tell the technician what to do for every possible sample result. In the settings of Chapter 10, no immediate action is contemplated; one merely assesses the strength of the evidence against H_0 in a particular sample. [Even so, when researchers use the (arbitrary) convention that $P < .05$ means "statistical significance," they imply that one should act as though H_0 were false.]

11.1	**Rejection Regions and Errors**

As in previous chapters, we assume throughout that data are obtained by random sampling (possibly without replacement). As in Chapter 10, we reduce the sample information to a *test statistic*. A test statistic should contain whatever information in the sample is relevant for deciding between H_0 and H_A. A *test* of H_0 is a rule that says which action to take depending on the results in a sample.

Example 11.1a   **Adjusting a Coffee Machine**

A coin-operated coffee machine dispenses a preset amount of coffee. Given any fixed setting, the amount dispensed varies from cup to cup, even if only slightly. Adjusting the setting increases or decreases the mean amount dispensed. Because settings can change with time, machines need to be checked periodically. The vending company wants to know if the average amount dispensed is what it is supposed to be.

Suppose the average amount dispensed is intended to be $\mu = 6$ fluid ounces. The null hypothesis is $H_0: \mu = 6$ (no adjustment needed). (We take this single value of μ as H_0, even though a setting that differs only slightly from 6 may be satisfactory in practice.) The alternative is $H_A: \mu \neq 6$ (requiring adjustment).

To check the adjustment, a technician takes a sample of n cupfuls and measures the amount in each. Let $\bar{X}$ denote the average amount dispensed.

If $\mu = 6$, then $\bar{X}$ should be near 6. A large discrepancy between $\bar{X}$ and 6 suggests that μ is not close to 6, calling for a readjustment. The vendor would want a readjustment made whether μ has drifted *above* 6 (in which case too much coffee would be given away, perhaps spilling some by overfilling), or *below* 6 (in which case the customer is short-changed). So consider the following type of decision rule:

Reject H_0 if $|\bar{X} - 6| > K$,

accept H_0 if $|\bar{X} - 6| \leq K$.

Each choice of K defines a rule; how to choose K will be addressed later. For the moment, suppose $K = .15$. If $\bar{X} = 5.7$ (for example), we would reject H_0 according to this rule, since $|5.7 - 6| > .15$. (To be continued.)   ■

Like the decision rule of this example, a test of H_0 versus H_A is characterized by a **rejection region** (sometimes called a *critical region*).

The **rejection region** R of a test of H_0 against H_A is the set of values of the test statistic that call for rejecting H_0.

Using a particular rejection region may lead to choosing the wrong action. There are two kinds of error:

Type I error: Rejecting H_0 when H_0 is true.
Type II error: Accepting H_0 when H_0 is false.

What are the chances that following a particular rule leads to the wrong decision? If we never encountered H_0, we'd never make a type I error. So to answer the question, we'd have to know "how often" we might encounter H_0—we'd need to assume probabilities for H_0 and H_A. (Compare Chapter 16.) Instead, we calculate the probabilities of taking the wrong action *given* H_0 and *given* H_A.

Whether a probability given H_0 or given H_A is even defined depends on the form of these hypotheses. If H_0 completely specifies a distribution for the test statistic, we can calculate the probability of rejecting H_0 when H_0 is true. This probability is called the **size** of the type I error, or the **significance level** of the test. It is generally denoted by α.

For a test with rejection region R, the size of the type I error or the **significance level** of the test is

$$\alpha = P(\text{reject } H_0 \,|\, H_0) = P(R \,|\, H_0).$$

Example 11.1b   **Adjusting the Coffee Machine (continued)**
The rule suggested in Example 11.1a has this rejection region:

$$R = \{|\bar{X} - 6| > .15\}.$$

It says to reject H_0 ($\mu = 6$) if $\bar{X} > 6.15$ or if $\bar{X} < 5.85$.

The null hypothesis $\mu = 6$ does not uniquely define the distribution of the test statistic $\bar{X}$. Suppose we assume $n = 10$ and $\sigma = .20$; this still does not define the distribution exactly, but the central limit theorem suggests that $\bar{X} \approx N(\mu, (.20)^2/10)$.

Thus, $\sigma_{\bar{X}} \doteq .20/\sqrt{10} = .0632$, and

$$\alpha = P(R \mid H_0) = P(|\bar{X} - 6| > .15 \mid \mu = 6)$$
$$= 1 - P(5.85 < \bar{X} < 6.15 \mid \mu = 6)$$
$$\doteq 1 - \Phi\left(\frac{6.15 - 6}{.0632}\right) + \Phi\left(\frac{5.85 - 6}{.0632}\right) = 2\Phi(-2.37) \doteq .018.$$

Figure 11.1 shows the sampling distribution of $\bar{X}$ under H_0, with α indicated as the sum of two (equal) tail-areas. (To be continued.)

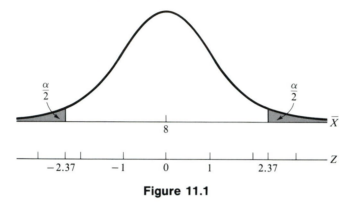

Figure 11.1 ■

Alternative hypotheses are often like the one in the preceding examples, stating only that a parameter θ is in some range of values. The probability of accepting H_0 when H_A is true depends on θ.

If H_A is specific enough, we may be able to calculate probabilities for the test statistic. When this is the case, the probability of accepting H_0 when H_A is true is well defined; it is called the *size* of the type II error and denoted by β. The quantity $1 - \beta$, the probability of rejecting H_0 when H_A is true, is called the **power** of the test; it measures the ability of the test to detect that H_0 is not true when it's not.

When H_A is not specific enough to determine probabilities for the test statistic, we can still calculate the probability of accepting H_0 for each particular distribution in H_A. This is a type II error size for that particular alternative.

For a test with rejection region R, the type II error size for a specific value of θ in H_A is

$$\beta = P(\text{accept } H_0 \mid \theta) = P(R^c \mid \theta).$$

The **power** of the test against that alternative is $1 - \beta$.

Example 11.1c   **Adjusting Lab Equipment (continued)**
We return to the setting of Example 11a. The rule the lab uses for deciding whether to proceed with the day's testing or to make an adjustment is defined by the rejection region

$$R = \{X > 15.9 \text{ or } X < 14.3\},$$

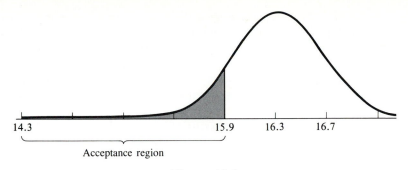

Acceptance region

Figure 11.2

where X is the result of a single measurement on the standard blood supply. The null and alternative hypotheses are

$$H_0: \mu = 15.1 \quad \text{and} \quad H_A: \mu \neq 15.1.$$

We assume (as does the lab) that $X \sim N(\mu, .16)$. We can then calculate the size of the type I error, since specifying $\mu = 15.1$ defines the distribution of X:

$$\alpha = P(R) = P(X > 15.9 \text{ or } X < 14.3 \,|\, \mu = 15.1)$$

$$= 1 - \Phi\left(\frac{15.9 - 15.1}{.40}\right) + \Phi\left(\frac{14.3 - 15.1}{.40}\right) = .0456.$$

The probability of accepting H_0 given H_A (a type II error) is *not* uniquely defined, since H_A does not specify μ and therefore does not determine a distribution for X. However, we can calculate a size of type II error for any particular value of μ in H_A. For example, when $\mu = 16.3$ (see Figure 11.2):

$$\beta = P(\text{accept } H_0 \,|\, \mu = 16.3) = P(14.3 < X < 15.9 \,|\, \mu = 16.3)$$

$$= \Phi\left(\frac{15.9 - 16.3}{.40}\right) - \Phi\left(\frac{14.3 - 16.3}{.40}\right) \doteq .1587.$$

The acceptance region for the test is the interval with endpoints $15.1 \pm 2 \times .40$; these are called 2σ-*limits*. State regulations actually call for using 3σ-limits: $15.1 \pm 3 \times .40$. Using these limits means rejecting H_0 if $X > 16.3$ or $X < 13.9$. (So the lab's standards are more stringent than required by the state.) The significance level with 3σ-limits is smaller:

$$\alpha = P(|Z| > 3) = .0026,$$

but now β is larger:

$$\beta = P(13.9 < X < 16.3 \,|\, \mu = 16.3)$$

$$= \Phi\left(\frac{16.3 - 16.3}{.40}\right) - \Phi\left(\frac{13.9 - 16.3}{.40}\right) \doteq .50.$$

(To be continued.)   ■

In this example, changing the rejection region to reduce α increased the size of the type II error. This is typical: Modifying a rejection region to reduce the size of one type of error increases the size of the other type of error.

P-Values and α

Significance levels (α's) and P-values are both tail-probabilities in the null distribution. A P-value is calculated from observed data, without regard to choices of action, whereas α is calculated for a given decision rule or test, independent of observed data. The P-value for a set of data can equal α; this happens when the observed value of the test statistic falls on the boundary of the rejection region. A P-value is sometimes called the *observed significance level*.

Example 11.1d   **Adjusting a Coffee Machine (continued)**

In Example 11.1b, we decided to reject H_0 when $|\bar{X} - 6| > .15$. Suppose we observe $\bar{X} = 5.85$—a value right on the boundary of the rejection region. The P-value is

$$P = 2 \cdot P(\bar{X} \le 5.85) = 2\Phi\left(\frac{5.85 - 6.00}{.0632}\right) \doteq .018.$$

This is precisely the α calculated in Example 11.b. So α is the P-value calculated from an observation that falls on the boundary of the rejection region. (To be continued.)

■

11.2   The Power Function

In Section 11.1, we saw that the size of the type II error depends on the particular distribution in H_A that we assume. When the distribution of the test statistic is determined by a parameter θ, the probability of rejecting H_0 is a function of θ, the **power function**, $\pi(\theta)$. Although the term *power* is especially appropriate for θ in H_A, the function $\pi(\theta)$ is defined for all θ, including θ in H_0. When H_0 is $\theta = \theta_0$, then $\pi(\theta_0) = \alpha$.

Example 11.2a   **Adjusting Lab Equipment (continued)**

In Examples 11a and 11.1c, we gave the rule used by a hospital lab for testing H_0: $\mu = 15.1$. It was defined by this rejection region:

$$R = \{X < 14.3 \text{ or } X > 15.9\}.$$

The power function of the test is the probability that X falls in the rejection region when the population mean is μ:

$$\pi(\mu) = P(\text{reject } H_0 \,|\, \mu) = P(R \,|\, \mu) = P(X < 14.3 \text{ or } X > 15.9 \,|\, \mu).$$

We reported in Example 11.1c that the lab assumes $X \sim N(\mu, .16)$. With this assumption,

$$\pi(\mu) = P(\text{reject } H_0 \,|\, \mu) = P(X < 14.3 \,|\, \mu) + P(X > 15.9 \,|\, \mu)$$

$$= \Phi\left(\frac{14.3 - \mu}{.40}\right) + 1 - \Phi\left(\frac{15.9 - \mu}{.40}\right)$$

$$= \Phi\left(\frac{14.3 - \mu}{.40}\right) + \Phi\left(\frac{\mu - 15.9}{.40}\right).$$

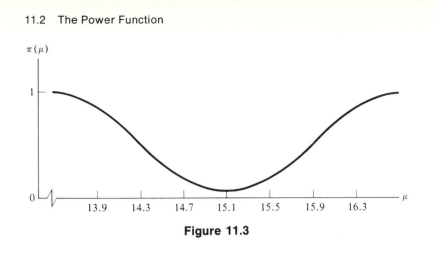

Figure 11.3

This is the sum of a decreasing function of μ (the first term) and an increasing function of μ (the second term). We have sketched the power function in Figure 11.3. It has the shape of an inverted bell with minimum value α at $\mu = 15.1$:

$$\alpha = P(R \mid \mu = 15.1) = \pi(15.1) = 2\Phi(-2) = .0456.$$

As μ moves away from 15.1 in either direction, the power increases. Thus, the greater the misadjustment, the more likely it is that the technician will be led to take corrective action. However, the power function is flat near its minimum at 15.1, so if the mean is only slightly out of adjustment, the test is not likely to call for a readjustment. ■

When the distribution of a test statistic depends on a parameter θ, the **power function** of a test with rejection region R is

$$\pi(\theta) = P(R \mid \theta). \qquad (1)$$

In particular, when H_0 specifies $\theta = \theta_0$,

$$\alpha = P(R \mid \theta_0) = \pi(\theta_0).$$

The power function for a given test or rejection region does not depend on how we've defined H_0 and H_A. However, it does suggest the type of partition of θ-values into an H_0 and an H_A for which the test is especially suited: Low power is desirable on H_0 and high power on H_A.

Example **11.2b** **One-Sided Regions for the Alternative $p > p_0$**

The campaign committee for a gubernatorial candidate commissions a poll to determine whether the candidate is already the choice of a majority of voters. If not, the committee will spend more on TV time and newspaper ads. Let p denote the population proportion who already favor the candidate and $\hat{p}$ the corresponding proportion in a random sample of 400. For the hypothesis $H_0: p = .5$, and the one-sided alternative $H_A: p > .5$, consider this test: Reject H_0 (buy more advertising) if $\hat{p} > .54$.

Since n is fairly large,

$$\hat{p} \approx N\left[p, \frac{p(1-p)}{n} \right],$$

according to Section 8.8. The significance level of the test is

$$\alpha = P(\hat{p} > .54 \,|\, p = .5) = 1 - \Phi\left(\frac{.54 - .50}{.5/20} \right) \doteq .055.$$

The power function is

$$\pi(p) = P(\hat{p} > .54 \,|\, p) = 1 - \Phi\left(\frac{.54 - p}{\sqrt{p(1-p)/20}} \right).$$

This is an increasing function of p (see Figure 11.4). Its value at $p = .5$ is $\alpha = .055$. The larger the actual value of p, the more likely it is that $p = .5$ will be rejected. If p is actually less than .5, the probability of rejecting H_0 is even smaller than when $p = .5$. Thus, the one-sided rejection region is quite appropriate for the one-sided alternative $p > .5$.

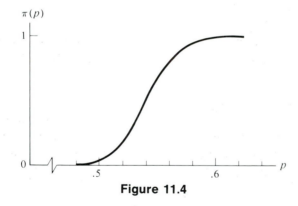

Figure 11.4

11.3   Choosing a Sample Size

In Chapter 10 and so far in this chapter, we have assumed a given, fixed sample size. This is appropriate when one has a particular data set to analyze. When setting up an experiment, one faces the question of sample size. Economic considerations are usually a constraint. Except for such constraints, the sample size should depend on how reliable we want our conclusions to be. In Section 9.2, we showed how, in estimating a parameter, we might determine a sample size from a specification of standard error. When testing hypotheses, the sample size can be determined by specifying allowable error sizes.

A finite amount of data seldom provides perfect information, so an experimenter who must make a decision based on those data must accept the possibility of error—both type I and type II. How large the sizes of these errors can be depends on the cost of wrong decisions and on the cost of gathering data. Here we'll assume that error sizes are specified.

Suppose the distribution of our test statistic is completely determined by the value of a parameter θ, and consider $H_0: \theta = \theta_0$. An investigator may be willing to tolerate $\alpha = .05$ but insists on $\beta = .20$ when $\theta = \theta_1$. We'll see that two such restrictions on the power function, $\pi(\theta_0) = \alpha$ and $\pi(\theta_1) = \beta$, serve to determine the sample size and rejection region.

Testing μ

Suppose the population mean μ is the unknown parameter and $\bar{X}$ is the test statistic. Suppose further that either the population is normal or the sample size is large enough that $\bar{X}$ is approximately normal. If σ is known, the power function for the rejection region $\bar{X} > K$ is

$$\pi(\mu) = P(\bar{X} > K) = \Phi\left(\frac{\mu - K}{\sigma/\sqrt{n}}\right). \tag{1}$$

This is symmetric about the point $(K, \frac{1}{2})$. The curve can be shifted to the right or left by increasing or decreasing K. For a given K, the slope of the curve at K can be increased by increasing n:

$$\pi'(K) = \frac{1}{\sigma}\sqrt{\frac{n}{2\pi}}. \tag{2}$$

Requiring $\pi(\mu)$ to pass through a specified point is a condition on n and K, and requiring it to pass through two specified points defines two simultaneous equations for n and K. Solving this system of equations produces n and K that meet the given requirements.

Example 11.3a   **Adjusting the Coffee Machine (conclusion)**
We return to the setting of Example 11.1d, and the rejection region

$$R = \{|\bar{X} - 6| > K\}.$$

Suppose, in deciding whether to adjust the machine, we want at most a 5% probability of calling for readjustment when none is needed (that is, when $\mu = 6$): $\pi(6) = .05$. Suppose further that we want an 80% probability of calling for readjustment when the mean μ has shifted to 6.2 oz or to 5.8 oz: $\pi(5.8) = \pi(6.2) = .80$. Assuming $\sigma = .20$ and making a calculation similar to that in Example 11.1b, we find

$$\pi(\mu) = P(|\bar{X} - 6| > K \mid \mu)$$
$$\doteq 1 - \Phi\left(\frac{K + 6 - \mu}{.2/\sqrt{n}}\right) + \Phi\left(\frac{-K + 6 - \mu}{.2/\sqrt{n}}\right). \tag{3}$$

Then

$$\alpha = \pi(6) \doteq 2\Phi\left(\frac{-K\sqrt{n}}{.2}\right) = .05.$$

This tells us that

$$\frac{-K\sqrt{n}}{.2} = -1.96, \qquad \text{or} \qquad K = \frac{.392}{\sqrt{n}}. \tag{4}$$

We then substitute this K and $\mu = 6.2$ in (3):

$$\pi(6.2) \doteq 1 - \Phi\left(\frac{(-.392/\sqrt{n}) - .2}{.2/\sqrt{n}}\right) + \Phi\left(\frac{(-.392/\sqrt{n}) - .2}{.2/\sqrt{n}}\right) = .80. \tag{5}$$

Simplifying the fractions and using $\Phi(z) = 1 - \Phi(-z)$, we obtain

$$\Phi(1.96 - \sqrt{n}) - \Phi(-1.96 - \sqrt{n}) = .20.$$

We could obtain an approximate solution of this equation by trial and error, but notice that the second term is negligible when $n > 1$. Thus,

$$1.96 - \sqrt{n} \doteq -.84.$$

The solution is $n \doteq 7.84$. We take the next larger integer for n and substitute in (4) to find $K = .1386$. With this K and $n = 8$ in (3), we get

$$\pi(6) = .05 \qquad \text{and} \qquad \pi(6.2) = .808.$$

The power at 6.2 is slightly greater than specified by (5) because we (necessarily) rounded n to an integer. ∎

The issues we discussed in Section 9.2 regarding the selection of a sample size when σ is unknown apply in the setting of testing hypotheses about μ. In particular, taking a pilot sample or monitoring accumulating data can be effective in estimating σ.

11.4   Quality Control

Example 11.4a   Quality Control

In Japan, the Deming medal is the highest and most prestigious award that Japanese companies can earn for the advancement of precision and dependability of products. It is named for W. Edwards Deming, a statistician who gave assistance to Japanese industry after World War II. He is given credit for helping to improve the quality of their products and increase the productivity of their plants. Many Japanese feel that the practical methods he taught led to the spectacular strengthening of their competitive position in international markets. These methods include sampling inspection and acceptance sampling, which are statistical tools for monitoring production quality. ∎

   The quality of manufactured items is important for consumers. It is equally important to manufacturers, who may well find it unprofitable to sell items of poor quality in a competitive market. In controlling quality it is important for a producer to monitor the quality of what is produced and for the customer to monitor the quality of what is received—to do **acceptance sampling**.

   Suppose that goods are shipped in lots. Both the manufacturer and the customer want to determine whether a lot is acceptable. Inspecting every item in a lot is costly, and complete inspection may not be totally reliable. In some cases, it may be necessary to destroy the items to assess reliability; in such cases, complete sampling is out of the question. In any case, suppose the manufacturer decides to inspect the items in a sample taken from the lot to help decide whether or not to pass the lot. This process is called **sampling inspection**. We deal here only with situations in which the items are classed as either good or defective.[1]

   We assume that sampling is at random and without replacement. Our notation is consistent with that of Chapter 4:

   N = lot size,

   n = sample size,

   p = lot fraction defective,

   $M = Np$ = number of defectives in the lot,

   Y = number of defectives in the sample,

   c = acceptable number of defectives in the sample.

We'll assume that N is fixed and take $p = M/N$ as the unknown parameter. Under sampling without replacement, Y is hypergeometric (see Section 4.3).

   In acceptance sampling, the choice of action is between rejecting and accepting the lot. The decision rule used in practice is a natural one: Reject the lot if the sample has too many defectives, and otherwise accept the lot. So the test statistic is Y, the number of defectives in the sample. The rejection region for such a rule is of the form

$$R = \{Y > c\}. \tag{1}$$

   The power function of the test (1) is a sum of hypergeometric probabilities:

$$\pi(p) = P(R \mid p) = 1 - P(Y \le c \mid p)$$
$$= 1 - \sum_{k=0}^{c} \frac{\binom{Np}{k}\binom{N-Np}{n-k}}{\binom{N}{n}}. \tag{2}$$

In industrial circles, the performance of a sampling plan is usually described by the **operating characteristic**, or OC curve. This is the probability of acceptance, or 1 minus the power function:

$$\text{OC}(p) = P(\text{lot is accepted} \mid p) = P(Y \le c \mid p) = 1 - \pi(p).$$

---

[1] For further reading on this and related topics, see A. J. Duncan, *Quality Control and Industrial Statistics*, 4th ed., (Homewood, Ill.: Irwin, 1974).

Example 11.4b   Suppose we take $c = 1, n = 3$, and $N = 10$. The power function of the sampling plan defined by these constants follows from (2):

$$\pi(p) = 1 - [P(Y = 0 | p) + P(Y = 1 | p)]$$

$$= 1 - \frac{\binom{10p}{0}\binom{10 - 10p}{3} + \binom{10p}{1}\binom{10 - 10p}{2}}{\binom{10}{3}}.$$

The power and OC functions are given in the following table:

p	0	.1	.2	.3	.4	.5	.6	.7	.8	.9	1
$\pi(p)$	0	0	.07	.18	.33	.50	.67	.83	.93	1	1
OC(p)	1	1	.93	.82	.67	.50	.33	.17	.07	0	0

With this plan, the chance of rejecting the lot increases as p increases—that is, as the lot quality worsens. ∎

When the sample size is small in comparison with the population or lot size ($n \ll N$), the hypergeometric probabilities in (2) can be approximated by binomial probabilities:

$$\pi(p) \doteq 1 - \sum_{k=0}^{c} \binom{n}{k} p^k (1 - p)^{n-k}. \tag{3}$$

These, in turn, can be approximated by Poisson probabilities when p is small and n is large and by normal probabilities when $npq > 5$ (see Sections 4.5 and 4.6).

The power function (2) and its binomial, Poisson, and normal approximations are increasing functions of p: The poorer the lot quality, the more likely the test is to reject the lot.

Example 11.4c   **Sampling a Large Lot**

Consider sampling 25 from a lot of 500 articles ($N = 500$). Consider the sampling plan defined by $c = 2$. The power function is the probability of more than two defectives in a sample of 25 from 500, of which $500p$ are defective. The power function (2) is

$$p(Y > 2 | p) = 1 - \frac{\binom{Np}{0}\binom{N - Np}{25} + \binom{Np}{1}\binom{N - Np}{24} + \binom{Np}{2}\binom{N - Np}{23}}{\binom{N}{25}}.$$

Because $n = 25$ is much smaller than $N = 500$, binomial probabilities (3) in place of the hypergeometric probabilities in (2) should provide sufficient accuracy:

$$\pi(p) \doteq 1 - \sum_{0}^{2} \binom{25}{k} p^k (1 - p)^{25-k}$$

$$= 1 - (1 - p)^{25} - 25p(1 - p)^{24} - 300p^2(1 - p)^{23}$$

$$= 1 - (1 - p)^{23}((1 - p)^2 + 25p(1 - p) + 300p^2).$$

For any particular value of p, this is easy to evaluate using a calculator with an "x^y" key. When p is small, we could use a Poisson approximation with $m = np = 25p$ (see Section 4.5):

$$\pi(p) = 1 - e^{-25p}\left[1 + 25p + \frac{(25p)^2}{2}\right].$$

The power function is given in Table 11.1 for selected values of p. The table also gives the binomial and Poisson approximations to $\pi(p)$. The difference between the power at $p = .05$ and $p = .20$ indicates that this test does a reasonably good job in discriminating between these two values.

Table 11.1 Power function for Example 11.4b

p	Hypergeometric	Binomial approximation	Poisson approximation
.05	.1224	.1271	.1315
.10	.4657	.4629	.4562
.15	.7534	.7463	.7229
.20	.9075	.9018	.8753

In the preceding example, we have analyzed the performance of a test without specifying a null hypothesis! Because the power function (2) increases with p, the sampling inspection procedure is suited to testing a hypothesis of the form $H_0: p \le p_0$ against the alternative $H_A: p > p_0$. Both H_0 and H_A involve many values of p, so the α and β are not uniquely defined. However, suppose the manufacturer specifies a fraction defective p_0 such that any lot with $p \le p_0$ is acceptable. The power at p_0 is called the *producer's risk*:

$$\alpha = \max_{p \le p_0} \pi(p) = \pi(p_0).$$

The consumer may specify a fraction defective p_1 such that any lot with $p \ge p_1$ is unacceptable. The probability of acceptance when $p = p_1$ is called the *consumer's risk*:

$$\beta = \max_{p \ge p_1}[1 - \pi(p)] = 1 - \pi(p_1).$$

Specifying these risks amounts to specifying that the power function pass through the two points (p_0, α) and $(p_1, 1 - \beta)$. As in Section 11.3, these two conditions determine an acceptance number c and sample size n for a test. (See Problem 18.)

11.5   **Most Powerful Tests**

Finding good tests is most straightforward when both the null and the alternative hypotheses completely specify a distribution for the test statistic. Let f denote the joint p.f. or p.d.f. of the sample $\mathbf{X} = (X_1, \ldots, X_n)$, and consider testing

$$H_0: f = f_0 \qquad \text{versus} \qquad H_1: f = f_1,$$

Simple and Composite Hypotheses

where f_0 and f_1 involve no unknown parameters. Such hypotheses are said to be **simple**. A hypothesis that is composed of more than one simple hypothesis is **composite**. To refer to one of the simple hypotheses that make up a composite H_A, we may say "a particular alternative in H_A."

A test is good if the sizes of both types of error are small. How do we find a rejection region that will have small error sizes of both types? One approach is to fix the size of α at some acceptable level and look for a rejection region whose β is as small as possible (with that α) or, equivalently, one whose power on the alternative is a maximum. Thus, in the continuous case, we seek a region R with a specified α:

$$\int_R f_0(\mathbf{x})\,d\mathbf{x} = \alpha,$$

such that

$$1 - \beta = \int_R f_1(\mathbf{x})\,d\mathbf{x}$$

is a maximum. A clue as to how to construct such a rejection region R from sample points $\mathbf{x}$ is found in the following example.

Example 11.5a **Maximizing the Ratio of Value to Volume**

A device that has been used in TV game shows is to have someone with a shopping cart dash madly through a grocery store, filling the cart with as much loot as possible, taking at most one of each type of item. The usual constraint is one of time, but suppose you have a cart and enough time to reflect on how to fill it most efficiently. How would you go about filling the cart to get the most total value?

Surely you would not take a large box of cornflakes, or a 12-pack of cola, or a package of cotton puffs—these all take up too much room for what they're worth. Rather, you would take things that are expensive but don't take up much room, things that have a high cost per unit volume—raspberries out of season, swordfish, caviar, and the like. Indeed, you might start by putting into your cart *first* those items that have the highest cost per unit volume. ∎

Likelihood Ratios

Think of $f_0(\mathbf{x})$ as the volume of the "item" $\mathbf{x}$, and $f_1(\mathbf{x})$ as its cost or value. The reasoning of the above example suggests that we get a region R of greatest value (integral of f_1 over R) for a given volume (integral of f_0 over R) by putting into it first those $\mathbf{x}$'s which have the smallest volume per unit price (smallest ratio of f_0 to f_1). The ratio of f_0 to f_1 is a **likelihood ratio**, Λ. We keep on putting points into R until it gets filled up—that is, until the total volume (integral of f_0 over R) reaches α, the capacity of the "cart" R.

Example 11.5b **Likelihood Ratio for Bin(p)**

Consider these simple hypotheses about the Bernoulli parameter p:

$$H_0: p = .50 \qquad \text{versus} \qquad H_1: p = .25.$$

Suppose we conduct five independent trials and base a test on the sufficient statistic Y, the number of successes. The possible values of Y and corresponding probabilities under these two models (from Table II in Appendix 1) are shown in Table 11.2. Also given is a column of values of the likelihood ratio statistic,

$$\Lambda(Y) = \frac{f_0(Y)}{f_1(Y)}.$$

(As is customary, our notation may drop the dependence of Λ on Y and refer to the likelihood ratio as Λ.)

Table 11.2

y	$f_0(y)$	$f_1(y)$	$\Lambda(y)$
0	.031	.2373	.13
1	.156	.3955	.39
2	.3125	.2637	1.2
3	.3125	.0879	3.6
4	.1562	.0146	10.7
5	.0312	.0010	31.2

From the column of likelihood ratios in Table 11.2, it is clear that Λ is an increasing function of y—the values of Y with the smallest Λ-values are those with smallest Y-values. So in forming a rejection region, we put into it first $y = 0$; then the next smallest, $y = 1$, and so on, until the sum of the f_0's for the next Y-value would exceed some specified α. The resulting rejection region will have the smallest β (greatest power) among all tests with the same α. Some of these **most powerful** rejection regions of Y-values are as follows:

R	α	β	$Power = 1 - \beta$
$\varnothing$	0	1	0
$\{0\}$	.031	.76	.24
$\{0, 1\}$	.187	.37	.63
$\{0, 1, 2\}$	.500	.10	.90

Of course, with $n = 5$ one could not hope that both α and β would turn out to be small, but these tests are better than others. For instance, both $R = \{0, 1\}$ and $R = \{4, 5\}$ have $\alpha = .187$, but $\beta = .985$ for the latter. (To be continued.) ∎

The mathematical theorem to the effect that rejection regions of the form $\Lambda < K$ are most powerful in the class of tests with no larger α is the Neyman–Pearson lemma:

Neyman–Pearson lemma: For testing $f_0(x)$ against $f_1(x)$ based on a sample **x**, a critical region of the form

$$\Lambda = \frac{L_0}{L_1} < K, \tag{1}$$

where L_i is the likelihood of f_i and K is a constant, has the greatest power (smallest β) in the class of tests with no larger α.

Proof of the Neyman– Pearson Lemma

To prove the Neyman–Pearson lemma, we consider the Neyman–Pearson region R defined by (1), a region in the n-dimensional sample space of **X**, and any competing region S whose type I error size is no larger than that of R:

$$\alpha_S \le \alpha_R. \tag{2}$$

We observe that for any function f (in abbreviated notation),

$$\left(\int_R - \int_S \right) f = \int_{RS^c} f - \int_{R^c S} f, \tag{3}$$

since the integral over the intersection RS cancels. Then, applying (3) to f_1, we have

$$\beta_S - \beta_R = \left(\int_R - \int_S \right) f_1 = \int_{RS^c} f_1 - \int_{R^c S} f_1.$$

By (1), $f_1 \ge f_0/K$ on R and $-f_1 \ge -f_0/K$ on R^c. Hence, with another application of (3), we see that

$$\beta_S - \beta_R \ge \frac{1}{K} \left(\int_{RS^c} f_0 - \int_{R^c S} f_0 \right) = \frac{1}{K} \left(\int_R - \int_S \right) f_0 = \frac{1}{K} (\alpha_R - \alpha_S).$$

In view of (2), the lemma is established.

α-β Plots

A graphical representation of the error sizes of tests is helpful. Each proposed rejection region R has an associated pair of error sizes, (α, β). Each such pair can be plotted as a point in a rectangular coordinate system. The collection of points corresponding to all rejection regions is contained in the unit square, since α and β are probabilities.

Example 11.5c   Neyman–Pearson Test for Bin(*p*)

We return to choosing between the two binomial models in Example 11.5b. Figure 11.5 shows the point (α, β) for each of the likelihood ratio tests (Neyman–Pearson tests). Also plotted are the points corresponding to the tests of the form $\Lambda > K$, which are the worst tests. (Reverse the argument for the best tests: Fill the shopping cart in Example 11.5a with voluminous items having little value.) All other possible tests, including randomized tests discussed below, are represented by points in the shaded region of the figure. This region is a convex set whose

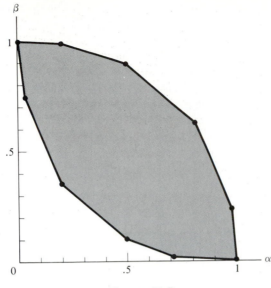

Figure 11.5

extreme or corner points are the best and worst nonrandomized tests. It is clear from the figure that not only do the Neyman–Pearson tests have the smallest β for a given α, they also have the smallest α for a given β.

An example of a **randomized test** is this: If $Y = 0$, reject H_0; if $Y > 1$, accept H_0; and if $Y = 1$, reject H_0 with probability γ, a number in the interval $(0, 1)$. Using the probabilities in Table 11.2, we find

$$\alpha = P(\text{rej. } H_0 \,|\, H_0), H_0$$
$$= P(\text{rej. } H_0 \,|\, Y = 0)P(Y = 0 \,|\, H_0) + P(\text{rej. } H_0 \,|\, Y = 1, H_0)P(Y = 1 \,|\, H_0)$$
$$= .031 + .156\gamma.$$

This is between $.031$ and $.031 + .156 = .187$, depending on γ. To achieve $\alpha = .05$, say, we set $\gamma = .019/.156$. Similarly, the type II error size is between $.76$ and $.37$. ∎

UMP Tests

A test that has the greatest power among tests with no larger α, not only against a particular alternative but against all the alternatives in H_A, is said to be **uniformly most powerful** (UMP). One way of constructing a test for a simple null against a composite alternative hypothesis, say $H_0\colon \theta = \theta_0$ versus $H_A\colon \theta < \theta_0$, is to select a particular $\theta_1 < \theta_0$ and construct a Neyman–Pearson test against $H_A\colon \theta = \theta_1$. If the form of this best test is the same for every $\theta_1 < \theta_0$, the test is said to be uniformly most powerful in testing θ_0 against $\theta < \theta_0$.

Example 11.5d   **UMP Test for Binomial p**

Consider again testing $H_0\colon p = p_0$ based on Y, the number of 1's in a random sample of size 5. We saw in Example 11.5b that a rejection region of the form $Y < K$ has the greatest power against the alternative $p = .25$ of all tests with size $\alpha = .187$. The result is the same when we use an arbitrary $p < .5$. The likelihood ratio is

$$\Lambda = \frac{(.5)^Y(.5)^{5-Y}}{p^Y(1-p)^{5-Y}} = \frac{(1/p - 1)^Y}{[2(1-p)]^5}.$$

This is an increasing function of Y, since $1/p > 2$ when $p < .5$. So Λ is less than a constant when Y is less than a constant. A test of the form $Y < K$ is therefore most powerful against every single $p < .5$ and, hence, UMP against the alternative $H_A: p < .5$. (To be continued.) ∎

Composite H_0

When the null hypothesis is composite, α may not be uniquely defined. In this case, we define α^* to be the maximum (or supremum) of the probabilities of rejection over all simple hypotheses in H_0. With α^* taking the place of α, we define uniformly most powerful tests as before.

Consider testing $\theta \le \theta_0$ versus $\theta > \theta_0$. Suppose we have found a rejection region R that is uniformly most powerful for testing $\theta = \theta_0$ against $\theta > \theta_0$. If its power function is increasing, the maximum power on H_0 occurs at θ_0, and this is then α^*. Any competing tests with extended size α^* are then included in the set of tests with $\alpha = \pi(\theta_0) = \alpha^*$. Therefore, R is also uniformly most powerful for $\theta \le \theta_0$ versus $\theta > \theta_0$.

Example 11.5e   **UMP Test for Binomial p (continued)**

In Example 11.5d, we showed that the rejection region $\{0, 1\}$ for the binomially distributed Y is UMP for $H_0: p = .5$ versus $H_A: p < .5$ in the class of tests with $\alpha = \pi(.5) = .187$. The power function of this test is

$$\pi(p) = P(Y = 0 \text{ or } 1 \,|\, p) = (1-p)^5 + 5(1-p)^4 p$$
$$= (1-p)^4(1+4p).$$

This is a decreasing function of p (for $0 < p < 1$). Consequently, the largest power over the range $p \ge .5$ is at the boundary, $p = .5$. So the extended size is

$$\alpha^* = \sup_{H_0} \pi(p) = \pi(.5) = .187.$$

Since the region $R = \{Y \le 1\}$ has greatest power at any alternative among tests with $\alpha = .187$, it has greatest power at any alternative among tests with $\alpha^* = .187$. (The latter class is a subclass of the former.) It is uniformly most powerful for $p \ge .5$ against $p < .5$. ∎

In view of the foregoing, it is useful to know when a power function is monotonic. A general class of distributions for which the power function of the likelihood ratio test for $\theta = \theta_0$ against $\theta = \theta_1$ is a monotonic function of the parameter θ is the exponential family defined in Example 8.3f. [This family includes $N(\mu, 1)$, $\text{Ber}(p)$, $\text{Geo}(p)$, and $\text{Exp}(\lambda)$, among others.][2]

---

[2] For a proof, see Lindgren [15].

> Suppose $\mathbf{X} = (X_1, \ldots, X_n)$ is a random sample from a population with p.d.f. $f(x \mid \theta) = B(\theta)h(x)e^{\theta x}$. The power function of the rejection region $\Sigma X > K$ is a nondecreasing function of θ.

11.6   Likelihood Ratio Tests

In Section 11.5, we considered the likelihood ratio statistic

$$\Lambda = \frac{L(\theta_0)}{L(\theta_1)}$$

for deciding between two simple hypotheses, θ_0 and θ_1. This statistic orders the sample points according to the value of Λ, rejection regions of the form $\Lambda(\mathbf{X}) < K$ being most powerful. We now extend the idea of using likelihood comparisons to a general method of constructing tests for cases in which either the null or the alternative hypothesis (or both) is composite.

The idea of the **generalized likelihood ratio** is a comparison of the best explanation of a given set of data among models in H_0 with the best explanation in H_A, where *best* means having maximum likelihood. A generalized likelihood ratio test often turns out to have good properties. Indeed, in many standard situations, the likelihood ratio approach leads to rejection regions that we've introduced already on other intuitive grounds. It also provides reasonable tests in cases where it is not so obvious how to proceed.

Consider a family of distributions indexed by θ, and take as H_0 a subset of the parameter space. The alternative is the complement of that subset. Let $\hat{\theta}$ denote the m.l.e. of θ, and let $\hat{\theta}_0$ denote the θ that maximizes the likelihood within H_0:

$$L(\hat{\theta}) = \sup_{H_0 \cup H_A} L(\theta), \qquad L(\hat{\theta}_0) = \sup_{H_0} L(\theta).$$

The generalized likelihood ratio is

$$\Lambda = \frac{L(\hat{\theta}_0)}{L(\hat{\theta})}. \tag{1}$$

Clearly, $\Lambda \leq 1$, and $\Lambda = 1$ if and only if $\hat{\theta} = \hat{\theta}_0$. When Λ is very small, the best explanation for the data in H_0 is much worse than the best available explanation — the one in H_A. So we take regions of the form $\Lambda < K$ as rejection regions for the likelihood ratio statistic.

Example 11.6a   **The *t*-test as a Likelihood Ratio Test**

Consider testing $H_0: \mu = \mu_0$ against $H_A: \mu \neq \mu_0$, where μ is the mean of a normal variable X with unknown variance σ^2. The population distributions are indexed by the parameter pair (μ, σ^2), and H_0 restricts μ but leaves σ^2 as a free parameter. The likelihood function is

$$L(\mu, \sigma^2) = (\sigma^2)^{-n/2} \exp\left\{-\frac{1}{2\sigma^2} \sum (X_i - \mu)^2\right\}.$$

This is maximized (see Solved Problem O in Chapter 9) at $(\bar{X}, V)$, where

$$V = \frac{1}{n} \sum (X_i - \bar{X})^2.$$

The maximum value of L is

$$L(\bar{X}, V) = V^{-n/2} \exp\left\{-\frac{1}{2V} \sum (X_i - \bar{X})^2\right\} = V^{-n/2} e^{-n/2}.$$

This is the denominator in (1).

To find the numerator of (1), we need to the maximum of L when μ is held fixed at μ_0. With $\mu = \mu_0$, the log of the likelihood is

$$\log L(\mu_0, \sigma^2) = -\frac{n}{2} \log \sigma^2 - \frac{1}{2\sigma^2} \sum (X_i - \mu_0)^2.$$

Setting the derivative with respect to σ^2 equal to zero, we find that the maximum likelihood within H_0 is attained when σ^2 equals

$$V_0 = \frac{1}{n} \sum (X_i - \mu_0)^2 = \frac{1}{n} \sum [(X_i - \bar{X})^2 + (\bar{X} - \mu_0)]^2$$

$$= V + (\bar{X} - \mu_0)^2.$$

(The cross-product term vanishes because the deviations of the X_i about $\bar{X}$ sum to zero.) The maximum value of L in H_0 is

$$L(\mu_0, V_0) = V_0^{-n/2} \exp\left\{-\frac{1}{2V_0} \sum (X_i - \mu_0)^2\right\} = V_0^{-n/2} e^{-n/2}.$$

So the likelihood ratio (1) is

$$\Lambda = \left(\frac{V_0}{V}\right)^{-n/2} = \left(1 + \frac{(\bar{X} - \mu_0)^2}{V}\right)^{-n/2}.$$

This is small when the second term in the braces is large. That term is proportional to T^2, the test statistic for a two-sided test of $\mu = \mu_0$ [see (1) of Section 10.4]:

$$T^2 = \frac{(\bar{X} - \mu_0)^2}{V/(n-1)}.$$

The likelihood ratio rejection region $\Lambda < K$ is therefore equivalent to a two-sided rejection region for $T: |T| > K$.   ■

Large Sample Distribution of the Λ

In this example, we have seen that Λ was expressible in terms of a statistic whose null distribution is familiar. This is not always the case. The null distribution of Λ can be quite complicated. However, when H_0 is formed by fixing the value of one or more components of the parameter vector θ, it has been shown[3] that as n becomes infinite, the statistic $-2 \log \Lambda$ is asymptotically chi-square under H_0. The number of degrees of freedom is equal to the number of parameters assigned specific values under H_0. (In the preceding example, H_0 assigns a value to the one parameter μ, so d.f. = 1.) The next example illustrates how this fact can be used.

Example 11.6b

A Two-Sided Test for $p = p_0$

Consider a random sample from Ber(p), and the problem of testing $p = .5$ against $p \neq .5$. Suppose we observe 100 independent trials and record 58 successes. The likelihood function (see Example 8.2b) is

$$L(p) = p^{58}(1 - p)^{42}.$$

This is a maximum (over all p) at $\hat{p} = .58$, the sample relative frequency of success. The maximum value of L is

$$L(\hat{p}) = (.58)^{58}(.42)^{42}.$$

When p is restricted to H_0, L has only one value:

$$L(.5) = (.5)^{58}(.5)^{42}.$$

The test statistic (1) is then

$$\Lambda = \frac{L(.5)}{L(\hat{p})} = \left(\frac{.5}{.58}\right)^{58}\left(\frac{.5}{.42}\right)^{42},$$

and

$$-2 \log \Lambda = -116 \log\left(\frac{.5}{.58}\right) - 84 \log\left(\frac{.5}{.42}\right) \doteq 2.57.$$

If we specify $\alpha = .05$, for example, the critical value of $-2 \log \Lambda$ is the 95th percentile of the chi-square distribution with one d.f., or 3.84. Since 2.57 < 3.84, we do not reject H_0.

Example 10.3 gave a Z-test for this same situation. The likelihood ratio test can be shown[4] to be asymptotically equivalent to the two-sided Z-test: Z^2 and χ^2 are asymptotically equal (under H_0), and their null distributions are asymptotically the same—namely, chi-square with one degree of freedom. For the present data, $Z^2 \doteq 2.89$, close to the value of $-2 \log \Lambda$. Thus, $Z \doteq 1.7$, and we do not reject H_0, since (for $\alpha = .05$) the critical value of Z is 1.96. (To be continued.) ∎

We have introduced the likelihood ratio test as a rule for deciding between a null and an alternative hypothesis. The statistic Λ is also useful in problems where a decision is not required: In the approach of Chapter 10, take small values of Λ as "extreme." If Λ is a monotonic function of some more familiar statistic whose

---

[3] See Wilks [20], pp. 419ff.
[4] Cf. Lindgren [15], p. 427.

distribution is known, we can find a P-value from that distribution. If n is large, so that $-2 \log \Lambda \approx \chi^2$, we can find a P-value in the chi-square table.

Example 11.6b   **A Two-Sided Test for $p = p_0$ (continued)**
In the preceding example, we found $Z^2 = 2.89$ and $-2 \log \Lambda = 2.57$. The (two-sided) P-value is .09 from Z and .11 from $-2 \log \Lambda$. For large samples, the Z-test and the likelihood ratio test will give similar results. ∎

The likelihood ratio approach in these examples has not given us new tests. However, the examples show that some of the standard tests considered earlier are in fact likelihood ratio tests.

Chapter Perspective

In this chapter, we have considered tests of hypotheses as decision rules—rules set up prior to the collection of data. The type of rejection region used is based on a consideration of alternative hypotheses prior to analysis of the data, and specific rejection limits are chosen to satisfy specifications of error sizes or power characteristics.

The notion of the power of a test is useful even when (as in the preceding chapter) an immediate decision is not required and we only report the degree of evidence against H_0. If a test is not powerful enough to notice a practically important alternative to H_0 (by giving us a "small" P-value), a larger sample is needed.

The likelihood ratio approach to test construction gives us a systematic method of devising tests for particular null and alternative hypotheses. This intuitive method usually gives good tests, at least for large samples.

We have introduced hypothesis testing in settings where we have a single sample for testing hypotheses about the population from which it was drawn. The next chapter takes up the important problem of testing hypotheses about two populations using a sample from each.

Solved Problems

Section 11.1

A. The sample space of a test statistic X has five values, $\{a, b, c, d, e\}$. Consider testing f_0 versus f_1, probability functions for X, defined as follows:

x	$f_0(x)$	$f_1(x)$
a	0	.3
b	.1	0
c	.2	.2
d	.3	.4
e	.4	.1

  (a)   Find α and β for the rejection region $\{b, c\}$.
  (b)   Find α and β for the rejection region $\{d\}$.

Solution:

(a) $\alpha = P(X = b \text{ or } c \mid H_0) = .1 + .2 = .3.$

$\beta = 1 - P(X = b \text{ or } c \mid H_1) = 1 - (0 + .2) = .8.$

(b) $\alpha = P(X = d \mid H_0) = .3.$

$\beta = 1 - P(X = d \mid H_1) = 1 - (.4) = .6.$

[Note that this rejection has the same α but a smaller β than the rejection region in (a), so it's a better test.]

B.   In a 1977 study[5] on sex and race discrimination, individuals visited nine car dealerships to request the best possible deal on a certain car. The prices offered a black female minus the corresponding prices offered a white male are as follows (in dollars):

$$111, -35, 37, 69, 86, 59, 155, -10, 253.$$

For testing the hypothesis that on average there is no mean difference in treatment between the two types of customer,

(a)   state suitable null and alternative hypotheses.

(b)   carry out a t-test, reporting the result as a P-value.

(c)   If you were asked to "test H_0 at the 5% significance level," what's your conclusion?

Solution:

(a)   No difference in treatment can be interpreted as no mean difference in the prices offered black females and white males. So H_0 is $\delta = 0$, where δ is the population mean difference in price ($\mu_{BF} - \mu_{WM}$). The alternative one usually has in mind in discrimination is that blacks and females are discriminated against; this would correspond to $\delta > 0$ ($\mu_{BF} > \mu_{WM}$).

(b)   The sample mean difference is $\bar{X} = 80.56$, and $S_X = 86.81$. Thus,

$$T = \frac{80.56 - 0}{86.81/\sqrt{9}} \doteq 2.78.$$

The one-sided P-value (8 d.f.) is .012.

(c)   Since $P < .05$ (*either* one- or two-sided), reject H_0.

C.   In Example 11.1a, suppose we decide to reject $H_0 : \mu = 8.0$, if a 90% confidence interval calculated from the amounts in n cupfuls does not include the value 8.0. (Assume that the sample mean is approximately normal.)

(a)   Express this rule as an inequality for the standardized sample mean.

(b)   Find α for this rule. (Can you generalize?)

Solution:

(a)   From Section 9.3, we have these confidence limits:

$$\bar{X} - 1.645\frac{\sigma}{\sqrt{n}} < \mu < \bar{X} + 1.645\frac{\sigma}{\sqrt{n}}.$$

---

[5] "Sex and race discrimination in the new-car showroom: A fact or myth," *J. Consumer Affairs* (1977), 107–113.

The rule considered here is to accept $\mu = 8.0$ if

$$\bar{X} - 1.645\frac{\sigma}{\sqrt{n}} < 8.0 < \bar{X} + 1.645\frac{\sigma}{\sqrt{n}}$$

or

$$\left|\frac{\bar{X} - 8.0}{\sigma/\sqrt{n}}\right| < 1.645.$$

(b) The probability of rejecting $\mu = 8.0$ under H_0 is the probability that $|Z| > 1.645$. This will be the case if a 90% confidence interval does not cover the true value of μ. So $\alpha = 1 - .90 = .10$. [The obvious generalization is that for a rule that rejects H_0: $\theta = \theta_0$ if a $100(1 - \alpha)\%$ confidence interval does not include θ_0, the size of the type I error is α.]

Sections 11.2–11.3

D. Find the power function for the test in Problem C, if $\sigma/\sqrt{n} = .1$.

Solution:
For large (or even only moderate) n, $\bar{X} \approx N(\mu, .01)$. So

$$\pi(\mu) = 1 - P(|\bar{X} - 8| < .1645) = P(\bar{X} < 7.8355) + P(\bar{X} > 8.1645)$$

$$\pi(\mu) = 1 - \Phi\left(\frac{8.1645 - \mu}{.10}\right) + \Phi\left(\frac{7.8355 - \mu}{.10}\right).$$

This curve has a trough at $\mu = 8$, where the power is $\alpha = .10$. The test is appropriate for the alternative $\mu \neq 8.0$.

E. In the setting of Problems C and D,
  (a) construct a one-sided confidence interval such that rejecting H_0 when the interval fails to cover $\mu = 8$ is equivalent to a critical region of the form $\bar{X} > K$.
  (b) determine the power function of the rule in (a).

Solution:
(a) We obtain a one-sided confidence interval as follows:

$$.90 = P\left(\frac{\bar{X} - \mu}{.10} > 1.28\right) = P(\bar{X} - \mu > .128) = P(\mu < \bar{X} - .128).$$

The rule that rejects $\mu = 8.0$ if it is not in the one-sided confidence region has this critical region: $\bar{X} > 8.0 - .128$.

(b) The power function is

$$\pi(\mu) = P(\bar{X} > 7.87 \,|\, \mu) = 1 - \Phi\left(\frac{7.87 - \mu}{.10}\right) = \Phi(10\mu - 78.7).$$

The shape of this is that of a normal c.d.f., with high power for large values of μ. The test would be appropriate for testing $\mu = 8.0$ versus $\mu > 8.0$.

F. Let a number $\varphi(\mathbf{x})$ be assigned to each $\mathbf{x}$ such that $0 \leq \varphi \leq 1$. Interpret φ as defining a test—a **randomized test**—as follows: If we observe $\mathbf{x}$, reject H_0 with probability $\varphi(\mathbf{x})$.

[That is, toss a "coin" with probability φ of heads and reject H_0 if heads turns up, otherwise accept H_0.]

(a) Find the power function of φ in testing a hypothesis about a parameter θ.

(b) Find a function φ that defines a test equivalent to that defined by a rejection region R.

Solution:

(a) $\pi_\varphi(\theta) = P(\text{rej. } H_0 \mid \theta) = E[P(\text{rej. } H_0 \mid \mathbf{x}, \theta)] = E[\varphi(\mathbf{x} \mid \theta)]$

(b) Reject H_0 with probability 1 when $\mathbf{x}$ is in R, and with probability 0 when $\mathbf{x}$ is in R^c:

$$\varphi(x) = \begin{cases} 1 & \text{if } \mathbf{x} \text{ is in } R, \\ 0 & \text{if } \mathbf{x} \text{ is not in } R. \end{cases}$$

So the class of tests defined by functions φ (with $0 \le \varphi \le 1$) includes those defined by critical regions R.

G. A study of adolescent suicides[6] points out the large sample size needed to recognize an effect of a treatment: "For example, let us assume that we have developed a preventive measure which, if successful, will reduce the suicide rate from 30 to 25 per 100,000. If we employ a probability value of .05 and a confidence interval of .95, we would need an experimental sample of almost 3 million ... and a control sample of the same size." To explain this would require a "two-sample" analysis, to be taken up in the next chapter. For now, consider testing $p = .00030$ against $p = .00025$, where p is the present suicide rate, and determine a sample size such that the type I and II error sizes are both .05.

Solution

The rejection region for the stated one-sided alternative is $\hat{p} < K$. Under the null hypothesis, $\sqrt{pq} = \sqrt{.0003 \times .9997} = .0173$, so s.d.$(\hat{p}) \doteq .0173/\sqrt{n}$. Then

$$\alpha = P(\hat{p} < K \mid H_0) = \Phi\left(\frac{K - .00030}{.0173/\sqrt{n}}\right) = .05,$$

$$1 - \beta = P(\hat{p} < K \mid H_1) = \Phi\left(\frac{K - .00025}{.0173/\sqrt{n}}\right) = .95.$$

So we have these simultaneous equations for K and n:

$$\begin{cases} K\sqrt{n} = .0003\sqrt{n} + (.0173)(-1.645) \\ K\sqrt{n} = .00025\sqrt{n} + (.0173)(1.645) \end{cases}$$

Solving, we find that n is about 1.3 million.

Section 11.4

H. An acceptance sampling scheme is to have probability .05 of rejecting a lot in which the lot fraction defective p is .04, and also a probability .05 of accepting a lot in which $p = .20$. Find an appropriate sample size and acceptance number.

---

[6] L. Eisenberg, "The epidemiology of suicide in adolescents," *Pediatric Annals 13* (1984), 47–54.

Solution:

The stated requirements are translated as these equations:

$$\sum_0^c f(k\,|\,.04) = 95, \qquad \sum_0^c f(k\,|\,.20) = .05.$$

Solving these simultaneously for a sample size n and an acceptance number c is difficult; indeed, with the requirement that these be integers, there may be no exact solution. One could obtain approximate solutions by trial and error or by using numerical methods. Rather, we'll simply show that $n = 36$ and $c = 3$ are approximate solutions.[7] When $p = .04$, $np = 1.44$, and we use a Poisson approximation for the hypergeometric probability $f(k\,|\,p)$:

$$\sum_4^\infty f(k\,|\,.04) = 1 - e^{-1.44}\left(1 + 1.44 + \frac{1.44^2}{2} + \frac{1.44^3}{6}\right) \doteq .06.$$

When $p = .20$ $np = 7.2$ and we use a normal approximation,

$$\sum_0^3 f(k\,|\,.20) \doteq \Phi\left(\frac{3.5 - 7.2}{2.4}\right) = \Phi(-1.54) = .062.$$

Section 11.5

I.   Referring to Problem A, find the most powerful rejection region with $\alpha = .3$.

Solution:

We saw that the region $\{d\}$ is more powerful than $\{b, c\}$. Is it most powerful? We repeat here the table of Problem A, with a column of values of Λ. In order of increasing Λ, the x-values are a, d, c, e, b. To form a Neyman–Pearson rejection region, we put x-values into it in that order. The Neyman–Pearson regions are $\varnothing$, $\{a\}$, $\{a, d\}$, $\{a, d, c\}$, $\{a, d, c, e\}$, $\{a, d, c, e, b\}$, with α's equal to 0, 0, .3, .5, .9, and 1, respectively. The one with $\alpha = .3$ is $\{a, d\}$. For this region, $\beta = 1 - P(X = a, d\,|\,H_1) = .3$.

x	$f_0(x)$	$f_1(x)$	$\Lambda(x)$
a	0	.3	0
b	.1	0	∞
c	.2	.2	1
d	.3	.4	$\dfrac{3/4}{4}$
e	.4	.1	4

J.   Consider a random sample of size $n = 5$ for testing H_0: $X \sim N(0, 1)$ against the alternative that X has this Cauchy p.d.f.:

$$f_1(x) = \frac{1/\pi}{1 + x^2}.$$

Find the Neyman–Pearson rejection regions.

---

[7] These values of c and n were obtained from a chart on p. 297 of Lindgren, McElrath, & Berry, *Intro. to Probability & Statistics*, 4th Ed., (New York: Macmillan, 1978).

Solution:

Under H_0, $\bar{X} \sim N(0, \frac{1}{5})$, but under H_A $\bar{X}$ has the same distribution as X. (See Solved Problem Q of Chapter 8.) The likelihood ratio Λ (except for a constant factor) is thus

$$\frac{e^{-5\bar{X}^2/2}}{(1 + \bar{X}^2)^{-1}} = \frac{1 + \bar{X}^2}{e^{5\bar{X}^2/2}},$$

so $\Lambda < K$ is equivalent to

$$\log \Lambda = \log(1 + X^2) - \frac{5}{2}X^2 < K'.$$

This is a decreasing function of X^2, so the Neyman–Pearson rejection region is of the form $X^2 > K$.

K. Find the uniformly most powerful tests for $\sigma = 1$ versus $\sigma > 1$ based on a random sample of size n from $N(0, \sigma^2)$.

Solution:

The Neyman–Pearson rejection regions for the alternative $\sigma = \sigma_0$ are of the form

$$\Lambda = \sigma_0^2 \exp\left\{ -\frac{1}{2}\left(1 - \frac{1}{\sigma^2}\right)\sum X_i^2 \right\} < K'.$$

Since Λ is a decreasing function of $\sum X_i^2$, this is equivalent to the region $R = \{\sum X_i^2 > K\}$. Since σ_0 can be any value of σ greater than 1, R is UMP for $\sigma > 1$.

Section 11.6

L. Construct a likelihood ratio test for $\sigma = \sigma_0$ versus $\sigma \neq \sigma_0$, based on a random sample of size n from $N(\mu, \sigma^2)$.

Solution:

The likelihood function is

$$L(\mu, \sigma^2) = (\sigma^2)^{-n/2} \exp\left\{ -\frac{1}{2\sigma^2}\sum(X_i - \mu)^2 \right\}.$$

The maximum of L over all (μ, σ^2) is attained at $(\bar{X}, V)$, where $V = \frac{n-1}{n}S^2$. With $\sigma = \sigma_0$, the maximum over μ is attained at $\bar{X}$. Thus,

$$\Lambda = \frac{L(\bar{X}, \sigma_0^2)}{L(\bar{X}, V)} = \frac{(\sigma_0^2 \, e^{V/(2\sigma_0^2)})^{-n/2}}{(eV)^{-n/2}} = \left(\frac{e^W}{We}\right)^{-n/2},$$

where $W = V/\sigma_0^2$. Thus, $\Lambda < K$ is equivalent to $W > K'$. The graph of e^W/W is concave up with a minimum at $W = 1$; the inequality $e^W/W > C$ will hold if $W > K_1$ or $W < K_2$, where K_1 and K_2 are two numbers such that $e^K/K = C$. Thus values of V that are much bigger or much smaller than σ_0^2 are extreme. The rejection region is two-sided, but it is not symmetric about σ_0^2, nor are the tail-areas under the null p.d.f. of V equal.

M. Apply the likelihood ratio method to testing $\lambda \leq 1$ against $\lambda > 1$ based on a random sample from $\text{Exp}(\lambda)$. Express the critical region in terms of the sample mean.

Solution:

The likelihood function is

$$L(\lambda) = \lambda^n e^{-\lambda \Sigma X} = (\lambda e^{-\lambda \bar{X}})^n.$$

This has a maximum at $\lambda = 1/\bar{X}$. A sketch of L shows that if $\bar{X} > 1$, the maximum of L over H_0 is the same as the maximum over all λ, so $\Lambda = 1$. If $\bar{X} < 1$, the maximum on H_0 is $L(1) = \exp(-n\bar{X})$. In this case

$$\Lambda = \frac{L(1)}{L(1/\bar{X})} = \bar{X} e^{-n(\bar{X}-1)}$$

This is an increasing function of $\bar{X}$. So $\Lambda < K$ is equivalent to a critical region for the sample mean of the form $\bar{X} < K'$.

Problems

For problems marked with an asterisk, answers are provided in Appendix 2.

Section 11.1

*1. Suppose Oscar (in Example 10a) decided to "accept" the hypothesis of ESP (i.e., reject the null hypothesis $p = \frac{1}{4}$) if Felix made seven or more correct identifications in ten trials. (We won't go into what actions might correspond to rejecting or accepting H_0.) Find the size of the type I error for this decision rule.

*2. Consider these two rejection regions: $R_1 = \{|\bar{X} - 10| > .5\}$, and $R_2 = \{|\bar{X} - 10| > .8\}$.
    (a) Which has the larger α?
    (b) Which has the larger β?

*3. Consider a random sample of size $n = 4$ from $N(\mu, 1)$, to be used in testing $H_0: \mu = 10$. For each of the rejection regions R_1 and R_2 of Problem 2,
    (a) find α.
    (b) find β for the alternative $\mu = 11$.
    (It is instructive to sketch the distribution of $\bar{X}$ under H_0 and under the alternative $\mu = 11$, identifying the areas that represent the error sizes you have calculated.)

4. The hospital lab mentioned in Example 11a also checked its equipment for measuring white blood cell count using standard supplies with low, average, and high counts. The white count of the "low" standard supply was 3,800, and the standard deviation of the measurements was assumed to be 400. The rule used to determine whether to proceed with the day's testing or to stop and look for an assignable cause was defined by a control chart. An upper control line was drawn at 4,600 and a lower control line at 3,000. A measurement of the standard supply outside those limits would call for rejecting the hypothesis that all was well and require a check of the equipment. Assume that the distribution of a single reading is $N(\mu, 400^2)$.
    (a) Find the size of the type I error for the lab's operating rule.
    (b) Find the probability that, following the rule, the lab would (incorrectly) fail to check the equipement if the mean of the measurements were actually (i) $\mu = 5,400$, and (ii) $\mu = 2,800$.

*5. Consider this rule: Reject the hypothesis that $\mu = 10$ if the usual 90% confidence interval for μ based on the mean of a random sample of size 100 does not include the value $\mu = 10$. Assume $\sigma = 1$.
  (a) Express the rule as a rejection region for the sample mean.
  (b) Find the α for this rule.

6. A panel of health experts will approve a certain drug for use, assuming no adverse side effects, if two studies comparing its effect with a placebo obtain statistically significant results (that is, if each of the two studies rejects the hypothesis of no treatment effect at $\alpha = .05$).
  (a) Find the probability that this rule serves to release a drug that has no effect whatever.
  (b) Suppose one wants the probability that an ineffective drug is released to be .01. How should α be set to accomplish this?

7. In the comic strip "Hi and Lois," the twins Dot and Ditto argued about the probability that a piece of buttered bread dropped on the floor would land buttered side down. Dot claimed it was .9 and Ditto, that it was .5. They conducted several trials (interrupted by the mother).
  *(a) Suppose they had decided to carry out five trials and would reject $H_0: p = .5$ and accept $H_1: p = .9$ if $Y > K$, where Y is the number of times the bread landed buttered side down in five trials. Calculate α and β for this rule, for each value of K: 0, 1, 2, 3, 4, 5, 6. Plot the seven pairs (α, β) obtained in this way, taking note of the inverse nature of the relationship.
  (b) Consider the rule of rejecting $p = .5$ if $Y < K$. Plot the seven (α, β)-pairs corresponding to $K = 0, 1, \ldots, 6$ and explain why these rules are foolish.

Sections 11.2–11.3

*8. Find the power function for the rule given in Problem 1.

*9. As in Problem 3, consider a random sample of size $n = 4$ from $N(\mu, 1)$ to be used in testing $H_0: \mu = 10$. For each of the rejection regions R_1 and R_2 in Problem 2, find the power function. (In the process of solving Problem 3, you would have actually calculated a couple of points on the power curve.)

10. Find and sketch the power function for the lab's operating rule in Problem 4. Suppose instead of 2σ-limits, the lab used the 3σ-limits specified in the state's regulations— rejecting H_0 if the observation fell outside $3,800 \pm 1,200$.
  (a) What would this do to the power curve?
  (b) Which rule is more protective of the patients' interests?

*11. Consider testing $\theta = 0$ on the basis of a single observation on a Cauchy variable X with median θ: $f(x \mid \theta) = \frac{1}{\pi}[1 + (x - \theta)^2]^{-1}$.
  (a) Find α and the power function for the rejection region $X > 2$.
  (b) Find α and the power function for the rejection region $|X| > 1.5$.

12. Suppose a treatment intended to increase a response. Consider the null hypothesis that the mean increase is $\mu = 0$. A study employing a sample of size n is to be conducted to determine the treatment's effect. A treatment effect is to be inferred if $\bar{X} > K$. The test is to have $\alpha = .05$ and an 80% chance of detecting a mean increase of 2 units. What sample size n and rejection value K should be used, given that $\sigma = 5$?

*13. Problem 5 considered this rule: Reject the hypothesis that $\mu = 10$ if a 90% confidence interval for μ based on a random sample of size 100 does not include the value $\mu = 10$. Find the power function. (Again, assume $\sigma = 1$.)

14. Find a function φ (see Solved Problem F) that defines the randomized test based on a statistic T that rejects H_0 if $T < K$, accepts H_0 if $T > K$, and rejects H_0 with probability γ if $T = K$.

15. Let φ be a randomized test [see Solved Problem F] and let T be sufficient for θ. Define $\varphi^*(t) = E[\varphi(\mathbf{X}) \mid T = t]$. Assume $\mathbf{X}$ is discrete.
   (a) Show that φ^* is a test. (In particular, you need to show that it does not depend on θ and that $0 \le \varphi^* \le 1$.)
   (b) Show that the power function for φ^* is the same as the power function for φ. (This shows that one can equal the performance of any test, in terms of power, with a test based on a sufficient statistic.)

Section 11.4

*16. A large shipment of disposable thermometers is to be accepted or rejected by a clinic, according as a sample of ten thermometers has no defectives or has at least one defective. Find and sketch the power function of this rule, as a function of p, the proportion of defectives in the shipment.

17. An acceptance sampling plan for lots of size 20 is to accept a lot if a random selection of five from the lot includes at most one defective item. Find the power function of this plan.

*18. Suppose a manufacturer of a battery-powered toy wants to buy batteries whose operating lives (X) average at least 40 hours. The purchase agreement with a certain supplier specifies that for a shipment to be acceptable, a sample of 25 batteries taken at random from the shipment must satisfy the condition $T > -1.5$, where

$$T = \frac{5(\bar{X} - 40)}{S}.$$

   (a) Find α for this rule, taking H_0 to be $\mu = 40$.
   (b) Power calculations for the given rule require a distribution for T when $\mu \ne 40$. This is a distribution (called *noncentral t*) that we have not studied. Instead, calculate the power function of the rejection region $\bar{X} - 40/.8 < -1.5$, assuming the population s.d. is known: $\sigma = 4$.
   (c) Suppose the manufacturer and supplier agree that a procedure with rejection region of the form $\bar{X} < K$ is acceptable to both if it has a 5% chance of rejecting a shipment in which $\mu = 42$ and a 1% chance of accepting a shipment in which $\mu = 39$. Again assuming $\sigma = 4$, find the sample size and rejection value K that will satisfy these criteria.

19. Find the OC-function for a plan that accepts a lot if a random selection of 15 items from the lot includes at most 1 defective. (The lot size is much larger than 15.)

Section 11.5

*20. Consider testing $\mu = 0$ against $\mu = 1$, based on a random sample of size n from $N(\mu, 1)$.
   (a) Find the most powerful rejection regions in terms of $\bar{X}$.

(b)   Find α and β as functions of the rejection boundary in (a) when $n = 4$.

(c)   Sketch a plot of β versus α as given by (b).

∗21.   Find the uniformly most powerful tests for $\mu = 0$ versus $\mu > 0$, based on a random sample of size n from $N(\mu, 1)$.

22.   For a random sample of size n from $\text{Exp}(\lambda)$,

(a)   find the most powerful tests for $\lambda = 1$ versus $\lambda = 2$.

(b)   find the form of uniformly most powerful tests for $H_0: \lambda = 1$ against $H_A: \lambda > 1$.

∗23.   Find the most powerful tests for $m = 1$ versus $m = 2$, based on a random sample of size n from $\text{Poi}(m)$.

24.   The random variable Z has one of these two distributions:

z	z_1	z_2	z_3	z_4	z_5
$f(z \mid H_0)$	.2	.3	.1	.3	.1
$f(z \mid H_A)$	.3	.1	.3	.2	.1

(a)   Find the most powerful tests.

(b)   Find the most powerful test of size $\alpha = .3$.

(c)   Find the most powerful (randomized) test of size $\alpha = .2$.

Section 11.6

∗25.   Given a random sample from $N(\mu, 1)$, find the likelihood ratio tests for $\mu = \mu_0$ against $\mu \neq \mu_0$, as rejection regions for the sample mean.

26.   Find the likelihood ratio rejection regions for testing $\mu = 0$ against $\mu = 1$, using a random sample of size n, when the population is normal with unknown variance.

∗27.   Find the likelihood ratio tests for $\sigma = \sigma_0$ against $\sigma \neq \sigma_0$, based on a random sample from $N(0, \sigma^2)$.

28.   Find the likelihood ratio tests for $\lambda = \lambda_0$ against $\lambda \neq \lambda_0$, based on a random sample from $\text{Exp}(\lambda)$.

∗29.   An observation Z takes on one of four values. Its distribution is one of three identified as θ_1, θ_2, and θ_3, shown in the following table:

	z_1	z_2	z_3	z_4
θ_1	.2	.3	.1	.4
θ_2	.6	.1	.1	.2
θ_3	.3	0	.4	.3

Find all the likelihood ratio rejection regions for testing $\theta = \theta_1$ against the alternative that $\theta = \theta_2$ or θ_3.

30.   Find the likelihood ratio test for $\theta = 0$ against $\theta \neq 0$ based on a single observation from a Cauchy distribution with median θ. (The likelihood function is $[1 + (X - \theta)^2]^{-1}$.)

31. Consider a random sample of size n from $N(\mu, 1)$ for testing $H_0: \mu \le 0$ against $H_A: \mu > 0$. The likelihood function [from (2) of Section 8.2] is

$$L(\mu) = \exp\left[-\frac{n}{2}(\mu - \bar{X})^2 \right].$$

    (a) Sketch $L(\mu)$ for an $\bar{X} < 0$ and for an $\bar{X} > 0$. In each case, deduce from the graphs the location of the maximum of L over $\mu \le 0$ and the maximum over $H_0 \cup H_A$.

  *(b) Calculate Λ for $\bar{X} < 0$ and for $\bar{X} > 0$.

    (c) Show that the rejection region $\Lambda < K$ is equivalent to one of the form $\bar{X} > K'$.

32. The function $f(z \mid \theta)$ is shown in the accompanying table. We want test $H_0: \theta = 0$ against $H_A: 0 < \theta < 1$.

    (a) Find the likelihood ratio test for which $\alpha = \frac{1}{6}$ and determine its power function.

    (b) For the test with rejection region $R = \{z_3\}$, find α and the power function. How does this test compare with the likelihood ratio test?

Z	$\theta = 0$	$0 < \theta < 1$
z_1	$\dfrac{1}{12}$	$\dfrac{\theta}{3}$
z_2	$\dfrac{1}{12}$	$\dfrac{1 - \theta}{3}$
z_3	$\dfrac{1}{6}$	$\dfrac{1}{2}$
z_4	$\dfrac{2}{3}$	$\dfrac{1}{6}$

CHAPTER 12

Comparing Two Populations

Suppose a new treatment is to be evaluated. Researchers continually propose new treatments (drugs, surgical techniques, chemical baths, fertilizers, teaching methods, and so on), as well as new formulas (for making bread, detergents, alloys, and the like). These are aimed at making some aspect of life less painful, more pleasant, more productive, or more profitable.

Sometimes a "treatment" is simply a set of circumstances that exists and may have an effect, although no treatment has been deliberately applied. For example, in some localities, being black may affect the punishment for a crime; being female may have an effect on income; smoking may affect health; and so on.

To assess the effects of a treatment or of a difference in treatments, we'd want to compare the population of responses of treated individuals or experimental materials with those that are not treated (or are given a second treatment). Generally speaking, we can compare populations only by comparing samples from them.

In comparing populations, the usual null hypothesis is that they are the same. Using samples to test the hypothesis that two populations are the same in every respect can be difficult, so we usually restrict consideration to a single population parameter. For example, we ask whether the two populations have the same mean, or have the same proportion of successes, or have the same variance. We employ the labels 1 and 2 to refer to the two populations—as subscripts or parameters (μ, p, and σ^2), sample sizes (n), and statistics ($\bar{X}$, $\hat{p}$, and S^2) to identify them with their respective populations.

We'll take up particular two-sample tests in later sections. First, we discuss some general aspects of evaluating treatments.

12.1   Treatment Effects

The response to a treatment in some settings is categorical—for example, life or death. For the success–failure type of response (as when a patient lives or dies), the treatment **effect** is the change in probability of success. When there are more than two categories (such as little, moderate, or complete relief from pain), the treatment effect is not as easily described.

In other settings, the response to a treatment is a numerical characteristic, such as blood pressure, hardness, yield, and the like. Responses will usually be different for different individuals or different pieces of experimental material. (Even the same individual may respond differently at different times.) So the treatment may increase some responses and decrease others; but if it increases the average response, we say there is a treatment effect. The *effect* is the average amount of increase.

Example **12.1a**   **Lowering Blood Pressure**

A biochemist designs a new drug that is intended to lower blood pressure. Once it has been shown to be safe when used on animals and then healthy human volunteers, it is used on patients with high blood pressure. The patients' blood pressures change after taking the drug; most fall, but some rise. The question is: Does the drug lower blood pressure on the average?   ■

In evaluating a treatment, one compares it with no treatment or with some standard treatment. There are two populations to consider: the **treatment population**—the population of responses of all individuals who might be treated— and the **control population**—the population of responses of individuals who are not treated or who are given the standard treatment.

Controls

Example **12.1b**   **Surgical Controls**

The following is quoted from an article in an American Airlines flight magazine[1] by the late Dr. William Nolen, author and surgeon:

> Before coronary by-passes became popular, another operation, called internal mammary artery ligation, was used in the treatment of angina pectoris. The internal mammary artery consists of two branches, one on either side of the breastbone; one of these goes toward the heart. The inventor of the procedure, a surgeon named Glover, would ligate, or tie off, one branch of the artery so that more blood would be shunted toward the heart. About 50 percent of his angina patients got better. It was a simple operation and could even be done under a local anesthesia. However, another surgeon decided to test it with some sham operations. He made a skin incision but did not tie off the internal mammary. He achieved about the same results as Dr. Glover—about 45 to 50 percent of his patients got better.

Without a control group, the real effect of a treatment cannot be assessed.   ■

When the distribution of responses in the control population is well known, it may be sufficient to sample only from the treatment population. Testing is then a one-sample problem and is treated using the methods discussed in Chapter 10. However, many statisticians recommend always using a control sample in addition to the treatment sample. It is a mistake to assume that the distribution of responses in the control population is completely known; the experimental units involved in

---

[1] *American Way*, August 1983.

the current experiment may be special in some important respect. The next example illustrates this point.

Example **12.1c** **Standard Treatment as a Control**

Suppose we're planning a clinical trial designed to evaluate a new therapy for cardiac rhythm disorders. An important question is how it compares with quinidine sulfate, a very old, standard treatment. The null hypothesis is that the new therapy is no more effective than quinidine. The literature contains many studies evaluating the effectiveness of quinidine. It is tempting to assume that its average effectiveness is known to be the average in these studies.

The results for quinidine in these studies vary greatly—from 20% of the patients showing marked improvement to 65% showing the same level of improvement. These studies had different entry criteria; and even when the criteria are the same, two different clinicians may admit different types of patients. For example, one study may involve much sicker patients than another. It is difficult or impossible to see which study group, if any, provides an appropriate comparison for the study being planned. The safest approach in a new study is to include a control group, assigning some patients (randomly) to the new treatment and the rest to quinidine. ∎

Randomizing some subjects to a treatment and the remainder to no treatment is the safest way to ascertain the effect of the treatment, but this may be impossible. For example, it could be considered unethical to assign patients who are terminally ill to a placebo when there is a treatment that might be effective. An alternative is to use the results for patients treated previously by the same clinicians. These are called **historical controls**. Another possibility is to cull patients from studies in the literature who match those in the current study—*literature controls*. The use of historical or literature controls is a controversial matter. Some statisticians claim that results based on such controls are invalid. In any case, they can serve an important function when randomized controls are not possible.

Studies that are planned before treatments are given and responses measured are called **prospective**. Studies that involve searching existing records of treatments and results are **retrospective**. A study may have prospective and retrospective aspects. For instance, two active treatments may be compared prospectively, but each may be compared to no treatment using historical controls.

An advantage of a prospective study is that subjects can be assigned to treatment or control independent of characteristics (such as severity of their disease) that may influence responses. The assignment should be made randomly so that these characteristics, including any that are unrecognized by the experimenter, tend to balance between treatment and control. A table of random digits can be used to make the assignment. Start at an arbitrary point in the table and assign each individual in sequence to treatment or control, depending on whether the next digit is even or odd. This does not guarantee equal sample sizes, but there are various ways of achieving equal sample sizes for the treatment and control groups if this is important. For example, assign each odd-numbered individual at random to treatment or control and assign the next (even-numbered) individual to the other

Historical Controls

Retrospective versus Prospective Studies

group. (There may be selection bias if the investigator knows that the treatment is used on the odd-numbered individuals and can choose the next individual to be treated.)

Example 12.1d   **Smoking and Health**
Many studies have investigated the effect of smoking on health. Some of these have compared smoking histories of diseased persons with those of healthy persons; others have compared disease histories of smokers with those of nonsmokers. Prospective studies are impossible, since one could not *assign* some people to the treatment (smoking) and others to the control. Tobacco interests and others argue that individuals prone to disease may also be apt to continue smoking. They may concede that there is a relationship, but they point out that this is not enough to establish smoking as the *cause* of the disease.   ■

An investigator needs to answer such questions as the following. What data are important? How expensive are the needed measurements? For a control population, do we want active controls (a standard treatment) or passive controls (no treatment)? What levels of treatment should be used—how large a dose? Or in the case of the intermittent feeding (Example 10b), how intermittent? How long should a treatment last?

The Placebo Effect   When the experimental units are humans, the design must avoid the possibility that results are confounded with a "placebo effect." A **placebo** is a nontherapeutic "treatment"—a *non*treatment that appears to the subject to be identical to a treatment—for example, a pill that contains no active ingredient but resembles the real thing. It is well established in medicine that people sometimes respond because they think they're being treated; this response is the placebo effect.

A 1980 newspaper article, dealing with the role of doctors' hands in healing—the "human touch"—stated:

> If you tell someone a drug will relieve pain—and that person believes it—the odds are very good that the pain will be relieved. Solid scientific studies show that placeboes relieve pain by releasing endorphins . . . .

Example 12.1e   **Nonsurgery**
In the article quoted in Example 12.1b, Dr. Nolen goes on to explain the success rate of the sham operation—a kind of placebo. "You might consider this an example of the placebo effect." Whether this is placebo effect is not clear. Many patients get better without *either* treatment or placebo. The placebo effect can be isolated only by using a control group of individuals who were given neither treatment nor placebo. (What is thought to be a placebo effect is sometimes a *regression effect*–see Section 15.6.)   ■

Using a placebo for the control group has become common practice in clinical trials. Of course, the subjects should not know who is getting the real thing; such trials are called **blind**. (The laws of many countries require that human subjects be informed that they *may* get placebo therapy.) It is also best that the clinician not know who is getting which treatment, lest the clinician subconsciously give better collateral treatment to those in one of the groups. Experiments are considered more

Double-Blind Studies

reliable when they are **double-blind**, with neither patient nor clinician aware of who is getting the actual treatment.

12.2 Large-Sample Comparison of Means

Suppose we have two large, independent random samples, one from population 1 and one from population 2. We'll use these to test $H_0: \mu_1 = \mu_2$. In terms of the single parameter $\delta = \mu_1 - \mu_2$ introduced in Section 9.6, this null hypothesis can be written as

$$H_0: \delta = 0. \tag{1}$$

In Section 9.6, we estimated δ using the difference between the corresponding sample means:

$$Y = \bar{X}_1 - \bar{X}_2. \tag{2}$$

We also used the fact that Y is approximately normal when n_1 and n_2 are large. We can exploit this approximate normality in a "Z-test" for H_0.

To standardize Y, we need its expected value and s.d. under H_0. These were given in Section 9.6 [as (1) and (2)]:

$$E(Y) = \delta, \quad \text{and} \quad \text{s.d.}(Y) = \sqrt{\frac{\sigma_1^2}{n_1} + \frac{\sigma_2^2}{n_2}}. \tag{3}$$

If the difference (2) is approximately normal, the Z-score

$$Z = \frac{Y - E(Y \mid H_0)}{\sqrt{\frac{\sigma_1^2}{n_1} + \frac{\sigma_2^2}{n_2}}} \tag{4}$$

is approximately $N(0, 1)$. When the variances are known, this is a statistic, and we again have a Z-test; but the population variances σ_1^2 and σ_2^2 are usually not known. So in place of the s.d. of Y, we use the s.e. [(4) of Section 9.4]:

$$\text{s.e.}(Y) = \sqrt{\frac{S_1^2}{n_1} + \frac{S_2^2}{n_2}} \tag{5}$$

in constructing an approximate standardized mean difference:

Z-test for testing $H_0: \delta = 0$. Calculate

$$Z = \frac{Y - E(Y \mid H_0)}{\text{s.e.}(Y)} = \frac{\bar{X}_1 - \bar{X}_2 - 0}{\sqrt{\frac{S_1^2}{n_1} + \frac{S_2^2}{n_2}}}. \tag{6}$$

When n_1 and n_2 are large, Z is approximately normal; find P-values in Table II of Appendix 1.

The Z-score (5) is the distance between the observed Y and its expected value under H_0, measured (as usual) in numbers of standard errors. How large the sample sizes have to be for (5) to be approximately standard normal depends on the shape of the population; in most cases, sample sizes of 25 or more would do. (The case of smaller samples will be taken up in Section 12.4.)

Example 12.2a

Platelet Counts in Cancer Diagnosis

In the study referred to in Example 9.2c, investigators were interested in whether blood platelet count could be useful in cancer diagnosis. They measured the platelet counts of 153 male cancer patients and of 25 healthy males. The results were as follows:

	n	$\bar{X}$	S
Cancer patients	153	395	170
Healthy patients	35	235	45.3

In the five-step outline of Section 10.2, we proceed as follows:

1. The investigators' null hypothesis was that the mean count is the same for the two types of males (cancer patient and healthy): $\delta = 0$. (This is not really the correct null hypothesis for deciding whether platelet count discriminates between healthy persons and cancer patients.)
2. Assuming that a difference either way would be useful, we adopt the two-sided alternative $H_A: \delta \neq 0$.
3. The test statistic is the two-sample Z-score given by (6):

$$Z = \frac{(395 - 235) - 0}{\sqrt{170^2/153 + 45.3^2/35}} \doteq 10.17.$$

4. In view of H_A, large values of $|Z|$ are extreme.
5. The P-value is the area under the standard normal curve outside $Z = \pm 10.17$: according to Table II, $P = .0000$. (In fact, it is about 10^{-22}.)

So the observed difference is "highly statistically significant": The very small P-value suggests that the evidence against the null hypothesis is overwhelming. (This conclusion substantiates what is evident from the two sample histograms shown in Figure 7.9.) ∎

Choosing Sample Sizes

In Section 11.3, we considered the problem of choosing the size of a single sample so that a deviation from H_0 deemed practically significant would be likely to be picked up as statistically significant. Here we have to choose *two* sample sizes, n_1 and n_2. As before, we need some knowledge of population variability, but now we also have to decide the relative sizes of n_1 and n_2. In many cases, if $n_1 + n_2$ is fixed as an even integer, then $n_1 = n_2$ provides the greatest power. Although $\frac{n_1}{n_2}$ is an important variable to be specified in advance, for simplicity we'll take $n_1 = n_2 = n$.

Example **12.2b** **Hypertension and Sodium Levels**

Consider planning an experiment to determine whether hypertensive patients have higher sodium levels than normotensive patients. We want to be fairly confident that our test will detect an average difference of 5 mEq/L. We take this to mean that the test should be powerful, having a high probability of rejecting H_0 if the mean difference is $\delta = 5$. Past experience suggests that the population s.d.'s are about $\sigma = 5$. Suppose we want $n_1 = n_2 = n$ and require $\alpha = .05$ and the power at $\delta = 5$ to be .80. These conditions determine a rejection region and a sample size, provided that n is large enough for a normal approximation to be appropriate.

Consider a rejection region of the form $Y > K$. Then

$$1 - \alpha = .95 = \Phi\left(\frac{K - 0}{5\sqrt{1/n + 1/n}}\right), \quad \text{or} \quad \frac{K\sqrt{n}}{5\sqrt{2}} = 1.645,$$

the 95th percentile of the standard normal distribution. So $K\sqrt{n} = 11.63$.

The requirement that the power at $\delta = 5$ be .80 is a second condition on K and n:

$$.80 = P(Y > K \,|\, \delta = 5) = 1 - \Phi\left(\frac{K - 5}{5\sqrt{\frac{2}{n}}}\right).$$

So the Z-score in parentheses must be $-.8416$, the 20th percentile of Z. Thus,

$$(K - 5)\sqrt{n} = 5\sqrt{2}(-.8416), \quad \text{or} \quad K\sqrt{n} = 5\sqrt{n} - 5.95. \tag{7}$$

However, we saw above that $K\sqrt{n} = 11.63$. Substituting this value in (7) and solving for n, we find $\sqrt{n} = 3.52$ or $n = 12.4$. Rounding to the nearest larger integer: $n = 13$. With this in (7), we find $K = 11.63/\sqrt{13} = 3.23$. In summary, we need to sample 13 hypertensive and 13 normotensive subjects, and we reject H_0 if the difference in sample means is greater than 3.23. [Should it turn out that the data are not consistent with the assumption $\sigma = 5$ (see Section 10.6), we may find that we had a larger ($\sigma < 5$) or smaller ($\sigma > 5$) sample than needed.] ■

Power calculations are most relevant before the study. People sometimes wrongly criticize a study that has found a statistically significant difference on the grounds that the sample sizes were not large enough to detect a clinically significant difference. However, if a result has turned out to be statistically significant, the sample sizes *were* adequate for rejecting the null hypothesis!

12.3   Comparing Proportions

We turn next to the case of Bernoulli populations, comparing two population proportions (or probabilities of success). The null hypothesis of no difference is H_0: $p_1 = p_2$. Equivalently, with $\delta = p_1 - p_2$, it is

$$H_0: \delta = 0. \tag{1}$$

[This is consistent with our use of δ as the difference in population means, since with the Bernoulli coding $(0 = 1)$, the population mean is the population proportion.]

The maximum likelihood estimator of δ is the difference between sample proportions:

$$\hat{\delta} = \hat{p}_1 - \hat{p}_2,$$

where $\hat{p}_i = Y_i/n_i$, and Y_i is the number of successes in sample i. We saw in Section 9.6 that the large sample distribution of $\hat{\delta}$ is approximately normal. To transform $\hat{\delta}$ into a Z-statistic for testing H_0, we need its mean and variance under H_0. The expected difference is the difference in proportions: $E(\hat{\delta}) = \delta$. So in view of (1),

$$E(\hat{\delta} \mid H_0) = 0. \tag{2}$$

Because the samples are independent, the variance of $\hat{\delta}$ is the sum of the variances of $\hat{p}_1$ and $\hat{p}_2$, each given by (7) of Section 8.7:

$$\begin{aligned}
\operatorname{var} \hat{\delta} &= \operatorname{var} \hat{p}_1 + \operatorname{var} \hat{p}_2 \\
&= \frac{p_1(1 - p_1)}{n_1} + \frac{p_2(1 - p_2)}{n_2}.
\end{aligned} \tag{3}$$

Under H_0, $p_1 = p_2$; call this common value p. Then

$$\operatorname{var}(\hat{\delta} \mid H_0) = \frac{p(1 - p)}{n_1} + \frac{p(1 - p)}{n_2} = p(1 - p)\left(\frac{1}{n_1} + \frac{1}{n_2}\right). \tag{4}$$

With this we can construct a standard score using (2) and (4):

$$Z = \frac{\hat{\delta} - 0}{\text{s.d.}(\hat{\delta} \mid H_0)} = \frac{\hat{p}_1 - \hat{p}_2}{\sqrt{p(1 - p)(1/n_1 + 1/n_2)}}. \tag{5}$$

Under H_0, this is approximately standard normal when n_1 and n_2 are large.

Estimating p

The value of p in (5) is unknown. It can be approximated from either of the two samples, or better still, by pooling the two samples. Under H_0, the two samples together constitute a single random sample of size $n_1 + n_2$ from a Bernoulli population with parameter p. The obvious (and maximum likelihood) estimate of this common p is the proportion $\hat{p}$ in the *combined* sample:

$$\hat{p} = \frac{Y_1 + Y_2}{n_1 + n_2} = \frac{n_1 \hat{p}_1 + n_2 \hat{p}_2}{n_1 + n_2}. \tag{6}$$

Thus, $\hat{p}$ is a weighted average of the two sample proportions, with weights proportional to the sample sizes. Replacing p in (5) by the estimate $\hat{p}$ from (6), we obtain an approximate Z-score as our test statistic:

$$Z = \frac{\hat{p}_1 - \hat{p}_2}{\sqrt{\hat{p}(1 - \hat{p})(1/n_1 + 1/n_2)}}. \tag{7}$$

Under H_0, the large sample distribution of (7) is approximately standard normal.

Test for the equality of population proportions using large, independent random samples: Calculate Z from (7), with $\hat{p}$ given by (6), and find the P-value in Table II of Appendix 1.

Example **12.3a**   **Windowlessness**

Researchers have hypothesized that visual decor in windowless offices would be more nature-oriented than in windowed offices. A study[2] of the visual material used to decorate offices at the University of Washington found nature-dominant themes in 134 of 195 windowless offices and in 45 out of 82 windowed offices.

1.   The null hypothesis is that there is no difference in choice of decorating themes between windowed and windowless offices: $\delta = 0$.
2.   Assuming that a difference either way is possible, we adopt the two-sided alternative H_A: $\delta \neq 0$,
3.   The test statistic is the standardized difference (7). For the given data,

$$\hat{p}_1 = \frac{134}{195} \doteq .69, \qquad \hat{p}_2 = \frac{45}{82} \doteq .55.$$

From (6), we calculate

$$\hat{p} = \frac{134 + 45}{195 + 82} = \frac{179}{277} = .646.$$

With this in (7), we obtain

$$Z = \frac{\frac{134}{195} - \frac{45}{82}}{\sqrt{\frac{179}{277} \cdot \frac{98}{277}\left(\frac{1}{195} + \frac{1}{82}\right)}} \doteq \frac{.1384}{.06293} \doteq 2.20.$$

4.   For our two-sided alternative, large values of $|Z|$ are extreme.
5.   The P-value is twice the area under the standard normal curve to the right of the $Z = 2.20$ or $2\Phi(-2.20) = .0278$.

The observed difference in sample proportions is seen to offer some evidence against equality of the population proportions.   ∎

Exact P-Value

When sample sizes are small, or when one or the other of the sample proportions is small, the normal approximation for Z is not very accurate. Actually, since the null hypothesis does not specify p, the common value of the population proportions, the P-value is net defined. In the normal approximation, we take p to be the m.l.e. $\hat{p}$, which amounts to assuming that $Y = Y_1 + Y_2$ is given. Taking the same approach, we can find an exact conditional P-value given Y, using the conditional

[2] J. Heerwager, and G. Orians, "Adaptations to windowlessness," *Environment and Behavior 18* (1986), 623–639.

distribution of Y_1 given Y. The joint probability function of (Y_1, Y_2) is the product of binomials:

$$f(y_1, y_2) = \binom{n_1}{y_1} p_1^{y_1} (1 - p_1)^{n-y_1} \cdot \binom{n_2}{y_2} p_2^{y_2} (1 - p_2)^{n-y_2}. \tag{8}$$

Under H_0, $Y \sim \text{Bin}(n_1 + n_2, p_1 + p_2)$. Dividing (8) by the binomial p.f., we obtain the conditional p.f. of Y_1 given $Y = y$:

$$f(y_1 | Y = y) = P(Y_1 = y_1 | Y_1 + Y_2 = y) = \frac{\binom{n_1}{y_1}\binom{n_2}{y_2}}{\binom{n_1+n_2}{y_1+y_2}}.$$

This is hypergeometric. The following example shows how we can use this distribution in finding exact *P*-values. Doing this is referred to as *Fisher's exact test*.

Example 12.3b **Pets in Therapy**

Example 2a referred to a study that considered the possibility that having a pet might be good therapy for patients with heart disease. The comparison was based on whether or not a patient was alive one year after a heart attack. The study included 92 patients hospitalized for a heart attack or other serious heart disease. It was found that 3 of the 53 patients who had a pet died within a year after their release, whereas 11 of 39 who did not have a pet died within a year.

The Z-statistic for testing the hypothesis of equal population proportions uses (6), the proportion of survivors in the combined sample:

$$\hat{p} = \frac{3 + 11}{53 + 39} = \frac{14}{92}.$$

With this in (7), we obtain

$$Z = \frac{\frac{3}{53} - \frac{11}{39}}{\sqrt{\frac{14}{92} \cdot \frac{78}{92}(\frac{1}{53} + \frac{1}{39})}} \doteq -2.98.$$

The approximating normal distribution yields a (one-sided) *P*-value of .0015 as the area under the standard normal curve to the left of the $Z = -2.98$.

An exact conditional *P*-value given $Y = 3 + 11 = 14$ is a hypergeometric probability:

$$P(Y_1 \leq 3 | Y = 14) = \frac{\binom{14}{3}\binom{78}{50} + \binom{14}{2}\binom{78}{51} + \binom{14}{1}\binom{78}{52} + \binom{14}{0}\binom{78}{53}}{\binom{92}{53}}.$$

$$\doteq .00358.$$

In this case, the normal approximation is poor because the observed relative frequency $\frac{3}{35}$ is very small.

The observed difference in proportions offers strong evidence against equality of the population proportions. However, we cannot conclude that having a pet is the *cause* of the higher survival proportion among pet owners—even if we reject $\delta = 0$. (This is the case in any retrospective study.) There may be factors that lead a person to have a pet which are also factors that lead to survival. For example, a healthier person may feel more capable of caring for a pet. A prospective randomized trial is not out of the question here, but we know of none that has been conducted. ■

We'll return to this problem of comparing proportions in Section 13.5, extending it there to the problem of comparing discrete populations with finitely many categories.

| 12.4 | ## Two-Sample *t*-Tests |

In Section 12.2, we based a large sample test of the hypothesis $\delta = 0$ (that is, $\mu_1 - \mu_2 = 0$) on $Y = \bar{X}_1 - \bar{X}_2$ using

$$\text{s.e.}(Y) = \sqrt{\frac{S_1^2}{n_1} + \frac{S_2^2}{n_2}} \tag{1}$$

as the denominator of a Z-score. However, when n_1 or n_2 is small, we encounter the same problem as in the one-sample case (Section 10.4): The variability in (1) introduces more variability into the Z-score than is represented by the standard normal curve. As in the one-sample case, we can use the t-distribution to take this into account.

Suppose we assume normal populations. Fashioning a t-test for $H_0: \delta = 0$ is difficult when population variances are not known. However, we can find appropriate null distributions if we include equality of population variances in the null hypothesis:

$$H_0: \delta = 0 \quad \text{and} \quad \sigma_1^2 = \sigma_2^2 = \sigma^2.$$

Assuming equal variances may be quite appropriate in many practical settings. In particular, when the null hypothesis is that a treatment has no effect at all, the treatment population is the same as the control population under H_0. In this case, both populations will have the same variance as well as the same mean under H_0.

The t-test to be presented is directed at detecting a difference in means. The alternative will be either $\delta \neq 0$ or one-sided ($\delta > 0$ or $\delta < 0$), depending on the context.

Estimating σ^2

When $\sigma_1^2 = \sigma_2^2 = \sigma^2$, either sample variance (S_1^2 or S_2^2) could serve as an estimate of σ^2, but an estimate that is better than either of these is the **pooled variance**, a weighted average of S_1^2 and S_2^2 in which the weights are proportional to degrees of freedom:

$$S_p^2 = \frac{(n_1 - 1)S_1^2 + (n_2 - 1)S_2^2}{n_1 + n_2 - 2}. \tag{2}$$

The coefficients of the sample variances are fractions that add to 1. If the sample sizes are equal, these fractions are $\frac{1}{2}$, and the pooled estimate is a simple average of the sample variances. If the sample sizes are unequal, the variance from the larger sample is weighted more heavily.

Substituting S_p^2 for each sample variance in (1) gives an approximation to the s.d. of Y for use in constructing a T-score:

$$T = \frac{Y - 0}{S_p\sqrt{1/n_1 + 1/n_2}}. \tag{3}$$

Under H_0 and the assumption of independent samples from normal populations, the quantity

$$(n_1 + n_2 - 1)\frac{S_p^2}{\sigma^2} = (n_1 - 1)\frac{S_1^2}{\sigma^2} + (n_2 - 1)\frac{S_2^2}{\sigma^2}$$

is the sum of independent chi-square variables with $n_1 - 1$ and $n_1 - 1$ degrees of freedom, respectively. So it is chi-square with $n_1 + n_2 - 2$ degrees of freedom. Thus, much as in the case of the one-sample t-statistic, we see that T has the t-distribution with $n_1 + n_2 - 2$ d.f. under H_0. Moreover, as in the case of the one-sample t-statistic, even if the populations are not normal but are close to normal, the null distribution of T is well approximated by the t-distribution.

Example 12.4a **Feeding Schedules and Blood Pressure (continued)**
Consider again the rat blood pressure data of Example 10b:

Regular feeding: 108, 133, 134, 145, 152, 155, 169.

Intermittent feeding: 115, 162, 162, 168, 170, 181, 199, 207.

We test $H_0: \delta = 0$ against the alternative $H_A: \delta \neq 0$. The difference in means is 28.2, and the two s.d.'s are 19.61 and 27.99. Using (2), we find the pooled variance to be

$$S_p^2 = \frac{6(19.61)^2 + 7(27.99)^2}{7 + 8 - 2} = 24.48^2.$$

From (3),

$$T = \frac{-28.2}{24.48\sqrt{\frac{1}{7} + \frac{1}{8}}} = -2.226.$$

The corresponding two-sided P-value (Table IIIb in Appendix 1) is .044.   ∎

Test of $H_0: \mu_1 = \mu_2$, assuming nearly normal populations with equal variances: Calculate

$$T = \frac{\bar{X}_1 - \bar{X}_2}{S_p\sqrt{1/n_1 + 1/n_2}}, \tag{3}$$

where S_p^2 is the pooled variance (2). Find P in Table III, d.f. $= n_1 + n_2 - 2$.

T or Z?   When should you use Z and when T? The simplest answer is to use T whenever population variances cannot be assumed known. If the sample sizes are large enough, the *t*-table takes you automatically to the Z-approximation in the row for d.f. $= \infty$. When population variances are known, T is *not* appropriate. (Whether Z is appropriate in this case hinges on whether the sample sizes are large enough that the difference in sample means is approximately normal.)

Example 12.4b   **Feminism and Authoritarianism**

A study[3] compared people's attitudes toward feminism with their degree of "authoritarianism." Two samples were used, one consisting of 30 subjects who were rated high in authoritarianism, and a second sample of 31 subjects who were rated low. Each subject was given an 18-item test, designed to reveal attitudes on feminism, with scores reported on a scale from 18 to 90. (High scores indicated pro-feminism.) Summary statistics from the study are as follows:

Authoritarianism	n	$\bar{X}$	S
High	30	67.7	11.8
Low	31	52.4	13.0

We proceed as usual: The null hypothesis in the study is that authoritarianism is not a factor in attitudes toward feminism. Under H_0, scores of the high and low authoritarianism types are equal on the average: $\delta = 0$. We'll take $\delta \neq 0$ as the alternative, so large values of $|T|$ in (3) are extreme.

To calculate (3), we first find the pooled variance using (2): $S_p^2 = 154.37$. So $S_p = 12.42$. From (3),

$$T = \frac{67.7 - 52.4}{12.42\sqrt{\frac{1}{30} + \frac{1}{31}}} = 4.81.$$

The *t*-distribution with 59 d.f. is close to normal. In Table IIIc of Appendix 1, we are driven to use the last row (d.f. $= \infty$), where we see that 4.81 is beyond the 99.5th percentile. So $P < .005$. (Table IIIb shows $P = .000$ for 40 d.f.)

Alternatively, since d.f. > 40, we could calculate Z using (6) of Section 12.2, with

$$\frac{S_1^2}{n_1} + \frac{S_2^2}{n_2} = \frac{11.8^2}{30} + \frac{13.0^2}{31} = 10.09.$$

The result is $Z = 4.82$, very close to the value of T. (Ordinarily, T and Z will not agree exactly, since they are based on different estimates of variance.) ∎

---

[3] G. Sarup, "Gender, authoritarianism, and attitude toward feminism," *Soc. Behav. Personality 4* (1976) 57–64.

Although the assumption of equal variances under H_0 is often reasonable, it may not be correct. If one fails to reject H_0, one cannot be certain that this is because δ is not 0 or because the variances are not equal. To handle the case of unequal population variances, various modifications of the basic two-sample t-test for equality of means have been proposed. One method uses the large sample Z-score of Section 12.2 in conjunction with the t-table and a modified number of d.f. (The two-sample test for equal means in most statistical computer packages has such an option.) However, there is not a consensus as to what approach is best.

In the next section, we consider a nonparametric alternative to the t-test—one that involves fewer assumptions.

12.5   Two-Sample Nonparametric Tests

The t-test is appropriate when the populations are "nearly normal." We now take up a rank test for comparing the locations of two continuous populations that are unrestricted as to shape. The null hypothesis is that the two populations are the same. The alternative hypothesis of interest, as in the case of the two-sample t-test of Section 12.4, is that the populations differ in location—that one population is *shifted* to the right or left of the other.

Two samples that come from the same population tend to be sprinkled over the axis of possible values in roughly the same way, so the observations will tend to be interspersed when the null hypothesis is true. Suppose we plot the observations from both populations on the same axis of values, using the symbol "1" for those from population 1 and the symbol "2" for those from population 2. This orders the $n_1 + n_2$ observations in the combined sample. If the populations are the same, the 1's and 2's will be interspersed in the combined order statistic. If population 1 is shifted to the right of population 2, however, the 1's will tend to be toward the right in the sequence. For example, if sample 1 is (10, 14, 17) and sample 2 is (6, 8, 9, 11), the combined sample is (6, 8, 9, *10*, 11, *14*, *17*), with sample 1 observations indicated here in italics. The corresponding sequence of 1's and 2's is 1 1 1 2 1 2 2.

The next step is to assign a rank to each observation in the *combined* sample, rank 1 to the smallest, rank 2 to the second smallest, and so on. For the sequence 1 1 1 2 1 2 2, the 1-ranks are 1, 2, 3, 5, and the 2-ranks are 4, 6, 7. If the observations from population 1 tend to be toward the right in the combined sample sequence, their average rank will be unusually large; if they tend to be toward the left, their average rank will be unusually small. So the average rank indicates the degree to which one population is shifted from the other. To avoid fractions, we use the *sum* of the ranks (which is proportional to the average) as our test statistic. This statistic was proposed by F. Wilcoxon.[4]

---

[4] Equivalent rank statistics were proposed more or less simultaneously by H. B. Mann and R. Whitney, and by J. B. S. Haldane and C. A. B. Smith.

> ### Wilcoxon's Rank-Sum Statistic:
>
> Given independent random samples of n_1 observations from population 1 and n_2 observations from population 2, combine them into a single ordered sequence and assign rank j to the jth smallest. Then
>
> R_i = sum of the ranks of the observations from population i.

Since $R_1 + R_2$ is the sum of the first $n_1 + n_2$ integers, it is constant, depending only on $n_1 + n_2$:

$$R_1 + R_2 = \frac{1}{2}(n_1 + n_2)(n_1 + n_2 + 1). \tag{1}$$

This means that we can use just one—either one—of the rank sums as the test statistic.

Null Distribution of R_1

As usual, we need to know the null distribution of the test statistic. Under H_0, the hypothesis that the two populations are the same, the two independent samples together constitute a single sample from the common population. The exchangeability of the $n_1 + n_2$ observations (under H_0) implies that all patterns of 1's and 2's are equally likely under H_0. To see how this leads to the null distribution of R_1, we consider an artificial example with small sample sizes.

Example 12.5a

Consider independent samples of sizes $n_1 = 2$ and $n_2 = 3$, and suppose the samples are as follows: (5, 10) and (2, 7, 9). Combining these into a single ordered sequence, we have $(2, 5, 7, 9, 10)$, where the observations from population 2 are shown in italics. With each observation replaced by its population of origin, the sequence becomes 2 1 2 2 1. This is one of $\binom{5}{3} = 10$ sequences of three 1's and two 2's that could have occurred. We list them here (the one obtained from the given sample is in boldface), along with the corresponding values of R_1 and R_2:

Sequence	R_1	R_2
2 2 2 1 1	9	6
2 2 1 2 1	8	7
2 1 2 2 1	**7**	**8**
2 2 1 1 2	7	8
2 1 2 1 2	6	9
1 2 2 2 1	6	9
1 2 2 1 2	5	10
2 1 1 2 2	5	10
1 2 1 2 2	4	11
1 1 2 2 2	3	12

[As indicated by (1), $R_1 + R_2 = 15 = (5 \cdot 6)/2$ in each case.] The ten possible sequences of two 1's and three 2's, shown in the above table, are equally likely under H_0. So we can find the null distribution of R_1 by counting. The probability table for R_1 is as follows:

r	$P(R_1 = r)$
3	1/10
4	1/10
5	2/10
6	2/10
7	2/10
8	1/10
9	1/10

The distribution of R_1 is symmetric—and this is true generally. So $E(R_1 \mid H_0) = 6$, the point of symmetry. The observed sequence (2 1 2 2 1) has rank-sum $R_1 = 7$. The corresponding one-sided P-value, for the alternative that population 2 is located to the left of population 1, is

$$P = P(R_1 \geq 7 \mid H_0) = 2/10 + 1/10 + 1/10 = 4/10.$$

In deriving the null distribution of R_1, we assumed nothing about the common population except that it is continuous. ∎

Counting sequences to determine the null distribution of a rank-sum statistic is of course much more tedious for larger sample sizes. However, as in the preceding example, these null distributions do *not* depend on the shape of the common distribution under H_0. The statistic is **distribution-free**. They do depend on the sample sizes, and Table VII of Appendix 1 gives tail-probabilities for the distribution of R_1, for sample sizes up to $n_1 = n_2 = 10$. To use this table, labels "1" and "2" must be assigned so that $n_1 \leq n_2$. The table gives only *left* tail-probabilities; these suffice because the distribution of R_1 is symmetric about its mean.

The rank-sum statistic is asymptotically normal as the sample sizes become infinite; it is approximately normal when both sample sizes exceed 10. (The proof of this is beyond the scope of this book.) To approximate probabilities for a rank-sum using a Z-score, we need formulas for the mean and variance. We won't derive these,[5] but you should check that they are correct for the case of Example 12.5a:

$$E(R_1) = \frac{1}{2} n_1 (n_1 + n_2 + 1), \tag{2}$$

$$\operatorname{var} R_1 = \frac{1}{2} n_1 n_2 (n_1 + n_2 + 1). \tag{3}$$

---

[5] See Lindgren [15], pp. 519–521.

> Rank-sum test for H_0: Populations 1 and 2 are the same, based on independent random samples of sizes n_1 and n_2. Right or left tail-values of R_1 are extreme according as H_A is that population 1 is to the right or left of population 2. For $n_1 \leq n_2 \leq 10$, P-values are in Table VII. For n's > 10, use a Z-score and Table II:
>
> $$Z = \frac{R_1 + \frac{1}{2} - \frac{1}{2}n_1(n_1 + n_2 + 1)}{\sqrt{\frac{1}{2}n_1 n_2 (n_1 + n_2 + 1)}}. \qquad (4)$$

Example 12.5b   Intermittent Feeding (continued)

Consider again the blood pressure data of Examples 10b, 10.4a, and 12.4a:

Regular feeding: 108, 133, 134, 145, 152, 155, 169.
Intermittent feeding: 115, 162, 162, 168, 170, 181, 199, 207.

To avoid assuming population normality, we use the rank-sum test.

The sample sizes are 7 and 8. So that $n_1 \leq n_2$, we label the regular feeding group as sample 1. The order statistic of the combined sample is

(*108*, 115, *133*, *134*, *145*, *152*, *155*, 162, 162, 168, *169*, 170, 181, 199, 207).

The sequence of 1's and 2's is

1 2 1 1 1 1 1 2 2 2 1 2 2 2 2.

For sample 1, the ranks are 1, 3, 4, 5, 6, 7, 11. The sum of these is $R_1 = 37$. For sample 2, the ranks are 2, 8, 9, 10, 12, 13, 14, 15, with sum $R_2 = 15(16)/2 - 37 = 83$. (Note: $R_1 + R_2 = 120 = (15 \cdot 16)/2$.) We proceed as usual:

1.   The null hypothesis is that the populations are the same.
2.   The alternative hypothesis is that they differ in location.
3.   The test statistic is the sum of the ranks of the observations in the smaller sample: $R_1 = 37$.
4.   Because H_A is two-sided, values of R_1 at either end of its distribution are extreme.
5.   The two-sided P-value is $2 \cdot P(R_1 \leq 37 \,|\, H_0)$. In Table VII, in the segment for sample sizes 7 and 8, we find .014 opposite $c = 37$, so $P = 2(.014) = .028$.

Although Table VII gives tail-probabilities of the exact distribution for these sample sizes, let's see what we get using the normal approximation (4). Using (2) and (3), we find that $E(R_1) = (7 \cdot 16)/2 = 56$, and var $R_1 = (7 \cdot 8 \cdot 16)/12 = 224/3$. So [with the continuity correction $\frac{1}{2}$ shown in (4)]

$$Z = \frac{37.5 - 56}{\sqrt{\frac{224}{3}}} = -2.14.$$

The two-sided P-value calculated from this Z-score is $2 \cdot \Phi(-2.14) = .032$, quite close to the exact value .028 found above. (The P-value from the t-test in Example 12.4a is .044.) ∎

Ranks versus t-Tests

Rank tests are appropriate when the assumptions needed for a t-test are in question, but they can be used even when a t-test seems appropriate. The matter of which test is preferred in such a case has received extensive study. The t-test is only slightly better when the populations are normal, the case for which the t-test is designed! When the populations are nonnormal, the rank test can be very much better than the t-test. Indeed, a t-test may be worthless if the populations are sufficiently far from normal. (See our comments after Example 10.4b.) Most researchers use t-tests (this is what they learned), but rank tests are better all-purpose procedures.

When data suggest population nonnormality, a transformation may make the data appear more like data from a normal population. Commonly used transformations are logarithms and powers—the square root, for example. One might select the transformation that makes the data look most like data from a normal population and carry out a t-test using the transformed data.

A **rank transformation** is an all-purpose transformation in which each observation is replaced by its rank in the combined order statistic.[6] We then apply the ordinary two-sample t-test using the two sets of ranks. This t-test on ranks is in fact approximately equivalent to the rank-sum test. One reason to use the T-score based on ranks is that it uses the commonly understood t-test and conveniently available t-table but does not require the assumption of normality of the original data.

t-Test on Ranks

Example 12.5c   Intermittent Feeding (continued)

In Example 12.5b, we replaced the data on intermittent versus regular feeding with their ranks in the combined, ordered sample. These ranks, and their respective means and s.d.'s, are as follows:

	Ranks	$\bar{X}$	S
Regular	1, 3, 4, 5, 6, 7, 11,	5.286	3.200
Intermittent	2, 8, 9, 10, 12, 13, 14, 15	10.375	4.173

---

[6] See W. J. Conover and R. L. Iman, "Rank transformations as a bridge between parametric and nonparametric statistics," *American Statistician 35* (1981), 124–129.

about 3.8. Since $2.04 < 3.8$, the P-value exceeds .05. There is little evidence against the hypothesis of equality of variances. ∎

It has been found that a test based on (1) in conjunction with the F-table is not robust with respect to nonnormality.

Chapter Perspective

This chapter has given a number of statistical tests, parametric and nonparametric, for comparing two populations with respect to location. One of the settings in which these are useful is determining whether a treatment is effective.

When a subject serves both as treatment and control, or when treatment and control are assigned at random in a pair of like subjects, we test the hypothesis of no mean treatment effect by applying a one-sample test to the sample of differences using the methods of Sections 10.3 and 10.4.

In Chapter 14, we'll extend the parametric methods of Sections 12.4 and 12.6 to the case of more than two populations.

Solved Problems

Sections 12.2–12.3

A. Example 10c raised the question, "Are boys better at math?" One study[8] gives SAT-M (math) scores for 3,674 intellectually gifted 7th and 8th graders, all of whom had been accelerated at least one grade level. Their scores are summarized as follows:

	Number	Mean	s.d.
Boys	2046	436	87
Girls	1628	404	77

Test the hypothesis that the corresponding population mean scores are equal.

Solution:
Let δ denote the difference in population means. We test $H_0: \delta = 0$ against $H_A: \delta > 0$ using the Z-score [(5) of Section 12.2]:

$$Z = \frac{(436 - 404) - 0}{\sqrt{87^2/2{,}046 + 77^2/1{,}628}} = 11.8.$$

---

[8] C. P. Benlow and J. C. Stanby, "Sex differences in mathematical ability: Fact or artifact?" *Science 210* (1980), 1262–1264.

Large values of Z are extreme, so $P = P(Z > 11.8) \doteq 2 \times 10^{-32}$, according to the asymptotic formula in the footnote to Table II in Appendix 1. This extremely small P-value seems overwhelming as evidence against the null hypothesis. (Even if H_A is true, this does not mean that boys are inherently better at math. For example, there may be sex-related social differences that mitigate against the most gifted girls advancing a grade level.)

B. A 1987 Gallup survey (reported April 12, 1987) found "no evidence of growing public intolerance towards gays." The 1987 poll included 1,015 adults in "scientifically selected localities across the nation" during the period March 14–18. One question asked was this: "Do you think homosexual relations between consenting adults should or should not be legal?" In 1986, the percentage who said "should not" was 54; in 1987 it was 55. Despite the increase from 54% to 55%, Gallup does not see it as evidence of growing intolerance. Is there a basis for this? Assume that the number polled in 1986 was also 1,015.

Solution:

We test the null hypothesis of no difference in population proportions, $H_0: p_1 = p_2$. The sample proportions are $\hat{p}_1 = .54$ and $\hat{p}_1 = .55$. Under H_0, the m.l.e. of the common proportion p [(6) of Section 12.3] is

$$\hat{p} = \frac{1015 \times .54 + 1015 \times .55}{2030} = .545.$$

The Z-score for the observed difference [(7) of Section 12.3] is

$$Z = \frac{.54 - .55 - 0}{\sqrt{.545 \times .455(\frac{1}{1015} + \frac{1}{1015})}} \doteq -.45.$$

So the observed difference is easily explainable as sampling variability and provides little evidence that the population proportions are different. (We note that even if the 1986 sample were much larger—say $n = \infty$—there is still little evidence that the population proportions are different: $Z \doteq -.64$.)

Sections 12.4–12.5

C. Stress-related factors in particular career aspirations are the subject of a study[9] that reported these data on systolic blood pressure for people in two types of careers:

	n	$\bar{X}$	S
(1) *Professional*	9	120.4	9.6
(2) *Semi-skilled*	14	110.1	8.4

Test $H_0: \delta = 0$ against $H_A: \delta > 0$—that professional people have higher blood pressures.

---

[9] "Contrasting patterns of blood pressure and related factors within a Maori and European population in New Zealand," *Social Science and Medicine 23* (1986), 439–444.

Solution:

With unknown population variances and small sample sizes, we use a t-statistic, assuming independent samples and equality of population variances. The pooled variance [(2) of Section 12.4] is

$$S_p^2 = \frac{1}{21}[9 \times 9.6^2 + 14 \times 8.4^2] = 78.79.$$

With this in (3) of Section 12.4, we obtain

$$T = \frac{120.4 - 110.1}{\sqrt{78.79(\frac{1}{9} + \frac{1}{14})}} = \frac{10.3}{3.792} \doteq 2.72.$$

The (one-sided) P-value from Table IIIb in Appendix 1 is $P = .007$.

D. Example 9.6a gave these amounts of water filtered with a new filtration device (in m^3/m^2 filter), using two methods of operation:

Method 1:	308	365	221	172	81	277	286	248	243	119
	243	216	196	182	157	194	148	182	105	98
Method 2:	221	249	271	187	99	222	161	379	307	294
	305	325	280	308	311	258	203	400	354	296
	253	415	344	329	282	356	323	248	91	

The summary statistics are as follows:

Method	n	Mean	s.d.
1	20	202.0	74.35
2	29	278.3	79.03

Test the hypothesis of no difference between the two methods in mean amount filtered.

Solution:

The data give no reason to question normality, so we could proceed with a two-sample t-test. Since d.f. $= 47$, the t-table leads us to a normal approximation, so we may as well treat this as a large-sample problem and use (5) of Section 12.2. For comparison purposes, however, we'll do both. First the t-test: The pooled variance [(2) of Section 12.4] is

$$S_p^2 = \frac{1}{47}[19 \times 74.35^2 + 28 \times 79.03^2] = 77.17^2.$$

Substituting in (6) of Section 12.4, we obtain

$$T = \frac{278.3 - 202.0}{77.17\sqrt{\frac{1}{20} + \frac{1}{29}}} = 3.40.$$

With d.f. $= 47$, we see from Table IIIa that $P < .001$. Substituting in (6) of Section 12.2, for a Z-test, we find

$$Z = \frac{278.3 - 202.1}{\sqrt{\dfrac{74.35^2}{20} + \dfrac{79.03^2}{29}}} = 3.44.$$

The P-value (from Table II) is again less than .001.

E.   An experiment with kittens compared pattern recognition of males and females. With the criterion 27 out of 30 correct responses to a visual stimulus, the number of trials to meet the criterion were recorded[10] as follows:

   Males: 120, 130, 155, 150, 40, 106, 382, 76, 89,

Females: 69, 117, 66, 391, 94, 103.

Test the hypothesis of no average difference.

Solution:

The dot diagram shown in Figure 12.1, showing two obvious outliers, suggests that the population distributions are not normal, so the rank-sum test is more appropriate than a t-test. The combined ordered sample is

40, *66*, *69*, 76, 89, *94*, *103*, 106, *117*, 120, 130, 150, 155, 382, *391*.

The ranks of the females are 2, 3, 6, 7, 9, 15, with sum 42. In Table VII of Appendix 1 ($m = 6$, $n = 9$), we find that 42 is not in the left tail of the distribution: $P > .164$. [The mean of R_F under H_0 is 48, from (1) of Section 12.5, so 42 is to the left of the middle and not in the right tail.] The given data do not provide convincing evidence of a difference.

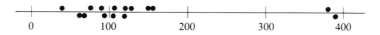

Figure 12.1 Dot diagram for Problem E

   An alternative method of analysis is to transform the data. Suppose we replace each observation by its rank in the combined order statistic and apply the two-sample t-test to these two sets of ranks. The result is $T = -.69$ (13 d.f.), which is not close to the "significant" range. (The mean difference is -1.667, and the pooled s.d. is 4.557.)

   If we (inappropriately) apply a t-test to the original data, we find $T = -.17$. The large standard deviations resulting from the outliers make the denominator of the T-statistic so large that the observed difference seems small by comparison.

F.   Verify the entry for $c = 24$, $m = 6$, $n = 7$ in Table VII of Appendix 1. Use the fact that the possible sequences of six X's and seven Y's are equally likely under the hypothesis that X and Y have the same distribution (and that the samples are independent).

---

[10] Dodwell, Wilkinson, and von Grunan, "Pattern recognition in kittens," *Perception 12* (1983), 393–410.

Solution:

The most extreme patterns are as follows, shown with R_X:

X X X X X X X Y Y Y Y Y Y Y   21

X X X X X Y X Y Y Y Y Y Y   22

X X X X Y X X Y Y Y Y Y Y   23

X X X X X Y Y X Y Y Y Y Y   23

X X X Y X X X Y Y Y Y Y Y   24

X X X X X Y Y Y Y X Y Y Y Y   24

X X X X Y X Y X Y Y Y Y Y   24.

So there are seven patterns with rank-sum of 24 or less, out of $\binom{13}{6} = 1{,}716$ equally likely patterns: $P = \frac{7}{1716} \doteq .0041$.

Sections 12.6–12.7

G. Plasma half-life data (in hours) of the drug verapamil for eight subjects, after administration of verapamil only (V) and after verapamil administered concurrently with another drug (V+), were obtained as follows:

Subject	1	2	3	4	5	6	7	8
V	2.55	1.81	1.99	2.37	3.03	2.25	1.89	1.83
V+	3.15	2.07	3.22	2.67	2.90	2.47	1.31	2.68
Differences	.60	.26	1.23	.40	−.13	.22	−.58	.85

Test the hypothesis that, on average, the second drug has no effect on the plasma half-life of verapamil.

Solution:

A dot diagram of the differences shows no obvious nonnormality, so we can use either a t-test or a signed-rank test. For the t-test, we need the mean difference $\bar{D} = .356$ and the s.d. of the differences $S_D = .562$. Substituting these in the one-sample T-statistic [(1) of Section 10.4], we find

$$T = \frac{.356}{.562/\sqrt{8}} \doteq 1.79.$$

The P-value from Table IIIa is .057.

    For the signed-rank test, we arrange the differences in order of magnitude:

$$-.13, .22, .26, .40, -.58, .60, .85, 1.23.$$

The first and fifth are negative, so $R_- = 1 + 5 = 6$. Table VIII ($n = 8$, $c = 6$) gives $P = .055$—very close to the P-value from the t-test.

H. Problem H of Chapter 10 gives decreases in VPB's for eleven heart patients after administration of a drug. The original data (VPB's/min) are as follows:

Patient No.	1	2	3	4	5	6	7	8	9	10	11	12
Before	6	9	17	22	7	5	5	14	9	7	9	51
After	5	2	0	0	2	1	0	0	0	0	13	0

Rank all 12 observations, replace each by its rank, and carry out a paired-sample t-test on the ranks. [This is another instance of rank-transformation (page 540).]

Solution:

The ranks (with tied observations assigned average ranks) are

Before	14	18	22	23	15.5	12	12	21	18	15.5	18	24
After	12	9.5	3.5	3.5	9.5	8	3.5	3.5	3.5	3.5	20	3.5
Difference	2	8.5	18.5	19.5	6	4	8.5	17.5	14.5	12	−2	20.5

The mean of these rank differences is 10.79, and the s.d. is 7.13. Using these in the one-sample T-statistics [(1) of Section 10.4], we find $T = 5.24$ (10 df), and $P < .001$.

I. Referring to Problem 16 (of this chapter), we calculate the s.d.'s of platelet counts as follows: 53.4 for normal individuals, 139.1 for thrombosis patients. Test the hypothesis of equality of population variances.

Solution:

We assume normality of the two populations and calculate the ratio of sample variances: $(139.1/53.4)^2 \doteq 6.8$. The sample sizes are 10 for normal individuals and 14 for patients, so under H_0 the ratio of sample variances is $F(13, 9)$. Table VIIIa does not include 9 d.f. for the numerator, so we consult Table VIIIb, which gives the 5% and 1% tail-probability values. Since

$$6.78 > 4.7 \doteq F_{.99}(13, 9),$$

we conclude that $P < .01$, fairly strong evidence against equality of population variances.

J. Referring to Problem A above, suppose we had wanted to test the hypothesis of equal population variances. The sample variances are different; is their difference statistically significant?

Solution:

The F-table runs out of entries at $n = 60$, and we have samples that are much larger. So we use instead the approximate normality of variances of samples from normal populations. Test scores tend to be close to normally distributed, so we'll proceed with this assumption and calculate a Z-score [see (2) and (5) in Section 9.7]:

$$Z = \frac{87^2 - 77^2 - 0}{\sqrt{2(87^4/2046 + 77^4/1628)}} \doteq 5.2$$

The evidence against H_0 is strong.

Problems

For problems marked with an asterisk, answers are provided in Appendix 2.

   Note: As in Chapter 10, many of these problems are to be addressed by phrasing them as tests of hypotheses and determining P-values. State whether your P-values are one- or two-sided. Assume random samples when necessary and independence of samples when appropriate.

Sections 12.2–12.3

*1. A research study on fluoride in toothpaste was conducted using Colgate's "MFP" formula, and a leading stannous fluoride (SF) toothpaste. Data[11] on the number of new cavities over a three-year period are summarized as follows:

	n	Mean	s.d.
MFP	208	19.98	10.6
SF	201	22.39	11.96

What conclusion can be drawn concerning the mean difference in number of cavities?

2. Two types of aggregate are tested for thermal conductivity, with 25 observations for each type:

   1 (low cost):   mean = .485,   s.d. = .180,

   2 (higher cost):   mean = .372,   s.d. = .160.

   Before choosing the higher-cost type, a certain buyer wants to be convinced that it has mean conductivity at least .05 less than that of the low-cost type. Do the data provide such evidence? (Test the null hypothesis $H_0: \mu_1 - \mu_2 = .05$ against the alternative $H_A: \mu_1 - \mu_2 > .05$.)

*3. In a comparison of pain and activity levels for good and poor sleepers, researchers[12] report these data on hours of activity:

	n	$\bar{X}$	S
Good sleepers	28	10.7	4.8
Poor sleepers	70	8.6	4.8

Test the hypothesis of no average difference in hours of activity between good and poor sleepers.

*4. Poll A uses a random sample of size 1,000 and reports that 42% are opposed to an amendment prohibiting abortions. Poll B uses the identically worded question in a

---

[11] S. F. Frankl and J. E. Alman, *J. Oral Therapeutics Pharmacol. 4* (1968), 443–449.
[12] I. Pilowsky, I. Crettenden, and M. Townley, "Sleep disturbance in pain clinic patients," *Pain 26* (1985), 27–33.

random sample of size 1,500; it reports 39% opposed. Assuming independent samples, test the hypothesis that the two populations have the same proportions opposing abortions.

5. Does the response rate on a questionnaire depend on whether the cover letter is a form letter or is semipersonal? In one study,[13] it was found that 225 among 1,022 receiving the form letter responded, and 325 of 1,018 who received the semipersonal letter responded. Test the hypothesis that the proportion of responders is the same for both types of cover letters.

*6. The proportion of smokers in the sample of male students listed in Table 7.1a is $\frac{8}{56}$, and the proportion in the sample of female students listed in Table 7.7a is $\frac{5}{48}$. Assuming these to be independent random samples, test the hypothesis that the proportion of smokers is the same for female students and male students.

7. Show that the pooled estimate of p given by (6) of Section 12.3 is the m.l.e. of p under $H_0: p_1 = p_2 = p$.

*8. A study[14] found that of 238 individuals with coronary heart disease (CHD), 145 smoked cigarettes; of 476 individuals in the same age group but with no CHD, 192 smoked cigarettes. Test the hypothesis that the proportion of cigarette smokers is the same in the population of individuals with CHD as in the population with no CHD.

9. The *New York Times* (February 15, 1983) carried a report of a study showing that a medicine commonly used to treat middle-ear infections in young children is no more effective than a placebo. Of 278 children who took the drug, 25% had no ear infection after four weeks, whereas 24% of the 275 who received a placebo had no infection. (On the other hand, those who took the drug experienced "significantly more" side effects— mild sedation and weakness.) In view of the sample proportions, is the study's conclusion warranted?

*10. In a study[15] on oral hygiene, 30 subjects used a test compound, and 34 subjects used a placebo. Improvements in an oral hygiene index were as follows:

Test: 10, 15, 6, 10, 11, 3, 8, 8, 3, 13, 10, 9, 8, 9, 8,
           4, 10, 15, 11, 5, 14, 7, 8, 8, 2, 13, 6, 2, 7, 3;

Placebo: 5, 6, 4, 3, 3, 5, 6, 4, 4, 2, 0, 7, 0, 3, 2, 2, 3, 6,
           0, 3, −1, 1, 6, 6, 8, 2, 12, 24, 5, 3, 3, 13, 4, 3.

Test the hypothesis that the test compound is no more effective, on average, than the placebo.

11. In the magazine *Natural History* (September 1988), an article on the burrowing owl in the Columbia River basin reports that of 25 nests lined with dung, only two were lost to badgers, whereas 13 of 24 unlined nests were destroyed by these carnivores. Find the *P*-

[13] M. J. Matteson, "Type of transmittal letter and questionnaire color as two variables influencing response in a mail survey," *J. Applied Psych. 59* (1974), 535–536.

[14] J. N. Morris et al., "Vigorous exercise in leisure-time and the incidence of coronary heart-disease," *The Lancet* (Feb. 17, 1973), 333–337.

[15] Zinner, Duany, and Chilton, *Pharmacology and Therapeutics in Dentistry 1* (1970), 7–15.

value for a test of the hypothesis that lining with dung does not affect destruction of owl nests by badgers,

(a) using a normal approximation.

(b) using an exact calculation based on the hypergeometric distribution.

Sections 12.4–12.5

*12. Measurements on anteroposterior chest diameter for 25 male pulmonary emphysema patients and 16 normal males yielded the following results:[16]

> Normal: Mean $= 20.2$,   s.d. $= 2.0$;
>
> Emphysema: Mean $= 23.0$,   s.d. $= 2.4$.

Test the hypothesis that there is no difference between average chest diameters of normal males and males with emphysema, assuming independent random samples. (The study is flawed because it confounds the difference of interest with both age and weight: The mean age in the normal sample was 30.9 years and in the other sample, 56 years; the mean weight in the normal sample was 74.5 kg and in the other, 62.1 kg. Taking weights into consideration suggests that the effect of emphysema is even greater, since the emphysema patients weigh less than normal but have larger chests.)

13. In a statistics class of 28 students, the 12 smokers determined their pulse rates to be

62, 66, 90, 92, 66, 70, 68, 70, 78, 100, 88, 62.

The 16 nonsmokers' pulse rates were

64, 58, 64, 74, 84, 68, 62, 76, 80, 68, 60, 62, 72, 70, 74, 66.

(a) Carry out a *t*-test for the hypothesis of no average difference between pulse rates of smokers and nonsmokers.

(b) Test the hypothesis of no difference, using a rank test.

*14. The following are summary statistics of plasma clearances in (l/sec) of a drug given intravenously to 8 smokers and 4 nonsmokers:

	n	*Mean*	*s.d.*
Smokers	8	12.12	4.18
Nonsmokers	4	15.14	5.06

Test the hypothesis of no difference between clearances of smokers and nonsmokers against a two-sided alternative.

15. Write out the sequences of four 1's and four 2's in which the 1's have rank-sums less than 14. Assume that the 70 possible patterns are equally likely to verify the entry in Table VII of Appendix 1 for $n_1 = n_2 = 4$ and $c = 13$.

---

[16] Kilburn and Asmundsson, "Anteroposterior chest diameter in emphysema," *Archiv. Internal Med. 123*, (1969) 379–382.

*16. Platelet counts (in 1,000's per mm³) for 10 normal males and for 14 patients with a recent thrombosis were obtained as follows:

Normal: 257, 185, 231, 220, 141, 237, 199, 295, 276, 319;

Patients: 597, 415, 264, 403, 681, 188, 364, 169, 426, 388, 364, 294, 368, 466.

Apply the rank-sum test for the hypothesis that the population distributions are the same against the alternative that the counts for thrombosis patients tend to be higher.

17. A study[17] addressed the relationship between the attitudes of children toward their fathers and their birth order. Fifteen first-born and 15 second-born males (independent samples) were given a questionnaire dealing with these attitudes. The scores:

1st: 40, 41, 44, 49, 53, 53, 54, 54, 56, 61, 62, 64, 65, 67, 67;

2nd: 23, 25, 38, 43, 44, 47, 49, 54, 55, 58, 58, 60, 66, 66, 72.

(A large score means that the child identifies with and supports the role of the father in the family.) Apply the rank-sum test for the hypothesis of no difference between attitudes of first- and second-born children against a two-sided alternative.

*18. For independent random samples of sizes $n_1 = 8$ and $n_2 = 10$, compare the (exact) P-value given in Table VIII for $R_1 = 57$ with the normal approximation given by (4) of Section 12.5.

19. Use a rank test for the null hypothesis in Problem D of this chapter.

*20. Obtain the m.l.e. of the common variance σ^2, for independent random samples from normal populations, and give its relationship to the pooled variance of Section 12.4.

21. In a study[18] on the effect of cod-liver oil, seven pigs were fed a diet containing large amounts of cod-liver oil. Eleven pigs were fed normally. The extent of artery blockage was determined for several arteries of each pig. For the right coronary artery, the results (measured in percentages) were as follows:

	n	Mean	s.d.
Oil fed	7	12.75	13.77
Control	11	53.46	22.80

Test $H_0: \delta = 0$, where δ is the population mean difference in blockage.

Sections 12.6–12.7

*22. In a diabetes study,[19] glycosylated hemoglobin was measured for each individual in four pairs of twins. One twin in each pair had diabetes and one did not:

Twin pair no.	1	2	3	4
Diabetic twin	9.6	8.5	8.9	7.0
Nondiabetic twin	4.9	5.1	5.2	5.8

[17] A. Roost, "A *Q*-sort analysis of family ordinal position," Ph.D. thesis, University of Minnesota, 1975.

[18] B. H. Weiner et al., "Inhibition of atherosclerosis by cod-liver oil in a hyperlipidemic swine model," *New Engl. J. Med. 315* (1986), 841–846.

[19] W. A. Kaye et al., "Acquired defect in Interleukin-2 production . . . ," *New Engl. J. Med 315* (1986), 920–924.

(The normal range for glycosylated hemoglobin is between 4 and 6.) Test the hypothesis of zero mean difference between diabetics and nondiabetics.

23. Darwin did an experiment[20] to learn whether self- or cross-pollinated seeds would produce more vigorous seeds (as indicated by plants of greater height). He obtained the following data, giving mature heights (in eighths of inches) for pairs of plants of genus *Zea mays* (Indian corn). Plants in a pair were matched genetically and assigned to self- and cross-pollinization at random, planted at opposite sides of a pot, and separated by a plane of glass parallel to the light rays.

Pair	1	2	3	4	5	6	7	8	9	10	11	12	13	14	15
Cross	188	96	168	176	153	172	177	163	146	171	186	168	177	184	96
Self	139	163	160	160	147	149	149	122	132	144	130	144	102	124	144

Assuming a two-sided alternative, test the hypothesis of no difference between self- and cross-pollinated seeds using
(a) the sign test.
(b) the signed-rank test.
(c) a t-test.

*24. Two pigs are selected from each of ten litters. One of each pair is fed a test diet and the other a standard diet. Weight gains over a given period are as shown in the following table:

Pair	1	2	3	4	5	6	7	8	9	10
Test	36.0	32.7	39.2	37.6	32.0	40.2	34.4	30.7	36.4	37.2
Standard	35.2	30.0	36.5	38.1	29.4	36.0	31.3	31.6	31.1	34.0

Test the hypothesis of no difference against the alternative that the test diet results in greater average weight gains.

25. Using the data of Problem 24 to illustrate the point discussed in the next to last paragraph of Section 12.6,
(a) calculate the correlation coefficient of weight gains.
(b) treat the test diet gains and the standard diet gains as though they were independent samples, and calculate the two-sample T-statistic for comparing means. (The variation from litter to litter makes the experimental error look larger and lowers the T-score. Elimination of litter-to-litter differences by pairing more than compensates for the loss of sensitivity associated with fewer degrees of freedom.)

*26. A prospective study was designed to see if honey in a diet would increase hemoglobin. Six pairs of twins in a children's home were the subjects. For a period of six weeks, each child was given a cup of milk at 9 P.M. One of the twins (always the same one) in each pair received a tablespoon of honey dissolved in the milk. The following increases in

---

[20] *Effect of Cross- and Self-Fertilization in the Vegetable Kingdom* (New York: Appleton, 1902).

hemoglobin were recorded:

Pair	1	2	3	4	5	6
Honey	19	12	9	17	24	24
No honey	14	8	4	4	11	11

(a) Apply the sign test and obtain a one-sided P-value.

(b) Apply the signed-rank test and obtain a one-sided P-value. Explain why this is the same as the P-value in (a).

(c) Apply a t-test on the ranks of the observations in the combined sample.

27. In the setting and using the data of Solved Problem D, test the hypothesis of no difference in mean amount filtered, using a t-test on ranks.

*28. Given the data in Solved Problem D, test the hypothesis that the two filtration procedures are the same with regard to *variance* of the amount filtered (per square unit of filter), against the hypothesis that the variances are different.

29. Use the data in Problem 13 to test the hypothesis that the variances of pulse rate are the same for smokers and nonsmokers.

*30. We know that for large n, the sample variance is approximately normal: $S^2 \approx N(\sigma^2, 2\sigma^4/n)$.

(a) Construct a large sample test for equal population variances based on the difference between sample variances.

(b) Apply the test in (a) to the situation of Solved Problem A. The sample statistics are as follows:

	Number	Mean	s.d.
Boys	2,046	436	87
Girls	1,628	404	77

Goodness of Fit

Tests for **goodness of fit** check the consistency between a set of data and a proposed model. The various tests presented in the preceding chapters are actually tests of fit. In formulating those tests, however, we usually phrased both null and alternative hypotheses in terms of a population parameter such as the population mean. Tests were designed to be sensitive to a particular type of parametric alternative such as $\mu > \mu_0$. In this chapter, we take up some tests that are not directed at such specific kinds of alternatives. Rather, the alternative hypothesis H_A is simply that the null hypothesis is not true. These are **nonparametric** tests.

We begin with testing a completely defined distribution as the null hypothesis, first in the discrete case and then in the continuous case. We then address the commonly occurring question of whether a data set fits—not a specific model, but a *class* of models—a type of probability distribution indexed by a small number of parameters. Thus, we want to know whether a population is normal, or binomial, or exponential, and so on, without specifying a particular member of the family. We'll show how to adapt the tests for fitting a particular distribution to this more general type of null hypothesis. Finally, we use tests of fit for hypotheses about bivariate categorical populations based on contingency table data.

13.1   Fitting a Distribution with Two Categories

As in earlier chapters, let p denote the probability or population proportion of "successes," and $1 - p$, the probability or proportion of "failures." In Section 10.3, we introduced a large-sample test for a hypothesis of the form $H_0: p = p_0$. The test was based on the sample relative frequency of successes ($\hat{p}$) and employed this Z-score:

$$Z = \frac{\hat{p} - p_0}{\sqrt{\dfrac{p_0(1 - p_0)}{n}}}.$$

This section generalizes (1) for the case of a distribution with more than two categories. To see how to generalize, we'll recast the statistic Z^2 in a form that treats the categories symmetrically. For this purpose, we change the notation: Let the two population categories be designated 1 and 2 and their probabilities p_1 and p_2. In a random sample of size n, let Y_j denote the frequency of category j. Thus, p_1 plays the role of p, p_2 plays the role of $1 - p$, and $Y_1/n = \hat{p}$.

Category	Probability	Frequency
1	p_1	Y_1
2	p_2	Y_2
Sums:	1	n

In Section 10.3, we called the specific value of p to be tested p_0. We now replace p_0 with π_1 and $1 - p_0$ with π_2. So the null hypothesis is

$$H_0: p_1 = \pi_1.$$

In this notation, the test statistic (1) is

$$Z = \frac{\dfrac{Y_1}{n} - \pi_1}{\sqrt{\dfrac{\pi_1 \pi_2}{n}}}. \tag{2}$$

In Section 10.3, for the alternative $H_A: p_1 \neq \pi_1$, we took large values of $|Z|$ to be extreme. This is the same as taking large Z^2 to be extreme. Squaring (2), we have

$$Z^2 = \frac{\left(\dfrac{Y_1}{n} - \pi_1\right)^2}{\dfrac{\pi_1 \pi_2}{n}} = \frac{(Y_1 - n\pi_1)^2}{n\pi_1\pi_2}. \tag{3}$$

Since

$$(Y_1 - n\pi_1)^2 = (n - Y_2 - n(1 - \pi_2))^2 = (Y_2 - n\pi_2)^2,$$

we may also write Z^2 as

$$Z^2 = \frac{(Y_2 - n\pi_2)^2}{n\pi_1\pi_2}. \tag{4}$$

To obtain a form more symmetrical than either (3) or (4), one that is easily generalized, we use the identity

$$\frac{1}{n\pi_1\pi_2} = \frac{\pi_1 + \pi_2}{n\pi_1\pi_2} = \frac{1}{n\pi_1} + \frac{1}{n\pi_2}. \tag{5}$$

Together with (3) and (4), this yields

$$Z^2 = \frac{(Y_1 - n\pi_1)^2}{n\pi_1} + \frac{(Y_2 - n\pi_2)^2}{n\pi_2}. \tag{6}$$

Example 13.1a   In Example 10.3c, we tested the hypothesis $p = .520$, where p is the probability of a male birth. The proportion of boys in a sample of size $n = 145$ was $\hat{p} = \frac{95}{145}$. In a Z-test, we found $Z = 3.26$. Here we'll calculate Z^2 using (6). To set the pattern for calculating extensions of (6) to several categories, we write the H_0 probabilities and the observed frequencies in table form.

Category	π_i	Y_i	$145\pi_i$	$Y_i - 145\pi_i$
Boy	.520	95	75.4	19.6
Girl	.480	50	69.6	-19.6
Sums:	1	145	145	0

Substituting in (6), we obtain

$$Z^2 = \frac{(19.6)^2}{75.4} + \frac{(-19.6)^2}{69.6} = 10.6 = (3.26)^2.$$

Under H_0, Z is nearly normal, so we use Table II in Appendix 1 to find the (two-sided) P-value:

$$P = P(Z^2 \geq 10.6) = P(|Z| \geq 3.26) = 2\Phi(-3.26) \doteq .0006. \tag{7}$$

Alternatively, since the square of a standard normal variable is $\chi^2(1)$ (see Section 6.5), we can find the P-value using Table Va in Appendix 1:

$$P(Z^2 \geq 10.6) = P[\chi^2(1) \geq 10.6] = .001.$$

[This differs from (7) only because the accuracy in Table Va is less than that in Table II.]   ∎

The two-sided Z-test and a one-sided test based on Z^2 are equivalent. Our reason for rephrasing the Z-test in terms of Z^2 [as given by (6)] is to suggest how to extend it to the case of more than two categories. This we do next.

13.2   The Chi-Square Test

Consider a discrete population with k categories. We label these $1, 2, \ldots, k$ and denote the corresponding probabilities by $p_1, p_2, \ldots, p_k$. We want to test the null hypothesis that these category probabilities are $\pi_1, \pi_2, \ldots, \pi_k$ (respectively), specific positive numbers that sum to 1:

$$H_0: p_j = \pi_j \quad \text{for} \quad j = 1, 2, \ldots, k.$$

Suppose we have a random sample of size n from such a population. We can summarize the sample observations in a frequency distribution (see Section 7.1). The category frequencies Y_j are sufficient (Section 8.3), and $Y_j \sim \text{Bin}(n, p_j)$ (Section 4.8). Under H_0, the expected frequency of category j is $n\pi_j$. The following table gives the population distribution, the expected frequencies under H_0, and the observed frequencies.

Category	Probability under H_0	Expected Frequency	Observed Frequency
1	π_1	$n\pi_1$	Y_1
2	π_2	$n\pi_2$	Y_2
$\vdots$	$\vdots$	$\vdots$	$\vdots$
k	π_k	$n\pi_k$	Y_k
Sums:	1	n	n

The Chi-Square Statistic

The chi-square statistic[1] is an overall measure of discrepancy between the observed frequencies and the expected frequencies under H_0. It is a natural generalization of the statistic Z^2 given by (6) of Section 13.1:

$$\chi^2 = \sum_1^k \frac{(Y_j - n\pi_j)^2}{n\pi_j}. \tag{1}$$

Since χ^2 is a weighted sum of *squared* deviations, it equals 0 if and only if every term on the right-hand side of (1) is zero. This happens only when each sample frequency Y_i is equal to the corresponding expected frequency $n\pi_i$—a perfect fit of the data to H_0. A poor fit is indicated by large discrepancies between observed and expected frequencies, which makes the value of χ^2 large. In such a case, it may be more reasonable to attribute the result to H_A than to sampling variability under H_0. So we take large values of χ^2 as extreme, and P-values are right tail-areas; but how large is "large"?

The null distribution of χ^2 is quite complicated and depends on the hypothesized probabilities π_j. However, when n is large, the distribution is approximately chi-square with $k - 1$ degrees of freedom (see Section 6.5), independent of the π_j.

The next example shows how one might test a genetic theory.

Example 13.2a   **Mendel's Pea Experiment**

In Example 8.2d, we gave Mendel's famous pea data, obtained by crossing round yellow pea plants with wrinkled green pea plants, obtaining plants bearing peas in one of four categories. Mendel's theory claimed that the expected frequencies of these characteristics are in the proportion $9:3:3:1$. This defines probabilities π_i, shown in Table 13.1 along with the sample frequencies Y_i and other ingredients of the chi-square statistic. These probabilities constitute our null hypothesis.

---

[1] The English statistician Karl Pearson proposed this statistic and gave its distribution in about 1900.

Table 13.1

i	Type	π_i	$556\pi_i$	Y_i	$Y_i - 556\pi_i$
1	R Y	$\dfrac{9}{16}$	312.75	315	2.25
2	R G	$\dfrac{3}{16}$	104.25	108	3.75
3	W Y	$\dfrac{3}{16}$	104.25	101	-3.25
4	W G	$\dfrac{1}{16}$	34.75	32	-2.75
Sums:		1	556.00	556	0

The fit of the data to the model of H_0 is not perfect, since the differences $Y_i - n\pi_i$ are not all zero. (Observe, however, that they sum to 0, as they always will.) Substituting in (1), we find

$$\chi^2 = \frac{2.25^2}{312.75} + \frac{3.75^2}{104.25} + \frac{(-3.25)^2}{104.25} + \frac{(-2.75)^2}{34.75} \doteq .47.$$

Because the fit is not perfect, χ^2 is positive; but this will generally be the case owing to sampling variability, even when H_0 is true. The question is whether χ^2 is so large as to suggest that H_0 is not true.

Under the null hypothesis, $\chi^2 \approx \chi^2(3)$. (There are four categories, so $k - 1 = 3$.) The graph of this distribution is shown in Figure 13.1, and tail-probabilities are given in Table Va in Appendix 1. For the pea data, the value of χ^2 is actually in the *left* tail of the distribution, whereas *large* values are extreme. The data are consistent with H_0.

(Very small values of χ^2 are also in a "tail" of the null distribution when d.f. > 4, but a very small χ^2 indicates a good fit—no reason to doubt H_0. A fit that is "too

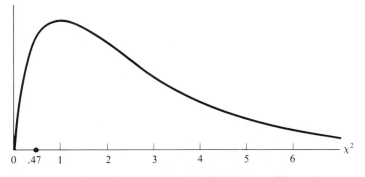

Figure 13.1 Chi-square p.d.f., 3 degrees of freedom

good to be true" may indicate something else. Some people have suggested that his gardener, having learned what Mendel was expecting, may have fudged the data to improve the fit.)   ■

Chi-square test for H_0: $p_j = \pi_j$, $j = 1, \ldots, k$. Given frequencies Y_j in a random sample of size n, calculate

$$\chi^2 = \sum_1^k \frac{(Y_j - n\pi_j)^2}{n\pi_j}. \tag{1}$$

For large n, find the P-value as a right tail-area of $\chi^2(k - 1)$ in Table Va in Appendix 1.

How large n must be for the chi-square approximation to work well depends on the cell probabilities. A conservative rule of thumb is that the expected frequencies $n\pi_i$ should be about 5 or more. When there are many cells, the approximation is good enough even if one or two expected frequencies are as small as 1. Cells with small expected frequencies can be combined to improve the approximation—even though in combining them, we'd be testing an approximation to H_0.

Testing a Distribution Type

The chi-square statistic can be used to test a distribution *type* without specifying a particular distribution of that type—for example, testing the hypothesis that a population distribution is binomial, or Poisson, or the like. Such a null hypothesis is a *family* of distributions indexed by unknown parameters. Specifying the values of these parameters specifies the category probabilities. If we replace the parameters by sample estimates, we can calculate estimated mean frequencies for use as "expected frequencies" in the chi-square statistic (1).

What are appropriate estimates of the unknown parameters, and what does using them in place of the parameters do to the null distribution of χ^2? With regard to the second question, using the sample to help define the null distribution seems like cheating, for it amounts to tailoring the null hypothesis to make it more like the sample. The fit then appears to be better than it really is. Since χ^2 is apt to be smaller, this suggests that its true null distribution is really shifted to the left.

With regard to the first question, we'll usually replace unknown parameters by their maximum likelihood estimates (Section 9.9). For the examples of this chapter, we'll use the sample mean, sample variance, and sample proportion as estimates of the corresponding population parameters. We let $\hat{p}_j$ denote the estimated probability of category j, calculated using m.l.e.'s of any unknown parameters. It can be shown[2] that under H_0, the large-sample distribution of the resulting value of the

---

[2] See Cramér [5], 417 ff. This result assumes that unknown parameters are replaced by m.l.e.'s, or by estimators asymptotically equivalent to m.l.e.'s.

chi-square statistic

$$\chi^2 = \sum_{j=1}^{k} \frac{(Y_j - n\hat{p}_j)^2}{n\hat{p}_j} \tag{2}$$

is again approximately chi-square. However, the number of degrees of freedom must be reduced to $k - 1 - r$, where r is the number of estimated parameters. Indeed, the distribution of $\chi^2(k - 1 - r)$ is shifted to the left as compared with that of $\chi^2(k - 1)$. So any particular percentile of χ^2 is smaller; the P-value for a calculated value of χ^2 is decreased accordingly.

Example 13.2b   **Boy–Girl Ratio in Families**

Does the sex of successive children in a family behave like independent Bernoulli trials, with the same probability p of having a boy? If so, the number of boys in a family of given size is binomially distributed. Consider families with eight children; we'll test the composite null hypothesis

$$H_0: p_j = \binom{8}{j} p^j (1 - p)^{8-j}, \qquad j = 0, 1, \ldots, 8, \tag{3}$$

where p_j is the probability that there will be j boys in a family of eight children. Table 13.2 summarizes data[3] on the numbers of boys in each of 1,000 families with eight children. The number of boys among the 8,000 children in these families is $\Sigma j Y_j = 4,040$. So our estimate of p is

$$\hat{p} = \frac{4040}{8000} = .505.$$

Table 13.2

j	Y_j	jY_j	$\hat{p}_j$	$n\hat{p}_j$
0	10	0	.0036	3.6
1	34	34	.0294	29.4
2	111	222	.1050	105.0
3	215	645	.2143	214.3
4	239	956	.2733	273.3
5	227	1135	.2231	223.1
6	115	690	.1138	113.8
7	34	238	.0332	33.2
8	15	120	.0042	4.2
	1000	4040	.9999	999.9

---

[3] Although the data are artificial, they follow the tendency observed in actual data. The real data are so extensive as to leave no room for doubt that male and female births do not behave as independent Bernoulli trials.

The probabilities (3) are thus estimated to be

$$\hat{p}_j = \binom{8}{j}(.505)^j(.495)^{8-j}, \qquad j = 0, 1, \ldots, 8.$$

Multiplying by $n = 1,000$ yields the expected frequencies $n\hat{p}_j$ shown in Table 13.2.

Substituting observed and expected frequencies from Table 13.2 into (2), we find $\chi^2 \doteq 50$. We estimated one parameter (p), so we look in Table Va in the column for 7 degrees of freedom $(k - 1 - r = 9 - 1 - 1 = 7)$ and find $P < .002$—strong evidence against the hypothesis that the population distribution is binomial.

(The sample distribution differs from binomial in a very special way: There are more families than expected in which one of the sexes predominates. Data such as these could arise if the sex distribution within each family is binomial, but different families have different p's. Estimating the distribution of p's from data such as given in this example is an interesting but difficult problem.) ∎

To test a discrete distribution whose probabilities p_j depend on one unknown parameter θ: $p_j = p_j(\theta)$, find an m.l.e. of θ and calculate χ^2, with π_i replaced by $\hat{p}_j = p_j(\hat{\theta})$:

$$\chi^2 = \sum_1^k \frac{(Y_j - n\hat{p}_j)^2}{n\hat{p}_j}$$

Under H_0, $\chi^2 \approx \chi^2(k - 2)$. When p_i depends on r parameters, the number of degrees of freedom is $k - r - 1$.

The chi-square test of fit is for a *discrete* distribution with a finite number of categories. However, it can be used in the continuous case by approximating the continuous distribution with one that is discrete. This is done by partitioning the axis of values into a finite number of class intervals and using these as categories. (See Problem B.) The null probability of an interval is the area above that interval under the null p.d.f. This procedure is not wholly satisfactory, since the test is only approximate—first because of the discrete approximation, and second because (as before) the chi-square distribution is only asymptotically correct. Moreover, the chi-square test does not take into account the ordering that is an essential aspect of a numerical variable (whether discrete or continuous).

13.3   Tests Based on c.d.f.'s

In this section, we consider testing a specific *continuous* distribution:

$$H_0: F(x) = F_0(x) \qquad \text{for all } x$$

against the general alternative

$$H_A: F(x) \neq F_0(x) \qquad \text{for some } x.$$

Various tests have been designed especially for continuous distributions. Some of these compare $F_0(x)$ with the **sample c.d.f.** The sample c.d.f., denoted by $F_n(x)$, is the c.d.f. of the discrete distribution that assigns probability $1/n$ to each sample value, as in Section 7.4. So F_n is a step function: It jumps an amount $1/n$ at each observation and is constant (horizontal) between successive observations in the ordered sample. If there are k observations with the *same* value, then the jump at that value is k/n. (Since the probability is zero that two observations will be exactly equal, multiple values should not occur, but they occur in practice because of round-off.)

Sample c.d.f.:

$$F_n(x) = \frac{1}{n}(\text{Number of observations} \leq x). \tag{1}$$

Kolmogorov–Smirnov Test

There are various ways of measuring the discrepancy between F_n and F_0. The *Cramér–von Mises* statistic measures "discrepancy" as an average of the squared difference; the **Kolmogorov–Smirnov** statistic measures discrepancy as the maximum[4] absolute difference. We'll treat only the Kolmogorov–Smirnov (K–S) statistic:

$$D_n = \max_x |F_n(x) - F_0(x)|. \tag{2}$$

The sample c.d.f. F_n is a step function, and F_0 is continuous. Examining the graphs, we see that the largest vertical distance between such functions occurs at one of the sample values—say, at the mth smallest, $X_{(m)}$. Then D_n is either

$$a_m = |F_0(X_{(m)}) - F_n(X_{(m)})| \quad \text{or} \quad b_m = |F_0(X_{(m)}) - F_n(X_{(m-1)})|.$$

These two cases are shown in the graphs of Figure 13.2.

If the data are not consistent with H_0, the statistic D_n will be large. So large values of D_n are extreme—evidence against H_0. To find P-values, we need the null distribution of D_n.

There are recursion formulas for calculating tail-probabilities, and these have been used in constructing Table IX in Appendix 1. It gives critical values of D_n ($n = 1$ to 20, 25, 30, 35) for $\alpha = .01, .05, .10, .15$, and .20. For large values of n ($n > 35$), we can use an approximation based on the asymptotic distribution of D_n:

$$P(D_n > v) \doteq 2e^{-2nv^2}. \tag{3}$$

---

[4] To be technically correct, *maximum* should read *supremum*. There may not be an actual maximum in the same sense that there is no largest number such that $x < 1$, although there is a least upper bound, or supremum (sup).

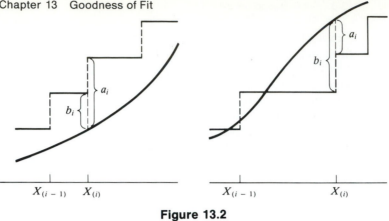

Figure 13.2

Example **13.3a**   **Hemoglobin Level**

Consider again the hospital lab's control chart described in Example 11a. It assumes that the results of tests on a particular blood sample are normally distributed, with mean μ and standard deviation $\sigma = .40$. On ten consecutive days, these measurements of hemoglobin were made on a standard blood sample:

15.36, 14.24, 15.69, 15.07, 16.89,
15.21, 15.09, 15.52, 16.28, 15.56.

Suppose we assume that the test equipment has not changed over the ten-day period and use the data to test the hypothesis that the population is normal, with $\mu = 15.5$ and $\sigma^2 = .16$. The two c.d.f.'s F_{10} and F_{10} are shown in Figure 13.3. To find D_n,

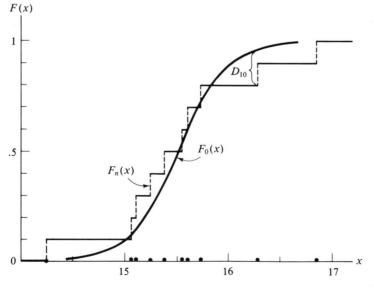

Figure 13.3

Table 13.3

$X_{(i)}$	$Z_{(i)}$	$F_0(X_{(i)})$	$F_n(X_{(i)})$	a_i	b_i
14.24	−3.15	.001	.1	.099	.001
15.07	−1.08	.140	.2	.060	.040
15.09	−1.02	.154	.3	.146	.046
15.21	−.72	.236	.4	.164	.064
15.36	−.35	.363	.5	.137	.037
15.52	.05	.520	.6	.080	.020
15.56	.15	.560	.7	.140	.040
15.69	.48	.684	.8	.116	.016
16.28	1.95	.974	.9	.074	.174
16.89	3.47	1.000	1.0	.000	.100

we need look only at the differences between F_0 and F_n at the ten sample values. Table 13.3 gives the ordered observations, the corresponding values of the two c.d.f.'s at those points, and the distances a_i and b_i.

The value of the K–S statistic (2)—the maximum distance between the two c.d.f.'s—is circled in the table: $D_{10} = .174$. How surprising is this result under the null hypothesis? Referring to Table IXb, we see that .174 is not in the tail of the distribution when $n = 10$. (On the graph in Table IXa, $10D_{10} = 1.74$ would have a P-value of at least .7.) There is no reason to question that F is F_0 on the basis of the given data.

Approximation (3) is not very successful for this small sample size; it gives $P = 2 \exp\{-2n(.174)^2\} = 1.09 > 1$! ∎

Kolmogorov-Smirnov test for a continuous c.d.f. $F_0(x)$: Calculate

$$D_n = \max_x |F_n(x) - F_0(x)|, \qquad (2)$$

where $F_n(x)$ is the sample c.d.f. Find the corresponding P-value in Table IX. For large n,

$$P(D_n > v) \doteq 2e^{-2nv^2}. \qquad (3)$$

The K–S test has good and bad points. On the good side, the small-sample distribution is known, so the test does not require a large sample. Also, it is distribution-free: The same table is used *no matter what F_0 is being tested.* On the bad side, it is designed with no very specific kind of alternative in mind. So we could not expect it to be as sensitive as the Z-test or the t-test in detecting a difference in mean value, an alternative for which the Z- and t-tests are especially designed. Thus, in the above example, if the alternative to $N(15.5, .16)$ were $N(\mu_1, .16)$, we'd use a Z-test.

When H_0 involves unknown parameters, we can calculate D_n by using sample estimates in their place, as we did in the case of chi-square (Section 13.2). And as with χ^2, this adjustment of the model toward the data affects the null distribution of D_n. The effect is not known in general, but an adjustment for testing normality will be given in the next section.

Although the sampling distribution of the K−S statistic as given by Table IX is based on the assumption of a continuous population variable, it can also be used when the population variable is discrete. Doing so is conservative, in the sense that the actual P-value does not exceed that given by the table, but it could be much smaller.

Discrete Models

Example 13.3b   In Example 13.2b, we tested the hypothesis that the number of boys in a family with eight children has a binomial distribution with $p = \frac{1}{2}$. Since the number of boys is a numerical variable, the K−S test may seem more appropriate than the chi-square test.

Table 13.4

x	Sample d.f. $F_n(x)$	Population d.f. $F_0(x)$	$\|F_n(x) - F_0(x)\|$
0	.010	.004	.006
1	.044	.035	.009
2	.155	.145	.011
3	.370	.363	.007
4	.609	.637	.028
5	.836	.856	.020
6	.951	.965	.014
7	.985	.996	.011
8	1.000	1.000	0

The data on 1,000 families ($n = 1000$) in Example 13.2b are given in Table 13.4 as cumulative relative frequencies F_n. The binomial distribution is given in terms of cumulative probabilities F_0. The last column gives the absolute differences between F_n and F_0 at $x = 0, 1, \ldots, 8$. These are the only differences that need be considered, since both the sample c.d.f. and the population c.d.f. jump only at these points. The maximum distance is $D_n = .028$, and the tail-area in the K−S limiting distribution [from (3)] is

$$P(D_n > .028) \doteq 2e^{-2000(.028)^2} = 2e^{-1.57} \doteq .42.$$

The chi-square test gave $P < .01$. In this case, the K−S statistic is not as sensitive to the heavy tails of F_n as is χ^2.   ■

The one-sample K−S statistic is based on the largest absolute difference between functions. This same measure of distance, applied to two sample c.d.f.'s, leads to the two-sample K–S statistic for testing the hypothesis that two populations are identical,

$$H_0: F(x) = G(x) \qquad \text{for all } x,$$

against the general alternative,

$$H_0: F(x) \neq G(x) \qquad \text{for some } x.$$

Two-Sample K–S Test

Given independent random samples from continuous populations, we define the two-sample K–S statistic as the maximum absolute difference between the sample c.d.f.'s. Under H_0, this largest absolute difference D is distribution-free: Its null distribution depends only on the two sample sizes m and n. So a single table (Table X) is applicable, no matter what the common population distribution (under H_0) may be. An asymptotic distribution is known; it provides the following approximation for large samples, which was used in constructing Table Xa:

$$P\left(D < y\sqrt{\frac{m+n}{mn}}\right) \doteq 1 - 2e^{-2y^2}.$$

Since the value of a sample distribution function at any point is a fraction with denominator equal to the sample size, the differences defining D are differences between two fractions. The value of D is then itself a fraction, one whose denominator is the least common multiple of the two sample sizes. Table Xb gives values of D corresponding to 5% and 1% in the right tail of the distribution for small samples, as well as asymptotic formulas for these percentiles for the case of large samples. Table Xa gives tail-probabilities for the null distribution of D for the case of equal sample sizes ($n = m = 4$ to 11, 15, and 20) and for large samples.

Test for $F = G$ based on the corresponding sample c.d.f.'s F_n and G_n: Find

$$D = \max|F_n(x) - G_n(x)| \tag{4}$$

and obtain P-values in Table X.

Example 13.3c   **Intermittent Feeding and Blood Pressure**

In an experiment to study the effect of intermittent feeding on blood pressure, the blood pressures of 15 rats were reported as follows (see Examples 10b, 12.4a, and 12.5b):

> Normal feeding: 169, 155, 134, 152, 133, 108, 145.
>
> Intermittent feeding: 170, 168, 115, 181, 162, 199, 207, 162.

The two sample c.d.f.'s are shown in Figure 13.4.

If there is no effect of the intermittent feeding, then the two populations are identical, so we test $H_0: F_N = F_I$. From the graphs, it is evident that

$$D = \frac{6}{7} - \frac{1}{8} = \frac{41}{56}.$$

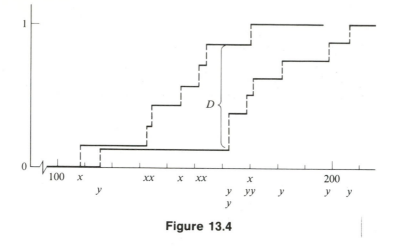

Figure 13.4

Table Xb shows the value of D for $P = .05$ to be $\frac{35}{56}$ and the value with $P = .01$ to be $\frac{42}{56}$. The P-value for $D = \frac{41}{56}$ is then between .01 and .05—actually, closer to .01.

The sample sizes are not especially large, but suppose we try the large sample approximation, which is conservative for smaller sizes. Referring to Table Xa (case 2), we have $Y \doteq 1.41$ and $P \doteq .038$.

For comparison, we note that the two-sample t-test (carried out in Example 12.4a) resulted in $P = .044$. The rank-sum test (performed in Example 12.5b) resulted in $P = .028$. These tests were directed toward shift alternatives—toward detecting a difference in location. The K–S test is a general-purpose test and is apt to be less sensitive against specific types of alternatives. In the present example, however, it happens to provide about the same degree of evidence against H_0 as the location tests. ■

13.4   Testing Normality

The assumption of population normality is fundamental to the t-tests of Chapters 10 and 12 and to various tests yet to come. Normality can be tested using the chi-square statistic (3) of Section 13.2, as illustrated in Problem B. However, a test in which we use the K–S statistic (Section 13.3), with unknown parameters estimated from sample data, is generally more powerful against a wider variety of alternatives than the chi-square test. So we define a c.d.f. F_0 for use in (2) of Section 13.3 by replacing μ and σ in the general normal c.d.f. with $\bar{X}$ and S, respectively.

The null distribution of this modified K–S statistic is known only approxi-

Adapting the K–S Statistic

mately, on the basis of simulations.[5] To simulate the null distribution of D_n (see Section 8.5), one obtains a great many samples from a normal distribution and calculates D_n for each. Table XI of Appendix 1 gives some percentiles of D_n obtained in this way.

Example 13.4a   **Sea Pollution**

A study of the effects of dumping wastes into the sea[6] gives the following measurements of zinc concentration (mg/kg) at 13 points within one mile of a dump site:

$$13.5, 23.8, 23.3, 20.9, 23.8, 29.0, 20.9, 24.4, 16.4, 18.3, 17.6, 25.4, 23.3$$

The mean and s.d. are $\bar{X} = 21.6$ and $S = 4.21$. To test normality, we use these as the mean and s.d. of a normal distribution and define

$$F_0(x) = \Phi\left(\frac{x - 21.6}{4.21}\right).$$

Figure 13.5 shows F_0 and F_n $(n = 13)$. Table 13.5 gives the values of F_0 and F_n at each observation x_i. In that table, a_i and b_i have the same interpretation as in Table 13.3.

The largest discrepancy between F_n and the estimated F_0 is $D_{13} = .186$. In Table XI, we find the area beyond $D_{13} = .193$ in the null distribution to be .20, so $P > .20$. According to this test, the data do not call the hypothesis of normality into serious question.   ■

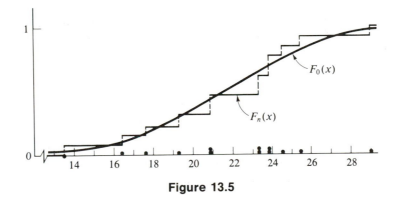

Figure 13.5

[5] H. W. Lilliefors, "On the Kolmogorov–Smirnov test for normality with mean and variance unknown," *J. Am. Stat. Assn. 64* (1967), 399–402. Slight adjustments of his tabled values based on more extensive sampling are given in G. Dallal and L. Wilkinson, "An analytic approximation to the distribution of Lilliefors' test statistic for normality," *Amer. Statistician 40* (1986), 294–296.

[6] K. J. Borwell, D. M. G. Kingston, and J. Webster, "Sludge disposal at sea—the Lothian experience," *Water Pollution Control 85* (1986), 269–276.

Table 13.5

$X_{(i)}$	$F_0[X_{(i)}]$	$F_n[X_{(i)}]$	Differences a_i	b_i
13.5	.0274	1/13	.049	.027
16.4	.0977	2/13	.056	.021
17.6	.1711	3/13	.060	.017
18.3	.2165	4/13	.091	.014
20.9	.4340	6/13	.028	.126
23.3	.6480	8/13	.033	.186
23.8	.6985	10/13	.071	.083
24.4	.7470	11/13	.099	.022
25.4	.8159	12/13	.107	.030
29.0	.9608	1	.039	.038

Comparison with Normal Scores

Another kind of test for normality is due to Shapiro and Wilk. It is based on a comparison of ordered sample values with their expected locations under the hypothesis of normality. Let $Z_{(i)}$ denote the ith smallest in a random sample from a *standard* normal distribution, and define $m_i = E[Z_{(i)}]$. The m_i are called **normal scores** or **rankits**. They are given in Table XII of Appendix 1 for sample sizes 2 to 20.

Denote the ith smallest sample value by $X_{(i)}$, as usual. Consider the pairs

$$(X_{(1)}, m_1), \quad (X_{(2)}, m_2), \ldots, \quad (X_{(n)}, m_n).$$

If the population is normal with mean μ and s.d. σ, then the Z-scores

$$Z_{(i)} = \frac{X_{(i)} - \mu}{\sigma}$$

are ordered observations from a standard normal distribution. So

$$E(X_{(i)}) = \sigma E(Z_{(i)}) + \mu = \sigma m_i + \mu.$$

Thus, when the population is normal, we "expect" the ordered sample values to be *linearly* related to the m_i's. Of course, the correlation with the *actual* ordered sample values will not be perfect because of sampling variability, even if the population is normal.

A scatterplot of the pairs $(X_{(i)}, m_i)$ is called a **normal-scores plot** or a **rankit plot**. With some experience in interpreting such a plot, one can make an informal judgment as to normality. We'll start you in this direction with some examples below.

The Shapiro–Wilk statistic W is approximately the square of the correlation coefficient of the pairs $(X_{(i)}, m_i)$. To use W, we need a table of tail-probabilities of its null distribution. Table 13.6[7] gives *left* tail-areas (α's). Because a small value of W

---

[7] The table is taken from S. Weisberg's *Multreg User's Manual*, which adapted it from S. Weisberg, "An empirical comparison of W and W'," *Biometrika 61* (1974), 645–646, and S. S. Shapiro, and R. S. Francia, "An approximate analysis of variance test for normality," *J. Amer. Stat. Assn. 67* (1972), 215–216.

Table 13.6 Selected percentage points of W

α n	.01	.05	.10	.50
5	.675	.777	.817	.922
10	.776	.842	.869	.940
15	.815	.878	.903	.954
20	.858	.902	.921	.962
35	.919	.943	.952	.976
50	.935	.953	.963	.981
75	.956	.969	.973	.986
99	.967	.976	.980	.989

means a poor fit of the scatterplot to a straight line, small values of W are "extreme," suggesting nonnormality of the population.

Example 13.4b **Sea Pollution (continued)**

Consider the 13 data points of Example 13.4a. The expected values of the normal order statistics for $n = 13$ are found in Table XI. We pair these with the ordered zinc concentrations $(X_{(i)}, m_i)$:

$(13.5, -1.668)$ $(16.4, -1.164)$ $(17.6, -.850)$ $(18.3, -.603)$
$(20.9, -.388)$ $(20.9, -.190)$ $(23.3, 0)$ $(23.3, .190)$
$(23.8, .388)$ $(23.8, .603)$ $(24.4, .850)$ $(25.4, 1.164)$
$(29.0, 1.668)$

The rankit plot is shown in Figure 13.6. The points lie very nearly on a straight line, so there appears to be little evidence against normality. The correlation coefficient is .979, so $W = .959$. This is not in the left half of the null distribution of W. (For $n = 10$ the median is .94, and for $n = 15$ it's about .95.) So there is little evidence against normality, consistent with the conclusion reached in the preceding example using the Lilliefors table. ∎

Some researchers use a test of normality to decide whether or not a t-test is valid. Using such a preliminary test with the same data as is used in the t-test can make the calculated P-value incorrect. However, it is usually a good idea to make at least an informal analysis (for example, inspecting a rankit plot) to see that there is not an obvious and drastic departure from normality. When there is such a departure, it may be appropriate to transform the observations to make their distribution more nearly normal. Alternatively, use a nonparametric procedure—one not sensitive to the assumption of normality (see Sections 10.5 and 12.5).

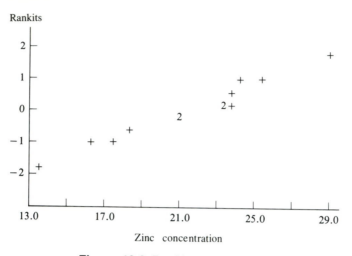

Figure 13.6 Rankits versus zinc

Testing Independence

A common problem in applied statistics is deciding whether two variables are related. In this section, we show how the chi-square statistic is useful in testing the independence of two categorical variables.

Example 13.5a   **Arrowhead Breakage**
Seventy-seven fractured arrowheads were found in central Tennessee.[8] The investigator considered as a null hypothesis that the location of fracture and the cause of fracture are unrelated. A cross-classification of 77 arrowheads yielded the following frequencies, laid out in a contingency table:

| | | *Fracture location:* | | | |
		Base	*Middle*	*Tip*	
Cause of	*Fire*	21	8	18	47
fracture	*Other means*	15	11	4	30
		36	19	22	77

The proportions fractured by fire are different for the different locations: $\frac{21}{36} \doteq .58$, $\frac{8}{19} \doteq .42$, $\frac{18}{22} \doteq .82$. The statistical question is whether the differences are explainable

---

[8] Reported by J. L. Hofman in "Eva projectile point breakage at Cave Spring: Pattern recognition and interpretive possibilities," *Midcont. J. Arch. 11*, (1986), 79–95.

as random variations or if they give evidence against the hypothesis of independence in the population from which these arrowheads were drawn. (To be continued.)   ∎

This example calls for a test of the hypothesis of independence in a discrete bivariate population. Consider discrete variables X with r categories and Y with c categories. According to Section 2.5, X and Y are independent if and only if their joint probabilities factor into products of marginal probabilities. Denote the probability of the pair (i, j) by p_{ij}, where i is a category of population X and j is a category of population Y. Marginal probabilities are the row and column sums in the two-way table of joint probabilities:

$$P(X = i) = p_{i1} + \cdots + p_{ic} = p_{i+},$$

and

$$P(Y = j) = p_{1j} + \cdots + p_{rj} = p_{+j}.$$

A "+" in a subscript indicates that a summation has been performed on the corresponding index. In this notation, the hypothesis of independence is

$$H_0: p_{ij} = p_{i+}p_{+j} \qquad \text{for} \qquad i = 1, \ldots, r, \qquad j = 1, \ldots, c. \tag{1}$$

The alternative is that H_0 is false—that X and Y are dependent.

The null hypothesis H_0 gives the rc cell probabilities in the joint distribution as functions of the $r - 1$ parameters p_{i+} and the $c - 1$ parameters p_{+j}. Thus, we may apply the chi-square statistic (2) of Section 13.2 to the case of a discrete distribution whose category probabilities depend on unknown parameters. The number of degrees of freedom of χ^2 is $rc - 1$ minus the number of estimated parameters:

$$\text{d.f.} = rc - 1 - [(r - 1) + (c - 1)] = (r - 1)(c - 1).$$

Consider a random sample of n observations on (X, Y). Let n_{ij} denote the frequency of the cell (i, j). Just as for the p's, let n_{i+} and n_{+j} indicate the sums of the cell frequencies over the corresponding index.

Under $H_0 \cup H_A$, the expected cell frequencies are np_{ij}. Under H_0, they are $np_{i+}p_{+j}$. We estimate the marginal probabilities by the corresponding marginal relative frequencies in the sample (their m.l.e.'s):

$$\hat{p}_{i+} = \frac{n_{i+}}{n}, \qquad \hat{p}_{+j} = \frac{n_{+j}}{n}. \tag{2}$$

So under H_0, the m.l.e. of $np_{i+}p_{+j}$ is

$$n\hat{p}_{i+}\hat{p}_{+j} = \frac{n_{i+}n_{+j}}{n} = e_{ij}.$$

(The e stands for expected frequency.)

Chi-square test for independence: To test $H_0: p_{ij} = p_{i+}p_{+j}$, calculate

$$\chi^2 = \sum_i \sum_j \frac{(n_{ij} - e_{ij})^2}{e_{ij}}, \tag{3}$$

where

$$e_{ij} = \frac{n_{i+}n_{+j}}{n}. \tag{4}$$

Under H_0, χ^2 has a chi-square distribution with $(r-1)(c-1)$ degrees of freedom.

Example 13.5b   **Arrowhead Breakage (continued)**

Returning to the arrowhead data of Example 13.5a, we calculate two of the expected cell frequencies:

$$e_{11} = \frac{36.47}{77} = 21.97, \qquad e_{12} = \frac{19.47}{77} = 11.60.$$

The others follow from these two by subtraction from marginal totals; this gives an interpretation for "degrees of freedom = 2." The complete table of expected frequencies e_{ij} is as follows:

| | | *Fracture location:* | | | |
		Base	*Middle*	*Tip*	
Cause of	*Fire*	21.97	11.60	13.43	47
fracture	*Other means*	14.03	7.40	8.57	30
		36	19	22	77

In this table, the rows are proportional and the columns are proportional, as in the table of joint probabilities for a pair of independent variables (see Section 2.5). Substituting the expected and observed frequencies from this table and the one given in Example 13.5a into (3) gives

$$\chi^2 = \frac{(21 - 21.97)^2}{21.97} + \frac{(8 - 11.60)^2}{11.60} + \frac{(18 - 13.43)^2}{13.43} + \frac{(15 - 14.03)^2}{14.03}$$

$$+ \frac{(11 - 7.40)^2}{7.40} + \frac{(4 - 8.57)^2}{8.57} = 6.97.$$

From Table Va (d.f. = 2) we find $P = .030$. So there is some evidence against H_0.

(In doing these calculations, there is a question of how much to round off. We kept two decimal places in the above table; if we round to one decimal place, we obtain $\chi^2 = 7.025$, not far from 6.97. If we round to the nearest integer, the result is $\chi^2 = 8.44$, appreciably in error. Rounding expected frequencies to the nearest tenth gives sufficient accuracy.) ■

The sampling scheme used in the above examples was to take a random sample of individuals from some population and record two characteristics (variables) for each. An alternative is to control one variable—say, x. For example, suppose individuals are to be given different cold remedies; "remedies" is a controlled variable. It is possible to control a variable when it defines identifiable sub-populations, such as males and females. In contrast to the uncontrolled case, we can always specify a sample size for each category of response of a controlled variable. This fixes the marginal frequencies of x. We can then compare the Y-samples for the various subpopulations identified by x.

Testing Homogeneity

A null hypothesis corresponding to "no difference among subpopulations" is that the probability distributions in those subpopulations are the same. This is the hypothesis of *homogeneity*. The hypothesis of independence of X and Y is the same as the hypothesis of homogeneity of the subpopulations defined by X. (See Section 2.5.) Since the null hypotheses are the same for the two sampling schemes, we use the same chi-square test in both cases. The next example illustrates a test for homogeneity.

Example 13.5c

Behavior Modification

A study of opinions on the usefulness of behavior modification[9] reported on opinions of psychologists and psychiatrists as follows:

	Useful?				
	Never	Occasionally	Often	Always	
Psychologists	12	26	29	18	85
Psychiatrists	5	13	7	0	25
	17	39	36	18	110

The psychologists and psychiatrists were two identified populations, and samples were selected from each. The null hypothesis is that psychologists and psychiatrists have the same opinions on behavior modification. The chi-square statistic is calculated by comparing the above observed frequencies with the expected

[9] G. P. Koocher and B. M. Pedulla, "Current practices in child psychotherapy," *Profess. Psychol. 8,* (1977), 275–286.

frequencies under H_0:

	Never	Occasionally	Often	Always
Psychologists	13.14	30.14	27.83	13.91
Psychiatrists	3.86	8.86	8.18	4.09

The value of χ^2 calculated from (3) is 8.45. Referring to Table Va under d.f. = $(4-1)(2-1) = 3$, we find that the P-value is about .037.

The authors report that 280 questionnaires were sent out. There were 138 responses, and of these only 110 were usable. Such a low response rate invalidates the study. (A rule of thumb says that the response rate should be at least 80%.) If these are random samples from some populations, they are the populations of responders, not of psychiatrists and psychologists generally.   ■

The preceding example is an instance typical of many in which the categories of one or both of the variables are naturally ordered. However, the chi-square test does not take order into account. For instance, if frequencies for "occasionally" and "always" were reversed, χ^2 would be unchanged. It might be better to use a test based on location. This can be done by coding the ordered categories, perhaps by rank, and applying a t-test, for example.

Example 13.5d   Behavior Modification—Another Approach
Ignoring problems with response rate, we consider again the data of the preceding example. Now code the categories of response by rank:

Never = 1,   occasionally = 2,   often = 3,   always = 4.

The data are then summarized in two frequency tables:

Code	Psychologists		Psychiatrists	
x	f	xf	f	xf
1	12	12	5	5
2	26	52	13	26
3	29	87	7	21
4	18	72	0	0
	85	223	25	52

The code means are $\frac{223}{85} = 2.62$ and $\frac{52}{25} = 2.08$, and the s.d.'s are .976 and .704, respectively. From (5) of Section 12.2,

$$Z = \frac{2.62 - 2.08}{\sqrt{.976^2/85 + .702^2/25}} \doteq 3.09.$$

(The sample sizes are large enough that Z is approximately normal.) The two-sided P-value is .002, so the evidence against the hypothesis of equal average ranks is strong—much stronger than the evidence that the two populations are different in some unspecified way. (Using χ^2, we found $P = .037$.)

Because this approach considers the ordering explicitly, we prefer it to a chi-square test, but we recognize that it depends on the assignment of numbers to categories. Thus, if "always" = 5 instead of 4, then $Z = 3.78$, whereas if "always" is combined with "often" and assigned the value 3, then $Z = 2.06$.   ■

We've considered only the case of two categorical variables and the corresponding two-dimensional contingency table. A considerable amount of research has been directed at understanding higher-dimensional contingency tables. For an introduction to this area, see Fienberg et al. [9].

13.6   A Likelihood Ratio Test for Goodness of Fit

The likelihood ratio method introduced in Section 11.6 is applicable in testing goodness of fit. In the notation of earlier sections, category i ($i = 1, \ldots, k$) has probability p_i and frequency Y_i in a random sample of size n. The likelihood function is

$$L(\mathbf{p}) = p_1^{Y_1} \cdots p_k^{Y_k}. \tag{1}$$

We saw in Section 9.9 that (1) is maximized when p_i is replaced by the corresponding sample relative frequency Y_i/n:

$$\sup L(\mathbf{p}) = \left[\frac{Y_1}{n}\right]^{Y_1} \cdots \left[\frac{Y_k}{n}\right]^{Y_k}. \tag{2}$$

Consider testing a null hypothesis of the form

$$H_0: p_1 = \pi_1(\theta), \ldots, p_k = \pi_k(\theta).$$

against H_A: H_0 is false. With θ unknown, the likelihood function under H_0 is

$$L_0(\theta) = [\pi_1(\theta)]^{Y_1} \cdots [\pi_k(\theta)]^{Y_k}. \tag{3}$$

This is a maximum at $\hat{\theta}$, the m.l.e. of θ. The likelihood ratio for testing H_0 against H_A is

$$\Lambda = \frac{\sup\limits_{\theta} L_0}{\sup\limits_{p} L} = \frac{(\pi_1(\hat{\theta}))^{Y_1} \cdots (\pi_k(\hat{\theta}))^{Y_k}}{(Y_1/n)^{Y_1} \cdots (Y_k/n)^{Y_k}}$$

$$= \left(\frac{\pi_k(\hat{\theta})}{Y_1/n}\right)^{Y_1} \cdots \left(\frac{\pi_k(\hat{\theta})}{Y_k/n}\right)^{Y_k}. \tag{4}$$

Small values of Λ are extreme; equivalently, large values of $-2 \log \Lambda$ are extreme. Under H_0 (according to Section 11.6),

$$-2 \log \Lambda \approx \chi^2(k - r - 1),$$

where r is the number of estimated parameters—the number of components of the vector parameter θ. (When the π_i are constants, $r = 0$.)

It is not just a coincidence that the large-sample null distribution of $-2 \log \Lambda$ is the same as the large-sample distribution of χ^2. It can be shown[10] that under the null hypothesis, these two test statistics are asymptotically equal.

Example 13.6a   **Kirby Puckett**

Consider the hypothesis that a baseball player's successive times at bat in a game are independent Bernoulli trials, where p is the player's "true" probability of getting a hit. Then the number of hits in four at-bats is Bin $(4, p)$ for some p. The following table gives the distribution of the number of hits by Kirby Puckett of the Minnesota Twins in the 1987 regular season, in the 78 games in which he had four official times at bat:

Number of hits, j	0	1	2	3	4
Frequency, Y_j	19	26	26	4	3

The null hypothesis is

$$H_0: p_j = \binom{4}{j} p^i (1 - p)^{4-j}, \qquad j = 0, 1, 2, 3, 4.$$

The alternative is that this isn't correct—that the p_j are not binomial for any p. The likelihood function is

$$L(p_0, \ldots, p_4) = p_0^{19} p_1^{26} p_2^{26} p_3^4 p_4^3,$$

where $p_0 + \cdots + p_4 = 1$. Under $H_0 \cup H_A$, this is a maximum when $p_i = Y_i/78$:

$$\sup L = \left(\frac{19}{78}\right)^{19} \left(\frac{26}{78}\right)^{26} \left(\frac{26}{78}\right)^{26} \left(\frac{4}{78}\right)^4 \left(\frac{3}{78}\right)^3.$$

Under the null hypothesis, the likelihood is

$$L_0(p) = [(1-p)^4]^{19} [p(1-p)^3]^{26} [p^2(1-p)^2]^{26} [p^3(1-p)]^4 [p^4]^3$$
$$= p^{102}(1-p)^{210}.$$

This is largest at $\hat{p} = \frac{102}{312} \doteq .3269$, Puckett's batting average in the 78 games. So

$$\log \Lambda = \log[\sup L_0(p)] - \log[\sup L(p_0, \ldots, p_4)]$$
$$= -3.409,$$

and $-2 \log \Lambda \doteq 6.82$. Since $k = 5$ and $r = 1$, the d.f. $= 3$. The P-value for $\chi^2 = 6.82$ (from Table Va) is about .078. There is some evidence against the binomial, but the data are not completely convincing.

---

[10] See Lindgren [15], p. 427.

(Applying the chi-square test of Section 13.2 yields $\chi^2 \doteq 8.7$ and $P \doteq .038$. The poor agreement of these P-values is the result of a small expected frequency for $j = 4$, which is less than 1. The .078 is more reliable.) ∎

The likelihood ratio statistic can also be applied to the data in a contingency table to test independence:

$$H_0: p_{ij} = p_{i+}p_{+j},$$

in terms of the notation of Section 13.5. The likelihood function is

$$L(p_{11}, \ldots, p_{rc}) = \prod_{i,j} p_{ij}^{n_{ij}}. \tag{5}$$

The m.l.e. of p_{ij} under $H_0 \cup H_A$ is just the cell relative frequency, n_{ij}/n. Under H_0, the m.l.e. of p_{ij} is the product of the m.l.e.'s of p_{i+} and p_{+j}, which is the product of the corresponding marginal relative frequencies. So

$$\Lambda = \frac{\prod_{i,j}\left(\dfrac{n_{i+}n_{+j}}{n^2}\right)^{n_{ij}}}{\prod_{i,j}\left(\dfrac{n_{ij}}{n}\right)^{n_{ij}}} = \frac{\prod_{i,j}(n_{i+}n_{+j})^{n_{ij}}}{n^n \prod_i \prod_j n_{ij}^{n_{ij}}}. \tag{6}$$

Taking logarithms, we have the following:

Large-sample likelihood ratio statistic for testing independence:

$$-2 \log \Lambda = 2n \log n + 2 \sum\sum n_{ij} \log n_{ij} - 2 \sum n_{i+} \log n_{i+}$$
$$- 2 \sum n_{+j} \log n_{+j}, \tag{7}$$

where $i = 1, \ldots, r$ and $j = 1, \ldots, c$. Under independence, $-2 \log \Lambda \approx \chi^2[(r-1)(c-1)]$.

Example 13.6b   **Coffee and Heart Disease**

Coffee consumption is suspected by many of contributing to the development of coronary heart disease (CHD). One investigation[11] into the possible relationship reported these data:

	Cups per day				
	0	1–2	3–4	5 or more	n_{i+}
CHD	4	17	17	9	47
No CHD	185	457	226	125	993
n_{+j}	189	474	243	134	1,040 = n

[11] A. Z. LaCroix et al., "Coffee consumption and the incidence of coronary heart disease," *New Engl. J. Med. 315* (1986), 977–982.

We'll test independence using the likelihood ratio statistic. Substituting in (7), we have

$$-2 \log \Lambda = 1040 \log 1040 + 4 \log 4 + 17 \log 17 + \cdots - 189 \log 189 - \cdots$$
$$-47 \log 47 - 993 \log 993 = 8.44.$$

Referring to Table Va, we find $P \doteq .038$.

The categories of "cups per day" are ordered. Another approach in investigating a relationship is to code these categories and test for a difference in means. Suppose we adopt this coding:

$$0 \text{ cups} = 0,$$
$$\text{One to two cups} = 1.5,$$
$$\text{Three to four cups} = 3.5,$$
$$\text{Five or more cups} = 6.$$

We find these summary statistics for the coded data:

	n	Mean	s.d.
CHD	47	2.96	1.86
No CHD	993	2.24	1.83

The Z-statistic is

$$Z = \frac{2.96 - 2.24}{\sqrt{1.86^2/47 + 1.83^2/993}} \doteq 2.6.$$

The two-sided P-value is .01, so the evidence against independence is somewhat stronger with this analysis. ∎

13.7   Testing Homogeneity Using Paired Data

Example 13.7a   Tonsillectomies and Hodgkin's Disease

A study[12] involved 85 patients with Hodgkin's disease. Each of these had a normal sibling (one who did not have the disease). In 26 of the pairs, both individuals had had tonsillectonies (T); in 37 pairs, both individuals had not had tonsillectomies (N); in 15 pairs, only the normal individual had had a tonsillectomy; in 7 pairs, only the one with Hodgkin's disease had had a tonsillectomy.

---

[12] S. Johnson and R. Johnson, "Tonsillectomy history in Hodgkin's disease," *New Engl. J. Med. 287* (1972), 1122–1125.

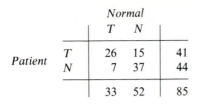

	Normal T	Normal N	
Patient T	26	15	41
Patient N	7	37	44
	33	52	85

A goal of the study was to determine whether there was a link between the disease and having had a tonsillectomy: Is the proportion of those who had tonsillectomies the same among those with Hodgkin's disease as among those who don't have it? (To be continued.) ∎

Consider a response variable with two categories, which we observe for each individual in n randomly selected matched pairs. In each pair, one is a "treatment" subject and the other a "control." We adopt the 0–1 coding for the variable of interest and also for treatment–control. Each pair will then be in one of these four cells: $(0, 0), (0, 1), (1, 0), (1, 1)$. Given a random sample of n pairs, each pair will fall in one of these categories, so the category frequencies are multinomial. The frequencies and corresponding probabilities can be organized in a contingency table, as follows.

Data:

	0	1	
0	n_{00}	n_{01}	n_{0+}
1	n_{10}	n_{11}	n_{1+}
	n_{+0}	n_{+1}	n

Model:

	0	1	
0	p_{00}	p_{01}	p_{0+}
1	p_{10}	p_{11}	p_{1+}
	p_{+0}	p_{+1}	1

The hypothesis we want to test is $H_0: p_{1+} = p_{+1}$. We can't use a standard comparison of proportions (Section 12.3), since we don't have independent samples (the data are paired).

The null hypothesis is equivalent to $p_{10} = p_{01}$. Intuition suggests that there is no information about this hypothesis in the cell frequencies n_{00} and n_{11}. Consider the conditional probability of $(0, 1)$ given the off-diagonal:

$$p_* = \frac{p_{01}}{p_{01} + p_{10}}.$$

In terms of p_*, the null hypothesis is $p_* = \frac{1}{2}$. The Z-score for testing $p_* = \frac{1}{2}$ based on n_{01} successes in $n_{01} + n_{10}$ independent trials (see Section 10.3) is

$$Z = \frac{n_{01} - \frac{1}{2}(n_{01} + n_{10})}{\sqrt{\frac{1}{4}(n_{01} + n_{10})}} = \frac{n_{01} - n_{10}}{\sqrt{n_{01} + n_{10}}}$$

For a two-sided test, large values of $|Z|$ or of Z^2 are extreme:

$$Z^2 = \frac{(n_{01} - n_{10})^2}{n_{01} + n_{10}}. \tag{1}$$

The null distribution of Z^2 is $\chi^2(1)$, and the test based on (1) in conjunction with the chi-square table (one d.f.) is called **McNemar's test**.

McNemar's test is actually a chi-square test. The four cell probabilities p_{ij} are functions of two parameters, p_{11} and p_*. The likelihood function under H_0 is a multinomial probability:

$$L(p_{11}, p_*) = p_{11}^{n_{11}} p_*^{n_{10}} p_*^{n_{01}} (1 - 2p_* - p_{11})^{n_{00}}, \tag{2}$$

and

$$\log L = n_{11} \log p_{11} + (n_{01} + n_{10}) \log p_* + n_{00} \log(1 - 2p_* - p_{11}).$$

Setting the derivatives of $\log L$ with respect to p_{11} and with respect to p_* equal to 0 and solving simultaneously, we find

$$\hat{p}_{11} = \frac{n_{11}}{n}, \qquad \hat{p}_* = \frac{n_{01} + n_{10}}{2n}, \qquad \hat{p}_{00} = \frac{n_{00}}{n}. \tag{3}$$

Using these estimates to form expected cell frequencies, one can show that Pearson's chi-square is identical with McNemar's statistic. The next example illustrates this equivalence.

Example 13.7b   **Tonsillectomies and Hodgkin's Disease (continued)**

Consider again the data in Example 13.7a.

		Normal	
		T	N
Patient	T	26	15
	N	7	37

Substituting $n_{01} = 15$ and $n_{10} = 7$ into (1), we find $Z^2 = 2.91$. From Table II in Appendix 1, we find $P \doteq .09$. (This can also be read from Table Va, with 1 d.f.)

To calculate the chi-square statistic, we arrange the cells in a column, as we did in Section 13.2, along with observed frequencies, null probabilities as estimated by maximum likelihood, and expected cell frequencies.

Category	f	$\hat{p}$	$n\hat{p}$
11	26	26/85	26
01	7	11/85	11
10	15	11/85	11
11	37	37/85	37

Then

$$\chi^2 = \frac{(7-11)^2}{11} + \frac{(15-11)^2}{11} = 2.91.$$

This is identical with the McNemar statistic (1). A continuity correction, subtracting 1 from the numerator of (1), results in $Z^2 = 2.86$, $P \doteq .09$. (To be continued.) ∎

We can also use a likelihood ratio test for $p_{01} = p_{10}$. Under H_0, the likelihood function (2) has a maximum when we substitute the m.l.e.'s (3). In the unrestricted model, the m.l.e.'s of the cell probabilities are the corresponding cell relative frequencies. So in the likelihood ratio statistic, the factors involving n_{00} and n_{11} cancel, leaving

$$\Lambda = \frac{\left(\dfrac{n_{01} + n_{10}}{2n}\right)^{n_{01} + n_{10}}}{\left(\dfrac{n_{01}}{n}\right)^{n_{01}} \left(\dfrac{n_{10}}{n}\right)^{n_{10}}}. \tag{4}$$

Example **13.7c** **Tonsillectomies and Hodgkin's Disease (continued)**

Substituting the data from the previous examples into (4), we obtain

$$\Lambda = \frac{\left(\frac{11}{85}\right)^{22}}{\left(\frac{7}{85}\right)^7 \left(\frac{15}{85}\right)^{15}}$$

and $-2 \log \Lambda \doteq 2.98$. The three parameters in the general likelihood were restricted to two under H_0, so d.f. = 1 for the approximating chi-square distribution. In Table Va, we find $P \doteq .08$. ∎

As in these examples, McNemar's test and the likelihood ratio test will usually give similar results. We gave the latter as yet another illustration of the likelihood ratio method, a method that is applicable in the analysis of higher-dimensional contingency tables where it is not clear how to formulate a chi-square test.

Chapter Perspective

Seeing how well data fit a proposed model is the idea behind most statistical tests, including the Z- and t-tests of the preceding chapters. Goodness-of-fit tests are special in that they are not directed at particular parametric alternatives. This makes them less sensitive to special classes of alternatives than are tests specifically designed to detect those alternatives. A goodness-of-fit test may detect gross deviations from a null hypothesis, but the lack of sensitivity of such tests limits their usefulness. However, we recommend normal-scores (rankit) plots and the Shapiro–Wilk test for normality because they are sensitive to the kind of nonnormality for which t-tests are especially inappropriate.

Goodness-of-fit tests, like many of the tests we've encountered, are tests of very precise null hypotheses. Making a null hypothesis precise allows for calculating a P-value. A small P-value is evidence against the null hypothesis, but a large P-value should not be taken as evidence that H_0 is precisely true. For instance, in a goodness-of-fit test of normality, failure to reject the hypothesis does not mean that the population is normal; there are many other distributions or types of distributions that would fit at least as well, and the normal distribution plays no special role. So a goodness-of-fit test can easily mislead.

Solved Problems

Sections 13.1–13.2

A. Prior to the introduction of a safe red coloring agent, the makers of M&M plain candies were mixed, according to the manufacturer, with colors in the proportions 4:2:2:1:1 for brown, yellow, orange, tan, and green, respectively. We counted the colors in a bag taken from the store shelf and found 234 brown, 101 yellow, 91 orange, 50 tan, and 42 green M&M's. Assuming this bag to contain a random selection from a supply of M&M's, test the hypothesis that the population proportions are those reported by the manufacturer.

Solution:

The sample size is 518. The observed and expected frequencies and the differences between them are as follows:

Color	Observed	Expected	Difference
Brown	234	207.2	26.8
Yellow	101	103.6	−2.6
Orange	91	103.6	−12.6
Tan	50	51.8	−1.8
Green	42	51.8	−9.8

So

$$\chi^2 = \frac{26.8^2}{207.2} + \frac{2.6^2}{103.6} + \frac{12.6^2}{103.6} + \frac{1.8^2}{51.8} + \frac{9.8^2}{51.8} \doteq 6.98.$$

Under H_0 and the assumption of random sampling, the test statistic is $\chi^2(4)$. (There are five categories: d.f. $= 5 - 1 = 4$.) Table Va gives the P-value as greater than .126. (Disposal of the experimental material posed a problem, but we managed.)

B. Use the birthweight data in Solved Problem B of Chapter 7 in a chi-square test of the hypothesis that the population is normal. The stem-leaf diagram from that problem is repeated here

Depth	Stem	Leaves (tenths)
1	3	1
1	3	
1	4	
2	4	5
5	5	443
10	5	66669
16	6	011114
33	6	56666666677889999
13	7	1111112233444
26	7	566688888899
14	8	033345
8	8	5668
4	9	22
1	9	9
1	10	
1	10	9

$n = 72$ brackets the rows with depths 33, 13, 26.

Solution:

We use (arbitrarily) the class intervals (2.95, 3.85), (3.85, 4.75), and so on. The frequency distribution is shown in Table 13.7. The mean is $\bar{X} = 7.125$, and the standard deviation

Table 13.7

Class interval	Class mark x	Frequency f	xf	Interval endpoint	z_i	p_i
				2.95	−3.42	
(2.95, 3.85)	3.4	1	3.4			.002
				3.85	−2.68	
(3.85, 4.75)	4.3	1	4.3			.016
				4.75	−2.05	
(4.75, 5.65)	5.2	7	36.4			.091
				5.65	−1.21	
(5.65, 6.55)	6.1	8	48.8			.206
				6.55	−.471	
(6.55, 7.45)	7.0	29	203.0			.286
				7.45	.266	
(7.45, 8.35)	7.9	16	126.4			.237
				8.35	1.00	
(8.35, 9.25)	8.8	8	70.4			.117
				9.25	1.74	
(9.25, 10.15)	9.7	1	9.7			.034
				10.15	2.48	
(10.15, 11.05)	10.6	1	10.6			.006
				11.05	3.21	

is $S = 1.2214$. To find normal curve areas between interval endpoints, we need Z-scores for each of those endpoints. These are given in Table 13.7 as z_i. This table also gives the areas under the normal curve for each class interval, calculated as $p_i = \Phi(z_i) - \Phi(z_{i-1})$. The p_i sum to .995—less than 1 because the normal curve extends beyond the range of the ten class intervals. Also, with $n = 72$, several of the expected frequencies np_i will be less than 5. For these reasons, we take class intervals at each end to be $(-\infty, 4.75)$ and $(9.25, \infty)$. This results in seven class intervals, with observed and expected frequencies as shown in Table 13.8.

Table 13.8

Interval	np_i	f_i
$(-\infty, 4.75)$	1.45	2
$(4.75, 5.65)$	6.71	7
$(5.65, 6.55)$	14.8	8
$(6.55, 7.45)$	20.6	29
$(7.45, 8.35)$	17.1	16
$(8.35, 9.25)$	8.41	8
$(9.25, \infty)$	2.94	2

The value of χ^2, calculated using (3) of Section 13.2, is 7.15. Two parameters were estimated: the mean and variance. So with seven intervals, we have d.f. $= 7 - 1 - 2 = 4$. The P-value from Table Va is about .13. (See Problem D below.)

Sections 13.3–13.4

C. The following data represent percentages of nitrogen oxides removed in a coal-burning facility, from Problem D, Chapter 7:

91	95	90	83	91	65	55	42	55
81	89	38	20	45	58	85	78	70.

For these data, $n = 18$, $\bar{X} = 68.39$, $S = 21.54$.
(a) Test the hypothesis that the distribution is $N(75, 15^2)$, using the K–S test.
(b) Test the hypothesis that the population is normal, using the Lilliefors test.
(c) Construct a rankit plot.

Solution:
(a) Figure 13.7 shows the sample c.d.f. with the hypothesized normal c.d.f. superimposed. The maximum deviation occurs at $x = 58$:

$$D_{18} = \frac{7}{18} - \Phi\left(\frac{58 - 75}{15}\right) \doteq .26.$$

Then $18D_{18} = 4.68$, and the graph in Table IXa shows $P \doteq .16$.
(b) The c.d.f. of $N(68.39, 21.54^2)$ is also shown in Figure 13.7. The maximum deviation is .17, at $x = 78$. The 20% critical value from Table XI is .167, so $P > .20$.
(c) The rankit plot, shown in Figure 13.8, is a scatterplot of the pairs $(X_{(i)}, E[Z_{(i)}])$, where the $E[Z_{(i)}]$ are found in Table XII: $(20, -1.82), (38, -1.35), (42, -1.07), \ldots$.

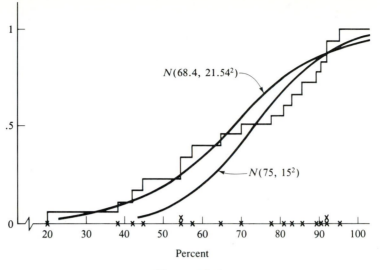

Figure 13.7

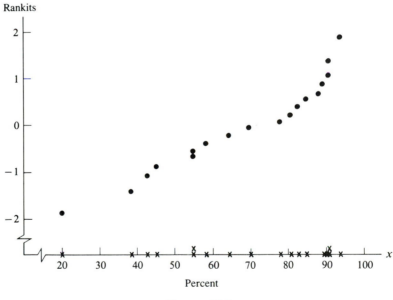

Figure 13.8

D. Test for normality of birthweight using the Shapiro–Wilk test on the data given in Problem B above.

Solution:

We entered the data in a computer statistics package and obtained the rankit plot shown in Figure 13.9. A "+" indicates a single data point; the number 3, for example, indicates

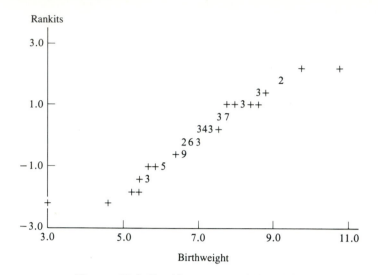

Figure 13.9 Rankits versus birthweight

three points superimposed (because of roundoff). The S–W statistic is shown as $W = .97$. Without a computer to do the work, calculating W and making the plot are quite tedious. (Indeed, Table XII doesn't give rankits for samples larger than $n = 20$.) According to Table 13.6 (page 573), the P-value is about .06. (Using a chi-square test in Problem B, we found $P \doteq .13$.)

Section 13.5

E. A study[13] of the dramatic increase in the number of working women gave these data:

	Husbands with employed wives	*Husbands with nonemployed wives*
n	341	141
Like wives working	74%	37%
Do not care	11%	26%
Dislike wives working	15%	37%

Is the pattern of attitudes different between the two types of husbands?

Solution:

We first convert percentages to frequencies (with some arbitrary round-off) and then

[13] D. V. Hiller and W. Philliber, "The division of labor in contemporary marriage: Expectations, perceptions, and performance," *Social Problems* (1986), 191–201.

calculate the expected frequencies [(4) of Section 13.5]:

Observed:				Expected:		
252	52	304		215.1	88.9	304
38	37	75		53.1	21.9	75
51	52	103		72.8	30.2	103
341	141	482		341	141	482

Using (3) of Section 13.5, we obtain

$$\chi^2 = \frac{36.9^2}{215.1} + \frac{36.9^2}{88.9} + \frac{15.1^2}{53.1} + \frac{15.1^2}{21.9} + \frac{21.8^2}{72.8} + \frac{21.8^2}{30.2} \doteq 58.6.$$

With 2 d.f., this is way off the table, so $P \ll .002$. The evidence against independence is strong.

F. A study[14] considered 85 female and 95 male Harvard medical students who contacted a mental health service between 1980 and 1984. Sixty-five of the 85 women (81%) and 62 of the 95 men (65%) selected a psychiatrist of the same sex. How strong is this evidence that men and women differ in the matter of selecting a psychiatrist of the same sex?

Solution:

Suppose these are independent random samples from the population of men and women medical students. We test the hypothesis of homogeneity:

$H_0: p_W = p_M = p,$

where p_W is the proportion of women who prefer a female psychiatrist and p_M is the proportion of men who prefer a male psychiatrist. The contingency table frequencies and expected frequencies are as follows:

	Observed:				Expected:		
	Prefer same sex	Prefer other sex			Prefer same sex	Prefer other sex	
Woman	69	16	85	Woman	61.9	23.1	85
Man	62	33	95	Man	69.1	25.9	95
	131	49	180		131	49	180

The chi-square statistic is

$$\chi^2 = \frac{(61.9 - 69)^2}{61.9} + \frac{(23.1 - 16)^2}{23.1} + \frac{(69.1 - 62)^2}{69.1} + \frac{(25.9 - 33)^2}{25.9},$$

or about 5.67. The P-value (1 d.f.) is .017.

---

[14] K. B. Kris and J. Silberger, "Medical students' gender preference when selecting a psychiatrist," *J. Amer. College Health 35* (1986), 5–10.

Let's check the equivalence of χ^2 and Z^2 in this 2×2 case by calculating Z. The (pooled) estimate of p is

$$\hat{p} = \frac{69 + 62}{85 + 95} = .728.$$

The Z-score is then

$$Z = \frac{\frac{69}{85} - \frac{62}{95}}{\sqrt{.728 \times .272(\frac{1}{85} + \frac{1}{95})}} = 2.39,$$

and $Z^2 = 5.71 \doteq \chi^2$. (This differs from $\chi^2 = 5.67$ only because of round-off.)

Sections 13.6–13.7

G. Do traditional female roles tend to conflict with out-of-town travel? A study[15] reported these data on willingness to travel for a job:

	Unwilling	Willing, a week or two	Willing, a month or two	
Males	74	78	339	491
Females	81	127	199	407
	155	205	538	898

Use a likelihood ratio test for homogeneity.

Solution:

Substituting the given data and $n = 898$ in (7) of Section 13.6, we find

$$\begin{aligned}
-\log \Lambda &= 898 \log 898 + 74 \log 74 + 78 \log 78 + 339 \log 339 \\
&\quad + 81 \log 81 + 127 \log 127 + 199 \log 199 - 155 \log 155 \\
&\quad - 205 \log 205 - 538 \log 538 - 491 \log 491 - 407 \log 407 \\
&= 20.56.
\end{aligned}$$

So $-2 \log \Lambda = 41.1$. The largest entry in Table Va for χ^2 is 13.0, for which P is .002; so $P \ll .002$.

Problems

For problems marked with an asterisk, answers are provided in Appendix 2.

Sections 13.1–13.2

*1. Continuing the experiment described in Solved Problem A, we obtained a two-pound

---

[15] W. T. Markham, C. M. Bonjean, and J. Corder, "Gender, out-of-town travel, and occupational advancement," *Sociology and Social Research* (1986), 156–160.

bag of M&M peanut candies and counted the colors. The results:

Color:	Brown	Orange	Yellow	Green
Frequency:	106	105	62	87

Use these data to test the manufacturer's claim that the four colors were mixed in equal proportions, assuming that the M&M's in this two-pound bag constitute a random sample.

2. We tossed a die 150 times and observed the results 1 to 6 with these respective frequencies: 28, 25, 30, 26, 22, and 19. Use these data to test the hypothesis that the six sides are equally likely.

∗3. Your friend believes that when tossing two coins, the outcomes 0 heads, 1 heads, and 2 heads are equally likely. You decide to test this hypothesis by tossing two coins 100 times. You get 0 heads 23 times, 1 head 52 times, and 2 heads 25 times.

(a) Carry out a chi-square test of the hypothesis that 0, 1, and 2 heads are equally probable.

(b) Carry out a chi-square test of the hypothesis that

$$P(0 \text{ heads}) = P(2 \text{ heads}) = \frac{1}{4}, \ P(1 \text{ heads}) = \frac{1}{2}.$$

(c) Toss two coins 100 times and repeat part (a) for *your* data.

4. In the 1987 baseball season, Gary Gaetti of the Minnesota Twins played in 80 games in which he was at bat four times, with these results:

Number of hits:	0	1	2	3	4
Number of games:	26	35	16	2	1

As in Example 13.6a, test the hypothesis that the number of hits in four at-bats has a binomial distribution (for this player).

∗5. An industrial plant compiled the following summary of accidents per month, over a period of 32 months:

No. of accidents:	0	1	2	3
No. of months:	22	6	2	2

Test the hypothesis that the number of accidents per month follows a Poisson distribution. (The chi-square test is for a model with a finite number of categories, whereas there are infinitely many possible values in a Poisson model. When calculating probabilities, combine the values 3, 4, . . . into the single category "3 or more" and test the resulting finite model.)

6. Table 7.3 gave a frequency distribution for 50 spot-weld strengths as follows:

Class mark	Frequency	Class mark	Frequency
366	2	421	7
377	2	432	3
388	10	443	2
399	15	454	1
410	8		

From Examples 7.4b and 7.5b, the mean and s.d. are $\bar{X} = 404.3$ and $S = 18.89$. Use χ^2 to test the hypothesis of normality.

*7. Some years ago, an extensive study[16] of times to failure reported data on a variety of units and components. Given the frequency distribution shown, in each case test the hypothesis that the time to failure is exponential, using a chi-square test. To approximate the sample mean, use the class marks shown in the table.

(a) Fifth bus motor failures:

Interval, thousands of miles	Observed number of failures	Class mark
0–20	29	8
20–40	27	27
40–60	14	45
60–80	8	68
80–up	7	120

(b) A type of radar indicator tube:

Hours	Frequency	Class mark
0–100	29	36
100–200	22	136
200–300	12	237
300–400	10	340
400–600	10	475
600–800	9	685
800–up	8	1100

8. The formula for the chi-square statistic [(1) of Section 13.2] is given in terms of *frequencies*. Show that if one uses proportions in place of frequencies (a common student error), the result is $1/n$ times the correct value of χ^2.

Sections 13.3–13.4

*9. The following are ten successive entries from the table of Gaussian deviates (Table XV):

.275, −.514, .982, −.071, −1.975, .027, .016, −.584, .669, .070.

The table was constructed deterministically, using a computer, but the source claims that they can serve as a random sample from a standard normal population. Use the K–S statistic to test the hypothesis that they constitute a random sample from $N(0, 1)$.

---

[16] From D. J. Davis, "An analysis of some failure data," *J. Amer. Stat. Assn. 47* (1952), 147.

10. Here are twenty successive sequences of three digits in the table of random digits (see Table XIV in Appendix 1). We have inserted decimal points to make them numbers in $(0, 1)$:

.633 .871 .026 .698 .276 .597 .465 .136 .828 .979,

.095 .216 .973 .165 .677 .914 .740 .245 .119 .612.

Test the hypothesis that this set of data was obtained as a random sample from $U(0, 1)$.

∗11. Consider the data in Problem 3 giving the number of heads in each of 100 tosses of two coins. Use the Kolmogorov–Smirnov statistic to test the hypothesis

(a) $f(0) = f(1) = f(2) = \dfrac{1}{3}$.

(b) $f(0) = f(2) = \dfrac{1}{4}$.

12. Use the K–S statistic to test the hypothesis that the following is a random sample from an exponential population with mean 1:

.890, .038, .014, .297, .077, 1.98, 1.94, .150, 1.39, 1.77.

[The population actually used was $U(0, 2)$.]

13. Use the K–S statistic to test the hypothesis that the data in Problem 5 are from a Poisson population with mean $\frac{1}{2}$.

∗14. Calculate the K–S statistic for comparing the scores for first- and second-born children given in Problem 17 of Chapter 12, which we repeat here:

1st: 40, 41, 44, 49, 53, 53, 54, 54, 56, 61, 62, 64, 65, 67, 67;

2nd: 23, 25, 38, 43, 44, 47, 49, 54, 55, 58, 58, 60, 66, 66, 72.

Use Table Xa to obtain an exact P-value.

15. The following are measures of zinc concentration (mg/kg) at various locations in an outer zone and in an inner zone around an ocean dump site.[17]

Inner:	13.5	23.8	23.3	20.9	23.8	20.0
	24.4	16.4	18.3	17.6	25.4	23.3;
Outer:	26.4	20.6	19.8	15.0	16.8	20.4	23.4	21.5.

Use the two-sample K–S statistic to test the hypothesis of no difference in distributions of concentrations in the two zones.

16. A pollution study[18] gives measures of biological oxygen demand (BOD) and chemical oxygen demand (COD) for 18 samples of the effluent from a factory manufacturing

---

[17] Data are from the reference cited in Example 13.4a.

[18] K. J. Shapland, "Industrial effluent treatability—a case study," *Water Pollution Control* 85 (1986), 75–80.

detergents and related materials:

BOD: 881 104   74 118   74   46   59   96   78
         59   78 101 128 144   82   77 111   87;

COD: 580 674 512 540 616 298 960 570 640
        588 556 588 582 844 574 420 696 620.

*(a)   Test BOD for normality.
*(b)   Test COD for normality.
 (c)   Construct rankit plots for BOD and for COD.

Section 13.5

*17.   A study[19] reports the following cross-classification of perceptions by 28 school superintendents and 43 school social workers as to the extent of social workers' involvement in the school entry process.

	Perceived by	
	Superintendents	*Social workers*
Much	15%	47%
Little	85%	53%

Convert the percentages to frequencies and test for homogeneity of superintendents and social workers with respect to perception,
(a)   using the chi-square test.
(b)   using a Z-test.

18.   An investigation[20] of the relationship between sex and field of study used questionnaires to classify individuals as masculine (M), feminine (F), androgynous (A), or undifferentiated (U). The data were reported as shown in the following cross-classification:

	M	F	A	U
Nonsci.	8	6	2	14
Biol. Sci.	11	10	2	7
Phys. Sci.	4	9	8	9

Test the independence of these two categorizations.

*19.   The sizes of a person's left and right feet are seldom exactly the same. The following table[21] cross-classifies 127 right-handed individuals according to sex and according to

---

[19] R. Constable and E. Montgomery, "Perception of the school social worker's role," *Social Work in Education 7*, (1985), 244–257.
[20] R. Baker, "Masculinity, femininity, and androgyny among male and female science and non-science majors," *School Science and Math 84* (1984), 459–467.
[21] J. Levy and M. M. Levy, "Human lateralization from head to foot: Sex-related factors," *Science 200* (1978), 1291.

whether the left-shoe size is larger, smaller, or the same as the right-shoe size.

	$L > R$	$L < R$	$L = R$
Males	2	28	10
Females	55	14	18

Test the hypothesis that sex and relative shoe size are independent in the population sampled.

20. Various research studies have dealt with the relationship between church attendance and truancy from school. A 1986 study[22] used data from a sample of teenagers (ages 13–18) throughout the United States in 1975.

<table>
<thead>
<tr><th rowspan="2"></th><th></th><th colspan="3">Truancy frequency:</th></tr>
<tr><th></th><th>Never</th><th>Occasional</th><th>Frequent</th></tr>
</thead>
<tbody>
<tr><td rowspan="3">Church attendance</td><td>Rare/never</td><td>91</td><td>68</td><td>136</td></tr>
<tr><td>Occasional</td><td>140</td><td>78</td><td>119</td></tr>
<tr><td>Frequent</td><td>296</td><td>106</td><td>90</td></tr>
</tbody>
</table>

Test for independence of truancy and church attendance.

*21. A survey of college students[23] reports these data on smoking and cardiovascular disease (CVD) in their immediate families:

<table>
<thead>
<tr><th></th><th colspan="4">Smokers</th></tr>
<tr><th></th><th>Both parents</th><th>Father only</th><th>Mother only</th><th>Neither parent</th></tr>
</thead>
<tbody>
<tr><td>CVD</td><td>22</td><td>18</td><td>6</td><td>23</td></tr>
<tr><td>No CVD</td><td>18</td><td>35</td><td>27</td><td>75</td></tr>
</tbody>
</table>

Test the hypothesis of independence of these classification variables in the reference population.

22. The data in Problem 21 can be used in testing an interesting hypothesis not involving CVD: Construct a contingency table with the categories S (smoke) and N (not smoke) for each of the parents, and test the hypothesis that the smoking habits of spouses are independent.

*23. One theory for the rise of Japanese companies in world markets centers on Japanese managers' ability to foster worker productivity. Seventy-two managers from Japan and 65 from the United States were asked[24] this question: "How would you attempt to

[22] D. M. Sloane and R. H. Potvin, "Religion and delinquency: Cutting through the maze," *Social Forces 65* (Sept. 1986), 87–105.

[23] "Cardiovascular risk factors and health knowledge among freshman college students...," *J. Am. College Health 34* (1986), 267–270.

[24] R. Hirokawa and A. Miyahara, "A comparison of influence strategies utilized by managers in American and Japanese organizations," *Comm. Quart.* (1986), 250–265.

convince a consistently tardy employee to report to work on time?" There were several categories of response; we repeat here only the totals for those involving punishment and those not:

	American	Japanese
Punishment	48	29
Nonpunishment	17	43

(a) Use χ^2 to test the hypothesis of homogeneity.

(b) Test homogeneity using a Z-test based on sample proportions.

24. Show that in testing homogeneity with a 2×2 contingency table, the value of Pearson's χ^2 is equal to the square of the Z-statistic used in comparing two population proportions.

*25. In Example 13.6b, we considered the following data concerning the possible relationship of coffee drinking and coronary heart disease (CHD):

	Cups per day				
	0	1–2	3–4	5 or more	n_{i+}
CHD	4	17	17	9	47
No CHD	185	457	226	125	993
n_{+j}	189	474	243	134	$1040 = n$

Test for independence using χ^2. (Compare the value of χ^2 with the value of $-2 \log \Lambda$ in Example 13.6b.)

26. Air Force lore says that fighter pilots are more likely to sire daughters than sons. A study[25] investigating the possibility that high G-force exposure may be a factor found the following data.

		G-force exposure level	
		High	Low
Sex of offspring	Male	295	66
	Female	287	100

Test the hypothesis of no relationship between G-force exposure and sex of offspring.

Sections 13.6–13.7

*27. Test the manufacturer's claim in Problem 1 using a likelihood ratio test.

28. Use the likelihood ratio method with the data in Problem 3 to test the hypothesis in Problem 3(a) against the alternative that the probabilities are $\frac{1}{4}$, $\frac{1}{2}$, and $\frac{1}{4}$.

---

[25] B. B. Little et al., "Pilot and astronaut offspring: Possible G-force effects on human sex ratio," *Aviation Space & Environmental Medicine 58* (1987), 707–709.

∗29. For the data of Problem 23, calculate $-2 \log \Lambda$, where Λ is the likelihood ratio statistic for testing independence. Find the approximate P-value, assuming $-2 \log \Lambda \approx \chi^2(1)$.

30. Four instructors, each teaching a section of the same statistics course, assigned final grades as follows:

Instructor:	1	2	3	4
A:	13	8	31	11
B:	21	20	45	18
C:	9	35	20	20
D:	2	16	8	4

Are the patterns of grade assignment different for the four instructors? Test using the likelihood ratio.

∗31. In the context of 2×2 contingency tables, consider the null hypothesis defined by the following table of joint probabilities:

	A_1	A_2
B_1	p^2	$p(1-p)$
B_2	$p(1-p)$	$(1-p)^2$

Under H_0, A and B are independent and identically distributed.

(a) Find the m.l.e.'s of the parameters in H_0 in terms of cell frequencies.

(b) Using the data in Example 13.7b, test H_0 against the hypothesis that A and B are merely exchangeable, using the likelihood ratio method.

(c) Use the likelihood ratio statistic with the data of Example 13.7b to test H_0 against the hypothesis that the cell probabilities are unrestricted (except that their sum is 1). [Observe the relationship among the values of $-2 \log \Lambda$ here, in (b), and in Example 13.7b.]

32. A multicenter trial[26] was conducted to assess the possibility of preventing blindness in premature babies by temporarily freezing the whites of their eyes. In the study, 135 babies with disease in both eyes were randomized—half of them had their left eye treated and other half, their right eye. The outcomes in both eyes were the same in all but 40 babies. In 34 of these, the outcome in the treated eye was favorable while that in the untreated eye was unfavorable, and in the other six the outcome in the treated eye was unfavorable while that in the untreated eye was favorable. Use McNemar's test to judge whether the treatment is effective.

---

[26] "Multicenter trial of cryotherapy for retinopathy of prematurity," *Archives of Ophthalmology 106* (1988), 471–479.

Analysis of Variance

In this chapter, we introduce tests for treatment effects, including tests that also take into account other factors that may affect responses. Suppose there are k treatments or levels of treatment; each defines a population of responses. The hypothesis that the population means are equal is the hypothesis that there is no treatment effect. The analytical method we use in such problems is called the **analysis of variance** (ANOVA).

Analysis of variance is a method for comparing the means of several populations using ratios of variance estimates. A **one-way ANOVA** generalizes the two-sample t-test of Chapter 12. A **two-way ANOVA** generalizes paired t-tests. Identifying important mean differences involves multiple comparisons of means, introduced in Section 14.3.

Like the t-tests for one and two samples, an F-test in ANOVA assumes that responses are normally distributed.

14.1   One-Way ANOVA: The Method

In Section 12.1, we discussed the running of an experiment to compare a treatment with a control, or to compare two treatments. Here we consider the case of more than two treatments. In designing an experiment to compare their effects, we first identify a population about which inferences are to be drawn. Then we randomly select **experimental units**—individuals or animals or pieces of experimental material—from that population. In a **completely randomized design**, we then randomly assign one of the treatments to each experimental unit. If there are k treatments, we get k samples, one from each treatment population.

Assumptions   In an extension of the two-sample t-test of Section 12.4, our test for the equality of k population means assumes k independent random samples, one from each treatment population. As in the two-sample case, we assume (as the simplest case) equality of the population variances.

600

Example 14.1a   Dial Coatings

To compare the effectiveness of three different types of phosphorescent coating of instrument dials, each type of coating was applied to eight dials. So $k = 3$. After a period of illumination by an ultraviolet light, the number of minutes each dial glowed was recorded, with these results:

Coating type	Sample size	Sample mean	Sample s.d.
1	8	57.1	4.250
2	8	59.175	3.185
3	8	68.45	4.098

The sample means are different, but we know that sample means are variable and that differences can arise solely because of sampling variability. The question is whether the differences in the sample means are larger than is typical of sampling variability. If they are, this suggests that the population means are different. (To be continued.)   ∎

Some notation: Let n_i denote the size of the sample from the ith population, $i = 1, \ldots, k$, and $n = n_1 + \cdots + n_k$. Let X_{ij} denote the jth observation in the ith sample, $j = 1, \ldots, n_i$. The following table introduces additional notation.

Population	Population mean	Population variance	Sample	Sample mean	Sample variance
1	μ_1	σ^2	$X_{11}, \ldots, X_{1n_1}$	$\bar{X}_1$	S_1^2
$\vdots$	$\vdots$	$\vdots$		$\vdots$	$\vdots$
k	μ_k	σ^2	$X_{k1}, \ldots, X_{kn_k}$	$\bar{X}_k$	S_k^2

We denote by $\bar{X}$ the mean of the combined sample—the set of all n observations. This is the **grand mean**:

$$\bar{X} = \frac{1}{n} \sum_1^k \sum_1^{n_i} X_{ij} = \frac{1}{n} \sum_1^k n_i \bar{X}_i. \tag{1}$$

The null hypothesis to be tested is that there are no treatment differences, on average:

$$H_0: \mu_1 = \cdots = \mu_k = \mu, \tag{2}$$

where μ is unknown. The alternative is that H_0 is not true.

When there are large differences among the several sample means, their deviations from the grand mean $\bar{X}$ are large. The squares of these deviations are the basis of the statistic we use in testing (2). When H_0 is true, the combined sample (the set of all n observations) is a random sample of size n from $N(\mu, \sigma^2)$. The sum of their squared deviations from the grand mean is the **total sum of squares**:

Sums of Squares

$$\text{SSTot} = \sum\sum (X_{ij} - \bar{X})^2. \tag{3}$$

We decompose this sum of squares as follows. First, we express the deviation of X_{ij} from $\bar{X}$ as the sum of its deviation from the mean of its sample plus the deviation of that sample mean from $\bar{X}$:

$$X_{ij} - \bar{X} = (X_{ij} - \bar{X}_i) + (\bar{X}_i - \bar{X}).$$

Squaring both sides, we obtain

$$(X_{ij} - \bar{X})^2 = (X_{ij} - \bar{X}_i)^2 + (\bar{X}_i - \bar{X})^2 + 2(X_{ij} - \bar{X}_i)(\bar{X}_i - \bar{X}). \tag{4}$$

The sum of the squared deviations (4) over all i and j is SSTot. Summing first on j within the ith sample, we obtain

$$\sum_{j=1}^{n_i} (X_{ij} - \bar{X})^2 = \sum_{j=1}^{n_i} (X_{ij} - \bar{X}_i)^2 + \sum_{j=1}^{n_i} (\bar{X}_i - \bar{X})^2 + \sum_{j=1}^{n_i} 2(X_{ij} - \bar{X}_i)(\bar{X}_i - \bar{X})$$

$$= \sum_{j=1}^{n_i} (X_{ij} - \bar{X}_i)^2 + n_i(\bar{X}_i - \bar{X})^2 + 2(\bar{X}_i - \bar{X}) \sum_{j=1}^{n_i} (X_{ij} - \bar{X}_i).$$

Since the sum of the deviations in the ith sample about its mean is zero for each i, the last term vanishes. Summing on i, we obtain this expression for SSTot:

$$\text{SSTot} = \sum\sum (X_{ij} - \bar{X})^2 = \sum\sum (X_{ij} - \bar{X}_i)^2 + \sum_i n_i(\bar{X}_i - \bar{X})^2. \tag{5}$$

(In each double sum here and in the following section, whether indicated or not, the inner summation is within the ith sample, from $j = 1$ to n_i, and the outer summation is across the samples, from $i = 1$ to k.)

We label the individual terms on the right of (5) as follows:

$$\sum\sum (X_{ij} - \bar{X}_i)^2 = \text{SSE} \qquad \text{(error sum of squares)}, \tag{6}$$

and

$$\sum n_i(\bar{X}_i - \bar{X})^2 = \text{SSTr} \qquad \text{(treatment sum of squares)}. \tag{7}$$

In this notation, the decomposition (5) is

$$\text{SSTot} = \text{SSE} + \text{SSTr}. \tag{8}$$

This is the essence of an analysis of variance. We have decomposed SSTot, which describes the overall variability in the combined sample, into a portion attributable to differences in means and a portion attributable to random error. We'll use the size of SSTr in judging whether there is a difference in means—a treatment effect. First, we examine the individual sums of squares more closely.

The error sum of squares (6) is an extension to k samples of the numerator of a pooled variance (from Section 12.4): Deviations of each observation in a sample about the mean of that sample are squared and summed over all observations. So SSE can be calculated from the individual sample variances:

$$\text{SSE} = \sum_{1}^{k} \left\{ \sum_{j=1}^{n_i} (X_{ij} - \bar{X}_i)^2 \right\} = \sum_{1}^{k} (n_i - 1)S_i^2. \tag{9}$$

Using (2) of Section 9.7, we can calculate the mean value of SSE:

$$E(\text{SSE}) = \sum E[(n_i - 1)S_1^2] = \sigma^2 \sum (n_i - 1) = (n - k)\sigma^2.$$

Thus, if we divide SSE by $\Sigma(n_i - 1) = n - k$, the result is an unbiased estimate of σ^2 (whether H_0 is true or not). It is called MSE, or **mean square for error**.

Mean square for error:

$$MSE = \frac{SSE}{n - k} = \frac{1}{n - k} \sum\sum(X_{ij} - \bar{X}_i)^2. \tag{10}$$

This is an unbiased estimate of σ^2:

$$E(MSE) = \sigma^2.$$

The treatment sum of squares SSTr is the sum of the squared deviations of the sample means from the grand mean. Problem 9 asks you to show that

$$E(SSTr) = (k - 1)\sigma^2 + \sum n_i(\mu_i - \mu)^2. \tag{11}$$

When there is no treatment effect ($\mu_i = \mu$), the second term on the right is 0. So under H_0,

$$MSTr = \frac{SSTr}{k - 1}$$

(**mean square for treatment**) is an unbiased estimate of σ^2.

Mean square for treatment:

$$MSTr = \frac{SSTr}{k - 1} = \frac{1}{k - 1} \sum_1^k (n_i - 1)S_i^2. \tag{12}$$

This is unbiased for σ^2 when $\mu_i = \mu$:

$$E(MSTr \mid H_0) = \sigma^2.$$

Under H_0, both MSE and MSTr are (unbiased) estimates of σ^2. However, when the population means are not all the same, (11) shows that SSTr (and hence MSTr) tends to be inflated, whereas MSE is the same under both H_0 and H_A. So the ratio of MSTr to MSE is useful as a test statistic for the hypothesis of equality of means:

$$F = \frac{SSTr/(k - 1)}{SSE/(n - k)} = \frac{MSTr}{MSE}. \tag{13}$$

Large values of F are considered extreme and taken as evidence against the null hypothesis (2).

In Section 14.2, we'll see that the divisors in numerator and denominator of (13) are in fact the degrees of freedom of the corresponding sums of squares. We'll also find that in the case of *normal* populations, the distribution of F under H_0 is $F(k - 1, n - k)$ [see Section 6.6], and that the likelihood ratio rejection region for testing H_0 is of the form $F > C$.

Example 14.1b   **Dial Coatings (continued)**
We return to the data of Example 14.1a. For those data, $k = 3$ and $n_1 = n_2 = n_3 = 8$. Using (1), we find the grand mean:

$$\bar{X} = \frac{1}{n} \sum n_i \bar{X}_i = \frac{8}{24}(57.1 + 59.175 + 68.45) = 61.575.$$

The sum of squared deviations of the sample means from the grand mean is

$$\text{SSTr} = \sum n_i(\bar{X}_i - \bar{X})^2$$
$$= 8[(57.1 - \bar{X})^2 + (59.175 - \bar{X})^2 + (68.45 - \bar{X})^2] = 584.41.$$

This is the sum of squares in the numerator of the statistic F in (13). The implied variance estimate (12) is

$$\text{MSTr} = 584.41/2 = 292.2.$$

To judge whether MSTr has been inflated by a treatment effect, we calculate MSE, an estimate of variance that is valid whether or not there is a treatment effect. Using (9) and the s.d.'s from Example 14.1a, we obtain

$$\text{SSE} = 7(4.250)^2 + 7(3.185)^2 + 7(4.098)^2 = 315.0.$$

Since $n - k = 24 - 3 = 21$, we find

$$\text{MSE} = \frac{315.0}{21} = 15.0.$$

This is much smaller than MSTr. The F-ratio (13) is

$$F = \frac{292.2}{15.0} \doteq 19.5.$$

ANOVA Table   The sums of squares, degrees of freedom, and ratios are conveniently laid out in an **ANOVA table**:

Source	*SS*	*d.f.*	$MS = \dfrac{SS}{d.f.}$	*F*
Treatment	584.41	2	292.2	19.5
Error	315.0	21	15.0	—
Total	899.41	23	—	

With 2 d.f. in the numerator and 21 in the denominator, the tail-area beyond $F = 19.5$ is less than .001 (from Table VIIIa in Appendix 1). So the P-value for the null hypothesis (2) is less than .001. The evidence against the hypothesis of equal population means is strong.   ∎

Test for the equality of means of k normal populations with equal variances: Calculate

$$F = \frac{\text{MSTr}}{\text{MSE}} = \frac{\sum n_i(\bar{X}_i - \bar{X})^2/k - 1}{\sum\sum(X_{ij} - \bar{X}_i)^2/n - k}.$$

Large values of F are extreme. Find critical values and P-values in Table VIII of Appendix 1. ∎

In the next section, we show that the two-sided, two-sample t-test of Section 12.4 is a special case of the F-test. Like the t-test, the F-test is moderately robust against the assumption of population normality. Roughly speaking, if there are no unusually large or unusually small observations, the P-value given by the F-distribution won't be far off. Also, the F-test is not very sensitive to the assumption of equal population variances.

In Example 14.1b, we concluded that the population means are different. How are they different? We see that $\bar{X}_3 > \bar{X}_2 > \bar{X}_1$, but can we conclude that $\mu_3 > \mu_2 > \mu_1$? Or that $\mu_3 > \mu_2$? Or $\mu_3 > \mu_1$? These kinds of questions are important in practice. Answering them is not easy. A number of methods for multiple comparisons have been proposed and are in use. We'll take up two of them in Section 14.3.

Multiple Comparisons

14.2   One-Way ANOVA: The Theory

We now give a theoretical basis and some further motivation for the method presented in Section 14.1. We assume that the X_{ij} are independent, with common variance σ^2. As in Section 14.1, μ_i denotes the mean of the ith population. Let $\varepsilon_{ij} = X_{ij} - \mu_i$, the random error component of X_{ij}, and [by analogy with (1) of Section 14.1] let μ denote the weighted average[1] of population means:

$$\mu = \frac{1}{n}\sum_{i=1}^{k} n_i\mu_i. \tag{1}$$

Under the null hypothesis that all μ_i are equal, μ is their common value.

The expected value of the ith sample mean is the mean of population i:

$$E(\bar{X}_i) = \frac{1}{n_i}\sum E(X_{ij}) = \mu_i.$$

The expected value of the grand mean is μ:

$$E(\bar{X}) = \frac{1}{n}\sum_i\sum_j E(X_{ij}) = \frac{1}{n}\sum_i\sum_j \mu_i = \frac{1}{n}\sum_i n_i\mu_i = \mu.$$

---

[1] It may seem odd that our definition of μ depends on the sample sizes n_i. This is for convenience; it does not affect our testing of H_0 versus H_A. (See also Problem 13.)

The expected deviation of $\bar{X}_i$ from the grand mean is

$$E(\bar{X}_i - \bar{X}) = \mu_i - \mu = \tau_i.$$

Treatment Effects

We call τ_i the ith *treatment effect*. The (weighted) average treatment effect is

$$\frac{1}{n}\sum n_i\tau_i = \frac{1}{n}\sum n_i(\mu_i - \mu) = \mu - \mu = 0.$$

Under the null hypothesis that all μ_i are equal, $\tau_i = 0$ for each i.

To complete the description of the model, we assume (as in Section 14.1) that the observations X_{ij} are normally distributed.

Model for one-way ANOVA:

$$X_{ij} = \mu + \tau_i + \varepsilon_{ij}, \tag{2}$$

where $\Sigma\, n_i\tau_i = 0$, and the errors ε_{ij} are independent, with $\varepsilon_{ij} \sim N(0, \sigma^2)$. The null hypothesis is

$$H_0\colon \tau_1 = \cdots = \tau_k = 0,$$

and the alternative is

$$H_A\colon \text{Not all } \tau_i \text{ are } 0.$$

Maximum Likelihood Estimates

We obtain next the m.l.e.'s of the parameters μ, μ_i, and σ^2. The p.d.f. of X_{ij} (with $\theta = \sigma^2$, for convenience in differentiating) is

$$f(x \mid \mu_i, \theta) \propto \frac{1}{\sqrt{\theta}} \exp\left(-\frac{1}{2\theta}(x - \mu_i)^2\right).$$

The likelihood function is therefore this product:

$$L(\mu_1, \ldots, \mu_k; \theta) = \prod\prod \frac{1}{\sqrt{\theta}} \exp\left(-\frac{1}{2\theta}(X_{ij} - \mu_i)^2\right).$$

$$= \theta^{-n/2} \exp\left(-\frac{1}{2\theta}\sum\sum(X_{ij} - \mu_i)^2\right),$$

with logarithm

$$\log L = -\frac{n}{2}\log\theta - \frac{1}{2\theta}\sum\sum(X_{ij} - \mu_i)^2. \tag{3}$$

Differentiating $\log L$ with respect to μ_i, we obtain

$$\frac{1}{2\theta}\sum_j(X_{ij} - \mu_i) = \frac{1}{2\theta}(n_i\bar{X}_i - n_i\mu_i).$$

Setting this equal to zero, we obtain $\hat{\mu}_i = \bar{X}_i$. Further,

$$\hat{\mu} = \frac{1}{n}\sum n_i\hat{\mu}_i = \frac{1}{n}\sum n_i\bar{X}_i = \bar{X}, \qquad \text{and} \qquad \hat{\tau}_i = \hat{\mu}_i - \hat{\mu} = \bar{X}_i - \bar{X}.$$

Thus, the differences identified in Section 14.1 as attributable to treatment differences are the m.l.e.'s of the treatment effects τ_i.

To find $\hat{\theta}$, we differentiate $\log L$ with respect to θ:

$$-\frac{n}{2\theta} + \frac{1}{2\theta^2} \sum\sum (X_{ij} - \mu_i)^2.$$

Setting this equal to zero and substituting $\mu_i = \bar{X}_i$, we get

$$\hat{\theta} = \hat{\sigma}^2 = \frac{1}{n} \sum\sum (X_{ij} - \bar{X}_i)^2 = \frac{\text{SSE}}{n}. \tag{4}$$

This is an average squared (within samples) error. It is smaller by the factor $\frac{n-k}{n}$ than the unbiased estimator of θ defined in Section 14.1: $\text{MSE} = \frac{\text{SSE}}{n-k}$.

Distribution of F

In Section 14.1, we stated that under the assumption of population normality, the null distribution of the test ratio [(13) of Section 14.1] is $F(n-k, k-1)$. We now demonstrate this fact. Assume $H_0: \mu_i = \mu$ for all i. The observations X_{ij} together constitute a single random sample from a normal population with mean μ and variance σ^2. The variance of this combined sample is

$$\frac{1}{n-1} \sum\sum (X_{ij} - \bar{X})^2 = \frac{\text{SSTot.}}{n-1}.$$

According to Section 8.10, $\frac{\text{SSTot}}{\sigma^2}$ is $\chi^2(n-1)$ under H_0.

The basic decomposition of SSTot [(8) of Section 14.1] is

$$\text{SSTot} = \text{SSE} + \text{SSTr}, \tag{5}$$

or

$$\sum\sum (X_{ij} - \bar{X})^2 = \sum\sum (X_{ij} - \bar{X}_i)^2 + \sum n_i (\bar{X}_i - \bar{X})^2. \tag{6}$$

The first term on the right of (6) divided by σ^2 is a sum of independent chi-square variables:

$$\sum_i \sum_j \frac{(X_{ij} - \bar{X}_i)^2}{\sigma^2} = \sum_i \frac{(n_i - 1)S_i^2}{\sigma^2}. \tag{7}$$

It is therefore chi-square with $\Sigma(n_i - 1) = n - k$ degrees of freedom. The two terms on the right of (6) are *independent*, since the first is a combination of sample variances and the second is calculated from sample means. (We know from Section 8.10 that in a random sample from a normal population, the sample mean and sample variance are independent.) It follows from Section 6.5 that

$$\frac{\text{SSTr}}{\sigma^2} \sim \chi^2(k-1),$$

with

$$\text{d.f.} = (n-1) - (n-k) = k - 1.$$

So under the null hypothesis, SSE/σ^2 and SSTr/σ^2 have independent chi-square distributions. This shows (see Section 6.6) that the null distribution of

$$F = \frac{\text{MSTr}}{\text{MSE}} = \frac{\dfrac{\text{SSTr}}{k-1}}{\dfrac{\text{SSE}}{n-k}} = \frac{\left.\sum \dfrac{n_i(\bar{X}_i - \bar{X})^2}{\sigma^2}\right/(k-1)}{\left.\sum\sum \dfrac{n_i(X_{ij} - \bar{X}_i)^2}{\sigma^2}\right/(n-k)} \tag{8}$$

is $F(k-1, n-k)$, as claimed in Section 14.1.

Likelihood Ratio Test

We said in Section 14.1 that the F-test is actually a likelihood ratio test for H_0: $\tau_1 = \cdots = \tau_k = 0$. For the denominator of the likelihood ratio statistic Λ (see Section 11.6), we need the maximum of (3) over all μ_i and σ^2. As usual, this occurs at the m.l.e.'s: $\hat{\mu}_i = \bar{X}_i$, $\hat{\sigma}^2 = \text{SSE}/n$. So

$$\max L = L(\bar{X}_1, \ldots, \bar{X}_k; \hat{\theta}) = \hat{\theta}^{-n/2} e^{-n/2}. \tag{9}$$

Under H_0, the likelihood function involves just two parameters:

$$L_0(\mu, \theta) = L(\mu, \ldots, \mu; \theta) = \theta^{-n/2} \exp\left(-\frac{1}{2\theta} \sum\sum (X_{ij} - \mu)^2\right).$$

Its logarithm is

$$\log L_0 = -\frac{n}{2} \log \theta - \frac{1}{2\theta} \sum\sum (X_{ij} - \mu)^2. \tag{10}$$

Setting the derivative of $\log L_0$ with respect to μ equal to 0, we find the maximizing value to be $\hat{\mu} = \bar{X}$. Setting the derivative with respect to θ equal to zero, substituting $\bar{X}$ for μ, and solving, we find the m.l.e. of θ under H_0:

$$\hat{\theta}_0 = \frac{1}{n} \sum\sum (X_{ij} - \bar{X})^2 = \frac{\text{SSTot}}{n}. \tag{11}$$

Then

$$L_0(\hat{\mu}, \hat{\theta}_0) = L_0\left(\bar{X}, \frac{\text{SSTot}}{n}\right) = (\hat{\theta}_0)^{-n/2} e^{-n/2}.$$

So the likelihood ratio is

$$\Lambda = \frac{(\hat{\theta})^{-n/2} e^{-n/2}}{(\hat{\theta})^{-n/2} e^{-n/2}} = \left(\frac{\hat{\theta}_0}{\hat{\theta}}\right)^{-n/2},$$

and

$$\Lambda^{-2/n} = \frac{\hat{\theta}_0}{\hat{\theta}} = \frac{\text{SSTot}}{\text{SSE}} = \frac{\text{SSE} + \text{SSTr}}{\text{SSE}} = 1 + \frac{\text{SSTr}}{\text{SSE}} = 1 + \left(\frac{n-1}{k-k}\right)F,$$

where F is given by (13) in Section 14.1. So the likelihood ratio critical region $\Lambda < C$ is equivalent to one of the form $F > C'$, as we set out to show.

When $k = 2$

To show that the F-test of this section is an extension of a two-sample t-test (Section 12.4), we'll show that when $k = 2$, the F-test for equality of means is exactly the same as a two-sided two-sample t-test. When $k = 2$, the grand mean is

$$\bar{X} = \frac{n_1 \bar{X}_1 + n_2 \bar{X}_2}{n_1 + n_2}.$$

We can then express the deviations of the sample means from the overall mean in terms of the difference in means:

$$\bar{X}_1 - \bar{X} = \frac{n_2}{n_1 + n_2}(\bar{X}_1 - \bar{X}_2), \quad \text{and} \quad \bar{X}_2 - \bar{X} = \frac{n_1}{n_1 + n_2}(\bar{X}_2 - \bar{X}_1).$$

Using these in MSTr [from (5)], we see after a little algebra that

$$\text{MSTr} = \frac{\text{SSTr}}{2 - 1} = \sum_{i=1}^{2} n_i(\bar{X}_i - \bar{X})^2 = \frac{(\bar{X}_1 - \bar{X}_2)^2}{1/n_1 + 1/n_2}.$$

The MSE in this case of two samples is just the pooled variance, (2) of Section 12.4:

$$\text{MSE} = \frac{1}{n - 2} \sum\sum(X_{ij} - \bar{X}_i)^2 = S_p^2.$$

Thus, the F-ratio (8) reduces to T^2 [(3) of Section 12.4]:

$$F = \frac{\text{MSTr}}{\text{MSE}} = \left\{ \frac{\bar{X}_1 - \bar{X}_2}{S_p\sqrt{1/n_1 + 1/n_2}} \right\}^2 = T^2. \tag{12}$$

According to Section 6.6, $F(1, n - 2)$ is the same distribution as $t^2(1, n - 2)$. So the P-value for F is the same as a two-sided P-value for T. (For this reason, we have not included P-values in Table VIII of Appendix 1 for numerator d.f. = 1.)

The equality of F and T^2 when $k = 2$ and the fact that $F(1, n - 2)$ and $t^2(1, n - 2)$ are the same distribution together establish the claim of Section 12.4 that the null distribution of the square of the two-sample t-statistic is $F(1, n - 2)$.

Example 14.2a   **Intermittent Feeding**

In Example 12.4a, we analyzed these blood pressure data:

Regular feeding: 108, 133, 134, 145, 152, 155, 169;
Intermittent feeding: 115, 162, 162, 168, 170, 181, 199, 207.

In testing the hypothesis of no treatment difference, we found

$$S_p^2 = \frac{6 \cdot 19.61^2 + 7 \cdot 27.99^2}{7 + 8 - 2} = \frac{7791.39}{13} = 599.34 = \text{MSE}.$$

To calculate SSTr, we first find the means:

$$\bar{X}_1 = 142.29, \quad \bar{X}_2 = 170.5, \quad \bar{X} = \frac{2360}{15} = 157.33,$$

and then

$$\text{SSTr} = 7(142.29 - 157.33)^2 + 8(170.5 - 157.33)^2 = 2971.9.$$

The ANOVA table is as follows:

Source	SS	d.f.	MS	F
Treatment	2971.9	1	2971.9	4.96
Error	7791.4	13	599.34	—

In the earlier example, we calculated $T = 2.226$. Since $2.226^2 = 4.96$, this verifies (12) in this particular case.

The two-sided P-value for $T = 2.226$ (13 d.f.) is .044, the same as the P-value for $F = 4.96$ on (1, 13) d.f.:

$$P(|T| > 2.226) = 2 \cdot P(T > 2.226) = .044 = P(F > 4.96).$$ ∎

14.3   Multiple Comparisons

The F-test we've given for equality of k population means may suggest that the means are not all the same, but it does not tell us which are different. The ordering of the sample means suggests the same ordering of the population means, but this ordering may be wrong because of sampling variability. One could do t-tests for the equality of the means in each pair of means, but the question of significance in such multiple testing is problematical, as the following example shows.

Example 14.3a   We generated ten independent random samples of size $n = 100$ from $N(20, 1)$. The summary statistics are as follows:

Sample # (Treatment #)	Mean	s.d.
1	19.83	1.0305
2	20.27	.9588
3	20.08	1.0045
4	19.97	.9086
5	19.89	.9764
6	19.78	.9814
7	20.01	1.0170
8	20.04	.9846
9	20.11	.9243
10	20.06	1.0071

There are $\binom{10}{2} = 45$ possible comparisons of two means. Applying the two-sample t-test to each such pair, we find differences in sample means that are significant ($P < .05$) in 6 of the 45 pairs: treatment 6 with treatment 2, 6 with 9, 1 with 2, 1 with 9, 5 with 2, 3 with 6, and 2 with 4. This is despite the fact that the population means were identical in every case! ∎

The samples in the preceding example are not especially unusual; when there are 45 opportunities to find significant differences, it is not surprising to find some! The expected number of pairs of sample means that are significantly different when the population means are actually equal is $45 \times .05 = 2.25$; the variance is quite large because the comparisons are highly correlated.

We saw in Chapter 11 (for instance, Solved Problem C) that constructing a confidence interval is equivalent to testing a hypothesis. The significance level α corresponds to a confidence level of $1 - \alpha$. In the remainder of this section, we focus on confidence limits, with the obvious implication of a corresponding test.

Consider now k populations, where population i is $N(\mu_i, \sigma^2)$, and (in the notation of the preceding sections) the ratio

$$T = \frac{\bar{X}_i - \bar{X}_j - (\mu_i - \mu_j)}{S_p \sqrt{1/n_i + 1/n_j}}, \tag{1}$$

for some pair (i, j). In Section 12.4, the S_p in the denominator of (3) was a pooled estimate of σ based on the two samples; here we have more information about σ and define the pooled variance as

$$S_p^2 = \text{MSE} = \frac{1}{n - k} \sum n_i S_i^2.$$

Then $(n - k)S_p^2 \sim \chi^2(n - k)$, and the distribution of (1) is $t(n - k)$. The ratio (1) is therefore pivotal (see Section 9.5). With it and an appropriate percentile of the t-distribution, we can construct a confidence interval for each difference $\mu_i - \mu_j$—but what is the appropriate percentile?

The Bonferroni Method

Suppose we want m confidence intervals at the level $1 - \alpha$. Let $t(m, \alpha)$ denote the value of $t(n - k)$ with area $\alpha/(2m)$ to its right:

$$P[|T| > t(m, \alpha)] = \frac{\alpha}{m}. \tag{2}$$

Let $E_{i,j}$ denote the event that the confidence interval formed from (1) using the multiplier $t(m, \alpha)$ covers $\mu_i - \mu_j$—that is, that

$$\bar{X}_i - \bar{X}_j - t(m, \alpha) \times \text{s.e.} < \mu_i - \mu_j < \bar{X}_i - \bar{X}_j + t(m, \alpha) \times \text{s.e.}, \tag{3}$$

where

$$\text{s.e.} = S_p \sqrt{\frac{1}{n_i} + \frac{1}{n_j}}.$$

By the choice of $t(m, \alpha)$, $P(E_{i,j}) = 1 - \frac{\alpha}{m}$. Then, using the Bonferroni inequality [Problem 55(b) of Chapter 1], it follows that

$$P(\text{union of all } E_{i,j}) \geq 1 - \sum P(E_{i,j}^c) = 1 - m\left(\frac{\alpha}{m}\right) = 1 - \alpha.$$

Thus, the probability that all of the m confidence intervals will cover the corresponding true mean differences is at least $1 - \alpha$. For simultaneous tests of the $m = \binom{k}{2}$ hypotheses of the form $\mu_i = \mu_j$, we construct confidence intervals for the corresponding differences in means as given by (3) and reject any for which the interval does not include 0. The probability is at most α that we'd incorrectly reject one or more of the hypotheses when all the means are actually equal.

Example 14.3b   Consider the data of Examples 14.1a and 14.1b. There were three populations ($k = 3$) and samples of size 8 from each. The mean square for error was found to be 15.0, with 21 d.f.

We might be interested in the three differences of two means ($m = 3$). To obtain simultaneous confidence limits for the three differences at level .95, we set $\alpha = .05$. Then $\alpha/m = .05/3 \doteq .0167$, so we look in Table IIIb of Appendix 1 for the value of $T(21)$ with about .0083 in each tail:

$$P(|T| > 2.6) = 2 \times .008 = .016.$$

Thus, $t(3,.05) \doteq 2.6$. So for confidence limits, we go either way from the point estimate by the amount

$$2.6\sqrt{15.0\left(\frac{1}{8} + \frac{1}{8}\right)} \doteq 2.6 \times 1.94 \doteq 5.04.$$

The sample means were 57.1, 59.2, 68.5, for populations 1, 2, and 3, respectively. So the three sets of confidence limits are as follows:

For $\mu_2 - \mu_1$: 2.1 ± 5.04.
For $\mu_3 - \mu_1$: 9.3 ± 5.04.
For $\mu_3 - \mu_2$: 11.4 ± 5.04.

The procedure used to obtain these limits has at least a 95% success rate. Since only the first interval includes the value 0, the differences $\bar{X}_3 - \bar{X}_1$ and $\bar{X}_3 - \bar{X}_2$ are significant, but the difference $\bar{X}_1 - \bar{X}_2$ is not.   ∎

Contrasts

A method of multiple comparisons due to H. Scheffé deals with all possible **contrasts**. A contrast is a linear combination of population means $\Sigma a_i \mu_i$ in which the coefficients sum to 0: $\Sigma a_i = 0$. For example, $\mu_1 - \mu_3$ is a contrast, and $\mu_1 + \mu_2 - 2\mu_3$ is another. The latter in effect compares the average mean of populations 1 and 2 with the mean of population 3. Similarly, the contrast

$$3(\mu_1 + \mu_2) - 2(\mu_3 + \mu_4 + \mu_5)$$

compares the average of the means of populations 1 and 2 with the average of the means of populations 3, 4, and 5.

Scheffé's Method

Scheffé's method constructs confidence limits for all possible contrasts of a set of means. It is based on the fact that the distribution of the ratio

$$F = \frac{\Sigma n_i(\bar{X}_i - \bar{X} - \tau_i)^2}{(k-1)S_p^2} \tag{4}$$

is $F(k - 1, n - k)$. (The ratio reduces to the test ratio in the one-way ANOVA test of Section 14.1 when $\tau_i = 0$.) If we choose f such that $P(F \le f)$ is the desired overall confidence level $1 - \alpha$, then

$$P(\sum n_i(\bar{X}_i - \bar{X} - \tau_i)^2 \le (k-1)fS_p^2) = 1 - \alpha.$$

Some further algebra (see Lindgren [15], p. 555) shows that then the probability is at least η that for every contrast $\sum a_i\mu_i$,

$$\sum a_i\bar{X}_i - S_p\sqrt{b} \le \sum a_i\mu_i \le \sum a_i\bar{X}_i + S_p\sqrt{b}, \tag{5}$$

where

$$b = (k-1)f\sum \frac{a_i^2}{n_i}. \tag{6}$$

The inequalities (5) define confidence intervals, and the probability is at least $1 - \alpha$ that they simultaneously cover the respective contrasts.

Example 14.3c   Consider again the data of the preceding examples. The distribution of (4) is $F(2, 21)$, and for an overall confidence level of .95 we use f defined as the 95th percentile: $f = 3.47$. For the contrast $\mu_i - \mu_j$, $a_i = 1$ and $a_j = -1$. The sample sizes n_i are all 8. Thus [from (6)],

$$b = 2 \times 3.47 \times \left(\frac{1}{8} + \frac{1}{8}\right) = 1.735.$$

We saw earlier that $S = \sqrt{15} = 3.87$, so $S\sqrt{b} \doteq 5.10$. So the confidence intervals for the three contrasts considered in Example 14.3a are 2.1 ± 5.10, 9.3 ± 5.10, 11.4 ± 5.10. Each is a little wider than the corresponding interval obtained in the earlier example. The Scheffé procedure gives simultaneous tests—not just for the pairwise comparisons, but for all possible contrasts. The extra width allows for this. ∎

For a detailed presentation of some other methods of multiple comparison see Chapter 11 of *An Introduction to Statistical Methods and Data Analysis* by Lyman Ott (Boston: Duxbury Press, 1984).

14.4   Two-Way ANOVA

In Section 12.6, in comparing two treatments, we explained how matching pairs of experimental subjects can help to eliminate the effect of irrelevant individual differences. In such a design, subjects are paired according to variables that may have an effect on the measured response. Then one member of the pair is assigned at random to each treatment. This matched pairs design is a special case of a **randomized block design** for comparing k treatments. In this design, we group $n = rc$ subjects or pieces of experimental material into r matched **blocks** of size c and randomly assign one member of each block to each treatment. The data consist of a rectangular array of responses X_{ij}—the response to treatment level i of the individual in block j receiving that treatment level, $i = 1, \ldots, r$ and $j = 1, \ldots, c$.

Example 14.4a   Abrasion Resistance

A wear-testing machine is used to determine resistance to abrasion. It has three weighted brushes, under which samples of fabrics are fixed. Resistance is measured in terms of loss of weight after a specified number of cycles. Four different fabrics were tested and the weight loss measured at each brush position. The data are as follows.

| | | *Fabric* | | | |
		A	B	C	D
	1	1.93	2.55	2.40	2.33
Brush position	2	2.38	2.72	2.68	2.40
	3	2.20	2.75	2.31	2.28

The main interest is in fabric differences, but measured wear may also depend on the brush position. It would be a mistake to ignore the brush position and compare the fabrics using a one-way ANOVA if there is appreciable variation due to brush position. In a one-way analysis, this variation would be part of experimental error and inflate the sum of squares in the denominator of the F-ratio. Fabric differences would be less apt to show up than they would if the brush effect could be removed. (To be continued.)   ■

As in a one-way ANOVA, we assume independent observations, normally distributed with constant variance. The mean response depends on the treatment level i and on the block index j. We define

$$\theta_i = \frac{1}{c} \sum_j E(X_{ij}), \qquad \varphi_j = \frac{1}{r} \sum_i E(X_{ij}), \qquad \mu = \frac{1}{rc} \sum_i \sum_j E(X_{ij})$$

and

$$\tau_i = \theta_i - \mu, \qquad \beta_j = \varphi_j - \mu.$$

This determines τ's and β's such that

$$\sum_i \tau_i = \sum_j \beta_j = 0. \tag{1}$$

Thus τ_i is an average effect of treatment i as compared with all the treatments, and β_j is the average effect of block j as compared with all the blocks. We then define

$$\psi_{ij} = \mu_{ij} - \mu - \tau_i - \beta_j.$$

When this is not zero, we take it as representing an *interaction* between level i of factor A and level j of factor B. With only one observation for each ij-combination, we can't get at this interaction and are forced to treat it (if it is not zero) as part of the "error," ε_{ij}.

Additive model for a two-way ANOVA:

$$X_{ij} = \mu + \tau_i + \beta_j + \varepsilon_{ij}, \tag{2}$$

where the ε_{ij} are independent and normally distributed with $E(\varepsilon_{ij}) = 0$ and var $\varepsilon_{ij} = \sigma^2$.

The null hypothesis we consider is that the average responses at the different levels of the treatment are the same. In terms of the τ_i:

$$H_0: \tau_i = 0 \qquad \text{for all } i.$$

(We could also test $\beta_j = 0$ for all j: simply interchange the roles of treatment and block in what follows.)

Maximum likelihood estimates of the overall mean μ and the blocking and treatment effects (Problem 25) are

$$\hat{\mu} = \bar{X} = \frac{1}{rs} \sum_i \sum_j X_{ij}, \qquad \hat{\tau}_i = \bar{X}_{i.} - \bar{X}, \qquad \hat{\beta}_j = \bar{X}_{.j} - \bar{X},$$

where

$$\bar{X}_{i.} = \frac{1}{c} \sum_{j=1}^{c} X_{ij}, \qquad \bar{X}_{.j} = \frac{1}{r} \sum_{i=1}^{r} X_{ij}.$$

An analysis of variance is a decomposition of the sum of squared deviations of responses about their overall average or grand mean $\bar{X}$, the total sum of squares, SSTot. For this decomposition, we write each deviation as the sum of three terms, one based on τ_i, related to the treatment effect; one based on β_j, related to the block effect; and a third based on what's left after subtracting out the estimates of treatment and block effects:

$$X_{ij} - \bar{X} = (\bar{X}_{i.} - \bar{X}) + (\bar{X}_{.j} - \bar{X}) + (X_{ij} - \bar{X}_{i.} - \bar{X}_{.j} + \bar{X}). \tag{3}$$

Squaring and summing, we find

$$\sum_i \sum_j (X_{ij} - \bar{X})^2 = \sum_i \sum_j (\bar{X}_{i.} - \bar{X})^2 + \sum_i \sum_j (\bar{X}_{.j} - \bar{X})^2$$
$$+ \sum_i \sum_j (X_{ij} - \bar{X}_{i.} - \bar{X}_{.j} + \bar{X})^2.$$

(All the cross-products are zero.) In sums of squares notation, this is, in corresponding order,

$$\text{SSTot} = \text{SSTr} + \text{SSB} + \text{SSE}. \tag{4}$$

The summands in SSTr and SSB are independent of the inner summation index. Thus, in SSTr the sum on j can be replaced by c, and in SSB the sum on i can be replaced by r:

$$\text{SSTr} = \sum c(\bar{X}_{i.} - \bar{X})^2, \qquad \text{SSB} = \sum r(\bar{X}_{.j} - \bar{X})^2. \tag{5}$$

When there is a treatment effect, SSTr will tend to be large; when there is a block effect, SSB will tend to be large. However, SSE is based on the residual when the overall mean, treatment, and block effects are removed:

$$(X_{ij} - \bar{X}_{i\cdot} - \bar{X}_{\cdot j} + \bar{X}) = X_{ij} - (\bar{X}_{i\cdot} - \bar{X}) - (\bar{X}_{\cdot j} - \bar{X}) - \bar{X}.$$

It measures the "error" and should be independent of treatment and block effects. Thus, a large ratio of SSTr to SSE suggests a treatment effect.

Under the hypotheses of no treatment and no block effect, the rc observations constitute a random sample from a population with mean μ. If we assume that the observations are *normally* distributed, then

$$\frac{\text{SSTot}}{\sigma^2} = \frac{\sum_i \sum_j (X_{ij} - \bar{X})^2}{\sigma^2}$$

has a chi-square distribution with $rc - 1$ degrees of freedom, since SSTot divided by $rc - 1$ is the variance of the sample consisting of all the observations. Dividing (4) through by σ^2, we have

$$\frac{\text{SSTot}}{\sigma^2} = \frac{\text{SSTr}}{\sigma^2} + \frac{\text{SSB}}{\sigma^2} + \frac{\text{SSE}}{\sigma^2}.$$

Under the hypothesis of no treatment and block effects, the terms on the right have independent chi-square distributions, with $r - 1$, $c - 1$, and $(r - 1)(c - 1)$ degrees of freedom, respectively, as in one-way ANOVA. (The proof of this is carried out along the lines of the derivation in Section 14.2.) It follows then that the corresponding test ratios have F-distributions, under the null hypotheses $\tau_i \equiv 0$ and $\beta_j \equiv 0$, respectively. (Although the numerator sums of squares are independent, the F-ratios for treatment and block are not independent, because SSE occurs in both denominators.)

F-ratio for H_0: $\tau_i = 0$, $i = 1, \ldots, r$:

$$F_{\text{Tr}} = \frac{\text{SSTr}/(r - 1)}{\text{SSE}/[(r - 1)(c - 1)]}. \tag{6}$$

Large values of F are extreme. Find P-values in Table VIIIa of Appendix 1, with $[r - 1, (r - 1)(c - 1)]$ d.f. For testing block effect, use

$$F_B = \frac{\text{SSB}/(s - 1)}{\text{SSE}/[(r - 1)(s - 1)]},$$

with $[c - 1, (r - 1)(c - 1)]$ d.f.

It can be shown (Problem 26) that the test defined by the rejection region $F > C$ (for either F-ratio) is equivalent to a likelihood ratio test.

Example **14.4b** **Abrasion Resistance (continued)**

We return to the data of Example 14.4a, which we repeat here with an added row and added column of averages.

Brush position

		1	2	3	$\bar{X}_{i\cdot}$
	A	1.93	2.38	2.20	2.17
Fabric type	B	2.55	2.72	2.75	2.673
	C	2.40	2.68	2.31	2.463
	D	2.33	2.40	2.28	2.337
	$\bar{X}_{\cdot j}$	2.3025	2.545	2.385	2.411

The ANOVA table is then as follows.

Source	SS	d.f.	MS	F
Fabric (treatment)	.405	3	.135	9.43
Brush position (block)	.122	2	.061	4.24
Error	.086	6	.0143	
Total	.613	11		

The P-value for $F(3, 6) = 9.43$ is about .011, so there is some evidence of a fabric effect. The P-value for $F(2, 6) = 4.24$ is about .071, so there is a suggestion of a brush position effect.

A way to see whether brush position matters is to ignore it. Suppose we treat the data as constituting four independent samples of size 3 in a one-way ANOVA. The ANOVA table would be as follows.

Source	SS	d.f.	MS	F
Fabric	.405	3	.135	5.19
Error	.208	8	.026	
Total	.613	11		

(The new error SS is the sum of SSE and SSB above.) The P-value for $F(3, 8) = 5.19$ (Table VIIIa) is .028. The case against H_0 is not as strong as when we took brush position into account. ■

We indicated earlier that two-way ANOVA generalizes the paired t-tests of Section 12.6. When $k = 2$, the F-test described at (6) is equivalent to a two-sided paired t-test. (See Solved Problem F.)

The analysis so far does not take into account any interaction—the possibility that treatment i has a different effect in block j from what it has in block j', while treatment i' has the same effect in both blocks. Although it is not possible to get at interactions with a single observation at each treatment/block combination when some cells have multiple observations, the variability within cells can be exploited to obtain an SSE (and an estimate of error variance) independent of any interactions. For the model and analysis, see any text on design of experiments; for the necessary distribution theory, see Lindgren [15], pp. 558ff.

Chapter Perspective

We have seen how a suitable decomposition of total variability can focus on the various factors affecting a response, for two simple experimental designs. It is perhaps apparent that such designs can be generalized in many ways; the basic ANOVA technique can be extended to apply to the more complicated designs. We refer the interested student to courses and textbooks in this extensively developed area of applied statistics.

The underlying probability structure we have assumed in this chapter is a special case of what is termed a *linear model*—one in which a response depends linearly on parameters identifying the various factors that affect the response. The next chapter takes up linear models in which the factors of interest are numerical variables.

Solved Problems

Section 14.1

A. Four instructors, each teaching a section of the same statistics course, assigned final grades as follows:

		Instructor			
		1	2	3	4
	A	13	8	31	11
Grade	B	21	20	45	18
	C	9	35	20	20
	D	2	16	8	4

[At the time, the university did not use the grade of F, but rather N (for "no credit"). This covered many types of cases, so we've omitted N's.] Are the instructors grading differently on average? Test the hypothesis of no difference and give the ANOVA table.

Solution:

In Chapter 13 (Problem 30) we treated this as a two-way contingency table and tested for homogeneity of instructors. We now exploit the fact that the grades are ordered and, indeed, are assigned numerical values for calculating grade point averages. Assuming we have random samples of the instructors' grades, we can use these numerical values

[A = 4, B = 3, C = 2, D = 1] to test the hypothesis that the instructors' true average grades are the same. We will assume normal populations; even though this is not exactly correct, the F-test is apt to be a good approximation. The relevant statistics are as follows:

Instructor	n	Mean	s.d.
1	45	3.00	.826
2	79	2.25	.898
3	104	2.95	.896
4	53	2.68	.894

The grand mean is 2.71. So $\text{SSTr} = \Sigma\, n_i(\bar{X}_i - 2.71)^2 = 26.54$. The error sum of squares is found from the given s.d.'s:

$$\text{SSE} = \sum(n_i - 1)S_i^2$$
$$= 44 \times .826^2 + 78 \times .898^2 + 103 \times .896^2 + 52 \times .894^2 = 217.2.$$

So

$$F = \frac{\text{MSTr}}{\text{MSE}} = \frac{26.54/3}{217.2/277} \doteq 11.3.$$

Table VIII in Appendix 1 does not extend to 277 denominator d.f., but it is clear that $P << .001$. [As the denominator d.f. tends to ∞, $(k-1)F$ tends to $\chi^2(k-1)$; so we get an approximate P-value in the chi-square table at $\chi^2 = 3F = 33.9$.] The ANOVA table is as follows:

Source	d.f.	SS	MS	F	P
Instructor	3	26.54	8.85	11.3	$< .001$
Error	277	217.2	.784		
Total	280	243.74			

Section 14.2

B. Show that SSTot can be calculated as follows:

$$\sum(X_{ij} - \bar{X})^2 = \sum X_{ij}^2 - \frac{\left(\sum X_{ij}\right)^2}{n}.$$

Solution:

Since $\bar{X}$ is the mean of all n observations ($n = \Sigma\, n_i$), this is just the parallel axis theorem applied to the second moment about the mean of that combined sample.

Section 14.3

C. To find statistically significant differences ($\alpha = .05$) for Problem A above, apply
(a) the Bonferroni method.
(b) the Scheffé method.

Solution:

(a) We consider the six possible differences among the four sample means; $m = 6$. Since $n - k = 281 - 4 = 277$, the quantity $t(m, \alpha)$ is found in the normal table as the value of Z with area $\alpha/m = .05/6$ to its left, or $t(6, .05) \doteq 2.6$. The quantity S_p is the square root of MSE, or .8855. For comparing means for instructors 1 and 2 we calculate the s.e. as

$$S_p\sqrt{\frac{1}{n_1} + \frac{1}{n_2}} = .8855\sqrt{\frac{1}{45} + \frac{1}{104}} = .158.$$

The limits for $\mu_1 - \mu_2$ are

$$\bar{X}_1 - \bar{X}_2 \pm t(6, .05) \times \text{s.e.} = .75 \pm .41.$$

Similar calculations yield these intervals:

1 versus 3: $.05 \pm .41$

1 versus 4: $.32 \pm .47$

2 versus 3: $.70 \pm .29$

2 versus 4: $.43 \pm .41$

3 versus 4: $.27 \pm .39$

So instructor 2 is judged different from the rest, but the differences between other pairs are not significant.

(b) First we find $f = $ 95th percentile of $F(3,277)$. As in Problem A, $3F \approx \chi^2(3)$, so $3f \doteq 7.81$. Then $b = 7.81(1/n_1 + 1/n_2)$. For instructors 1 and 2, $\sqrt{b} = .522$, and the interval for $\mu_1 - \mu_2$ is $.75 \pm .46$. Similarly, for the other pairs:

1 versus 3: $.05 \pm .44$

1 versus 4: $.32 \pm .50$

2 versus 3: $.70 \pm .37$

2 versus 4: $.43 \pm .44$

3 versus 4: $.27 \pm .42$

Section 14.4

D. Twelve patients with cardiac arrhythmias were treated with three active drugs, A, B, and C, in a double-blind, three-period crossover trial. Each period consisted of one week of treatment followed by a 24-hour ambulatory EKG recording. (These periods were separated by two long periods with no drug.) The accompanying table gives the responses (measured as a mean number of PVC's per hour).[2]

(a) Carry out a two-way ANOVA to test for treatment differences.

(b) Rank all 36 observations and carry out an ANOVA test on the ranks. [This is an extension of the rank-transformation idea introduced in Chapter 12. It is more appropriate than the test in (a) because the population distributions seem quite skewed.]

Patient	1	2	3	4	5	6	7	8	9	10	11	12
A	170	19	187	10	216	49	7	474	.4	1.4	27	29
B	7	1.4	205	.3	.2	33	37	9	.6	63	145	0
C	0	6	18	1	22	30	3	5	0	36	26	0

Solution:

Tables for a two-way ANOVA are as follows.

		d.f.	SS	MS	F	P
	Patient	11	92801	8436.5	1.03	.452
Original data	Drug	2	46857	23429	2.87	.078
	Error	22	179520	8159.8		
	Total:	35	319170			

		d.f.	SS	MS	F	P
	Patient	11	1679.7	152.7	2.03	.076
Transformed data	Drug	2	545.79	272.7	2.63	.043
	Error	22	1653.5	7516		
	Total:	35	38790			

[A rankit plot of the original data shows considerable skewing; a plot of the transformed data yields a Wilk–Shapiro statistic for testing normality of about .96.]

E. In the notation and context of Section 14.4, show the following:

$$E\left(\sum n_i(\bar{X}_{i\cdot} - \bar{X})^2\right) = \frac{n(r-1)\sigma^2}{rc} + \sum n_i\tau_i^2. \tag{1}$$

(This shows that SSTr tends to be larger when the τ_i are not all 0 than when they are.)

---

[2] Reported in D. A. Berry, "Logarithmic transformations in ANOVA," *Biometrics 43* (1987), 439–456.

Solution:

We expand the following square, where $\mu_i - \mu = \tau_i$:

$$(\bar{X}_{i.} - \bar{X} - (\mu_i - \mu))^2 = (X_{i.} - \bar{X})^2 + \tau_i^2 - 2\tau_i(\bar{X}_{i.} - \bar{X}).$$

Multiplying by n_i and summing on i, we obtain

$$\sum n_i(\bar{X}_{i.} - \bar{X} - \tau_i)^2 = \sum n_i(\bar{X}_{i.} - \bar{X})^2 + \sum n_i\tau_i^2 - 2\sum n_i\tau_i(\bar{X}_{i.} - \bar{X}).$$

Upon taking expected values, we find that the last term is just twice the next to the last term, so they can be combined:

$$\sum n_i E(\bar{X}_{i.} - \bar{X} - \tau_i)^2) = E(\sum n_i(\bar{X}_{i.} - \bar{X})^2) - \sum n_i\tau_i^2,$$

or

$$E(\sum n_i(\bar{X}_{i.} - \bar{X})^2) = \sum n_i E((\bar{X}_{i.} - \bar{X} - \tau_i)^2) + \sum n_i\tau_i^2. \tag{2}$$

The expected squares in the first sum on the right-hand side of (2) are

$$\text{var}(\bar{X}_{i.} - \bar{X}) = \text{var } \bar{X}_{i.} + \text{var } \bar{X} - 2\,\text{cov}(\bar{X}_{i.}, \bar{X}).$$

Then, since

$$\text{cov}(\bar{X}_{i.}, \bar{X}) = \text{cov}\left(\bar{X}_{i.}, \frac{1}{r}\sum \bar{X}_{i.}\right) = \frac{1}{r}\,\text{var } \bar{X}_{i.} = \frac{1}{r}\left(\frac{\sigma^2}{c}\right),$$

it follows that

$$E((\bar{X}_{i.} - \bar{X} - \tau_i)^2) = \text{var}(\bar{X}_{i.} - \bar{X}) = \frac{\sigma^2}{c} + \frac{\sigma^2}{rc} - 2\frac{\sigma^2}{rc} = \frac{(r-1)\sigma^2}{rc}. \tag{3}$$

Substituting (3) in (2), we obtain (1).

F. Show that the F-test described at (6) in Section 14.4 is equivalent to a two-sided paired t-test when $r = 2$.

Solution:

Let $c = n$ and $r = 2$. Then

$$\bar{X} = \frac{\bar{X}_1 + \bar{X}_2}{2},$$

and the treatment sum of squares is

$$\text{SSTr} = n((\bar{X}_1 - \bar{X})^2 + (\bar{X}_2 - \bar{X})^2)$$

$$= \frac{n}{2}(\bar{X}_1 - \bar{X}_2)^2.$$

The error sum of squares is the sum of two terms:

$$\text{SSE} = \sum(X_{1j} - \bar{X}_1 - \bar{X}_{.j} + \bar{X})^2 + \sum(X_{2j} - \bar{X}_2 - \bar{X}_{.j} + \bar{X})^2$$

$$= \frac{1}{2}\sum(X_{1j} - X_{2j} - [\bar{X}_1 - \bar{X}_2])^2 = \frac{n-1}{2}S_D^2,$$

where S_D is the standard deviation of the set of n differences $\{X_{1j} - X_{2j}\}$. Then

$$F = \frac{\text{SSTr}/(2-1)}{\text{SSE}/[(2-1)(n-1)]} = \frac{(\bar{X}_1 - \bar{X}_2)^2}{S_D^2/n} = T^2,$$

where T is the paired t-statistic of Section 12.6. The rejection region $F > C$ is equivalent to a two-sided rejection region for $T: |T| > C'$. Since $F(1, n-1)$ is the same distribution as $t^2(n-1)$, the P-values will agree.

Problems

For problems marked with an asterisk, answers are provided in Appendix 2.

Section 14.1

*1. Calculate the F-statistic for testing the hypothesis of no treatment effect, given these data:

Treatment	Sample
A	2, 6, 4
B	12, 9, 7, 4, 8
C	4, 6, 8, 2

2. Find the within samples estimate of error variance for these data:

Treatment	Sample
A	4, 4, 4
B	6, 6, 6, 6, 6
C	7, 7

What is the implication of this result for testing for a treatment effect?

*3. Three plots each of five varieties of clover were planted at the Rosemount Experiment Station in Minnesota. Yields in tons per acre were as follows:

Variety	Yield	Mean	s.d.
Spanish	2.79, 2.26, 3.09	2.713	.420
Evergreen	1.93, 2.07, 2.45	2.150	.269
Commercial yellow	2.76, 2.34, 1.87	2.323	.445
Madrid	2.31, 2.30, 2.49	2.367	.107
Wisconsin A46	2.39, 2.05, 2.68	2.373	.315

(a) Construct the ANOVA table and find the F-statistic for testing the hypothesis of no difference in mean yield among the five varieties.

(b) Construct a 90% confidence interval for the mean difference in yield between the Spanish and Evergreen varieties.

4. In some large computers, printed circuit boards have several layers. Each layer has many small holes. A substantial clearance between holes in different layers is necessary to avoid shorting. For six holes selected from each of five locations on a board, the clearances were measured as follows (in thousandths of an inch):

Location	Clearance	Mean	s.d.
A	6.9, 6.9, 7.5	7.10	.346
B	10.1, 10.1, 7.3, 7.3, 9.1, 9.9	8.97	1.343
C	5.0, 8.0, 6.4, 6.8, 7.2, 8.8	7.03	1.317
D	8.8, 8.0, 7.3, 8.5, 7.8, 8.9	8.22	.624
E	7.8, 8.8, 8.9, 7.2, 8.1, 7.1	7.98	.768

(Three holes at location A were accidentally ground through in preparing the board for measurement.) Test the hypothesis of no difference in mean clearance at the various locations on the board.

*5. Use a one-way ANOVA to redo Problem 13 of Chapter 12. (Does the answer agree with that obtained using the two-sample t-test?)

*6. The following data[3] were obtained (in part) to examine the variation in strength between and within bobbins for a type of worsted yarn.

Bobbin	1	2	3	4	5	6
	18.2	17.2	15.3	15.6	19.2	16.2
	16.8	18.5	15.9	16.0	18.0	15.9
	18.1	15.0	14.5	15.2	17.0	14.9
	17.0	16.2	14.2	14.9	16.9	15.5
Mean	17.52	16.72	14.97	15.42	17.77	15.62
S^2	.5292	2.209	.5958	.2292	1.149	.3158

Test the hypothesis of no average difference among bobbins.

7. The following data[4] are amounts filled by a 24-head machine for filling bottles of vegetable oil; the five groups of bottles reflect in part the variability among heads.

---

[3] Reported in E. J. Snell, *Applied Statistics* (New York: Chapman and Hall, 1987), p. 104.
[4] Reported in W. H. Swallow and S. R. Searle, "Minimum variance quadratic unbiased estimation of variance components," *Technometrics 20* (1978), 265–272.

1	2	3	4	5
15.70	15.69	15.75	15.61	15.65
15.68	15.71	15.82	15.66	15.60
15.64		15.75	15.59	
15.60		15.71		
		15.84		

Test the hypothesis of no difference among group means.

Section 14.2

*8. In the setting and notation of Sections 14.1 and 14.2, calculate

$$E(\bar{X}_i - \bar{X}) \qquad \text{and} \qquad E(X_{ij} - \bar{X}).$$

9. In the setting and notation of Sections 14.1 and 14.2, show (11) of Section 14.1, namely:

$$E\left(\sum n_i(\bar{X}_i - \bar{X})^2\right) = (k - 1)\sigma^2 + \sum n_i\tau_i^2.$$

[Hint: Start with $(\bar{X}_i - \mu_i) = (\bar{X}_i - \bar{X}) + (\bar{X} - \mu) + (\mu - \mu_i)$, square both sides, multiply by n_i, sum, and take expected values.]

10. Show that MSE in Sections 14.1 and 14.2 reduces to the pooled variance S_p^2 of Section 12.4 when $k = 2$.

11. Show that

$$\text{SSTr} = \sum n_i \bar{X}_i^2 - \frac{\left(\sum\sum X_{ij}\right)^2}{n},$$

where the term subtracted is the same as the term subtracted in Solved Problem B in finding SSTot. Use this formula for calculating SSTr in Problem 1.

12. Carry out the details of the maximization of the likelihood function L_0 leading to (11) of Section 14.2 and to the maximum value, $L_0(\hat{\mu}, \hat{\theta}_0)$.

13. Suppose we define μ as the *unweighted* average $(\sum \mu_i)/k$ [rather than as (1) of Section 14.2]. Then define $\alpha_i = \mu_i - \mu$, so that the model is $X_{ij} = \mu + \alpha_i + \varepsilon_{ij}$, where $\varepsilon_{ij} \sim N(\mu_i, \sigma^2)$. Show that the m.l.e. of μ is the unweighted average of the sample means. [This m.l.e. is the same as the m.l.e. of the μ defined by (1) of Section 14.2 *if* the sample sizes are all the same. In general, it is not the $\bar{X}$ used in forming SSTot, and not the $\bar{X}$ in the treatment sum of squares that appears in the natural F-ratio.]

Section 14.3

*14. A study[5] of the effects of exercise reports the following summary statistics for physical fitness scores.

---

[5] Ph.D. dissertation by D. Lobstein, Purdue University, 1983, as recorded in Moore and McCabe, *Introduction to the Practice of Statistics* (New York: Freeman, 1989).

	n	Mean	s.d.
Treatment	10	291.9	38.17
Control	5	309.0	32.07
Joggers	11	366.9	41.19
Sedentary	10	226.1	63.53

(The treatment subjects were participants in an exercise program; control subjects were volunteers for the program who were unable to attend; subjects in the other two groups were similar to those in the first two groups in age and other characteristics.)

(a) Test the hypothesis of no average difference among the types of subjects.

(b) Apply the Bonferroni and Scheffé methods for multiple comparisons.

15. Apply the Bonferroni and Scheffé methods for multiple comparisons to the data in Problem 6. Use $\alpha = .05$

*16. Compare the ten pairs of means in the case of the data in Problem 4 at $\alpha = .10$, using

(a) the Bonferroni method.

(b) the Scheffé method, given $f_{.90}(4, 22) = 2.5$.

Section 14.4

*17. Construct the ANOVA table and calculate F-ratios for the following set of artificial data.

		Level of Factor A		
		1	2	3
	1	3	5	4
Level of Factor B	2	11	10	12
	3	16	21	17

18. Calculate SSE for the following array.

		Level of Factor T			
		1	2	3	4
	1	16	12	9	11
Level of Factor B	2	11	7	4	6
	3	15	11	8	10

(a) What do the F-ratios suggest?

(b) Give estimates of the treatment and block effects.

*19. The whiteness of clothes after washing may depend on both water temperature and brand of detergent. With three water temperatures and three brands of detergent, whiteness was measured for each combination of temperature and detergent brand, with results as follows.

	Detergent		
	A	B	C
Warm	45	47	55
Temperature Cold	36	41	55
Hot	42	47	46

Give the ANOVA table and calculate the F-statistic and corresponding P-value for testing
(a)   the hypothesis of no difference among detergent types.
(b)   the hypothesis of no water temperature effect.

20.   A study of methods of freezing meat loaf investigated the effect of oven position on drip loss.[6] Three batches of eight loaves each were baked and analyzed. The drip losses were recorded as follows.

		Oven position								Means
		1	2	3	4	5	6	7	8	
Batch	1	7.33	3.22	3.28	6.44	3.83	3.28	5.06	4.44	4.610
	2	8.11	3.72	5.11	5.78	6.50	5.11	5.11	4.28	5.465
	3	8.06	4.28	4.56	8.61	7.72	5.56	7.83	6.33	6.619
Means		7.833	3.740	4.317	6.943	6.017	4.650	6.000	5.017	5.565

Given SSE = 9.290,
(a)   test the hypothesis of no oven position effect.
(b)   test the hypothesis of no batch effect.

*21.   In the preceding problem, suppose the data had been analyzed as consisting of three random samples of size 8, ignoring the possibility of a batch effect. Combine the batch and error sums squares in that problem to obtain an SSE for a one-way ANOVA, and test the hypothesis of no batch effect. (Does taking oven position into account make a difference?)

22.   The data in Problem 6 were for yarn of type B. Strengths for type A yarn (with the same six bobbins) were also measured. Mean strengths for the two types are as follows.

Bobbin:	1	2	3	4	5	6
Type A	15.33	16.50	16.20	14.22	15.10	15.50
Type B	17.52	16.72	14.97	15.42	17.77	15.62

Use these means for a two-way analysis of variance to test the hypothesis of no mean difference between yarn types.

---

[6] Study by B. Bobeng, and B. David, described by T. J. Ryan, Jr., et al., in *Minitab: A Student Handbook* (North Scituate, Mass.: Duxbury Press, 1976).

∗23. The detailed data for yarn A, whose means are given in the preceding problem, are as follows.

Bobbin	1	2	3	4	5	6
	15.0	15.7	14.8	14.9	13.0	15.9
	17.0	15.6	15.8	14.2	16.2	15.6
	13.8	17.6	18.2	15.0	16.4	15.0
	15.5	17.1	16.0	12.8	14.8	15.5
S^2	1.756	1.007	2.053	1.029	2.467	.1400

Think of each combination of bobbin and yarn as a cell, and calculate an error sum of squares based on the variation of strengths within the cells—the sum of squared deviations of strengths about the respective cell means. With 3 d.f. within each cell, this sum of squares involves 36 d.f. Find the corresponding mean square and use this as a denominator for an F-test of no difference between yarns A and B. (The numerator will be $\Sigma\Sigma\Sigma(\bar{X}_{i..} - \bar{X})^2 = \Sigma\Sigma\, 4(\bar{X}_{i..} - \bar{X})$, or four times the SSBobbin based on cell means in Problem 22.)

24. Along the lines of Problem 10, obtain these calculation formulas for the two-way classification:

$$\text{SSTr} = \frac{1}{r}\sum_i\left(\sum_j X_{ij}\right)^2 - C, \qquad \text{SSB} = \frac{1}{c}\sum_j\left(\sum_i X_{ij}\right)^2 - C,$$

where

$$C = \frac{1}{rc}\left(\sum\sum X_{ij}\right)^2.$$

25. Derive the m.l.e.'s of τ_i, β_j, and σ^2 given in Section 14.4, assuming the normal model defined at (2) of Section 14.4. (See page 616.)

26. Derive the tests based on the F-ratios at (6) in Section 14.4 as likelihood ratio tests.

Regression

Example **15a**   The height (Y) that a cake rises depends on the oven temperature (x_1) and on the amounts of various ingredients—the amount of flour (x_2), the amount of soda (x_3), and so on. Thus,

$$Y = g(x_1, x_2, \ldots, x_k).$$

We'd want to know the importance of each of the explanatory variables x_i and to be able to predict Y from the values of the x's.

Because of variations in the ingredient materials and factors not taken into account, the height Y is not completely determined by the x's: It includes a *random error*. This randomness makes the appropriate analysis and prediction statistical. One way to infer the contributions of the different x's is to bake several cakes using various combinations of values of those variables.   ■

Regression models are used in describing the dependence of a random numerical response Y on one or more explanatory or predictor variables ($x_1, \ldots, x_k$). The **regression function** is the locus of the (conditional) mean of the response variable, given the values of the explanatory or predictor variables. It is usually assumed to have a specific form or type, involving unknown parameters. The type considered in this chapter is linear:

$$Y = \beta_0 + \beta_1 x_1 + \cdots + \beta_k x_k,$$

although certain nonlinear functions can be thought of as subsumed within this type. Regression analysis involves inferences about those parameters β_j based on data.

We begin by studying the case of a single predictor or explanatory variable.

15.1   Models with a Single Explanatory Variable

When X Is
Controlled

We consider two types of sampling scheme. Both generate pairs (X, Y) in which Y is the **response** and X is the **explanatory** or **predictor variable**. These schemes are analogous to the two sampling schemes considered in Section 13.5 for categorical data. In one scheme, X can be controlled—set at a particular value. To stress that the explanatory variable is nonrandom, we'll use the lower case x. The variable $Y = Y(x)$ is the response for a given x. For instance, a reactor yield (Y) may depend on temperature (x); sales (Y) on amount of advertising (x); change in blood pressure (Y), on dosage of an antihypertensive drug (x); and so on. The relationship between response and controlled variable may be completely predictable. But more often than not there are unavoidable, uncontrolled factors that result in different responses for the same value of x. Such factors include measurement errors and changes in experimental conditions such as temperature, humidity, power surges, and vibrations. Their effects on the response are thought of as random components of the measured response. The aim of a statistical analysis is to sort out the underlying relationship from the "noise," the random component in the observed Y-values.

In such contexts, the data one obtains for investigating the relationship of a response Y to a controlled variable x consist of pairs (x_i, Y_i), where the x_i are selected (and fixed) values of x and the Y_i $(i = 1, \ldots, n)$ are the corresponding responses.

Example **15.1a**   **A Welding Experiment**

The data in Table 15.1 are from the report of an inertia welding experiment.[1] The controlled variable x is the angular velocity of a rotating part, and the response y is the breaking strength of the weld. Seven welds were made using four different velocities of rotation.

Table 15.1

Velocity (x) $(10^2 \ ft/min)$	Breaking strength (Y) (ksi)
2.00	89
2.5	97, 91
2.75	98
3.00	100, 104, 97

The data are plotted in Figure 15.1. This plot shows a nearly linear relationship between velocity and breaking strength. Deviations from linearity might be attributed to random errors.

---

[1] See Example d of the Prologue.

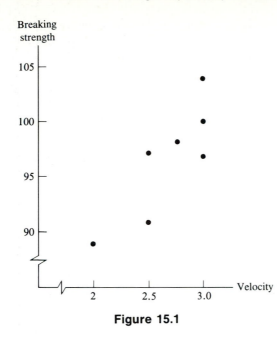

Figure 15.1   ■

Let x denote a fixed value of the controlled variable and $Y = Y(x)$ the corresponding response. We assume this response to be the sum of a deterministic function $g(x)$ and a random error ε:

$$Y(x) = g(x) + \varepsilon. \tag{1}$$

We assume $E(\varepsilon) = 0$, so $E[Y \mid x] = g(x)$. [Even though x is the value of a controlled rather than a random variable, we use the bar notation, consistent with our earlier use in such expressions as $E(X \mid \theta)$.] We call $E(Y \mid x)$ the **regression function of Y on x**. It specifies the underlying relationship between x and Y. Researchers are usually interested in the form of g. Learning about g from data pairs (x, y) is complicated by the presence of the random error ε in each response.

Another sampling scheme is to select experimental subjects or materials at random, and make two measurements (X, Y) on each one. In this scheme, both X and Y are random, and the basic model is a bivariate distribution for (X, Y). In such settings, one is usually interested in predicting the value of one, given a value of the other. We'll denote the variable to be predicted by Y and the predictor variable by X.

X and *Y*
Both Random

Example 15.1b   **GPA Versus SAT Scores**
Many high school students take college aptitude tests such as the SAT (Scholastic Aptitude Test) in their junior or senior year. Such a test is intended to indicate a student's readiness for college and to predict how well the student will do in college. Let X denote a student's SAT score and Y the student's GPA (grade point average) at the end of the first year in college. Both X and Y are random when a student is selected at random from some population. They are correlated, but not perfectly

so. (Typically, ρ is about .6.) The relationship between the variables can aid in the prediction of GPA for a student whose SAT score is known. But any prediction is subject to error when $|\rho| < 1$.   ■

To predict Y given $X = x$, we'll use the conditional mean $E(Y \mid x)$. This choice is explained in Section 15.5. The conditional mean is a function of x called the **regression function of Y on X** [see (5) of Section 6.9]. The source of the term *regression* will be explained in Section 15.6.

The data for estimating the regression function consist of pairs (x_i, y_i)—just like the data one obtains when X is controlled. The methods of analysis are the same for both types of sampling scheme. We now introduce these methods in the context in which x is controlled.

The error ε is generally different for each pair (x, y). We denote it by ε rather than, for example, by $\varepsilon(x)$ because we'll consider only the case in which its distribution is the same at all values of x. We denote the variance of ε by σ^2. Further, we assume that the errors involved in measured responses are *independent* random variables. This in turn means that those responses are independent.

Regression model:

$$Y(x) = g(x) + \varepsilon, \tag{1}$$

where $E(\varepsilon) = 0$, var $\varepsilon = \sigma^2$. The measured responses Y are independent.

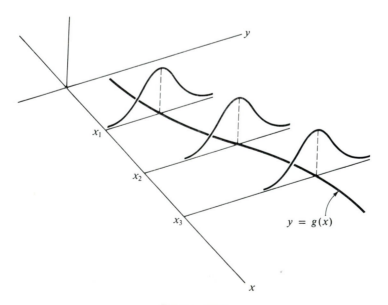

Figure 15.2

The response $Y(x)$ is distributed about its mean $g(x)$ in the same way that ε is distributed about its mean 0. This error distribution is usually assumed to be *normal*. Figure 15.2 represents the situation schematically; it shows the pattern of variation about the regression function as a normal p.d.f. centered at $g(x)$.

In applications, one usually assumes a particular form for the function g, such as linear ($\alpha + \beta x$), or trigonometric ($\alpha \sin \beta x$), or exponential ($\alpha e^{\beta x}$). The unspecified coefficients ($\alpha, \beta, \ldots$) are parameters of the model. The data (x, y) can be used to estimate these parameters.

The simplest nontrivial function is *linear*. In the next few sections, this is the only form of regression function we'll consider: $g(x) = \alpha + \beta x$. In this case, (1) becomes

$$Y(x) = \alpha + \beta x + \varepsilon. \tag{2}$$

When g is not linear, a linear function is often a useful approximation, even if only over a limited range of values of x.

15.2   Least Squares

We turn now to finding estimates of the slope β and intercept α of the linear regression function $\alpha + \beta x$ based on data pairs $(x_i, y_i), i = 1, 2, \ldots, n$. Finding a line that comes close to most of the data points in some overall sense is an instance of *curve fitting*. Suppose we try the line $y = a + bx$. The amount by which a particular y_i deviates from the value of the function $a + bx$ at $x = x_i$ is called a **residual** (see Figure 15.3):

$$y_i - (a + bx_i). \tag{1}$$

The n residuals are all zero when the data points all lie on the line $y = a + bx$, but these points will not ordinarily lie on a line. So we choose a and b to make the residuals (1) *close* to zero in some overall sense. The **method of least squares**, devised

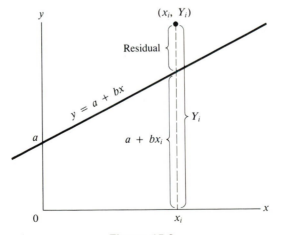

Figure 15.3

by K. F. Gauss late in the eighteenth century, measures closeness in terms of the sum of the squared residuals. This is a generalization of the method used at (5) of Section 7.5, which gives $\bar{X}$ as an estimate for a population mean.

In the method of least squares, we fit a line by minimizing the sum of the squared residuals:

$$\sum_{i=1}^{n} [y_i - (a + bx_i)]^2. \tag{2}$$

The linear function that minimizes (2) defines the **least-squares line**.

Example 15.2a   Consider these three data points, (0, 1), (1, 0), (2, 2), chosen to make the arithmetic simple. For any particular values a and b, the residuals are

$$1 - (a + 0b), \qquad 0 - (a + b), \qquad 2 - (a + 2b),$$

respectively. The sum of their squares (2) is

$$(1 - 2a + a^2) + (a^2 + 2ab + b^2) + (4 + a^2 + 4b^2 - 4a + 4ab - 8b)$$
$$= 3a^2 - 6a + 5b^2 - 8b + 6ab + 5.$$

To find the a and b that minimize this, we can use either algebra or calculus. According to the method of calculus, the derivative with respect to a (holding b fixed) and the derivative with respect to b (holding a fixed) must vanish at a minimizing pair (a, b):

$$\begin{cases} 6a + 6b - 6 = 0 \\ 6a + 10b - 8 = 0. \end{cases}$$

Solving these equations simultaneously yields $a = b = \frac{1}{2}$ as the only critical point. This must produce a minimum, because the sum of squared residuals is bounded below (by 0) and so *has* a minimum. So the least-squares line is $y = \frac{1}{2}(1 + x)$, shown along with the three data points in Figure 15.4.

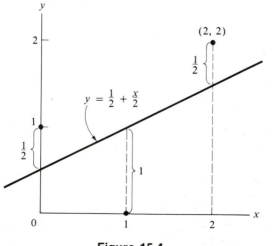

Figure 15.4

The residuals for the three data points are easy to calculate, or they can be read from the graph: $\frac{1}{2}$, -1, and $\frac{1}{2}$. (These sum to 0, as is true generally; see Problem 8. This fact provides a convenient check on the arithmetic.) The sum of their squares is $\frac{1}{4} + 1 + \frac{1}{4} = \frac{3}{2}$, and this is the minimum value of (2). The sum of squared residuals about any other line is larger. ∎

The detailed calculation of the last example was tedious, even with relatively simple arithmetic. We will now use an algebraic method to derive general formulas for the least-squares coefficients. As a by-product, the derivation will yield a convenient formula for the smallest sum of squared residuals. To minimize (2), we first rewrite each residual as follows:

$$y_i - a - bx_i = (y_i - \bar{y}) + (\bar{y} - a - b\bar{x}) - b(x_i - \bar{x}).$$

After squaring this as a trinomial, we sum on i:

$$\sum_{i=1}^{n} [y_i - (a + bx_i)]^2 = \sum_{i=1}^{n} (y_i - \bar{y})^2 + n(\bar{y} - a - b\bar{x})^2$$

$$+ b^2 \sum_{i=1}^{n} (x_i - \bar{x})^2 - 2b \sum_{i=1}^{n} (x_i - \bar{x})(y_i - \bar{y}). \tag{3}$$

(The other cross-product terms drop out because the sum of the deviations of the x_i's about their mean and the sum of the deviations of the y_i's about their mean are both 0.)

Some notation will be helpful at this point. The sum of squared deviations of any set of numbers x about their mean $\bar{x}$ is

$$SS_{xx} = \sum (x - \bar{x})^2 = \sum x^2 - \frac{(\sum x)^2}{n}. \tag{4}$$

(Here and in the remainder of this chapter, all sums extend from $i = 1$ to $i = n$.) Similarly, the sum of products of deviations of x's and y's, each about the respective mean, is

$$SS_{xy} = \sum (x - \bar{x})(y - \bar{y}) = \sum xy - \frac{(\sum x)(\sum y)}{n}. \tag{5}$$

In this notation, we rewrite the right-hand side of (3) as

$$SS_{yy} + n(\bar{y} - a - b\bar{x})^2 + b^2 SS_{xx} - 2b SS_{xy}. \tag{6}$$

Next we complete the square in the last two terms of (6):

$$b^2 SS_{xx} - 2b SS_{xy} = SS_{xx} \left\{ b^2 - 2b \frac{SS_{xy}}{SS_{xx}} + \left(\frac{SS_{xy}}{SS_{xx}} \right)^2 \right\} - \frac{SS_{xy}^2}{SS_{xx}}$$

$$= SS_{xx} \left(b - \frac{SS_{xy}}{SS_{xx}} \right)^2 - r^2 SS_{yy},$$

where r is the correlation coefficient of the pairs (x_i, y_i):

$$r = \frac{SS_{xy}}{\sqrt{SS_{xx} SS_{yy}}} \tag{7}$$

(see Section 7.6). Clearly, there is a relationship between r and $\hat\beta$:

$$\hat\beta = \sqrt{\frac{SS_{yy}}{SS_{xx}}} \cdot r = \frac{S_y}{S_x} \cdot r,$$

where S_y and S_x are sample s.d.'s. Returning to (6), we see that

$$\sum[y_i - (a + bx_i)]^2 = n(\bar y - a - b\bar x)^2 + SS_{xx}\left(b - \frac{SS_{xy}}{SS_{xx}}\right)^2 + SS_{yy}(1 - r^2). \quad (8)$$

Since the first two terms on the right are nonnegative, the sum of squared residuals is minimized when a and b are chosen to make those terms equal to 0. We denote the minimizing values of a and b by $\hat\alpha$ and $\hat\beta$, because we'll use them as estimates of the parameters α and β of the true regression line [(2) in Section 15.1]. With these minimizing values in (8), we obtain the *smallest* sum of squared residuals:

$$SSRes = \sum[y_i - (\hat\alpha + \hat\beta x_i)]^2 = SS_{yy}(1 - r^2).$$

Define

$$\hat\beta = \frac{SS_{xy}}{SS_{xx}} = \frac{S_y}{S_x} \cdot r, \qquad \hat\alpha = \bar y - \hat\beta\bar x. \tag{9}$$

The least-squares line (empirical regression line) is given by

$$y = \hat\alpha + \hat\beta x. \tag{10}$$

The sum of the squared residuals about the least-squares line is

$$SSRes = SS_{yy}(1 - r^2). \tag{11}$$

Equation (11) provides another demonstration of the fact that r^2 cannot exceed 1, for SSRes is always nonnegative, being the sum of squared residuals. Moreover, (11) shows that SSRes $= 0$ when $r^2 = 1$; in this case, all the data points lie on the least-squares line.

Example 15.2b   Consider again the data of Example 15.2a. For hand calculations, it is convenient to lay out the work like this:

x	y	x^2	xy	y^2
1	0	1	0	0
0	1	0	0	1
2	2	4	4	4
Sums: 3	3	5	4	5

$$SS_{xy} = 4 - \frac{3.3}{3} = 1,$$

$$SS_{xx} = 5 - \frac{3^2}{3} = 2 = SS_{yy},$$

$$\hat\beta = \frac{1}{2}, \qquad \hat\alpha = 1 - \frac{1}{2} = \frac{1}{2}.$$

$$r = \frac{1}{\sqrt{2 \cdot 2}} = \frac{1}{2}.$$

As in Example 15.2a, we obtain $y = \frac{1}{2}(1 + x)$ as the least-squares line. To find the minimum sum of squared residuals, we substitute into (11):

$$\text{SSRes} = 2\left(1 - \frac{1}{4}\right) = \frac{3}{2}.$$

This also agrees with the result obtained earlier.   ■

Example 15.2c   A Dose–Response Curve

A study in surgery[2] examined the increase (Y) in pancreatic intraductal pressure (PIP) in response to doses of a potent cholinesterase inhibitor (x). Six different doses were administered to one dog, with the following results (x, Y): (0, 14.6), (5, 24.5), (10, 21.8), (15, 34.5), (20, 35.1), (25, 43). We assume that the dose–response curve is linear—that the increment in PIP is the same for each unit increment in dose, regardless of the dose level. This assumption is clearly not correct for some dose levels; in particular, PIP is bounded. However, it may be reasonable to assume linearity over the range of doses considered. (In practice, one should check the assumption of linearity, at least by informal inspection of the scatter plot. A more formal way is to introduce, for example, a quadratic term in the model to see if the resulting model gives an appreciably better fit.)

For hand calculation, we lay out the work as we did in the preceding example.

x(dose)	y(pressure)	xy	x^2
0	14.6	0	0
5	24.5	122.5	25
10	21.8	218	100
15	34.5	517.5	225
20	35.1	702	400
25	43.0	1075	625
75	173.5	2635	1375

We find

$$\bar{x} = 12.5, \qquad \bar{y} = 28.92, \qquad \text{SS}_{xx} = 437.5, \qquad \text{SS}_{xy} = 466.25.$$

Substituting these into (9) yields

$$\hat{\beta} = \frac{466.25}{437.5} = 1.065, \qquad \hat{\alpha} = 28.92 - 1.065 \times 12.5 = 15.6.$$

The least-squares regression line (10) is thus

$$y = 15.6 + 1.065x.$$

Its graph is shown in Figure 15.5, along with a scatter plot of the data points.

---

[2] T. D. Dressel et al., "Sensitivity of the canine pancreatic intraductal pressure to subclinical reduction in cholinesterase activity," *Annals of Surgery 190*, No. 1 (July 1979).

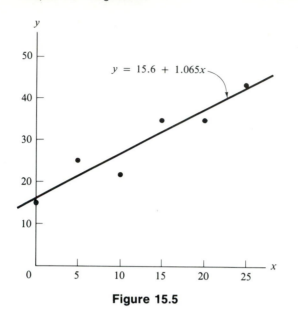

Figure 15.5

We calculate the minimum mean squared residual using (11) with $SS_{yy} = 542.868$ and $r^2 = (466.25)^2/(542.868 \times 437.5) = .9153$:

$$SSRes = 542.868(1 - .9153) = 45.98. \qquad \blacksquare$$

We've applied the method of least squares to the special case of fitting a straight line, but the method can be used in fitting other types of curves (Solved Problem C) and generalized to the case of several explanatory variables (Section 15.7).

15.3   ## Distribution of the Least-Squares Estimators

The least-squares coefficients $\hat{\alpha}$ and $\hat{\beta}$ [(9) of Section 15.2] are functions of the data. They vary from sample to sample: They are random variables—functions of the random responses Y_i. We'll need the sampling distributions of these coefficients when using them to construct confidence intervals and test hypotheses about the true regression coefficients α and β.

Any particular data set is the result of sampling the populations defined by the regression model (2) of Section 15.1, in which

$$E[Y(x)] = \alpha + \beta x, \qquad \text{var}[Y(x)] = \sigma^2.$$

Let $Y_i = Y(x_i)$, $i = 1, \ldots, n$, and assume further that $Y_1, \ldots, Y_n$ are independent.

In deriving the distributions of $\hat{\alpha}$ and $\hat{\beta}$, it helps to rewrite SS_{xy} in a way that shows clearly that it is a *linear* combination of the Y_i's:

$$SS_{xy} = \sum(x_i - \bar{x})(Y_i - \bar{Y}) = \sum(x_i - \bar{x})Y_i. \qquad (1)$$

These two expressions for SS_{xy} are equal, since $\Sigma(x_i - \bar{x})\bar{Y} = 0$. Similarly,

$$SS_{xx} = \sum x_i(x_i - \bar{x}). \tag{2}$$

Now

$$\hat{\beta} = \frac{S_{xy}}{S_{xx}},$$

Mean and Variance of $\hat{\alpha}$ and $\hat{\beta}$

so in view of (1), $\hat{\beta}$ is also a linear function of the Y_i's. This makes it easy to find the mean and variance of $\hat{\beta}$. Since $E(Y_i) = \alpha + \beta x_i$, we have

$$E(\hat{\beta}) = \frac{1}{SS_{xx}}\sum(x_i - \bar{x})E(Y_i) = \frac{1}{SS_{xx}}\sum(x_i - \bar{x})(\alpha + \beta x_i).$$

Multiplying out and using (2), we get

$$E(\hat{\beta}) = \frac{\alpha}{SS_{xx}}\sum(x_i - \bar{x}) + \frac{\beta}{SS_{xx}}\sum x_i(x_i - \bar{x}) = 0 + \beta = \beta.$$

So $\hat{\beta}$ is an unbiased estimate of β. Problem 9 asks you to show that $\hat{\alpha}$ is an unbiased estimate of α.

Because $\hat{\alpha}$ and $\hat{\beta}$ are unbiased, their mean squared errors are simply their variances [see (3) of Section 9.1]. To obtain these variances, we again exploit the fact that $\hat{\alpha}$ and $\hat{\beta}$ are linear in the Y's. We have assumed that var $Y_i = \sigma^2$, so it follows, from independence and the formula for the variance of a linear combination, that

$$\text{var } \hat{\beta} = \frac{1}{(SS_{xx})^2}\sum(x_i - \bar{x})^2 \text{ var } Y_i = \frac{\sigma^2}{SS_{xx}}. \tag{3}$$

[See (11) of Section 5.6 and (8) of Section 5.8.] Thus, the variability in the slope estimate is proportional to the error variance—the larger the component of error in the response variable, the larger the typical error in estimating β.

For the estimator $\hat{\alpha}$ of α, we have

$$\text{var } \hat{\alpha} = \text{var}(\bar{Y} - \hat{\beta}\bar{x}) = \text{var } \bar{Y} + (\bar{x})^2 \text{ var } \hat{\beta} - 2(\bar{x})\text{cov}(\bar{Y}, \hat{\beta}).$$

Problem 11 asks you to show that the covariance of $\bar{Y}$ and $\hat{\beta}$ is 0. Using this result and (3), we obtain the variance of $\hat{\alpha}$:

When $Y_1, \ldots, Y_n$ are independent, with

$$E(Y_i) = \alpha + \beta x_i, \qquad \text{var } Y_i = \sigma^2,$$

the least-squares estimates $\hat{\alpha}$ and $\hat{\beta}$ are unbiased, and

$$\text{var } \hat{\alpha} = \sigma^2\left(\frac{1}{n} + \frac{(\bar{x})^2}{SS_{xx}}\right), \qquad \text{var } \hat{\beta} = \frac{\sigma^2}{SS_{xx}}. \tag{4}$$

**Selecting
x-Values**

Suppose we can set the values of the x's and have the goal of estimating α and β as precisely as possible. The variance of α is minimized by choosing $\bar{x}$ as close to zero as possible. Since var β is inversely proportional to the variance of the x's, they should be spread out as much as possible. There are usually practical limits for the choice of the x's, and SS_{xx} will be largest if half the x-values are at the lower limit and half at the upper limit. However, unless one is certain of the linearity of the regression relation, it's a good idea to choose some intermediate values as well.

For making inferences about $\hat{\alpha}$ and $\hat{\beta}$, we need a further assumption about the distribution of responses. In the remainder of this section and throughout the next, we assume them to be *normally* distributed:

$$Y(x) \sim N(\alpha + \beta x, \sigma^2)$$

(see Figure 15.2).

**Maximum
Likelihood
Estimates**

When responses are normally distributed, the least-squares estimates $\hat{\alpha}$ and $\hat{\beta}$ are maximum likelihood estimates, as we show next. Because of the assumed independence of responses, the likelihood function is proportional to the product of the p.d.f.'s of the individual observations. Thus, since $Y_i \sim N(\alpha + \beta x_i, \sigma^2)$,

$$L(\alpha, \beta, \sigma^2) = \prod \sigma^{-1} \exp\left\{ -\frac{1}{2\sigma^2}(Y_i - \alpha - \beta x_i)^2 \right\}$$

$$= \sigma^{-n} \exp\left\{ -\frac{1}{2\sigma^2} \sum(Y_i - \alpha - \beta x_i)^2 \right\}. \tag{5}$$

To maximize (5), α and β must be chosen to minimize the sum in the exponent. Minimizing this sum of squares is precisely what led us to the least-squares coefficients in Section 15.2! Thus, the m.l.e.'s of α and β are their least-squares estimates. To find the m.l.e. of σ^2, we set the derivative with respect to σ^2 equal to zero, solve for σ^2, and replace α and β with their least-squares estimates:

$$\hat{\sigma}^2 = \frac{1}{n} \sum(Y_i - \hat{\alpha} - \hat{\beta} x_i)^2 = \frac{SSRes}{n} \tag{6}$$

[see (11) of Section 15.2].

We've already pointed out that the estimates $\hat{\alpha}$ and $\hat{\beta}$ depend linearly on the responses Y_i. With a normal distribution for each Y_i, this linearity implies that the

**Sampling
Distributions**

slope $\hat{\beta}$ and intercept $\hat{\alpha}$ have a *bivariate normal* distribution (see Section 6.8). As we explain next,

$$\frac{n\hat{\sigma}^2}{\sigma^2} = \frac{SSRes}{\sigma^2}$$

has a chi-square distribution with $n - 2$ degrees of freedom.

The following identity follows upon multiplying out the product on the right and using the definition of $\hat{\alpha}$:

$$Y_i - \alpha - \beta x_i = (Y_i - \hat{\alpha} - \hat{\beta} x_i) + [\bar{Y} - (\alpha + \beta\bar{x})] + (\hat{\beta} - \beta)(x_i - \bar{x}). \tag{7}$$

When we square the trinomial on the right and sum on i, the three cross-product terms drop out (Solved Problem D). Thus,

$$\sum_{1}^{n}\left(\frac{Y_i - \alpha - \beta x_i}{\sigma}\right)^2 = \frac{n\hat{\sigma}^2}{\sigma^2} + \frac{[\bar{Y} - (\alpha + \beta\bar{x})]^2}{\sigma^2/n} + \frac{(\hat{\beta} - \beta)^2}{\sigma^2/SS_{xx}}. \qquad (8)$$

The left-hand side of (8) is the sum of squares of n independent, standard normal variables, so its distribution is $\chi^2(n)$. Consider the right-hand side of (8). The second term is $\chi^2(1)$—the square of a single standard normal variable. The third term is also $\chi^2(1)$, since $\hat{\beta}$ is normal with mean β and variance σ^2/SS_{xx}. From Problem 11, we know that $\bar{Y}$ and $\hat{\beta}$ are uncorrelated and hence (because they have a bivariate normal distribution) independent. So the second and third terms are independent. Actually, all three terms on the right-hand side of (8) are independent,[3] and this implies [according to Section 6.5] that the first term on the right is $\chi^2(n-2)$.

When $Y_1, \ldots, Y_n$ are independent, with $Y_i \sim N(\alpha + \beta x_i, \sigma^2)$, $i = 1, \ldots, n$, the m.l.e.'s $\hat{\alpha}$ and $\hat{\beta}$ are the least-squares estimates and are normally distributed. Also, $\hat{\sigma}^2$ and $\hat{\alpha} + \hat{\beta}x$ are independent, and

$$\frac{n\hat{\sigma}^2}{\sigma^2} = \frac{SSRes}{\sigma^2} \sim \chi^2(n-2).$$

The m.l.e. $\hat{\sigma}^2$ given by (6) is an ordinary average of the squared residuals. In standard terminology, the **mean squared residual** (MSRes) is the average obtained when we divide SSRes instead by its degrees of freedom:

$$S_\varepsilon^2 = \frac{SSRes}{n-2} = \frac{n\hat{\sigma}^2}{n-2}. \qquad (9)$$

This is an unbiased estimate of σ^2, because the mean of SSRes is $(n-2)\sigma^2$ [from (5) of Section 6.5]:

$$E(S_\varepsilon^2) = E\left(\frac{SSRes}{n-2}\right) = \frac{n-2}{n-2} \cdot \sigma^2 = \sigma^2. \qquad (10)$$

Under the assumption of normality, the variance of S_ε^2 is

$$\operatorname{var}(S_\varepsilon^2) = \operatorname{var}\left(\frac{SSRes}{\sigma^2} \cdot \frac{\sigma^2}{n-2}\right) = \frac{\sigma^4}{(n-2)^2} \cdot (n-2) = \frac{\sigma^4}{n-2} \qquad (11)$$

[from (6) of Section 6.5]. This tends to zero as n becomes infinite, so S_ε^2 is a consistent estimator of σ^2 (as is $\hat{\sigma}^2$).

---

[3] The independence is a consequence of a general decomposition theorem (for sampling normal populations), of which the independence of mean and variance shown in Section 8.10 is another special case. See for example Lindgren [15], pp. 525ff.

15.4   Inference for Regression Parameters

Standard Error of $\hat{\beta}$

In applications, the slope parameter β is usually of special interest. It gives the increase in the response Y when x is increased by one unit. In particular, when $\beta = 0$, the response Y is not affected by the controlled variable x. Estimating β to be $\hat{\beta}$ involves a random error. The standard deviation of $\hat{\beta}$ is

$$\sigma_{\hat{\beta}} = \frac{\sigma}{\sqrt{SS_{xx}}}. \tag{1}$$

This depends on σ, the population s.d., which is usually unknown. Replacing σ in (1) with S_ε [from (9) in Section 15.3] yields a standard error:

$$\text{s.e.}(\hat{\beta}) = \frac{S_\varepsilon}{\sqrt{SS_{xx}}}. \tag{2}$$

Confidence Limits for β

We construct confidence limits for the slope β in the usual way, by taking the point estimate $\hat{\beta}$ plus or minus a multiple of the standard error:

$$\hat{\beta} \pm k \cdot \text{s.e.}(\hat{\beta}). \tag{3}$$

When n is large, $k = 1.96$ for 95% confidence, 2.58 for 99% confidence, and so on. (See Table II in Appendix 1.) When the sample size is *small*, the large-sample confidence interval is too narrow, owing to the added uncertainty in replacing σ with S_ε. The t-distribution comes into play, as in Section 9.4. Its validity again depends on assumption of nearly normal errors.

In Section 15.3, we gave a decomposition (8) in which each term has a chi-square distribution; the ratio of the last term to the first term on the right-hand side of (8) in Section 15.3, multiplied by $n - 2$, is

$$F = \frac{(\hat{\beta} - \beta)^2}{[\text{s.e.}(\hat{\beta})]^2} = \frac{\dfrac{(\hat{\beta} - \beta)^2}{\sigma^2/SS_{xx}}}{\dfrac{SSRes}{\sigma^2}\Big/(n-2)} = \frac{A/1}{B/(n-2)}, \tag{4}$$

where A and B are independent chi-square variables with 1 and $n - 2$ degrees of freedom, respectively. Thus, F has the F-distribution with $(1, n - 2)$ d.f. [see Section 6.6], and its symmetric square root

$$T = \frac{\hat{\beta} - \beta}{\text{s.e.}(\hat{\beta})} \tag{5}$$

has the t-distribution with $n - 2$ degrees of freedom (see Section 6.6). So k in (3) is found in Table IIIc of Appendix 1, in the row labeled $n - 2$ degrees of freedom. For a confidence level of $1 - \alpha$, we use the $100(1 - \alpha/2)$ percentile.

Example 15.4a   Dose Response (continued)

The estimate of the slope of the regression line in Example 15.2c is $\hat{\beta} = 1.065$. From that example, SSRes = 45.98. So

$$S_\varepsilon^2 = \frac{45.98}{4} = 11.495, \qquad SS_{xx} = 437.5,$$

and

$$\text{s.e.}(\hat{\beta}) = \sqrt{\frac{11.495}{437.5}} = .1621.$$

In view of the small sample size ($n = 6$) and the need to estimate σ, we turn to the t-table. For 95% confidence, we set $k = 2.45$ in (3)—this is the 97.5th percentile of $t(4)$ as given in Table III of Appendix 1. The 95% confidence limits are thus

$$\hat{\beta} \pm 2.45[\text{s.e.}(\hat{\beta})] = 1.065 \pm .397.$$

Using the t-table assumes that the error distribution is normal, or close to normal. In practice, some consideration should be given to checking this assumption, or at least worrying about it. ∎

Testing
$\beta = \beta_0$

To test $H_0: \beta = \beta_0$ against $H_A: \beta \neq \beta_0$, we use the likelihood ratio statistic Λ. The maximum value of the likelihood function [(5) of Section 15.3] over $H_0 \cup H_A$ is

$$L(\hat{\alpha}, \hat{\beta}, \hat{\sigma}^2) = (\hat{\sigma}^2)^{-n/2} e^{-n/2}.$$

Under H_0, the m.l.e. of α is $\bar{Y} - \beta_0 \bar{x}$, so

$$\max_{H_0} L = L(\bar{Y} - \beta_0 \bar{x}, \beta_0, \hat{\sigma}_0^2) = (\hat{\sigma}_0^2)^{-n/2} e^{-n/2}, \tag{6}$$

where (see Solved Problem E)

$$n\hat{\sigma}_0^2 = SS\text{Res} + (\hat{\beta} - \beta_0)^2 SS_{xx}. \tag{7}$$

Then

$$\Lambda^{-2/n} = \frac{\hat{\sigma}_0^2}{\hat{\sigma}^2} = \frac{SS\text{Res} + (\hat{\beta} - \beta_0)^2 SS_{xx}}{SS\text{Res}} = 1 + (n-2)T^2, \tag{8}$$

where

$$T = \frac{\hat{\beta} - \beta_0}{\text{s.e.}(\hat{\beta})}. \tag{9}$$

For a likelihood ratio test, the rejection region is $\Lambda < C$; in view of (8), this region is equivalent to $|T| > C$.

Likelihood ratio test of $H_0: \beta = \beta_0$ versus $H_A: \beta \neq \beta_0$: Reject H_0 if

$$|T| = \frac{|\hat{\beta} - \beta_0|}{\text{s.e.}(\hat{\beta})} > C. \tag{10}$$

Under H_0, $T \sim t(n-2)$.

The rejection region (10) for T is intuitively appealing, since T measures the discrepancy between the hypothesized and estimated values of β, as compared with s.e.$(\hat{\beta})$, the "typical" deviation of the estimate from the true value. So it has the same form as the t-statistic for testing a population mean.

For the *one*-sided alternative H_A: $\beta > \beta_0$, the likelihood ratio critical region is $T > C$, for an appropriately chosen C. For example, if $n = 10$ and $\alpha = .05$, then d.f. $= 8$ and $C = 1.86$.

Equation (11) of Section 15.2 expresses the residual sum of squares in terms of the SS_{yy}, often called the *total sum of squares*:

$$\text{SSTot} = SS_{yy} = r^2 SS_{yy} + \text{SSRes}.$$

ANOVA in Regression

The first term on the right is termed the *sum of squares for regression*:

$$r^2 SS_{yy} = \hat{\beta}^2 SS_{xx} = \text{SSReg},$$

and we write

$$\text{SSTot} = \text{SSReg} + \text{SSRes} = r^2 SS_{yy} + (1 - r^2) SS_{yy}.$$

This decomposition of SS_{yy} is an analysis of variance. It expresses the total variation in Y-values as a fraction r^2 attributable to regression (that is, to the term βx) plus a fraction $(1 - r^2)$ attributable to error. (The residuals about the regression line are estimates of the error component ε in the model of Section 15.1.) The fraction r^2 attributed to regression is the **coefficient of determination**.

Example 15.4b   Dose Response (continued)

In Example 15.2c, we found $r^2 = .915$. We say then that 91.5% of the total variation among the Y-values is attributed to dose and 8.5% to random error.

We now arrange the various sums of squares (taken from Example 15.2c) in an ANOVA table:

Source	SS	d.f.	MS	F
Regression	496.89	1	496.89	43.22
Residual	45.98	4	11.495	—
Total:	542.87	5	—	—

The t-statistic for testing $\beta = 0$ is $\sqrt{F} = \sqrt{43.22} = 6.57$, highly statistically significant (the P-value is less than .001). ∎

X and Y Both Random

So far, we've assumed that x is controlled. Now suppose that the n pairs (X_i, Y_i) constitute a random sample from a bivariate distribution. Suppose further that the conditional mean given $X = x$—the regression function—is linear in x: $E(Y|x) = \alpha + \beta x$, and that the conditional variance is constant: $\text{var}(Y|x) = \sigma^2$. The distribution of the Y's given the X's is then exactly the same as it is when x is controlled. So conditional on the values of X in the sample, the estimates $\hat{\alpha}$, $\hat{\beta}$, and $\hat{\sigma}^2$ have the same distributions as before.

Testing $\beta = 0$ is of special interest for bivariate normal populations. In this case, $\beta = 0$ is equivalent to $\rho = 0$, which in turn is equivalent to independence of X and Y. With $\beta_0 = 0$, (9) becomes

$$T = \frac{\hat{\beta}}{\text{s.e.}(\hat{\beta})} = \frac{\dfrac{SS_{xy}}{SS_{xx}}}{\sqrt{\dfrac{(1 - r^2)SS_{yy}}{(n - 2)SS_{xx}}}} = \sqrt{n - 2}\,\frac{r}{\sqrt{1 - r^2}}. \tag{11}$$

When $\beta = 0$, this is $t(n - 2)$, conditional on the x's. Since the x's are not involved in the distribution, the unconditional distribution of T is also $t(n - 2)$.

Example 15.4c   Suppose a sample of size $n = 25$ from a bivariate normal population shows a correlation .30 between X and Y. Is this evidence against $\rho = 0$? (That is, against the hypothesis that X and Y are independent?) Even if $\rho = 0$, the *sample* correlation r will generally be different from zero. Is $r = .30$ so far from 0 as to suggest dependence?

To test $H_0: \rho = 0$ (or $\beta = 0$), we use the t-statistic (11):

$$T = \sqrt{n - 2}\,\frac{r}{\sqrt{1 - r^2}} = \sqrt{23}\,\frac{.30}{\sqrt{1 - .09}} = 1.51.$$

With 23 d.f., the two-sided P-value (from Table IIIb in Appendix 1) is found to be about $2(.073) \doteq .15$. So $r = .30$ is not statistically significant.   ∎

15.5   Predicting Y from X

Suppose you want to predict the demand for a new product or the response to a drug. You have information X that may aid in the prediction. In the first instance, X may come from a market survey; in the second, it may come from a dose–response clinical trial. How should you use the information, and how helpful is it?

Predicting Without X

First consider a random demand or response Y in the *absence* of correlative information X. The error in predicting that Y will be c is the difference $Y - c$. The absolute error is $|Y - c|$, a random variable. In choosing a predictor c, we'd like to make the absolute error small in some average sense. In the spirit of least squares, we choose c to minimize the **mean squared prediction error**:

$$\text{m.s.p.e.} = E[(Y - c)^2].$$

The m.s.p.e. is a second moment of Y about c. We saw in Sections 3.3 and 5.5 that the second moment of a random variable about its mean value is the smallest second moment. So to minimize the mean squared error in predicting Y, we use $c = E(Y)$. With this choice of c, the m.s.p.e. is the variance of Y, and the root-mean-square (r.m.s.) prediction error is σ_Y.

Prediction Using X

When X and Y are related and it becomes known that $X = x$, the appropriate distribution for predicting Y is conditional—the distribution of $Y \,|\, x$. Applying the conclusion of the preceding paragraph, we use the *conditional* mean $E(Y \,|\, x)$ to predict Y. This is the regression function of Y on x [see (4) of Section 5.9]. When we use it to predict, the m.s.p.e. is the conditional variance, $\text{var}(Y|x)$.

When $E(Y|x)$
Is Linear

The regression function $E(Y|x)$ may be linear in x. Suppose it is: $E(Y|x) = \alpha + \beta x$. Using iterated expectations (Section 5.9), we have

$$E(Y) = E[E(Y|X)] = \alpha + \beta\, E(X), \qquad \text{or} \qquad \mu_Y = \alpha + \beta\mu_X.$$

The regression equation $y = \alpha + \beta x$ can therefore be written

$$y = \mu_Y + \beta(x - \mu_X).$$

Now

$$E(XY|X = x)] = E(xY|X = x) = x \cdot E(Y|x) = \alpha x + \beta x^2.$$

Again using iterated expectations, we have

$$E(XY) = E(\alpha X + \beta X^2) = \alpha\mu_X + \beta E(X^2).$$

Subtracting

$$E(X)E(Y) = \mu_X(\alpha + \beta\mu_X),$$

we obtain

$$\sigma_{X,Y} = \beta\sigma_X^2, \qquad \text{or} \qquad \beta = \frac{\sigma_{X,Y}}{\sigma_X^2}.$$

The equation of the regression line is thus

$$y = \mu_Y + \frac{\sigma_{X,Y}}{\sigma_X^2}(x - \mu_X), \tag{1}$$

or

$$\frac{y - \mu_Y}{\sigma_Y} = \rho \cdot \frac{x - \mu_X}{\sigma_X}. \tag{2}$$

Suppose next that the regression function is linear *and* the conditional variance is constant. We can find the value of that constant as follows:

$$\sigma_{Y|x}^2 = E(\sigma_{Y|x}^2) = \sigma_Y^2 - \text{var}[E(Y|X)] = \sigma_Y^2 - \text{var}(\beta X) = \sigma_Y^2 - \beta^2\sigma_X^2,$$

or

$$\sigma_{Y|x}^2 = \sigma_Y^2(1 - \rho^2). \tag{3}$$

So the mean squared prediction error using the predictor (1) is $\sigma_Y^2(1 - \rho^2)$. This decreases as $|\rho|$ increases. Indeed, when $|\rho| = 1$, there is no prediction error. However, if we predict *without* using X, the best predictor is $E(Y)$, with average squared error σ_Y^2. So (3) shows that by using X, we reduce the m.s. error by the factor $(1 - \rho^2)$. For example, if $\rho = \sqrt{\frac{3}{4}} = .886$, the r.m.s.p.e. can be cut in half by using the empirical regression function to predict Y. When $\rho = 0$, the r.m.s. error is just σ_Y, whether we use X or not.

Example 15.5a   **Predicting Achievement from Aptitude**

In educational testing, a typical correlation between an aptitude score (X) and an achievement score (Y) is .6. The reduction in error for predicting achievement using (1) is about 20%: $\sqrt{1 - .36} = .80$.

Suppose $\rho = .6$, $\mu_X = 22$, $\sigma_X = 2.4$, $\mu_Y = 540$, and $\sigma_Y = 50$. For any given x, (1) gives the least-squares prediction for achievement:

$$y = 540 + (.6) \cdot \frac{50}{2.4}(x - 22) = 265 + 12.5x.$$

The predicted achievement score for an individual with an aptitude score of 28 is $265 + 12.5 \times 28 = 615$. From (3), m.s.p.e. $= .64\sigma_Y^2$, so the r.m.s.p.e. is $.8\sigma_Y = 40$ (as compared with 50 when x is ignored). ∎

**Linear
Predictors**

When (X, Y) is bivariate normal, the regression function *is* linear; so it must have the form (1), which agrees with (8) of Section 6.8. Moreover, in this case, the conditional variance *is* constant and so is given by (3), which agrees with (9) of Section 6.8.

In general, the regression function may not be linear. Suppose it's not. The regression function is still the best predictor in the sense of mean square, but a linear function may serve almost as well. Consider the linear predictor $a + bx$. In terms of mean square, we find the best choice of the coefficients a and b by minimizing

$$\text{m.s.p.e.} = E\{[Y - (a + bX)]^2\}. \tag{4}$$

Expanding the square, averaging with respect to the joint distribution of X and Y, and rearranging terms [as in obtaining (8) of Section 15.2], we obtain

$$\text{m.s.p.e.} = (a - [\mu_Y - b\mu_X])^2 + \sigma_X^2\left(b - \frac{\sigma_{X,Y}}{\sigma_X^2}\right)^2 + \sigma_Y^2(1 - \rho^2). \tag{5}$$

The equality of (4) and (5) is easily checked (Problem 27) by verifying that the corresponding coefficients of a, b, ab, a^2, and b^2 in the two expressions agree.

It is clear that the m.s.p.e. (5) is smallest when

$$a = \mu_Y - b\mu_X \qquad \text{and} \qquad b = \frac{\sigma_{X,Y}}{\sigma_X^2}. \tag{6}$$

So the best **straight-line predictor** of Y given x is

$$y = \mu_Y + \frac{\sigma_{X,Y}}{\sigma_X^2}(x - \mu_X), \tag{7}$$

where *best* means smallest m.s.p.e. When we use (7), this smallest m.s.p.e. is the last term in (5):

$$\text{m.s.p.e.} = \sigma_Y^2(1 - \rho^2). \tag{8}$$

We'll illustrate with an example in which the linear predictor happens not to do very well.

Example 15.5b   Consider a uniform distribution on the triangle bounded by $y = 0$, $y = 1 - x$, and $y = 1 + x$. (See Figure 15.6.) From the symmetry, it is apparent that $\rho = 0$, so the best *linear* predicting function (7) is $y = \mu_Y = \frac{1}{3}$—a horizontal line. The corresponding m.s. error (8), the average (over X as well as Y) is $\sigma_Y^2 = \frac{1}{18}$.

The best predicting function for given x is the conditional mean:

$$E(Y \mid x) = \tfrac{1}{2}(1 - |x|), \qquad -1 < x < 1,$$

which is not linear in x. (See Figure 15.6.) The m.s.p.e. is the conditional variance:

$$\mathrm{var}(Y \mid x) = \tfrac{1}{12}(1 - |x|)^2, \qquad -1 < x < 1.$$

Although this is actually greater than $\sigma_Y^2 = \frac{1}{18}$ when x is near 0, its average with respect to the distribution of X is smaller than σ_Y^2, as it must always be:

$$E[\tfrac{1}{12}(1 - |X|)^2] = \tfrac{1}{24} < \tfrac{1}{18}.$$

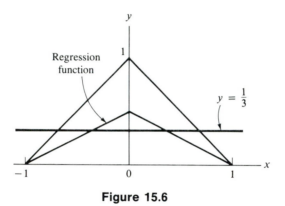

Figure 15.6   ■

Using Data to Predict

The true regression function is usually unknown, but suppose we have data (x_i, y_i), $i = 1, \ldots, n$. We can use this information to estimate $E(Y \mid x_0)$. The obvious estimator of the best linear predicting function (7) is

$$\hat{\alpha} + \hat{\beta} x_0, \tag{9}$$

with $\hat{\alpha}$ and $\hat{\beta}$ defined by (9) of Section 15.2. If we use this to predict the value of Y when $X = x_0$, how accurate is the prediction?

As in Section 15.3, suppose we assume that $Y \mid x \sim N(\alpha + \beta x, \sigma^2)$ and that the observations are independent random variables. When we use (9) with $X = x_0$ to predict $Y = Y(x_0)$, the prediction error is $Y - \hat{\alpha} - \hat{\beta} x_0$. In this expression, Y, $\hat{\alpha}$, and $\hat{\beta}$ are random variables, and Y is independent of $\hat{\alpha}$ and $\hat{\beta}$ because it is independent of the observations from which they are calculated. The distribution of the estimates $\hat{\alpha}$ and $\hat{\beta}$ (conditional on the x_i's) was given in Section 15.3. With it, we can calculate the mean squared prediction error:

$$\begin{aligned} \text{m.s.p.e.}(Y) = E[(Y - \hat{\alpha} - \hat{\beta} x_0)^2] &= \mathrm{var}(Y - \hat{\alpha} - \hat{\beta} x_0) \\ &= \mathrm{var}\, Y + \mathrm{var}\, \hat{\alpha} + x_0^2 \, \mathrm{var}\, \hat{\beta} + 2x_0 \, \mathrm{cov}(\hat{\alpha}, \hat{\beta}). \end{aligned}$$

Substituting from (4) of Section 15.3 and using the expression for $\text{cov}(\hat{\alpha}, \hat{\beta})$ from Problem 12, we obtain

$$\text{m.s.p.e.}(Y) = \sigma^2 + \sigma^2\left(\frac{1}{n} + \frac{(\bar{x})^2}{\text{SS}_{xx}}\right) + (x_0^2 - 2x_0\bar{x})\frac{\sigma^2}{\text{SS}_{xx}}.$$

In using $\hat{\alpha} + \hat{\beta}x_0$ to predict $Y(x_0)$, the mean squared prediction error is

$$\text{m.s.p.e.} = \sigma^2\left(1 + \frac{1}{n} + \frac{(x_0 - \bar{x})^2}{\text{SS}_{xx}}\right). \tag{10}$$

We estimate σ^2 [from (9) of Section 15.3] to be

$$S_\varepsilon^2 = \frac{\text{SS}_{yy}(1 - r^2)}{n - 2} = \frac{\text{SS}_{yy} - \hat{\beta}^2\text{SS}_{xx}}{n - 2}. \tag{11}$$

Substituting (11) for σ^2 in (10) yields an approximate m.s.p.e.

Example 15.5c   **Arthropodal Thermometer**

The frequency of chirping of a cricket is thought to be related to temperature. This suggests the possibility that temperature can be estimated (predicted) from the chirp frequency. The following data[4] give frequency–temperature pairs observed for the striped ground cricket:

Chirps/sec. (x)	20	16	20	18	17	16	15	17	15	16
Temperature, °F (y)	89	72	93	84	81	75	70	82	69	83

These data are plotted in Figure 15.7.

To find the correlation and the least-squares line, we need these sample moments:

$$\bar{X} = 17, \qquad \bar{Y} = 79.8, \qquad \text{SS}_{xx} = 30, \qquad \text{SS}_{yy} = 589.6, \qquad \text{SS}_{xy} = 122.$$

Substituting these in (7) and (9) of Section 15.2, we obtain

$$\hat{\beta} = \tfrac{1}{3} \times 12.2, \qquad \hat{\alpha} = 79.8 - \tfrac{1}{3} \times 17 \times 12.2 = \tfrac{1}{3} \times 22.$$

A correlation this high, even with only ten data points, would seem to show that $\rho \neq 0$, and a formal test yields $T = 6.5$ [(11) of Section 15.4] with 8 d.f., so $P < .001$.

The empirical regression line is

$$y = \tfrac{1}{3}(32 + 12.2x).$$

Given a chirp frequency of 19 per second, we'd predict the temperature to be

$$\tfrac{1}{3}(32 + 12.2 \times 19) = 87.9.$$

---

[4] G. W. Pierce, *The Songs of Insects* (Cambridge, Mass: Harvard University Press, 1949), pp. 12–21.

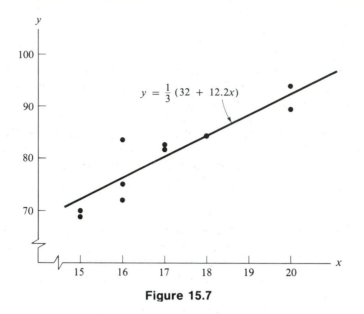

$$y = \tfrac{1}{3}(32 + 12.2x)$$

Figure 15.7

To calculate the m.s.p.e. at $x = 19$, we first find S_ε^2 using SSRes = 94.47 [from (11) of Section 15.2]:

$$S_\varepsilon^2 = \tfrac{1}{8} \times 94.47 = 11.81.$$

From (10) we have

$$\text{m.s.p.e.} \doteq 11.8[1 + 1/10 + (19 - 17)^2/30] \doteq 14.57.$$

The r.m.s. prediction error is the square root, 3.82 degrees.

Had we ignored x and estimated the temperature to be $\bar{Y} = 79.8$ instead of 87.8, our r.m.s.p.e. would be $S_Y = 7.68$. Although using x does not eliminate prediction error, it substantially reduces the typical error. ∎

The term involving $(x_0 - \bar{x})^2$ in (10) shows that the prediction error is large when predicting at an x_0 that is far from the mean of the x's in the data. This is because of the uncertainty in estimating the slope β. Moreover, the true regression function may be nonlinear outside the range of x-values in the data set, so using a linear predictor beyond the range where one has sample information is doubly risky.

15.6   The Regression Effect

Example 15.6a   Training Teachers

A psychology professor, D. Kahneman of the University of British Columbia, related this anecdote.[5] He was teaching a course in the psychology of training to air

---

[5] Quoted in *Discover* (June 1985).

force flight instructors at Hebrew University in the 1960s. He cited studies showing that in teaching, rewards are more effective than punishment. One of his students objected, saying, "I've often praised people warmly for beautifully executed maneuvers, and the next time they almost always do worse. And I've screamed at people for badly executed maneuvers, and by and large the next time they improve. Don't tell me that reward works and punishment doesn't. My experience contradicts it." Other students agreed.

Professor Kahneman said, "I suddenly realized that this was an example of the statistical principle of regression to the mean, and that nobody else had ever seen this before. I think this was one of the most exciting moments of my career." "Once you become sensitized to it," he remembered saying that day, "you see regression everywhere." Elaborating further on this ubiquity, Professor Kahneman pointed out that great movies have disappointing sequels, and disastrous presidents have better successors. ∎

Example 15.6b   **Regression in Baseball**
Astute baseball analysts have noticed the regression effect in American baseball. They call it the *law of competitive balance*. It works like this. Teams or players who do extremely well one year *tend* to do worse in the following year—better than the rest of the league but worse than their own previous high. For example, a "Rookie of the Year" will usually do worse in the second year. (Conversely, teams or players that do very poorly one year tend to do better the next, although usually worse than average.)

The reason for this is really quite simple. To do very well in a particular year requires some combination of two things: skill and luck. The fact that a particular team did well suggests that it had both. In the following year, the skill may still be there (barring extensive changes in personnel), but the luck is apt not to be—at least not in the same degree. So the team will *tend* to do well (the skill part), but not as well (the luck part) as before.

If you want to predict a player's batting average (BA) next year, average his previous year's BA with that year's BA of the entire league. This assumes that skill and luck contribute to BA in equal measure; they don't, but the formula works quite well. In particular, it does much better than using the player's previous year's BA as a predictor. ∎

The nineteenth-century geneticist Francis Galton measured the sizes of the seeds of mother and daughter sweet pea plants. He observed that the sizes of daughter plants seemed to revert (or regress) to the mean. From scatter plots of the heights of fathers and sons, he again noted a regression to average: Sons of tall men tended to be tall, but not as tall as their fathers, while sons of short men tended to be short, but not as short as their fathers.

In Section 15.5, we saw that when the regression function is linear (as it is, in particular, in the bivariate normal case), the equation of the regression line is

$$\frac{y - \mu_y}{\sigma_y} = \rho \frac{x - \mu_x}{\sigma_x}, \qquad \text{or} \qquad z_y = \rho z_x,$$

where z_x and z_y are the standard scores corresponding to x and y. Suppose we observe a value of X, two s.d.'s above average: $Z_x = 2$. The best prediction for the corresponding Y (mean-square sense) is a value of Y *less* than two s.d.'s above the average Y by the factor ρ. Similarly, if we observe an X that is three s.d.'s below average, the best prediction for Y is a value *less* than three s.d.'s below average, by the factor ρ. This is the regression effect. So even though there may be as many tall sons as tall fathers, a father that is very tall will have a son that is on average not quite so tall.

Figure 15.8 shows a contour curve for a bivariate normal distribution with $\rho = \frac{1}{2}$, drawn on the (z_x, z_y) axes. The equation of this curve is

$$z_x^2 - z_x z_y + z_y^2 = c^2,$$

where c^2 is a positive constant [see (7) in Section 6.8, with $2\rho = 1$]. Since a rotation through 45 degrees would result in an equation of the form $u^2 + 3v^2 = $ constant, the major axis of the ellipse is the 45-degree line, $z_y = z_x$. However, the line we use for predicting, the regression line, is $z_y = \frac{1}{2}z_x$ a line that bisects the vertical cords of the ellipse.

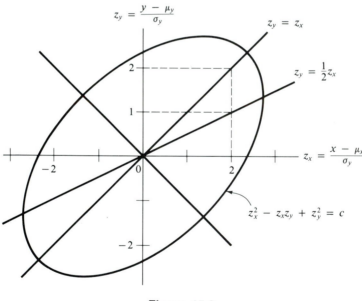

Figure 15.8

When $|\rho| = 1$, of course, there is no regression effect, and when $|\rho|$ is close to 1, the effect is slight.

Example **15.6c**   **Regression in Golf**
First- and second-round scores in the 1988 LPGA championship match in Mason, Ohio, are shown in the scatter diagram of Figure 15.9. The correlation between first-

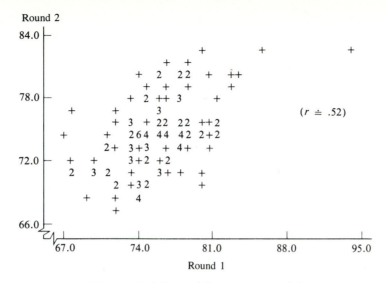

Figure 15.9 Round 2 versus round 1

and second-round scores is about .52. It is evident that among those with very low first-round scores, some got high and some got low second-round scores, the average being even a bit above the overall average. So the second-round score of a player with a low first-round score should increase.

The regression effect is shown somewhat more clearly in the plot of Figure 15.10, in which the *change* in score from first to second round is plotted against

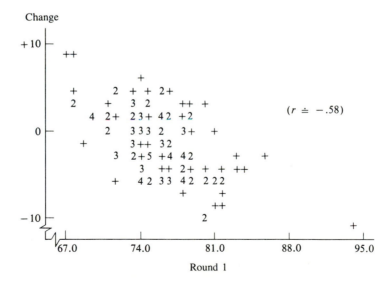

Figure 15.10 Change versus round 1

first-round scores. The correlation here is about $-.58$. Golfers who had higher-than-average scores for the first round tended to go down in the second; those whose first-round scores were lower than average tended to go up in the second round. ■

The regression effect shows up practically everywhere, but in the case of clinical trials, its misinterpretation is particularly unfortunate. A common criterion for entry in a clinical trial is that the patient be "very sick." Thus, in a trial of an antihypertensive drug, only those with "high blood pressure" are included. Blood pressure measurements are quite variable within each individual. Just as in baseball, where a result is a combination of skill and luck, a diagnosis of high blood pressure is a combination of a real propensity to have high blood pressure and a random component. Those selected tend to have a high random component at entry, so their blood pressures tend to decrease—even if the drug has no effect whatever! The regression effect can easily be confused with a treatment effect. One solution is to use separate measurements for admission to the trial and for data analysis. Another is to compare the drug group with a control group only, paying no heed to the apparent effect in the drug group alone.

15.7 Multiple Regression

Example 15.7a Data on water absorption (y), flour protein percentage (x_1), and starch damage (x_2) in wheat flour were obtained in a study reported in *Cereal Chemistry*.[6] They are shown in Table 15.2.

Table 15.2

y	x_1	x_2	y	x_1	x_2	y	x_1	x_2	y	x_1	x_2
30.9	8.5	2	47.6	12.0	32	47.0	12.9	24	48.3	12.1	34
32.7	8.9	3	47.2	12.5	31	46.8	12.0	25	48.6	11.3	35
36.7	10.6	3	44.0	10.9	28	45.9	12.9	28	50.2	11.1	40
41.9	10.2	20	47.7	12.2	36	48.8	13.1	28	49.6	11.5	45
40.9	9.8	22	43.9	11.9	28	46.2	11.4	32	53.2	11.6	50
42.9	10.8	20	46.8	11.3	30	47.8	13.2	28	54.3	11.7	55
46.3	11.6	31	46.2	13.0	27	49.2	11.6	35	55.8	11.7	57

Linear correlations between y and x_1 and between y and x_2 are .69 and .95, respectively. Perhaps it would be useful to use both x_1 and x_2 in predicting y. (To be continued.) ■

---

[6] "An ultracentrifuge flour absorption method," *Cereal Chemistry* (1978), 96–101.

In a simple extension of the linear regression in Section 15.1 to the case of two predictors, we assume a regression function of this form:

$$g(x_1, x_2) = \beta_0 + \beta_1 x_1 + \beta_2 x_2. \tag{1}$$

As in the case of a single predictor variable, β_i is the amount of increase in response caused by a unit increase in x_i.

With subscripts 1 and 2 referring to the predictors, we need a second subscript for observations. Data consist of triples: (x_{1j}, x_{2j}, y_j), $j = 1, \ldots, n$, represented graphically as points in three dimensions. Each triple is referred to in computer packages as a *case*. The regression function (1) is represented graphically as a plane, but the data points are usually not coplanar. The method of least squares provides estimates of the coefficients β_i by finding the plane that best fits the n data points (in the sense of least squares).

The residual of the jth data point about (1) is

$$y_j - \beta_0 - \beta_1 x_{1j} - \beta_2 x_{2j}.$$

To minimize the sum of the squares of the residuals,

$$\sum_j (y_j - \beta_0 - \beta_1 x_{1j} - \beta_2 x_{2j})^2,$$

we differentiate with respect to each β in turn, obtaining three *linear* equations in the three β's:

$$\begin{cases} \sum y_j = n\beta_0 + \left(\sum x_{1j}\right)\beta_1 + \left(\sum x_{2j}\right)\beta_2 \\ \sum x_{1j} y_j = \left(\sum x_{1j}\right)\beta_0 + \left(\sum x_{1j}^2\right)\beta_1 + \left(\sum x_{1j} x_{2j}\right)\beta_2 \\ \sum x_{2j} y_j = \left(\sum x_{2j}\right)\beta_0 + \left(\sum x_{1j} x_{2j}\right)\beta_1 + \left(\sum x_{2j}^2\right)\beta_2. \end{cases}$$

These three equations usually have a unique solution $(\hat{\beta}_0, \hat{\beta}_1, \hat{\beta}_2)$, which must be a minimizing point.

Example 15.7b   **Moisture in Flour (continued)**

Calculation of the necessary sums from the data in Example 15.7a yields the following equations for the β's:

$$\begin{cases} 1287.4 = 28\beta_0 + 322.3\beta_1 + 829.0\beta_2 \\ 14940 = 322.3\beta_0 + 3746.4\beta_1 + 9746.6\beta_2 \\ 40016 = 829\beta_0 + 9746.6\beta_1 + 29327\beta_2 \end{cases}$$

Solving these simultaneously yields $\hat{\beta}_0 = 19.44$, $\hat{\beta}_1 = 1.448$, $\hat{\beta}_2 = .3356$. So the least-squares plane is

$$y = 19.44 + 1.442 x_1 + .3356 x_2. \qquad\blacksquare$$

When there are k predictor variables, there are $k + 1$ β's:

$$g(x_0, x_1, \ldots, x_k) = \beta_0 x_0 + \beta_1 x_1 + \cdots + \beta_k x_k.$$

The minimization process yields $k + 1$ linear equations in the β's. Solution by hand calculation is at best tedious (perhaps prohibitive). Statistical packages for microcomputers will handle the equations easily. (In some applications more than 100 predictors are used, and special software or a larger computer may be required in such a case.)

In the algebra of least squares for multiple regression, it does not usually matter that some predictor variables may depend on others. Thus, one can take certain types of interaction into account by including a predictor variable that is a product of two others. For instance, one might assume a regression function of this form:

$$\beta_0 + \beta_1 x_1 + \beta_2 x_2 + \beta_3 x_1^2 + \beta_4 x_1 x_2.$$

However, if one includes a variable twice (for example, $\beta_1 x_1 + \beta_2 x_1$), it is clear that the β's are not uniquely determined, and the least-squares equations are indeterminate. The computer program will balk—as it will if there are dependencies among the x's that are sufficiently close to linear.

Usefulness of a Predictor

Whenever one includes several terms in a regression function, the obvious question is whether they are all needed. An ANOVA is particularly useful in considering a hierarchy of regression functions. To illustrate, assume that the regression function depends linearly on two predictors:

$$g(x_1, x_2) = \beta_0 + \beta_1 x_1 + \beta_2 x_2. \tag{2}$$

Testing $H_0: \beta_2 = 0$ against $H_A: \beta_2 \neq 0$ is a way of determining whether x_2 makes a useful contribution to the model. This tests whether the full model (2) provides an appreciably better fit than a model with x_1 as the sole predictor variable.

Let $\hat{\beta}_{00}$ and $\hat{\beta}_{10}$ denote the m.l.e.'s of β_0 and β_1 when $\beta_2 = 0$, and $\hat{\beta}_0, \hat{\beta}_1$, and $\hat{\beta}_2$, the m.l.e.'s under $H_0 \cup H_A$. Given n data pairs (x_{1i}, x_{2i}, y_i), the likelihood ratio statistic for testing H_0 versus H_A (Problem 32) is

$$\Lambda = \left(\frac{\hat{\sigma}_1^2}{\hat{\sigma}_2^2} \right)^{-n/2},$$

where

$$n\hat{\sigma}_1^2 = \sum (y_i - \hat{\beta}_{00} - \hat{\beta}_{10} x_{1i})^2 = \text{SS1},$$

and

$$n\hat{\sigma}_2^2 = \sum (y_i - \hat{\beta}_0 - \hat{\beta}_1 x_{1i} - \hat{\beta}_2 x_{2i})^2 = \text{SS2},$$

where SS1 is the residual sum of squares for testing $\beta_2 = 0$, and SS2 is a residual sum of squares we identify as "error" after fitting a quadratic regression function. The inequality $\Lambda < C$ is equivalent to $\text{SS}_1/\text{SS}_2 > C'$. It will be convenient below to notice that this is equivalent to

$$\frac{\text{SS1} - \text{SS2}}{\text{SS2}} > C''.$$

With $\text{SSTot} = \text{SS}_{yy}$ (as in Section 15.2), the ANOVA for testing H_0 is based on this identity:

$$\text{SSTot} = (\text{SSTot} - \text{SS1}) + (\text{SS1} - \text{SS2}) + \text{SS2}. \tag{3}$$

When the terms on the right are divided by σ^2, their distributions are chi-square with d.f.'s 1, 1, and $n - 3$, respectively, and they are independent. (See Lindgren [15], p. 525.) The d.f.'s on the right sum to the d.f. on the left:

$$n - 1 = 1 + 1 + (n - 3).$$

So under H_0, the test ratio

$$F = \frac{(SS1 - SS2)/1}{SS2/(n - 3)} \tag{4}$$

has an F-distribution with $(1, n - 3)$ d.f.

An F-ratio for testing the hypothesis that $\beta_1 = \beta_2 = 0$ is called an *overall F* in many computer packages. It addresses the question of whether the regression function is useful in prediction. Combining the first two terms on the right-hand side of (3), we obtain

Usefulness of the Model

$$SSTot = SSReg + SSRes, \tag{5}$$

where

$$SSReg = SSTot - SS2 \quad \text{and} \quad SSRes = SS2.$$

The test statistic is then

$$\text{Overall } F = \frac{SSReg}{SSRes}, \tag{6}$$

whose null distribution is $F(2, n - 3)$. The ratios

$$R^2 = \frac{SSReg}{SSTot} \quad \text{and} \quad 1 - R^2 = \frac{SSRes}{SSTot} \tag{7}$$

give fractions of the total variation (of the Y's about their mean) attributable, respectively, to regression and to error. The ratio R^2 is called the **coefficient of determination**.

A somewhat more useful measure of the worth of the regression function in predicting Y is

$$\text{Adjusted } R^2 = 1 - \frac{(n - 1)SSRes}{(n - p - 1)SSTot}, \tag{8}$$

where p is the number of predictor variables. Adding a term to the regression function may result in a better fit, but at the cost of a more complicated model and the loss of a degree of freedom from SSRes. Choosing a model with a large adjusted R^2 (rather than simply a large R^2) is apt to give a more useful model.

Example 15.7c   Metabolism Versus Weight

The data in the following table give body weight (x) and metabolic clearance rate per body weight (y) for 14 cattle.

x	y
110	235, 198, 173
230	174, 149, 124
360	115, 130, 102, 95
505	122, 112, 98, 96

Suppose we are interested in the possibility of nonlinearity of the regression function and introduce a quadratic term: $x_1 = x$ and $x_2 = x^2$:

$$g(x_1, x_2) = \beta_0 + \beta_1 x + \beta_2 x^2.$$

The computer printout shows that the least-squares quadratic regression function is

$$275.3 - .748x + .000820x^2.$$

```
UNWEIGHTED  LEAST  SQUARES  LINEAR  REGRESSION  FOR  RATE

PREDICTOR
VARIABLES    COEFFICIENT    STD ERROR    STUDENT'S T      P

CONSTANT       275.3         26.71           10.31      0.0000
WT           -7.481E-01      1.954E-01       -3.83      0.0028
SQWT          8.197E-04      3.070E-04        2.67      0.0218

CASES INCLUDED          14       MISSING CASES  0
DEGREES OF FREEDOM      11
OVERALL F              24.39     P VALUE    0.0000
ADJUSTED R SQUARED    0.7825
R SQUARED             0.8160
MEAN SQUARED ERROR     399.4


STEPWISE  ANALYSIS  OF  VARIANCE  OF  RATE

           INDIVIDUAL CUM CUMULATIVE CUMULATIVE ADJUSTED
SOURCE        SS       DF     SS         MS      R-SQUARED

CONSTANT 2.6414E+05
WT       1.6634E+04   1  1.6634E+04 1.6634E+04    0.6714
SQWT     2847.6       2  1.9482E+04 9740.9        0.7825
RESIDUAL 4393.4      13  2.3875E+04 1836.6

CASES INCLUDED          14       MISSING CASES  0
DEGREES OF FREEDOM      11
OVERALL F              24.39     P VALUE    0.0000
ADJUSTED R SQUARED    0.7825
R SQUARED             0.8160
MEAN SQUARED ERROR     399.4
```

It also gives SS2 = 4393.4 and SS1 − SS2 = 2847.6. With these, we calculate the F-statistic (4) for testing $\beta_2 = 0$ versus $\beta_2 \neq 0$:

$$F = \frac{2847.6/1}{4393.4/11} = 7.13.$$

This is the square of the student's T in Table 15.2. Entering Table IIIa of Appendix 1 with $T = \sqrt{7.13} = 2.67$, 11 d.f., we find the single tail area to be .022, so $P = .044$. (The t-statistic shown opposite "*wt*" in Table 15.2 is for testing $\beta_1 = 0$ when β_2 is *not* assumed to be zero.)

The overall F in Table 15.2 tests the hypothesis that neither predictor variable is useful ($\beta_1 = \beta_2 = 0$): $F = \frac{9740.9}{399.4} \doteq 24.4$, with (2, 11) d.f. ($P < .001$). The printout also gives the value of R^2:

$$R^2 = 1 - \frac{\text{SSRes}}{\text{SSTot}} = 1 - \frac{7241}{23875} \doteq .8160,$$

as well as the adjusted R^2 (.7825, obtained from (8) with $p = 2$). ∎

The mathematics of linear multiple regression models is much simplified with the use of matrix algebra. Because we do not assume linear algebra as a prerequisite, we'll only set up the notation. Suppose the regression function is

$$y = \beta_0 + \beta_1 x_1 + \cdots + \beta_k x_k.$$

Matrix Formulation

Denote by **Y** the vector of responses, by **β** the vector of parameters, and by **ε** the vector of errors:

$$\mathbf{Y} = \begin{pmatrix} Y_1 \\ \vdots \\ Y_n \end{pmatrix}, \qquad \boldsymbol{\beta} = \begin{pmatrix} \beta_0 \\ \vdots \\ \beta_k \end{pmatrix}, \qquad \boldsymbol{\varepsilon} = \begin{pmatrix} \varepsilon_1 \\ \vdots \\ \varepsilon_n \end{pmatrix}.$$

The *design matrix* **X** is the following $n \times (k + 1)$ matrix of predictor variables:

$$\mathbf{X} = \begin{pmatrix} 1 & x_{11} & \cdots & x_{k1} \\ 1 & x_{12} & \cdots & x_{21} \\ \vdots & \vdots & & \vdots \\ 1 & x_{1n} & \cdots & x_{kn} \end{pmatrix}.$$

Together, **X** and **Y** constitute the data. The model can now be written

$$\mathbf{Y} = \mathbf{X}\boldsymbol{\beta} + \boldsymbol{\varepsilon}.$$

The least-squares criterion is to choose **β** to minimize the quadratic form

$$(\mathbf{Y} - \mathbf{X}\boldsymbol{\beta})'(\mathbf{Y} - \mathbf{X}\boldsymbol{\beta}).$$

The solution can be shown to be

$$\hat{\boldsymbol{\beta}} = (\mathbf{X}'\mathbf{X})^{-1}\mathbf{X}'\mathbf{Y}.$$

Those interested in pursuing this approach are referred to Draper and Smith [6] or Weisberg [19].

Chapter Perspective

The regression models we've assumed so far are sometimes overly restrictive. For example, it may be necessary to allow for the error variance to depend on x. Also, errors may be correlated, and the assumption of *normally* distributed random errors may not be reasonable. When normality is questionable, inferences based on t- and F-distributions are also questionable.

Many regression models fall in the category of *linear models*. The term *linear* refers not to the dependence of the response on the predictor variable, but rather to its dependence on the model parameters. With these models, the equations resulting from the minimizing squared residuals are linear and can be solved by methods of linear algebra to yield parameter estimates.

When the appropriate regression function g involves parameters in a nonlinear way, transformations of response and/or predictor variables may produce a linear model. Transformations are also sometimes used in attempting to make the error component closer to normal.

Solved Problems

Sections 15.1–15.2

A.   For the data of Example 15.1a, find
  (a)   the correlation coefficient.
  (b)   the least-squares regression line.

Solution:

The data are shown in the accompanying table, along with columns for finding needed sums.

x	y	x^2	xy	y^2
2	89	4	178	7,921
2.5	97	6.25	242.5	9,409
2.5	91	6.25	227.5	8,281
2.75	98	7.5625	269.5	9,604
3	100	9	300	10,000
3	104	9	312	10,816
3	97	9	291	9,409
18.75	676	51.0625	1820.5	65,440

(a)   To find r, we use (3) of Section 7.6:

$$r = \frac{1{,}820.5 - \dfrac{18.75 \times 676}{7}}{\sqrt{51.0625 - \dfrac{(18.75)^2}{7}}\sqrt{65{,}440 - \dfrac{676^2}{7}}} = \frac{9.7857}{\sqrt{.8393} \times 157.714} \doteq .85.$$

(b) The slope is $\hat{\beta} = 9.7857/.8393 = 11.66$. The intercept is $\hat{\alpha} = \bar{Y} - \hat{\beta}\bar{x} = 96.57 - 11.66 \times 2.679 = 65.34$. The least-squares line is $y = 65.34 + 11.66x$.

B. Let $\tilde{Y}_i = \hat{\alpha} + \hat{\beta}x_i$, the *fitted value* at x_i. Show that in the n pairs $(Y_i, \tilde{Y}_i)$, the correlation coefficient is $|r|$, where r is the correlation coefficient of the (x_i, Y_i).

Solution:

The average of the fitted values $\tilde{Y}_i$ is $\bar{Y}$:

$$\frac{1}{n}\sum \tilde{Y}_i = \frac{1}{n}\sum(\hat{\alpha} + \hat{\beta}x_i) = \alpha + \hat{\beta}\bar{x} = \bar{Y}.$$

Also, $\tilde{Y}_i - \bar{Y} = \hat{\beta}(x_i - \bar{x})$. Hence,

$$SS_{\tilde{y}\tilde{y}} = \sum(\tilde{Y}_i - \bar{Y})^2 = \sum \hat{\beta}(x_i - \bar{x})^2 = \hat{\beta}^2 SS_{xx},$$

and

$$SS_{y\tilde{y}} = \sum Y_i(\tilde{Y}_i - \bar{Y}) = \sum Y_i\hat{\beta}(x_i - \bar{x}) = \hat{\beta}SS_{xy}.$$

So

$$r_{y\tilde{y}} = \frac{SS_{y\tilde{y}}}{\sqrt{SS_{yy}SS_{\tilde{y}\tilde{y}}}} = \frac{\hat{\beta}SS_{xy}}{|\hat{\beta}|\sqrt{SS_{yy}SS_{xx}}} = |r|.$$

C. Use the method of least squares to fit a quadratic function through the origin to these data points: $(1, 0), (2, 2), (3, 2)$.

Solution:

For the quadratic function $\alpha x + \beta x^2$, the sum of squared residuals is

$$\sum(Y_i - \alpha x - \beta x^2)^2 = (\alpha + \beta)^2 + (2 - 2\alpha - 4\beta)^2 + (2 - 3\alpha - 9\beta)^2.$$

Differentiate with respect to α and to β and set the results equal to 0:

$$2(\alpha + \beta) + 2(2 - 2\alpha - 4\beta)\cdot(-2) + 2(2 - 3\alpha - 9\beta)\cdot(-3) = 0.$$
$$2(\alpha - \beta) + 2(2 - 2\alpha - 4\beta)(-4) + 2(2 - 3\alpha - 9\beta)\cdot(-9) = 0.$$

Collecting terms, we have

$$\begin{cases} 14\alpha + 36\beta = 10 \\ 36\alpha + 98\beta = 26. \end{cases}$$

Solving simultaneously, we find $\alpha = \frac{11}{19}$, $\beta = \frac{1}{19}$. So the least-squares regression function is

$$y = \frac{11x + x^2}{19}.$$

Section 15.3

D. In Section 15.3, we say that the cross-product terms in (8) vanish. Show that they do.

Solution:

We need to show

$$\sum(Y_i - \hat{\alpha} - \hat{\beta}x_i) = 0 \qquad \text{and} \qquad \sum(x_i - \bar{x})(Y_i - \hat{\alpha} - \hat{\beta}x_i) = 0.$$

With $\hat{\alpha}$ substituted from (7) of Section 15.2, the first becomes

$$\sum Y_i - n(\bar{Y} - \hat{\beta}\bar{x}) - \hat{\beta}\sum x_i = n\bar{Y} - n\bar{Y} + n\hat{\beta}\bar{x} - n\hat{\beta}\bar{x} = 0.$$

Substituting for $\hat{\alpha}$ and $\hat{\beta}$ in the second yields

$$\sum(x_i - \bar{x})Y_i - \bar{Y}\sum(x_i - \bar{x}) - \hat{\beta}\sum(x_i - \bar{x})^2 = SS_{xy} + 0 - \hat{\beta}\cdot SS_{xx}.$$

Since $\hat{\beta} = SS_{xy}/SS_{xx}$, the first term cancels the last. (In Problem 5, you'll see that when you use calculus to minimize the mean squared residual, the necessary conditions for a minimum are precisely the equalities we've just established.)

Section 15.4

E.   Show (7) of Section 15.4:

$$n\hat{\sigma}_0^2 = SSRes + (\hat{\beta} - \beta_0)^2 SS_{xx},$$

where

$$n\hat{\sigma}_0^2 = \sum[Y_i - \hat{\alpha}_0 - \beta_0 x_i]^2 \quad \text{and} \quad \hat{\alpha}_0 = \bar{Y} - \beta_0\bar{x}.$$

Solution:

Write $Y_i - \hat{\alpha}_0 - \beta_0 x_i = [Y_i - \bar{Y} - \hat{\beta}(x_i - \bar{x})] + (\hat{\beta} - \beta_0)(x_i - \bar{x}).$

Then

$$n\hat{\sigma}_0^2 = \sum[Y_i - \hat{\alpha}_0 - \beta_0 x_i]^2 = \sum[Y_i - \hat{\alpha} - \hat{\beta}x_i]^2 + (\hat{\beta} - \beta_0)^2 SS_{xx},$$

where the cross-product has dropped out (in view of Problem D). The first term on the right is SSRes.

F.   Problem 13 deals with the estimate $\tilde{Y} = \hat{\alpha} + \hat{\beta}x_0$ for the value of the regression function at a point x_0. The mean and variance of $\tilde{Y}$ (from Problem 13) are

$$E(\tilde{Y}) = \alpha + \beta x_0, \qquad \text{var } \tilde{Y} = \sigma^2\left[\frac{1}{n} + \frac{(x_0 - \bar{x})^2}{SS_{xx}}\right].$$

Obtain confidence limits for $\alpha + \beta x_0$, assuming the normal model of Section 15.1.

Solution:

Since $\hat{Y}$ is normal, the square of the standardized $\hat{Y}$,

$$\frac{[\tilde{Y} - (\alpha + \beta x_0)]^2}{\sigma^2\left[\dfrac{1}{n} + \dfrac{(x_0 - \bar{x})^2}{SS_{xx}}\right]},$$

is $\chi^2(1)$. From Section 15.3, we know that

$$\frac{(n-2)\hat{\sigma}^2}{\sigma^2} \sim \chi^2(n-2)$$

and is independent of $\tilde{Y}$. Then

$$T = \frac{\tilde{Y} - (\alpha + \beta x_0)}{\hat{\sigma}\sqrt{\left[\dfrac{1}{n} + \dfrac{(x_0 - \bar{x})^2}{SS_{xx}}\right]}} = \frac{\tilde{Y} - (\alpha + \beta x_0)}{\text{s.e.}(\tilde{Y})}$$

has a t-distribution with $n - 2$ d.f. The desired confidence limits are $\tilde{Y} \pm k \cdot \text{s.e.}(\tilde{Y})$, where k is determined from Table III ($n - 2$ d.f.).

Sections 15.5–15.6

G. It has been found that the correlation between first and second exam scores in a certain course is .90. The mean scores are 65, and the s.d.'s are both 12. Predict the second exam score of a student whose first exam score is

(a) 45 (b) 95 (c) unknown,

and give the r.m.s. prediction error in each case.

Solution:
We use the linear predictor (1) of Section 15.5 to obtain

(a) $65 + .9(45 - 65) = 47,$ (b) $65 + .9(95 - 65) = 92.$

For these, the r.m.s.p.e. is $12\sqrt{1 - .81} = 5.23$.

(c) With first exam score unknown, we predict the mean, 65, with r.m.s. error equal to the s.d., 12.

H. Given the data in Example 15.1a, repeated in Problem A above, give a predicted value of Y for $x = 3.5$, together with an estimate of r.m.s. prediction error and a 95% prediction interval.

Solution:
Substituting $x = 3.5$ in the least-squares function found in Problem A, we predict $y = 65.34 + 11.66 \times 3.5 = 106.15$. The estimate of σ_e given by (11) of Section 15.5 is

$$S_e = \sqrt{\frac{157.714(1 - .72343)}{5}} = 2.954,$$

and the r.m.s. prediction error is obtained using (10) of Section 15.5:

$$2.954 \sqrt{1 + \frac{1}{7} + \frac{(3.5 - 2.679)^2}{.8393}} \doteq 4.12.$$

For a 95% prediction interval, we use the multiplier $t_{.975}(5) = 2.57$, to obtain the limits 106.15 ± 10.59.

Section 15.7

I. Eye weight in grams and corneal thickness in micrometers were recorded for nine randomly selected calves:[7]

Calf	1	2	3	4	5	6	7	8	9
Weight (x)	.2	1.4	2.2	2.7	4.9	5.3	8.0	8.8	9.6
Thickness (y)	416	673	733	801	967	1036	883	736	567

Test the hypothesis that the regression function for y on x is quadratic against the hypothesis that it is linear.

[7] Reproduced by Devore and Peck in *Statistics* (St. Paul, Minn.: West, 1986) in a problem calling for a test of $\rho = 0$. The data are taken from "The collagens of the developing bovine cornea," *Exper. Eye Res.* (1984), 639–652.

Solution:

Assuming the model $y = \alpha + \beta_1 x + \beta_2 x^2$, we test the hypothesis $H_0: \beta_2 = 0$ against $H_A: \beta_2 \neq 0$. We did this using a microcomputer statistics package, which printed these results:

Predictor	Coefficient	s.e.	t	P
Const.	354.4	34.14	10.38	.000
x	237.2	17.29	13.72	.000
x^2	-22.05	1.668	-13.22	.000

So the regression coefficients are $\hat{\alpha} = 354.4$, $\hat{\beta}_1 = 237.2$, and $\hat{\beta}_2 = -22.05$. The ANOVA table is as follows:

Source	SS	df	MS
Weight	20,688	1	
Sq. weight	265,720	1	265,720
Residual	9,122	6	1,520
Total:	295,530	8	

The F-statistic for testing $\beta_2 = 0$ is $F = \frac{265,720}{1,520} = 174.8$, which is the square of the t shown in the above table for the coefficient of x^2. The evidence against $\beta_2 = 0$ is strong.

Problems

For problems marked with an asterisk, answers are provided in Appendix 2.

Sections 15.1–15.2

*1. Given the data points

   (0, 1), (1, 0), (1, 1), (0, 0), (2, 2),

   (a) determine the least-squares line.
   (b) find the sum of squared residuals about the least-squares line, (i) directly and (ii) using (11) of Section 15.2.

*2. Decrease in surface tension of liquid copper (y) appears to be a linear function of the logarithm of the percentage of sulfur (x). Given these data,[8] find the least-squares line (y on x) and the correlation coefficient:

Log percent sulfur	-3.38	-2.38	-1.20	$-.92$	$-.49$	$-.19$
Decrease (deg/cm)	308	426	590	624	649	727

---

[8] Modified from data of Baer and Kellogg, *Journal of Metals 5* (1953), 634–648.

3. The following data[9] are age in years (x) and annual trunk diameter growth increment in mm (y), for a sample of trees:

(17, 2.20)   (30, .85)   (54, .50)   (50, .75)   (93, .70)

(55, 1.00)   (40, 1.70)   (23, 1.25)   (46.5, .75).

Make a scatter plot and find
(a) the least-squares line, graphing it on your scatter plot.
(b) the correlation coefficient.
(c) the residual sum of squares.

*__4.__ In an investigation[10] of the damage to residential structures caused by blasting, the following data on frequency $(x,$ in cps$)$ and particle displacement $(y,$ in inches$)$ were obtained for 13 blasts.

x	y
2.5	.390, .250, .200
3.0	.360
3.5	.250
9.3	.077
9.9	.140
11.0	.180, .093, .080
16.0	.052
25.0	.051, .150

Find the least-squares regression line of Y on $\log x$.

5. Use calculus to minimize (2) of Section 15.2 to find the least-squares line $y = \hat{\alpha} + \hat{\beta}x$. (Differentiate with respect to a, holding b fixed, and with respect to b, holding a fixed. Then set these derivatives equal to 0 and solve simultaneously. The resulting values of a and b are the estimates $\hat{\alpha}$ and $\hat{\beta}x$, respectively.)

*__6.__ Apply the method of least squares, following the pattern of Example 15.2a, to find the *parabola* $y = a + bx + cx^2$ that best fits these four points: (0,4), (1,0), (2,0), (3,3).

7. Suppose there are m observations at each x_i, $i = 1, \ldots, k$.
(a) Show that the least-squares line for the mk data points can be found by applying (11) of Section 15.2 to the k pairs $(x_i, \bar{y}_i)$.
(b) Explain why the method in (a) does not work when the numbers of observations at the various x_i are not all equal.

8. Show that the sum of the residuals about the least-squares regression line is zero.

---

[9] "Pedunculate oak woodland severe environment," *J. Ecology* (1978), 707–740.
[10] H. Nicholls, C. Johnson, and W. Duvall, "Blasting vibrations and their effects on structures," U.S. Dept. of the Interior, Bureau of Mines *Bulletin 656* (1971), 17, Fig. 3.3.

Section 15.3

9. Show that the y-intercept ($\hat{\alpha}$) of the least-squares line $y = \hat{\alpha} + \hat{\beta}x$ [see (9) of Section 15.2] is an unbiased estimate of α in the regression function $\alpha + \beta x$, with the assumptions at (1) of Section 15.1.

*10. For estimating the intercept α as $\hat{\alpha}$ [see (9) of Section 15.2],
   (a) obtain a formula for the standard error [see (4) of Section 15.3].
   (b) give the form of a confidence interval for α.

11. Assuming the regression model defined by (1) and (2) of Section 15.1, show that $\bar{Y}$ and $\hat{\beta}$ [in (9) of Section 15.2] are uncorrelated.

12. Use the result of Problem 11 to show that

$$\operatorname{cov}(\hat{\alpha}, \hat{\beta}) = -\frac{\sigma^2 \bar{x}}{SS_{xx}}.$$

13. Suppose we use the least-squares line based on n data points [(10) of Section 15.2] to predict a new value of Y when $x = x_0$ as $\tilde{Y} = \hat{\alpha} + \hat{\beta}x_0$.
   (a) Show that $\tilde{Y}$ is an unbiased estimate of $\alpha + \beta x_0$.
   (b) Show $\operatorname{var}(\tilde{Y}) = \sigma^2 \left[\dfrac{1}{n} + \dfrac{(x_0 - \bar{x})^2}{SS_{xx}} \right].$

Section 15.4

*14. Find a P-value for testing the hypothesis $\beta = 0$ against $\beta \neq 0$ with the data in Problem 3 above.

15. Give 90% confidence limits for the slope of the regression line in Problem 4 above.

*16. For the data in Problem 3 and $\tilde{Y}$ as defined in Problem 13 with $x_0 = 60$,
   (a) construct the ANOVA table for Problem 3.
   (b) find the standard error for $\tilde{Y}$ as an estimate of $\alpha + 60\beta$.
   (c) obtain 90% confidence limits for $\alpha + 60\beta$. [See Solved Problem F.]
   (d) for a new observation at $x = 60$, find 90% prediction limits when we use $\tilde{Y}$ as the predicted value.

17. In the development of the likelihood ratio test for $H_0: \beta = \beta_0$ in Section 15.4, verify the calculation of max L under H_0 [(6) of Section 15.4].

*18. Consider the regression model (1) of Section 15.1 with $g(x) = \beta x$. (This is appropriate if the response for $x = 0$ must be $Y = 0$.)
   (a) Determine the least-squares line.
   For (b) and (d), assume a normally distributed error component.
   (b) Find the distribution of the m.l.e. of β.
   (c) Show: $\sum (Y_i - \beta_0 x)^2 = \sum (Y_i - \hat{\beta}X_i)^2 + (\hat{\beta} - \beta_0)^2 \sum x_i^2$.
   (d) Find the likelihood ratio test for $\beta = \beta_0$ versus $\beta \neq \beta_0$, using (c) to cast it in terms of the difference $\hat{\beta} - \beta_0$.

19. Thinking that students who sit toward the front of a class tend to do better on exams than those who sit toward the back, we collected the following data: y is an exam score (50-point scale), and x is the number of the row in which the student usually sat (1 = front row).

x	y	x	y	x	y	x	y	x	y	x	y
7	41	9	36	7	41	7	38	10	24	7	37
4	42	1	35	6	22	3	41	11	13	8	34
9	38	10	28	7	43	11	40	9	15	7	40
8	35	8	40	4	11	6	37	10	30	9	40
3	14	8	31	2	19	3	41	5	17	2	35
10	29	2	23	2	30	9	42	1	41	7	42
5	38	9	20	10	25	9	44	10	31	4	20
10	28	5	46	10	29	5	42	6	24	2	39
8	27	2	46	9	44	5	28	8	38	8	40
6	19	4	32	9	32	9	19	6	29	11	37
2	45	10	25	5	43	3	46	3	24		

(a) Find the sample correlation coefficient.

(b) Assuming normality of Y given X, test $H_0 : \rho = 0$.

Sections 15.5–15.6

*20. The director of graduate studies in a statistics department finds that for 25 entering students with GRE scores, the correlation with end-of-first-year GPA's is .75. For GPA, the mean is 3.4 and the s.d. is 0.22. For GRE, the mean is 700 and the s.d. is 30.

(a) Find the best fitting line (in the sense of least squares) for predicting GPA from GRE.

(b) What GPA would be predicting for a student with $GRE = 780$?

(c) What is the r.m.s. prediction error?

(d) How does the Z-score for the predicted GPA in (b) compare with the Z-score for $X = 780$?

*21. Consider the data in Table 15.3.[11] The researchers suggest that a linear relation between log of runoff volume and log of peak discharge is best for predicting peak discharge.

(a) Use statistical computer software to find the correlations between peak discharge and each of the other four variables. (You won't need to enter the last two columns, since you can obtain them by transforming other columns.)

(b) Obtain a least-squares line for peak discharge on runoff volume.

(c) Use the line in (b) to predict the peak discharge for a runoff volume of 1,000 and give an estimate of prediction error.

(d) Obtain a least-squares line for log peak discharge on log of runoff volume.

(e) Use the line in (d) to predict the peak discharge for a runoff volume of 1,000 and give an estimate of prediction error.

22. Given the data of Problem 4, predict the particle displacement (Y) for a frequency of 6.0, and give the r.m.s. prediction error.

23. Verify that (4) and (5) in Section 15.5 are equal.

---

[11] V. P. Singh, and H. Ainian, "An empirical relation between volume and peak of direct runoff," *Water Resources Bulletin 22* (1986), 725–730.

Table 15.3

Rain volume (cm)	Runoff volume (cm)	Peak discharge (cm/hr)	log of runoff volume	log of peak discharge
4,042	3,341	2.0100	8.1140	0.6981
3,480	1,102	0.3710	7.0049	−0.9916
5,544	1,832	0.7980	7.5132	−0.2256
4,677	3,531	1.5880	8.1693	0.4625
3,503	2,259	1.3000	7.7227	0.2624
4,064	430	0.0150	6.0638	−4.1997
8,890	2,441	0.0380	7.8002	−3.2702
3,032	866	0.0160	6.7639	−4.1352
7,874	106	0.0020	4.6634	−6.2146
2,794	287	0.0097	5.6595	−4.6356
2,292	3,426	0.9270	8.1391	−0.0758
2,652	1,410	0.4320	7.2513	−0.8393
7,057	3,274	1.3700	8.0938	0.3148
6,867	1,826	0.8620	7.5099	−0.1485
4,370	1,548	0.5870	7.3447	−0.5327

24. Batting averages (BA's) during the first half and during the second half of 1984 for most of the regulars in the American League East are summarized as follows.

	Mean	s.d.	
1st half	.2714	.03403	$n = 66$
2nd half	.2702	.03944	$r = .4719$

(a) Find the least-squares line in a regression of second-half averages on first-half averages.

(b) Find SSRes.

(c) Find S_ϵ, the estimate of σ defined by (9) of Section 15.3.

(d) Use the line in (a) to predict the second-half BA for a batter whose first-half BA is (i) .300, (ii) .200.

(e) Determine the (approximate) r.m.s. prediction errors for your predictions in (d).

∗25. Given the joint p.d.f. $f(x, y) = 24xy$ in the triangle bounded by $x = 0$, $y = 0$, and $x + y = 1$, find

(a) the regression function of Y on X, $E(Y \mid x)$.

(b) the best *linear* predictor of Y given $X = x$.

26. Given the joint p.d.f. $f(x, y) = e^{-y}$ for $0 < x < y$, find

(a) the regression function of Y on X, $E(Y \mid x)$.

(b) the best *linear* predictor of Y given $X = x$.

27. Show the asserted equality of (4) and (5) in Section 15.5 (page 647).

28. Let (X, Y) have a uniform distribution in the triangle with vertices $(0, 0)$, $(0, 1)$, $(1, 1)$.
    (a) Find the least-squares regression line of Y on X.
    (b) Find the upper quartiles of X and Y. Given an X at its upper quartile, would you predict Y to be equal to, larger than, or smaller than the upper quartile of Y?

*29. In a large data set, we find $r = .7$ and use the least-squares line to predict the Y-value for another individual whose X-score is $\bar{x} - 2S_x$. Will the predicted value be equal to, greater than, or less than $\bar{y} - 2S_y$? Why?

30. Find a data set that you think may show the regression effect. Sports data are particularly accessible. Make a scatter plot of (x, y) and a second scatter plot of $(x, y - x)$. Comment on the strength of the effect for your data set—no calculations are necessary.

Section 15.7

*31. The following data[12] are power plant cost (C), date of construction permit (D), and power plant net capacity (S) for nine light water reactor power plants in the United States, outside the northeast region.

Plant	C	D	S
1	452.99	67.33	1065
2	443.22	69.33	1065
3	412.18	68.42	530
4	289.66	68.42	530
5	567.79	68.75	913
6	621.45	69.67	786
7	473.64	70.42	538
8	697.14	71.80	1130
9	288.48	67.17	821

(a) Obtain the least-squares linear regression function of C on D and S.
(b) Calculate the overall F-ratio and corresponding P-value.
(c) Find R^2.
(d) Calculate the F-ratio for testing the hypothesis that the coefficient of S is zero.

32. In studies of the reaction rate of a synthetase of a bovine lens, it has appeared that the reciprocal of this rate (Y) is linearly related to the reciprocal of the substrate concentration (x). The following data were obtained.[13]

---

[12] From W. E. Mooz, "Cost analysis of light water reactor power plants," Report R-2304-DOE, Rand Corporation, Santa Monica, Calif.
[13] Private communication.

x	Y
24	.429, .444
20	.293, .293
16	.251, .268
12	.207, .216
8	.239, .218
6	.180, .199
2	.156, .167

   (a) Testing the hypothesis that the coefficient of x^2 in a quadratic regression function is 0.

   (b) Assuming linearity (and combining SS's for a new SSE), obtain estimates of α and β, together with standard errors.

33. Derive the likelihood ratio statistic for $\beta_2 = 0$ in the regression function (1) of Section 15.7.

*34. Using the data of Problem 4,

   (a) fit a quadratic regression function: $y = \alpha + \beta x + \gamma x^2$.

   (b) test the hypothesis $\gamma = 0$.

CHAPTER 16

Bayesian Methods and Making Decisions

So far in this text, we have dealt with probabilities in the *data space*. For example, a P-value is the probability under H_0 of data at least as extreme as that observed. Many researchers and others who use statistical reasoning incorrectly attribute P-values to sets in the *hypothesis space*. The desire to have probabilities of hypotheses seems natural; researchers want answers to such questions as "How likely is the null hypothesis in view of these data?"

Representing uncertainties about hypotheses and parameters as probabilities is fundamental to the **Bayesian** approach to inference. Given such representations, we can calculate the probability that a parameter lies in any given interval, or the probability of a hypothesis about a parameter, using the current distribution of the parameter. (Neither of these has meaning in the classical statistical approach of earlier chapters.) Faced with an inference problem or a decision problem, the statistician can make an inference or a decision based on the current probability distribution of any unknowns and an assessment of the possible penalties for wrong decisions.

Analyzing data yields information that is apt to change the description of one's uncertainty. Bayes' theorem is the fundamental tool for a learning process that involves changing one's probabilities *prior* to gathering the data to those *posterior* to the data.

The connection (the only mathematical connection) between probabilities in the data space given a hypothesis and probabilities in the hypothesis space given the data is Bayes' theorem. Referring to (1) of Section 2.4, we substitute hypothesis H for E and the observed data for F, and write Bayes' theorem as

Bayes'
Theorem

$$P(H \mid \text{data}) = \frac{P(\text{data} \mid H)P(H)}{P(\text{data})}. \tag{1}$$

We frequently write such an expression in this form:

$$P(H \mid \text{data}) \propto P(\text{data} \mid H) \cdot P(H), \tag{2}$$

because $P(\text{data})$ is unconditional and thus does not depend on H.

The probability of H given the data, $P(H\mid \text{data})$, is called the **posterior probability** of H—posterior to the data. The unconditional probability of H, $P(H)$, is the **prior probability** of H. The probability of the data given H, $P(\text{data}\mid H)$, is the **likelihood** of H as defined in Section 8.2. We remember (2) in words as "Posterior is proportional to likelihood times prior."

An alternative way of expressing Bayes's theorem (1) is in terms of odds. For any hypothesis H,

$$\frac{P(H\mid \text{data})}{P(H^c\mid \text{data})} = \frac{P(\text{data}\mid H)}{P(\text{data}\mid H^c)} \cdot \frac{P(H)}{P(H^c)}. \tag{3}$$

Or, in words: The posterior odds ratio equals the likelihood ratio times the prior odds ratio.

Notice in (1) that the probability of H given the data depends on probability in the data space only through the likelihood function, which in turn depends only on the observed data. In particular, the probability in a tail of the null distribution (as in calculating a P-value) is irrelevant.

Using (1) to find a posterior probability requires having a prior probability $P(H)$. Where does $P(H)$ come from?

To take a concrete example, suppose H is the hypothesis that a particular drug has no effect in treating AIDS. Knowing the chemical structure and being familiar with similar chemicals and their biological effect, one scientist might say $P(H) = .3$, another might say $P(H) = .02$, and still a third might say $P(H) = 0$. Who's right? Some statisticians say none of them is right; others say that they're all right! The former feel that $P(H)$ is something that cannot be known by anyone; the latter consider probabilities as subjective, depending on the experience and information possessed by the person making the assessment. This second view is called *Bayesian* because Bayes's theorem is the tool used to update probabilities. The first view disallows using Bayes's theorem because probabilities in the hypothesis space are not even defined; it requires another approach, such as the one described in the preceding chapters of this text.

The Bayesian approach has much to recommend it. Direct probability statements about hypotheses are permitted, so it is ideal for weighing pros and cons when making decisions. Bayesian inference may well be the wave of the future in statistics generally, but it is already important in fields as diverse as industrial decision making and medical diagnosis.

16.1   Assessing Prior Probabilities

A probability distribution for an unknown parameter is intended to embody one's uncertainty, based on beliefs, convictions, experience, hunches, and past data. It is *prior* (to the data) if it describes this uncertainty as it exists prior to collecting the data. It is *posterior* if it describes uncertainty after the data are obtained. In any case, it is subjective or personal because it involves beliefs and opinions that are peculiar to an individual.

Example **16.1a** **Disputed Paternity**

In many countries and in many of the United States, a critical issue in cases of disputed paternity is the probability that the alleged father is the actual father, calculated on the basis of genetic data. Typically, samples of blood are taken from the mother, the child, and the alleged father. These samples are analyzed to determine various genetic traits, including blood group. An expert on blood testing testifies in court as to this information and converts it into a "probability of paternity." For this, one needs prior probabilities or prior odds.

Denote by H the hypothesis that the alleged father is actually the father. Blood banks always use $P(H) = \frac{1}{2}$. The corresponding prior odds ratio is 1, and [according to (3) above] the posterior odds ratio is then equal to the likelihood ratio. Suppose the likelihood ratio is 5:1. The probability of paternity is then $P(H \mid \text{data}) = \frac{5}{6}$. The report given to the court would say that the alleged father has an 83% chance of being the father.

How reasonable is $P(H) = \frac{1}{2}$ in a paternity case? Suppose the mother names three men as possible fathers. According to standard blood bank procedures, each would have probability $\frac{1}{2}$. This makes the total probability greater than one—$\frac{3}{2}$ for these three men plus whatever probability is associated with the actual father's being none of these three. The postulates of probability are clearly violated.

Is $P(H) = \frac{1}{2}$ ever reasonable? Suppose that two men (including the alleged father) had intercourse with the mother during the time she could have become pregnant, both the same number of times, and that the two men were equally fertile. Then, prior to blood typing, $P(H) = \frac{1}{2}$ is appropriate. However, these conditions are unlikely to hold in any particular case. Assuming $P(H) = \frac{1}{2}$ in every case, regardless of testimony given, is not appropriate.

In assuming $P(H) = \frac{1}{2}$, a blood bank takes none of the other evidence in a case into account. To make an informed decision, a juror in a paternity case must listen to the other testimony: What do the witnesses say about the possible fathers, about the frequency and timing of intercourse, and so on. The juror must also take into consideration the possibility of lying by the witnesses (even subconscious lying)— testimony from the two sides is almost always contradictory. Weighing the testimony, deciding whom to believe and how much, the juror forms an opinion about H which is the basis for the juror's $P(H)$. This probability may change in the course of subsequent deliberations, but at any stage it embodies just the personal view of that juror. (To be continued.) ∎

Consider a situation in which you are to give a prior probability for some hypothesis H. We ask you what $P(H)$ is, and you say you don't know. Not accepting this answer, we proceed as follows to elicit *your* $P(H)$. All we require is that you value money and that you agree that there is a fair coin—one that results in heads with probability $\frac{1}{2}$.

We offer you this proposition: You may choose to receive \$1 if H is true or to receive \$1 if the coin comes up heads. If you prefer the former, then $P(H) > \frac{1}{2}$; if the latter, then $P(H) < \frac{1}{2}$; and if you are indifferent, then $P(H) = \frac{1}{2}$. If $P(H) > \frac{1}{2}$, we offer you a second proposition (not forgetting the first, which we will honor in any case): You may choose to receive another \$1 if H is true or to receive \$1 if a second (independent) coin toss also comes up heads. If you prefer the former, then $P(H) > \frac{3}{4}$; if the latter, then $P(H) < \frac{3}{4}$; and if you are indifferent, then $P(H) = \frac{3}{4}$. [It should be clear how the second proposition must be modified in case the first proposition shows $P(H) < \frac{1}{2}$.]

We proceed along these lines, offering you similar propositions, always splitting the interval of uncertainty in two. In this way, we can elicit your $P(H)$ to any degree of accuracy. For example, it will cost us a maximum of $10 to identify an interval of width $\frac{1}{1024}$ containing your $P(H)$.

In dealing with hypotheses about a parameter θ, we summarize a person's beliefs, opinions, past experience, etc., in a probability distribution for θ. Thus, mathematically, θ is a random variable.[1] In many cases, the range of θ-values is an interval, and the appropriate distribution is continuous. We can elicit the probability of any given set of parameter values just as we did for a hypothesis H. Or we can use the same technique to find any specified percentile of your distribution. A judicious selection of percentiles reveals the shape of your distribution for θ, at least approximately.

Example 16.1b   Suppose we want your distribution for N, the population of Wales in millions. We offer a choice between $1 if a coin falls heads and $1 if $N < 1$. Suppose you choose the latter; then your median value is at least 1. Next we offer $1 for heads or $1 if $N < 5$; if you take the coin toss, we know that the median is less than 5. Proceeding in this way, we can home in on your median; suppose it is 3.

To determine your first quartile or 25th percentile, we repeat the process but replacing the coin by a random selection from a bowl with four chips, one red and three blue. The first choice is between $1 if the chip selected is red and $1 if, say, $N < 2$; and so on. For other percentiles, we change the contents of the bowl accordingly. ■

Our discussion has taken for granted an issue of some practical importance. Suppose H is a hypothesis that no one will ever know to be true or false. (For example, H may be that of all brontosauruses that ever lived, more than half were female.) In such a case, you'd have to be willing to take part in the elicitation we've just described, imagining that there will be payoffs even though you know there will be none. If you are not willing to do this, we cannot discover your $P(H)$. Another issue of some importance is that one dollar may not be enough for you to take the experiment seriously. In this case, we'd have to increase the payoff.

---

[1] We are abandoning the convention of using uppercase letters for random variables and corresponding lowercase letters for their particular values. Consistency is too awkward—capital μ, for instance, looks like capital m. Following a rather common practice, we'll denote the random variable by θ and use this same symbol for a particular value that the random variable can take on and a variable of integration.

16.2   ## Using Bayes' Theorem to Update Probabilities

We gave several examples using Bayes' theorem in Section 2.4. Although mathematically similar, the examples we'll give here are concerned with applying the theorem to problems of statistical inference.

Example 16.2a   ### Disputed Paternity (continued)

Taking up the thread of Example 16.1a, we'll keep things simple and consider calculating the probability of paternity using only ABO–blood-group data. Suppose the mother is type O, the child is type B, and the alleged father is type AB. If the alleged father were type O or type A, he could not be the father. The fact that he is not one of these two types is *positive* evidence of paternity—not *completely* convincing, because there are many other men who are not excluded, but positive nonetheless. Just how convincing is it?

Let H be the hypothesis that the alleged father is actually the father, and take his blood type and that of the mother as given throughout. Let "data" refer to the child's blood type. Then according to Bayes' theorem [(6) of Section 2.4],

$$P(H \mid \text{data}) = \frac{P(\text{data} \mid H)P(H)}{P(\text{data} \mid H)P(H) + P(\text{data} \mid H^c)P(H^c)}, \tag{1}$$

or in terms of odds [see (3) in the chapter introduction],

$$\frac{P(H \mid \text{data})}{P(H^c \mid \text{data})} = \frac{P(\text{data} \mid H)}{P(\text{data} \mid H^c)} \cdot \frac{P(H)}{1 - P(H)}. \tag{2}$$

The child must have an O-allele from the mother and so must have got a B-allele from the father. Given the data, the likelihood of H follows from Mendelian genetic theory. Assuming H, the father is the alleged father and thus has an A-allele and a B-allele; the B-allele is selected in mating with probability $\frac{1}{2}$. So

$$P(\text{data} \mid H) = \frac{1}{2}.$$

The likelihood of H^c, given the data and assuming random mating, is the proportion of B-alleles in the population. Human populations differ according to this characteristic, but a typical proportion is 10%:

$$P(\text{data} \mid H^c) = \frac{1}{10}.$$

So the likelihood ratio of H to H^c, needed for (2), is

$$\frac{P(\text{data} \mid H)P(H)}{P(\text{data} \mid H^c)P(H^c)} = \frac{1/2}{1/10} = 5. \tag{3}$$

This means that the data serve to increase the odds in favor of H by a factor of 5.

The other factor needed in (2) is the prior odds ratio,

$$\frac{P(H)}{P(H^c)}.$$

Blood banks always use $P(H) = \frac{1}{2}$, and this is generally accepted by courts (although it shouldn't be). (See Example 16.1a for a discussion of the appropriateness of this prior.) The corresponding odds ratio is 1, so according to (2) and (3), the posterior odds are 5:1 that the alleged father is in fact the father. The blood bank would report to the court that the man has an 83.3% chance of being the father.   ∎

Example 16.2b   **A Crossover Clinical Trial**

A clinical trial is designed to test the effectiveness of a new therapy for high blood pressure. Twenty patients are administered a placebo therapy for one week and their diastolic blood pressures (dbp) measured; then they are given the new therapy, and a week later their changes in dbp are recorded. Twenty different patients are treated in the same scheme except that the order of treatment is reversed: new therapy and then placebo. This is called a *crossover* or *changeover* design. If there is a differential residual effect between two treatments, such a design can discover it. However, for our present purposes, we assume that there is no such effect—the crossover accomplished its purpose, and we'll ignore the crossover aspect of the trial in what follows.

It turns out that of the 40 patients, 30 have a better response on therapy and 10 have a better response on the placebo. (Diastolic blood pressure is measured on a numerical scale, and the amounts of improvement are relevant, but here we'll consider only their signs.) Let p denote the probability that the therapy is better than the placebo on a randomly selected individual. For the given data, the likelihood of p is

$$L(p) = P(\text{data} \mid p) \propto p^{30}(1 - p)^{10}.$$

Now consider testing $H_0: p = \frac{1}{2}$ against $H_A: p = \frac{2}{3}$. Under H_0, the therapy has no effect, and under H_A, it is effective two-thirds of the time. (The alternative $p = \frac{2}{3}$ is somewhat artificial; in Section 16.7, we'll consider the more realistic alternative $p > \frac{1}{2}$.)

The likelihood ratio of H_0 to H_A is

$$\frac{P(\text{data} \mid H_0)}{P(\text{data} \mid H_A)} = \frac{L\left(\frac{1}{2}\right)}{L\left(\frac{2}{3}\right)} = \frac{\left(\frac{1}{2}\right)^{30}\left(\frac{1}{2}\right)^{10}}{\left(\frac{2}{3}\right)^{30}\left(\frac{1}{3}\right)^{10}} = \frac{3^{40}}{2^{70}} \doteq .010.$$

This means that the posterior odds on H_0 are about 1% of the prior odds:

$$\frac{P(H_0 \mid \text{data})}{P(H_A \mid \text{data})} = (.010) \times \frac{P(H_0)}{P(H_A)}.$$

If our prior odds are even $[P(H_0) = P(H_A) = \frac{1}{2}]$, the posterior odds on H_0 are 1:100.

So

$$P(H_0 \,|\, \text{data}) = \frac{.010}{1.010} \doteq .01, \qquad P(H_A \,|\, \text{data}) \doteq .99.$$

Even if the prior odds seem heavily against the new therapy, this evidence can tip the balance the other way; for instance, if

$$\frac{P(H_0)}{P(H_A)} = 9,$$

that is, $P(H_A) = .1$), then

$$\frac{P(H_0 \,|\, \text{data})}{P(H_A \,|\, \text{data})} = .09$$

and $P(H_A \,|\, \text{data}) = .92.$ ■

In the preceding example, we allowed for just two possible values of a population parameter ($p = \frac{1}{2}$ and $p = \frac{2}{3}$). More generally, suppose a parameter θ is real-valued. In most problems, θ is a number on the interval $(0, 1)$, on the semi-infinite interval $(0, \infty)$ or on the whole real line $(-\infty, \infty)$. In a Bayesian analysis, θ is a random variable. Let $g(\theta)$ be your p.f. or p.d.f. of θ prior to obtaining data and $h(\theta \,|\, \text{data})$, your p.f. or p.d.f. of θ posterior to the data. In terms of the likelihood function $L(\theta)$, Bayes's theorem says

$$h(\theta \,|\, \text{data}) \propto L(\theta)g(\theta). \tag{4}$$

Since the integral (or sum, in the discrete case) of h must equal 1, we can write (4) as

$$h(\theta \,|\, \text{data}) = \frac{L(\theta)g(\theta)}{\displaystyle\int L(\theta)g(\theta)d\theta}. \tag{5}$$

The Mean of a Normal Population

The normal distribution is important in modeling populations, so inference concerning the mean μ of a normal population is an important statistical problem. We now derive the posterior distribution of μ when the prior for μ is normal and the variance σ^2 is assumed known. Given a random sample of size n, the likelihood function [see (2) of Section 8.2] is

$$L(\mu) \propto \exp\left(\frac{-(\mu - \bar{X})^2}{2\sigma^2/n}\right). \tag{6}$$

A normal prior p.d.f. combines neatly with this likelihood in using Bayes's theorem. Suppose the moments of the normal prior are $E(\mu) = v$ and $\text{var}(\mu) = \tau^2$. The prior p.d.f. is then

$$g(\mu) \propto \exp\left(\frac{-(\mu - v)^2}{2\tau^2}\right). \tag{7}$$

The posterior p.d.f. is proportional to the product of the likelihood (6) and the prior (7):

$$h(\mu \mid \bar{X}) \propto \exp\left(\frac{-(\mu - \bar{X})^2}{2\sigma^2/n} + \frac{-(\mu - v)^2}{2\tau^2}\right). \tag{8}$$

The exponent in (8) is quadratic in μ. This means that h is a *normal* p.d.f. [see Section 6.1]:

$$h(\mu \mid \bar{X}) \propto \exp\left\{-\frac{[\mu - E(\mu \mid \bar{X})]^2}{2 \, \text{var}(\mu \mid \bar{X})}\right\}. \tag{9}$$

We can find the mean and variance of this distribution by completing the square in μ in the exponent of (8). Except for the constant factor $-\frac{1}{2}$, that exponent is

$$\frac{(\mu - v)^2}{\tau^2} + \frac{(\mu - \bar{X})^2}{\sigma^2/n} = \left(\frac{1}{\tau^2} + \frac{n}{\sigma^2}\right)\mu^2 - 2\mu\left(\frac{v}{\tau^2} + \frac{n\bar{X}}{\sigma^2}\right) + \frac{v^2}{\tau^2} + \frac{n\mu^2}{\sigma^2}$$

$$= \left(\frac{1}{\tau^2} + \frac{n}{\sigma^2}\right)\left\{\mu - \frac{\dfrac{v}{\tau^2} + \dfrac{n\bar{X}}{\sigma^2}}{\dfrac{1}{\tau^2} + \dfrac{n}{\sigma^2}}\right\}^2 + \text{(terms free of } \mu\text{).} \tag{10}$$

We read the posterior mean and variance from (10):

$$E(\mu \mid \bar{X}) = \frac{\dfrac{v}{\tau^2} + \dfrac{n\bar{X}}{\sigma^2}}{\dfrac{1}{\tau^2} + \dfrac{n}{\sigma^2}} = \frac{\dfrac{1}{\tau^2}}{\dfrac{1}{\tau^2} + \dfrac{n}{\sigma^2}} v + \frac{\dfrac{n}{\sigma^2}}{\dfrac{1}{\tau^2} + \dfrac{n}{\sigma^2}} \bar{X} \tag{11}$$

and

$$\text{var}(\mu \mid \bar{X}) = \frac{1}{\dfrac{1}{\tau^2} + \dfrac{n}{\sigma^2}} = \frac{\sigma^2\tau^2}{\sigma^2 + n\tau^2}. \tag{12}$$

Equation (12) says that the reciprocal of the posterior variance of μ is the sum of the reciprocal variances of the prior and the sample mean. The reciprocal variance is sometimes called *precision*, which we denote by π. Thus, for prior, data, and posterior, we define

$$\pi_{\text{pr}} = \frac{1}{\tau^2}, \qquad \pi_{\text{data}} = \frac{n}{\sigma^2}, \qquad \pi_{\text{po}} = \frac{1}{\text{var}(\mu \mid \bar{X})}, \tag{13}$$

and rewrite (12) as

$$\pi_{\text{po}} = \pi_{\text{pr}} + \pi_{\text{data}}. \tag{14}$$

The precision π indicates how well the random variable in question is known. A very small τ^2 (high prior precision) indicates a strong prior belief that μ is very likely to be close to v. On the other hand, a very large τ^2 (low prior precision) means that the prior is quite diffuse. If one's prior information about μ is about as good as a single observation, τ^2 should equal σ^2. More generally, if the prior information is worth about the same as k observations, then τ^2 should equal σ^2/k.

In terms of precisions, the posterior mean is

$$E(\mu \mid \bar{X}) = \left(\frac{\pi_{\text{pr}}}{\pi_{\text{po}}}\right) v + \left(\frac{\pi_{\text{data}}}{\pi_{\text{po}}}\right) \bar{X}. \tag{15}$$

This has intuitive appeal. The posterior mean is a weighted average of the sample mean $\bar{X}$ and the prior mean; the weights are the corresponding precisions. If π_{pr} is much less than π_{data}, the posterior mean will be very close to $\bar{X}$; in this case, the sample information dominates the prior. This will be the case when n is very large. It is clear from (13) and (15) that the posterior mean $E(\mu \mid \bar{X})$ tends to the sample mean $\bar{X}$ as n becomes infinite. Since $\bar{X}$ is a consistent estimate of μ (see Section 9.10), the posterior mean is a consistent estimate of μ. (This may not be true in other settings.)

Example 16.2c   We want to establish the precise weight of a particular block of metal whose nominal weight is "10 grams." We'll use a precision balance known to be unbiased, in that it gives the correct weight on the average. The results of individual weighings are variable, distributed approximately normally with standard deviation 6 micrograms (μg). Our experience with similar blocks suggests that their weights are usually a bit different from the nominal weight, sometimes lower and sometimes higher. Let X denote the amount (in μg) by which the measured weight exceeds the actual weight, so that $\mu = E(X) = 0$.

Suppose we summarize our prior notions about μ in a normal distribution with mean $v = 0$ and s.d. $\tau = 4$. We then weigh the block five times and find $\bar{X} = 7$; how does this information change our distribution for μ?

The likelihood is given by (6) with $\sigma = 6$ and $n = 5$. The posterior distribution of μ is normal with p.d.f. (9). The precision of the prior and of the data [from (13)] are

$$\pi_{\text{pr}} = \frac{1}{16} = \frac{9}{144} \quad \text{and} \quad \pi_{\text{data}} = \frac{5}{36} = \frac{20}{144}.$$

The posterior precision, from (14), is the sum:

$$\pi_{\text{po}} = \pi_{\text{pr}} + \pi_{\text{data}} = \frac{29}{144},$$

from which we find the posterior standard deviation to be $12/\sqrt{29} \doteq 2.23$. The posterior mean [from (15)] is a weighted average of the prior mean and the sample mean, with weights defined by the precisions:

$$E(\mu \mid \bar{X}) = \left(\frac{9}{29}\right) \cdot 0 + \left(\frac{20}{29}\right) \cdot 7 = \frac{140}{29} \doteq 4.83.$$

(Figure 16.1 shows graphs of the likelihood, prior, and posterior.) This is an average of 0, our best guess from past experience, and 7, our estimate from the data at hand. In a way, it is an adjustment for the regression effect (Section 15.6).

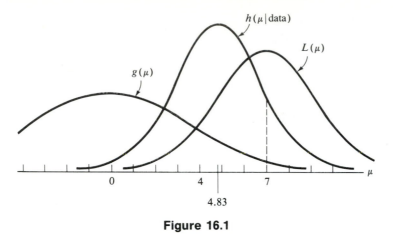

Figure 16.1

The data are rather convincing that the block is heavier than its nominal weight:

$$P(\mu > 0 \mid \bar{X}) = 1 - \Phi\left(\frac{0 - 140/29}{12/\sqrt{29}}\right) \doteq 1 - \Phi(-2.17) \doteq .98,$$

but our past experience with such blocks suggests that μ is not as large as $\bar{X}$:

$$P(\mu < 7 \mid \bar{X}) = 1 - \Phi\left(\frac{7 - 140/29}{12/\sqrt{29}}\right) \doteq \Phi(.97) \doteq .83. \qquad \blacksquare$$

Conjugate Priors

In the preceding discussion and example, the normal prior and likelihood fit neatly together to yield a normal posterior. When the posterior is a member of the same family of distributions as the prior, the prior is said to be **conjugate** for that likelihood. Another population for which there is a conjugate prior is the exponential, with p.d.f.

$$f(x \mid \lambda) = \lambda e^{-\lambda x}, \qquad \text{for } x > 0.$$

Given a random sample of size n, the likelihood is a function of the sample mean [which is sufficient for λ (Section 8.3)]:

$$L(\lambda) = \lambda^n e^{-n\lambda x}, \qquad \lambda > 0. \tag{16}$$

Suppose the prior is $\text{Gam}(\alpha, \beta)$, with p.d.f.

$$g(\lambda) \propto \lambda^{\alpha - 1} e^{-\beta \lambda}, \qquad \lambda > 0. \tag{17}$$

[see (Section 6.3)]. The posterior p.d.f. is

$$h(\lambda \mid \bar{X}) \propto g(\lambda)L(\lambda) = \lambda^{n + \alpha - 1} e^{-\lambda(n\bar{X} + \beta)}, \qquad \lambda > 0. \tag{18}$$

This is $\text{Gam}(\alpha + n, \beta + n\bar{X})$, with mean

$$E(\lambda \mid \bar{X}) = \frac{\alpha + n}{\beta + n\bar{X}} = \frac{\alpha/n + 1}{\beta/n + \bar{X}}, \tag{19}$$

and variance

$$\text{var}(\lambda \mid \bar{X}) = \frac{\alpha + n}{(\beta + n\bar{X})^2}. \tag{20}$$

As n becomes infinite, the posterior mean tends to the m.l.e. $\hat{\lambda} = 1/\bar{X}$, and the posterior variance tends to 0. As in the earlier normal case, this in turn implies that the posterior mean is a consistent estimate of the population mean.

Example 16.2d Consider estimating a Bernoulli parameter p, with quadratic loss, using a random sample of size n. The number of successes in n independent trials is a sufficient statistic. Given this number Y, the likelihood function is

$$L(p) = p^Y(1 - p)^{n-Y}, \qquad 0 < p < 1.$$

Suppose we assume that the prior for p is Beta(r, s) [see (6) of Section 6.4], with p.d.f.

$$g(p) \propto p^{r-1}(1 - p)^{s-1}, \qquad 0 < p < 1. \tag{21}$$

The posterior p.d.f. is then

$$h(p) \propto g(p)L(p) \propto p^{r+Y-1}(1 - p)^{s+n-Y-1}, \qquad 0 < p < 1. \tag{22}$$

This is Beta($r + Y$, $s + n - Y$). So the beta distribution is a conjugate prior for the Bernoulli likelihood. ■

16.3   **Loss Functions**

In Chapter 12, we considered questions such as these: Does fertilizer A give a higher yield than fertilizer B? Is a certain drug and diet combination more effective than the diet alone in lowering blood pressure? Is formulation A of a product more acceptable to consumers than is formulation B? Such questions are often preliminary to making **decisions**.

In the case of the fertilizer, a farmer needs to decide which one to buy. Fertilizer A may tend to give a higher yield, but it may be more expensive or may result in a crop of poorer quality. In the case of treating high blood pressure, a physician must decide whether to prescribe the drug in addition to the diet.

The U.S. Food and Drug Administration faces many decision problems concerning experimental drugs. A typical scenario is this: A pharmaceutical company submits data to the FDA, hoping to obtain approval for marketing the drug. Perhaps the drug has been shown to be significantly more effective than a standard treatment, both statistically and clinically. However, a benefit is usually accompanied by some risks, known or unknown. Questions to be answered include: Are the incidence and severity of adverse events and side effects adequately identified? Are there some patients for whom use of the drug should be restricted? Does the drug interact with other drugs or chemicals that might be used concurrently? Should usage recommendations depend on a patient's age? The ultimate questions are these: Will approving the drug do more good than harm? If the drug is approved, what restrictions should be placed on its use? Some guesswork is involved, but if there is too much uncertainty, the FDA can ask for more data.

All decision problems involve questions like those facing the FDA. First, there is a set of possible actions (such as "approve," "disapprove," and "ask for further studies"). Second, there is information for assessing the consequences of the various possible actions. Third, there is a probability distribution (possibly posterior to some data) for the various possible states of nature. A decision is to be made, weighing the various uncertainties and costs.

The scale for assessing consequences may or may not be monetary. Even when it is, the money scale is usually not a good one to use. (An increment of $10,000 may not mean the same to two different people.) We shall assume that there is a numerical scale on which costs of wrong decisions are measured.[2] These costs or losses depend on the action taken and on unknown population characteristics—on one or more parameters θ. We denote by $\ell(\theta, a)$ the loss that would be incurred upon taking a particular action a when the parameter value is θ. The function ℓ is called the *loss function*. If we know the value of θ, an obvious choice of action is one that minimizes $\ell(\theta, a)$; but the value of θ is usually not known, and therein lies the decision "problem."

Assessing losses is a personal matter, in that the assessment depends on who is making it. So decision making based on such evaluations of loss and on personal probabilities is a subjective process.

In this chapter, we adopt the view that (for the decision maker) θ is a random variable. Thus, for each possible action, the loss $\ell(\theta, a)$ is random. Averaging it with respect to the decision maker's distribution of θ, we define the **Bayes loss** for action a as

$$B(a) = E[\ell(\theta, a)]. \tag{1}$$

The distribution of θ to be used in averaging ℓ in (1) is the decision maker's current (subjective) distribution. We say that an action a is *best* if it is minimizes (1) and call it a **Bayes action**.

The following example illustrates the calculations leading to a Bayes decision. We've kept the numbers small so that the calculations do not obscure the concepts.

Example 16.3a   You're a quality control engineer in a company that produces machine parts. Each batch contains five of a certain part and sells for $180. Any defective part in the batch will be returned by the customer, to be replaced with a good part at a cost (to your company) of $60.

You select one part at random from each batch. After inspecting that part, you must decide whether to reject the batch or pass the batch, with a good part replacing the inspected part if it was bad. (For simplicity, we suppose that these are your only options.) Let M denote the number of defective parts in a batch that is passed. The loss in dollars is then

$$\ell(M, a) = \begin{cases} 60M - 180 & \text{if } a = \text{accept the lot,} \\ 0 & \text{if } a = \text{reject the lot.} \end{cases}$$

---

[2] The theory of *utility* shows that "ideal" decision makers have such a scale. See Berger [1].)

Suppose the part you inspect is *good*. Should you accept or reject the lot? To answer this, we need a prior distribution for the parameter M. Assume M is uniform on $\{0, 1, 2, 3, 4, 5\}$: $g(M) = \frac{1}{6}$. The likelihood function is the probability of a good article: $L(M) = (5 - M)/5$. The posterior p.f. for M is then proportional to the product of g and L; normalizing so that $\Sigma\, h = 1$ yields

$$h(M \,|\, \text{good}) = \frac{5 - M}{15}, \qquad M = 0, 1, \ldots, 5,$$

with mean $E(M \,|\, \text{good}) = \frac{4}{3}$. The Bayes loss for the action "reject" is

$$B(\text{reject} \,|\, \text{good}) = E[\ell(M, \text{reject})] = 0.$$

For the action "accept," it is

$$B(\text{accept} \,|\, \text{good}) = E[\ell(M, \text{accept})] = 60 \cdot E(M \,|\, \text{good}) - 180 = -100.$$

So $B(\text{accept} \,|\, \text{good}) < B(\text{reject} \,|\, \text{good})$, and the better of the two possible actions, if the inspected part is good, is to accept the lot.

Next, suppose that the inspected part is defective. The likelihood is $L(M) = M/5$. With the same prior, the posterior is

$$h(M \,|\, \text{def.}) = \frac{M}{15}, \qquad M = 0, 1, 2, 3, 4, 5,$$

with mean $E(M \,|\, \text{def.}) = \frac{11}{3}$. As before, $B(\text{reject} \,|\, \text{def}) = 0$, but now

$$B(\text{accept} \,|\, \text{def}) = 60 \cdot E[M - 1 \,|\, \text{def}] - 180 + 60 = 40.$$

(This includes an extra \$60 for the part that replaces the one found to be defective, but not the extra cost for making sure that it is actually good.) So, if the sampled part is defective, the better action is to reject the batch. (To be continued.) ∎

In the preceding example, we found the Bayes action and the Bayes loss for both possible outcomes of the sample. We next address the question of how much the sample is worth and ask whether it might be better not to sample at all.

Example 16.3b   In the setting of Example 16.3a, we now determine the Bayes action for the case of no data. We use the prior p.f. of that example: $g(M) = \frac{1}{6}$, for $M = 0, 1, \ldots . 5$, which has mean $E(M) = 2.5$. Since

$$B(\text{accept}) = 60 E(M) - 180 = -30,$$

and

$$B(\text{reject}) = 0,$$

the Bayes action is to accept the batch, with Bayes loss of -30 (that is, an expected profit of \$30).

If instead we sample one part and take the appropriate Bayes action, as in Example 16.3a, the loss is the cost of the sample plus the minimum Bayes loss. The latter depends on whether the part selected is good or defective. The probability of getting a defective is

$$P(\text{def}) = E[P(\text{def} \mid M) = E\left(\frac{M}{5}\right) = \frac{2.5}{5} = \frac{1}{2}$$

(which should be clear by symmetry). Then, of course, $P(\text{good}) = \frac{1}{2}$. Recall that the Bayes action is to reject the batch if the article selected is defective and accept it if the article is good. The minimum Bayes loss depends on the sample result:

$$B(\text{reject} \mid \text{def}) = 0, \qquad B(\text{accept} \mid \text{good}) = -100.$$

The Bayes loss (in dollars) including the sampling cost c is

$$B(\text{take sample}) = \frac{0 \cdot 1}{2} + (-100) \cdot \frac{1}{2} + c = -50 + c.$$

The sample is worth taking if

$$-50 + c < -30 = B(\text{don't sample}),$$

or if $c < \$20$.   ∎

16.4   Bayesian Point Estimates

Estimating the value of a parameter θ can be viewed as a decision problem in which the action a is an estimate of θ. One possible loss function assigns no loss ($\ell = 0$) when $a = \theta$, and a constant loss ($\ell = 1$) if $a \neq \theta$. (Such a loss structure seems reasonable only when θ is discrete.)

Example 16.4a   The parameter M in Example 16.3a, the number of defectives in a batch of five parts, is discrete. Suppose we sample two parts at random from a batch (without replacement), and let X denote the number of defectives in the sample. With a uniform prior for M, the posterior p.f. of M is proportional to the likelihood. If we observe $X = 0$,

$$h(M \mid X = 0) \propto (5 - M)(4 - M), \qquad M = 0, 1, \dots, 5.$$

With $\ell(M, a) = 0$ or 1 depending on whether $a = M$ or $a \neq M$, the Bayes loss for the estimated value a is

$$B(a \mid X = 0) = E[\ell(M, a) \mid x = 0] = P(M \neq a \mid X = 0) = 1 - h(a \mid X = 0).$$

This is smallest when $a = 0$, so the Bayes estimate of the parameter M is 0.

If the sample contains one defective ($X = 1$), the posterior p.f. is

$$h(M \mid X = 1) \propto M(5 - M), \qquad M = 0, 1, \dots, 5.$$

The Bayes loss is smallest when $a = 2$ or $a = 3$. Both of these are Bayes estimates of M when $X = 1$. If $X = 2$ (two defectives in the sample), the posterior p.f. is

$$h(M \mid X = 2) \propto M(M - 1), \qquad M = 0, 1, \dots, 5.$$

The Bayes loss is smallest at $M = 5$.   ∎

The "all or nothing" loss of this last example is usually unrealistic—very unrealistic in the case of continuous θ. A more natural kind of loss is one that depends continuously on the magnitude of the error $\theta - a$ and increases as a moves away from θ in either direction. In particular, consider a loss that is proportional to the magnitude of the error:

$$\ell(\theta, a) = k \cdot |\theta - a|. \tag{1}$$

For an estimate a, the Bayes loss is

$$B(a) = E[\ell(\theta, a)] = kE(|\theta - a|).$$

This is a minimum when a is the *median* of the distribution of θ, so the median of the current distribution of θ is a Bayes estimate.

> In estimating θ with loss (1), the *median* of the distribution of θ is the Bayes estimate of θ.

Example 16.4b   Consider estimating a population mean when we assume the loss function $\ell(\theta, a) = |\theta - a|$. Suppose the population distributions is $N(\mu, \sigma^2)$, where σ^2 is known, and suppose further that the prior for μ is $N(v, \tau^2)$. The posterior distribution of μ is then also normal [see (9) of Section 16.2]. Before collecting data, the Bayes estimate of μ is the median of the prior distribution, which is its mean v. (A normal distribution is symmetric about its mean.) Given some data, the Bayes estimate of μ is the posterior median, equal to the posterior mean, given by (11) of Section 16.2. ∎

In our discussion of estimation in Section 9.1, we took the average of the *squared* error over the data space as a measure of inaccuracy of an estimator. Suppose we take as a loss function the squared error in the parameter space:

$$\ell(\theta, a) = (\theta - a)^2. \tag{2}$$

The corresponding Bayes loss is the average of (2) with respect to the distribution of θ:

$$B(a) = E[\ell(\theta, a)] = E[(\theta - a)^2]. \tag{3}$$

For any given a, $B(a)$ is a second moment of the distribution of θ. It is minimized by taking $a = E(\theta)$ [see (7) of Section 3.3]. Thus, the Bayes estimate of θ, with the quadratic loss (2), is the mean of the current distribution of θ.

> In estimating θ with quadratic loss (2), the *mean* of the distribution of θ is the Bayes estimate of θ.

Example 16.4c   **Estimating p**

Consider estimating a Bernoulli parameter p using a random sample of size n, as in Example 16.2e. When the prior is Beta(r, s), and Y is the number of successes in n independent trials, the posterior is Beta$(r + Y, s + n - Y)$. The posterior mean is then

$$E(p \mid Y) = \frac{r + Y}{r + s + n}, \tag{4}$$

[from (8) of Section 7.4]. This is the Bayes estimate of p for quadratic loss. In particular, when $r = s = 1$, the prior is $U(0, 1)$. The posterior mean (4) is then

$$E(p \mid Y) = \frac{Y + 1}{n + 2} = \left(\frac{n}{n + 2}\right)\hat{p} + \left(1 - \frac{n}{n + 2}\right)\frac{1}{2}. \tag{5}$$

[The second expression for $E(p \mid Y)$ shows that the estimate is a weighted average of the sample proportion and the mean of the prior; the larger the sample, the closer it is to the sample proportion.] This estimate of p has an extensive history and is known as *Laplace's rule of succession*.   ∎

16.5   **Probability of an Interval**

In introducing this chapter, we pointed out that the classical statistical inferences presented in earlier chapters are often misinterpreted. For example, it's hard *not* to view a confidence level as the *probability* that the parameter value is in the corresponding confidence interval. With a probability distribution for θ, we can actually calculate the probability that θ is in any specified interval. (Indeed, in Example 16.2c, we did such a calculation.) So, given a confidence interval for θ based on a particular set of data, we can calculate the posterior probability that θ is in that interval. This probability may or may not be close to the confidence level.

Example 16.5a   **Confidence interval probability**

In Example 9.3c, we worked with data giving blood platelet counts of 35 healthy males, with mean $\bar{X} = 234$ and s.d. $S = 45.05$. We found a 95% confidence interval for μ to be $219.1 < \mu < 248.9$, taking the sample s.d. as the population s.d. σ and assuming $\bar{X}$ to be normally distributed.

The precision of the sample mean is $n/\sigma^2 = .0173$. Suppose our prior for μ is normal with mean $v = 250$ and s.d. $\tau = 5$; the prior precision is $1/\tau^2 = .04$. The posterior, given $\bar{X} = 234$, is normal with precision

$$\pi_{\text{po}} = .0173 + .04 = .0573 = \frac{1}{(4.18)^2}$$

and mean

$$E(\mu \mid \bar{X}) = \frac{.0173\bar{X} + .04v}{.0573} = 245.2.$$

The posterior probability of the interval (219.1, 248.9) is then

$$P(219.1 < \mu < 248.9) = \Phi\left(\frac{248.9 - 245.2}{4.18}\right) - \Phi\left(\frac{219.1 - 245.2}{4.18}\right)$$

$$\doteq \Phi(.89) - \Phi(-6.24) = .8133.$$

Now suppose instead that the normal prior has parameters $v = 240$ and $\tau = 5$. With similar calculations, we find the posterior mean to be 238.2 and the posterior s.d. 4.18. The 95% confidence interval then has posterior probability

$$P(219.1 < \mu < 248.9) = \Phi(2.56) - \Phi(-4.57) = .995.$$

Thus, the probability of the confidence interval can be either greater or smaller than the confidence level, depending on the prior.

For a sequence of normal priors in which (for any fixed mean) the prior variance becomes infinite, the posterior mean and variance tend toward the sample mean $\bar{X}$ and its variance σ^2/n. This means that the probability of the confidence interval tends to the confidence level. So in the limit, as one's prior information becomes ever more diffuse, the popular interpretation of a confidence level becomes correct. ■

Example 16.5b   **A Duck Census**

A U.S. Fish and Wildlife report[3] applies the Bayesian method to a census of breeding pairs of mallard ducks. The parameter p of interest is the proportion of quarter-sections in which breeding mallards are present.

For the assumption of ignorance about p, the report uses the prior density

$$g(p) \propto p^{-.5}(1 - p)^{-.5}. \tag{1}$$

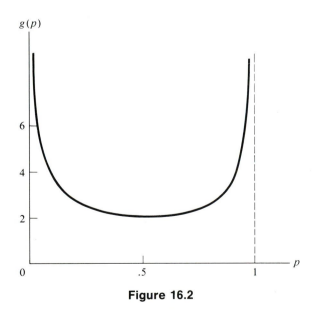

Figure 16.2

---

[3] U. S. Department of the Interior, *Special Scientific Report—Wildlife*, No. 203, by D. H. Johnson.

This is quite flat over the interval $.2 < p < .8$ but rises sharply to infinity at 0 and at 1 (see Figure 16.2 and Problem 26).

A 1972 study by Stewart and Kantrud[4] found mallards breeding in 79 of 130 randomly selected quarter-sections; thus the m.l.e. of p is

$$\hat{p} = \frac{Y}{n} = \frac{79}{130} \doteq .608.$$

Assuming the binomial model to be applicable, the likelihood function (which is proportional to the probability that $Y = 79$ given p) is

$$L(p) \propto p^{79}(1 - p)^{51}. \tag{2}$$

The posterior density is then proportional to the product of (1) and (2):

$$h(p \mid 79) \propto p^{78.5}(1 - p)^{50.5}.$$

The posterior mean [from (7) of Section 6.4] is $79.5/131 \doteq .607$. This is then the Bayes estimate of p with the quadratic loss function (2) of Section 16.3. The posterior s.d. [from (7) of Section 6.4] is .0425, so the distribution is concentrated quite near the posterior mean, which in turn is quite close to the sample mean.

A 95% confidence interval, found as in Section 9.2, has limits $.608 \pm .084$ (that is, $\hat{p} \pm 1.96$ times its standard error). The Bayesian approach permits us to calculate the posterior probability of this interval:

$$P(.524 < p < .692) = \frac{1}{B(79.5, 51.5)} \int_{.524}^{.692} p^{78.5}(1 - p)^{50.5} \, dp = .952.$$

(The value of the integral can be calculated using numerical methods or a normal approximation.) With the rather diffuse prior (1), the posterior probability of the confidence interval turns out to be very close to the confidence level.   ■

16.6   Probability of the Null Hypothesis

In Section 10.7, we mentioned the tendency for researchers to view a P-value as the probability of the null hypothesis. It is natural to think in terms of probabilities of hypotheses. Although $P(H_0)$ and $P(H_0 \mid \text{data})$ do not make sense in classical significance testing, they have meaning when parameters are random variables. Consider testing $H_0: \theta \leq \theta_0$ against $H_A: \theta > \theta_0$, where θ is a continuous parameter. In this case, the probability of H_0—prior or posterior—is just the integral of the appropriate p.d.f. of θ over the interval $(-\infty, \theta_0)$.

Example **16.6a**   Because of the role of college aptitude test scores in college entrance decisions, there are minicourses that purport to teach students how to take these tests. A particular aptitude test has been found to produce scores that are normally distributed, with

---

[4] "Population estimates of breeding birds in North Dakota." *Auk* 89(4), 766–788.

mean 500. If the minicourse directed at this test is effective (on average), the mean score μ of students who take the course is larger than 500; otherwise it is not. We want to test

$$H_0: \mu \le 500 \qquad \text{versus} \qquad H_A: \mu > 500,$$

and our prior for μ is $N(520, 20^2)$. The prior probability of the null hypothesis is

$$P(H_0) = P(\mu \le 500) = \Phi\left(\frac{500 - 520}{20}\right) \doteq .16.$$

Suppose we take a random sample of 50 students who took the course and find that their test scores have mean $\bar{X} = 513$ and s.d. $S = 80$. The precisions of the prior and the data are

$$\pi_{\text{pr}} = \frac{1}{\tau^2} = \frac{1}{400},$$

$$\pi_{\text{data}} = \frac{n}{\sigma^2} \doteq \frac{n}{S^2} = \frac{50}{6400} = \frac{1}{128}.$$

Using these in (11) and (12) of Section 16.2, we find that the posterior is normal with mean

$$E(\mu \,|\, \bar{X}) = \frac{\left(\frac{1}{128}\right)\bar{X} + \left(\frac{1}{400}\right)v}{\frac{1}{128} + \frac{1}{400}} = .758\bar{X} + .242v = 514.7.$$

and variance

$$\text{var}(\mu \,|\, \bar{X}) = \frac{1}{\frac{1}{128} + \frac{1}{400}} = (9.85)^2.$$

The posterior probability of H_0 is then

$$P(H_0 \,|\, \text{data}) = P(\mu \le 500 \,|\, \text{data}) = \Phi\left(\frac{500 - 514.7}{9.85}\right) \doteq \Phi(-1.49) \doteq .068.$$

For comparison, the "P-value" is

$$P = P(\bar{X} > 513 \,|\, \mu = 500) = 1 - \Phi\left(\frac{513 - 500}{\sqrt{128}}\right) \doteq \Phi(-1.15) \doteq .125.$$

Suppose we decide that to be worth our recommending the special training, the increase in mean score should be at least 10 points. What is the probability that $\mu \ge 510$? To calculate this, we integrate the posterior from 510 to ∞:

$$P(\theta \ge 510 \,|\, \text{data}) = 1 - \Phi\left(\frac{510 - 514.7}{9.85}\right) \doteq \Phi(.32) \doteq .37. \qquad \blacksquare$$

Consider testing $\theta = \theta_0$ against the two-sided alternative $\theta \ne \theta_0$. The point null hypothesis $\theta = \theta_0$ is usually only a convenient idealization of the hypothesis that θ is in some small interval including θ_0, so as an approximation, we assign a positive weight to H_0, regarding the distribution of θ as a mixture of a singular distribution (at θ_0) and a continuous distribution over the alternative.

Example 16.6b   The null hypothesis in a sign test for a pairwise comparison of two treatments is H_0: $p = \frac{1}{2}$, where p is the probability that treatment A wins over treatment B in a single paired comparison. Suppose the alternative is $p \neq \frac{1}{2}$. Let Y denote the number of pairs in which A wins in ten independent trials. Suppose A wins in nine trials; the two-sided P-value is then

$$\frac{2(11)}{2^{10}} \doteq .021.$$

To find the probability of H_0, we need a distribution for p. Suppose our prior probability that the treatments are equally effective[5] is

$$P\left(p = \frac{1}{2}\right) = \gamma.$$

Given that $p \neq \frac{1}{2}$, suppose that p has this symmetric beta p.d.f.

$$g\left(p \,\middle|\, p \neq \frac{1}{2}\right) = 6p(1 - p), \qquad 0 < p < 1.$$

For $Y = 9$ and $n = 10$, the likelihood function is

$$L(p) = P(Y = 9 \,|\, p) = 10p^9(1 - p).$$

The unconditional probability of nine wins for A is

$$P(Y = 9) = P\left(Y = 9 \,\middle|\, p = \frac{1}{2}\right) \cdot P\left(p = \frac{1}{2}\right) + P\left(Y = 9 \,\middle|\, p \neq \frac{1}{2}\right) \cdot P\left(p \neq \frac{1}{2}\right)$$

$$= \gamma \cdot 10(2^{-10}) + (1 - \gamma) \cdot 60 \int_0^1 p^{10}(1 - p)^2 \, dp$$

$$= 10\left[\frac{\gamma}{1024} + \frac{1 - \gamma}{143}\right].$$

The posterior odds on H_0 is the ratio of these two terms [see (3) in the introduction to this chapter]:

$$\frac{P(H_0 \,|\, Y = 9)}{P(H_A \,|\, Y = 9)} = \frac{\gamma}{1 - \gamma} \cdot \frac{143}{1024}.$$

The probability of H_0 is

$$P(H_0 \,|\, \text{data}) = \frac{143\gamma}{143\gamma + 1024(1 - \gamma)}.$$

Figure 16.3 shows this as a function of γ ($0 \leq \gamma \leq 1$). When $\gamma = \frac{1}{2}$, $P(H_0 \,|\, \text{data}) = .123$, but if $\gamma = .1$, $P(H_0 \,|\, \text{data}) \doteq .015$.   ∎

---

[5] If one is a control and the other is a new cancer therapy, γ would be quite large: There are countless new therapies, most of which are worthless.

$P(H_0|\text{data})$

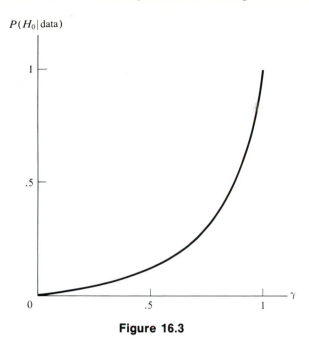

Figure 16.3

16.7   Hypothesis Testing as Decision Making

In Chapter 11, we presented hypothesis testing as a process of deciding between a null and an alternative hypothesis. The choice of significance level α is usually left to the investigator, who typically conforms to tradition and chooses $\alpha = .05$. Such an arbitrary choice of significance level does not take into consideration the costs of wrong decisions.

Consider first the case of a simple null hypothesis H_0 and a simple alternative hypothesis H_1. Let f_0 and f_1 denote the p.d.f.'s of a sample $\mathbf{X} = (X_1, \ldots, X_n)$ under H_0 and H_1, respectively. The corresponding likelihoods are then

$$L(H_0) = f_0(\mathbf{X}) \quad \text{and} \quad L(H_1) = f_1(\mathbf{X}).$$

Suppose there is no loss for a correct choice of action; for an incorrect choice, let the losses be denoted

$$A = \ell(H_0, \text{rej. } H_0) \quad \text{and} \quad B = \ell(H_1, \text{acc. } H_0).$$

To find the Bayes losses, we need prior probabilities for the two competing hypotheses:

$$g_0 = P(H_0) \quad \text{and} \quad g_1 = P(H_1) = 1 - g_0.$$

The posterior probabilities are then

$$h(H_0|\mathbf{X}) = \frac{g_0 f_0(\mathbf{X})}{g_0 f_0(\mathbf{X}) + g_1 f_1(\mathbf{X})}, \qquad h(H_1|\mathbf{X}) = \frac{g_1 f_1(\mathbf{X})}{g_0 f_0(\mathbf{X}) + g_1 f_1(\mathbf{X})}.$$

The ratio of Bayes losses for accepting and for rejecting H_0 is

$$\frac{B(\text{reject} \mid \mathbf{X})}{B(\text{accept} \mid \mathbf{X})} = \frac{E[\ell(H, \text{reject}) \mid \mathbf{X}]}{E[\ell(H, \text{accept}) \mid \mathbf{X}]} = \frac{A \cdot g_0 f_0(\mathbf{X})}{B \cdot g_1 f_1(\mathbf{X})}.$$

Rejecting H_0 is the better action when this ratio is less than 1, or when

$$\frac{f_0(\mathbf{X})}{f_1(\mathbf{X})} < K = \frac{Bg_1}{Ag_0}.$$

Thus, we are led to a test of the Neyman–Pearson type (see Section 11.5). Moreover, a Neyman–Pearson test is a Bayes test for some choice of losses A and B and prior probabilities g_0.

Suppose next that H_0 and H_A are composite hypotheses about a parameter θ—that is, about sets of θ-values. If we assume no penalty for correct decisions, the most general type of loss function for testing H_0 against H_A is

$$\ell(\theta, \text{accept}) = \begin{cases} b(\theta) & \text{if } \theta \text{ is in } H_A, \\ 0 & \text{if } \theta \text{ is in } H_0, \end{cases}$$

$$\ell(\theta, \text{reject}) = \begin{cases} 0 & \text{if } \theta \text{ is in } H_A, \\ a(\theta) & \text{if } \theta \text{ is in } H_0, \end{cases}$$

where a and b are nonnegative functions. (An especially simple special case is that in which the functions a and b are constant.) If the current distribution for θ has p.d.f. g, the Bayes losses are

$$B(\text{accept}) = \int_{H_A} b(\theta)g(\theta)\,d\theta, \qquad B(\text{reject}) = \int_{H_0} a(\theta)g(\theta)\,d\theta.$$

Example 16.7a In Section 11.1, we considered testing the hypothesis that a coffee machine is dispensing an amount whose mean μ is 8 oz. If $\mu < 8$ and we do not adjust, the customer is short-changed (on average) and may either stop using the machine (causing loss of future income) or ask for a full or partial refund, depending on how short the measure. If $\mu > 8$ and we do not adjust, the machine will run out of coffee sooner than otherwise, and the company will again suffer a loss. Suppose that if we accept H_0 ($\mu = 8$), the loss over the period until we next check the machine is

$$\ell(\mu, \text{accept}) = \begin{cases} 2(8 - \mu), & 0 \le \mu \le 8, \\ \mu - 8, & \mu > 8. \end{cases}$$

On the other hand, if we reject H_0, the cost to the company is

$$\ell(\mu, \text{reject}) = a.$$

The Bayes loss for accepting H_0 and not making an adjustment is

$$B(\text{accept}) = E[\ell(\mu, \text{accept})] = 2\int_0^8 (8 - \mu)g(\mu)\,d\mu + \int_8^\infty (\mu - 8)g(\mu)\,d\mu,$$

where $g(\mu)$ is the current p.d.f. for μ. Suppose g is approximately $N(7.3, .16)$. Then $B(\text{accept}) \doteq 1.57$ (see Problem 24). So if $a < 1.57$, it is better to reset the machine. ∎

16.8   **Bayesian Predictive Distributions**

Our discussion of Bayesian inference thus far has focused on population parameters. It is often more natural and important to predict the next observation
or sequence of observations. Consider a random sample from $f(x\,|\,\theta)$: $\mathbf{X} =$
$(X_1, \ldots, X_n)$ and the problem of predicting the next m independent observations
from the same population: $\mathbf{Y} = (Y_1, \ldots, Y_m)$. The likelihood function, given the
data, is

$$L(\theta) = f(X_1\,|\,\theta) \cdots f(X_n\,|\,\theta),$$

and the joint p.f. of the new observations is

$$f^*(y_1, \ldots, y_m\,|\,\theta) = f(y_1\,|\,\theta) \cdots f(y_m\,|\,\theta). \tag{1}$$

Given a distribution for the parameter θ, we define the Bayesian predictive
distribution of the Y_i to be the average of f^* with respect to the distribution of θ:

$$f^*(y_1, \ldots, y_m) = E[f^*(y_1, \ldots, y_m\,|\,\theta)]. \tag{2}$$

As a special case, consider the setting of a Bernoulli population. We might ask,
"What is the probability that the next observation is a success?" The *conditional*
probability of success at a single trial is p. The unconditional probability (2) is

$$E[P(\text{success}\,|\,p)] = E(p), \tag{3}$$

the average with respect to the current distribution of p. More generally, the
probability of exactly k successes in the next m trials is

$$E\left\{\binom{m}{k}p^k(1-p)^{m-k}\right\}. \tag{4}$$

Example 16.8a   The following situation arose in an actual legal case, although we have modified the
specifics. Twenty windows in a high-rise office building broke in the first year after it
was built. The question at issue is the number of these that were caused by a
particular type D of defect in the glass. Four windows were selected at random from
the 20 and analyzed; all four were found to have been caused by D. Of the other 16,
how many were caused by D?

   The windows on the building were all from a particular lot. A glass expert
testifies that the lot proportion p of windows with defect D among all windows with
defects varies, with a J-shaped distribution with p.d.f.

$$g(p) \propto p^{-3/4}(1-p)^{-1/4}, \qquad 0 < p < 1.$$

This is a beta density, shown in Figure 16.4, with mean $E(p) = \tfrac{1}{4}$. In particular, D is
the predominant defect in only 22% of the lots:

$$P\left(p > \frac{1}{2}\right) = \frac{\displaystyle\int_{1/2}^1 p^{-3/4}(1-p)^{-1/4}\,dp}{\displaystyle\int_0^1 p^{-3/4}(1-p)^{-1/4}\,dp} \doteq .22.$$

$g(p)$

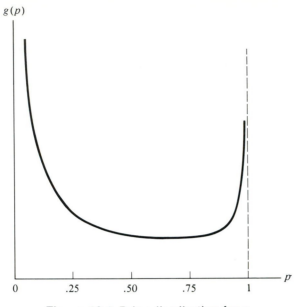

Figure 16.4 Prior distribution for p

Consider then 20 independent observations X_i on Ber(p), where p has the above prior distribution. Let $Y = X_1 + \cdots + X_4$, and $Z = X_5 + \cdots + X_{20}$. We have observed $Y = 4$ and want to know the predictive distribution of Z.

The likelihood function is $L(p) = p^4$, so the posterior p.d.f. is

$$h(p \,|\, Y = 4) \propto p^4 p^{-3/4}(1 - p)^{-1/4}, \qquad 0 < p < 1.$$

Given p, the p.f. of Z is binomial:

$$f(z \,|\, p) = \binom{16}{k} p^z (1 - p)^{16 - z}, \qquad z = 0, 1, \ldots, 16.$$

The predictive p.f. is then

$$f^*(z) = E[f(z \,|\, p)] = \binom{16}{z} \frac{\displaystyle\int_0^1 p^{z + 3.25}(1 - p)^{15.75 - z}\,dp}{\displaystyle\int_0^1 p^{3.25}(1 - p)^{-.25}\,dp}$$

$$= \binom{16}{z} \frac{\dfrac{\Gamma(z + 4.25)\Gamma(16.75 - z)}{\Gamma(21)}}{\dfrac{\Gamma(4.25)\Gamma(.75)}{\Gamma(5)}}.$$

These probabilities are shown in Table 16.1 and pictured in Figure 16.5. We found

Table 16.1

z	$f(z)$
0	.00008
1	.00036
2	.00096
3	.00205
4	.00378
5	.00637
6	.01006
7	.01510
8	.02184
9	.03069
10	.04217
11	.05701
12	.07626
13	.10168
14	.13667
15	.19004
16	.30486

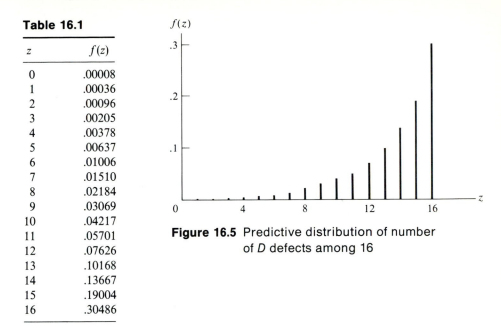

Figure 16.5 Predictive distribution of number of D defects among 16

them by first calculating $f^*(16)$ and then using the recursion relation

$$\frac{f^*(k)}{f^*(k+1)} = \frac{k+1}{k+4.25} \cdot \frac{15.75-k}{16-k}.$$

The table and figure are appropriate for presentation in court. The court may also be interested in $E(Z)$. Since Z, given p, is Bin(16, p), we have $E(Z\,|\,p) = 16p$, so

$$E(Z) = E[E(Z\,|\,p)] = 16E(p) = \frac{16 \times 4.25}{5} = 13.6,$$

from the formula $r/(r+s)$ for the mean of Beta(r, s), with $r = 4.75$ and $s = .75$. (You can verify this from Table 16.1.) Among the 20 broken windows, the expected number caused by defect D, given $Y = 4$, is then 17.6.   ■

Example 16.8b   In Problem 20 we considered an exponential distribution for life-length X, with p.d.f.

$$f(x\,|\,\lambda) = \lambda e^{-\lambda x}, \qquad x > 0.$$

Given n observations, the likelihood function is

$$L(\lambda) = \prod f(x_i\,|\,\theta) = \lambda^n e^{-n\lambda\bar{x}}, \qquad \lambda > 0.$$

Assuming a gamma prior (as in Problem 20) with p.d.f.

$$g(\lambda) \propto \lambda^{\alpha-1}e^{-\beta\lambda}, \qquad \lambda > 0,$$

we find the posterior p.d.f. to be

$$h(\lambda \mid \bar{x}) \propto \lambda^{n+\alpha-1} e^{-(n\bar{x}+\beta)}, \qquad \lambda > 0.$$

Suppose now we take a new unit from the same population. The life-length Y has p.d.f. $f(y \mid \lambda) = \lambda e^{-\lambda y}$, $y > 0$. The predictive p.d.f. of Y is then

$$f^*(y) = E_h[f(y \mid \lambda)] = \int_0^\infty \lambda e^{-\lambda y} h(\lambda \mid \bar{x}) \, d\lambda$$

$$= \frac{\displaystyle\int_0^\infty \lambda^{n+\alpha} e^{-(n\bar{x}+\beta+y)\lambda} \, d\lambda}{\displaystyle\int_0^\infty \lambda^{n+\alpha-1} e^{-(n\bar{x}+\beta)\lambda} \, d\lambda} = \frac{(n+\alpha)(n\bar{x}+\beta)^{n+\alpha}}{(n\bar{x}+\beta+y)^{n+\alpha+1}},$$

for $y > 0$. The mean of this predictive distribution is

$$\int_0^\infty y f^*(y) \, dy = \frac{n\bar{x}+\beta}{n+\alpha-1},$$

which tends to $\bar{x}$ as n becomes infinite. ∎

Chapter Perspective

This chapter's approach to statistical inference is very different from that of the rest of the book. Here we do what those who use statistics quite naturally want to do; namely, attach probabilities to hypotheses and intervals of parameter values. Bayes's theorem makes this possible. However, there is a price to pay; one must assume a prior distribution.

Prior distributions are necessarily subjective or personal, and many statisticians regard subjectivity as unscientific. So the traditional approach presented in Chapters 1 through 15 has prevailed in most statistical teaching and practice. As we have pointed out on several occasions, the traditional methods also involve subjectivity: for instance, in the interpretation of P-values, and in the setting of significance and confidence levels. Thus, *both* the traditional and the Bayesian approaches involve subjective elements; the Bayesian puts personal judgments and opinions clearly on display whereas those of the traditionalist are hidden.

Bayesian methods have been used for a number of years in the making of business decisions. Decision making under uncertainty plays an increasingly important role in many other fields of application and the approach of this chapter is ideally suited to such applications.

Solved Problems

Sections 16.1–16.2

A. Show that if the distribution of an observation X (which could be the value of a sample statistic) does not depend on the parameter θ, the posterior distribution for θ given X is the same as the prior (or unconditional) distribution.

Solution:

The likelihood function, which is the p.f. or p.d.f. of X taken as a function of θ, is in this case constant in θ. Hence, the posterior (prior times likelihood) is proportional to the prior:

$$h(\theta \mid x) \propto g(\theta) f(x \mid \theta) \propto g(\theta).$$

This can be so only if the prior and the posterior are the same.

B. Consider inference about the parameter θ in a *Rayleigh distribution*:

$$f(x \mid \theta) = (\theta x) \exp\left(\frac{-\theta x^2}{2}\right), \qquad x > 0,$$

based on a random sample of size n. Find a conjugate prior and the corresponding posterior distribution for θ.

Solution:

The sample sum of squares $Y = \Sigma x^2$ is sufficient. Given this number Y, the likelihood function is

$$L(\theta) \propto \theta^n e^{-\theta Y/2}.$$

What's needed is a family of distributions on the interval $(0, \theta)$ with the same form as L but with parameters in place of n and Y. The gamma distributions constitute such a family:

$$g(\theta) \propto \theta^{\alpha - 1} e^{-\lambda \theta}, \qquad \theta > 0.$$

With this prior, the *posterior* p.d.f. for θ is

$$h(\theta \mid Y) \propto g(\theta) L(\theta) \propto \theta^{n + \alpha - 1} e^{-\theta(\lambda + Y/2)}, \qquad \theta > 0.$$

Since this is also a gamma density, the gamma distribution is a conjugate prior for the Rayleigh likelihood.

C. You are a military intelligence expert sent on a secret mission to an enemy training area to count enemy tanks. You know that the enemy has (not too wisely) numbered its tanks in the area from 1 to N. You observe tanks individually (perhaps through a telescope), randomly and independently (with replacement). You observe six tanks, whose numbers are 2, 7, 13, 5, 8. Assume a uniform prior distribution for N and find the posterior.

Solution:

The number of positive integers is infinite, so we can't actually assign equal probabilities to them. However, suppose we simply assign some constant weight to each positive integer and see where treating these as probabilities leads. We know that $X_{(5)}$ is sufficient for N (see Problem 15 of Chapter 8), so the likelihood function depends only on $X_{(5)}$ (and the sample size):

$$L(N) = \begin{cases} \dfrac{1}{N^5}, & \text{if } N \geq X_{(5)}, \\ 0, & \text{otherwise.} \end{cases}$$

Using Bayes's theorem with a constant prior gives

$$h(N \mid X_{(5)} = 13) \propto \frac{1}{N^5}, \qquad N = 13, 14, \ldots .$$

We take care of the proportionality constant by dividing by the sum:

$$h(N \mid X_{(5)} = 13) = \frac{\dfrac{1}{N^5}}{\displaystyle\sum_{k=13}^{\infty} \dfrac{1}{k^5}}, \qquad N = 13, 14, \ldots .$$

So we obtain a proper p.f. for h even though what we used for g is not a p.f. The sum in the denominator can be evaluated numerically; it is approximately 6.5×10^{-7}. A partial table of posterior probabilities is as follows:

k	13	14	15	16	17	$\cdots$
$h(k \mid 13)$	.319	.204	.135	.092	.064	$\cdots$

(Although this was obtained with an improper prior, it is in fact the limit of posterior distributions obtained using a sequence of proper priors—say, $g(N) = 1/M$ for $N = 1, 2, \ldots, M$.)

Sections 16.3–16.4

D. Suppose that a parameter θ has three possible values, and consider the problem of choosing one of three possible actions when the loss function is defined by the following table:

	a_1	a_2	a_3
θ_1	0	2	4
θ_2	3	5	0
θ_3	6	0	2

Given the prior probabilities $(\frac{3}{6}, \frac{2}{6}, \frac{1}{6})$ for $(\theta_1, \theta_2, \theta_3)$,
(a) find the Bayes action.
(b) suppose Z is a discrete random variable with distributions as follows under the various states:

	θ_1	θ_2	θ_3
z_1	.4	.2	.7
z_2	.6	.8	.3

Find the Bayes action given $Z = z_1$.

Solution:

(a) The expected value of the loss with respect to the given distribution for θ is

$$\frac{1}{6}(0 \times 3 + 3 \times 2 + 6 \times 1) = \frac{12}{6} \qquad \text{for } a_1,$$

$$\frac{1}{6}(2 \times 3 + 5 \times 2 + 0 \times 1) = \frac{16}{6} \qquad \text{for } a_2,$$

$$\frac{1}{6}(4 \times 3 + 0 \times 2 + 2 \times 1) = \frac{14}{6} \qquad \text{for } a_3.$$

The Bayes action is a_1, since it gives the smallest mean loss.

(b) Given $Z = z_1$, we first find the new distribution for θ:

$$h(\theta \mid z_1) = \frac{1}{6}(3 \times .4, 2 \times .2, 1 \times .7) = \left(\frac{12}{23}, \frac{4}{23}, \frac{7}{23}\right)$$

The expected losses are then

$$\frac{1}{2.3}(0 \times 12 + 3 \times 4 + 6 \times 7) = \frac{54}{23} \qquad \text{for } a_1,$$

$$\frac{1}{2.3}(2 \times 12 + 5 \times 4 + 0 \times 7) = \frac{44}{23} \qquad \text{for } a_2,$$

$$\frac{1}{2.3}(4 \times 12 + 0 \times 4 + 2 \times 7) = \frac{62}{23} \qquad \text{for } a_3,$$

The Bayes action is therefore a_2.

E. Consider a random sample from $f(x \mid \beta) = \beta x e^{-\beta x^2/2}$ for $x > 0$ and $\beta > 0$. Find the Bayes estimate of β assuming a gamma prior: $g(\beta) \propto \beta^{\alpha-1} e^{-\lambda \beta}$, and a quadratic loss function.

Solution:

Given the sample $(X_1, \ldots, X_n)$, the likelihood function is

$$L(\beta) = \beta^n e^{-\beta \Sigma X^2/2}, \qquad \beta > 0.$$

Multiplying g and L, we obtain the posterior as

$$h(\beta \mid X) = \beta^{n+\alpha-1} e^{-\beta[\Sigma X^2/2 + \lambda]}.$$

With quadratic loss, the estimate is the mean of this posterior distribution, which is

$$\text{Gam}\left(n + \alpha, \frac{\Sigma X^2}{2} + \lambda\right).$$

From Section 6.3, we know the mean to be the ratio of these parameters:

$$\frac{n + \alpha}{\dfrac{\Sigma X^2}{2} + \lambda}.$$

Sections 16.5–16.8

F. A patient needs an organ transplant; to judge histocompatibility with possible donors, the patient's human leukocyte antigen (HLA) is determined. National blood bank records show that in a random sample of 5,000 individuals, only one has the patient's HLA type. Let λ denote the population rate per 5,000 for this HLA type and assume a uniform (improper) prior for λ: $g(\lambda) = 1$. Find a 95% probability interval for λ of the form $(\lambda_{.025}, \lambda_{.975})$, where λ_p is the $100p$th percentile of the distribution of λ given the data (that is, given one HLA in the sample of 5,000).

Solution:

The distribution of the number of compatible HLA's in the sample of 5,000 is $\text{Bin}(5,000, p)$; but we'll approximate it with $\text{Poi}(\lambda)$, where $\lambda = 5,000p$, the population number of HLA's per 5,000. Then, given one HLA in 5,000.

$$L(\lambda) = P(\text{one HLA} \mid \lambda) = f_{\text{Poi}(\lambda)}(1) = \lambda e^{-\lambda}.$$

With this and the given prior, the posterior p.d.f. for λ is

$$h(\lambda \mid \text{data}) \propto \lambda e^{-\lambda}, \qquad \lambda > 0,$$

and c.d.f.

$$H(\lambda \mid \text{data}) = \int_0^\lambda x e^{-x} dx = 1 - e^{-\lambda}(1 + \lambda) = 1 - F_{\text{Poi}(\lambda)}(1).$$

Since $e^{-\lambda}(1 + \lambda)$ is the probability of 0 or 1 in $\text{Poi}(\lambda)$, we look in the Poisson table, finding $e^{-\lambda}(1 + \lambda) = .974$ for $c = .25$ and $.982$ when $c = .20$. Interpolating, we find $\lambda_{.025} \doteq .243$. Similarly, $\lambda_{.975} \doteq 5.57$. So a 95% probability interval for the population rate λ is from about one in 20,000 to one in 900.

G. A treatment for respiratory failure in newborn infants termed ECMO (extracorporeal membrane oxygenation) was the object of a randomized study in 1985.[6] The result was that eleven patients given ECMO survived, the only patient in the control group (standard care) survived, and the only patient in the control group (standard care) died. (The therapy assignment scheme favored a therapy if it had been more successful.) Let p_T and p_C denote the probabilities of survival under the treatment and control, respectively. Assume a uniform distribution in the unit square for (p_T, p_C) and find the probability that $p_T > p_C$.

Solution:

The likelihood function is the product of likelihoods (assuming independence of the treatment and control samples):

$$L \propto p_T^{11}(1 - p_T)^0 \cdot p_C^0(1 - p_C)^1.$$

---

[6] R. H. Bartlett et al., "Extra-corporeal circulation in neonatal respiratory failure: A prospective randomized study," *Pediatrics 76* (1985), 479–487.

Given that the prior is uniform, the posterior is proportional to the likelihood:

$$h(p_T, p_C \,|\, \text{data}) = \frac{p_T^{11}(1 - p_C)^1}{\displaystyle\int_0^1 \int_0^1 p_T^{11}(1 - p_C)^1 \, dp_T \, dp_C} = 24 p_T^{11}(1 - p_C)^1,$$

for p_T and p_C between 0 and 1. Then

$$P(p_T > p_C) = 24 \int_0^1 \int_0^x x^{11}(1 - y) \, dy \, dx = 24 \int_0^1 x^{11}\left(x - \frac{x^2}{2}\right) dx = \frac{90}{91}.$$

H. A couple has five children, all boys. Suppose that p, the probability p of a boy for this couple, is constant. Assume further that the sexes of successive children are independent given p. Find the probability that their next child is a boy when

(a) the prior for p is uniform on $(0, 1)$.

(b) the prior is more concentrated near $p = \frac{1}{2}$, with p.d.f.

$$g(p) \propto p^{12}(1 - p)^{11}.$$

Solution:

(a) According to Bayes's theorem,

$$h(p \,|\, 5 \text{ boys}) \propto p^5, \qquad 0 < p < 1.$$

The unconditional probability of a boy at the next trial, from (3) of Section 16.8, is the posterior mean, $\frac{6}{7}$. [This is another instance of Laplace's rule of succession (see Example 16.4c).

(b) With this prior, the probability of a boy in the absence of data [from (3) of Section 16.8] is $E(p) = \frac{13}{25} \doteq .52$. (The proportion of boys varies among different populations, but .52 is typical.) Given the data (first five children were male), the posterior p.d.f. of p is

$$h(p \,|\, \text{data}) \propto p^{17}(1 - p)^{11}, \qquad 0 < p < 1,$$

[according to (22) of Section 16.2 with $r = 12$, $s = 11$, and $Y = n = 5$]. Now the probability that the sixth child is a boy is

$$E(p \,|\, 5 \text{ boys}) = \frac{18}{30} = .6,$$

an instance of (4) of Section 16.4.

Problems

For problems marked with an asterisk, answers are provided in Appendix 2.

Sections 16.1–16.2

1. Let μ denote the mean weight of adult male emperor penguins. Use the technique of Section 16.1 to elicit a friend's prior distribution for μ, to the extent of determining the median and quartiles, and the first and 99th percentiles. (Since this distribution is personal, there is no single correct answer.)

*2.   Example 10a described Oscar's test of Felix's claim of ESP. Let p denote the probability that Felix correctly identifies a card shown to him. Given six successes in ten trials, find the posterior p.d.f. of p when
   (a)   the prior is uniform on $(0, 1)$.
   (b)   the prior p.d.f. is proportional to $p(1 - p)^4$.
   (c)   the prior is proportional to $p^5(1 - p)$.

3.   Repeat the calculations of Problem 2, but with six successes in ten trials replaced by ten out of ten.

4.   Referring to Example 16.2a (page 675), find a juror's posterior probability that the accused is the father, in view of the blood-type data, if his prior probability from *other* evidence and testimony is
   (a)   .8   (b)   .2.

5.   Referring to Example 16.2b, suppose only 25 of the 40 persons involved are better on therapy. Find $P(H_0 \mid \text{data})$ given that the prior probabilities are equal, as in the example.

*6.   Given a single observation X from Poi(λ), find the posterior distribution for λ when the prior has p.d.f. $g(\lambda) = 3e^{-3\lambda}$, $\lambda > 0$.

7.   Suppose X is geometric: $f(x \mid p) = p(1 - p)^x$ for $x = 0, 1, \ldots$. Given a random sample of size n, find the posterior distribution for p when the prior is Beta(r, s).

8.   For given θ, observations (X_1, X_2) on X are independent with p.d.f. $f(x \mid \theta)$, and a prior for θ with p.d.f. $g(\theta)$, find the posterior p.d.f. for θ in two ways: (i) Find the posterior after observing X_1 and use this as a prior with the likelihood defined by X_2. (ii) Find the posterior using the prior g, with the likelihood defined by the pair (X_1, X_2). Show that the results are the same in (i) and (ii).

9.   Find the posterior distribution of μ given a random sample of size n from $N(\mu, 1)$, assuming a uniform prior on $(-\infty, \infty)$. (A uniform distribution on an infinite interval is improper, but take $g(\mu) \equiv 1$ and proceed as though it were proper.) Verify that the result is the same as the limit of the posterior found using a proper normal prior for μ as the prior precision tends to 0.

10.   Show that if all of the prior probability for θ is concentrated on a single value θ_0, then (with one exceptional case) the posterior distribution is the same as the prior. That is, no amount of data can change the mind of a "know-it-all" scientist.

Sections 16.3–16.4

*11.   Refer to Solved Problem D(b).
   (a)   Find the Bayes action when one observes $Z = z_2$, assuming the prior given in that problem. If the losses are in dollars, how much should one be willing to pay to observe Z?
      (*Hint*: Find the difference between the Bayes loss with no data and the expected Bayes loss when Z is observed.)
   (b)   Find the Bayes actions assuming a uniform prior for θ
      (i)   given $Z = z_1$ and (ii) given $Z = z_2$.

*12.   Consider estimating a mean μ whose distribution is $N(\mu_0, 1)$. Suppose the loss function $\ell(\mu, a)$ is 0 or 1 according as $|\mu - a| \leq 1$ or $|\mu - a| > 1$. Find the Bayes estimate.

13. Consider a two-action decision problem with unknown parameter μ and losses as follows:

$$\ell(\mu, a_1) = A \quad \text{when } \mu > \mu^* \quad \text{and } 0 \text{ when } \mu \leq \mu^*,$$
$$\ell(\mu, a_2) = B \quad \text{when } \mu < \mu^* \quad \text{and } 0 \text{ when } \mu \geq \mu^*$$

Assume that $\mu \sim N(\mu_0, 1)$ and find the Bayes action (which will depend on the relation between μ^* and μ_0).

*14. Obtain the Bayes estimate of p
  (a) for each case in Problem 2, assuming quadratic loss.
  (b) for ten successes in ten trials, with a uniform prior and assuming the loss $\ell(p, a) = |p - a|$.

15. Obtain a general formula for the posterior mean of the Poisson parameter λ, assuming a random sample of size n, a gamma prior with $g(\lambda) \propto \lambda^{\alpha-1}e^{-\beta\lambda}$ for $\lambda > 0$, and a quadratic loss function.

16. For the case of a continuous distribution, show that $E|X - a|$ is a minimum when a is a population median.
  [*Hint*: Write the integral for $E|X - a|$ as the sum of two terms, one an integral over $(-\infty, a)$ where $|x - a| = a - x$, and the other an integral over (a, ∞), where $|x - a| = x - a$. Then calculate the derivative with respect to a, using the fundamental theorem of calculus.]

*17. Consider a random sample of size n from $N(0, 1/\theta)$. (So θ is the precision of each observation.) Assume a quadratic loss function for estimating θ.
  (a) Assuming a prior for θ that is Gam(α, λ), find the (no data) Bayes estimate of θ.
  (b) If the prior for θ is Gam(5, 5), the sample size is $n = 8$, and the observations are $(3, -1, 1, -1, -1, 0, 2, 0)$, find the Bayes estimate of θ.

18. Suppose X has p.d.f. $\theta e^{-\theta x}$, $x > 0$. Given a random sample of size $n = 5$ in which $\Sigma X_i = 10$, find the Bayes estimate of θ, assuming a quadratic loss and prior density $g(\theta) = \theta e^{-\theta}$, $\theta > 0$.

*19. Given Y successes in n independent trials of a Bernoulli (p) experiment, find the Bayes estimate of p with quadratic loss and (improper) prior $g(p) = [p(1 - p)]^{-1}$, $0 < p < 1$.

20. Problem 7 of Chapter 13 gave data on length of life in hours for a certain type of radar indicator tube. It was found there that an exponential distribution provides an adequate description of life length X. Assume that in fact X is exponential, with p.d.f.

$$f(x\,|\,\lambda) = \lambda e^{-\lambda x}, \quad x > 0,$$

for some $\lambda > 0$ that is not precisely known. Suppose that prior to obtaining the data, we had a gamma distribution for the failure rate λ with mean $\mu_\lambda = \sigma_\lambda = .0005$. From the earlier problem, $\bar{X} = 297.5$ and $n = 100$. Find the Bayes estimate of λ.

*21. In Problem 14, you found Bayes estimates (with quadratic loss) of the Bernoulli parameter p, given six successes in ten independent trials. Suppose the sampling had been planned to continue until six successes were obtained, and that it happened to take ten trials. How would this affect the Bayes estimates?

Sections 16.5–16.8

*22. In the dental study quoted in Problem 1 of Chapter 12, the mean difference in number of new cavities between the MFP and SF samples is approximately $N(\theta, 1.252)$. The null hypothesis is $\theta \leq 0$. Assume that the prior for θ is $N(4, 4)$ [that is, $v = 4$ and $\tau^2 = 4$ in (7) of Section 16.2].

   (a) Find the posterior distribution of θ given the observed mean difference 2.41 (from the study), and from this find the probability of the null hypothesis.

   (b) Find a 95% confidence interval for θ and calculate the posterior probability of this interval.

23. Repeat (a) and (b) of the preceding problem, using a normal prior with mean 1 and variance 16.

24. Verify the value $B(\text{accept}) = 1.57$ in Example 16.7a for the normal prior with mean 7.3 and s.d. .40.

*25. Consider testing $H_0: p \leq .5$ against $H_A: p > .5$, where p is the parameter of a Bernoulli distribution. Given the losses

$$\ell(p, a) = \begin{cases} 0 & \text{if } a \text{ is the correct action,} \\ 3 & \text{if } a = \text{reject } H_0 \text{ when } p \leq .5, \\ 2 & \text{if } a = \text{accept } H_0 \text{ when } p > .5, \end{cases}$$

   find the Bayes action if our prior p.d.f. for p is Beta(7, 3), and in ten independent trials of Ber(p), we obtain three successes.

26. The prior distribution used in Example 16.5b is known as *Jeffreys' prior*, defined by the p.d.f.

$$g(p) = C \cdot [p(1 - p)]^{-1/2}, 0 < p < 1.$$

   (a) Find C, the constant of proportionality.

   (b) Find the (unconditional) probability of success, assuming Jeffreys' prior.

   (c) Suppose you observe a success. What is the new probability of success at the next trial? (That is, given that one trial results in success, find the predictive distribution for success/failure at a second trial.)

*27. Referring to Problem 2, suppose the prior puts probability .9 on the value $p = \frac{1}{4}$ and spreads the remaining .1 uniformly over the interval (0, 1).

   (a) Calculate $P(Y = 6) = E[P(Y = 6 \mid p)]$ as a sum of two terms, one the product of $P(Y = 6 \mid \frac{1}{4})$ and $P(\frac{1}{4})$, and the other as an integral of $P(Y = 6 \mid p)$ with respect to the probability element $.1\, dp$.

   (b) Find $P(p = \frac{1}{4} \mid Y = 6)$ using Bayes's theorem.

   (c) Repeat (a) and (b) with a prior that puts only .5 probability on $p = \frac{1}{4}$ and spreads the rest uniformly over (0, 1).

28. Referring to Solved Problem F, find a 90% probability interval for λ if the 5,000 in the sample had included no one of the patient's HLA type, assuming the same prior as in Problem F.

∗29. Refer to Solved Problem G.

    (a) Show that p_T and p_C are independent, given the data in that problem.

    (b) Find the probability that the next patient survives (i) if given the new treatment, and (ii) if given the control treatment.

    (c) The research report cited in Problem G says that past experience shows $p_C \doteq .2$. With a uniform prior for p_T, find the posterior probability that p_T exceeds this value .2.

30. A university professor (who was trying to convince a U.S. senator to vote against the MX missile program) posed this question: If an enemy missile has been fired successfully ten times in ten trials, what is the probability that in the next five firings, there are no failures? If one takes $\hat{p} = \frac{10}{10} = 1$ as the value of p, the answer is $(1)^5 \doteq 1$. However, this is unrealistic. Suppose (prior to learning the data) we'd judge that p may be near 1 and assume a beta prior for p with p.d.f.

$$g(p) \propto p^4, \qquad 0 < p < 1,$$

which peaks at $p = 1$, with $E(p) = \frac{5}{6}$. Given this prior and the assumed data, find the predictive distribution of the number of failures in the next five firings.

References & Further Readings

[1] Berger, J. O., *Statistical Decision Theory and Bayesian Analysis*, 2nd Ed., New York: Springer Verlag, 1985.

[2] Bickel, P., and K. Doksum, *Mathematical Statistics*, Oakland, Calif.: Holden-Day, 1977.

[3] Cleveland, W. S., *The Elements of Graphing Data*, Pacific Grove, Calif.: Wadsworth Advanced Books, 1985.

[4] Cox, D. R., and D. V. Hinkley, *Theoretical Statistics*, London: Chapman & Hall, Ltd., 1974.

[5] Cramér, H., *Mathematical Methods of Statistics*, Princeton, N. J.: Princeton University Press, 1946.

[6] Draper N., and H. Smith, *Applied Regression Analysis*, New York: Wiley, 1981.

[7] Feller, W., *An Introduction to Probability Theory and Its Applications*, Vol. 1, 3rd Ed., New York: Wiley, 1968.

[8] Ferguson, T., *Mathematical Statistics*, New York: Academic Press, 1967.

[9] Fienberg, S., Y. Bishop, and P. Holland, *Categorical Data Analysis*, Cambridge, Mass.: M. I. T. Press, 1975.

[10] Fisher, R. A., *Statistical Methods and Scientific Inference*, 3rd Ed, New York: Hafner Press, 1973.

[11] Hicks, C. R., *Fundamental Concepts in the Design of Experiments*, 2nd Ed., New York: Holt, Rinehart & Winston, 1973.

[12] Hollander, M., and D. A. Wolfe, *Nonparametric Statistical Methods*, New York: Wiley, 1973.

[13] Lehmann, E., *Theory of Point Estimation*, New York: Wiley, 1983.

[14] Lehmann, E., *Nonparametrics: Statistical Methods Based on Ranks*, San Francisco, Calif.: Holden-Day, 1975.

[15] Lindgren, B. W., *Statistical Theory*, 3rd Ed., New York: Macmillan, 1976.

[16] Rao, C. R., *Linear Statistical Inference and Its Applications*, 2nd Ed., New York: Wiley, 1973.

[17] Snedecor, G. W., and W. G. Cochran, *Statistical Methods*, 7th Ed., Ames, Iowa: Iowa State University Press, 1980.

[18] Tukey, J., *Exploratory Data Analysis*, Reading, Mass: Addison-Wesley, 1977.

[19] Weisberg, S., *Applied Regression Analysis*, 2nd Ed., New York: Wiley, 1985.

[20] Wilks, S. S., *Mathematical Statistics*, New York: Wiley, 1962.

Tables

Ia.   Binomial probabilities   708

Ib.   Cumulative binomial probabilities   710

II.   Standard normal c.d.f.   712

IIa.   Standard normal percentiles   714

IIb.   Two-tailed standard normal probabilities   714

IIc.   Confidence interval multipliers, large n 714

IIIa.   Confidence interval multipliers, σ unknown   715

IIIb.   Tail-probabilities of student's t-distribution   716

IIIc.   Percentiles of the t-distribution   719

IV.   Cumulative Poisson probabilities   720

Va.   Tail-probabilities of the chi-square distribution   723

Vb.   Percentiles of the chi-square distribution   726

VI.   Tail-probabilities of the one-sample Wilcoxon signed-rank statistic   727

VII.   Tail-probabilities of the two-sample Wilcoxon rank-sum statistic   728

VIIIa.   Tail-probabilities of the F-distribution   730

VIIIb.   Percentiles of the F-distribution   735

IXa.   Tail-probabilities for the one-sample Kolmogorov–Smirnov statistic   737

IXb.   Critical values for the one-sample Kolmogorov–Smirnov statistic   738

Xa.   Tail-probabilities for the two-sample Smirnov statistic   739

Xb.   Critical values for the two-sample Smirnov statistic   740

XI.   Critical values for the Lilliefors distribution   741

XII.   Expected values of normal order statistics   742

XIII.   Distribution of the standardized range   743

XIV.   Random digits   744

XV.   Normal random numbers   749

Table Ia Binomial Probabilities

$$P(k \text{ successes in } n \text{ trials}) = \binom{n}{k} p^k (1-p)^{n-k}$$

							p							
n	k	.01	.05	.10	.15	1/6	.20	.25	.30	1/3	.35	.40	.45	.50
5	0	.9510	.7738	.5905	.4437	.4019	.3277	.2373	.1681	.1317	.1160	.0778	.0503	.0312
	1	.0480	.2036	.3280	.3915	.4019	.4096	.3955	.3601	.3292	.3124	.2592	.2059	.1562
	2	.0010	.0214	.0729	.1382	.1608	.2048	.2637	.3087	.3292	.3364	.3456	.3369	.3125
	3	.0000	.0011	.0081	.0244	.0322	.0512	.0879	.1323	.1646	.1811	.2304	.2757	.3125
	4	.0000	.0000	.0004	.0022	.0032	.0064	.0146	.0283	.0412	.0488	.0768	.1128	.1562
	5	.0000	.0000	.0000	.0001	.0001	.0003	.0010	.0024	.0041	.0053	.0102	.0185	.0312
6	0	.9415	.7351	.5314	.3771	.3349	.2621	.1780	.1176	.0878	.0754	.0467	.0277	.0156
	1	.0571	.2321	.3543	.3993	.4019	.3932	.3560	.3025	.2634	.2437	.1866	.1359	.0938
	2	.0014	.0305	.0984	.1762	.2009	.2458	.2966	.3241	.3293	.3280	.3110	.2780	.2344
	3	.0000	.0021	.0146	.0415	.0536	.0819	.1318	.1852	.2195	.2355	.2765	.3032	.3125
	4	.0000	.0001	.0012	.0055	.0080	.0154	.0330	.0595	.0823	.0951	.1382	.1861	.2344
	5	.0000	.0000	.0001	.0004	.0006	.0015	.0044	.0102	.0165	.0205	.0369	.0609	.0938
	6	.0000	.0000	.0000	.0000	.0000	.0001	.0002	.0007	.0014	.0018	.0041	.0083	.0156
7	0	.9321	.6983	.4783	.3206	.2791	.2097	.1335	.0824	.0585	.0490	.0280	.0152	.0078
	1	.0659	.2573	.3720	.3960	.3907	.3670	.3115	.2471	.2049	.1848	.1306	.0872	.0547
	2	.0020	.0406	.1240	.2097	.2344	.2753	.3115	.3177	.3073	.2985	.2613	.2140	.1641
	3	.0000	.0036	.0230	.0617	.0781	.1147	.1730	.2269	.2561	.2679	.2903	.2918	.2734
	4	.0000	.0002	.0026	.0109	.0156	.0287	.0577	.0972	.1280	.1442	.1935	.2388	.2734
	5	.0000	.0000	.0002	.0012	.0018	.0043	.0115	.0250	.0384	.0466	.0774	.1172	.1641
	6	.0000	.0000	.0000	.0001	.0001	.0004	.0013	.0036	.0064	.0084	.0172	.0320	.0547
	7	.0000	.0000	.0000	.0000	.0000	.0000	.0001	.0002	.0005	.0006	.0016	.0037	.0078
8	0	.9227	.6634	.4305	.2725	.2326	.1678	.1001	.0576	.0390	.0319	.0168	.0084	.0039
	1	.0746	.2793	.3826	.3847	.3721	.3355	.2670	.1977	.1561	.1373	.0896	.0548	.0312
	2	.0026	.0515	.1488	.2376	.2605	.2936	.3115	.2965	.2731	.2587	.2090	.1569	.1094
	3	.0001	.0054	.0331	.0839	.1042	.1468	.2076	.2541	.2731	.2786	.2787	.2568	.2187
	4	.0000	.0004	.0046	.0185	.0260	.0459	.0865	.1361	.1707	.1875	.2322	.2627	.2734
	5	.0000	.0000	.0004	.0026	.0042	.0092	.0231	.0467	.0683	.0808	.1239	.1719	.2187
	6	.0000	.0000	.0000	.0002	.0004	.0011	.0038	.0100	.0171	.0217	.0413	.0703	.1094
	7	.0000	.0000	.0000	.0000	.0000	.0001	.0004	.0012	.0024	.0033	.0079	.0164	.0312
	8	.0000	.0000	.0000	.0000	.0000	.0000	.0000	.0001	.0002	.0002	.0007	.0017	.0039
9	0	.9135	.6302	.3874	.2316	.1938	.1342	.0751	.0404	.0260	.0207	.0101	.0046	.0020
	1	.0830	.2985	.3874	.3679	.3489	.3020	.2253	.1556	.1171	.1004	.0605	.0339	.0176
	2	.0034	.0629	.1722	.2597	.2791	.3020	.3003	.2668	.2341	.2162	.1612	.1110	.0703
	3	.0001	.0077	.0446	.1069	.1302	.1762	.2336	.2668	.2731	.2716	.2508	.2119	.1641
	4	.0000	.0006	.0074	.0283	.0391	.0661	.1168	.1715	.2048	.2194	.2508	.2600	.2461
	5	.0000	.0000	.0008	.0050	.0078	.0165	.0389	.0735	.1024	.1181	.1672	.2128	.2461
	6	.0000	.0000	.0001	.0006	.0013	.0028	.0087	.0210	.0341	.0424	.0743	.1160	.1641
	7	.0000	.0000	.0000	.0000	.0001	.0003	.0012	.0039	.0073	.0098	.0212	.0407	.0703
	8	.0000	.0000	.0000	.0000	.0000	.0000	.0001	.0004	.0009	.0013	.0035	.0083	.0176
	9	.0000	.0000	.0000	.0000	.0000	.0000	.0000	.0000	.0001	.0001	.0003	.0008	.0020

Table Ia (continued)

							p							
n	k	.01	.05	.10	.15	1/6	.20	.25	.30	1/3	.35	.40	.45	.50
10	0	.9044	.5987	.3487	.1969	.1615	.1074	.0563	.0282	.0173	.0135	.0060	.0025	.0010
	1	.0914	.3151	.3874	.3474	.3230	.2684	.1877	.1211	.0867	.0725	.0403	.0207	.0098
	2	.0042	.0746	.1937	.2759	.2907	.3020	.2816	.2335	.1951	.1757	.1209	.0763	.0439
	3	.0001	.0105	.0574	.1298	.1550	.2013	.2503	.2668	.2601	.2522	.2150	.1665	.1172
	4	.0000	.0010	.0112	.0401	.0543	.0881	.1460	.2001	.2276	.2377	.2508	.2384	.2051
	5	.0000	.0001	.0015	.0085	.0130	.0264	.0584	.1029	.1366	.1536	.2007	.2340	.2461
	6	.0000	.0000	.0001	.0012	.0022	.0055	.0162	.0368	.0569	.0689	.1115	.1596	.2051
	7	.0000	.0000	.0000	.0001	.0002	.0008	.0031	.0090	.0163	.0212	.0425	.0746	.1172
	8	.0000	.0000	.0000	.0000	.0000	.0001	.0004	.0014	.0030	.0043	.0106	.0229	.0439
	9	.0000	.0000	.0000	.0000	.0000	.0000	.0000	.0001	.0003	.0005	.0016	.0042	.0098
	10	.0000	.0000	.0000	.0000	.0000	.0000	.0000	.0000	.0000	.0000	.0001	.0003	.0010
11	0	.8953	.5688	.3138	.1673	.1346	.0859	.0422	.0198	.0116	.0088	.0036	.0014	.0005
	1	.0995	.3293	.3835	.3248	.2961	.2362	.1549	.0932	.0636	.0518	.0266	.0125	.0054
	2	.0050	.0867	.2131	.2866	.2961	.2953	.2581	.1998	.1590	.1395	.0887	.0513	.0054
	3	.0002	.0137	.0710	.1517	.1777	.2215	.2581	.2568	.2384	.2254	.1774	.1259	.0806
	4	.0000	.0014	.0158	.0536	.0711	.1107	.1721	.2201	.2384	.2428	.2365	.2060	.1611
	5	.0000	.0001	.0025	.0132	.0199	.0388	.0803	.1321	.1669	.1830	.2207	.2360	.2256
	6	.0000	.0000	.0003	.0023	.0040	.0097	.0268	.0566	.0835	.0985	.1471	.1931	.2256
	7	.0000	.0000	.0000	.0003	.0006	.0017	.0064	.0173	.0298	.0379	.0701	.1128	.1611
	8	.0000	.0000	.0000	.0000	.0001	.0002	.0011	.0037	.0075	.0102	.0234	.0462	.0806
	9	.0000	.0000	.0000	.0000	.0000	.0000	.0001	.0005	.0012	.0018	.0052	.0126	.0269
	10		.0000	.0000	.0000	.0000	.0000	.0000	.0000	.0001	.0002	.0007	.0021	.0054
	11		.0000	.0000	.0000	.0000	.0000	.0000	.0000	.0000	.0000	.0000	.0000	.0005
12	0	.8864	.5404	.2824	.1322	.1122	.0687	.0317	.0138	.0077	.0057	.0022	.0008	.0002
	1	.1074	.3413	.3766	.3012	.2692	.2062	.1267	.0712	.0462	.0368	.0174	.0075	.0029
	2	.0060	.0988	.2301	.2924	.2961	.2835	.2323	.1678	.1272	.1088	.0639	.0339	.0161
	3	.0002	.0173	.0852	.1720	.1974	.2362	.2581	.2397	.2120	.1954	.1419	.0923	.0537
	4	.0000	.0021	.0213	.0683	.0888	.1329	.1936	.2311	.2384	.2367	.2128	.1700	.1208
	5	.0000	.0002	.0038	.0193	.0284	.0532	.1032	.1585	.1908	.2039	.2270	.2225	.1934
	6	.0000	.0000	.0005	.0040	.0066	.0155	.0401	.0792	.1113	.1281	.1655	.2124	.2256
	7	.0000	.0000	.0000	.0006	.0011	.0033	.0115	.0291	.0477	.0591	.1009	.1489	.1934
	8	.0000	.0000	.0000	.0001	.0001	.0005	.0024	.0078	.0149	.0199	.0420	.0762	.1208
	9	.0000	.0000	.0000	.0000	.0000	.0001	.0004	.0015	.0033	.0048	.0125	.0277	.0537
	10	.0000	.0000	.0000	.0000	.0000	.0000	.0000	.0002	.0005	.0008	.0025	.0068	.0161
	11	.0000	.0000	.0000	.0000	.0000	.0000	.0000	.0000	.0000	.0001	.0003	.0010	.0029
	12	.0000	.0000	.0000	.0000	.0000	.0000	.0000	.0000	.0000	.0000	.0000	.0001	.0002

NOTE: For $p > .5$, reverse the roles of p and $q = 1 - p$. (For example, the probability of k successes when $p = .7$ is found as the entry for $n - k$ under $p = .3$.)

Table Ib Cumulative Binomial Probabilities

$$P(\text{at least } k \text{ successes in } n \text{ trials}) = \sum_{i=k}^{n} \binom{n}{i} p^i (1-p)^{n-i}$$

n	k	.01	.05	.10	.15	1/6	.20	.25	.30	1/3	.35	.40	.45	.50
5	1	.0490	.2262	.4095	.5563	.5981	.6723	.7627	.8319	.8683	.8840	.9222	.9497	.9688
	2	.0010	.0226	.0815	.1648	.1962	.2627	.3672	.4718	.5391	.5716	.6630	.7538	.8125
	3	.0000	.0012	.0086	.0266	.0355	.0579	.1035	.1631	.2099	.2352	.3174	.4069	.5000
	4	.0000	.0000	.0005	.0022	.0033	.0067	.0156	.0308	.0453	.0540	.0870	.1312	.1875
	5	.0000	.0000	.0000	.0001	.0001	.0003	.0010	.0024	.0041	.0053	.0102	.0185	.0313
6	1	.0585	.2649	.4686	.6229	.6651	.7379	.8220	.8824	.9122	.9246	.9533	.9723	.9844
	2	.0015	.0328	.1143	.2235	.2632	.3446	.4661	.5798	.6488	.6809	.7667	.8364	.8906
	3	.0000	.0022	.0159	.0473	.0623	.0989	.1694	.2557	.3196	.3529	.4557	.5585	.6563
	4	.0000	.0001	.0013	.0059	.0087	.0170	.0379	.0705	.1001	.1174	.1792	.2553	.3438
	5	.0000	.0000	.0001	.0004	.0007	.0016	.0046	.0109	.0178	.0223	.0410	.0692	.1094
	6	.0000	.0000	.0000	.0000	.0000	.0001	.0002	.0007	.0014	.0018	.0041	.0083	.0156
7	1	.0679	.3017	.5217	.6794	.7209	.7903	.8665	.9176	.9415	.9510	.9720	.9848	.9922
	2	.0020	.0444	.1497	.2834	.3302	.4233	.5551	.6706	.7366	.7662	.8414	.8976	.9375
	3	.0000	.0038	.0257	.0738	.0958	.1480	.2436	.3529	.4294	.4677	.5801	.6836	.7734
	4	.0000	.0002	.0027	.0121	.0176	.0333	.0706	.1260	.1733	.1998	.2898	.3917	.5000
	5	.0000	.0000	.0002	.0012	.0020	.0047	.0129	.0288	.0453	.0556	.0963	.1529	.2266
	6	.0000	.0000	.0000	.0001	.0001	.0004	.0013	.0038	.0069	.0090	.0188	.0357	.0625
	7	.0000	.0000	.0000	.0000	.0000	.0000	.0001	.0002	.0005	.0006	.0016	.0037	.0078
8	1	.0773	.3366	.5695	.7275	.7674	.8322	.8999	.9424	.9610	.9681	.9832	.9916	.9961
	2	.0027	.0572	.1869	.3428	.3953	.4967	.6329	.7447	.8049	.8309	.8936	.9368	.9648
	3	.0001	.0058	.0381	.1052	.1348	.2031	.3215	.4482	.5318	.5722	.6846	.7799	.8555
	4	.0000	.0004	.0050	.0214	.0307	.0563	.1138	.1941	.2587	.2936	.4059	.5230	.6367
	5	.0000	.0000	.0004	.0029	.0046	.0104	.0273	.0580	.0879	.1061	.1737	.2604	.3633
	6	.0000	.0000	.0000	.0002	.0004	.0012	.0042	.0113	.0197	.0253	.0498	.0885	.1445
	7	.0000	.0000	.0000	.0000	.0000	.0001	.0004	.0013	.0026	.0036	.0085	.0181	.0352
	8	.0000	.0000	.0000	.0000	.0000	.0000	.0000	.0001	.0002	.0002	.0007	.0017	.0039
9	1	.0865	.3698	.6126	.7684	.8062	.8658	.9249	.9596	.9740	.9793	.9899	.9954	.9980
	2	.0034	.0712	.2252	.4005	.4573	.5638	.6997	.8040	.8569	.8789	.9295	.9615	.9805
	3	.0001	.0084	.0530	.1409	.1783	.2618	.3993	.5372	.6228	.6627	.7682	.8505	.9102
	4	.0000	.0006	.0083	.0339	.0480	.0856	.1657	.2703	.3497	.3911	.5174	.6386	.7461
	5	.0000	.0000	.0009	.0056	.0090	.0196	.0489	.0988	.1448	.1717	.2666	.3786	.5000
	6	.0000	.0000	.0001	.0006	.0011	.0031	.0100	.0253	.0424	.0536	.0994	.1658	.2539
	7	.0000	.0000	.0000	.0000	.0001	.0003	.0013	.0043	.0083	.0112	.0250	.0498	.0898
	8	.0000	.0000	.0000	.0000	.0000	.0000	.0001	.0004	.0010	.0014	.0038	.0091	.0195
	9	.0000	.0000	.0000	.0000	.0000	.0000	.0000	.0000	.0001	.0001	.0003	.0008	.0020

Table Ib (*continued*)

n	k	.01	.05	.10	.15	1/6	.20	.25	.30	1/3	.35	.40	.45	.50
10	1	.0956	.4013	.6513	.8031	.8385	.8926	.9437	.9718	.9827	.9865	.9940	.9975	.9990
	2	.0043	.0861	.2639	.4557	.5155	.6242	.7560	.8507	.8960	.9140	.9536	.9767	.9893
	3	.0001	.0115	.0702	.1798	.2248	.3222	.4744	.6172	.7009	.7384	.8327	.9004	.9453
	4	.0000	.0010	.0128	.0500	.0697	.1209	.2241	.3504	.4407	.4862	.6177	.7340	.8281
	5	.0000	.0001	.0016	.0099	.0155	.0328	.0781	.1503	.2131	.2485	.3669	.4956	.6230
	6	.0000	.0000	.0001	.0014	.0024	.0064	.0197	.0473	.0766	.0949	.1662	.2616	.3770
	7	.0000	.0000	.0000	.0001	.0003	.0009	.0035	.0106	.0197	.0260	.0548	.1020	.1719
	8	.0000	.0000	.0000	.0000	.0000	.0001	.0004	.0016	.0034	.0048	.0123	.0274	.0547
	9	.0000	.0000	.0000	.0000	.0000	.0000	.0000	.0001	.0004	.0005	.0017	.0045	.0107
	10	.0000	.0000	.0000	.0000	.0000	.0000	.0000	.0000	.0000	.0000	.0001	.0003	.0010
11	1	.1047	.4312	.6862	.8327	.8654	.9141	.9578	.9802	.9884	.9912	.9964	.9986	.9995
	2	.0052	.1019	.3026	.5078	.5693	.6779	.8029	.8870	.9249	.9394	.9698	.9861	.9941
	3	.0002	.0152	.0896	.2212	.2732	.3826	.5448	.6873	.7659	.7999	.8811	.9348	.9673
	4	.0000	.0016	.0185	.0694	.0956	.1611	.2867	.4304	.5274	.5745	.7037	.8089	.8867
	5	.0000	.0001	.0028	.0159	.0245	.0504	.1146	.2103	.2890	.3317	.4672	.6029	.7256
	6	.0000	.0000	.0003	.0027	.0046	.0117	.0343	.0782	.1221	.1487	.2465	.3669	.5000
	7	.0000	.0000	.0000	.0003	.0006	.0020	.0076	.0216	.0386	.0501	.0994	.1738	.2744
	8	.0000	.0000	.0000	.0000	.0001	.0002	.0012	.0043	.0088	.0122	.0293	.0610	.1133
	9	.0000	.0000	.0000	.0000	.0000	.0000	.0001	.0006	.0014	.0020	.0059	.0148	.0327
	10	.0000	.0000	.0000	.0000	.0000	.0000	.0000	.0000	.0001	.0002	.0007	.0022	.0059
	11	.0000	.0000	.0000	.0000	.0000	.0000	.0000	.0000	.0000	.0000	.0000	.0002	.0005
12	1	.1136	.4596	.7176	.8578	.8878	.9313	.9683	.9862	.9923	.9943	.9978	.9992	.9998
	2	.0062	.1184	.3410	.5565	.6187	.7251	.8416	.9150	.9460	.9576	.9804	.9917	.9968
	3	.0002	.0196	.1109	.2642	.3226	.4417	.6093	.7472	.8189	.8487	.9166	.9579	.9807
	4	.0000	.0022	.0256	.0922	.1252	.2054	.3512	.5075	.6069	.6533	.7747	.8655	.9270
	5	.0000	.0002	.0043	.0239	.0364	.0726	.1576	.2763	.3685	.4167	.5618	.6956	.8062
	6	.0000	.0000	.0005	.0046	.0079	.0194	.0544	.1178	.1777	.2127	.3348	.4731	.6128
	7	.0000	.0000	.0000	.0007	.0013	.0039	.0143	.0386	.0664	.0846	.1582	.2607	.3872
	8	.0000	.0000	.0000	.0001	.0002	.0006	.0028	.0095	.0188	.0255	.0573	.1117	.1938
	9	.0000	.0000	.0000	.0000	.0000	.0001	.0004	.0017	.0039	.0056	.0153	.0356	.0730
	10	.0000	.0000	.0000	.0000	.0000	.0000	.0000	.0002	.0005	.0008	.0028	.0079	.0193
	11	.0000	.0000	.0000	.0000	.0000	.0000	.0000	.0000	.0000	.0001	.0003	.0011	.0032
	12	.0000	.0000	.0000	.0000	.0000	.0000	.0000	.0000	.0000	.0000	.0000	.0001	.0002

Table II Standard Normal c.d.f.

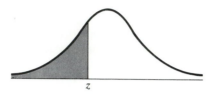

z	0	1	2	3	4	5	6	7	8	9
−3.	.0013	.0010	.0007	.0005	.0003	.0002	.0002	.0001	.0001	.0000
−2.9	.0019	.0018	.0017	.0017	.0016	.0016	.0015	.0015	.0014	.0014
−2.8	.0026	.0025	.0024	.0023	.0023	.0022	.0021	.0021	.0020	.0019
−2.7	.0035	.0034	.0033	.0032	.0031	.0030	.0029	.0028	.0027	.0026
−2.6	.0047	.0045	.0044	.0043	.0041	.0040	.0039	.0038	.0037	.0036
−2.5	.0062	.0060	.0059	.0057	.0055	.0054	.0052	.0051	.0049	.0048
−2.4	.0082	.0080	.0078	.0075	.0073	.0071	.0069	.0068	.0066	.0064
−2.3	.0107	.0104	.0102	.0099	.0096	.0094	.0091	.0089	.0087	.0084
−2.2	.0139	.0136	.0132	.0129	.0126	.0122	.0119	.0116	.0113	.0110
−2.1	.0179	.0174	.0170	.0166	.0162	.0158	.0154	.0150	.0146	.0143
−2.0	.0228	.0222	.0217	.0212	.0207	.0202	.0197	.0192	.0188	.0183
−1.9	.0287	.0281	.0274	.0268	.0262	.0256	.0250	.0244	.0238	.0233
−1.8	.0359	.0352	.0344	.0336	.0329	.0322	.0314	.0307	.0300	.0294
−1.7	.0446	.0436	.0427	.0418	.0409	.0401	.0392	.0384	.0375	.0367
−1.6	.0548	.0537	.0526	.0516	.0505	.0495	.0485	.0475	.0465	.0455
−1.5	.0668	.0655	.0643	.0630	.0618	.0606	.0594	.0582	.0570	.0559
−1.4	.0808	.0793	.0778	.0764	.0749	.0735	.0722	.0708	.0694	.0681
−1.3	.0968	.0951	.0934	.0918	.0901	.0885	.0869	.0853	.0838	.0823
−1.2	.1151	.1131	.1112	.1093	.1075	.1056	.1038	.1020	.1003	.0985
−1.1	.1357	.1335	.1314	.1292	.1271	.1251	.1230	.1210	.1190	.1170
−1.0	.1587	.1562	.1539	.1515	.1492	.1469	.1446	.1423	.1401	.1379
−.9	.1841	.1814	.1788	.1762	.1736	.1711	.1685	.1660	.1635	.1611
−.8	.2119	.2090	.2061	.2033	.2005	.1977	.1949	.1922	.1894	.1867
−.7	.2420	.2389	.2358	.2327	.2297	.2266	.2236	.2206	.2177	.2148
−.6	.2743	.2709	.2676	.2643	.2611	.2578	.2546	.2514	.2483	.2451
−.5	.3085	.3050	.3015	.2981	.2946	.2912	.2877	.2843	.2810	.2776
−.4	.3446	.3409	.3372	.3336	.3300	.3264	.3228	.3192	.3156	.3121
−.3	.3821	.3783	.3745	.3707	.3669	.3632	.3594	.3557	.3520	.3483
−.2	.4207	.4168	.4129	.4090	.4052	.4013	.3974	.3936	.3897	.3859
−.1	.4602	.4562	.4522	.4483	.4443	.4404	.4364	.4325	.4286	.4247
−.0	.5000	.4960	.4920	.4880	.4840	.4801	.4761	.4721	.4681	.4641

Table II (*continued*)

z	0	1	2	3	4	5	6	7	8	9
.0	.5000	.5040	.5080	.5120	.5160	.5199	.5239	.5279	.5319	.5359
.1	.5398	.5438	.5478	.5517	.5557	.5596	.5636	.5675	.5714	.5753
.2	.5793	.5832	.5871	.5910	.5948	.5987	.6026	.6064	.6103	.6141
.3	.6179	.6217	.6255	.6293	.6331	.6368	.6406	.6443	.6480	.6517
.4	.6554	.6591	.6628	.6664	.6700	.6736	.6772	.6808	.6844	.6879
.5	.6915	.6950	.6985	.7019	.7054	.7088	.7123	.7157	.7190	.7224
.6	.7257	.7291	.7324	.7357	.7389	.7422	.7454	.7486	.7517	.7549
.7	.7580	.7611	.7642	.7673	.7703	.7734	.7764	.7794	.7823	.7852
.8	.7881	.7910	.7939	.7967	.7995	.8023	.8051	.8078	.8106	.8133
.9	.8159	.8186	.8212	.8238	.8264	.8289	.8315	.8340	.8365	.8389
1.0	.8413	.8438	.8461	.8485	.8508	.8531	.8554	.8577	.8599	.8621
1.1	.8643	.8665	.8686	.8708	.8729	.8749	.8770	.8790	.8810	.8830
1.2	.8849	.8869	.8888	.8907	.8925	.8944	.8962	.8980	.8997	.9015
1.3	.9032	.9049	.9066	.9082	.9099	.9115	.9131	.9147	.9162	.9177
1.4	.9192	.9207	.9222	.9236	.9251	.9265	.9278	.9292	.9306	.9319
1.5	.9332	.9345	.9357	.9370	.9382	.9384	.9406	.9418	.9430	.9441
1.6	.9452	.9463	.9474	.9484	.9495	.9505	.9515	.9525	.9535	.9545
1.7	.9554	.9564	.9573	.9582	.9591	.9599	.9608	.9616	.9625	.9633
1.8	.9641	.9648	.9656	.9664	.9671	.9678	.9686	.9693	.9700	.9706
1.9	.9713	.9719	.9726	.9732	.9738	.9744	.9750	.9756	.9762	.9767
2.0	.9772	.9778	.9783	.9788	.9793	.9798	.9803	.9808	.9812	.9817
2.1	.9821	.9826	.9830	.9834	.9838	.9842	.9846	.9850	.9854	.9857
2.2	.9861	.9864	.9868	.9871	.9874	.9878	.9881	.9884	.9887	.9890
2.3	.9893	.9896	.9898	.9901	.9904	.9906	.9909	.9911	.9913	.9916
2.4	.9918	.9920	.9922	.9925	.9927	.9929	.9931	.9932	.9934	.9936
2.5	.9938	.9940	.9941	.9943	.9945	.9946	.9948	.9949	.9951	.9952
2.6	.9953	.9955	.9956	.9957	.9959	.9960	.9961	.9962	.9963	.9964
2.7	.9965	.9966	.9967	.9968	.9969	.9970	.9971	.9972	.9973	.9974
2.8	.9974	.9975	.9976	.9977	.9977	.9978	.9979	.9979	.9980	.9981
2.9	.9981	.9982	.9982	.9983	.9984	.9984	.9985	.9985	.9986	.9986
3.	.9987	.9990	.9993	.9995	.9997	.9998	.9998	.9999	.9999	1.0000

Notes: 1. Enter table at Z, read out $P(Z \leq z)$, the shaded area.
  2. For a general normal X, enter table at $z = (x - \mu)/\sigma$ to read $P(X \leq x)$.
  3. Entries opposite 3 are for 3.0, 3.1, 3.2 . . . , 3.9.
  4. For $z \geq 4$, $P(Z > z) = P(Z < -z) \doteq \dfrac{1}{\sigma\sqrt{2\pi}}e^{-z^2/2}$.

Table IIa Standard Normal Percentiles

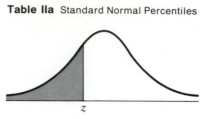

$P(Z \leq z)$	z
.001	-3.0902
.005	-2.5758
.01	-2.3263
.02	-2.0537
.03	-1.8808
.04	-1.7507
.05	-1.6449
.10	-1.2816
.15	-1.0364
.20	$-.8416$
.25	$-.6745$
.30	$-.5244$
.35	$-.3853$
.40	$-.2533$
.45	$-.1257$
.50	0
.55	.1257
.60	.2533
.65	.3853
.70	.5244
.75	.6745
.80	.8416
.85	1.0364
.90	1.2816
.95	1.6449
.96	1.7507
.97	1.8808
.98	2.0537
.99	2.3263
.995	2.5758
.999	3.0902

Table IIb Two-Tailed Probabilities for the Standard Normal Distribution

$P(|Z| > K)$	K
.001	3.2905
.002	3.0902
.005	2.8070
.01	2.5758
.02	2.3263
.03	2.1701
.04	2.0537
.05	1.9600
.06	1.8808
.08	1.7507
.10	1.6449
.15	1.4395
.20	1.2816
.30	1.0364

Table IIc Multipliers for Large Sample Confidence Intervals

Confidence level	Multiplier
.68	1
.80	1.28
.90	1.645
.95	1.96
.99	2.58

Table IIIa Multipliers of Standard Error

Two-Sided Confidence Intervals for μ (σ unknown)

Sample size	Confidence level .90	Confidence level .95	Confidence level .99	Sample size	Confidence level .90	Confidence level .95	Confidence level .99
2	6.31	12.7	63.7	26	1.71	2.06	2.79
3	2.92	4.30	9.92	27	1.71	2.06	2.78
4	2.35	3.18	5.84	28	1.70	2.05	2.77
5	2.13	2.78	4.60	29	1.70	2.05	2.76
				30	1.70	2.04	2.76
6	2.01	2.57	4.03				
7	1.94	2.45	3.71	31	1.70	2.04	2.75
8	1.90	2.36	3.50	32	1.70	2.04	2.74
9	1.86	2.31	3.36	33	1.69	2.04	2.74
10	1.83	2.26	3.25	34	1.69	2.03	2.73
				35	1.69	2.03	2.73
11	1.81	2.23	3.17				
12	1.80	2.20	3.11	36	1.69	2.03	2.72
13	1.78	2.18	3.06	37	1.69	2.03	2.72
14	1.77	2.16	3.01	38	1.69	2.02	2.71
15	1.76	2.14	2.98	39	1.69	2.02	2.71
				40	1.68	2.02	2.70
16	1.75	2.13	2.95				
17	1.75	2.12	2.92	60	1.65	1.98	2.65
18	1.74	2.11	2.90				
19	1.73	2.10	2.88				
20	1.73	2.09	2.86				
21	1.72	2.09	2.84				
22	1.72	2.08	2.83				
23	1.72	2.07	2.82				
24	1.71	2.07	2.81				
25	1.71	2.06	2.80				
∞	1.645	1.96	2.58	∞	1.645	1.96	2.58

The entries are percentiles of the t-distribution ($n - 1$ d.f.). For example, for $n = 35$, 2.03 is the 97.5th percentile of $t(34)$.

Table IIIb Tail-Probabilities of Student's *t*-Distribution

t	Degrees of freedom													
	4	5	6	7	8	9	10	11	12	13	14	15	16	17
1.0	.186	.181	.178	.175	.173	.172	.170	.169	.169	.168	.167	.167	.166	.166
1.1	.166	.160	.157	.154	.152	.150	.149	.147	.146	.146	.145	.144	.144	.143
1.2	.145	.142	.138	.135	.132	.130	.129	.128	.127	.126	.125	.124	.124	.123
1.3	.131	.125	.121	.117	.115	.113	.111	.110	.109	.108	.107	.107	.106	.105
1.4	.117	.110	.105	.102	.100	.098	.096	.095	.093	.092	.092	.091	.090	.090
1.5	.104	.097	.092	.089	.086	.084	.082	.081	.080	.079	.078	.077	.077	.076
1.6	.092	.085	.080	.077	.074	.072	.070	.069	.068	.067	.066	.065	.065	.064
1.7	.082	.075	.070	.066	.064	.062	.060	.059	.057	.056	.056	.055	.054	.054
1.8	.073	.066	.061	.057	.055	.053	.051	.050	.049	.048	.047	.046	.045	.045
1.9	.065	.058	.053	.050	.047	.045	.043	.042	.041	.040	.039	.038	.038	.037
2.0	.058	.051	.046	.043	.040	.038	.037	.035	.034	.033	.033	.032	.031	.031
2.1	.052	.045	.040	.037	.034	.033	.031	.030	.029	.028	.027	.027	.026	.025
2.2	.046	.040	.035	.032	.029	.028	.026	.025	.024	.023	.023	.022	.021	.021
2.3	.042	.035	.031	.027	.025	.023	.022	.021	.020	.019	.019	.018	.018	.017
2.4	.037	.031	.027	.024	.022	.020	.019	.018	.017	.016	.015	.015	.014	.014
2.5	.033	.027	.023	.020	.018	.017	.016	.015	.014	.013	.013	.012	.012	.011
2.6	.030	.024	.020	.018	.016	.014	.013	.012	.012	.011	.010	.010	.010	.009
2.7	.027	.021	.018	.015	.014	.012	.011	.010	.010	.009	.009	.008	.008	.008
2.8	.025	.019	.016	.013	.012	.010	.009	.009	.008	.008	.007	.007	.006	.006
2.9	.022	.017	.014	.011	.010	.009	.008	.007	.007	.006	.006	.005	.005	.005
3.0	.020	.015	.012	.010	.009	.007	.007	.006	.006	.005	.005	.004	.004	.004
3.1	.018	.013	.011	.009	.007	.006	.006	.005	.005	.004	.004	.004	.003	.003
3.2	.017	.012	.009	.008	.006	.005	.005	.004	.004	.003	.003	.003	.003	.003
3.3	.015	.011	.008	.007	.005	.005	.004	.004	.003	.003	.003	.002	.002	.002
3.4	.014	.010	.007	.006	.005	.004	.003	.003	.002	.002	.002	.002	.002	.002
3.5	.013	.009	.006	.005	.004	.003	.003	.002	.002	.002	.002	.002	.001	.001
3.6	.012	.008	.006	.004	.003	.003	.002	.002	.002	.002	.001	.001	.001	.001
3.7	.011	.007	.005	.004	.003	.002	.002	.002	.002	.001	.001	.001	.001	.001
3.8	.010	.006	.004	.003	.003	.002	.002	.001	.001	.001	.001	.001	.001	.001
3.9	.009	.006	.004	.003	.002	.002	.001	.001	.001	.001	.001	.001	.001	.001
4.0	.008	.005	.004	.003	.002	.002	.001	.001	.001	.001	.001	.001	.001	.000
4.5	.006	.003	.002	.001	.001	.001	.001	.000	.000	.000	.000	.000	.000	.000
5.0	.004	.002	.001	.001	.000	.000	.000	.000	.000	.000	.000	.000	.000	.000

Table IIIb Tail-Probabilities of Student's *t*-Distribution (*continued*)

t	18	19	20	21	22	23	24	25	26	27	28	29	30	35	40
1.0	.165	.165	.165	.164	.164	.164	.164	.163	.163	.163	.163	.163	.163	.162	.162
1.1	.143	.143	.142	.142	.142	.141	.141	.141	1.41	.141	.140	.140	.140	.139	.139
1.2	.123	.122	.122	.122	.121	.121	.121	.121	.120	.120	.120	.120	.120	.119	.119
1.3	.105	.105	.104	.104	.104	.103	.103	.103	.103	.102	.102	.102	.102	.101	.101
1.4	.089	.089	.088	.088	.088	.087	.087	.087	.087	.086	.086	.086	.086	.085	.085
1.5	.075	.075	.075	.074	.074	.074	.073	.073	.073	.073	.072	.072	.072	.071	.071
1.6	.064	.063	.063	.062	.062	.062	.061	.061	.061	.061	.060	.060	.060	.059	.059
1.7	.053	.053	.052	.052	.052	.051	.051	.051	.051	.050	.050	.050	.050	.049	.048
1.8	.044	.044	.043	.043	.043	.042	.042	.042	.042	.042	.041	.041	.041	.040	.040
1.9	.037	.036	.036	.036	.035	.035	.035	.035	.034	.034	.034	.034	.034	.033	.032
2.0	.030	.030	.030	.029	.029	.029	.028	.028	.028	.028	.028	.027	.027	.027	.026
2.1	.025	.025	.024	.024	.024	.023	.023	.023	.023	.023	.022	.022	.022	.022	.021
2.2	.021	.020	.020	.020	.019	.019	.019	.019	.018	.018	.018	.018	.018	.017	.017
2.3	.017	.016	.016	.016	.016	.015	.015	.015	.015	.015	.015	.014	.014	.014	.013
2.4	.014	.013	.013	.013	.013	.012	.012	.012	.012	.012	.012	.012	.011	.011	.011
2.5	.011	.011	.011	.010	.010	.010	.010	.010	.010	.009	.009	.009	.009	.009	.008
2.6	.009	.009	.009	.008	.008	.008	.008	.008	.008	.007	.007	.007	.007	.007	.006
2.7	.007	.007	.007	.007	.007	.006	.006	.006	.006	.006	.006	.006	.006	.005	.005
2.8	.006	.006	.006	.005	.005	.005	.005	.005	.005	.005	.005	.004	.004	.004	.004
2.9	.005	.005	.004	.004	.004	.004	.004	.004	.004	.004	.004	.004	.003	.003	.003
3.0	.004	.004	.004	.003	.003	.003	.003	.003	.003	.003	.003	.003	.003	.002	.002
3.1	.003	.003	.003	.003	.003	.003	.002	.002	.002	.002	.002	.002	.002	.002	.002
3.2	.002	.002	.002	.002	.002	.002	.002	.002	.002	.002	.002	.002	.002	.001	.001
3.3	.002	.002	.001	.001	.001	.001	.001	.001	.001	.001	.001	.001	.001	.001	.001
3.4	.001	.001	.001	.001	.001	.001	.001	.001	.001	.001	.001	.001	.001	.001	.001
3.5	.001	.001	.001	.001	.001	.001	.001	.001	.001	.001	.001	.001	.001	.000	.000
3.6	.001	.001	.001	.001	.001	.001	.001	.001	.001	.000	.000	.000	.000	.000	.000
3.7	.001	.001	.001	.001	.001	.000	.000	.000	.000	.000	.000	.000	.000	.000	.000
3.8	.001	.001	.000	.000	.000	.000	.000	.000	.000	.000	.000	.000	.000	.000	.000
3.9	.001	.000	.000	.000	.000	.000	.000	.000	.000	.000	.000	.000	.000	.000	.000
4.0	.000	.000	.000	.000	.000	.000	.000	.000	.000	.000	.000	.000	.000	.000	.000
4.5	.000	.000	.000	.000	.000	.000	.000	.000	.000	.000	.000	.000	.000	.000	.000
5.0	.000	.000	.000	.000	.000	.000	.000	.000	.000	.000	.000	.000	.000	.000	.000

Degrees of freedom

Table IIIc Percentiles of the *t*-Distribution

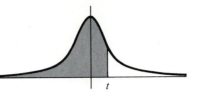

Degrees of Freedom	p								
	.60	.70	.80	.85	.90	.95	.975	.99	.995
1	.325	.727	1.38	1.96	3.08	6.31	12.7	31.8	63.7
2	.289	.617	1.06	1.39	1.89	2.92	4.30	6.96	9.92
3	.277	.584	.978	1.25	1.64	2.35	3.18	4.54	5.84
4	.271	.569	.941	1.19	1.53	2.13	2.78	3.75	4.60
5	.267	.559	.920	1.16	1.48	2.01	2.57	3.36	4.03
6	.265	.553	.906	1.13	1.44	1.94	2.45	3.14	3.71
7	.263	.549	.896	1.12	1.42	1.90	2.36	3.00	3.50
8	.262	.546	.889	1.11	1.40	1.86	2.31	2.90	3.36
9	.261	.543	.883	1.10	1.38	1.83	2.26	2.82	3.25
10	.260	.542	.879	1.09	1.37	1.81	2.23	2.76	3.17
11	.260	.540	.876	1.09	1.36	1.80	2.20	2.72	3.11
12	.259	.539	.873	1.08	1.36	1.78	2.18	2.68	3.06
13	.259	.538	.870	1.08	1.35	1.77	2.16	2.65	3.01
14	.258	.537	.868	1.08	1.34	1.76	2.14	2.62	2.98
15	.258	.536	.866	1.07	1.34	1.75	2.13	2.60	2.95
16	.258	.535	.865	1.07	1.34	1.75	2.12	2.58	2.92
17	.257	.534	.863	1.07	1.33	1.74	2.11	2.57	2.90
18	.257	.534	.862	1.07	1.33	1.73	2.10	2.55	2.88
19	.257	.533	.861	1.07	1.33	1.73	2.09	2.54	2.86
20	.257	.533	.860	1.07	1.32	1.72	2.09	2.53	2.84
21	.257	.532	.859	1.06	1.32	1.72	2.08	2.52	2.83
22	.256	.532	.858	1.06	1.32	1.72	2.07	2.51	2.82
23	.256	.532	.858	1.06	1.32	1.71	2.07	2.50	2.81
24	.256	.531	.857	1.06	1.32	1.71	2.06	2.49	2.80
25	.256	.531	.856	1.06	1.32	1.71	2.06	2.48	2.79
26	.256	.531	.856	1.06	1.32	1.71	2.06	2.48	2.78
27	.256	.531	.855	1.06	1.31	1.70	2.05	2.47	2.77
28	.256	.530	.855	1.06	1.31	1.70	2.05	2.47	2.76
29	.256	.530	.854	1.06	1.31	1.70	2.04	2.46	2.76
30	.256	.530	.854	1.05	1.31	1.70	2.04	2.46	2.75
40	.255	.529	.851	1.05	1.30	1.68	2.02	2.42	2.70
∞	.253	.524	.842	1.04	1.28	1.64	1.96	2.33	2.58

Notes: 1. The area to the right of the table entry is $1 - p$.
2. The distribution is symmetric. For example, for 10 degrees of freedom, $P(-1.37 < t < 1.37) = .9 - .1 = .8$.

Table IV Cumulative Poisson Probabilities

$$P(X \le c \mid m) = \sum_{0}^{c} \frac{m^k}{k!} e^{-m}$$

m (expected value)

c	.02	.04	.06	.08	.10	.15	.20	.25	.30	.35	.40
0	.980	.961	.942	.923	.905	.861	.819	.779	.741	.705	.670
1	1.000	.999	.998	.997	.995	.990	.982	.974	.963	.951	.938
2		1.000	1.000	1.000	1.000	.999	.999	.998	.996	.994	.992
3							1.000	1.000	1.000	1.000	1.000	.999
4											1.000

c	.45	.50	.55	.60	.65	.70	.75	.80	.85	.90	.95
0	.638	.607	.577	.549	.522	.497	.472	.449	.427	.407	.387
1	.925	.910	.894	.878	.861	.844	.827	.809	.791	.772	.754
2	.989	.986	.982	.977	.972	.966	.959	.953	.945	.937	.929
3	.999	.998	.998	.997	.996	.994	.993	.991	.989	.987	.984
4	1.000	1.000	1.000	1.000	.999	.999	.999	.999	.998	.998	.997
5					1.000	1.000	1.000	1.000	1.000	1.000	1.000

c	1.0	1.1	1.2	1.3	1.4	1.5	1.6	1.7	1.8	1.9	2.0
0	.368	.333	.301	.273	.247	.223	.202	.183	.165	.150	.135
1	.736	.699	.663	.627	.592	.558	.525	.493	.463	.434	.406
2	.920	.900	.879	.857	.833	.809	.783	.757	.731	.704	.677
3	.981	.974	.966	.957	.946	.934	.921	.907	.891	.875	.857
4	.996	.996	.992	.989	.986	.981	.976	.970	.964	.956	.947
5	.999	.999	.998	.998	.997	.996	.994	.992	.990	.987	.983
6	1.000	1.000	1.000	1.000	.999	.999	.999	.998	.997	.997	.995
7					1.000	1.000	1.000	1.000	.999	.999	.999
8									.000	1.000	1.000

c	2.2	2.4	2.6	2.8	3.0	3.2	3.4	3.6	3.8	4.0	4.2
0	.111	.091	.074	.061	.050	.041	.033	.027	.022	.018	.015
1	.355	.308	.267	.231	.199	.171	.147	.126	.107	.092	.078
2	.623	.570	.518	.469	.423	.380	.340	.303	.269	.238	.210
3	.819	.779	.736	.692	.647	.603	.558	.515	.473	.433	.395
4	.928	.904	.877	.848	.815	.781	.744	.706	.668	.629	.590
5	.975	.964	.951	.935	.916	.895	.871	.844	.816	.785	.753
6	.993	.988	.983	.976	.966	.955	.942	.927	.909	.889	.867
7	.998	.997	.995	.992	.988	.983	.977	.969	.960	.949	.936
8	1.000	.999	.999	.998	.996	.994	.992	.988	.984	.979	.972
9		1.000	1.000	.999	.999	.998	.997	.996	.994	.992	.989
10				1.000	1.000	1.000	.999	.999	.998	.997	.996
11							1.000	1.000	.999	.999	.999
12									1.000	1.000	1.000

Table IV (*continued*)

						m					
c	4.4	4.6	4.8	5.0	5.2	5.4	5.6	5.8	6.0	6.2	6.4
0	.012	.010	.008	.007	.006	.005	.004	.003	.002	.002	.002
1	.066	.056	.048	.040	.034	.029	.024	.021	.017	.015	.012
2	.185	.163	.143	.125	.109	.095	.082	.072	.062	.054	.046
3	.359	.326	.294	.265	.238	.213	.191	.170	.151	.134	.119
4	.551	.513	.476	.440	.406	.373	.342	.313	.285	.259	.235
5	.720	.686	.651	.616	.581	.546	.512	.478	.446	.414	.384
6	.844	.818	.791	.762	.732	.702	.670	.638	.606	.574	.542
7	.921	.905	.887	.867	.845	.822	.797	.771	.744	.716	.687
8	.964	.955	.944	.932	.918	.903	.886	.867	.847	.826	.803
9	.985	.980	.975	.968	.960	.951	.941	.929	.916	.902	.886
10	.994	.992	.990	.986	.982	.977	.972	.965	.957	.949	.939
11	.998	.997	.996	.995	.993	.990	.988	.984	.980	.975	.969
12	.999	.999	.999	.998	.997	.996	.995	.993	.991	.989	.986
13	1.000	1.000	1.000	.999	.999	.999	.998	.997	.996	.995	.994
14				1.000	1.000	1.000	.999	.999	.999	.998	.997
15							1.000	1.000	.999	.999	.999
16									1.000	1.000	1.000

						m					
c	6.6	6.8	7.0	7.2	7.4	7.6	7.8	8.0	8.5	9.0	9.5
0	.001	.001	.001	.001	.001	.001	.000	.000	.000	.000	.000
1	.010	.009	.007	.006	.005	.004	.004	.003	.002	.001	.001
2	.040	.034	.030	.025	.022	.019	.016	.014	.009	.006	.004
3	.105	.093	.082	.072	.063	.055	.048	.042	.030	.021	.015
4	.213	.192	.173	.156	.140	.125	.112	.100	.074	.055	.040
5	.355	.327	.301	.276	.253	.231	.210	.191	.150	.116	.089
6	.511	.480	.450	.420	.392	.365	.338	.313	.256	.207	.165
7	.658	.628	.599	.569	.539	.510	.481	.453	.386	.324	.269
8	.780	.755	.729	.703	.676	.648	.620	.593	.523	.456	.392
9	.869	.850	.830	.810	.788	.765	.741	.717	.653	.587	.522
10	.927	.915	.901	.887	.871	.854	.835	.816	.763	.706	.645
11	.963	.955	.947	.937	.926	.915	.902	.888	.849	.803	.752
12	.982	.978	.973	.967	.961	.954	.945	.936	.909	.876	.836
13	.992	.990	.987	.984	.980	.976	.971	.966	.949	.926	.898
14	.997	.996	.994	.993	.991	.989	.986	.983	.973	.959	.940
15	.999	.998	.998	.997	.996	.995	.993	.992	.986	.978	.967
16	.999	.999	.999	.999	.998	.998	.997	.996	.993	.989	.982
17	1.000	1.000	1.000	.999	.999	.999	.999	.998	.997	.995	.991
18				1.000	1.000	1.000	1.000	.999	.999	.998	.996
19								1.000	.999	.999	.998
20									1.000	1.000	.999
21											1.000

Table IV (continued)

c	10.0	10.5	11.0	11.5	12.0	m 12.5	13.0	13.5	14.0	14.5	15.0
2	.003	.002	.001	.001	.001	.000					
3	.010	.007	.005	.003	.002	.002	.001	.001	.000		
4	.029	.021	.015	.011	.008	.005	.004	.003	.002	.001	.001
5	.067	.050	.038	.028	.020	.015	.011	.008	.006	.004	.003
6	.130	.102	.079	.060	.046	.035	.026	.019	.014	.010	.008
7	.220	.179	.143	.114	.090	.070	.054	.041	.032	.024	.018
8	.333	.279	.232	.191	.155	.125	.100	.079	.062	.048	.037
9	.458	.397	.341	.289	.242	.201	.166	.135	.109	.088	.070
10	.583	.521	.460	.402	.347	.297	.252	.211	.176	.145	.118
11	.697	.629	.579	.520	.462	.406	.353	.304	.260	.220	.185
12	.792	.742	.689	.633	.576	.519	.463	.409	.358	.311	.268
13	.864	.825	.781	.733	.682	.628	.573	.518	.464	.413	.363
14	.917	.888	.854	.815	.772	.725	.675	.623	.570	.518	.466
15	.951	.932	.907	.878	.844	.806	.764	.718	.669	.619	.568
16	.973	.960	.944	.924	.899	.869	.835	.798	.756	.711	.664
17	.986	.978	.968	.954	.937	.916	.890	.861	.827	.790	.749
18	.993	.988	.982	.974	.963	.948	.930	.908	.883	.853	.819
19	.997	.994	.991	.986	.979	.969	.957	.942	.923	.901	.875
20	.998	.997	.995	.992	.988	.983	.975	.965	.952	.936	.917
21	.999	.999	.998	.996	.994	.991	.986	.980	.971	.960	.947
22	1.000	.999	.999	.998	.997	.995	.992	.989	.983	.976	.967
23		1.000	1.000	.999	.999	.998	.996	.994	.991	.986	.981
24				1.000	.999	.999	.998	.997	.995	.992	.989
25					1.000	.999	.999	.998	.997	.996	.994
26						1.000	1.000	.999	.999	.998	.997
27								1.000	.999	.999	.998
28									1.000	.999	.999
29										1.000	1.000

Table Va   Tail-probabilities of the Chi-square Distribution

χ^2	1	d.f. 2	3	χ^2	4	d.f. 5	6	χ^2	7	d.f. 8	9
2.2	.138	.333	.532	7.2	.126	.206	.303	12.2	.094	.143	.202
2.4	.121	.301	.494	7.4	.116	.192	.285	12.4	.088	.134	.192
2.6	.107	.273	.457	7.6	.107	.179	.269	12.6	.082	.126	.182
2.8	.094	.247	.423	7.8	.099	.167	.253	12.8	.077	.119	.172
3.0	.083	.223	.391	8.0	.092	.156	.238	13.0	.072	.112	.163
3.2	0.73	.202	362	8.2	.085	.145	.224	13.2	.067	.105	.154
3.4	.065	.183	.334	8.4	.078	.135	.210	13.4	.063	.099	.145
3.6	.058	.165	.308	8.6	.072	.126	.197	13.6	.059	.093	.137
3.8	.051	.150	.284	8.8	.066	.117	.185	13.8	.055	.087	.130
4.0	.045	.135	.261	9.0	.061	.109	.174	14.0	.051	.082	.122
4.2	.040	.122	.240	9.2	.056	.101	.163	14.2	.048	.077	.115
4.4	.036	.111	.221	9.4	.052	.094	.152	14.4	.044	.072	.109
4.6	.032	.105	.203	9.6	.048	.087	.143	14.6	.041	.067	.103
4.8	.028	.100	.187	9.8	.044	.081	.133	14.8	.039	.063	.097
5.0	.025	.091	.171	10.0	.040	.075	.125	15.0	.036	.059	.091
5.2	.022	.082	.157	10.2	.037	.070	.116	15.2	.034	.055	.086
5.4	.020	.074	.144	10.4	.034	.065	.109	15.4	.031	.052	.081
5.6	.018	.061	.132	10.6	.031	.060	.102	15.6	.029	.048	.076
5.8	.016	.055	.121	10.8	.029	.055	.095	15.8	.027	.045	.071
6.0	.014	.050	.111	11.0	.027	.051	.088	16.0	.025	.042	.067
6.2	.012	.045	.102	11.2	.024	.047	.082	16.2	.023	.040	.063
6.4	.011	.041	.093	11.4	.022	.044	.077	16.4	.022	.037	.059
6.6	.010	.037	.086	11.6	.021	.041	.072	16.6	.020	.035	.055
6.8	.009	.033	.078	11.8	.019	.038	.067	16.8	.019	.032	.052
7.0	.008	.030	.072	12.0	.017	.035	.062	17.0	.017	.030	.049
7.2	.007	.027	.066	12.2	.016	.032	.058	17.2	.016	.028	.046
7.4	.006	.025	.060	12.4	.015	.030	.054	17.4	.015	.026	.043
7.6	.006	.022	.055	12.6	.013	.027	.050	17.6	.014	.024	.040
7.8	.005	.020	.050	12.8	.012	.025	.046	17.8	.013	.023	.038
8.0	.004	.018	.046	13.0	.011	.023	.043	18.0	.012	.021	.035
8.2	.004	.017	.042	13.2	.010	.022	.040	18.2	.011	.020	.033
8.4	.004	.015	.038	13.4	.009	.020	.037	18.4	.010	.018	.031
8.6	.003	.014	.035	13.6	.009	.018	.034	18.6	.010	.017	.029
8.8	.003	.012	.032	13.8	.008	.017	.032	18.8	.009	.016	.027
9.0	.003	.011	.029	14.0	.007	.016	.030	19.0	.008	.015	.025
9.2	.002	.010	.027	14.2	.007	.014	.027	19.2	.008	.014	.024
9.4	.002	.009	.024	14.4	.006	.013	.025	19.4	.007	.013	.022
9.6	.002	.008	.022	14.6	.006	.012	.024	19.6	.006	.012	.021
9.8	.002	.007	.020	14.8	.005	.011	.022	19.8	.008	.011	.019
10.0	.001	.007	.018	15.0	.005	.010	.020	20.0	.006	.010	.018

(continues)

Table Va (*continued*)

χ^2	1	d.f. 2	3	χ^2	4	d.f. 5	6	χ^2	7	d.f. 8	9
10.2	.001	.006	.017	15.2	.004	.010	.019	20.2	.005	.010	.017
10.4	.001	.006	.015	15.4	.004	.009	.017	20.4	.005	.009	.016
10.6	.001	.005	.014	15.6	.004	.008	.016	20.6	.004	.008	.015
10.8	.001	.005	.013	15.8	.003	.007	.015	20.8	.004	.008	.014
11.0	.001	.004	.012	16.0	.003	.007	.014	21.0	.004	.007	.013
11.2	.001	.004	.011	16.2	.003	.006	.013	21.2	.003	.007	.012
11.4	.001	.003	.010	16.4	.003	.006	.012	21.4	.003	.006	.011
11.6	.001	.003	.009	16.6	.002	.005	.011	21.6	.003	.006	.010
11.8	.001	.003	.008	16.8	.002	.005	.010	21.8	.003	.005	.010
12.0	.000	.002	.007	17.0	.002	.004	.009	22.0	.003	.005	.009
12.2	.000	.002	.007	17.2	.002	.004	.009	22.2	.002	.005	.008
12.4	.000	.002	.006	17.4	.002	.004	.008	22.4	.002	.004	.008
12.6	.000	.002	.006	17.6	.001	.003	.007	22.6	.002	.004	.007
12.8	.000	.002	.005	17.8	.001	.003	.007	22.8	.002	.004	.007
13.0	.000	.002	.005	18.0	.001	.003	.006	23.0	.002	.003	.006
15.2	.125	.174	.231	20.2	.090	.124	.164	24.2	.085	.114	.149
15.4	.118	.165	.220	20.4	.086	.118	.157	24.4	.081	.109	.142
15.6	.112	.157	.210	20.6	.081	.112	.150	24.6	.077	.104	.136
15.8	.106	.149	.267	20.8	.077	.107	.143	24.8	.073	.099	.131
16.0	.100	.141	.191	21.0	.073	.102	.137	25.0	.070	.095	.125
16.2	.094	.134	.182	21.2	.071	.097	.131	25.2	.066	.090	.120
16.4	.089	.127	.174	21.4	.065	.092	.125	25.4	.063	.086	.114
16.6	.084	.120	.165	21.6	.062	.087	.119	25.6	.060	.082	.109
16.8	.079	.114	.157	21.8	.059	.083	.113	25.8	.057	.078	.104
17.0	.074	.108	.150	22.0	.055	.079	.108	26.0	.054	.074	.100
17.2	.070	.102	.142	22.2	.052	.075	.103	26.2	.051	.071	.095
17.4	.066	.097	.135	22.4	.049	.071	.098	26.4	.049	.067	.091
17.6	.062	.091	.128	22.6	.047	.067	.093	26.6	.046	.064	.087
17.8	.058	.086	.122	22.8	.044	.064	.088	26.8	.044	.061	.083
18.0	.055	.082	.116	23.0	.042	.060	.084	27.0	.041	.058	.079
18.2	.052	.077	.110	23.2	.039	.057	.080	27.2	.039	.055	.075
18.4	.049	.073	.104	23.4	.037	.054	.076	27.4	.037	.052	.072
18.6	.046	.069	.099	23.6	.035	.051	.072	27.6	.035	.050	.068
18.8	.043	.065	.093	23.8	.033	.048	.069	27.8	.033	.047	.065
19.0	.040	.061	.089	24.0	.031	.046	.065	28.0	.020	.045	.062
19.2	.038	.058	.084	24.2	.029	.043	.062	28.2	.030	.043	.059
19.4	.035	.054	.079	24.4	.028	.041	.059	28.4	.028	.040	.056
19.6	.033	.051	.075	24.6	.026	.039	.056	28.6	.027	.038	.053
19.8	.031	.048	.071	24.8	.025	.037	.053	28.8	.025	.036	.051
20.0	.029	.045	.067	25.0	.023	.035	.050	29.0	.024	.035	.048

Table Va (*continued*)

χ^2	10	d.f. 11	12	χ^2	13	d.f. 14	15	χ^2	16	d.f. 17	18
20.2	.027	.043	.063	25.2	.022	.033	.047	29.2	.023	.033	.046
20.4	.026	.040	.060	25.4	.020	.031	.045	29.4	.021	.031	.044
20.6	.024	.038	.057	25.6	.019	.029	.042	29.6	.020	.029	.042
20.8	.023	.036	.053	25.8	.018	.027	.040	29.8	.019	.028	.039
21.0	.021	.033	.050	26.0	.017	.026	.038	30.0	.018	.026	.037
21.2	.020	.031	.048	26.2	.016	.024	.036	30.2	.017	.025	.036
21.4	.018	.029	.045	26.4	.015	.023	.034	30.4	.016	.024	.034
21.6	.017	.028	.042	26.6	.014	.022	.032	30.6	.015	.022	.032
21.8	.016	.026	.040	26.8	.013	.020	.030	30.8	.014	.021	.030
22.0	.014	.024	.038	27.0	.012	.019	.029	31.0	.013	.020	.029
22.2	.013	.023	.035	27.2	.012	.018	.027	31.2	.013	.019	.027
22.4	.012	.021	.033	27.4	.011	.017	.026	31.4	.012	.018	.026
22.6	.012	.020	.031	27.6	.010	.016	.024	31.6	.011	.017	.025
22.8	.011	.019	.029	27.8	.010	.015	.023	31.8	.011	.016	.023
23.0	.010	.018	.028	28.0	.009	.014	.022	32.0	.010	.015	.022
23.2	.009	.017	.026	28.2	.008	.013	.020	32.2	.009	.014	.021
23.4	.009	.016	.025	28.4	.008	.013	.019	32.4	.009	.013	.020
23.6	.008	.015	.023	28.6	.007	.012	.018	32.6	.008	.013	.019
23.8	.008	.014	.022	28.8	.007	.011	.017	32.8	.008	.012	.018
24.0	.004	.013	.020	29.0	.007	.010	.016	33.0	.007	.011	.017
24.2	.004	.012	.019	29.2	.006	.010	.015	33.2	.007	.011	.016
24.4	.004	.011	.018	29.4	.006	.009	.014	33.4	.007	.010	.015
24.6	.003	.010	.017	29.6	.005	.009	.013	33.6	.006	.009	.014
24.8	.003	.010	.016	29.8	.005	.008	.013	33.8	.006	.009	.013
25.0	.003	.009	.015	30.0	.005	.008	.012	34.0	.005	.008	.013
25.2	.003	.009	.014	30.2	.004	.007	.011	34.2	.005	.008	.012
25.4	.002	.008	.013	30.4	.004	.007	.010	34.4	.005	.007	.011
25.6	.002	.007	.012	30.6	.004	.006	.010	34.6	.005	.007	.011
25.8	.002	.007	.011	30.8	.004	.006	.009	34.8	.004	.007	.010
26.0	.002	.006	.011	31.0	.003	.006	.009	35.0	.004	.006	.009

Table Vb Percentiles of the Chi-Square Distribution

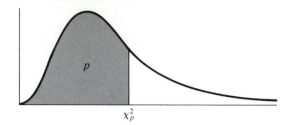

$$\chi^2_p$$

Degrees of freedom	.01	.025	.05	.10	.20	.30	.50	.70	.80	.90	.95	.975	.99
1	.000	.001	.004	.016	.064	.148	.455	1.07	1.64	2.71	3.84	5.02	6.63
2	.020	.051	.103	.211	.446	.713	1.39	2.41	3.22	4.61	5.99	7.38	9.21
3	.115	.216	.352	.584	1.01	1.42	2.37	3.66	4.64	6.25	7.81	9.35	11.3
4	.297	.484	.711	1.06	1.65	2.19	3.36	4.88	5.99	7.78	9.49	11.1	13.3
5	.554	.831	1.15	1.61	2.34	3.00	4.35	6.06	7.29	9.24	11.1	12.8	15.1
6	.872	1.24	1.64	2.20	3.07	3.83	5.35	7.23	8.56	10.6	12.6	14.4	16.8
7	1.24	1.69	2.17	2.83	3.82	4.67	6.35	8.38	9.80	12.0	14.1	16.0	18.5
8	1.65	2.18	2.73	3.49	4.59	5.53	7.34	9.52	11.0	13.4	15.5	17.5	20.1
9	2.09	2.70	3.33	4.17	5.38	6.39	8.34	10.7	12.2	14.7	16.9	19.0	21.7
10	2.56	3.25	3.94	4.87	6.18	7.27	9.34	11.8	13.4	16.0	18.3	20.5	23.2
11	3.05	3.82	4.57	5.58	6.99	8.15	10.3	12.9	14.6	17.3	19.7	21.9	24.7
12	3.57	4.40	5.23	6.30	7.81	9.03	11.3	14.0	15.8	18.5	21.0	23.3	26.2
13	4.11	5.01	5.89	7.04	8.63	9.93	12.3	15.1	17.0	19.8	22.4	24.7	27.7
14	4.66	5.63	6.57	7.79	9.47	10.8	13.3	16.2	18.2	21.1	23.7	26.1	29.1
15	5.23	6.26	7.26	8.55	10.3	11.7	14.3	17.3	19.3	22.3	25.0	27.5	30.6
16	5.81	6.91	7.96	9.31	11.2	12.6	15.3	18.4	20.5	23.5	26.3	28.8	32.0
17	6.41	7.56	8.67	10.1	12.0	13.5	16.3	19.5	21.6	24.8	27.6	30.2	33.4
18	7.01	8.23	9.39	10.9	12.9	14.4	17.3	20.6	22.8	26.0	28.9	31.5	34.8
19	7.63	8.91	10.1	11.7	13.7	15.4	18.3	21.7	23.9	27.2	30.1	32.9	36.2
20	8.26	9.59	10.9	12.4	14.6	16.3	19.3	22.8	25.0	28.4	31.4	34.8	37.6
21	8.90	10.3	11.6	13.2	15.4	17.2	20.3	23.9	26.2	29.6	32.7	35.5	38.9
22	9.54	11.0	12.3	14.0	16.3	18.1	21.3	24.9	27.3	30.8	33.9	36.8	40.3
23	10.2	11.7	13.1	14.8	17.2	19.0	22.3	26.0	28.4	32.0	35.2	38.1	41.6
24	10.9	12.4	13.8	15.7	18.1	19.9	23.3	27.1	29.6	33.2	36.4	39.4	43.0
25	11.5	13.1	14.6	16.5	18.9	20.9	24.3	28.2	30.7	34.4	37.7	40.6	44.3
26	12.2	13.8	15.4	17.3	19.8	21.8	25.3	29.2	31.8	35.6	38.9	41.9	45.6
27	12.9	14.6	16.2	18.1	20.7	22.7	26.3	30.3	32.9	36.7	40.1	43.2	47.0
28	13.6	15.3	16.9	18.9	21.6	23.6	27.3	31.4	34.0	37.9	41.3	44.5	48.3
29	14.3	16.0	17.7	19.8	22.5	24.6	28.3	32.5	35.1	39.1	42.6	45.7	49.6
30	15.0	16.8	18.5	20.6	23.4	25.5	29.3	33.5	36.2	40.3	43.8	47.0	50.9
40	22.1	24.4	26.5	29.0	32.4	35.0	39.3	44.2	47.3	51.8	55.8	59.3	63.7
50	29.7	32.3	34.8	37.7	41.5	44.4	49.3	54.7	58.2	63.2	67.5	71.4	76.2
60	37.5	40.5	43.2	46.5	50.7	53.9	59.3	65.2	69.0	74.4	79.1	83.3	88.4

Note: For degrees of freedom $k > 30$, use $\chi^2_p = \frac{1}{2}(z_p + \sqrt{2k-1})^2$, where z_p is the corresponding percentile of the standard normal distribution.

Table VI Tail-Probabilities of the One-Sample Wilcoxon Signed-Rank Statistic

n / c	4	5	6	7	8	9	10	11	12	13	14	15	n / c
0	.062	.031	.016	.008	.004	.002	.001	.000	.000	.000	.000	.000	0
1	.125	.062	.031	.016	.008	.004	.002	.001	.000	.000	.000	.000	1
2	.188	.094	.047	.023	.012	.006	.003	.001	.001	.000	.000	.000	2
3	.312	.156	.078	.039	.020	.010	.005	.002	.001	.001	.000	.000	3
4		.219	.109	.055	.027	.014	.007	.003	.002	.001	.000	.000	4
5			.156	.078	.039	.020	.010	.005	.002	.001	.001	.000	5
6			.219	.109	.055	.027	.014	.007	.003	.002	.001	.000	6
7				.148	.074	.037	.019	.009	.005	.002	.001	.001	7
8				.188	.098	.049	.024	.012	.006	.003	.002	.001	8
9				.234	.125	.064	.032	.016	.008	.004	.002	.001	9
10					.156	.082	.042	.021	.010	.005	.003	.001	10
11					.191	.102	.053	.027	.013	.007	.003	.002	11
12					.230	.125	.065	.034	.017	.009	.004	.002	12
13						.150	.080	.042	.021	.011	.005	.003	13
14						.180	.097	.051	.026	.013	.007	.003	14
15						.213	.116	.062	.032	.016	.008	.004	15
16							.138	.074	.039	.020	.010	.005	16
17							.161	.087	.046	.024	.012	.006	17
18							.188	.103	.055	.029	.015	.008	18
19							.216	.120	.065	.034	.018	.009	19
20								.139	.076	.040	.021	.011	20
21								.160	.088	.047	.025	.013	21
22								.183	.102	.055	.029	.015	22
23								.207	.117	.064	.034	.018	23
24									.113	.073	.039	.021	24
25									.151	.084	.045	.024	25
26									.170	.095	.052	.028	26
27									.190	.108	.059	.032	27
28									.212	.122	.068	.036	28
29										.137	.077	.042	29
30										.153	.086	.047	30
31										.170	.097	.053	31
32										.188	.108	.060	32
33										.207	.121	.068	33
34											.134	.076	34
35											.148	.084	35
36											.163	.094	36
37											.179	.104	37
38											.196	.115	38
$\dfrac{n(n+1)}{2}$	10	15	21	28	36	45	55	66	78	91	105	120	

Notes: 1. Table entries are $P(R_+ \leq c) = P[R_+ \geq \frac{1}{2}n(n+1) - c]$.

2. $R_- = \frac{1}{2}n(n+1) - R_+$.

3. For $n > 15$, use a normal approximation (see Section 10.5).

Table VII Tail-Probabilities of the Two-Sample Wilcoxon Rank-Sum Statistic

m n c	4 4	4 5	4 6	4 7	4 8	4 9	4 10
10	.014	.008	.005	.003	.002	.001	.001
11	.029	.016	.010	.006	.004	.003	.002
12	.057	.032	.019	.012	.008	.006	.004
13	.100	.056	.033	.021	.014	.010	.007
14	.171	.096	.057	.036	.024	.017	.012
15	.243	.143	.086	.055	.036	.025	.018
16		.206	.129	.082	.055	.038	.027
17			.176	.115	.077	.053	.038
18				.158	.107	.074	.053
19				.206	.141	.099	.071
20					.184	.130	.094
21						.165	.120
M	36	40	44	48	52	56	60

m n c	5 5	5 6	5 7	5 8	5 9	5 10
15	.004	.002	.001	.001	.000	.000
16	.008	.004	.003	.002	.001	.001
17	.016	.009	.005	.003	.002	.001
18	.028	.015	.009	.005	.003	.002
19	.048	.026	.015	.009	.006	.004
20	.075	.041	.024	.015	.009	.006
21	.111	.063	.037	.023	.014	.010
22	.155	.089	.053	.033	.021	.014
23	.210	.123	.074	.047	.030	.020
24		.165	.101	.064	.041	.028
25		.214	.134	.085	.056	.038
26			.172	.111	.073	.050
27			.216	.142	.095	.065
28				.177	.120	.082
28					.149	.103
M	55	60	65	70	75	80

Notes: 1. m is the size of the smaller sample.

2. The entry opposite c is the cumulative tail probability:

$$P(R \leq c) = P(R \geq M - c),$$

where R is the rank sum for the smaller sample, and M is the sum of the minimum and maximum values of R. [Note that $\frac{M}{2} = \text{m.v.}(R)$.]

3. For m or $n > 10$, use a normal approximation (see Section 12.5).

Table VII (*continued*)

m / n / c	6 6	6 7	6 8	6 9	6 10		m / n / c	7 7	7 8	7 9	7 10
24	.008	.004	.002	.001	.001		34	.009	.005	.003	.002
25	.013	.007	.004	.002	.001		35	.013	.007	.004	.002
26	.021	.011	.006	.004	.002		36	.019	.010	.006	.003
27	.032	.017	.010	.006	.004		37	.027	.014	.008	.005
28	.047	.026	.015	.009	.005		38	.036	.020	.011	.007
29	.066	.037	.021	.013	.008		39	.049	.027	.016	.009
30	.090	.051	.030	.018	.011		40	.064	.036	.001	.012
31	.120	.069	.041	.025	.016		41	.082	.047	.027	.017
32	.155	.090	.054	.033	.021		42	.104	.060	.036	.022
33	.197	.117	.071	.044	.028		43	.130	.076	.045	.028
34		.147	.091	.057	.036		44	.159	.095	.057	.035
35		.183	.114	.072	.047		45	.191	.116	.071	.044
36			.141	.091	.059		46		.140	.087	.054
37			.172	.112	.074		47		.168	.105	.067
38				136	.090		48		.198	.126	.081
39				.164	.110		49			.150	.097
							50			.176	.115
M	78	84	90	96	102		M	105	112	119	126

m / n / c	8 8	8 9	8 10		m / n / c	9 9	9 10		m / n / c	10 10
46	.010	.006	.003		59	.009	.005		74	.009
47	.014	.008	.004		60	.012	.007		75	.012
48	.019	.010	.006		61	.016	.009		76	.014
49	.025	.014	.008		62	.020	.011		77	.018
50	.032	.018	.010		63	.025	.014		78	.022
51	.041	.023	.013		64	.031	.017		79	.026
52	.052	.030	.017		65	.039	.022		80	.032
53	.065	.037	.022		66	.047	.027		81	.038
54	.080	.046	.027		67	.057	.033		82	.045
55	.097	.057	.034		68	.068	.039		83	.053
56	.117	.069	.042		69	.081	.047		84	.062
57	.139	.084	.051		70	.095	.056		85	.072
58	.164	.100	.061		71	.111	.067		86	.083
59	.191	.118	.073		72	.129	.078		87	.095
60		.138	.086		73	.149	.091		88	.109
61		.161	.102		74	.170	.106		89	.124
M	136	144	152		M	171	180		M	210

Table VIIIa Tail-Probabilities of the *F*-distribution

2 d.f. in numerator

F	5	6	7	8	9	10	11	12	13	14	15	20	30	40
2.0	.230	.216	.206	.198	.191	.186	.182	.178	.175	.172	.170	.162	.153	.149
2.2	.206	.192	.181	.173	.167	.162	.157	.153	.150	.148	.145	.137	.128	.124
2.4	.186	.171	.161	.153	.146	.141	.136	.133	.130	.127	.125	.116	.108	.104
2.6	.168	.154	.143	.135	.128	.123	.119	.115	.112	.110	.107	.099	.091	.087
2.8	.153	.138	.128	.120	.113	.108	.104	.100	.099	.095	.093	.085	.077	.073
3.0	.139	.125	.115	.107	.100	.095	.091	.088	.085	.082	.080	.073	.065	.061
3.2	.127	.113	.103	.095	.089	.084	.080	.077	.074	.072	.070	.062	.055	.051
3.4	.117	.103	.093	.085	.079	.075	.071	.068	.065	.063	.061	.054	.047	.043
3.6	.108	.094	.084	.077	.071	.066	.063	.060	.057	.055	.053	.046	.040	.037
3.8	.099	.086	.076	.069	.064	.059	.056	.053	.050	.048	.046	.040	.034	.031
4.0	.092	.079	.069	.063	.057	.053	.049	.047	.044	.042	.041	.035	.029	.026
4.2	.085	.072	.063	.057	.051	.047	.044	.041	.039	.037	.036	.030	.025	.022
4.4	.079	.067	.058	.051	.046	.043	.039	.037	.035	.033	.031	.026	.021	.019
4.6	.074	.062	.053	.047	.042	.038	.035	.033	.031	.029	.028	.023	.018	.016
4.8	.069	.057	.049	.043	.038	.035	.032	.029	.027	.026	.024	.020	.016	.014
5.0	.064	.053	.045	.039	.035	.031	.029	.026	.025	.023	.022	.017	.013	.012
5.2	.060	.049	.041	.036	.032	.028	.026	.024	.022	.020	.019	.015	.012	.010
5.4	.056	.046	.038	.033	.029	.026	.023	.021	.020	.018	.017	.013	.010	.008
5.6	.053	.042	.035	.030	.026	.023	.021	.019	.018	.016	.015	.012	.009	.007
5.8	.050	.040	.033	.028	.024	.021	.019	.017	.016	.014	.013	.012	.009	.006
6.0	.047	.037	.030	.026	.022	.019	.017	.016	.014	.013	.012	.009	.006	.005
6.2	.044	.035	.028	.024	.020	.018	.016	.014	.013	.012	.011	.008	.006	.005
6.4	.042	.033	.026	.022	.019	.016	.014	.013	.012	.011	.010	.007	.005	.004
6.6	.040	.031	.024	.020	.017	.015	.013	.012	.011	.010	.009	.006	.004	.003
6.8	.037	.029	.023	.019	.016	.014	.012	.011	.010	.009	.008	.006	.004	.003
7.0	.036	.027	.021	.017	.015	.013	.011	.010	.009	.008	.007	.005	.003	.002
7.2	.034	.025	.020	.016	.014	.012	.010	.009	.008	.007	.006	.004	.003	.002
7.4	.032	.024	.019	.015	.013	.011	.009	.008	.007	.006	.006	.004	.002	.002
7.6	.030	.023	.018	.014	.012	.010	.008	.007	.007	.006	.005	.004	.002	.002
7.8	.029	.021	.017	.013	.011	.009	.008	.007	.006	.005	.005	.003	.002	.001
8.0	.028	.020	.016	.012	.010	.008	.007	.006	.005	.005	.004	.003	.002	.001
8.2	.026	.019	.015	.012	.009	.008	.007	.006	.005	.004	.004	.003	.001	.001
8.4	.025	.018	.014	.011	.009	.007	.006	.005	.005	.004	.004	.002	.001	.001
8.6	.024	.017	.013	.010	.008	.007	.006	.005	.004	.004	.003	.002	.001	.001
8.8	.023	.016	.012	.010	.008	.006	.005	.004	.004	.003	.003	.002	.001	.001
9.0	.022	.016	.012	.009	.007	.006	.005	.004	.004	.003	.003	.002	.001	.001
9.5	.020	.014	.010	.008	.006	.005	.004	.003	.003	.002	.002	.001	.001	.000
10.0	.018	.012	.009	.007	.005	.004	.003	.003	.002	.002	.002	.001	.000	.000

Table VIIIa (*continued*)

2 d.f. in numerator

F	5	6	7	8	9	Denominator degrees of freedom 10	11	12	13	14	15	20	30	40
10.5	.016	.011	.008	.006	.004	.003	.003	.002	.002	.002	.001	.001	.000	.000
11.0	.015	.010	.007	.005	.004	.003	.002	.002	.002	.001	.001	.001	.000	.000
11.5	.013	.009	.006	.005	.004	.004	.003	.003	.002	.001	.001	.000	.000	.000
12.0	.012	.008	.005	.004	.003	.002	.002	.001	.001	.001	.001	.000	.000	.000
13.0	.010	.007	.004	.003	.002	.002	.001	.001	.001	.001	.001	.000	.000	.000
15.0	.008	.005	.003	.002	.001	.001	.001	.001	.000	.000	.000	.000	.000	.000
20.0	.004	.002	.001	.001	.000	.000	.000	.000	.000	.000	.000	.000	.000	.000

3 d.f. in numerator

F	5	6	7	8	9	10	11	12	13	14	15	20	30	40
2.0	.233	.216	.203	.193	.185	.178	.172	.168	.164	.160	.157	.146	.135	.129
2.2	.206	.189	.176	.166	.158	.151	.146	.141	.137	.133	.130	.120	.109	.103
2.4	.184	.166	.153	.143	.135	.129	.123	.119	.115	.111	.108	.098	.087	.082
2.6	.158	.147	.134	.124	.117	.110	.105	.100	.097	.093	.091	.081	.070	.065
2.8	.148	.131	.118	.109	.101	.095	.090	.085	.082	.079	.076	.066	.057	.052
3.0	.134	.117	.105	.095	.085	.082	.077	.073	.069	.066	.064	.055	.046	.042
3.2	.121	.105	.093	.084	.077	.071	.066	.062	.059	.056	.054	.045	.037	.033
3.4	.110	.094	.083	.074	.067	.062	.057	.053	.050	.048	.046	.038	.030	.027
3.6	.101	.085	.074	.065	.059	.054	.050	.046	.043	.041	.039	.032	.025	.022
3.8	.092	.077	.066	.058	.052	.047	.043	.040	.037	.035	.033	.026	.020	.017
4.0	.085	.070	.060	.052	.046	.041	.038	.035	.032	.030	.028	.022	.017	.014
4.2	.078	.064	.054	.046	.041	.036	.033	.030	.028	.026	.024	.019	.014	.011
4.4	.072	.058	.049	.042	.036	.032	.029	.026	.024	.022	.021	.016	.011	.009
4.6	.067	.053	.044	.037	.032	.029	.025	.023	.021	.019	.018	.013	.009	.007
4.8	.062	.049	.040	.034	.029	.025	.022	.020	.018	.017	.015	.011	.008	.006
5.0	.058	.045	.037	.031	.026	.023	.020	.018	.016	.015	.013	.010	.006	.005
5.2	.054	.042	.034	.028	.023	.020	.018	.016	.014	.013	.012	.008	.005	.004
5.4	.050	.039	.031	.025	.021	.018	.016	.014	.012	.011	.010	.007	.004	.003
5.6	.047	.036	.028	.023	.019	.016	.014	.012	.011	.010	.009	.006	.004	.003
5.8	.044	.033	.026	.021	.017	.015	.013	.011	.110	.009	.008	.005	.003	.002
6.0	.041	.031	.024	.019	.016	.014	.011	.010	.009	.008	.007	.004	.002	.002
6.2	.039	.029	.022	.018	.014	.012	.010	.009	.008	.007	.006	.004	.002	.001
6.4	.036	.027	.020	.016	.013	.011	.009	.008	.007	.006	.005	.003	.002	.001
6.6	.034	.025	.019	.015	.012	.010	.008	.007	.006	.005	.005	.003	.001	.001
6.8	.032	.023	.018	.014	.011	.009	.007	.006	.005	.005	.004	.002	.001	.001

(*continues*)

Table VIIIa (*continued*)

3 d.f. in numerator

F	5	6	7	8	9	10	11	12	13	14	15	20	30	40
						Denominator degrees of freedom								
7.0	.031	.022	.016	.013	.010	.008	.007	.006	.005	.004	.004	.002	.001	.001
7.2	.029	.021	.015	.012	.009	.007	.006	.005	.004	.004	.003	.002	.001	.001
7.4	.028	.019	.014	.011	.008	.007	.005	.004	.004	.003	.003	.002	.001	.000
7.6	.026	.018	.013	.010	.008	.006	.005	.004	.003	.003	.003	.001	.001	.001
7.8	.025	.017	.012	.009	.007	.006	.005	.004	.003	.003	.002	.001	.001	.000
8.0	.024	.016	.012	.009	.007	.005	.004	.003	.003	.002	.002	.001	.001	.000
8.2	.022	.015	.011	.008	.006	.005	.004	.003	.003	.002	.002	.001	.001	.000
8.4	.021	.014	.010	.007	.006	.005	.004	.003	.003	.002	.002	.001	.000	.001
8.6	.020	.014	.010	.007	.005	.004	.003	.003	.002	.002	.001	.001	.000	.000
8.8	.019	.013	.009	.006	.005	.004	.003	.002	.002	.002	.001	.001	.000	.000
9.0	.019	.012	.008	.006	.005	.003	.003	.002	.002	.001	.001	.001	.000	.000
9.5	.017	.011	.007	.005	.004	.003	.002	.002	.001	.001	.001	.000	.000	.000
10.0	.015	.009	.006	.004	.003	.002	.002	.001	.001	.001	.001	.000	.000	.000
10.5	.013	.008	.006	.004	.003	.002	.001	.001	.001	.001	.001	.000	.000	.000
11.0	.012	.007	.005	.003	.002	.002	.001	.001	.001	.001	.000	.000	.000	.000
11.5	.011	.007	.004	.003	.002	.001	.001	.001	.001	.000	.000	.000	.000	.000
12.0	.010	.006	.003	.002	.002	.001	.001	.001	.000	.000	.000	.000	.000	.000
13.0	.008	.005	.003	.002	.001	.001	.001	.000	.000	.000	.000	.000	.000	.000
15.0	.006	.003	.002	.001	.001	.000	.000	.000	.000	.000	.000	.000	.000	.000
20.0	.003	.002	.001	.000	.000	.000	.000	.000	.000	.000	.000	.000	.000	.000

4 d.f. in numerator

F	5	6	7	8	9	10	11	12	13	14	15	20	30	40
2.0	.233	.214	.199	.188	.178	.171	.164	.159	.154	.150	.146	.133	.120	.113
2.2	.205	.185	.171	.159	.150	.142	.136	.130	.126	.122	.118	.106	.093	.086
2.4	.181	.162	.147	.136	.127	.119	.113	.108	.103	.099	.096	.074	.072	.066
2.6	.161	.142	.128	.117	.108	.100	.094	.089	.085	.082	.078	.067	.056	.050
2.8	.144	.125	.111	.100	.092	.085	.079	.075	.071	.067	.064	.054	.044	.039
3.0	.134	.111	.097	.087	.079	.072	.067	.063	.059	.056	.053	.043	.034	.030
3.2	.117	.099	.086	.076	.068	.062	.057	.053	.049	.046	.044	.035	.027	.023
3.4	.106	.088	.076	.066	.059	.053	.048	.046	.041	.038	.036	.028	.021	.017
3.6	.096	.079	.067	.058	.051	.046	.041	.038	.035	.032	.030	.023	.016	.013
3.8	.088	.071	.060	.051	.045	.040	.035	.032	.029	.027	.025	.019	.013	.010
4.0	.080	.065	.053	.045	.039	.034	.031	.027	.025	.023	.021	.015	.010	.008
4.2	.074	.059	.048	.040	.034	.030	.026	.024	.021	.019	.018	.013	.008	.006
4.4	.068	.053	.043	.036	.030	.026	.023	.020	.018	.016	.015	.010	.006	.005
4.6	.063	.049	.039	.032	.027	.023	.020	.018	.016	.014	.013	.009	.005	.004
4.8	.058	.044	.035	.029	.024	.020	.017	.015	.013	.012	.011	.007	.004	.003

Table VIIIa (*continued*)

4 d.f. in numerator

F	5	6	7	8	9	10	11	12	13	14	15	20	30	40
						Denominator degrees of freedom								
5.0	.054	.041	.032	.026	.021	.018	.015	.013	.012	.010	.009	.006	.003	.002
5.2	.050	.037	.029	.023	.019	.016	.013	.012	.010	.009	.008	.005	.003	.002
5.4	.046	.034	.026	.021	.017	.014	.013	.010	.009	.008	.007	.004	.002	.001
5.6	.043	.032	.024	.019	.015	.012	.010	.009	.008	.007	.006	.003	.002	.001
5.8	.040	.029	.022	.017	.014	.011	.009	.008	.107	.006	.005	.003	.001	.001
6.0	.038	.027	.020	.016	.012	.010	.008	.007	.006	.005	.004	.002	.001	.001
6.2	.036	.025	.019	.014	.011	.009	.007	.006	.005	.004	.004	.002	.001	.001
6.4	.033	.023	.017	.013	.010	.008	.007	.005	.004	.004	.003	.002	.001	.000
6.6	.031	.022	.016	.012	.009	.007	.006	.005	.004	.003	.003	.001	.001	.000
6.8	.030	.020	.015	.011	.008	.007	.005	.004	.004	.003	.002	.001	.001	.000
7.0	.028	.019	.014	.010	.008	.006	.005	.004	.003	.003	.002	.001	.000	.000
7.2	.026	.018	.013	.009	.007	.005	.004	.003	.003	.002	.002	.001	.000	.000
7.4	.025	.017	.012	.009	.006	.005	.004	.003	.002	.002	.002	.001	.000	.000
7.6	.024	.016	.011	.008	.006	.004	.003	.003	.002	.002	.001	.001	.000	.000
7.8	.022	.015	.010	.006	.005	.004	.003	.002	.002	.002	.001	.001	.000	.000
8.0	.021	.014	.009	.007	.005	.004	.003	.002	.002	.001	.001	.001	.000	.000
8.2	.020	.013	.009	.006	.005	.003	.003	.002	.002	.001	.001	.000	.000	.000
8.4	.019	.012	.008	.006	.004	.003	.002	.002	.001	.001	.001	.000	.000	.000
8.6	.018	.012	.008	.005	.004	.003	.002	.002	.001	.001	.001	.000	.000	.000
8.8	.017	.011	.007	.006	.004	.003	.002	.001	.001	.001	.001	.000	.000	.000
9.0	.017	.010	.007	.005	.003	.002	.002	.001	.001	.001	.001	.000	.000	.000
9.5	.015	.009	.006	.004	.003	.002	.001	.001	.001	.001	.000	.000	.000	.000
10.0	.013	.008	.005	.003	.002	.002	.001	.001	.001	.000	.000	.000	.000	.000
10.5	.012	.007	.004	.003	.002	.001	.001	.001	.001	.000	.000	.000	.000	.000
11.0	.011	.006	.004	.002	.002	.001	.001	.001	.000	.000	.000	.000	.000	.000
11.5	.010	.006	.003	.002	.001	.001	.001	.000	.000	.000	.000	.000	.000	.000
12.0	.009	.005	.003	.002	.001	.001	.001	.000	.000	.000	.000	.000	.000	.000
13.0	.007	.004	.002	.001	.001	.001	.000	.000	.000	.000	.000	.000	.000	.000
15.0	.005	.003	.002	.001	.001	.000	.000	.000	.000	.000	.000	.000	.000	.000
20.0	.003	.001	.001	.000	.000	.000	.000	.000	.000	.000	.000	.000	.000	.000

5 d.f. in numerator

F	5	6	7	8	9	10	11	12	13	14	15	20	30	40
2.0	.233	.212	.196	.183	.173	.164	.157	.151	.146	.141	.137	.123	.107	.100
2.2	.204	.182	.166	.154	.144	.135	.128	.122	.117	.113	.109	.095	.081	.073
2.4	.179	.158	.142	.130	.120	.112	.105	.099	.095	.090	.087	.074	.060	.004
2.6	.159	.138	.123	.110	.101	.093	.087	.081	.077	.073	.069	.057	.045	.040
2.8	.141	.121	.106	.094	.085	.078	.072	.067	.063	.059	.056	.045	.034	.029

(*continues*)

Table VIIIa (*continued*)

5 d.f. in numerator

F	\multicolumn{14}{c}{Denominator degrees of freedom}
	5	6	7	8	9	10	11	12	13	14	15	20	30	40
3.0	.127	.107	.092	.081	.072	.066	.060	.055	.051	.048	.045	.035	.026	.022
3.2	.114	.095	.081	.070	.062	.055	.050	.046	.042	.039	.037	.028	.020	.016
3.4	.103	.084	.071	.061	.053	.047	.042	.038	.034	.032	.030	.022	.015	.012
3.6	.093	.075	.062	.053	.046	.040	.036	.032	.029	.027	.024	.017	.011	.009
3.8	.085	.067	.055	.046	.040	.034	.030	.027	.024	.022	.020	.014	.009	.007
4.0	.077	.061	.049	.041	.035	.030	.026	.023	.020	.018	.017	.011	.007	.005
4.2	.071	.055	.044	.036	.030	.026	.022	.019	.017	.015	.014	.009	.005	.004
4.4	.065	.050	.039	.032	.026	.022	.019	.017	.015	.013	.011	.007	.004	.003
4.6	.060	.045	.035	.028	.023	.019	.016	.014	.012	.011	.010	.006	.003	.002
4.8	.055	.041	.032	.025	.020	.017	.014	.012	.010	.009	.008	.005	.002	.002
5.0	.051	.038	.029	.023	.018	.015	.012	.010	.009	.008	.007	.004	.002	.001
5.2	.047	.034	.026	.020	.016	.013	.011	.009	.008	.007	.006	.003	.001	.001
5.4	.044	.032	.024	.018	.014	.012	.009	.008	.007	.006	.005	.003	.001	.001
5.6	.041	.029	.022	.016	.013	.010	.008	.007	.006	.005	.004	.002	.001	.001
5.8	.038	.027	.020	.015	.011	.009	.007	.006	.005	.004	.004	.002	.001	.001
6.0	.036	.025	.018	.013	.010	.008	.006	.005	.004	.004	.003	.001	.000	.000
6.2	.033	.023	.017	.012	.009	.007	.006	.005	.004	.003	.003	.001	.000	.000
6.4	.031	.021	.015	.011	.008	.006	.005	.004	.003	.003	.002	.001	.000	.000
6.6	.029	.020	.014	.010	.008	.006	.005	.004	.003	.002	.002	.001	.000	.000
6.8	.028	.019	.013	.009	.007	.005	.004	.003	.003	.002	.002	.001	.000	.000
7.0	.026	.017	.012	.008	.006	.005	.004	.003	.002	.002	.001	.001	.000	.000
7.2	.025	.016	.011	.008	.006	.004	.003	.002	.002	.002	.001	.001	.000	.000
7.4	.023	.015	.010	.007	.005	.004	.003	.002	.002	.001	.001	.000	.000	.000
7.6	.022	.014	.009	.007	.005	.003	.003	.002	.002	.001	.001	.000	.000	.000
7.8	.021	.013	.009	.006	.004	.003	.002	.002	.001	.001	.001	.000	.000	.000
8.0	.020	.012	.008	.006	.004	.003	.002	.002	.001	.001	.001	.000	.000	.000
8.2	.019	.012	.008	.005	.004	.003	.002	.001	.001	.001	.001	.000	.000	.000
8.4	.018	.011	.007	.005	.003	.002	.002	.001	.001	.001	.001	.000	.000	.000
8.6	.017	.010	.007	.004	.003	.002	.002	.001	.001	.001	.001	.000	.000	.000
8.8	.016	.010	.006	.004	.003	.002	.001	.001	.001	.001	.001	.000	.000	.000
9.0	.015	.009	.006	.004	.003	.002	.001	.001	.001	.001	.000	.000	.000	.000
9.5	.014	.008	.005	.003	.002	.001	.001	.001	.001	.000	.000	.000	.000	.000
10.0	.012	.007	.004	.003	.002	.001	.001	.001	.000	.000	.000	.000	.000	.000
10.5	.011	.006	.004	.002	.001	.001	.001	.000	.000	.000	.000	.000	.000	.000
11.0	.010	.006	.003	.002	.001	.001	.001	.000	.000	.000	.000	.000	.000	.000
11.5	.009	.005	.003	.002	.001	.001	.000	.000	.000	.000	.000	.000	.000	.000
12.0	.008	.004	.003	.001	.001	.000	.000	.000	.000	.000	.000	.000	.000	.000
13.0	.007	.004	.002	.001	.001	.000	.000	.000	.000	.000	.000	.000	.000	.000
15.0	.005	.002	.001	.001	.000	.000	.000	.000	.000	.000	.000	.000	.000	.000
20.0	.004	.002	.001	.001	.000	.000	.000	.000	.000	.000	.000	.000	.000	.000

Table VIIIb Ninety-fifth Percentiles of the F-Distribution

| | Numerator degrees of freedom | | | | | | | | | | | | |
	1	2	3	4	5	6	8	10	12	15	20	24	30
1	161	200	216	225	230	234	239	242	244	246	248	249	250
2	18.5	19.0	19.2	19.2	19.3	19.3	19.4	19.4	19.4	19.4	19.4	19.5	19.5
3	10.1	9.55	9.28	9.12	9.01	8.94	8.85	8.79	8.74	8.70	8.66	8.64	8.62
4	7.71	6.94	6.59	6.39	6.26	6.16	6.04	5.96	5.91	5.86	5.80	5.77	5.75
5	6.61	5.79	5.41	5.19	5.05	4.95	4.82	4.74	4.68	4.62	4.56	4.53	4.50
6	5.99	5.14	4.76	4.53	4.39	4.28	4.15	4.06	4.00	3.94	3.87	3.84	3.81
7	5.59	4.74	4.35	4.12	3.97	3.87	3.73	3.64	3.57	3.51	3.44	3.41	3.38
8	5.32	4.46	4.07	3.84	3.69	3.58	3.44	3.35	3.28	3.22	3.15	3.12	3.08
9	5.12	4.26	3.86	3.63	3.48	3.37	3.23	3.14	3.07	3.01	2.94	2.90	2.86
10	4.96	4.10	3.71	3.48	3.33	3.22	3.07	2.98	2.91	2.85	2.77	2.74	2.70
11	4.84	3.98	3.59	3.36	3.20	3.09	2.95	2.85	2.79	2.72	2.65	2.61	2.57
12	4.75	3.89	3.49	3.26	3.11	3.00	2.85	2.75	2.69	2.62	2.54	2.51	2.47
13	4.67	3.81	3.41	3.18	3.03	2.92	2.77	2.67	2.60	2.53	2.46	2.42	2.38
14	4.60	3.74	3.34	3.11	2.96	2.85	2.70	2.60	2.53	2.46	2.39	2.35	2.31
15	4.54	3.68	3.29	3.06	2.90	2.79	2.64	2.54	2.48	2.40	2.33	2.29	2.25
16	4.49	3.63	3.24	3.01	2.85	2.74	2.59	2.49	2.42	2.35	2.28	2.24	2.19
17	4.45	3.59	3.20	2.96	2.81	2.70	2.55	2.45	2.38	2.31	2.23	2.19	2.15
18	4.41	3.55	3.16	2.93	2.77	2.66	2.51	2.41	2.34	2.27	2.19	2.15	2.11
19	4.38	3.52	3.13	2.90	2.74	2.63	2.48	2.38	2.31	2.23	2.16	2.11	2.07
20	4.35	3.49	3.10	2.87	2.71	2.60	2.45	2.35	2.28	2.20	2.12	2.08	2.04
21	4.32	3.47	3.07	2.84	2.68	2.57	2.42	2.32	2.25	2.18	2.10	2.05	2.01
22	4.30	3.44	3.05	2.82	2.66	2.55	2.40	2.30	2.23	2.15	2.07	2.03	1.98
23	4.28	3.42	3.03	2.80	2.64	2.53	2.37	2.27	2.20	2.13	2.05	2.01	1.96
24	4.26	3.40	3.01	2.78	2.62	2.51	2.36	2.25	2.18	2.11	2.03	1.98	1.94
25	4.24	3.39	2.99	2.76	2.60	2.49	2.34	2.24	2.16	2.09	2.01	1.96	1.92
30	4.17	3.32	2.92	2.69	2.53	2.42	2.27	2.16	2.09	2.01	1.93	1.89	1.84
40	4.08	3.23	2.84	2.61	2.45	2.34	2.18	2.08	2.00	1.92	1.84	1.79	1.74
60	4.00	3.15	2.76	2.53	2.37	2.25	2.10	1.99	1.92	1.84	1.75	1.70	1.65

Denominator degrees of freedom (left axis label)

Note: The fifth percentiles are obtainable as follows:

$$F_{.05}(r, s) = \frac{1}{F_{.95}(s, r)}.$$

$$\left[\text{For example, } F_{.05}(3, 6) = \frac{1}{F_{.95}(6, 3)} = \frac{1}{8.94}. \right]$$

Table VIIIb Ninety-ninth Percentiles of the F-Distribution

| | \multicolumn{13}{c}{Numerator degrees of freedom} |
	1	2	3	4	5	6	8	10	12	15	20	24	30
1	4050	5000	5400	5620	5760	5860	5980	6060	6110	6160	6210	6235	6260
2	98.5	99.0	99.2	99.2	99.3	99.3	99.4	99.4	99.4	99.4	99.4	99.5	99.5
3	34.1	30.8	29.5	28.7	28.2	27.9	27.5	27.3	27.1	26.9	26.7	26.6	26.5
4	21.2	18.0	16.7	16.0	15.5	15.2	14.8	14.5	14.4	14.2	14.0	13.9	13.8
5	16.3	13.3	12.1	11.4	11.0	10.7	10.3	10.1	9.89	9.72	9.55	9.47	9.38
6	13.7	10.9	9.78	9.15	8.75	8.47	8.10	7.87	7.72	7.56	7.40	7.31	7.23
7	12.2	9.55	8.45	7.85	7.46	7.19	6.84	6.62	6.47	6.31	6.16	6.07	5.99
8	11.3	8.65	7.59	7.01	6.63	6.37	6.03	5.81	5.67	5.52	5.36	5.28	5.20
9	10.6	8.02	6.99	6.42	6.06	5.80	5.47	5.26	5.11	4.96	4.81	4.73	4.65
10	10.0	7.56	6.55	5.99	5.64	5.39	5.06	4.85	4.71	4.56	4.41	4.33	4.25
11	9.65	7.21	6.22	5.67	5.32	5.07	4.74	4.54	4.40	4.25	4.10	4.02	3.94
12	9.33	6.93	5.95	5.41	5.06	4.82	4.50	4.30	4.16	4.01	3.86	3.78	3.70
13	9.07	6.70	5.74	5.21	4.86	4.62	4.30	4.10	3.96	3.82	3.66	3.59	3.51
14	8.86	6.51	5.56	5.04	4.69	4.46	4.14	3.94	3.80	3.66	3.51	3.43	3.35
15	8.68	6.36	5.42	4.89	4.56	4.32	4.00	3.80	3.67	3.52	3.37	3.29	3.21
16	8.53	6.23	5.29	4.77	4.44	4.20	3.89	3.69	3.55	3.41	3.26	3.18	3.10
17	8.40	6.11	5.18	4.67	4.34	4.10	3.79	3.59	3.46	3.31	3.16	3.08	3.00
18	8.29	6.01	5.09	4.58	4.25	4.01	3.71	3.51	3.37	3.23	3.08	3.00	2.92
19	8.18	5.93	5.01	4.50	4.17	3.94	3.63	3.43	3.30	3.15	3.00	2.92	2.84
20	8.10	5.85	4.94	4.43	4.10	3.87	3.56	3.37	3.23	3.09	2.94	2.86	2.78
21	8.02	5.78	4.87	4.37	4.04	3.81	3.51	3.31	3.17	3.03	2.88	2.80	2.72
22	7.95	5.72	4.82	4.31	3.99	3.76	3.45	3.26	3.12	2.98	2.83	2.75	2.67
23	7.88	5.66	4.76	4.26	3.94	3.71	3.41	3.21	3.07	2.93	2.87	2.70	2.62
24	7.82	5.61	4.72	4.22	3.90	3.67	3.36	3.17	3.03	2.89	2.74	2.66	2.58
25	7.77	5.57	4.68	4.18	3.86	3.63	3.32	3.13	2.99	2.85	2.70	2.62	2.54
30	7.56	5.39	4.51	4.02	3.70	3.47	3.17	2.98	2.84	2.70	2.55	2.47	2.39
40	7.31	5.18	4.31	3.83	3.51	3.29	2.99	2.80	2.66	2.52	2.37	2.29	2.20
60	7.08	4.98	4.13	3.65	3.34	3.12	2.82	2.63	2.50	2.35	2.20	2.12	2.03

Denominator degrees of freedom

Table IXa Tail-Probabilities for the One-Sample Kolmogorov–Smirnov Statistic

$$nD_n = n \sup_x |F_n(x) - F_0(x)|$$

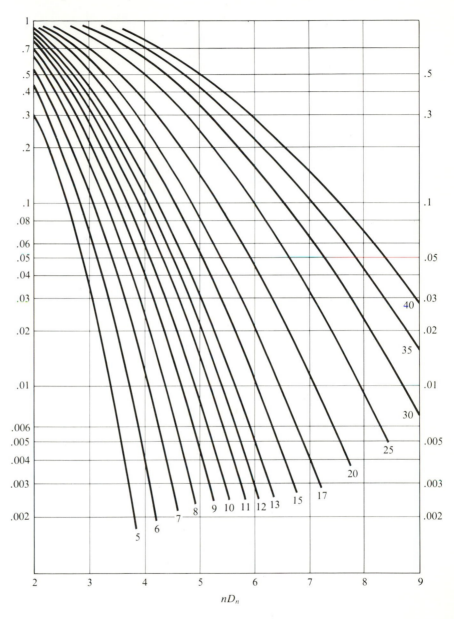

Table IXb Critical Values for the One-Sample
Kolmogorov–Smirnov Statistic

Sample size (n)	Significance level				
	.20	.15	.10	.05	.01
1	.900	.925	.950	.975	.995
2	.684	.726	.776	.842	.929
3	.565	.597	.642	.708	.829
4	.494	.525	.564	.624	.734
5	.446	.474	.510	.563	.669
6	.410	.436	.470	.521	.618
7	.381	.405	.438	.486	.577
8	.358	.381	.411	.457	.543
9	.339	.360	.388	.432	.514
10	.322	.342	.368	.409	.486
11	.307	.326	.352	.391	.468
12	.295	.313	.338	.375	.450
13	.284	.302	.325	.361	.433
14	.274	.292	.314	.349	.418
15	.266	.283	.304	.338	.404
16	.258	.274	.295	.328	.391
17	.250	.266	.286	.318	.380
18	.244	.259	.278	.309	.370
19	.237	.252	.272	.301	.361
20	.231	.246	.264	.294	.352
25	.21	.22	.24	.264	.32
30	.19	.20	.22	.242	.29
35	.18	.19	.21	.23	.27
40				.21	.25
50				.19	.23
60				.17	.21
70				.16	.19
80				.15	.18
90				.14	
100				.14	
Asymptotic formula:	$\dfrac{1.07}{\sqrt{n}}$	$\dfrac{1.14}{\sqrt{n}}$	$\dfrac{1.22}{\sqrt{n}}$	$\dfrac{1.36}{\sqrt{n}}$	$\dfrac{1.63}{\sqrt{n}}$

Reject the hypothetical distribution if $D_n = \max|F_n(x) - F(x)|$ exceeds the tabulated value.

(For $\alpha = .01$ and .05, asymptotic formulas give values that are too high—by 1.5% for $n = 80$.)

This table is taken from F. J. Massey, Jr., "The Kolmogorov–Smirnov test for goodness of fit," *J. Am. Stat. Assn. 46* (1951), 68–78 except that certain corrections and additional entries are from Z. W. Birnbaum, "Numerical tabulation of the distribution of Kolmogorov's statistic for finite sample size," *J. Am. Stat. Assn.* 47 (1952), 425–441 with the kind permission of the authors and the *J. Am. Stat. Assn.*

Table Xa Tail-Probabilities for the Two-Sample Kolmogorov–Smirnov Statistic

Case 1: *P-values when sample sizes are equal.*
Independent, random samples from F and G, each of size n.

Statistic: $X = n \cdot \sup|F_n - G_n|$.

x / n	4	5	6	7	8	9	10	11	12	15	20	x
2	.771	.873	.931	.963	.980	.989	.994	.997	.998	1.00	1.00	2
3	.229	.357	.474	.575	.660	.730	.787	.833	.869	.938	.983	3
4	.029	.079	.143	.212	.283	.352	.418	.479	.536	.678	.832	4
5		.008	.026	.053	.087	.126	.168	.211	.256	.386	.571	5
6			.002	.008	.019	.034	.052	.075	.100	.184	.336	6
7				.001	.002	.006	.012	.021	.031	.075	.175	7
8					.000	.001	.002	.004	.008	.026	.081	8
						.000	.000	.001	.001	.008	.034	9
							.000	.000	.000	.002	.012	10
								.000	.000	.000	.004	11
									.000	.000	.001	12
										.000	.000	13

Case 2: *P-values for large samples.*
Independent random samples: size m from F, size n from G.

Statistic: $Y = \sqrt{\dfrac{mn}{m+n}} \cdot \sup|F_m - G_n|$.

y	P	y	P	y	P	y	P	y	P
1.00	.270	1.20	.112	1.40	.040	1.60	.012	1.80	.003
1.01	.259	1.21	.107	1.41	.038	1.61	.011	1.81	.003
1.02	.249	1.22	.102	1.42	.035	1.62	.011	1.82	.003
1.03	.239	1.23	.097	1.43	.033	1.63	.010	1.83	.002
1.04	.230	1.24	.092	1.44	.032	1.64	.009	1.84	.002
1.05	.220	1.25	.088	1.45	.030	1.65	.009	1.85	.002
1.06	.211	1.26	.084	1.46	.028	1.66	.008	1.86	.002
1.07	.202	1.27	.079	1.47	.027	1.67	.008	1.87	.002
1.08	.194	1.28	.075	1.48	.025	1.68	.007	1.88	.002
1.09	.186	1.29	.072	1.49	.024	1.69	.007	1.89	.002
1.10	.178	1.30	.068	1.50	.022	1.70	.006	1.90	.001
1.11	.170	1.31	.065	1.51	.021	1.71	.006	1.91	.001
1.12	.163	1.32	.061	1.52	.020	1.72	.005	1.92	.001
1.13	.155	1.33	.058	1.53	.019	1.73	.005	1.93	.001
1.14	.149	1.34	.055	1.54	.017	1.74	.005	1.94	.001
1.15	.142	1.35	.052	1.55	.016	1.75	.004	1.95	.001
1.16	.136	1.36	.049	1.56	.015	1.76	.004	1.96	.001
1.17	.129	1.37	.047	1.57	.014	1.77	.004	1.97	.001
1.18	.123	1.38	.044	1.58	.014	1.78	.004	1.98	.001
1.19	.118	1.39	.038	1.59	.013	1.79	.003	1.99	.001

Table Xb Critical Values for the Two-Sample Kolmogorov–Smirnov Statistic

		Sample size n_1											
Sample size n_2		1	2	3	4	5	6	7	8	9	10	12	15
1		*	*	*	*	*	*	*	*	*	*		
		*	*	*	*	*	*	*	*	*	*		
2			*	*	*	*	*	*	7/8	16/18	9/10		
			*	*	*	*	*	*	*	*	*		
3				*	*	12/15	5/6	18/21	18/24	7/9		9/12	
				*	*	*	*	*	*	8/9		11/12	
4					3/4	16/20	9/12	21/28	6/8	27/36	14/20	8/12	
					*	*	10/12	24/28	7/8	32/36	16/20	10/12	
5						4/5	20/30	25/35	27/40	31/45	7/10		10/15
						4/5	25/30	30/35	32/40	36/45	8/10		11/15
6							4/6	29/42	16/24	12/18	19/30	7/12	
							5/6	35/42	18/24	14/18	22/30	9/12	
7								5/7	35/56	40/63	43/70		
								5/7	42/56	47/63	53/70		
8									5/8	45/72	23/40	14/24	
									6/8	54/72	28/40	16/24	
9										5/9	52/90	20/36	
										6/9	62/90	24/36	
10											6/10		15/30
											7/10		19/30
12												6/12	30/60
												7/12	35/60
15													7/15
													8/15

Notes: 1. Reject H_0 if $D = \max|F_{n_2}(x) - F_{n_1}(x)|$ exceeds the tabulated value. The upper value gives a level at most .05 and the lower at most .01.

2. Where * appears, do not reject H_0 at the given level.

3. For large values of n_1 and n_2, the following approximate formulas may be used:

$$\alpha = .05: \ 1.36\sqrt{\frac{n_1 + n_2}{n_1 n_2}}.$$

$$\alpha = .01: \ 1.63\sqrt{\frac{n_1 + n_2}{n_1 n_2}}.$$

Table XI Critical Values for the Lilliefors
Distribution (for testing normality)

			Right tail-probability			
n	.20	.15	.10	.05	.01	.001
5	.289	.303	.319	.343	.397	.439
6	.269	.281	.297	.323	.371	.424
7	.252	.264	.280	.304	.351	.402
8	.239	.250	.265	.288	.333	.384
9	.227	.238	.252	.274	.317	.365
10	.217	.228	.241	.262	.304	.352
11	.208	.218	.231	.251	.291	.338
12	.200	.210	.222	.242	.281	.325
13	.193	.202	.215	.234	.271	.314
14	.187	.196	.208	.226	.262	.305
15	.181	.190	.201	.219	.254	.296
16	.176	.184	.195	.213	.247	.287
17	.171	.179	.190	.207	.240	.279
18	.167	.175	.185	.202	.234	.273
19	.163	.170	.181	.197	.228	.266
20	.159	.166	.176	.192	.223	.260
25	.143	.150	.159	.173	.201	.236
30	.131	.138	.146	.159	.185	.217
40	.115	.120	.128	.139	.162	.189
100	.074	.077	.082	.089	.104	.122
400	.037	.039	.041	.045	.052	.061

Assumes that in calculating the K–S statistic, the population
mean and variance are replaced by the sample mean and
variance, respectively.

Table XII Expected Values of Order Statistics from a Standard Normal Population

n	$n-9$	$n-8$	$n-7$	$n-6$	$n-5$	$n-4$	$n-3$	$n-2$	$n-1$	n	i / n
2										.564	2
3										.846	3
4									.297	1.029	4
5									.495	1.163	5
6								.202	.642	1.267	6
7								.353	.757	1.352	7
8							.153	.473	.852	1.424	8
9							.275	.572	.932	1.485	9
10						.123	.376	.656	1.001	1.539	10
11						.225	.462	.729	1.062	1.586	11
12					.103	.312	.537	.793	1.116	1.629	12
13					.190	.388	.603	.850	1.164	1.668	13
14				.088	.267	.456	.662	.901	1.208	1.703	14
15				.165	.335	.516	.715	.948	1.248	1.736	15
16			.077	.234	.396	.570	.763	.990	1.285	1.766	16
17			.146	.295	.451	.619	.807	1.030	1.319	1.794	17
18		.069	.208	.351	.502	.665	.848	1.066	1.350	1.820	18
19		.131	.264	.402	.548	.707	.886	1.099	1.380	1.844	19
20	.062	.187	.315	.448	.590	.745	.921	1.131	1.408	1.867	20

Notes: 1. Assume random sample of size n.

2. Entries are expected locations of $X_{(i)}$, the ith smallest observation.

3. Entries for the lower half are the negatives of the entries for the corresponding upper half. For example, for $n = 5$, expected locations are -1.163, $-.495$, 0, $.495$, 1.163.

Table XIII Distribution of the Standardized Range $W = R/\sigma$ (assuming a normal population)

	Sample size											
	2	3	4	5	6	7	8	9	10	12	15
$E(W)$	1.128	1.693	2.059	2.326	2.534	2.704	2.847	2.970	3.078	3.258	3.472
σ_W	.853	.888	.880	.864	.848	.833	.820	.808	.797	.778	.755
$W_{.005}$	.01	.13	.34	.55	.75	.92	1.08	1.21	1.33	1.55	1.80
$W_{.01}$	.02	.19	.43	.66	.87	1.05	1.20	1.34	1.47	1.68	1.93
$W_{.025}$	.04	.30	.59	.85	1.06	1.25	1.41	1.55	1.67	1.88	2.14
$W_{.05}$	.09	.43	.76	1.03	1.25	1.44	1.60	1.74	1.86	2.07	2.32
$W_{.1}$	.18	.62	.98	1.26	1.49	1.68	1.83	1.97	2.09	2.30	2.54
$W_{.2}$	.36	.90	1.29	1.57	1.80	1.99	2.14	2.28	2.39	2.59	2.83
$W_{.3}$	.55	1.14	1.53	1.82	2.04	2.22	2.38	2.51	2.62	2.82	3.04
$W_{.4}$	.74	1.36	1.76	2.04	2.26	2.44	2.59	2.71	2.83	3.01	3.23
$W_{.5}$	.95	1.59	1.98	2.26	2.47	2.65	2.79	2.92	3.02	3.21	3.42
$W_{.6}$	1.20	1.83	2.21	2.48	2.69	2.86	3.00	3.12	3.23	3.41	3.62
$W_{.7}$	1.47	2.09	2.47	2.73	2.94	3.10	3.24	3.35	3.46	3.63	3.83
$W_{.8}$	1.81	2.42	2.78	3.04	3.23	3.39	3.52	3.63	3.73	3.90	4.09
$W_{.9}$	2.33	2.90	3.24	3.48	3.66	3.81	3.93	4.04	4.13	4.29	4.47
$W_{.95}$	2.77	3.31	3.63	3.86	4.03	4.17	4.29	4.39	4.47	4.62	4.80
$W_{.975}$	3.17	3.68	3.98	4.20	4.36	4.49	4.61	4.70	4.79	4.92	5.09
$W_{.99}$	3.64	4.12	4.40	4.60	4.76	4.88	4.99	5.08	5.16	5.29	5.45
$W_{.995}$	3.97	4.42	4.69	4.89	5.03	5.15	5.26	5.34	5.42	5.54	5.70

Table XIV Random Digits

42916	50199	26435	97117	77100	62919	74498	14252	11052	70038
49019	02101	14580	14421	58592	30885	60248	29783	39125	97534
04421	62261	52644	36493	53146	31906	00208	98915	27613	58180
74606	07765	21788	03093	69158	44498	51540	61267	70550	90599
76288	24031	13826	61989	54283	95614	20378	35853	86644	68259
42866	46273	43621	93636	23582	59351	29828	53006	06004	00427
99017	74447	14581	32223	89571	38437	43037	17654	32705	02726
67245	80759	07378	06307	51311	52458	57898	15213	72105	18792
29317	02377	60654	51918	97109	38972	71750	81431	69776	00892
56457	56692	88071	93055	31559	77054	33921	24189	47537	18470
66908	96815	00106	47915	72072	34460	04085	74036	99640	88672
83966	92418	68500	70046	30009	99166	49224	68804	34733	69265
53196	82252	58476	40657	09612	15380	70717	33052	93954	14642
63291	73919	67613	81329	27561	97499	79346	28385	20829	73829
62205	91166	04127	19669	17699	31072	16918	81168	72908	00561
83502	34546	70327	79999	26659	68085	43541	69983	09041	05677
20293	65765	45954	12799	49028	44691	19957	40928	81503	07030
72932	94622	89404	69024	73518	29828	35482	83798	92363	13918
69803	06247	23872	32055	36776	77634	01444	88377	50827	83716
01155	81380	11691	18090	13236	34313	13390	31223	64796	40116
44290	82296	81987	09423	44272	24414	43248	50536	52161	18884
16980	43552	32970	87214	99340	79058	70912	03514	87351	05102
73249	52463	51467	18602	28336	41484	49543	74121	04575	78007
06050	29975	60715	02040	12974	02831	52032	69726	67679	13772
34216	50564	74588	70102	62585	25511	38134	13802	98334	76947
04607	52269	21767	98347	69224	44987	31255	00344	60841	53970
92738	66714	58465	83216	95109	31032	99817	18844	31514	44004
76152	98002	84257	47518	53932	46337	96349	17004	81135	26247
24405	52117	41434	82281	02756	40000	26893	71507	55783	78195
99046	61444	59911	58255	45299	60971	72833	61883	52645	60945
26312	73154	21070	90104	42013	27302	55283	13166	14051	81929
36315	59502	91215	86654	44578	04159	63389	43516	48971	40922
52467	19775	71391	63601	84377	63350	59557	74397	06289	74426
66790	72193	63999	20307	47423	55164	93870	43783	06851	90065
16427	71681	64661	59249	74118	46257	69308	31035	64498	19592
63988	01319	15012	95770	82029	99778	81793	73836	11528	81863
67468	22553	71756	30281	28244	58696	72161	46240	63452	56485
60477	14463	49722	95808	73193	37865	84147	46004	43753	92444
95384	28822	12047	59393	14588	22723	64262	93653	00284	05594
51396	45671	08283	96848	27039	20852	38008	65531	65322	51775

Table XIV (*continued*)

70321	26394	01403	77390	52111	27816	33570	28064	41906	81867
98710	50639	43559	34442	25514	32178	83688	31018	11232	70459
61664	16238	04228	33224	18550	02255	34597	64773	97872	28450
12906	19628	77265	38578	00958	67476	92199	70519	32591	80452
07633	02489	78236	70986	74294	29591	31175	20817	64727	70957
35933	31203	16796	66581	55006	90733	07198	65126	54346	42214
57652	46065	59420	33920	44589	70899	41795	86683	27317	74817
86860	69306	49382	48964	92022	98252	47414	05190	66648	35104
54447	02332	11406	27021	60064	70307	42155	15810	08324	36194
69865	39302	09057	46982	14177	94534	90536	44442	43337	16371
40500	21406	00571	87320	81683	42788	86367	44686	22159	67015
35892	49668	83991	72088	30210	74009	86370	97956	02132	93512
54819	26094	51409	21485	94764	85806	13393	48543	07042	76538
64224	47909	09994	23750	17351	52141	30486	60380	86546	66606
36913	58173	45709	83679	82617	23381	09603	61107	00566	06572
64745	10614	86371	43244	97154	10397	50975	68006	20045	16942
25536	74031	31807	70133	78790	40341	68730	39635	39013	66841
44043	96215	21270	59427	25034	40645	84741	52083	54503	36861
27659	95463	53847	40921	70116	61536	56756	08967	31079	20097
76014	99818	16606	19713	66904	27106	24874	96701	73287	76772
06073	57343	51428	91171	28299	17520	64903	04177	36071	94952
59008	28543	11576	74547	13260	20688	41261	02780	06633	37536
08844	95774	49323	30448	14154	83379	71259	23302	68402	43750
88505	15575	44927	06584	29867	21541	65763	12154	86616	79877
73259	68626	98962	68548	86576	48046	51755	64995	03661	64585
81550	46798	49319	50206	22024	05175	12923	23427	55915	91723
55831	83784	81034	86779	34622	84570	18960	48798	42970	95789
39465	82353	68905	44234	18244	54345	05592	89361	14644	67924
66415	89349	88530	72096	44459	05258	48317	48866	56886	90458
75889	04514	37227	11302	04667	02129	80414	86289	15887	87380
50749	83220	50529	20619	11606	36531	23409	78122	19566	76564
33045	66703	30017	35347	35038	12952	13971	03922	98702	11786
38388	69556	76728	60535	59961	23634	42211	98387	34880	27755
93182	99040	96390	65989	38375	03652	59657	57431	24666	11061
64713	85185	72849	58611	31220	26657	77056	24553	24993	05210
89024	32054	46997	92652	28363	98992	22593	97710	47766	37646
93573	95502	33790	92973	27766	62671	89698	10877	73893	41004
96035	18795	48080	59666	30241	35233	87353	43647	13404	41982
19264	29229	61369	08309	39383	42305	25944	13577	51545	68990
69801	37145	79189	55897	57793	66816	21930	56771	79296	73793

(*continues*)

Table XIV (*continued*)

21632	42301	23693	72641	56310	85576	03004	25669	69221	32996
23040	65782	23712	13414	10758	15590	97298	74246	51511	46900
36795	38292	03852	06384	84421	03446	91670	45312	27609	87034
06683	83891	88991	16533	09197	31427	60384	48525	90978	46107
21693	12956	21804	46558	37682	81207	85840	53238	35026	04835
53264	41376	17783	64756	39278	25403	33042	20954	31193	24247
45911	92453	25370	86602	48574	57865	26436	16122	76614	17028
21262	59718	77821	14036	31033	90563	45410	15158	90209	84089
38053	60780	54166	14255	33120	27171	71798	91214	80040	56699
12475	40193	59415	04769	75920	01036	02692	75862	16612	73670
61182	03305	90334	00187	91659	28063	75684	50017	82643	09282
77376	85469	08164	05584	36623	82597	83859	03435	98460	70095
80257	04381	06501	08924	35514	14297	54373	71369	05172	15955
82441	04636	48215	06821	03385	17663	40107	55679	30366	42390
95895	16083	58499	17176	55993	51034	49296	04010	78974	35930
02019	96226	27167	68245	53109	59037	37843	79243	10262	58797
61490	82590	52411	54783	29447	94551	30026	97959	93939	73217
82573	62154	78291	33728	39102	11484	86210	43794	73553	87435
01110	77108	56521	78610	08254	01842	43068	70415	79195	26136
49786	47279	38471	20379	54704	86614	91138	51595	50818	80186
95809	54837	55978	10534	46194	00273	03659	57186	73342	95949
76742	70505	64773	48334	00869	80439	69374	35279	99952	85860
03880	30798	40515	66819	40691	72678	17590	76085	62741	93844
81045	58617	37788	64693	50968	31853	95733	08068	21988	60613
54802	96997	52909	14310	08726	09630	49081	66952	05603	08950
63119	68055	76641	87635	64835	06121	86006	76257	51695	44571
85397	39692	66765	50318	33763	45429	30943	20128	14439	68279
06618	14101	87706	77153	54866	00025	34092	92939	12528	24763
19525	93122	11658	06188	43735	43104	18115	28815	21863	22218
49494	16854	95248	00045	40357	73893	50732	11319	38804	15121
51694	84242	33341	77153	71970	66070	80879	10293	70875	61168
01838	72046	40042	59287	78115	74332	36858	17687	14357	58846
32697	10332	92643	06454	39300	20099	61461	15730	29333	95548
75128	04855	96418	26636	53328	69758	17597	56658	81043	40374
15675	20566	45945	10123	21679	38139	95843	76372	78669	14598
78142	72095	47327	43718	48286	88374	65046	27199	50484	03834
72869	28546	22578	94059	56817	88443	65557	75239	38101	17180
13896	62838	09470	31133	65941	84219	23017	34539	77391	52502
95808	40466	39870	79974	71187	02420	23124	15714	91874	86307
52320	60822	38657	81962	32388	50425	53231	62797	95490	87063

Table XIV (*continued*)

03564	46703	30528	28041	86108	99297	31593	21021	12451	90445
89432	52921	28068	71091	12944	06524	42605	02606	69417	81733
88573	55150	01443	97336	79910	49014	02237	85000	32344	45649
32920	01391	01105	15435	10918	92181	03839	92364	84229	83989
52704	39386	81791	35616	97616	64947	60456	16196	79527	43770
02696	86377	34209	38850	43712	58088	58490	42162	16423	79089
83961	43893	81108	79331	27601	30995	25447	05835	26029	01069
21914	31443	85624	29878	97401	66466	88421	76385	65526	93134
60215	43656	42638	13774	87380	32166	44914	57637	95151	08573
17644	23867	35765	75634	22484	86921	29597	94523	39661	15403
69499	38275	93129	99455	06429	10947	62748	09375	53925	65096
96761	14313	79554	48204	56142	39889	98293	07233	25422	43510
42115	20104	10771	93968	76480	82630	09458	50774	00461	17435
96628	79221	70360	47978	68880	91249	42500	92943	89942	94929
98056	88721	38743	37395	23774	29013	39877	56221	08293	74795
62011	34646	99276	37811	30494	51236	30385	22514	77077	68381
97562	82265	27078	02950	45701	53691	27376	17196	53122	40779
81485	72983	95838	93212	68260	21176	33964	32478	98334	81713
53465	74671	88519	84254	65937	56020	45728	52449	17785	75868
14640	29533	35425	50917	85742	38691	31928	68477	70081	31907
58410	98236	76474	16076	17250	91650	83632	54718	16705	22827
26780	38018	96714	32836	11929	56912	71592	29622	71248	49260
74439	29381	62148	13205	68606	03817	35829	21987	40162	35558
51015	49183	15384	28173	53705	96163	72306	10015	63078	95319
28516	56674	30562	96465	17886	00360	11265	05653	51383	85153
79501	69898	55076	54853	66742	70410	44434	15140	88331	75362
35468	52850	17797	78112	42126	33055	99776	40129	70370	27342
29763	07791	20976	69285	32965	95201	96582	71055	16511	13122
76537	32386	84442	97095	31922	39406	56418	76857	51158	43193
74519	89378	58353	83848	02802	06046	74264	76358	08642	31973
94988	12022	77021	60277	39048	03087	18920	98682	26756	05107
72363	40974	09594	10276	09631	43203	13227	90021	35899	21515
74967	66480	83894	82989	24784	42757	24447	44970	60048	26514
26236	32399	81419	47377	93952	89101	79748	03446	23212	10489
05632	68465	67842	85597	02094	42059	86912	23145	43060	94694
67352	41392	17545	30949	87565	83820	19827	34043	00575	23260
92727	35027	03117	80848	74559	96797	08118	72948	91838	00281
18223	91136	39695	39943	77413	48937	32672	94704	99738	50907
80723	91394	02992	11530	67845	05881	76173	18594	91937	74215
75007	85671	88211	55080	15581	02685	07889	26594	47083	76723

(continues)

Table XIV (continued)

60050	80463	30926	74970	38951	14928	81875	61424	62060	96004
69715	28522	73974	99491	50647	20252	44455	66593	23255	94807
92096	43555	48882	60717	07963	39375	21441	66090	80430	68380
35482	63353	08086	66635	71009	95777	70335	65808	69105	42800
24879	78061	38949	21123	28430	72627	04565	14741	85781	20795
68985	60486	58133	07709	25899	68531	10370	46536	46506	86675
96601	96785	20850	70389	74637	34020	61780	33461	32496	51247
66706	67664	93292	05934	71050	68192	51898	18872	01371	95558
39273	41912	40198	36441	89472	38835	41709	85397	57429	33822
42539	21771	58672	71421	92528	67229	57837	89729	97494	40943
78209	77315	08393	95809	15832	31381	83170	32933	90911	49431
31279	02627	93411	96192	88570	88861	79463	64823	93331	97785
76915	11168	58452	30237	43211	88094	40120	44502	89799	09877
31714	15972	03620	07957	84828	01328	66806	45975	01001	72953
23131	71925	82240	57451	86216	08900	42868	39434	74956	93714
95186	13811	57341	15008	70542	51583	01563	50348	01403	32881
69801	54360	53265	70858	65549	02535	43657	75928	00109	59990
22293	14758	37440	49589	96421	74696	46442	27647	63950	60872
91834	04476	12776	34486	64027	15943	12307	36791	13600	32570
18615	53505	27034	34479	81642	20618	19992	12006	37023	48116
97224	65695	75788	47328	86654	52342	31619	97675	40129	57606
26686	27899	76013	33828	00554	35016	56081	74251	51761	20546
71559	57727	59162	88726	75150	37162	90733	79311	48085	67398
46858	08197	42531	34583	46155	77480	13712	15607	61265	81522
68979	78824	23553	16776	72208	12765	33682	68422	53269	99473
82578	98058	70623	49929	23105	14041	96639	89657	53337	31527
26830	21649	96350	76934	95421	23398	24309	97283	55317	06715
66780	42434	36053	11518	10114	29978	35558	73063	47919	41275
26850	83856	07550	60184	71509	33950	19410	57673	82122	98268
83552	46929	39407	19738	08959	95800	66562	70100	16777	17829
69664	34188	44958	12585	14608	48201	21973	64338	56105	01633
91846	59409	44903	39135	09758	79483	93450	39965	19354	55190
94766	30727	36436	87104	45035	27194	40273	71694	57000	40477
06482	16026	77039	23656	00584	01539	43904	13910	79229	91014
38098	51630	27394	80495	28942	12426	84937	10365	44686	73746
42624	08957	40485	75135	35223	18951	86245	86562	52403	37080
75639	11741	34530	96298	61180	04574	41471	76100	49195	68552
58681	84924	85898	94144	40948	88720	92349	75081	72752	03225
49384	67921	07641	03287	85245	06555	59403	71346	25280	03515
02176	23783	96594	05593	94006	41335	81326	28049	25784	22043

Table XV Normal Random **Numbers**

.638	.158	1.136	2.145−	.602−	.474	.185−	1.007	.049	1.181
.398	1.779	.632−	.372−	.849	.410−	.117−	1.389	.100−	.821−
2.310	1.964−	.929	.478−	.856	.322	.614	.298−	.862−	1.367
.391−	1.466−	.583	.073−	1.875	1.580−	.902−	.256	1.656	2.209
1.320	.080−	1.675−	.610−	.971−	.039−	1.672−	.056	.372−	.867
.489−	.737	.303−	.495	.563−	.075−	.036	.585−	.051−	.915
.124−	1.409	.502−	.651−	.301	1.621−	1.700−	1.605	1.583−	.452−
1.342−	1.696−	2.305−	.158	.222	.334	.380	.219−	.477−	.551−
.321	.239−	.851−	1.322−	.535	1.233−	.812−	.887−	.581−	2.939−
3.402−	.013−	.628	1.286−	.540	.354−	.254	.543−	.330	1.573
.069	.150	.095	.269	.430−	.253	1.514	1.467	.196−	1.790−
2.029	.204−	2.024	.419	.718	.231	.940−	1.030−	.815	.453−
2.507	.662−	1.370	.175−	1.938	.508−	.131−	1.441	.783	.799−
.388	.894	1.628	1.216−	.826	.124−	2.307−	.595−	.490−	.214−
.195−	.404−	1.770	2.368	.360−	.836	.522−	1.241−	1.111−	1.621
1.172−	.049−	.215	.469−	.614	.059	1.031−	1.488	.649	1.509−
1.380	.642−	.814−	.086−	1.419−	1.164	.316−	.139	.542−	.411−
1.128	.437	.272−	1.296−	1.947−	1.477−	.425−	.508	.752	.143−
.551	1.116−	.383	1.194−	.347	.188	1.355	.594	.096−	.198
.359−	.301−	1.725	1.369−	.179	.558−	.255	1.673	1.604	1.904−
1.288	1.262−	.753−	1.467−	.784−	.621−	.732	1.054−	.320−	1.041
1.683	1.997	.319	1.129	1.094	1.457−	.060−	.363−	.480−	1.270−
.723−	.833−	.579	.042−	.147	1.370	.234	.404	1.522	1.406
.413	1.672	.843−	2.219	.525	.957−	.054−	.111−	.763−	1.406
1.390−	.975	1.151	.702	.017	.585−	1.277	1.270−	.679−	.876
.856−	.506	.419−	.234	.112−	.065−	.369−	.503−	.927−	.516
.078−	.401−	1.334	1.799−	2.176−	.693	1.630	.652−	.258−	.062
.611	1.061	.686	.291	.057−	1.084−	1.513−	.048−	1.136	.523−
.135	.543	1.130−	.012	.161	.085	.864	2.715	.081	1.582
.333−	.317−	2.097−	.782	.868−	.462−	.820−	.147−	.405	.616
.450−	.756−	.096−	.247−	.741	.311	.670−	1.028−	1.749−	1.502
.216−	.134−	.260−	2.319−	.734−	.471−	.164−	1.856−	.524−	.098−
.003−	.736−	.572	2.223	1.561−	2.532	.490	.339	.807−	.276
1.327−	.355−	.495	.333	.559	.386	.339	.380−	2.039−	1.089−
1.214−	.864−	.048	.645	.046	.198	1.003−	.173−	.988	.063−
.395−	1.049−	1.060−	.440	.641−	.405	.345−	.735−	1.656	.458
.309−	.541−	.181	1.161−	.226	.837−	1.347	2.651	.389−	.875−
.453	.746	.286	1.387	1.878	1.165	2.725−	.010	1.589	2.842−
1.333	1.053−	.096−	.802	1.078−	.714	.027	.242−	.123	.758−
1.023−	1.580−	.341−	.081−	.713−	.247	.069	.317−	.122	.889−

(continues)

Table XV (*continued*)

1.293	.910−	.952	.271−	1.912	1.902	.470	.087−	2.270−	.226
1.480−	.738−	1.939−	1.407−	2.615−	2.257−	.878−	.318−	1.788	1.812−
.382	.229−	.199	.311−	.154−	1.743−	.412−	.249	.370−	1.838
.715	.379	.623	1.520−	.772	.164	.075−	.038−	.294	.173−
.079−	2.231	.669−	.924−	.365	.707	.203	.879	.813	.674−
1.522	.210−	1.111	.669−	.160	.648−	.931	1.122	.492−	.894
.938−	.133−	.600−	.808	1.063−	1.022−	2.292	.705−	.284−	.915
1.705	.669−	2.209	1.280−	2.380−	.568	.635−	1.524−	.382−	1.096
.573−	.796	.680−	1.393	1.196	1.677−	1.409	.789	.529	.106−
1.511−	2.557−	.242−	.800	.483−	.167−	.458−	.773−	.847−	1.401−
.670−	.518	.387	.523	.641	1.243	.322	2.607−	1.097	.012−
2.912	1.448	1.343	.122−	.726	.617−	.609	2.319	.450−	1.197−
.028−	.790−	.057	1.425	1.940	1.161	.878−	.716−	.244−	1.151−
1.257−	.774	.003	.388	1.060	1.028	.236−	1.172	.442	.157−
2.372	1.376−	1.318−	1.236	.738	.337	.534−	.090	.886	.676
.970−	.438	.672−	.180−	.667	1.370	.481−	.329	.842	.449
1.228−	.129	.426−	.165−	.028	2.696	1.201	1.351−	.724	1.017−
.369−	.310	.432	.237	.884	1.224−	.539	.852	.497	.283−
1.161	1.219	1.615	.336	1.100	.528−	.161	.278	.675	1.143−
.284−	2.609	.792	1.825	.249−	1.654	.621	.979	1.472−	1.173−
.578−	.789−	.106	.832	.597−	.496	.561−	1.033−	.578−	.378−
.074	.261	.766−	1.046−	.361	.043−	1.927−	1.527	.605	1.475
.230	.046	.978	1.901−	1.162	.545−	.697	1.151	2.033	.080
2.162	.562−	1.190	.925	1.057−	.015	1.371−	1.067	1.080−	1.129−
1.020−	1.130−	.315−	.628	.140−	2.050	.030−	.629−	.128	1.221−
1.323	.836−	.284−	.249−	.768−	1.242	.879−	.417−	.013	.502−
2.329	1.884	.033	.598	.217−	.260	.431	1.914−	.205	1.155
2.761	1.800	.562−	.714	.407−	.009	.724−	1.168−	.247	1.166
.232−	.605	.023−	.531−	.542	.155−	.697	1.037	.316−	.003−
.742−	.210	.741−	1.099−	.158	2.112	.765−	.319−	.247−	.345
1.410−	.419	.705	1.444	1.057	.843−	.043	.571−	.001−	.203
2.272	.719−	.679	2.007	.180−	.698	1.137−	.688	.571−	.100−
2.832	.925	1.350−	1.529	.260−	1.007−	2.350−	1.501−	.289	1.522
1.086−	.558−	.973	1.285−	.021−	.077	.915	.241−	.249−	.529−
.134	1.815	.313	1.571−	.216−	2.261	.696	.130−	.393	.017
.783	.600	.745−	1.127	.684−	.519−	.125	.499−	1.543	.082−
..174−	.897−	.575	.751−	.694	2.959−	.529	1.587	.339	.813−
1.319−	.556	2.963	1.218	1.199	1.746−	1.611	.467	.490−	.202
1.298	.940−	1.143−	1.136−	1.516−	.548	.629	.250	1.087−	.322
.676−	1.107−	1.483−	.278	.493	.442−	1.078	.336−	.177−	.057−

Table XV (*continued*)

1.287–	.775	1.095–	1.161	1.877–	1.874	1.703	1.619–	.725–	1.407–
.260	.028–	1.982–	.811	.999	1.662	.908	1.476	1.137–	.945–
.481	1.060	1.441	.163	.720	1.490	.026–	.502–	.427	.351–
.794	.725	1.971	.384	.579–	1.079–	1.440–	.859–	.346–	.077
.584	.554–	1.460	.791	.426–	.682–	.430	1.922	2.099–	.221
.114–	.379	.698–	1.570	.511–	.725–	.680	.591–	1.091–	.357
1.128–	1.707–	.921	.859–	1.566–	1.523	.900–	.988–	.264	.282
.691	.153	.076	1.691	.553	.457	1.107–	.322	.633	.007
1.115	.777	.738–	.868	1.484	1.792–	.950	.842–	.192–	.620
.389–	.559	.670	.315–	1.234	.475	1.117	1.286	.649–	1.880–
.330	.750	.642–	.148	.608–	.866	1.720–	.653	.210–	.959–
.333—	.084–	1.239	.049–	.095–	.197–	.213–	1.420–	.491–	.102
1.718	1.111	.548–	.653–	1.534	.456–	.395–	1.614	.531–	.785–
.182–	.620	1.178	1.071–	.444	.072–	1.001–	1.325	.302–	1.119–
1.260	1.192–	.182	.397–	.705–	1.085–	1.492–	1.642	.673	.707–
1.204–	1.725–	1.695	1.473	.665	.489–	.020	.267	1.230	.865
.619–	.307	.226–	.096–	.987	1.195–	1.412–	.433	2.052	.022
.272–	.096–	.137	.361–	.653	.156–	1.309	.480–	.397–	1.302
.245	.690–	.493	1.123–	1.465	.132	.582	.429–	.225	.125
.101	.855–	.782	1.040–	2.113	1.423–	1.010–	.158	.106	1.232–

Reprinted with permission from *A Million Random Digits with 100,000 Normal Deviates,* Rand Corporation, Santa Monica, Calif.

Answers to Selected Problems

Chapter 1

1. **(a)** $\{6, N6, NN6, NN6, \ldots\}$, where $N =$ not 6   **(b)** $\{0, 1, 2, \ldots, 30\}$
    (c) {Democrat, Republican, Other, None]
    (d) An interval of real numbers from 50 to 400 ought to suffice.
    (e) The set of nonnegative real numbers
2. **(a)** $E = \{Ht, Th, Tt\}$, $F = \{Hh, Ht, Th\}$, $G = \{Hh, Th\}$
    (b) $EFG = \{th\}$; $E^c \cup F^c = \{Hh, Tt\}$; $(EF)^c = \{Hh, Tt\}$; $E \cup FG = \Omega = \{Hh, Ht, Tt, Th\}$; $EG^c = \{Ht, Tt\}$.
    (c) 16
5. **(a)** E **(b)** ϕ **(c)** Ω **(d)** E **(e)** E **(f)** E **(g)** Ω **(h)** ϕ
8. **(a)** $\frac{3}{4}$ **(b)** $\frac{5}{13}$ **(c)** $\frac{5}{26}$ **(d)** $\frac{1}{2}$ **(e)** $\frac{2}{13}$ **(f)** $\frac{21}{26}$ **10.** $\frac{2}{5}$ **12.** $\frac{3}{10}$
14. 12   **16.** $(15)_4 = 32,760$ **18.** **(a)** $4^{20} \doteq 1.0995 \times 10^{12}$ **(b)** $4 \times 3^{19} \doteq 4.649 \times 10^9$
20. **(a)** $2^{10} = 1,024$ **(b)** 210   **(c)** $\frac{1}{2^9} = \frac{1}{512}$ **22.** $(12)_8 = 19,958,400$ **23.** **(a)** 420
24. **(a)** 5,040   **(b)** 720   **(c)** 288   **25.** $n(n-1)$
26. **(a)** 20,358,520   **(b)** Approximately 1.9×10^{14} **28.** **(a)** 120   **(b)** 112   **(c)** 64
29. 1,023   **31.** 15   **34.** $\frac{1}{56}$ **35.** **(a)** $\frac{1}{1,352,000}$ for each   **(b)** $\frac{20}{1,352,000}$
36. **(a)** $\frac{13}{45}$ **(b)** $\frac{11}{15}$ **(c)** 1   **(d)** $\frac{4}{10}$ **38.** **(a)** $\frac{5}{14}$ **(b)** $\frac{2}{7}$
40. Approximately 6.44×10^{-7} **42.** .504   **44.** **(a)** 56   **(b)** $\frac{1}{28}$ **(c)** $\frac{1}{14}$ **(d)** $\frac{3}{7}$ **(e)** $\frac{3}{7}$
45. **(a)** .0211 $\left(\frac{54,912}{2,598,960}\right)$ **(b)** .00144 $\left(\frac{3,744}{2,598,960}\right)$
47. **(a)** $y^5 + 5xy^4 + 10x^2y^3 + 10x^3y^2 + 5x^4y + y^5$
    (b) $z^3 + 3z^2y + 3z^2x + 6xyz + 3zy^2 + 3zx^2 + 3x^2y + 3xy^2 + y^3 + x^3$ **49.** 1, 6, 15, 20, 15, 6, 1
50. **(a)** 56   **(b)** 105   **51.** **(a)** $\frac{1}{6}$ **(c)** $\frac{25}{36}$ **52.** **(a)** .38   **(b)** .91

Chapter 2

1.

x	2	3	4	5	6	7	8	9	10	11	12
$f(x)$	$\frac{1}{36}$	$\frac{1}{36}$	$\frac{3}{36}$	$\frac{4}{36}$	$\frac{5}{36}$	$\frac{6}{36}$	$\frac{5}{36}$	$\frac{4}{36}$	$\frac{3}{36}$	$\frac{2}{36}$	$\frac{1}{36}$

2.

y	-5	-4	-3	-2	-1	0	1	2	3	4	5
$f(y)$	$\frac{1}{36}$	$\frac{2}{36}$	$\frac{3}{36}$	$\frac{4}{36}$	$\frac{5}{36}$	$\frac{6}{36}$	$\frac{5}{36}$	$\frac{4}{36}$	$\frac{3}{36}$	$\frac{2}{36}$	$\frac{1}{36}$

4. $\omega = 12, 13, 14, 15, 23, 24, 25, 34, 35, 45$

(a)

s	3	4	5	6	7	8	9
$f_S(s)$	$\frac{1}{10}$	$\frac{1}{10}$	$\frac{2}{10}$	$\frac{2}{10}$	$\frac{2}{10}$	$\frac{1}{10}$	$\frac{1}{10}$

(b)

d	1	2	3	4
$f_D(d)$	$\frac{4}{10}$	$\frac{3}{10}$	$\frac{2}{10}$	$\frac{1}{10}$

(c)

u	2	3	4	5
$f_U(u)$	$\frac{1}{10}$	$\frac{2}{10}$	$\frac{3}{10}$	$\frac{4}{10}$

5. (a) .5161 (b) .2577 (c) .5625 (d) .1538

7. (a)

Education	Grade	High	College
Probability	.2	.3	.5

(b)

Opinion	Favor	Oppose
Probability	.6	.4

9. (b) .8 (c)

z	2	4	5
$f(z)$	.2	.7	.1

11. (a) $\frac{1}{13}$ (b) $\frac{1}{4}$ (c) $\frac{1}{2}$ (d) 1 (e) $\frac{1}{13}$ (f) $\frac{1}{2}$ **13.** (a) $.2845 = \frac{600}{2,109}$ (b) .24

15. (a) $\frac{11}{100}$ (b) $\frac{357}{494}$ (c) $\frac{1}{4}$ **17.** (a) $\frac{1}{5}$ (b) $\frac{1}{9}$ **18.** $\frac{1}{25}$ **20.** $\frac{3}{5}$

23. (b)

y \ x	1	2	3	4
0	$\frac{1}{8}$	$\frac{1}{24}$	0	0
1	$\frac{1}{8}$	$\frac{1}{6}$	$\frac{1}{8}$	0
2	0	$\frac{1}{24}$	$\frac{1}{8}$	$\frac{1}{4}$

(c)

y	0	1	2
$f_Y(y)$	$\frac{1}{6}$	$\frac{5}{12}$	$\frac{5}{12}$

(d) $\frac{3}{5}$

25. $\frac{2}{47}$ **28.** .2266   **30.** (a) .276   (b) .446   **31.** (a) $\frac{1}{36}$ (b) $\frac{1}{6}$ (c) $\frac{1}{12}$ **32.** .2015

33. $P(\text{at least one hit in 20 games}) = .0111$ **35.** No: If $W = 13$, $Z = 10$.   **38.** (a) $\frac{25}{36}$ (b) $\frac{2}{3}$

40. $\frac{91}{216}$ **42.** A and S, A and B **44.** .24   **49.** (a) $\frac{8}{13}$ (b) $\frac{2}{3}$ (c) $\frac{10}{39}$

Chapter 3

1. $EX = .5,$ $EY = 2$ **2.** $7, \frac{7}{3}, \frac{7}{2}, \frac{14}{3}$ **3.** (a) $\frac{1}{2}$ (b) $\frac{13}{22}$

5. (a)

ω	TTT	TTH	THT	HTT	THH	HTH	HHT	HHH
$X(\omega)$	0	1	1	1	2	2	2	3
$Y(\omega)$	0	0	0	0	1	1	1	0

(b) $\frac{3}{2}, \frac{3}{8}$

7. (a) $f(k) = \frac{\binom{k-1}{2}}{\binom{8}{3}},$ $k = 3, \ldots, 8$ (b) $\frac{13}{14}$ (c) $\frac{27}{4}$ **9.** (a) $\frac{1}{4}$ (b) $\frac{1}{4}$ (c) 1

13. 4   **15.** .866   **17.** 1.31   **19.** .584

20. (a) $f(k) = \frac{1}{4},$ $k = 1, 2, 3, 4$ (b) 2.5, 1.118   (c) $f(k) = \frac{3^{k-1}}{4^k},$ $k = 1, 2, \ldots$ **22.** $\frac{4}{3}, \frac{5}{9}$

24. $\frac{3}{4}$ **25.** (a) .08   (b) .6   (c) 1.01   (d) $\frac{29}{36}$

26. (a) 1   (b) 4   (c) 0   (d) 1   (e) 2   (f) 25   **29.** (a) $\frac{1}{48}$ (b) 1   **30.** .209

32. $1 - p + pt$ **33.** $(1 - p - pt)^3,$ $3(1 - p)p^2$ **35.** (a) $\frac{t}{2 - t}$ (b) $\left[\frac{t}{2 - t}\right]^2;$ $\frac{1}{4}$

Chapter 4

1. (a) .2461   (b) .1719   (c) $5, \sqrt{\frac{10}{4}}$ **2.** (a) .3826   (b) .8131   (c) .5695   (d) .8

4. (a) 5, 25   (b) 9.68   (c) .176   (d) .0059   **6.** 10, 40, 60, 40, 10, respectively   **8.** $n \geq 7$

10. (a) 1.25   (b) .86   (c) .0677   (d) .0726   **12.** (a) .7636   (b) 1   (c) .1697

14. **(a)** $\frac{1}{1/94\frac{1}{7},792}$ **(b)** .000092   **(c)** .0033   **(d)** .3048   **(e)** $5.30

15. **(a)** .0229   **(b)** 10   **(c)** 30   **17.** **(a)** 6   **(b)** 3.5

19. **(a)** 17.26   **(b)** .2734   **(c)** 24   **21.** $\frac{25}{3}$ **23.** .1151   **25.** .0521   **27.** .0736

29. **(a)** .223   **(b)** .3353   **(c)** .934

31. **(a)** $\dfrac{[\binom{950}{20} + 50\binom{950}{19}]}{\binom{1,000}{20}}(\doteq .736)$ **(b)** .73584   **(c)** .736   **(d)** .70

33. **(a)** .593   **(b)** .135   **(c)** .008   **(d)** .385

35. **(a)** 1.8   **(b)** .0273   **(c)** .025   **(d)** .884

37. 44.33   **40.** **(a)** .0110   **(b)** .01344   **(c)** .251   **(d)** .086   **(e)** .090

42. **(a)** .0335   **(b)** .0209   **(c)** .114

44. **(a)** $\text{Bin}(n, p_1)$ **(b)** $\text{Bin}(n, p_1 + p_2)$ **(c)** $\text{Bin}\left[n - k, \dfrac{p_1}{p_1 + p_3}\right]$ **47.** $\{p[1 - (1 - p)t]^{-1}\}^r$

Chapter 5

---

1. **(a)** $\frac{1}{8}$ **(b)** $\frac{1}{2}$ **(c)** $\frac{1}{2}$ **(d)** $\frac{1}{2}$ **3.** **(a)** 0, 1   **(b)** $1/\sqrt{2}$

5. **(a)** $x^2, \quad 0 < x < 1$ **(b)** $y^2/144, \quad 0 < y < 12$

6. **(a)** $\frac{1}{2}$ **(b)** $\frac{1}{2}(y + 1), \quad -1 < y < 1$ **(c)** $z^2, \quad 0 < z < 1$

8. **(a)** $\frac{1}{6}$ **(b)** $\frac{1}{3}$ **(c)** $\frac{2}{3}$ **(d)** $\frac{1}{2}$ **(e)** $\frac{5}{6}$ **(f)** 0   **10.** $\frac{2}{5}$ **11.** $f(x) = \frac{1}{2}, \quad 0 < x < 2$

13. **(a)** $f(x) = \begin{cases} \frac{1}{2}, & 0 < x < 1, \\ \frac{1}{4}, & 1 < x < 3. \end{cases}$ **(b)** $\frac{1}{2}$

15. **(a)** 1   **(b)** $\frac{1}{4}$ **(c)** Median = 0, $Q_1 = \sqrt{\frac{1}{2}} - 1$, $Q_3 = 1 - \sqrt{\frac{1}{2}}$

   (d) $F(x) = \begin{cases} \frac{1}{2}(1 + x)^2, & -1 < x < 0, \\ 1 - \frac{1}{2}(1 - x)^2, & 0 < x < 1. \end{cases}$

17. **(a)** $\frac{1}{3}$ **(b)** $\frac{1}{5}$ **(c)** $g(y) = \begin{cases} \frac{2}{3}(2 - y), & 1 < y < 2 \\ \frac{2}{3}, & 0 < y < 1. \end{cases}$

19. .0315

21. **(a)** $f(x) = \frac{1}{2}, \quad -1 < x < 1$ **(b)** $f(y) = 1, \quad 0 < y < 1$ **(c)** $f(z) = \sqrt{z}, \quad 0 < z < 1$

23. **(a)** $\sqrt{2}$ **(b)** $\frac{4}{3}$ **(c)** .9   **25.** **(a)** $\frac{1}{6}$ **(b)** $\frac{1}{3}$ **27.** **(a)** $\frac{1}{\sqrt{2}}$ **(b)** $\frac{-1}{6}$ **(c)** .45

29. **(a)** $\frac{1}{2}$ **(b)** 1   **31.** $2\sqrt{2/\pi}$ **33.** $\frac{1}{\sqrt{5}}$ **35.** $\frac{3}{8}$ **37.** $\frac{1}{18}$ **38.** **(a)** 1.25   **(b)** -1.58

39. **(a)** $5x^4 - 4x^5, \quad 0 < x < 1$ **(b)** $\frac{2}{3}$ **(c)** $\frac{2}{63}$ **(d)** $\frac{1}{12}$

42. **(a)** $\frac{3}{4}$ **(b)** $\frac{1}{2}$ **(c)** $f_X(u) = f_Y(u) = 2(1 - u), \quad 0 < u < 1$ **(d)** $\frac{1}{4}$ **43.** $-\frac{1}{12}$

45. $-\frac{1}{144}, \; -\frac{1}{11}$ **47.** **(a)** $F(z) = z^2$ and $f(z) = 2z, \quad 0 < z < 1$ **51.** $-[\frac{1}{2}(1 - p_{XY})]^{1/2}$

52. **(a)** No; support is not rectangular.   **(b)** $\frac{1}{4}$ **(c)** $f_X(u) = f_Y(u) = \frac{1}{2}, \quad -1 < u < 1$ **(d)** $\frac{1}{4}$

54. **(a)** Triangular on $(0, 2)$ **(b)** Triangular on $(-1, 1)$ **(c)** Same as for Y **(d)** Same as (b)
   (e) Uniform on $(-1, 1)$ **(f)** $f(u) = 1 - u/2, \quad 0 < u < 2$

55. **(a)** $f_X(u) = f_Y(u) = 1 - |u|, \quad -1 < u < 1$
   (b) No: Support is a rectangle, but its sides are not parallel to the coordinate axes.   **(c)** 0   **(d)** $\frac{1}{3}$
   (e) $\frac{1}{2}(1 + \lambda), \quad -1 < \lambda < 1$

57. **(a)** $\exp(-\Sigma X_i), \quad x_i > 0$ for all i **59.** **(a)** $\frac{8}{5}$ **(b)** 0   **(c)** $\frac{4}{75}$ **(d)** $\frac{4}{75}$ **61.** $\frac{1}{\sqrt{2}}$

62. $\frac{1}{2}(1 - y)$ **64.** $\frac{1}{2}(1 - x)$ **66.** $\psi(t) = \dfrac{e^t - 1}{t}, \qquad E(X^k) = \dfrac{1}{k + 1}$

67. **(a)** k odd: $E(X^k) = 0$; k even: $E(X^k) = k!$ **(b)** 2   **69.** $\exp\left(\dfrac{-t}{2}\right) \cdot \psi_X(t)$

71. **(a)** $\frac{2}{3}$ **(b)** $\frac{1}{18}$ **73.** No.   **74.** $e^{m(e^t - 1)}$

Chapter 6

1. **(a)** 4   **(b)** -1.5 2. **(a)** .1587   **(b)** .6826
4. **(a)** .089   **(b)** 416.34, 450, 483.7   **(c)** $(\tfrac{1}{2})$ **(d)** 433
6. **(a)** $-6, 64$ **(b)** $\exp[-6t + 32t^2]$
8. 14   10. $\sqrt{\pi\theta/2}$ 11. **(a)** $N(11, 7), N(-1, 5), N(1, 10)$ 12. .8944
13. Normal with mean $\Sigma\, a_i\mu_i$, variance $\Sigma\, a_i^2\sigma_i^2$ 15. **(a)** .135   **(b)** .393   **(c)** .323
17. **(a)** $\tfrac{5}{3}$ days   **(b)** $\tfrac{5}{6}$ 19. $\lambda^n\exp\{-\lambda \Sigma x_i\},\ x_i > 0$
20. **(a)** $135{,}135\sqrt{\pi}$ **(b)** $\frac{1}{18{,}018}$ **(c)** $\tfrac{3}{8}\pi$ 21. **(a)** $9!/2^{10}$ 22. **(a)** .00595   23. 1.68
24. $\text{Gam}(\alpha_1 + \alpha_2, \lambda)$ 26. $\dfrac{(2n-1)(2n-3)\cdots 5\cdot 3\cdot 1}{2^n}\sqrt{\pi}$ 29. $\dfrac{rs}{(r+s)^2(r+s+1)}$

30. **(a)** .042   32. .963   35. 1   37. $C = \dfrac{\Gamma[\tfrac{1}{2}(m+1)]}{\sqrt{m\pi}\,\Gamma(\tfrac{1}{2}m)}$

39. **(a)** $2[1 - \Phi(x/2)]$ **(b)** .317, .0026   **(c)** Batteries wear out.   40. $\tfrac{1}{2}$ 42. $3/\pi$
44. $R_1(R_2 + R_3 - R_2 R_3)$ 46. **(a)** $\alpha/(\alpha - 1)$ $2^{1/\alpha}$ 47. $N(2, 2), N(-2, 4)$
49. $\dfrac{1}{2\pi\sqrt{5}}\exp\left\{-\dfrac{9}{10}\left[x^2 - \dfrac{4x(y-4)}{27} + \dfrac{(y-4)^2}{9}\right]\right\}$
51. **(a)** $E(Y\,|\,x) = \tfrac{1}{2}(x + 2)$, $E(X\,|\,y) = y - 1$
    (b) $\mu_X = 0$, $\mu_Y = 1$,   var $X = 2$,   var $Y = 1$, $\sigma_{X,Y} = -1$, $\rho = \tfrac{-1}{\sqrt{2}}$
52. Bivariate normal: $\mu^X = \mu_Y = 0$, $\sigma_X^2 = 5$, $\sigma_Y^2 = 10$, $\sigma_{X,Y} = 1$
53. **(a)** Means: $-8, 6$;   variances: 8, 18; $\rho = \tfrac{-1}{3}$ **(b)** $-\tfrac{2}{9}y - \tfrac{20}{3}$.

Chapter 7

1. **(a)**

Class	1	2	3	4	5
Freq.	3	12	19	9	2

Marijuana:	
Yes	15
No	30

# Sibs.	0	1	2	3	4	5	6	7	8	9	13
Freq.	1	10	8	10	1	5	2	3	2	1	2

(b)

	1	2	3	4	5
Y	1	4	6	4	0
N	2	8	13	5	2

3. **(a)**

	1	2	3	4	5
Y	0	4	11	7	3
N	0	3	11	5	4

(b)

1:	Y	N
M	0	0
F	1	2

2:	Y	N
M	4	3
F	4	8

3:	Y	N
M	11	11
F	6	13

4:	Y	N
M	7	5
F	4	5

5:	Y	N
M	3	4
F	0	2

5. (a) Male:

		Below	At	Above
	B	4	12	13
	W	11	16	14

Female:

		Below	At	Above
	B	6	14	9
	W	15	9	6

9. (a)

8	846
9	432
10	4878768
11	24466
12	084083
13	13322
14	4

13. 8.42, 7.35, 7.425, 14.55   **15. (a)** 114 oz   **(b)** 113.5 oz   **16.** 405.08   **18.** 79
20. 5-number summaries: **(a)** 3.1, 5.75, 7.35, 9.8, 26   **(b)** 2.3, 6.3, 7.05, 7.7, 10.5   **22.** 1.155
23. (a) 4.473   **(b)** 3.05   **24.** 15.83 oz   **27.** 184, 570   **31.** 10, 5   **33.** .798
35. Same   **37. (a)** .833   **(b)** 0

Chapter 8

1. (a) $f(x_1, \ldots, x_4) = \frac{1}{16}$, $-1 < x_i < 1$, $i = 1, \ldots, 4$
     (b) $f(x_1, \ldots, x_4) = 16 \exp\{-2 \Sigma X_i\}$, $x_i > 0$ for $i = 1, \ldots, 4$

2. (a) $f(x_1, \ldots, x_4) = \prod_1^5 \binom{3}{x_i}(\frac{1}{4})^y(\frac{3}{4})^{5-y}$, where $y = \sum_1^5 x_i$, $x_i = 0, 1, 2, 3$, $i = 1, \ldots, 5$

3. (a) $\lambda^3 e^{-13\lambda}$, $\lambda > 0$

5. (a) $f(x_1, \ldots, x_n) = (1 - p^{\Sigma x_i - n} p^n)$, $x_i = 1, 2, \ldots$ **(b)** $L(p) = (1 - p)^{27} p^3$

7. (a), (b) $(M^2 - M)(12 - M)(11 - M)$ **(c)** $M = 6$

9. (a) $\sum(X_i)^2$ **(b), (c)** $\sum X_i$ **11.** $(\sum X_i, \sum Y_i)$ **13.** $[X_{(1)}, X_{(n)}]$ **15.** $X_{(n)}$

17. Uniform on the $\binom{t-1}{n-1}$ sample points with $\Sigma x = t$.   **18. (c)** $f(s) = \begin{cases} \frac{1}{10}, & s = 6, 7, 11, 12 \\ \frac{2}{10}, & s = 8, 9, 10 \end{cases}$

21. (a) $N(n\mu, n\theta)$ **(b)** $\text{Gam}(n, \lambda)$ **(c)** $\text{Negbin}(n, p)$ **22.** $\dfrac{nk}{nk + 1}$

23. (a) $\text{Beta}(3, 3)$ **(b)** $\frac{1}{2}$, $\frac{1}{28}$ **24.** .891
26. $f(u, v) = (n)_4 F(u)[F(v) - F(u)]^{n-4}[1 - F(v)]f(u)f(v)$, $u < v$
27. $f(u, v) = \dfrac{n(n-1)}{(b-a)^n} \cdot (v - u)^{n-2}$, $a < u < v < b$

29. (a) $f_{\bar{X}}(y) = \begin{cases} \frac{1}{15}, & y = 24, 25, 27, 30, 39, 41 \\ \frac{2}{15}, & y = 23, 26, 28, 37 \end{cases}$ **(b)** 30, 36.27

30. $70, \frac{3}{20}$ **32.** .0439   **34. (a)** .003   **(b)** .058   **36.** .23   **38.** 190

Chapter 9

1. $.42 \pm .0127$ **3.** 603.2 ± 33.6 **4. (a)** $\dfrac{-\theta}{n+1}$ **(b)** $\dfrac{2\theta^2}{(n+1)(n+2)}$

6. .0114   **8.** $\Sigma X_i/n$ **9. (a)** 25   **(b)** 100   **10. (a)** 2,500   **(b)** 625   **13.** $18.31 \pm .29$

15. 89.14 ± 12.53, 89.14 ± 17.22 **17. (a)** 19   **(b)** .358   **19.** $1.69 < \lambda < 3.69$
21. .35 (or .38 without continuity correction)   **22. (a)** .048   **(b)** $.113 \pm .124$ **24.** 30 ± 2.1

26. .0616   **28.** $2.84 < \sigma < 8.06$ **29.** For 95% confidence: $\dfrac{\sum X^2}{\chi^2_{.975}} < \sigma^2 < \dfrac{\sum X^2}{\chi^2_{.025}}$

30. (a) 245.7 ± 9.065 **(b)** $40.5 < \sigma < 53.6$
32. (a) $\sigma^2(\beta_n^2 + (\alpha_n - 1)^2) = \sigma^2(2(1 - \alpha_n) - 1/n)$ **(b)** $\frac{n}{n-1}\beta_n^2\sigma^2$ **35.** $n/\sum X^2$ (in both cases)

36. $\sum X_i^2/n$ (in both cases)   **38.** $\bar{X}$ **39.** $(X_{(1)} \cdot X_{(n)})$ **41. (a)** $\sqrt{\dfrac{\sum Y_i}{\sum X_i}}$ **(c)** $1.22 < \theta < 3.28$

43. $\hat{\beta} = \dfrac{\sum cy - \dfrac{\sum c \sum y}{n}}{\sum c^2 - \dfrac{(\sum c)^2}{n}}$, $\hat{\alpha} = \bar{Y} - \hat{\alpha}\bar{c}$,

$\hat{\sigma}^2 = \dfrac{1}{n}\sum(y - \hat{\alpha} - \hat{\beta}c)^2$

45. $2\Phi\left(-\dfrac{\sqrt{n}\varepsilon}{\sigma}\right)$ **50. (a)** $I(p) = \dfrac{n}{p(1-p)}$ **53.** $\dfrac{\lambda}{n-1}$, $\dfrac{\lambda^2}{(n-1)^2(n-2)}$

Chapter 10

1. .10 (.12 with continuity correction)   **2. (a)** .0197   **(b)** 1.18   **(c)** .303
3. Some evidence that $\mu < 500$ $(Z = 1.77, P = .038)$
5. $Z = 2.97$, $P = .0015$—hard to explain as sampling variability   **7.** 2-sided $P = .09$
9. $Z = 3$, 1-sided $P = .001$ (strong evidence against H_0)   **10. (a)** .0227   **(b)** .0227
12. Strong evidence against H_0: $Z \doteq 6$
14. $T = -1.27$ (6 d.f.), 2-sided $P = .242$ (little evidence of difference)
16. $T = 1.97$ (7 d.f.), 2-sided $P = .086$ (marginal evidence against no difference)
18. $T = 1.64$ (7 d.f.), 2-sided $P = .146$ (little evidence against H_0)   **20.** $R_- = 6$, 2-sided $P = .11$
22. (a) (1-sided) $P = .29$ **(b)** $P \doteq .07$ $(R_- = 23.5)$ **(c)** $P \doteq .04$ $(T = 1.87, 12$ d.f.$)$
24. (a) 256   **(b)** $\frac{7}{256}$ **26. (a)** Outlier suggests heavy tail.   **(b) and (c)** 2-sided $P = 1/256$

Chapter 11

1. .0035   **2. (a)** R_1 **(b)** R_2
3. (a) $\alpha_1 = .3174$, $\alpha_2 = .1096$ **(b)** $\beta_1 = .1574$, $\beta_2 = .3444$
5. (a) $R = \{\bar{X} > 10.164 \text{ or } \bar{X} < 9.836\}$ **(b)** .10
7. (a)

K	α	β
0	1	0
1	.9688	.0000
2	.8125	.0005
3	.5000	.0086
4	.1875	.0815
5	.0313	.4095
6	0	1

[For given α, β is larger in (b) except for $\varnothing$, Ω]

8. $p^7(120q^3 + 45pq^2 + 10p^2q + p^3)$, $q = 1 - p$

9. R_1: $\Phi(2\mu - 21) + \Phi(19 - 2\mu)$
     R_2: $\Phi(2\mu - 21.6) + \Phi(18.4 - 2\mu)$

11. **(a)** $15; \frac{1}{2} + \frac{1}{\pi}$ Arctan$(\theta - 2)$ **(b)** $.37;$ $1 + \frac{1}{\pi}[$Arctan$(\theta - 1.5) - $Arctan$(\theta + 1.5)]$

13. $\Phi(10\mu - 101.645) + \Phi(98.355 - 10\mu)$ **16.** $1 - (1 - p)^{10}$

18. **(a)** .073   **(b)** $\Phi(48.5 - 1.25\mu)$ **(c)** $n = 28$, $K = 40.76$

20. **(a)** $\bar{X} > K$ **(b)** $\alpha = 1 - \Phi(2K)$, $\beta = \Phi(2K - 2)$ **21.** $\bar{X} > K$ **23.** $\bar{X} > K$

25. $|\bar{X} - \mu_0| > K$ **27.** $\dfrac{e^u}{u} > K$,   where $u = \hat{\theta}/\theta_0$ **29.** $\phi, \{z_3\}, \{z_3, z_1\}, \{z_3, z_1, z_4\}, \Omega$

31. **(b)** $\Lambda = e^{-n\bar{X}^2/2}$, $\bar{X} > 0$, and $\Lambda = 1$, $\bar{X} < 0$.

Chapter 12

1. $Z = 2.15$, $P = .016$ (evidence of a difference)   **3.** $Z = 1.96$, 2-sided $P = .05$ (marginal)

4. $Z = 1.5$, 2-sided $P = .134$ (little evidence against H_0)   **6.** $Z = .59$ (little evidence of a difference)

8. $Z = 5.2$, strong evidence against H_0 **10.** $Z = 3.44$, 1-sided $P \doteq .0002$ (strong evidence against H_0)

12. $T = 3.9$ (39 D.F.), or $Z \doteq 4$ $(P < .001)$ **14.** $T = 1.1$ (10 d.f.), 2-sided $P = .3$

16. $R = 77$, $Z = -2.78$, one-sided $P = .0027$ **18.** Exact: .051; Normal approximation: .0501.

20. $\dfrac{n_1 + n_2 - 2}{n_1 + n_2} S_p^2$ **22.** $T = 4.4$ (3 d.f.), 1-sided $P \doteq .01$ **24.** $R_- = 4$, 1-sided $P = .007$

26. **(a)** $P = .0156$ **(b)** $R_- = 0$, $P = .0156$ **(c)** $T \doteq 3.3$, $P = .011$.

28. $F = 1.13$ (28, 19) d.f., $P > .05$ (not significant)   **30.** **(b)** $Z = 3.5$, $P = .0002$

Chapter 13

1. $\chi^2 \doteq 14.2$ (3 d.f.), $P < .005$.

3. **(a)** $\chi^2 \doteq 15.7$ (2 d.f.), $P < .002$.   **(b)** $\chi^2 \doteq .24$ (2 d.f.) (data are consistent with H_0)

5. Grouping two cells: $\chi^2 \doteq 2$ (1 d.f.), $P > .138$

7. **(a)** $\chi^2 \doteq 4.4$ (3 d.f.), $P = .22$ **(b)** $\chi^2 \doteq 2.4$ (5 d.f.), $P = .8$

9. $D_{10} \doteq .19$, $P \doteq .7$ (no reason to reject normality)

11. **(a)** $D_n \doteq .103$, $P \doteq .20$ (from asymptotic formula in Table IXb)
     (b) $D_n \doteq .02$ (data are consistent with H_0)   **14.** $D = .20$, $P > .5$

16. **(a)** $W = .30$ (strong evidence against normality)   **(b)** $W = .88$, $.01 < P < .05$

17. **(a)** $\chi^2 \doteq 7.9$ (1 d.f.), $P \doteq .005$ ("two-sided")   **(b)** $Z \doteq 2.8$, $P \doteq .003$ (one-sided)

19. $\chi^2 = 45$ (2 d.f.), $P < .001$ **21.** $\chi^2 = 16.17$ (3 d.f.), $P < .001$

23. **(a)** $\chi^2 = 15.64$ (1 d.f.), $P < .005$ **(b)** $Z^2 = 15.64$ (same test as in (a))

25. $\chi^2 = 8.44$ (3 d.f.), $P = .038$ **27.** $-2 \log \Lambda \doteq 15$ (3 d.f.), $P < .005$

29. $-2 \log \Lambda = 16.03$ (1 d.f.), $P < .001$

31. **(a)** $\hat{p} = \dfrac{2n_{11} + n_{12} + n_{21}}{2n}$ **(b)** $-2 \log \Lambda = 19.70$ (1 d.f.)

     (c) $-2 \log \Lambda = 22.68$ $(= 19.70 + 2.98)$, 2 d.f.

Chapter 14

1. $F = 2.61$

3. (a)

Source	d.f.	SS	MS	F	P
Variety	4	.502	.126	1.12	.40
Error	10	1.12	.112		
Total:	14	1.62			

(b) $(-.043, 1.17)$

5. $SSE = .00272$, $F = 3.33$ $(1, 26)$ d.f., $P \doteq .08$

6. $F = 6.51$ $(5, 18$ d.f.$)$, $.001 < P < .002$

8. Both $= \tau_i$

14. (a) $F = 14.3$ $(3, 32$ d.f.$)$, $P < .001$

   (b) (i) Using $t(6, .05) \doteq 2.55$, T versus C: 17.1 ± 65.6,
                  T versus S: 65.8 ± 53.5, C versus J: 57.9 ± 64.6
                  [S different from rest, T, C, J not distinguished
                  However, with $\alpha = .10$, C and J are different.]

       (ii) With $f \doteq 2.9$, T versus C: 17.1 ± 76.0
                                  T versus S: 65.8 ± 62.0
                                  C versus J: 57.9 ± 74.6
                                  S versus C: 82.9 ± 76.0 (diff.)
                                  T versus J: 75.0 ± 60.5 (diff.)

16. (a) Confidence limits:
       CA: $.20 \pm 2.23$ AB: 1.87 ± 2.23
       AE: $.88 \pm 2.23$ AD: 1.12 ± 2.23 (C, A, E, D not diff., A, E, D, B not diff., but C, B diff.)
       ED: $.24 \pm 1.82$ BC: 2.07 ± 1.82
       DB: $.75 \pm 1.82$ BE: $.99 \pm 1.82$
       CD: 1.32 ± 1.82 CE: 1.08 ± 1.82

   (b) Confidence limits for adjacent pair differences:
       CA: $.20 \pm .68$ (not sig.)
       AE: $.88 \pm .56$ (sig.)   Grouping: $C\&A$, $E\&D$, B
       ED: $.24 \pm .56$ (not sig.)
       DB: $.75 \pm .56$ (sg.)

17.

Source	d.f.	SS	MS	F	P
A	2	6	3	1	.44
B	2	294	147	49	.0015
Error	4	12	3		
Total:	8	312			

19.

Source	d.f.	SS	MS	F	P
Detergent	2	186	93	4.77	.087
Temperature	2	42	2.1	1.08	.42
Error	4	78	19.5		
Total:	8	306			

21. $F = 3.61$ $(7, 16)$ d.f., $.01 < P < .05$ **23.** $F = 1.98$ $(1, 36)$ d.f., $P = .085$

Chapter 15

1. (a) $y = \frac{2}{7} + \frac{9}{14}x$ (b) $\frac{23}{14}$ **2.** $y = 736 + 127.6x, \quad r = .996$ **4.** $y = .383 - .104(\log x)$

6. $y = 3.95 - 5.55x + 1.75x^2$

10. (a) $\text{s.e.}(\hat{\alpha}) = S_\varepsilon \sqrt{\dfrac{1}{n} + \dfrac{\bar{x}^2}{SS_{xx}}}$

    (b) $\bar{a} \pm k$ s.e. $(\hat{\alpha})$, where k is an appropriate percentile of $t(n-2)$ **14.** $t = -2.09(7 \text{ d.f.}), \qquad P = .076$

16. (a)

Source	SS	d.f.	MS	F
Age	.928	1	.928	4.34
Resid	1.497	7	.214	

    (b) .187   (c) (.500, 1.212)   (d) $(-.092, 1.80)$

18. (a) $y = \hat{\beta}x,$ where $\hat{\beta} = \dfrac{\sum x_i y_i}{\sum x_i^2}$ (b) $\hat{\beta} \sim N\left(\beta, \dfrac{\sigma^2}{\sum x^2}\right)$ (c) Rej. H_0 when $\dfrac{(\hat{\beta} - \beta_0)^2}{\hat{\sigma}^2/\sum x^2} > C.$

20. (a) $-.45 + .0055x$ (b) 3.84   (c) .15   (d) Smaller: $2 < 2.67$

21. (a)

	Rain	Runoff	Peakdisch	Logrunoff
Runoff	.07			
Peakdisch	−.066	.834		
Logrunoff	−.052	.906	.731	
Logpeakdisch	−.174	.796	.84	.885

    (b) $y = -.164 + .00046x$ (c) $.298 \pm .391$ (d) $y = -15.63 + 1.956x$ (e) 1940 ± 283.8

25. (a) and (b) $\frac{2}{3}(1-x)$ **29.** Greater: predicted Y is $\bar{Y} - 1.4S_Y > \bar{Y} - S_Y$

30. (a) $C = -4889 + 75.59D + .3232S$ (b) 13.46, $P < .001$ (c) .8177   (d) 10.5, $P = .018$

34. (a) $.380 - .0359x + .000989x^2$ (b) $T = 3.16 \quad (P \doteq .01)$

Chapter 16

2. (a) Beta(7, 5)   (b) Beta(8, 9)   (c) Beta(12, 6)   **4.** (a) $\frac{20}{21}$ (b) $\frac{5}{9}$

6. $h(\lambda \,|\, x) = \dfrac{4^{x+1}}{\Gamma(x+1)} \lambda^x e^{-4\lambda}; \qquad 13\lambda^7 e^{-4\lambda}, \quad \lambda > 0$ **11.** (a) a_1 (b) (i) a_2 (ii) a_3

12. μ_0 **14.** (a) $\frac{11}{12} = .917$, $\frac{12}{17}$, $\frac{8}{9}$ (b) .939   **17.** (a) $\dfrac{n/2 + \alpha}{\lambda + \sum X_i^2}$ (b) $\frac{2}{3}$

19. Y/n (same as the m.l.e.)   **21.** They'd be the same.   **22.** (a) .0026   (b) .964   **25.** Accept H_0

27. (a) .027   (b) .616   (c) .151   **29.** (b) $\frac{90}{91}$ (c) $\frac{12}{13}, \frac{1}{3}$ (d) Approx. $1 - 4 \times 10^{-9}$

Index

Acceptance sampling 501
Addition rule 14, 29
Additivity of expectations 113
Additivity of variances
Additivity of probabilities 29
Alternative hypothesis 461
Analysis of variance 608–618
ANOVA 600–618
ANOVA table 604
Asymptotic efficiency 444
Averages 93–120

Bayes action 682
Bayes loss 682
Bayes' theorem 60, 64
Bayesian estimation 684
Bayesian methods 671–696
Bernoulli process 137
Bernoulli variable 135
Beta distribution 282
Beta function 282
Bias 414
Binomial approximation 145
Binomial coefficient 25
Binomial distribution 139, 194, 708
Binomial theorem 25
Bivariate density function 219
Bivariate normal distribution 294
Bivariate probability function 51
Blocking 613
Bonferroni inequality 55, 611
Bonferroni method 611
Box–Muller transformation 381
Box plot 332

Categorical data 319
Cauchy distribution 200, 208, 212
C.d.f. 188
Central limit theorem 392
Central moments 96
Chebyshev inequality 437

Chi-square distribution 284, 722
Chi-square test 559–564
Class interval 384
Class mark 324
Coefficient of determination 644, 657
Combinations 20–21
Complement 8
Composite experiments 15
Composite hypothesis 504
Conditional density function 234
Conditional mean 101, 235
Conditional probability function 59
Confidence interval for μ 421, 425
    for p 423
Conjugate priors 680
Consistent estimate 436
Contingency tables 319, 575, 581
Continuity correction 151, 539
Continuous random variable 187–190
Contrasts 612
Controlled variable 630
Controls 524, 525
Convolution 240
Correlation coefficient
    population 111, 226
    sample 343
Covariance
    continuous variables 226
    discrete variables 109
    sample 344
Cramér–Rao inequality
Critical region 493
Cumulative distribution function 188

Decisions 691
Degrees of freedom
    chi-square 284
    t 286
    F 288
Density 196

Disjoint events 9
Distribution-free statistic 476, 538
Distribution function 188
Distributive laws 11, 12
Dot diagram 322
Double blind experiment 527

Efficient estimator 442
Empty set 9
Equality of means, testing 527, 534,
    600, 605
Equally likely outcomes 12
Errors in estimation 414
    in testing 493
    type I and type II 493
Estimating a mean 416
    a difference of μ's 427
    a difference of p's 429
    a proportion 415, 434
    a variance 430
Estimator 413
Event 6
Event algebra 7–12
Exchangeable random variables 70,
    91, 115, 226
Expected value
    discrete 95
    continuous 210
Explanatory variable 630
Exponential distribution 275
Exponential family 374, 442

F-distribution 288, 729
Factorial moment 165
Factorial moment generating function
    165
Fisher's exact test 532
Frequency distribution 317, 324
Functions of random variables 99,
    212, 224

Gamma distribution 278
Gamma function 279
Gaussian distribution 269
Generating functions 117
Geometric distribution 147
Geometric series 147
Gompertz distribution 293
Goodness of fit 557, 579
Grand mean 601
Grouped data 335

Hazard 290
Histogram 304, 324, 327
Historical control 525
Horse race 26–27
Hypergeometric distribution 143–145
Hypothesis 461

Independence of mean and variance
    397
Independence, test for 574, 581
Independent events 69
Independent random variables 67, 69,
    229
Indicator function 54, 120
Information inequality 440
Interquartile range 331
Intersection 9
Inverse sampling 147, 150
IQR 331
Iterated expectation 101, 235

Jeffreys' prior 704
Joint density function 219, 225
Joint frequency distribution 319
Joint probability function 51

Khintchine's theorem 438
Kolmogorov–Smirnov tests 565–570,
    736–739

Law of averages 161, 438
Law of large numbers 161, 438
Law of total probability 14–15, 29
Layering a three-way table 320
Least squares 633
Least squares line 634
Likelihood 368
Likelihood function 369
Likelihood ratio 504, 509
Likelihood ratio test 509, 579, 608
Lilliefors test 570, 740
Linear predictor 647
Linear transformations 108, 111, 191,
    213, 228, 295, 341, 346
Loss function 681
Lottery 13

M.a.d. 214, 339
Marginal density function 222
Marginal frequency distribution 319

Marginal probability function 53
Mass analogy 98
Mathematical expectation 94
Maximum likelihood estimation 433
McNemar's test 584
Mean deviation 214, 339
Mean value
    continuous variables 210
    discrete variables 95, 104
Mean, sample 333
Mean square
    for error 603
    for treatment 603
Mean squared error 414
Mean squared prediction error 645
Means, comparison of 527
Median
    sample 329
    population 204
Memoryless distribution 277
Method of moments 432
Midhinge 332
Midmean 336
Midrange 332
Models 13
Moment generating function 236–242
Moments
    continuous variables 241
    of $\bar{X}$ 387
    of $\hat{p}$ 389
    discrete variables 105–106
Most powerful test 505–506
Multinomial coefficient 25
Multinomial distribution 162–164
Multinomial sampling 371
Multiple comparisons 605, 610–617
Multiple regression 654
Multiplication principle 16
Multiplication rule 57, 58

Negative binomial distribution
    148–149
Negative exponential distribution 275
Neyman–Pearson lemma 506
Nonparametric tests 474, 476, 536
Normal approximation to the
    binomial 151
Normal distribution 269–275, 712
Normal scores 572, 741
Normality testing 570–574
Null distribution 462
Null hypothesis 461, 688–690

Odds 13
One-way ANOVA 600, 605
Operating characteristic 501

P-value 464, 467, 496
Paired comparison 541, 582
Parallel axis theorem 106
Partition 14, 372

Partitioning 21
Paternity 62, 673, 675
P.d.f. 196
Percentiles
    sample 331
    population 205
Permutations 19
Pilot sample 419
Pivotal quantity 426
Placebo effect 526
Poisson approximation to the
    binomial 154
Poisson distribution 155–158, 719
Poisson process 158
Poker 21, 72
Pooled variance 533
Population distribution 367
Population parameter 365
Posterior probability 672
Power function 496
Prediction 645–650, 656
Predictive distribution 693
Prior probability 672–674
Probability 12–14, 28
Probability axioms 29
Probability density 196
Probability element 195, 217
Probability function 48, 50
Probability generating function 117,
    164
Probability model 28
Product moment 108
Product set 229–233, 240
Prospective study 525

Quality control 366, 376, 381, 500
Quartiles
    sample 331
    population 205

Random numbers 24, 375, 744
Random sample 70, 367
Random sampling 365
Random selection 13, 22
Random variable 46
Random vector 51
Randomized block design 613
Range, sample 331
Rank sum test 537
Rank transformation 540
Rankits 572
Rayleigh distribution 697
Regression effect 650–654
Regression function 235, 298, 629,
    631–632
Regression model 632
Regression sum of squares 644
Rejection region 493
Relative efficiency 439
Relative frequency 317
Reliability function 290

Residual mean square 641
Residual sum of squares 641
Retrospective study 525
R.m.s. error
    of estimation 415
    of prediction 645
Robustness 472
Roulette 7

Sample correlation 343
Sample mean 333, 383, 395
Sample median 329
Sample range 331
Sample size 417, 498, 528
Sample space 6
Sample standard deviation 338
Sample variance 337–338
Sampling distribution 365, 375, 378
Sampling inspection 501
Sampling with replacement 137
Sampling without replacement 23, 24,
    137, 390
Scatter diagram 327
Scheffé's method 612
Schwarz inequality 112, 228
Set equality 8
Set inclusion 8
Shapiro–Wilk statistic 573
Sign tests 475
Signed-rank tests 476
Significance 464
Significance level 493
Simple hypothesis 504
Simulation 378
Spock, Dr. 5

Standard deviation
    continuous variables 214
    discrete variables 107
    sample 338
Standard error
    of a proportion 416
    of the mean 416
Standard normal distribution 270
Standard score 216
Standardization 216
Star plot 322
Statistic 329
Stem–leaf diagram 323
Students t-distribution 286
Sufficient partition 372
Sufficient statistic 372
Sums of random variables 113, 239
Support of a distribution 197
Symmetric distribution 206–207, 211
Symmetric probability function 71, 98

t-distribution 287, 716
Tchebycheff inequality 437
Test of significance 464
Tests for
    a mean 470, 473
    a proportion 470
    a variance 481
    equality of means 527, 534, 600
    equality of proportions 529, 577
    equality of variances
    homogeneity 577
    independence 574, 581
    normality 570
Two-way ANOVA 613–617

Transformations
    general 191, 203
    linear 191, 213, 228, 295, 341, 346
Treatment effect 523, 606
Tree diagram 17
Trimmed mean 336
Type I and II errors 493

UMP tests 507
Unbiased estimate 414
Uniform distribution 189
Union of events 9
Utility 100

Variance
    continuous variable 213
    discrete variable 105
    sample 337–338
Variance of a sum 113, 115
Variances, comparison of 543
Venn diagram 8
Voluntary response sample 364

Waiting time distribution 190, 197,
    201–202, 210, 237
Weibull distribution 291
Wilcoxon tests 476, 537, 726–727

Z-score 216
Z-test 468

Table II Standard Normal c.d.f.

z	0	1	2	3	4	5	6	7	8	9
−3.	.0013	.0010	.0007	.0005	.0003	.0002	.0002	.0001	.0001	.0000
−2.9	.0019	.0018	.0017	.0017	.0016	.0016	.0015	.0015	.0014	.0014
−2.8	.0026	.0025	.0024	.0023	.0023	.0022	.0021	.0021	.0020	.0019
−2.7	.0035	.0034	.0033	.0032	.0031	.0030	.0029	.0028	.0027	.0026
−2.6	.0047	.0045	.0044	.0043	.0041	.0040	.0039	.0038	.0037	.0036
−2.5	.0062	.0060	.0059	.0057	.0055	.0054	.0052	.0051	.0049	.0048
−2.4	.0082	.0080	.0078	.0075	.0073	.0071	.0069	.0068	.0066	.0064
−2.3	.0107	.0104	.0102	.0099	.0096	.0094	.0091	.0089	.0087	.0084
−2.2	.0139	.0136	.0132	.0129	.0126	.0122	.0119	.0116	.0113	.0110
−2.1	.0179	.0174	.0170	.0166	.0162	.0158	.0154	.0150	.0146	.0143
−2.0	.0228	.0222	.0217	.0212	.0207	.0202	.0197	.0192	.0188	.0183
−1.9	.0287	.0281	.0274	.0268	.0262	.0256	.0250	.0244	.0238	.0233
−1.8	.0359	.0352	.0344	.0336	.0329	.0322	.0314	.0307	.0300	.0294
−1.7	.0446	.0436	.0427	.0418	.0409	.0401	.0392	.0384	.0375	.0367
−1.6	.0548	.0537	.0526	.0516	.0505	.0495	.0485	.0475	.0465	.0455
−1.5	.0668	.0655	.0643	.0630	.0618	.0606	.0594	.0582	.0570	.0559
−1.4	.0808	.0793	.0778	.0764	.0749	.0735	.0722	.0708	.0694	.0681
−1.3	.0968	.0951	.0934	.0918	.0901	.0885	.0869	.0853	.0838	.0823
−1.2	.1151	.1131	.1112	.1093	.1075	.1056	.1038	.1020	.1003	.0985
−1.1	.1357	.1335	.1314	.1292	.1271	.1251	.1230	.1210	.1190	.1170
−1.0	.1587	.1562	.1539	.1515	.1492	.1469	.1446	.1423	.1401	.1379
−.9	.1841	.1814	.1788	.1762	.1736	.1711	.1685	.1660	.1635	.1611
−.8	.2119	.2090	.2061	.2033	.2005	.1977	.1949	.1922	.1894	.1867
−.7	.2420	.2389	.2358	.2327	.2297	.2266	.2236	.2206	.2177	.2148
−.6	.2743	.2709	.2676	.2643	.2611	.2578	.2546	.2514	.2483	.2451
−.5	.3085	.3050	.3015	.2981	.2946	.2912	.2877	.2843	.2810	.2776
−.4	.3446	.3409	.3372	.3336	.3300	.3264	.3228	.3192	.3156	.3121
−.3	.3821	.3783	.3745	.3707	.3669	.3632	.3594	.3557	.3520	.3483
−.2	.4207	.4168	.4129	.4090	.4052	.4013	.3974	.3936	.3897	.3859
−.1	.4602	.4562	.4522	.4483	.4443	.4404	.4364	.4325	.4286	.4247
−.0	.5000	.4960	.4920	.4880	.4840	.4801	.4761	.4721	.4681	.4641

(*continues*)